T0273832

PHYSICAL MATHEMATICS

Second edition

Unique in its clarity, examples, and range, *Physical Mathematics* explains simply and succinctly the mathematics that graduate students and professional physicists need to succeed in their courses and research. The book illustrates the mathematics with numerous physical examples drawn from contemporary research. This second edition has new chapters on vector calculus, special relativity and artificial intelligence, and many new sections and examples. In addition to basic subjects such as linear algebra, Fourier analysis, complex variables, differential equations, Bessel functions, and spherical harmonics, the book explains topics such as the singular value decomposition, Lie algebras and group theory, tensors and general relativity, the central limit theorem and Kolmogorov's theorems, Monte Carlo methods of experimental and theoretical physics, Feynman's path integrals, and the standard model of cosmology.

KEVIN CAHILL is Professor of Physics and Astronomy at the University of New Mexico. He has carried out research at NIST, Saclay, Ecole Polytechnique, Orsay, Harvard University, NIH, LBL, and SLAC, and has worked in quantum optics, quantum field theory, lattice gauge theory, and biophysics. *Physical Mathematics* is based on courses taught by the author at the University of New Mexico and at Fudan University in Shanghai.

PHYSICAL MATHEMATICS

Second edition

KEVIN CAHILL
University of New Mexico

CAMBRIDGE
UNIVERSITY PRESS

CAMBRIDGE
UNIVERSITY PRESS

University Printing House, Cambridge CB2 8BS, United Kingdom

One Liberty Plaza, 20th Floor, New York, NY 10006, USA

477 Williamstown Road, Port Melbourne, VIC 3207, Australia

314-321, 3rd Floor, Plot 3, Splendor Forum, Jasola District Centre, New Delhi - 110025, India

79 Anson Road, #06-04/06, Singapore 079906

Cambridge University Press is part of the University of Cambridge.

It furthers the University's mission by disseminating knowledge in the pursuit of education, learning and research at the highest international levels of excellence.

www.cambridge.org
Information on this title: www.cambridge.org/9781108470032
DOI: 10.1017/9781108555814

First published 2013
Reprinted with corrections 2014
Second edition 2019

A catalogue record for this publication is available from the British Library

Library of Congress Cataloging in Publication data
Names: Cahill, Kevin, 1941– author.
Title: Physical mathematics / Kevin Cahill (University of New Mexico).
Description: Second edition. | Cambridge ; New York, NY : Cambridge University Press, 2019. | Includes bibliographical references and index.
Identifiers: LCCN 2019008214 | ISBN 9781108470032 (alk. paper)
Subjects: LCSH: Mathematical physics. | Mathematical physics – Textbooks. | Mathematics – Study and teaching (Higher)
Classification: LCC QC20 .C24 2019 | DDC 530.15–dc23
LC record available at https://lccn.loc.gov/2019008214

ISBN 978-1-108-47003-2 Hardback

Additional resources for this publication at www.cambridge.org/Cahill2ed

To Ginette, Michael, Sean, Peter, Micheon,
Danielle, Rachel, Mia, James, Dylan,
Christopher, and Liam

Brief Contents

Contents

Preface

To the student

You will find some physics crammed in amongst the mathematics. Don't let the physics bother you. As you study the math, you'll learn some physics without extra effort. The physics is a freebie. I have tried to explain the math you need for physics and have left out the rest.

You can find codes and scripts for the simulations and figures of the book in the repositories of its chapters at `github.com/kevinecahill`.

To the professor

The book is for students who also are taking mechanics, electrodynamics, quantum mechanics, and statistical mechanics nearly simultaneously and who soon may use probability or path integrals in their research. Linear algebra and Fourier analysis are the keys to physics, so the book starts with them, but you may prefer to skip the algebra or postpone the Fourier analysis. The book is intended to support a one- or two-semester course for graduate students or advanced undergraduates.

My purpose in this book is to cover the mathematics that graduate students and working physicists need to know. I began the book with linear algebra, vector calculus, and Fourier analysis because these ideas are assumed without proof and often without mention in many courses on quantum mechanics. The chapter on infinite series lays the basis for the one that follows on the theory of functions of a complex variable, which is another set of ideas that students are assumed to know. The chapters on differential equations and on special functions cover concepts used in courses on quantum mechanics and on electrodynamics. These nine chapters make up material for a one-semester course on the basic mathematics of physics for undergraduates and beginning graduate students.

The second half of the book is about more advanced and more interesting topics. Most theoretical physicists use group theory, tensors and general relativity, and probability and statistics. Some use forms, Monte Carlo methods, path integrals, and the renormalization group. Most experimental physicists use probability and statistics and Monte Carlo methods. Chapters 10–21 provide material for a second-semester course on physical mathematics.

I have been guided by two principles. The first is to write simply so as to add no extra mental work to what the students need to do to understand the mathematics. My guide here is George Orwell's six rules for writing English.

The second principle is to describe the simple ideas that are the essence of the mathematics of physics. These ideas are so simple that their descriptions may serve as their proofs.

A list of errata is maintained at panda.unm.edu/cahill, and solutions to all the exercises are available for instructors at www.cambridge.org/Cahill2ed.

Acknowledgments

Several friends and colleagues – Rouzbeh Allahverdi, Susan Atlas, Bernard Becker, Steven Boyd, Charles Boyer, Robert Burckel, Marie Cahill, Sean Cahill, Colston Chandler, Vageli Coutsias, David Dunlap, Daniel Finley, Franco Giuliani, Roy Glauber, Pablo Gondolo, André de Gouvêa, Igor Gorelov, Kurt Hinterbichler, Jiaxing Hong, Fang Huang, Dinesh Loomba, Don Lichtenberg, Yin Luo, Lei Ma, Michael Malik, Richard Matzner, Kent Morrison, Sudhakar Prasad, Randy Reeder, Zhixiang Ren, Dmitri Sergatskov, Shashank Shalgar, David Spergel, Dimiter Vassilev, David Waxman, Edward Witten, and James Yorke – have helped me.

Students have supplied questions, ideas, and corrections, notably David Amdahl, Thomas Beechem, Chris Cesare, Yihong Cheng, Charles Cherqui, Robert Cordwell, Austin Daniel, Amo-Kwao Godwin, Aram Gragossian, Aaron Hankin, Kangbo Hao, Tiffany Hayes, Yiran Hu, Shanshan Huang, Tyler Keating, Joshua Koch, Zilong Li, Miao Lin, ZuMou Lin, Sheng Liu, Yue Liu, Manuel Munoz Arias, Ben Oliker, Boleszek Osinski, Ravi Raghunathan, Akash Rakholia, Christian Roberts, Xingyue Tian, Toby Tolley, Jiqun Tu, Christopher Vergien, Weizhen Wang, James Wendelberger, Xukun Xu, Huimin Yang, Zhou Yang, Changhao Yi, Daniel Young, Mengzhen Zhang, Lu Zheng, Lingjun Zhou, and Daniel Zirzow.

I should also like to thank the copy editor, Jon Billam, for his excellent work.

1

Linear Algebra

1.1 Numbers

The **natural** numbers are the positive integers and zero. **Rational** numbers are ratios of integers. **Irrational** numbers have decimal digits d_n

$$x = \sum_{n=m_x}^{\infty} \frac{d_n}{10^n} \tag{1.1}$$

that do not repeat. Thus the repeating decimals $1/2 = 0.50000\ldots$ and $1/3 = 0.\bar{3} \equiv 0.33333\ldots$ are rational, while $\pi = 3.141592654\ldots$ is irrational. Decimal arithmetic was invented in India over 1500 years ago but was not widely adopted in Europe until the seventeenth century.

The **real** numbers $\mathbb{R}$ include the rational numbers and the irrational numbers; they correspond to all the points on an infinite line called the **real line**.

The **complex** numbers $\mathbb{C}$ are the real numbers with one new number i whose square is -1. A complex number z is a linear combination of a real number x and a real multiple iy of i

$$z = x + iy. \tag{1.2}$$

Here $x = \text{Re}z$ is the **real part** of z, and $y = \text{Im}z$ is its **imaginary part**. One adds complex numbers by adding their real and imaginary parts

$$z_1 + z_2 = x_1 + iy_1 + x_2 + iy_2 = x_1 + x_2 + i(y_1 + y_2). \tag{1.3}$$

Since $i^2 = -1$, the product of two complex numbers is

$$z_1 z_2 = (x_1 + iy_1)(x_2 + iy_2) = x_1 x_2 - y_1 y_2 + i(x_1 y_2 + y_1 x_2). \tag{1.4}$$

The polar representation of $z = x + iy$ is

$$z = re^{i\theta} = r(\cos\theta + i\sin\theta) \tag{1.5}$$

in which r is the **modulus** or **absolute value** of z

$$r = |z| = \sqrt{x^2 + y^2} \tag{1.6}$$

and θ is its **phase** or **argument**

$$\theta = \arctan{(y/x)}. \tag{1.7}$$

Since $\exp(2\pi i) = 1$, there is an inevitable ambiguity in the definition of the phase of any complex number $z = re^{i\theta}$: for any integer n, the phase $\theta + 2\pi n$ gives the same z as θ. In various computer languages, the function atan2(y, x) returns the angle θ in the interval $-\pi < \theta \le \pi$ for which $(x, y) = r(\cos\theta, \sin\theta)$.

There are two common notations z^* and $\bar{z}$ for the **complex conjugate** of a complex number $z = x + iy$

$$z^* = \bar{z} = x - iy. \tag{1.8}$$

The square of the modulus of a complex number $z = x + iy$ is

$$|z|^2 = x^2 + y^2 = (x + iy)(x - iy) = \bar{z}z = z^*z. \tag{1.9}$$

The inverse of a complex number $z = x + iy$ is

$$z^{-1} = (x + iy)^{-1} = \frac{x - iy}{(x - iy)(x + iy)} = \frac{x - iy}{x^2 + y^2} = \frac{z^*}{z^*z} = \frac{z^*}{|z|^2}. \tag{1.10}$$

Grassmann numbers θ_i are **anticommuting** numbers, that is, the **anticommutator** of any two Grassmann numbers vanishes

$$\{\theta_i, \theta_j\} \equiv [\theta_i, \theta_j]_+ \equiv \theta_i\theta_j + \theta_j\theta_i = 0. \tag{1.11}$$

So the square of any Grassmann number is zero, $\theta_i^2 = 0$. These numbers have amusing properties (used in Chapter 20). For example, because $\theta_1\theta_2 = -\theta_2\theta_1$ and $\theta_1^2 = \theta_2^2 = 0$, the most general function of two Grassmann numbers is

$$f(\theta_1, \theta_2) = a + b\,\theta_1 + c\,\theta_2 + d\,\theta_1\theta_2 \tag{1.12}$$

and $1/(1 + a\,\theta_i) = 1 - a\,\theta_i$ in which a, b, c, d are complex numbers (Hermann Grassmann, 1809–1877).

1.2 Arrays

An **array** is an **ordered set** of numbers. Arrays play big roles in computer science, physics, and mathematics. They can be of any (integral) dimension.

A 1-dimensional array $(a_1, a_2, \ldots, a_n)$ is variously called an ***n*-tuple**, a **row vector** when written horizontally, a **column vector** when written vertically, or an ***n*-vector**. The numbers a_k are its **entries** or **components**.

A 2-dimensional array a_{ik} with i running from 1 to n and k from 1 to m is an $n \times m$ **matrix**. The numbers a_{ik} are its **entries, elements**, or **matrix elements**. One can think of a matrix as a stack of row vectors or as a queue of column vectors. The entry a_{ik} is in the ith row and the kth column.

One can add together arrays of the same dimension and shape by adding their entries. Two n-tuples add as

$$(a_1, \ldots, a_n) + (b_1, \ldots, b_n) = (a_1 + b_1, \ldots, a_n + b_n) \tag{1.13}$$

and two $n \times m$ matrices a and b add as

$$(a + b)_{ik} = a_{ik} + b_{ik}. \tag{1.14}$$

One can multiply arrays by numbers: Thus z times the 3-dimensional array a_{ijk} is the array with entries $z\, a_{ijk}$. One can multiply two arrays together no matter what their shapes and dimensions. The **outer product** of an n-tuple a and an m-tuple b is an $n \times m$ matrix with elements

$$(a\, b)_{ik} = a_i\, b_k \tag{1.15}$$

or an $m \times n$ matrix with entries $(ba)_{ki} = b_k a_i$. If a and b are complex, then one also can form the outer products $(\overline{a}\, b)_{ik} = \overline{a_i}\, b_k$, $(\overline{b}\, a)_{ki} = \overline{b_k}\, a_i$, and $(\overline{b}\, \overline{a})_{ki} = \overline{b_k}\, \overline{a_i}$. The outer product of a matrix a_{ik} and a 3-dimensional array $b_{j\ell m}$ is a five-dimensional array

$$(a\, b)_{ikj\ell m} = a_{ik}\, b_{j\ell m}. \tag{1.16}$$

An **inner product** is possible when two arrays are of the same size in one of their dimensions. Thus the **inner product** $(a, b) \equiv \langle a|b\rangle$ or **dot product** $a \cdot b$ of two real n-tuples a and b is

$$(a, b) = \langle a|b\rangle = a \cdot b = (a_1, \ldots, a_n) \cdot (b_1, \ldots, b_n) = a_1 b_1 + \cdots + a_n b_n. \tag{1.17}$$

The inner product of two complex n-tuples often is defined as

$$(a, b) = \langle a|b\rangle = \overline{a} \cdot b = (\overline{a_1}, \ldots, \overline{a_n}) \cdot (b_1, \ldots, b_n) = \overline{a_1}\, b_1 + \cdots + \overline{a_n}\, b_n \tag{1.18}$$

or as its complex conjugate

$$(a, b)^* = \langle a|b\rangle^* = (\overline{a} \cdot b)^* = (b, a) = \langle b|a\rangle = \overline{b} \cdot a. \tag{1.19}$$

The inner product of a vector with itself is nonnegative $(a, a) \geq 0$.

The product of an $m \times n$ matrix a_{ik} times an n-tuple b_k is the m-tuple b' whose ith component is

$$b'_i = a_{i1} b_1 + a_{i2} b_2 + \cdots + a_{in} b_n = \sum_{k=1}^{n} a_{ik} b_k. \tag{1.20}$$

This product is $b' = a\,b$ in matrix notation.

If the size n of the second dimension of a matrix a matches that of the first dimension of a matrix b, then their product $a\,b$ is a matrix with entries

$$(a\,b)_{i\ell} = a_{i1}\,b_{1\ell} + \cdots + a_{in}\,b_{n\ell} = \sum_{k=1}^{n} a_{ik}\,b_{k\ell}. \tag{1.21}$$

1.3 Matrices

Matrices are 2-dimensional arrays.

The **trace** of a square $n \times n$ matrix a is the sum of its diagonal elements

$$\mathrm{Tr}\, a = \mathrm{tr}\, a = a_{11} + a_{22} + \cdots + a_{nn} = \sum_{i=1}^{n} a_{ii}. \tag{1.22}$$

The trace of the product of two matrices is independent of their order

$$\mathrm{Tr}\,(a\,b) = \sum_{i=1}^{n}\sum_{k=1}^{n} a_{ik} b_{ki} = \sum_{k=1}^{n}\sum_{i=1}^{n} b_{ki} a_{ik} = \mathrm{Tr}\,(b\,a) \tag{1.23}$$

as long as the matrix elements are numbers that commute with each other. It follows that the trace is **cyclic**

$$\mathrm{Tr}\,(a\,b\,c \ldots z) = \mathrm{Tr}\,(b\,c \ldots z\,a) = \mathrm{Tr}\,(c \ldots z\,a\,b) = \ldots \tag{1.24}$$

The **transpose** of an $n \times \ell$ matrix a is an $\ell \times n$ matrix a^{T} with entries

$$\left(a^{\mathsf{T}}\right)_{ij} = a_{ji}. \tag{1.25}$$

Mathematicians often use a prime to mean transpose, as in $a' = a^{\mathsf{T}}$, but physicists tend to use primes to label different objects or to indicate differentiation. One may show that transposition inverts the order of multiplication

$$(a\,b)^{\mathsf{T}} = b^{\mathsf{T}}\,a^{\mathsf{T}}. \tag{1.26}$$

A matrix that is equal to its transpose

$$a = a^{\mathsf{T}} \tag{1.27}$$

is **symmetric,** $a_{ij} = a_{ji}$.

The (hermitian) **adjoint** of a matrix is the complex conjugate of its transpose. That is, the (hermitian) adjoint $a^{\dagger}$ of an $N \times L$ complex matrix a is the $L \times N$ matrix with entries

$$(a^{\dagger})_{ij} = a^{*}_{ji}. \tag{1.28}$$

One may show that

$$(a\,b)^\dagger = b^\dagger\, a^\dagger. \tag{1.29}$$

A matrix that is equal to its adjoint

$$a_{ij} = (a^\dagger)_{ij} = a_{ji}^* \tag{1.30}$$

(and which must be a square matrix) is **hermitian** or **self adjoint**

$$a = a^\dagger \tag{1.31}$$

(Charles Hermite 1822–1901).

Example 1.1 (The Pauli matrices) All three of Pauli's matrices

$$\sigma_1 = \begin{pmatrix} 0 & 1 \\ 1 & 0 \end{pmatrix}, \quad \sigma_2 = \begin{pmatrix} 0 & -i \\ i & 0 \end{pmatrix}, \quad \text{and} \quad \sigma_3 = \begin{pmatrix} 1 & 0 \\ 0 & -1 \end{pmatrix} \tag{1.32}$$

are hermitian (Wolfgang Pauli 1900–1958).

A real hermitian matrix is symmetric. If a matrix a is hermitian, then the quadratic form

$$\langle v|a|v\rangle = \sum_{i=1}^{N}\sum_{j=1}^{N} v_i^* a_{ij} v_j \in \mathbb{R} \tag{1.33}$$

is real for all complex n-tuples v.

The **Kronecker delta δ_{ik}** is defined to be unity if $i = k$ and zero if $i \neq k$

$$\delta_{ik} = \begin{cases} 1 & \text{if } i = k \\ 0 & \text{if } i \neq k \end{cases} \tag{1.34}$$

(Leopold Kronecker 1823–1891). The **identity matrix** I has entries $I_{ik} = \delta_{ik}$.

The **inverse** a^{-1} of an $n \times n$ matrix a is a square matrix that satisfies

$$a^{-1}\,a = a\,a^{-1} = I \tag{1.35}$$

in which I is the $n \times n$ identity matrix.

So far we have been writing n-tuples and matrices and their elements with lower-case letters. It is equally common to use capital letters, and we will do so for the rest of this section.

A matrix U whose adjoint $U^\dagger$ is its inverse

$$U^\dagger U = U U^\dagger = I \tag{1.36}$$

is **unitary**. Unitary matrices are square.

A real unitary matrix O is **orthogonal** and obeys the rule

$$O^{\mathsf{T}} O = O O^{\mathsf{T}} = I. \tag{1.37}$$

Orthogonal matrices are square.

An $N \times N$ hermitian matrix A is **nonnegative**

$$A \geq 0 \tag{1.38}$$

if for all complex vectors V the quadratic form

$$\langle V|A|V\rangle = \sum_{i=1}^{N} \sum_{j=1}^{N} V_i^* A_{ij} V_j \geq 0 \tag{1.39}$$

is nonnegative. It is **positive** or **positive definite** if

$$\langle V|A|V\rangle > 0 \tag{1.40}$$

for all nonzero vectors $|V\rangle$.

Example 1.2 (Kinds of positivity) The nonsymmetric, nonhermitian 2×2 matrix

$$\begin{pmatrix} 1 & 1 \\ -1 & 1 \end{pmatrix} \tag{1.41}$$

is positive on the space of all real 2-vectors but not on the space of all complex 2-vectors.

Example 1.3 (Representations of imaginary and grassmann numbers) The 2×2 matrix

$$\begin{pmatrix} 0 & -1 \\ 1 & 0 \end{pmatrix} \tag{1.42}$$

can represent the number i since

$$\begin{pmatrix} 0 & -1 \\ 1 & 0 \end{pmatrix} \begin{pmatrix} 0 & -1 \\ 1 & 0 \end{pmatrix} = \begin{pmatrix} -1 & 0 \\ 0 & -1 \end{pmatrix} = -I. \tag{1.43}$$

The 2×2 matrix

$$\begin{pmatrix} 0 & 0 \\ 1 & 0 \end{pmatrix} \tag{1.44}$$

can represent a Grassmann number since

$$\begin{pmatrix} 0 & 0 \\ 1 & 0 \end{pmatrix} \begin{pmatrix} 0 & 0 \\ 1 & 0 \end{pmatrix} = \begin{pmatrix} 0 & 0 \\ 0 & 0 \end{pmatrix} = 0. \tag{1.45}$$

To represent two Grassmann numbers, one needs 4×4 matrices, such as

$$\theta_1 = \begin{pmatrix} 0 & 0 & 1 & 0 \\ 0 & 0 & 0 & -1 \\ 0 & 0 & 0 & 0 \\ 0 & 0 & 0 & 0 \end{pmatrix} \quad \text{and} \quad \theta_2 = \begin{pmatrix} 0 & 1 & 0 & 0 \\ 0 & 0 & 0 & 0 \\ 0 & 0 & 0 & 1 \\ 0 & 0 & 0 & 0 \end{pmatrix}. \tag{1.46}$$

The matrices that represent n Grassmann numbers are $2^n \times 2^n$ and have 2^n rows and 2^n columns.

Example 1.4 (Fermions) The matrices (1.46) also can represent lowering or annihilation operators for a system of two fermionic states. For $a_1 = \theta_1$ and $a_2 = \theta_2$ and their adjoints $a_1^\dagger$ and $a_2^\dagger$, the creation operaors, satisfy the anticommutation relations

$$\{a_i, a_k^\dagger\} = \delta_{ik} \quad \text{and} \quad \{a_i, a_k\} = \{a_i^\dagger, a_k^\dagger\} = 0 \tag{1.47}$$

where i and k take the values 1 or 2. In particular, the relation $(a_i^\dagger)^2 = 0$ implements **Pauli's exclusion principle**, the rule that no state of a fermion can be doubly occupied.

1.4 Vectors

Vectors are things that can be multiplied by numbers and added together to form other vectors in the same **vector space**. So if U and V are vectors in a vector space S over a set F of numbers x and y and so forth, then

$$W = x\,U + y\,V \tag{1.48}$$

also is a vector in the vector space S.

A **basis** for a vector space S is a set B of vectors B_k for $k = 1, \ldots, n$ in terms of which every vector U in S can be expressed as a linear combination

$$U = u_1 B_1 + u_2 B_2 + \cdots + u_n B_n \tag{1.49}$$

with numbers u_k in F. The numbers u_k are the **components** of the vector U in the basis B. If the **basis vectors** B_k are **orthonormal**, that is, if their inner products are $(B_k, B_\ell) = \langle B_k | B_\ell \rangle = \bar{B}_k \cdot B_\ell = \delta_{k\ell}$, then we might represent the vector U as the n-tuple $(u_1, u_2, \ldots, u_n)$ with $u_k = \langle B_k | U \rangle$ or as the corresponding column vector.

Example 1.5 (Hardware store) Suppose the vector W represents a certain kind of washer and the vector N represents a certain kind of nail. Then if n and m are natural numbers, the vector

$$H = nW + mN \tag{1.50}$$

would represent a possible inventory of a very simple hardware store. The vector space of all such vectors H would include all possible inventories of the store. That space is a 2-dimensional vector space over the natural numbers, and the two vectors W and N form a basis for it.

Example 1.6 (Complex numbers) The complex numbers are a vector space. Two of its vectors are the number 1 and the number i; the vector space of complex numbers is then the set of all linear combinations

$$z = x1 + yi = x + iy. \tag{1.51}$$

The complex numbers are a 2-dimensional vector space over the real numbers, and the vectors 1 and i are a basis for it.

The complex numbers also form a 1-dimensional vector space over the complex numbers. Here any nonzero real or complex number, for instance the number 1 can be a basis consisting of the single vector 1. This 1-dimensional vector space is the set of all $z = z1$ for arbitrary complex z.

Example 1.7 (2-space) Ordinary flat 2-dimensional space is the set of all linear combinations

$$\boldsymbol{r} = x\hat{\mathbf{x}} + y\hat{\mathbf{y}} \tag{1.52}$$

in which x and y are real numbers and $\hat{\mathbf{x}}$ and $\hat{\mathbf{y}}$ are perpendicular vectors of unit length (unit vectors with $\hat{x} \cdot \hat{x} = 1 = \hat{y} \cdot \hat{y}$ and $\hat{x} \cdot \hat{y} = 0$). This vector space, called $\mathbb{R}^2$, is a 2-d space over the reals.

The vector $\boldsymbol{r}$ can be described by the basis vectors $\hat{\mathbf{x}}$ and $\hat{\mathbf{y}}$ and also by any other set of basis vectors, such as $-\hat{\mathbf{y}}$ and $\hat{\mathbf{x}}$

$$\boldsymbol{r} = x\hat{\mathbf{x}} + y\hat{\mathbf{y}} = -y(-\hat{\mathbf{y}}) + x\hat{\mathbf{x}}. \tag{1.53}$$

The components of the vector $\boldsymbol{r}$ are (x, y) in the $\{\hat{\mathbf{x}}, \hat{\mathbf{y}}\}$ basis and $(-y, x)$ in the $\{-\hat{\mathbf{y}}, \hat{\mathbf{x}}\}$ basis. **Each vector is unique, but its components depend upon the basis.**

Example 1.8 (3-space) Ordinary flat 3-dimensional space is the set of all linear combinations

$$\mathbf{r} = x\hat{\mathbf{x}} + y\hat{\mathbf{y}} + z\hat{\mathbf{z}} \tag{1.54}$$

in which x, y, and z are real numbers. It is a 3-d space over the reals.

Example 1.9 (Matrices) Arrays of a given dimension and size can be added and multiplied by numbers, and so they form a vector space. For instance, all complex 3-dimensional arrays a_{ijk} in which $1 \le i \le 3$, $1 \le j \le 4$, and $1 \le k \le 5$ form a vector space over the complex numbers.

Example 1.10 (Partial derivatives) Derivatives are vectors; so are partial derivatives. For instance, the linear combinations of x and y partial derivatives taken at $x = y = 0$

$$a\,\frac{\partial}{\partial x} + b\,\frac{\partial}{\partial y} \tag{1.55}$$

form a vector space.

Example 1.11 (Functions) The space of all linear combinations of a set of functions $f_i(x)$ defined on an interval $[a, b]$

$$f(x) = \sum_i z_i\, f_i(x) \tag{1.56}$$

is a vector space over the natural $\mathbb{N}$, real $\mathbb{R}$, or complex $\mathbb{C}$ numbers $\{z_i\}$.

Example 1.12 (States in quantum mechanics) In quantum mechanics, if the properties of a system have been measured as completely as possible, then the system (or our knowledge of it) is said to be in a **state**, often called a **pure state**, and is represented by a vector ψ or $|\psi\rangle$ in Dirac's notation. If the properties of a system have not been measured as completely as possible, then the system (or our knowledge of it) is said to be in a **mixture** or a **mixed state**, and is represented by a density operator (section 1.35).

If c_1 and c_2 are complex numbers, and $|\psi_1\rangle$ and $|\psi_2\rangle$ are any two states, then the linear combination

$$|\psi\rangle = c_1|\psi_1\rangle + c_2|\psi_2\rangle \tag{1.57}$$

also is a possible state of the system.

A harmonic oscillator in its kth excited state is in a state described by a vector $|k\rangle$. A particle exactly at position q is in a state described by a vector $|q\rangle$. An electron moving with momentum $\boldsymbol{p}$ and spin σ is in a state represented by a vector $|\boldsymbol{p}, \sigma\rangle$. A hydrogen atom at rest in its ground state is in a state $|E_0\rangle$.

Example 1.13 (Polarization of photons and gravitons) The general state of a photon of momentum $\vec{k}$ is one of elliptical polarization

$$|\vec{k}, \theta, \phi\rangle = \cos\theta\, e^{i\phi}|\vec{k}, +\rangle + \sin\theta\, e^{-i\phi}|\vec{k}, -\rangle \tag{1.58}$$

in which the states of positive and negative helicity $|\vec{k}, \pm\rangle$ represent a photon whose angular momentum $\pm\hbar$ is parallel or antiparallel to its momentum $\vec{k}$. If $\theta = \pi/4 + n\pi$, the polarization is linear, and the electric field is parallel to an axis that depends upon ϕ and is perpendicular to $\vec{k}$.

The general state of a graviton of momentum $\vec{k}$ also is one of elliptical polarization (1.58), but now the states of positive and negative helicity $|\vec{k}, \pm\rangle$ have angular momentum $\pm 2\hbar$ parallel or antiparallel to the momentum $\vec{k}$. Linear polarization again is $\theta = \pi/4 + n\pi$. The state $|\vec{k}, +\rangle$ represents space being stretched and squeezed along one axis while being squeezed and stretched along another axis, both axes perpendicular to each other and to $\vec{k}$. In the state $|\vec{k}, \times\rangle$, the stretching and squeezing axes are rotated by 45° about $\vec{k}$ relative to those of $|\vec{k}, +\rangle$.

1.5 Linear Operators

A **linear operator** A maps each vector V in its **domain** into a vector $V' = A(V) \equiv A\,V$ in its **range** in a way that is linear. So if V and W are two vectors in its domain and b and c are numbers, then

$$A(bV + cW) = bA(V) + cA(W) = bA\,V + cA\,W. \tag{1.59}$$

If the domain and the range are the same vector space S, then A maps each basis vector B_i of S into a linear combination of the basis vectors B_k

$$A\,B_i = a_{1i}B_1 + a_{2i}B_2 + \cdots + a_{ni}B_n = \sum_{k=1}^{n} a_{ki}\,B_k \tag{1.60}$$

a formula that is clearer in Dirac's notation (Section 1.12). The square matrix a_{ki} **represents** the linear operator A in the B_k basis. The effect of A on any vector $V = u_1B_1 + u_2B_2 + \cdots + u_nB_n$ in S then is

$$A\,V = A\sum_{i=1}^{n} u_iB_i = \sum_{i=1}^{n} u_iAB_i = \sum_{i,k=1}^{n} u_ia_{ki}B_k = \sum_{i,k=1}^{n} a_{ki}u_iB_k. \tag{1.61}$$

So the kth component u'_k of the vector $V' = A\,V$ is

$$u'_k = a_{k1}u_1 + a_{k2}u_2 + \cdots + a_{kn}u_n = \sum_{i=1}^{n} a_{ki}\,u_i. \tag{1.62}$$

Thus the column vector u' of the components u'_k of the vector $V' = A\,V$ is the product $u' = a\,u$ of the matrix with elements a_{ki} that represents the linear operator A in the B_k basis and the column vector with components u_i that represents the vector V in that basis. In each basis, vectors and linear operators are represented by column vectors and matrices.

Each linear operator is unique, but its matrix depends upon the basis. If we change from the B_k basis to another basis B'_i

$$B'_i = \sum_{\ell=1}^{n} u_{ki}\,B_k \tag{1.63}$$

in which the $n \times n$ matrix $u_{\ell k}$ has an inverse matrix u^{-1}_{ki} so that

$$\sum_{k=1}^{n} u^{-1}_{ki}B'_k = \sum_{k=1}^{n} u^{-1}_{ki}\sum_{\ell=1}^{n} u_{\ell k}B_\ell = \sum_{\ell=1}^{n}\left(\sum_{k=1}^{n} u_{\ell k}u^{-1}_{ki}\right)B_\ell = \sum_{\ell=1}^{n} \delta_{\ell i}B_\ell = B_i \tag{1.64}$$

then the old basis vectors B_i are given by

$$B_i = \sum_{k=1}^{n} u_{ki}^{-1} B'_k. \tag{1.65}$$

Thus (Exercise 1.9) the linear operator A maps the basis vector B'_i to

$$A\, B'_i = \sum_{k=1}^{n} u_{ki}\, A\, B_k = \sum_{j,k=1}^{n} u_{ki} a_{jk} B_j = \sum_{j,k,\ell=1}^{n} u_{ki} a_{jk} u_{\ell j}^{-1}\, B'_\ell. \tag{1.66}$$

So the matrix a' that represents A in the B' basis is related to the matrix a that represents it in the B basis by a **similarity transformation**

$$a'_{\ell i} = \sum_{jk=1}^{n} u_{\ell j}^{-1} a_{jk} u_{ki} \quad \text{or} \quad a' = u^{-1}\, a\, u \tag{1.67}$$

in matrix notation. If the matrix u is **unitary**, then its inverse is its hermitian adjoint

$$u^{-1} = u^{\dagger} \tag{1.68}$$

and the similarity transformation (1.67) is

$$a'_{\ell i} = \sum_{jk=1}^{n} u_{\ell j}^{\dagger} a_{jk} u_{ki} = \sum_{jk=1}^{n} u_{j\ell}^{*} a_{jk} u_{ik} \quad \text{or} \quad a' = u^{\dagger}\, a\, u. \tag{1.69}$$

Because traces are cyclic, they are invariant under similarity transformations

$$\mathrm{Tr}(a') = \mathrm{Tr}(u\, a\, u^{-1}) = \mathrm{Tr}(a\, u^{-1}\, u) = \mathrm{Tr}(a). \tag{1.70}$$

Example 1.14 (Change of basis) Let the action of the linear operator A on the basis vectors $\{B_1, B_2\}$ be $AB_1 = B_2$ and $AB_2 = 0$. If the column vectors

$$b_1 = \begin{pmatrix} 1 \\ 0 \end{pmatrix} \quad \text{and} \quad b_2 = \begin{pmatrix} 0 \\ 1 \end{pmatrix} \tag{1.71}$$

represent the basis vectors B_1 and B_2, then the matrix

$$a = \begin{pmatrix} 0 & 0 \\ 1 & 0 \end{pmatrix} \tag{1.72}$$

represents the linear operator A. But if we use the basis vectors

$$B'_1 = \frac{1}{\sqrt{2}}(B_1 + B_2) \quad \text{and} \quad B'_2 = \frac{1}{\sqrt{2}}(B_1 - B_2) \tag{1.73}$$

then the vectors

$$b'_1 = \frac{1}{\sqrt{2}}\begin{pmatrix} 1 \\ 1 \end{pmatrix} \quad \text{and} \quad b'_2 = \frac{1}{\sqrt{2}}\begin{pmatrix} 1 \\ -1 \end{pmatrix} \tag{1.74}$$

would represent B_1 and B_2, and the matrix

$$a' = \frac{1}{2}\begin{pmatrix} 1 & 1 \\ -1 & -1 \end{pmatrix} \tag{1.75}$$

would represent the linear operator A (Exercise 1.10).

A linear operator A also may map a vector space S with basis B_k into a different vector space T with its own basis C_k.

$$A\,B_i = \sum_{k=1}^{M} a_{ki}\,C_k. \tag{1.76}$$

It then maps an arbitrary vector $V = u_1 B_1 + \cdots + u_n B_n$ in S into the vector

$$A\,V = \sum_{k=1}^{M}\left(\sum_{i=1}^{n} a_{ki}\,u_i\right) C_k \tag{1.77}$$

in T.

1.6 Inner Products

Most of the vector spaces used by physicists have an inner product. A **positive-definite inner product** associates a number (f, g) with every ordered pair of vectors f and g in the vector space V and obeys the rules

$$(f, g) = (g, f)^* \tag{1.78}$$

$$(f, z\,g + w\,h) = z\,(f, g) + w\,(f, h) \tag{1.79}$$

$$(f, f) \ge 0 \quad \text{and} \quad (f, f) = 0 \iff f = 0 \tag{1.80}$$

in which f, g, and h are vectors, and z and w are numbers. The first rule says that the inner product is **hermitian**; the second rule says that it is **linear** in the second vector $z\,g + w\,h$ of the pair; and the third rule says that it is **positive definite**. The first two rules imply that (Exercise 1.11) the inner product is **antilinear** in the first vector of the pair

$$(z\,g + w\,h, f) = z^*(g, f) + w^*(h, f). \tag{1.81}$$

A **Schwarz inner product** obeys the first two rules (1.78, 1.79) for an inner product and the fourth (1.81) but only the first part of the third (1.80)

$$(f, f) \ge 0. \tag{1.82}$$

This condition of **nonnegativity** implies (Exercise 1.15) that a vector f of zero length must be orthogonal to all vectors g in the vector space V

$$(f, f) = 0 \implies (g, f) = 0 \ \text{for all} \ g \in V. \tag{1.83}$$

So a Schwarz inner product is *almost* positive definite.

Inner products of 4-vectors can be negative. To accomodate them we define an **indefinite** inner product without regard to positivity as one that satisfies the first two rules (1.78 and 1.79) and therefore also the fourth rule (1.81) and that instead of being positive definite is **nondegenerate**

$$(f, g) = 0 \text{ for all } f \in V \implies g = 0. \tag{1.84}$$

This rule says that only the zero vector is orthogonal to all the vectors of the space. The positive-definite condition (1.80) is stronger than and implies nondegeneracy (1.84) (Exercise 1.14).

Apart from the indefinite inner products of 4-vectors in special and general relativity, most of the inner products physicists use are Schwarz inner products or positive-definite inner products. For such inner products, we can define the **norm** $|f| = \| f \|$ of a vector f as the square root of the nonnegative inner product (f, f)

$$\| f \| = \sqrt{(f, f)}. \tag{1.85}$$

A vector $\hat{f} = f / \| f \|$ has unit norm and is said to be **normalized**. Two measures of the distance between two normalized vectors f and g are the **norm of their difference** and the **Bures distance**

$$D(f, g) = \| f - g \| \text{ and } D_B(f, g) = \arccos(|(f, g)|). \tag{1.86}$$

Example 1.15 (Euclidian space) The space of real vectors U, V with n components U_i, V_i forms an n-dimensional vector space over the real numbers with an inner product

$$(U, V) = \sum_{i=1}^{n} U_i \, V_i \tag{1.87}$$

that is nonnegative when the two vectors are the same

$$(U, U) = \sum_{i=1}^{n} U_i \, U_i = \sum_{i=1}^{n} U_i^2 \ge 0 \tag{1.88}$$

and vanishes only if all the components U_i are zero, that is, if the vector $U = 0$. Thus the inner product (1.87) is positive definite. When (U, V) is zero, the vectors U and V are **orthogonal**.

Example 1.16 (Complex euclidian space) The space of complex vectors with n components U_i, V_i forms an n-dimensional vector space over the complex numbers with inner product

$$(U, V) = \sum_{i=1}^{n} U_i^* V_i = (V, U)^*. \tag{1.89}$$

The inner product (U, U) is nonnegative and vanishes

$$(U, U) = \sum_{i=1}^{n} U_i^* U_i = \sum_{i=1}^{n} |U_i|^2 \ge 0 \tag{1.90}$$

only if $U = 0$. So the inner product (1.89) is positive definite. If (U, V) is zero, then U and V are orthogonal.

Example 1.17 (Complex matrices) For the vector space of $n \times m$ complex matrices A, B, ..., the trace of the adjoint (1.28) of A multiplied by B is an inner product

$$(A, B) = \text{Tr}A^\dagger B = \sum_{i=1}^{n}\sum_{j=1}^{m} (A^\dagger)_{ji} B_{ij} = \sum_{i=1}^{n}\sum_{j=1}^{m} A_{ij}^* B_{ij} \tag{1.91}$$

that is nonnegative when the matrices are the same

$$(A, A) = \text{Tr}A^\dagger A = \sum_{i=1}^{n}\sum_{j=1}^{m} A_{ij}^* A_{ij} = \sum_{i=1}^{n}\sum_{j=1}^{m} |A_{ij}|^2 \ge 0 \tag{1.92}$$

and zero only when $A = 0$. So this inner product is positive definite.

A vector space with a positive-definite inner product (1.78–1.81) is called an **inner-product space**, a **metric space**, or a **pre-Hilbert space**.

A sequence of vectors f_n is a **Cauchy sequence** if for every $\epsilon > 0$ there is an integer $N(\epsilon)$ such that $\|f_n - f_m\| < \epsilon$ whenever both n and m exceed $N(\epsilon)$. A sequence of vectors f_n **converges** to a vector f if for every $\epsilon > 0$ there is an integer $N(\epsilon)$ such that $\|f - f_n\| < \epsilon$ whenever n exceeds $N(\epsilon)$. An inner-product space with a norm defined as in (1.85) is **complete** if each of its Cauchy sequences converges to a vector in that space. A **Hilbert space** is a complete inner-product space. Every finite-dimensional inner-product space is complete and so is a Hilbert space. An infinite-dimensional complete inner-product space, such as the space of all square-integrable functions, also is a Hilbert space (David Hilbert, 1862–1943).

Example 1.18 (Hilbert space of square-integrable functions) For the vector space of functions (1.56), a natural inner product is

$$(f, g) = \int_a^b dx\, f^*(x)\, g(x). \tag{1.93}$$

The squared norm $\| f \|$ of a function $f(x)$ is

$$\| f \|^2 = \int_a^b dx\, |f(x)|^2. \tag{1.94}$$

A function is **square integrable** if its norm is finite. The space of all square-integrable functions is an inner-product space; it also is complete and so is a Hilbert space.

Example 1.19 (Minkowski inner product) The Minkowski or Lorentz inner product (p, x) of two 4-vectors $p = (E/c, p_1, p_2, p_3)$ and $x = (ct, x_1, x_2, x_3)$ is $\boldsymbol{p} \cdot \boldsymbol{x} - Et$. It is indefinite, nondegenerate (1.84), and invariant under Lorentz transformations, and often is written as $p \cdot x$ or as $p\,x$. If p is the 4-momentum of a freely moving physical particle of mass m, then

$$p \cdot p = \boldsymbol{p} \cdot \boldsymbol{p} - E^2/c^2 = -\,c^2 m^2 \le 0. \tag{1.95}$$

The Minkowski inner product satisfies the rules (1.78, 1.79, and 1.84), but it is **not positive definite**, and it does not satisfy the Schwarz inequality (Hermann Minkowski 1864–1909, Hendrik Lorentz 1853–1928).

Example 1.20 (Inner products in quantum mechanics) The probability $P(\phi|\psi)$ that a system in the state $|\psi\rangle$ will be measured to be in the state $|\phi\rangle$ is the absolute value squared of the inner product $\langle\phi|\psi\rangle$ divided by the squared norms of the two states

$$P(\phi|\psi) = \frac{|\langle\phi|\psi\rangle|^2}{\langle\phi|\phi\rangle\langle\psi|\psi\rangle}. \tag{1.96}$$

If the two states are normalized, then the probability is just the absolute value squared of their inner product, $P(\phi|\psi) = |\langle\phi|\psi\rangle|^2$.

1.7 Cauchy–Schwarz Inequalities

For any two vectors f and g, the Schwarz inequality

$$(f, f)\,(g, g) \ge |(f, g)|^2 \tag{1.97}$$

holds for any Schwarz inner product (and so for any positive-definite inner product). The condition (1.82) of **nonnegativity** ensures that for any complex number λ the inner product of the vector $f - \lambda g$ with itself is nonnegative

$$(f - \lambda g, f - \lambda g) = (f, f) - \lambda^*(g, f) - \lambda(f, g) + |\lambda|^2(g, g) \ge 0. \tag{1.98}$$

Now if $(g, g) = 0$, then for $(f - \lambda g, f - \lambda g)$ to remain nonnegative for all complex values of λ it is necessary that $(f, g) = 0$ also vanish (Exercise 1.15). Thus if $(g, g) = 0$, then the Schwarz inequality (1.97) is trivially true because both sides of it vanish. So we assume that $(g, g) > 0$ and set $\lambda = (g, f)/(g, g)$. The inequality (1.98) then gives us

$$(f - \lambda g,\ f - \lambda g) = \left(f - \frac{(g, f)}{(g, g)}\,g,\ f - \frac{(g, f)}{(g, g)}\,g\right) = (f, f) - \frac{(f, g)(g, f)}{(g, g)} \ge 0$$

which is the Schwarz inequality (1.97)

$$(f, f)(g, g) \geq |(f, g)|^2. \tag{1.99}$$

Taking the square root of each side, we have

$$\| f \| \| g \| \geq |(f, g)| \tag{1.100}$$

(Hermann Schwarz 1843–1921).

Example 1.21 (Some Schwarz inequalities) For the dot product of two real 3-vectors $\boldsymbol{r}$ and $\boldsymbol{R}$, the Cauchy–Schwarz inequality is

$$(\boldsymbol{r} \cdot \boldsymbol{r})\ (\boldsymbol{R} \cdot \boldsymbol{R}) \geq (\boldsymbol{r} \cdot \boldsymbol{R})^2 = (\boldsymbol{r} \cdot \boldsymbol{r})\ (\boldsymbol{R} \cdot \boldsymbol{R})\ \cos^2\theta \tag{1.101}$$

where θ is the angle between $\boldsymbol{r}$ and $\boldsymbol{R}$.

The Schwarz inequality for two real n-vectors $\boldsymbol{x}$ is

$$(\boldsymbol{x} \cdot \boldsymbol{x})\ (\boldsymbol{y} \cdot \boldsymbol{y}) \geq (\boldsymbol{x} \cdot \boldsymbol{y})^2 = (\boldsymbol{x} \cdot \boldsymbol{x})\ (\boldsymbol{y} \cdot \boldsymbol{y})\ \cos^2\theta \tag{1.102}$$

and it implies (Exercise 1.16) that

$$\|\boldsymbol{x}\| + \|\boldsymbol{y}\| \geq \|\boldsymbol{x} + \boldsymbol{y}\|. \tag{1.103}$$

For two complex n-vectors $\boldsymbol{u}$ and $\boldsymbol{v}$, the Schwarz inequality is

$$\left(\boldsymbol{u}^* \cdot \boldsymbol{u}\right)\ \left(\boldsymbol{v}^* \cdot \boldsymbol{v}\right) \geq \left|\boldsymbol{u}^* \cdot \boldsymbol{v}\right|^2 = \left(\boldsymbol{u}^* \cdot \boldsymbol{u}\right)\ \left(\boldsymbol{v}^* \cdot \boldsymbol{v}\right)\ \cos^2\theta \tag{1.104}$$

and it implies (exercise 1.17) that

$$\|\boldsymbol{u}\| + \|\boldsymbol{v}\| \geq \|\boldsymbol{u} + \boldsymbol{v}\|. \tag{1.105}$$

The inner product (1.93) of two complex functions f and g provides another example

$$\int_a^b dx\, |f(x)|^2 \int_a^b dx\, |g(x)|^2 \geq \left| \int_a^b dx\, f^*(x)\, g(x) \right|^2 \tag{1.106}$$

of the Schwarz inequality.

1.8 Linear Independence and Completeness

A set of n vectors $V_1, V_2, \ldots, V_n$ is **linearly dependent** if there exist numbers c_i, *not all zero*, such that the linear combination

$$c_1 V_1 + \cdots + c_n V_n = 0 \tag{1.107}$$

vanishes. A set of vectors is **linearly independent** if it is not linearly dependent.

A set $\{V_i\}$ of linearly independent vectors is **maximal** in a vector space S if the addition of any other vector U in S to the set $\{V_i\}$ makes the enlarged set $\{U, V_i\}$ linearly dependent.

A set of n linearly independent vectors $V_1, V_2, \dots, V_n$ that is maximal in a vector space S can represent any vector U in the space S as a linear combination of its vectors, $U = u_1 V_1 + \cdots + u_n V_n$. For if we enlarge the maximal set $\{V_i\}$ by including in it any vector U not already in it, then the bigger set $\{U, V_i\}$ will be linearly dependent. Thus there will be numbers $c_0, c_1, \dots, c_n$, not all zero, that make the sum

$$c_0 U + c_1 V_1 + \cdots + c_n V_n = 0 \tag{1.108}$$

vanish. Now if c_0 were 0, then the set $\{V_i\}$ would be linearly dependent. Thus $c_0 \neq 0$, and so we may divide by c_0 and express the arbitrary vector U as a linear combination of the vectors V_i

$$U = -\frac{1}{c_0}(c_1 V_1 + \cdots + c_n V_n) = u_1 V_1 + \cdots + u_n V_n \tag{1.109}$$

with $u_k = -c_k/c_0$. Thus a set of linearly independent vectors $\{V_i\}$ that is maximal in a space S can represent every vector U in S as a linear combination $U = u_1 V_1 + \cdots + u_n V_n$ of its vectors. Such a set $\{V_i\}$ of linearly independent vectors that is maximal in a space S is called a **basis** for S; it **spans** S; it is a **complete** set of vectors in S.

1.9 Dimension of a Vector Space

If $V_1, \dots, V_n$ and $W_1, \dots, W_m$ are any two bases for a vector space S, then $n = m$.

To see why, suppose that the n vectors $C_1, C_2, \dots, C_n$ are complete in a vector space S, and that the m vectors $L_1, L_2, \dots, L_m$ in S are linearly independent (Halmos, 1958, sec. 1.8). Since the C's are complete, the set of vectors $L_m, C_1, \dots, C_n$ is linearly dependent. So we can omit one of the C's and the remaining set $L_m, C_1, \dots, C_{i-1}, C_{i+1}, \dots, C_n$ still spans S. Repeating this argument, we find that the vectors

$$L_{m-1}, L_m, C_1, \dots, C_{i-1}, C_{i+1}, \dots, C_n \tag{1.110}$$

are linearly dependent, and that the vectors

$$L_{m-1}, L_m, C_1, \dots, C_{i-1}, C_{i+1}, \dots, C_{j-1}, C_{j+1}, \dots, C_n \tag{1.111}$$

still span S. We continue to repeat these steps until we run out of L's or C's. If n were less than m, then we'd end up with a set of vectors $L_k, \dots, L_m$ that would be complete and therefore each of the vectors $L_1, \dots, L_{k-1}$ would have to be linear combinations of the vectors $L_k, \dots, L_m$. But the L's by assumption are linearly independent. So $n \geq m$. Thus if both the C's and the L's are bases for the same space S, and so are both complete and linearly independent in it, then both $n \geq m$ and $m \geq n$. So all the bases of a vector space consist of the same number of vectors. This number is the **dimension** of the space.

The steps of the above demonstration stop for $n = m$ when the m linearly independent L's have replaced the n complete C's leaving us with $n = m$ linearly independent L's that are complete. Thus in a vector space of n dimensions, every set of n linearly independent vectors is complete and so forms a basis for the space.

1.10 Orthonormal Vectors

Suppose the vectors $V_1, V_2, \ldots, V_n$ are linearly independent. Then we can make out of them a set of n vectors U_i that are orthonormal

$$(U_i, U_j) = \delta_{ij}. \tag{1.112}$$

There are many ways to do this, because there are many such sets of orthonormal vectors. We will use the Gram–Schmidt method. We set

$$U_1 = \frac{V_1}{\sqrt{(V_1, V_1)}} \tag{1.113}$$

so the first vector U_1 is normalized. Next we set $u_2 = V_2 + c_{12} U_1$ and require that u_2 be orthogonal to U_1

$$0 = (U_1, u_2) = (U_1, c_{12} U_1 + V_2) = c_{12} + (U_1, V_2). \tag{1.114}$$

Thus $c_{12} = -(U_1, V_2)$, and so

$$u_2 = V_2 - (U_1, V_2)\, U_1. \tag{1.115}$$

The normalized vector U_2 then is

$$U_2 = \frac{u_2}{\sqrt{(u_2, u_2)}}. \tag{1.116}$$

We next set $u_3 = V_3 + c_{13} U_1 + c_{23} U_2$ and ask that u_3 be orthogonal to U_1

$$0 = (U_1, u_3) = (U_1, c_{13} U_1 + c_{23} U_2 + V_3) = c_{13} + (U_1, V_3) \tag{1.117}$$

and also to U_2

$$0 = (U_2, u_3) = (U_2, c_{13} U_1 + c_{23} U_2 + V_3) = c_{23} + (U_2, V_3). \tag{1.118}$$

So $c_{13} = -(U_1, V_3)$ and $c_{23} = -(U_2, V_3)$, and we have

$$u_3 = V_3 - (U_1, V_3)\, U_1 - (U_2, V_3)\, U_2. \tag{1.119}$$

The normalized vector U_3 then is

$$U_3 = \frac{u_3}{\sqrt{(u_3, u_3)}}. \tag{1.120}$$

We may continue in this way until we reach the last of the n linearly independent vectors. We require the kth unnormalized vector u_k

$$u_k = V_k + \sum_{i=1}^{k-1} c_{ik}\, U_i \tag{1.121}$$

to be orthogonal to the $k-1$ vectors U_i and find that $c_{ik} = -(U_i, V_k)$ so that

$$u_k = V_k - \sum_{i=1}^{k-1} (U_i, V_k)\, U_i. \tag{1.122}$$

The normalized vector then is

$$U_k = \frac{u_k}{\sqrt{(u_k, u_k)}}. \tag{1.123}$$

A basis is more convenient if its vectors are orthonormal.

1.11 Outer Products

From any two vectors f and g, we may make an **outer-product** operator A that maps any vector h into the vector f multiplied by the inner product (g, h)

$$A\, h = f\,(g, h) = (g, h)\, f. \tag{1.124}$$

The operator A is linear because for any vectors e, h and numbers z, w

$$A\,(z\,h + w\,e) = (g, z\,h + w\,e)\, f = z\,(g, h)\, f + w\,(g, e)\, f = z\,A\,h + w\,A\,e. \tag{1.125}$$

If f, g, and h are vectors with components f_i, g_i, and h_i in some basis, then the linear transformation is

$$(Ah)_i = \sum_{j=1}^{n} A_{ij}\, h_j = f_i \sum_{j=1}^{n} g_j^*\, h_j \tag{1.126}$$

and in that basis A is the matrix with entries

$$A_{ij} = f_i\, g_j^*. \tag{1.127}$$

It is the **outer product** of the vectors f and g^*. The outer product of g and f^* is different, $B_{ij} = g_i\, f_j^*$.

Example 1.22 (Outer product) If in some basis the vectors f and g are

$$f = \begin{pmatrix} 2 \\ 3i \end{pmatrix} \quad \text{and} \quad g = \begin{pmatrix} i \\ 1 \\ 3i \end{pmatrix} \tag{1.128}$$

then their outer products are the matrices

$$A = \begin{pmatrix} 2 \\ 3i \end{pmatrix} \begin{pmatrix} -i & 1 & -3i \end{pmatrix} = \begin{pmatrix} -2i & 2 & -6i \\ 3 & 3i & 9 \end{pmatrix} \tag{1.129}$$

and

$$B = \begin{pmatrix} i \\ 1 \\ 3i \end{pmatrix} \begin{pmatrix} 2 & -3i \end{pmatrix} = \begin{pmatrix} 2i & 3 \\ 2 & -3i \\ 6i & 9 \end{pmatrix}. \tag{1.130}$$

Example 1.23 (Dirac's outer products) Dirac's notation for outer products is neat. If the vectors $f = |f\rangle$ and $g = |g\rangle$ are

$$|f\rangle = \begin{pmatrix} a \\ b \\ c \end{pmatrix} \quad \text{and} \quad |g\rangle = \begin{pmatrix} z \\ w \end{pmatrix} \tag{1.131}$$

then their outer products are

$$|f\rangle\langle g| = \begin{pmatrix} az^* & aw^* \\ bz^* & bw^* \\ cz^* & cw^* \end{pmatrix} \quad \text{and} \quad |g\rangle\langle f| = \begin{pmatrix} za^* & zb^* & zc^* \\ wa^* & wb^* & wc^* \end{pmatrix} \tag{1.132}$$

as well as

$$|f\rangle\langle f| = \begin{pmatrix} aa^* & ab^* & ac^* \\ ba^* & bb^* & bc^* \\ ca^* & cb^* & cc^* \end{pmatrix} \quad \text{and} \quad |g\rangle\langle g| = \begin{pmatrix} zz^* & zw^* \\ wz^* & ww^* \end{pmatrix}. \tag{1.133}$$

1.12 Dirac Notation

Outer products are important in quantum mechanics, and so Dirac invented a notation for linear algebra that makes them easy to write. In his notation, a vector f is a **ket** $f = |f\rangle$. The new thing in his notation is the **bra** $\langle g|$. The inner product of two vectors (g, f) is the **bracket** $(g, f) = \langle g|f\rangle$. A matrix element (g, cf) of an operator c then is $(g, cf) = \langle g|c|f\rangle$ in which the bra and ket bracket the operator c.

In Dirac notation, an outer product like (1.124) $A\,h = (g, h)\,f = f\,(g, h)$ reads $A\,|h\rangle = |f\rangle\langle g|h\rangle$, and the outer product A itself is $A = |f\rangle\langle g|$.

The bra $\langle g|$ is the **adjoint** of the ket $|g\rangle$, and the ket $|f\rangle$ is the adjoint of the bra $\langle f|$

$$\langle g| = (|g\rangle)^\dagger \text{ and } |f\rangle = (\langle f|)^\dagger, \quad \text{so } \langle g|^{\dagger\dagger} = \langle g| \text{ and } |f\rangle^{\dagger\dagger} = |f\rangle. \tag{1.134}$$

The adjoint of an outer product is

$$(z\,|f\rangle\langle g|)^\dagger = z^*\,|g\rangle\langle f|. \tag{1.135}$$

In Dirac's notation, the most general linear operator is an arbitrary linear combination of outer products

$$A = \sum_{k\ell} z_{k\ell}\,|k\rangle\langle\ell|. \tag{1.136}$$

Its adjoint is

$$A^\dagger = \sum_{k\ell} z_{k\ell}^*\,|\ell\rangle\langle k|. \tag{1.137}$$

The adjoint of a ket $|h\rangle = A|f\rangle$ is

$$(|h\rangle)^\dagger = (A|f\rangle)^\dagger = \left(\sum_{k\ell} z_{k\ell}\,|k\rangle\langle\ell|f\rangle\right)^\dagger = \sum_{k\ell} z_{k\ell}^*\,\langle f|\ell\rangle\langle k| = \langle f|A^\dagger. \tag{1.138}$$

Before Dirac, bras were implicit in the definition of the inner product, but they did not appear explicitly; there was no simple way to write the bra $\langle g|$ or the outer product $|f\rangle\langle g|$.

If the kets $|k\rangle$ form an orthonormal basis in an n-dimensional vector space, then we can expand an arbitrary ket in the space as

$$|f\rangle = \sum_{k=1}^{n} c_k|k\rangle. \tag{1.139}$$

Since the basis vectors are orthonormal $\langle\ell|k\rangle = \delta_{\ell k}$, we can identify the coefficients c_k by forming the inner product

$$\langle\ell|f\rangle = \sum_{k=1}^{n} c_k\,\langle\ell|k\rangle = \sum_{k=1}^{n} c_k\,\delta_{\ell,k} = c_\ell. \tag{1.140}$$

The original expasion (1.139) then must be

$$|f\rangle = \sum_{k=1}^{n} c_k|k\rangle = \sum_{k=1}^{n}\langle k|f\rangle\,|k\rangle = \sum_{k=1}^{n}|k\rangle\,\langle k|f\rangle = \left(\sum_{k=1}^{n}|k\rangle\,\langle k|\right)|f\rangle. \tag{1.141}$$

Since this equation must hold for every vector $|f\rangle$ in the space, it follows that the sum of outer products within the parentheses is the identity operator for the space

$$I = \sum_{k=1}^{n} |k\rangle \langle k|. \tag{1.142}$$

Every set of kets $|\alpha_j\rangle$ that forms an orthonormal basis $\langle \alpha_j | \alpha_\ell \rangle = \delta_{j\ell}$ for the space gives us an equivalent representation of the identity operator

$$I = \sum_{j=1}^{n} |\alpha_j\rangle \langle \alpha_j| = \sum_{k=1}^{n} |k\rangle \langle k|. \tag{1.143}$$

These resolutions of the identity operator give every vector $|f\rangle$ in the space the expansions

$$|f\rangle = \sum_{j=1}^{n} |\alpha_j\rangle \langle \alpha_j | f\rangle = \sum_{k=1}^{n} |k\rangle \langle k|f\rangle. \tag{1.144}$$

Example 1.24 (Linear operators represented as matrices) The equations (1.60–1.67) that relate linear operators to the matrices that represent them are much clearer in Dirac's notation. If the kets $|B_k\rangle$ are n orthonormal basis vectors, that is, if $\langle B_k|B_\ell\rangle = \delta_{k\ell}$, for a vector space S, then a linear operator A acting on S maps the basis vector $|B_i\rangle$ into (1.60)

$$A|B_i\rangle = \sum_{k=1}^{n} |B_k\rangle\langle B_k|A|B_i\rangle = \sum_{k=1}^{n} a_{ki}\, |B_k\rangle, \tag{1.145}$$

and the matrix that represents the linear operator A in the $|B_k\rangle$ basis is $a_{ki} = \langle B_k|A|B_i\rangle$. If a unitary operator U maps these basis vectors into $|B'_k\rangle = U|B_k\rangle$, then in this new basis the matrix that represents A as in (1.138) is

$$\begin{aligned} a'_{\ell i} &= \langle B'_\ell|A|B'_i\rangle = \langle B_\ell|U^\dagger\, A\, U|B_i\rangle \\ &= \sum_{j=1}^{n}\sum_{k=1}^{n} \langle B_\ell|U^\dagger|B_j\rangle\langle B_j|A|B_k\rangle\langle B_k|U|B_i\rangle = \sum_{j=1}^{n}\sum_{k=1}^{n} u^\dagger_{\ell j} a_{jk} u_{ki} \end{aligned} \tag{1.146}$$

or $a' = u^\dagger\, a\, u$ in matrix notation.

Example 1.25 (Inner-product rules) In Dirac's notation, the rules (1.78—1.81), of a positive-definite inner product are

$$\begin{aligned} \langle f|g\rangle &= \langle g|f\rangle^* \\ \langle f|z_1 g_1 + z_2 g_2\rangle &= z_1\langle f|g_1\rangle + z_2\langle f|g_2\rangle \\ \langle z_1 f_1 + z_2 f_2|g\rangle &= z_1^*\langle f_1|g\rangle + z_2^*\langle f_2|g\rangle \\ \langle f|f\rangle \ge 0 \ \text{ and } \ \langle f|f\rangle &= 0 \iff f = 0. \end{aligned} \tag{1.147}$$

States in Dirac notation often are labeled $|\psi\rangle$ or by their quantum numbers $|n, l, m\rangle$, and one rarely sees plus signs or complex numbers or operators inside bras or kets. But one should.

Example 1.26 (Gram–Schmidt) In Dirac notation, the formula (1.122) for the kth orthogonal linear combination of the vectors $|V_\ell\rangle$ is

$$|u_k\rangle = |V_k\rangle - \sum_{i=1}^{k-1} |U_i\rangle\langle U_i|V_k\rangle = \left(I - \sum_{i=1}^{k-1} |U_i\rangle\langle U_i|\right)|V_k\rangle \tag{1.148}$$

and the formula (1.123) for the kth orthonormal linear combination of the vectors $|V_\ell\rangle$ is

$$|U_k\rangle = \frac{|u_k\rangle}{\sqrt{\langle u_k|u_k\rangle}}. \tag{1.149}$$

The vectors $|U_k\rangle$ are not unique; they vary with the order of the $|V_k\rangle$.

Vectors and linear operators are abstract. The numbers we compute with are inner products like $\langle g|f\rangle$ and $\langle g|A|f\rangle$. In terms of n orthonormal basis vectors $|j\rangle$ with $f_j = \langle j|f\rangle$ and $g_j^* = \langle g|j\rangle$, we can use the expansion (1.142) of the identity operator to write these inner products as

$$\begin{aligned}\langle g|f\rangle &= \langle g|I|f\rangle = \sum_{j=1}^{n}\langle g|j\rangle\langle j|f\rangle = \sum_{j=1}^{n} g_j^* f_j \\ \langle g|A|f\rangle &= \langle g|IAI|f\rangle = \sum_{j,\ell=1}^{n}\langle g|j\rangle\langle j|A|\ell\rangle\langle \ell|f\rangle = \sum_{j,\ell=1}^{n} g_j^*\, A_{j\ell}\, f_\ell\end{aligned} \tag{1.150}$$

in which $A_{j\ell} = \langle j|A|\ell\rangle$. We often gather the inner products $f_\ell = \langle \ell|f\rangle$ into a column vector f with components $f_\ell = \langle \ell|f\rangle$

$$f = \begin{pmatrix} \langle 1|f\rangle \\ \langle 2|f\rangle \\ \vdots \\ \langle n|f\rangle \end{pmatrix} = \begin{pmatrix} f_1 \\ f_2 \\ \vdots \\ f_n \end{pmatrix} \tag{1.151}$$

and the $\langle j|A|\ell\rangle$ into a matrix A with matrix elements $A_{j\ell} = \langle j|A|\ell\rangle$. If we also line up the inner products $\langle g|j\rangle = \langle j|g\rangle^*$ in a row vector that is the transpose of the complex conjugate of the column vector g

$$g^\dagger = \left(\langle 1|g\rangle^*,\ \langle 2|g\rangle^*, \ldots, \langle n|g\rangle^*\right) = \left(g_1^*, g_2^*, \ldots, g_n^*\right) \tag{1.152}$$

then we can write inner products in matrix notation as $\langle g|f\rangle = g^\dagger f$ and as $\langle g|A|f\rangle = g^\dagger A f$.

One can compute the inner product $\langle g, f\rangle$ of two vectors f and g by doing the sum (1.150) of $g_j^* f_j$ over the index j only if one knows their components

f_j and g_j which are their inner products $f_j = \langle j|f\rangle$ and $g_j = \langle j|g\rangle$ with the orthonormal states $|j\rangle$ of some basis. Thus an inner product implies the existence of an orthonormal basis and a representation of the identity operator

$$I = \sum_{j=1}^{n} |j\rangle\langle j|. \tag{1.153}$$

If we switch to a different basis, say from $|k\rangle$'s to $|\alpha_k\rangle$'s, then the components of the column vectors change from $f_k = \langle k|f\rangle$ to $f'_k = \langle \alpha_k|f\rangle$, and similarly those of the row vectors $g^\dagger$ and of the matrix A change, but the bras, the kets, the linear operators, and the inner products $\langle g|f\rangle$ and $\langle g|A|f\rangle$ do not change because the identity operator is basis independent (1.143)

$$\begin{aligned} \langle g|f\rangle &= \sum_{k=1}^{n} \langle g|k\rangle\langle k|f\rangle = \sum_{k=1}^{n} \langle g|\alpha_k\rangle\langle \alpha_k|f\rangle \\ \langle g|A|f\rangle &= \sum_{k,\ell=1}^{n} \langle g|k\rangle\langle k|A|\ell\rangle\langle \ell|f\rangle = \sum_{k,\ell=1}^{n} \langle g|\alpha_k\rangle\langle \alpha_k|A|\alpha_\ell\rangle\langle \alpha_\ell|f\rangle. \end{aligned} \tag{1.154}$$

Dirac's outer products show how to change from one basis to another. The sum of outer products

$$U = \sum_{k=1}^{n} |\alpha_k\rangle\langle k| \tag{1.155}$$

maps the ket $|\ell\rangle$ of one orthonormal basis into that $|\alpha_\ell\rangle$ of another

$$U|\ell\rangle = \sum_{k=1}^{n} |\alpha_k\rangle\langle k|\ell\rangle = \sum_{k=1}^{n} |\alpha_k\rangle\,\delta_{k\ell} = |\alpha_\ell\rangle. \tag{1.156}$$

Example 1.27 (Simple change of basis) If the ket $|\alpha_k\rangle$ of the new basis is simply $|\alpha_k\rangle = |k+1\rangle$ with $|\alpha_n\rangle = |n+1\rangle \equiv |1\rangle$, then the operator that maps the n kets $|k\rangle$ into the kets $|\alpha_k\rangle$ is

$$U = \sum_{k=1}^{n} |\alpha_k\rangle\langle k| = \sum_{k=1}^{n} |k+1\rangle\langle k|. \tag{1.157}$$

The square U^2 of U also changes the basis; it sends $|k\rangle$ to $|k+2\rangle$. The set of operators U^ℓ for $\ell = 1, 2, \ldots, n$ forms a group known as Z_n.

To compute the inner product (U, V) of two vectors U and V, one needs the components U_i and V_i of these vectors in order to do the sum (1.89) of $U_i^* V_i$ over the index i.

1.13 Adjoints of Operators

In Dirac's notation, the most general linear operator (1.136) on an n-dimensional vector space is a sum of outer products $z\,|k\rangle\langle\ell|$ in which z is a complex number and the kets $|k\rangle$ and $|\ell\rangle$ are two of the n orthonormal kets that make up a basis for the space. The **adjoint** (1.135) of this basic linear operator is

$$(z\,|k\rangle\langle\ell|)^\dagger = z^*\,|\ell\rangle\langle k|. \tag{1.158}$$

Thus with $z = \langle k|A|\ell\rangle$, the most general linear operator on the space is

$$A = IAI = \sum_{k,\ell=1}^{n} |k\rangle\langle k|A|\ell\rangle\langle\ell| \tag{1.159}$$

and its adjoint $A^\dagger$ is the operator $I A^\dagger I$

$$A^\dagger = \sum_{k,\ell=1}^{n} |\ell\rangle\langle\ell|A^\dagger|k\rangle\langle k| = \sum_{k,\ell=1}^{n} |\ell\rangle\langle k|A|\ell\rangle^*\langle k|. \tag{1.160}$$

It follows that $\langle\ell|A^\dagger|k\rangle = \langle k|A|\ell\rangle^*$ so that the matrix $A^\dagger_{k\ell}$ that represents $A^\dagger$ in this basis is

$$A^\dagger_{\ell k} = \langle\ell|A^\dagger|k\rangle = \langle k|A|\ell\rangle^* = A^*_{\ell k} = A^{*\mathsf{T}}_{k\ell} \tag{1.161}$$

in agreement with our definition (1.28) of the adjoint of a matrix as the transpose of its complex conjugate, $A^\dagger = A^{*\mathsf{T}}$. We also have

$$\langle g|A^\dagger f\rangle = \langle g|A^\dagger|f\rangle = \langle f|A|g\rangle^* = \langle f|Ag\rangle^* = \langle Ag|f\rangle. \tag{1.162}$$

Taking the adjoint of the adjoint is by (1.158)

$$\left[(z\,|k\rangle\langle\ell|)^\dagger\right]^\dagger = \left[z^*\,|\ell\rangle\langle k|\right]^\dagger = z\,|k\rangle\langle\ell| \tag{1.163}$$

the same as doing nothing at all. This also follows from the matrix formula (1.161) because both $(A^*)^* = A$ and $(A^\mathsf{T})^\mathsf{T} = A$, and so

$$\left(A^\dagger\right)^\dagger = \left(A^{*\mathsf{T}}\right)^{*\,\mathsf{T}} = A \tag{1.164}$$

the adjoint of the adjoint of a matrix is the original matrix.

Before Dirac, the adjoint $A^\dagger$ of a linear operator A was defined by

$$(g, A^\dagger f) = (A\,g, f) = (f, A\,g)^*. \tag{1.165}$$

This definition also implies that $A^{\dagger\dagger} = A$ since

$$(g, A^{\dagger\dagger} f) = (A^\dagger g, f) = (f, A^\dagger g)^* = (Af, g)^* = (g, Af). \tag{1.166}$$

We also have $(g, Af) = (g, A^{\dagger\dagger} f) = (A^\dagger g, f)$.

1.14 Self-Adjoint or Hermitian Linear Operators

An operator A that is equal to its adjoint $A^\dagger = A$ is **self adjoint** or **hermitian**. In view of (1.161), the matrix elements of a self-adjoint linear operator A satisfy $\langle k|A^\dagger|\ell\rangle = \langle \ell|A|k\rangle^* = \langle k|A|\ell\rangle$ in any orthonormal basis. So a matrix that represents a hermitian operator is equal to the transpose of its complex conjugate

$$A_{k\ell} = \langle k|A|\ell\rangle = \langle k|A^\dagger|\ell\rangle = \langle \ell|A|k\rangle^* = A^{*\mathsf{T}}_{k\ell} = A^\dagger_{k\ell}. \tag{1.167}$$

We also have

$$\langle g|\, A\, |f\rangle = \langle A\, g|\, f\rangle = \langle f|A\, g\rangle^* = \langle f|\, A\, |g\rangle^* \tag{1.168}$$

and in pre-Dirac notation

$$(g, A\, f) = (A\, g, f) = (f, A\, g)^*. \tag{1.169}$$

A matrix A_{ij} that is **real and symmetric** or **imaginary and antisymmetric** is hermitian. But a self-adjoint linear operator A that is represented by a matrix A_{ij} that is real and symmetric (or imaginary and antisymmetric) in one orthonormal basis will not in general be represented by a matrix that is real and symmetric (or imaginary and antisymmetric) in a different orthonormal basis, but it will be represented by a hermitian matrix in every orthonormal basis.

A ket $|a'\rangle$ is an **eigenvector** of a linear operator A with **eigenvalue** a' if $A|a'\rangle = a'|a'\rangle$. As we'll see in Section 1.29, hermitian matrices have real eigenvalues and complete sets of orthonormal eigenvectors. Hermitian operators and matrices represent physical variables in quantum mechanics.

Example 1.28 (Fierz identities for $n \times n$ hermitian matrices) The n^2 $n \times n$ hermitian matrices t^a form a vector space with an inner product $\langle a|b\rangle$ (Section 1.6) defined by the trace (1.22) $\langle a|b\rangle = \mathrm{Tr}(t^a\, t^b)$. One can use the Gram–Schmidt method (Section 1.10) to make them orthonormal, so that

$$\langle a|b\rangle = \mathrm{Tr}(t^a\, t^b) = \sum_{i,k=1}^{n} t^a_{ik}\, t^b_{ki} = \delta_{ab}. \tag{1.170}$$

Then the sum of their n^2 outer products (1.22) is the identity matrix of the n^2-dimensional vector space

$$\left(\sum_{a=1}^{n^2} |a\rangle\langle a|\right)_{ij,k\ell} = \sum_{a=1}^{n^2} t^a_{ij}\, t^a_{k\ell} = I_{ik,\ell j} = \delta_{i\ell}\, \delta_{kj} \tag{1.171}$$

because

$$t_{ij}^b = (|b\rangle)_{ij} = \sum_{a=1}^{n^2} (|a\rangle)_{ij} \langle a|b\rangle = \sum_{a=1}^{n^2} t_{ij}^a \mathrm{Tr}(t^a\, t^b) = \sum_{a=1}^{n^2} \sum_{k,\ell=1}^{n} t_{ij}^a t_{k\ell}^a\, t_{\ell k}^b. \tag{1.172}$$

(Markus Fierz, 1912–2006)

1.15 Real, Symmetric Linear Operators

In quantum mechanics, we usually consider complex vector spaces, that is, spaces in which the vectors $|f\rangle$ are complex linear combinations

$$|f\rangle = \sum_{k=1}^{n} z_k\, |k\rangle \tag{1.173}$$

of complex orthonormal basis vectors $|i\rangle$.

But real vector spaces also are of interest. A real vector space is a vector space in which the vectors $|f\rangle$ are real linear combinations

$$|f\rangle = \sum_{k=1}^{n} x_k\, |k\rangle \tag{1.174}$$

of real orthonormal basis vectors, $x_k^* = x_k$ and $|k\rangle^* = |k\rangle$.

A real linear operator A on a real vector space

$$A = \sum_{k,\ell=1}^{n} |k\rangle\langle k|A|\ell\rangle\langle\ell| = \sum_{k,\ell=1}^{n} |k\rangle A_{k\ell}\langle\ell| \tag{1.175}$$

is represented by a real matrix $A_{k\ell}^* = A_{k\ell}$. A real linear operator A that is self adjoint on a real vector space satisfies the condition (1.169) of hermiticity but with the understanding that complex conjugation has no effect

$$(g, A\, f) = (A\, g, f) = (f, A\, g)^* = (f, A\, g). \tag{1.176}$$

Thus its matrix elements are symmetric, $\langle g|A|f\rangle = \langle f|A|g\rangle$. Since A is hermitian as well as real, the matrix $A_{k\ell}$ that represents it (in a real basis) is real and hermitian, and so is symmetric $A_{k\ell} = A_{\ell k}^* = A_{\ell k}$.

1.16 Unitary Operators

A **unitary operator** U is one whose adjoint is its inverse

$$U\, U^\dagger = U^\dagger\, U = I. \tag{1.177}$$

Any operator that maps one orthonormal basis $|k\rangle$ to another $|\alpha_k\rangle$

$$U = \sum_{k=1}^{n} |\alpha_k\rangle\langle k| \tag{1.178}$$

is unitary since

$$\begin{aligned} UU^\dagger &= \sum_{k=1}^{n} |\alpha_k\rangle\langle k| \sum_{\ell=1}^{n} |\ell\rangle\langle\alpha_\ell| = \sum_{k,\ell=1}^{n} |\alpha_k\rangle\langle k|\ell\rangle\langle\alpha_\ell| \\ &= \sum_{k,\ell=1}^{n} |\alpha_k\rangle\delta_{k,\ell}\langle\alpha_\ell| = \sum_{k=1}^{n} |\alpha_k\rangle\langle\alpha_k| = I \end{aligned} \tag{1.179}$$

as well as

$$U^\dagger U = \sum_{\ell=1}^{n} |\ell\rangle\langle\alpha_\ell| \sum_{k=1}^{n} |\alpha_k\rangle\langle k| = \sum_{k=1}^{n} |k\rangle\langle k| = I. \tag{1.180}$$

A unitary operator maps every orthonormal basis $|k\rangle$ into another orthonormal basis $|\alpha_k\rangle$. For if $|\alpha_k\rangle = U|k\rangle$, then the vectors $|\alpha_k\rangle$ are orthonormal $\langle\alpha_k|\alpha_\ell\rangle = \delta_{k,\ell}$ (Exercise 1.22). They also are complete because they provide a resolution of the identity operator

$$\sum_{k=1}^{n} |\alpha_k\rangle\langle\alpha_k| = \sum_{k=1}^{n} U|k\rangle\langle k|U^\dagger = U\,I\,U^\dagger = U\,U^\dagger = I. \tag{1.181}$$

If we multiply the relation $|\alpha_k\rangle = U|k\rangle$ by the bra $\langle k|$ and then sum over the index k, we get

$$\sum_{k=1}^{n} |\alpha_k\rangle\langle k| = \sum_{k=1}^{n} U|k\rangle\langle k| = U\sum_{k=1}^{n} |k\rangle\langle k| = U. \tag{1.182}$$

Every unitary operator maps every orthonormal basis into another orthonormal basis or into itself.

Inner products do not change under unitary transformations because $\langle g|f\rangle = \langle g|U^\dagger\,U|f\rangle = \langle Ug|U|f\rangle = \langle Ug|Uf\rangle$ which in pre-Dirac notation is $(g, f) = (g, U^\dagger\,Uf) = (Ug, Uf)$.

Unitary matrices have unimodular determinants, $\det U = 1$, because the determinant of the product of two matrices is the product of their determinants (1.222) and because transposition doesn't change the value of a determinant (1.205)

$$1 = \det I = \det(UU^\dagger) = \det U \det U^\dagger = \det U(\det(U^{\mathsf{T}}))^* = |\det U|^2. \tag{1.183}$$

A unitary matrix that is real is **orthogonal** and satsfies

$$OO^{\mathsf{T}} = O^{\mathsf{T}}O = I. \tag{1.184}$$

1.17 Hilbert Spaces

We have mainly been talking about linear operators that act on finite-dimensional vector spaces and that can be represented by matrices. But infinite-dimensional vector spaces and the linear operators that act on them play central roles in electrodynamics and quantum mechanics. For instance, the Hilbert space $\mathcal{H}$ of all "wave" functions $\psi(\boldsymbol{x}, t)$ that are square integrable over 3-dimensional space at all times t is of infinite dimension.

In one space dimension, the state $|x'\rangle$ represents a particle at position x' and is an eigenstate of the hermitian position operator x with eigenvalue x', that is, $x|x'\rangle = x'|x'\rangle$. These states form a basis that is orthogonal in the sense that $\langle x|x'\rangle = 0$ for $x \neq x'$ and normalized in the sense that $\langle x|x'\rangle = \delta(x - x')$ in which $\delta(x - x')$ is Dirac's delta function. The delta function $\delta(x - x')$ actually is a **functional** $\delta_{x'}$ that maps any suitably smooth function f into its value at x'

$$\delta_{x'}[f] = \int \delta(x - x')\, f(x)\, dx = f(x'). \tag{1.185}$$

Another basis for the Hilbert space of 1-dimensional quantum mechanics is made of the states $|p\rangle$ of well-defined momentum. The state $|p'\rangle$ represents a particle or system with momentum p'. It is an eigenstate of the hermitian momentum operator p with eigenvalue p', that is, $p|p'\rangle = p'|p'\rangle$. The momentum states also are orthonormal in Dirac's sense, $\langle p|p'\rangle = \delta(p - p')$.

The operator that translates a system in space by a distance a is

$$U(a) = \int |x + a\rangle\langle x|\, dx. \tag{1.186}$$

It maps the state $|x'\rangle$ to the state $|x' + a\rangle$ and is unitary (Exercise 1.23). Remarkably, this translation operator is an exponential of the momentum operator $U(a) = \exp(-i\, p\, a/\hbar)$ in which $\hbar = h/2\pi = 1.054 \times 10^{-34}$ Js is Planck's constant divided by 2π.

In 2 dimensions, with basis states $|x, y\rangle$ that are orthonormal in Dirac's sense, $\langle x, y|x', y'\rangle = \delta(x - x')\delta(y - y')$, the unitary operator

$$U(\theta) = \int |x \cos\theta - y \sin\theta, x \sin\theta + y \cos\theta\rangle\langle x, y|\, dx dy \tag{1.187}$$

rotates a system in space by the angle θ. This rotation operator is the exponential $U(\theta) = \exp(-i\,\theta\, L_z/\hbar)$ in which the z component of the angular momentum is $L_z = x\, p_y - y\, p_x$.

We may carry most of our intuition about matrices over to these unitary transformations that change from one infinite basis to another. But we must use common sense and keep in mind that infinite sums and integrals do not always converge.

1.18 Antiunitary, Antilinear Operators

Certain maps on states $|\psi\rangle \to |\psi'\rangle$, such as those involving time reversal, are implemented by operators K that are **antilinear**

$$K(z\psi + w\phi) = K(z|\psi\rangle + w|\phi\rangle) = z^* K|\psi\rangle + w^* K|\phi\rangle = z^* K\psi + w^* K\phi \tag{1.188}$$

and **antiunitary**

$$(K\phi, K\psi) = \langle K\phi|K\psi\rangle = (\phi, \psi)^* = \langle\phi|\psi\rangle^* = \langle\psi|\phi\rangle = (\psi, \phi). \tag{1.189}$$

The adjoint $K^\dagger$ of an antiunitary operator K is defined by $\langle K^\dagger\phi|\psi\rangle = \langle\phi|K|\psi\rangle^*$ so that $\langle K^\dagger K\phi|\psi\rangle = \langle K\phi|K\psi\rangle^* = \langle\phi|\psi\rangle^{**} = \langle\phi|\psi\rangle$.

1.19 Symmetry in Quantum Mechanics

In quantum mechanics, a symmetry is a map of states $|\psi\rangle \to |\psi'\rangle$ and $|\phi\rangle \to |\phi'\rangle$ that preserves probabilities

$$|\langle\phi'|\psi'\rangle|^2 = |\langle\phi|\psi\rangle|^2. \tag{1.190}$$

Eugene Wigner (1902–1995) showed that every symmetry in quantum mechanics can be represented either by an operator U that is linear and unitary or by an operator K that is antilinear and antiunitary. The antilinear, antiunitary case occurs when a symmetry involves time reversal. Most symmetries are represented by operators that are linear and unitary. Unitary operators are of great importance in quantum mechanics. We use them to represent rotations, translations, Lorentz transformations, and internal-symmetry transformations.

1.20 Determinants

The **determinant** of a 2×2 matrix A is

$$\det A = |A| = A_{11}A_{22} - A_{21}A_{12}. \tag{1.191}$$

In terms of the 2×2 antisymmetric ($e_{ij} = -e_{ji}$) matrix $e_{12} = 1 = -e_{21}$ with $e_{11} = e_{22} = 0$, this determinant is

$$\det A = \sum_{i=1}^{2}\sum_{j=1}^{2} e_{ij} A_{i1} A_{j2} = \sum_{i=1}^{2}\sum_{j=1}^{2} e_{ij} A_{1i} A_{2j}. \tag{1.192}$$

It's also true that

$$e_{k\ell} \det A = \sum_{i=1}^{2}\sum_{j=1}^{2} e_{ij} A_{ik} A_{j\ell}. \tag{1.193}$$

Example 1.29 (Area of a parallelogram) Two 2-vectors $V = (V_1, V_2)$ and $W = (W_1, W_2)$ define a parallelogram whose area is the absolute value of a 2×2 determinant

$$\text{area}(V, W) = \left| \det \begin{pmatrix} V_1 & V_2 \\ W_1 & W_2 \end{pmatrix} \right| = |\, V_1 W_2 - V_2 W_1 \,| . \tag{1.194}$$

To check this formula, rotate the coordinates so that the 2-vector V runs from the origin along the x-axis. Then $V_2 = 0$, and the determinant is $V_1 W_2$ which is the base V_1 of the parallelogram times its height W_2.

These definitions (1.191–1.193) extend to any square matrix. If A is a 3×3 matrix, then its determinant is

$$\det A = \sum_{i,j,k=1}^{3} e_{ijk} A_{i1} A_{j2} A_{k3} = \sum_{i,j,k=1}^{3} e_{ijk} A_{1i} A_{2j} A_{3k} \tag{1.195}$$

in which e_{ijk} is the totally antisymmetric Levi-Civita symbol whose nonzero values are

$$e_{123} = e_{231} = e_{312} = 1, \quad \text{and} \quad e_{213} = e_{132} = e_{321} = -1. \tag{1.196}$$

The symbol vanishes whenever an index appears twice, thus

$$e_{111} = e_{112} = e_{113} = e_{222} = e_{221} = e_{223} = e_{333} = e_{331} = e_{332} = 0 \tag{1.197}$$

and so forth. The sums over i, j, and k run from 1 to 3

$$\begin{aligned} \det A &= \sum_{i=1}^{3} A_{i1} \sum_{j,k=1}^{3} e_{ijk} A_{j2} A_{k3} \\ &= A_{11} (A_{22}A_{33} - A_{32}A_{23}) + A_{21} (A_{32}A_{13} - A_{12}A_{33}) \\ &\quad + A_{31} (A_{12}A_{23} - A_{22}A_{13}) . \end{aligned} \tag{1.198}$$

The **minor** $M_{i\ell}$ of the matrix A is the 2×2 determinant of the matrix A without row i and column ℓ, and the **cofactor** $C_{i\ell}$ is the minor $M_{i\ell}$ multiplied by $(-1)^{i+\ell}$. Thus $\det A$ is the sum

$$\begin{aligned} \det A &= A_{11}(-1)^2 (A_{22}A_{33} - A_{32}A_{23}) + A_{21}(-1)^3 (A_{12}A_{33} - A_{32}A_{13}) \\ &\quad + A_{31}(-1)^4 (A_{12}A_{23} - A_{22}A_{13}) \\ &= A_{11}C_{11} + A_{21}C_{21} + A_{31}C_{31} \end{aligned} \tag{1.199}$$

of the products $A_{i1}C_{i1} = A_{i1}(-1)^{i+1}M_{i1}$ where

$$
\begin{aligned}
C_{11} &= (-1)^2 M_{11} = A_{22}A_{33} - A_{23}A_{32} \\
C_{21} &= (-1)^3 M_{21} = A_{32}A_{13} - A_{12}A_{33} \\
C_{31} &= (-1)^4 M_{31} = A_{12}A_{23} - A_{22}A_{13}.
\end{aligned} \tag{1.200}
$$

Example 1.30 (Volume of a parallelepiped) The determinant of a 3×3 matrix is the dot product of the vector of its first row with the cross-product of the vectors of its second and third rows

$$
\begin{vmatrix} U_1 & U_2 & U_3 \\ V_1 & V_2 & V_3 \\ W_1 & W_2 & W_3 \end{vmatrix} = \sum_{ijk=1}^{3} e_{ijk}\, U_i V_j W_k = \sum_{i=1}^{3} U_i \,(\boldsymbol{V} \times \boldsymbol{W})_i = \boldsymbol{U} \cdot (\boldsymbol{V} \times \boldsymbol{W}). \tag{1.201}
$$

The absolute value of this **scalar triple product** is the volume of the parallelepiped defined by U, V, and W as one can see by placing the parallelepiped so the vector U runs from the origin along the x-axis. The 3×3 determinant (1.201) then is $U_1(V_2 W_3 - V_3 W_2)$ which is the height of the parallelepiped times the area (1.194) of its base.

Laplace used the totally antisymmetric symbol $e_{i_1 i_2 \ldots i_n}$ with n indices and with $e_{123\ldots n} = 1$ to define the determinant of an $n \times n$ matrix A as

$$
\det A = \sum_{i_1 i_2 \ldots i_n = 1}^{n} e_{i_1 i_2 \ldots i_n} A_{i_1 1} A_{i_2 2} \ldots A_{i_n n} \tag{1.202}
$$

in which the sums over $i_1 \ldots i_n$ run from 1 to n. In terms of cofactors, two forms of his expansion of this determinant are

$$
\det A = \sum_{i=1}^{n} A_{ik} C_{ik} = \sum_{k=1}^{n} A_{ik} C_{ik} \tag{1.203}
$$

in which the first sum is over the row index i but not the (arbitrary) column index k, and the second sum is over the column index k but not the (arbitrary) row index i. The cofactor C_{ik} is $(-1)^{i+k} M_{ik}$ in which the minor M_{ik} is the determinant of the $(n-1) \times (n-1)$ matrix A without its ith row and kth column. It's also true that

$$
\begin{aligned}
e_{k_1 k_2 \ldots k_n} \det A &= \sum_{i_1 i_2 \ldots i_n = 1}^{n} e_{i_1 i_2 \ldots i_n} A_{i_1 k_1} A_{i_2 k_2} \ldots A_{i_n k_n} \\
&= \sum_{i_1 i_2 \ldots i_n = 1}^{n} e_{i_1 i_2 \ldots i_n} A_{k_1 i_1} A_{k_2 i_2} \ldots A_{k_n i_n}.
\end{aligned} \tag{1.204}
$$

In particular, since $e_{12\ldots n} = 1$, the determinant of the transpose of a matrix is equal to the determinant (1.202) of the matrix

$$\det A^{\mathsf{T}} = \sum_{i_1 i_2 \ldots i_n = 1}^{n} e_{i_1 i_2 \ldots i_n} A_{1 i_1} A_{2 i_2} \ldots A_{n i_n} = \det A. \tag{1.205}$$

The interchange $A \to A^{\mathsf{T}}$ of the rows and columns of a matrix has no effect on its determinant.

The key feature of a determinant is that it is an *antisymmetric* combination of products of the elements A_{ik} of a matrix A. One implication of this antisymmetry is that the interchange of any two rows or any two columns changes the sign of the determinant. Another is that if one adds a multiple of one column to another column, for example a multiple $x A_{i2}$ of column 2 to column 1, then the determinant

$$\det A' = \sum_{i_1 i_2 \ldots i_n = 1}^{n} e_{i_1 i_2 \ldots i_n} \left(A_{i_1 1} + x A_{i_1 2} \right) A_{i_2 2} \ldots A_{i_n n} \tag{1.206}$$

is unchanged. The reason is that the extra term $\delta \det A$ vanishes

$$\delta \det A = \sum_{i_1 i_2 \ldots i_n = 1}^{n} x \, e_{i_1 i_2 \ldots i_n} \, A_{i_1 2} A_{i_2 2} \ldots A_{i_n n} = 0 \tag{1.207}$$

because it is proportional to a sum of products of a factor $e_{i_1 i_2 \ldots i_n}$ that is antisymmetric in i_1 and i_2 and a factor $A_{i_1 2} A_{i_2 2}$ that is symmetric in these indices. For instance, when i_1 and i_2 are 5 and 7 and 7 and 5, the two terms cancel

$$e_{57 \ldots i_n} A_{52} A_{72} \ldots A_{i_n n} + e_{75 \ldots i_n} A_{72} A_{52} \ldots A_{i_n n} = 0 \tag{1.208}$$

because $e_{57 \ldots i_n} = -e_{75 \ldots i_n}$.

By repeated additions of $x_2 A_{i2}$, $x_3 A_{i3}$, and so forth to A_{i1}, we can change the first column of the matrix A to a linear combination of all the columns

$$A_{i1} \longrightarrow A_{i1} + \sum_{k=2}^{n} x_k A_{ik} \tag{1.209}$$

without changing $\det A$. In this linear combination, the coefficients x_k are arbitrary. The analogous operation with arbitrary y_k

$$A_{i\ell} \longrightarrow A_{i\ell} + \sum_{k=1, k \neq \ell}^{n} y_k A_{ik} \tag{1.210}$$

replaces the ℓth column by a linear combination of all the columns without changing $\det A$.

Suppose that the columns of an $n \times n$ matrix A are linearly dependent (Section 1.8), so that the linear combination of columns

$$\sum_{k=1}^{n} y_k A_{ik} = 0 \quad \text{for } i = 1, \ldots, n \tag{1.211}$$

vanishes for some coefficients y_k not all zero. Suppose $y_1 \neq 0$. Then by adding suitable linear combinations of columns 2 through n to column 1, we could make all the modified elements A'_{i1} of column 1 vanish without changing $\det A$. But then $\det A$ as given by (1.202) would vanish. **Thus the determinant of any matrix whose columns are linearly dependent must vanish.**

Now suppose that the columns of an $n \times n$ matrix are linearly independent. Then the determinant of the matrix cannot vanish because any linearly independent set of n vectors in a vector space of n dimensions is complete (Section 1.8). Thus if the columns of a matrix A are linearly independent and therefore complete, some linear combination of all columns 2 through n when added to column 1 will convert column 1 into a nonzero multiple of the n-dimensional column vector $(1, 0, 0, \ldots, 0)$, say $(c_1, 0, 0, \ldots, 0)$. Similar operations will convert column 2 into a nonzero multiple of the column vector $(0, 1, 0, \ldots, 0)$, say $(0, c_2, 0, \ldots, 0)$. Continuing in this way, we may convert the matrix A to a matrix with nonzero entries c_i along the main diagonal and zeros everywhere else. The determinant $\det A$ then is the product $c_1 c_2 \ldots c_n$ of the nonzero diagonal entries c_i's, and so $\det A$ cannot vanish.

We may extend these arguments to the rows of a matrix. The addition to row k of a linear combination of the other rows

$$A_{ki} \longrightarrow A_{ki} + \sum_{\ell=1,\ell\neq k}^{n} z_\ell A_{\ell i} \tag{1.212}$$

does not change the value of the determinant. In this way, one may show that the determinant of a matrix vanishes if and only if its rows are linearly dependent. The reason why these results apply to the rows as well as to the columns is that the determinant of a matrix A may be defined either in terms of the columns or in terms of the rows as in the definitions (1.202 and 1.204). These and other properties of determinants follow from a study of **permutations** (Section 11.13). Detailed proofs are in (Aitken, 1959).

Let us return for a moment to Laplace's expansion (1.203) of the determinant $\det A$ of an $n \times n$ matrix A as a sum of $A_{ik} C_{ik}$ over the row index i with the column index k held fixed

$$\det A = \sum_{i=1}^{n} A_{ik} C_{ik} = \sum_{i=1}^{n} A_{ki} C_{ki} \tag{1.213}$$

in order to prove that

$$\delta_{k\ell} \det A = \sum_{i=1}^{n} A_{ik} C_{i\ell} = \sum_{i=1}^{n} A_{ki} C_{\ell i}. \tag{1.214}$$

For $k = \ell$, this formula just repeats Laplace's expansion (1.213). But for $k \neq \ell$, it is Laplace's expansion for the determinant of a matrix that has two copies of its kth column. Since the determinant of a matrix with two identical columns vanishes, the rule (1.214) also is true for $k \neq \ell$.

The rule (1.214) provides a formula for the inverse of a matrix A whose determinant does not vanish. Such matrices are said to be **nonsingular**. The inverse A^{-1} of an $n \times n$ nonsingular matrix A is the transpose of the matrix of cofactors divided by the determinant of the matrix

$$\left(A^{-1}\right)_{\ell i} = \frac{C_{i\ell}}{\det A} \quad \text{or} \quad A^{-1} = \frac{C^{\mathsf{T}}}{\det A}. \tag{1.215}$$

To verify this formula, we use it for A^{-1} in the product $A^{-1}A$ and note that by (1.214) the ℓkth entry of the product $A^{-1}A$ is just $\delta_{\ell k}$

$$\left(A^{-1}A\right)_{\ell k} = \sum_{i=1}^{n} \left(A^{-1}\right)_{\ell i} A_{ik} = \sum_{i=1}^{n} \frac{C_{i\ell}}{\det A} A_{ik} = \delta_{\ell k}. \tag{1.216}$$

Example 1.31 (Inverting a 2×2 matrix) Our formula (1.215) for the inverse of the general 2×2 matrix

$$A = \begin{pmatrix} a & b \\ c & d \end{pmatrix} \tag{1.217}$$

gives

$$A^{-1} = \frac{1}{ad - bc} \begin{pmatrix} d & -b \\ -c & a \end{pmatrix} \tag{1.218}$$

which is the correct inverse as long as $ad \neq bc$.

The simple example of matrix multiplication

$$\begin{pmatrix} a & b & c \\ d & e & f \\ g & h & i \end{pmatrix} \begin{pmatrix} 1 & x & y \\ 0 & 1 & z \\ 0 & 0 & 1 \end{pmatrix} = \begin{pmatrix} a & xa + b & ya + zb + c \\ d & xd + e & yd + ze + f \\ g & xg + h & yg + zh + i \end{pmatrix} \tag{1.219}$$

shows that the operations (1.210) on columns that don't change the value of the determinant can be written as matrix multiplication from the right by a matrix that has unity on its main diagonal and zeros below. Now consider the matrix product

$$\begin{pmatrix} A & 0 \\ -I & B \end{pmatrix} \begin{pmatrix} I & B \\ 0 & I \end{pmatrix} = \begin{pmatrix} A & AB \\ -I & 0 \end{pmatrix} \tag{1.220}$$

in which A and B are $n \times n$ matrices, I is the $n \times n$ identity matrix, and 0 is the $n \times n$ matrix of all zeros. The second matrix on the left-hand side has unity on its main diagonal and zeros below, and so it does not change the value of the determinant of the matrix to its left, which then must equal that of the matrix on the right-hand side:

$$\det\begin{pmatrix} A & 0 \\ -I & B \end{pmatrix} = \det\begin{pmatrix} A & AB \\ -I & 0 \end{pmatrix}. \tag{1.221}$$

By using Laplace's expansion (1.203) along the first column to evaluate the determinant on the left-hand side and his expansion along the last row to compute the determinant on the right-hand side, one finds that **the determinant of the product of two matrices is the product of the determinants**

$$\det A \, \det B = \det AB. \tag{1.222}$$

Example 1.32 (Two 2×2 matrices) When the matrices A and B are both 2×2, the two sides of (1.221) are

$$\det\begin{pmatrix} A & 0 \\ -I & B \end{pmatrix} = \det\begin{pmatrix} a_{11} & a_{12} & 0 & 0 \\ a_{21} & a_{22} & 0 & 0 \\ -1 & 0 & b_{11} & b_{12} \\ 0 & -1 & b_{21} & b_{22} \end{pmatrix} \tag{1.223}$$
$$= a_{11}a_{22} \det B - a_{21}a_{12} \det B = \det A \det B$$

and

$$\det\begin{pmatrix} A & AB \\ -I & 0 \end{pmatrix} = \det\begin{pmatrix} a_{11} & a_{12} & (ab)_{11} & (ab)_{12} \\ a_{21} & a_{22} & (ab)_{21} & (ab)_{22} \\ -1 & 0 & 0 & 0 \\ 0 & -1 & 0 & 0 \end{pmatrix} \tag{1.224}$$
$$= (-1)C_{42} = (-1)(-1) \det AB = \det AB$$

and so they give the product rule $\det A \det B = \det AB$.

Often one uses the notation $|A| = \det A$ to denote a determinant. In this more compact notation, the obvious generalization of the product rule is

$$|ABC \ldots Z| = |A||B| \ldots |Z|. \tag{1.225}$$

The product rule (1.222) implies that $\det\left(A^{-1}\right)$ is $1/\det A$ since

$$1 = \det I = \det\left(AA^{-1}\right) = \det A \det\left(A^{-1}\right). \tag{1.226}$$

Example 1.33 (Derivative of the logarithm of a determinant) We see from our formula (1.213) for $\det A$ that its derivative with respect to any given element A_{ik} is the corresponding cofactor C_{ik}

$$\frac{\partial \det A}{\partial A_{ik}} = C_{ik} \tag{1.227}$$

because the cofactors C_{ij} and C_{jk} for all j are independent of A_{ik}. Thus the derivative of the logarithm of this determinant with respect to any parameter β is

$$\begin{aligned} \frac{\partial \ln \det A}{\partial \beta} &= \frac{1}{\det A} \sum_{ik} \frac{\partial \det A}{\partial A_{ik}} \frac{\partial A_{ik}}{\partial \beta} = \sum_{ik} \frac{C_{ik}}{\det A} \frac{\partial A_{ik}}{\partial \beta} \\ &= \sum_{ik} A^{-1}_{ki} \frac{\partial A_{ik}}{\partial \beta} = \mathrm{Tr}\left(A^{-1} \frac{\partial A}{\partial \beta} \right). \end{aligned} \tag{1.228}$$

Example 1.34 (Numerical tricks) Adding multiples of rows to other rows does not change the value of a determinant, and interchanging two rows only changes a determinant by a minus sign. So we can use these operations, which leave the absolute values of determinants invariant, to make a matrix **upper triangular**, a form in which its determinant is just the product of the factors on its diagonal. Thus to make the matrix

$$A = \begin{pmatrix} 1 & 2 & 1 \\ -2 & -6 & 3 \\ 4 & 2 & -5 \end{pmatrix} \tag{1.229}$$

upper triangular, we add twice the first row to the second row

$$\begin{pmatrix} 1 & 2 & 1 \\ 0 & -2 & 5 \\ 4 & 2 & -5 \end{pmatrix}$$

and then subtract four times the first row from the third

$$\begin{pmatrix} 1 & 2 & 1 \\ 0 & -2 & 5 \\ 0 & -6 & -9 \end{pmatrix}. \tag{1.230}$$

Next, we subtract three times the second row from the third

$$\begin{pmatrix} 1 & 2 & 1 \\ 0 & -2 & 5 \\ 0 & 0 & -24 \end{pmatrix}.$$

We now find as the determinant of A the product of its diagonal elements:

$$|A| = 1(-2)(-24) = 48. \tag{1.231}$$

Incidentally, Gauss, Jordan, and modern mathematicians have developed much faster ways of computing determinants and matrix inverses than those (1.203 & 1.215) due to Laplace. Sage, Octave, Matlab, Maple, Mathematica, and Python use these modern techniques, which are freely available as programs in C and FORTRAN from www.netlib.org/lapack.

Example 1.35 (Using Matlab) The Matlab command to make the matrix (1.229) is A = [1 2 1; -2 -6 3; 4 2 -5] . The command d = det(A) gives its determinant, d = 48, and Ainv = A$^{(-1)}$ gives its inverse

```
Ainv =     0.5000     0.2500     0.2500
           0.0417    -0.1875    -0.1042
           0.4167     0.1250    -0.0417 .
```

The **permanent** of a square $n \times n$ matrix A_{ik} is the sum over all permutations $1, 2, \ldots, n \to s_1, s_2, \ldots, s_n$ of the products $A_{s_1 1} A_{s_2 2} \cdots A_{s_n n}$

$$\mathrm{perm}(A) = \sum_s A_{s_1 1} A_{s_2 2} \cdots A_{s_n n}. \tag{1.232}$$

1.21 Jacobians

When one changes variables in a multiple integral from coordinates x_1, x_2 and area element $dx_1 dx_2$, one must find the new element of area in terms of the new variables y_1, y_2. If $\hat{\boldsymbol{x}}_1$ and $\hat{\boldsymbol{x}}_2$ are unit vectors in the x_1 and x_2 directions, then as the new coordinates (y_1, y_2) change by dy_1 and dy_2, the point they represent moves by

$$\boldsymbol{dy}^1 = \left(\frac{\partial x_1}{\partial y_1} \hat{\boldsymbol{x}}_1 + \frac{\partial x_2}{\partial y_1} \hat{\boldsymbol{x}}_2 \right) dy_1 \quad \text{and by} \quad \boldsymbol{dy}^2 = \left(\frac{\partial x_1}{\partial y_2} \hat{\boldsymbol{x}}_1 + \frac{\partial x_2}{\partial y_2} \hat{\boldsymbol{x}}_2 \right) dy_2. \tag{1.233}$$

These vectors, $\boldsymbol{dy}^1$ and $\boldsymbol{dy}^2$ define a parallelogram whose area (1.194) is the absolute value of a determinant

$$\mathrm{area}(\boldsymbol{dy}^1, \boldsymbol{dy}^2) = \left| \det \begin{pmatrix} \dfrac{\partial x_1}{\partial y_1} & \dfrac{\partial x_2}{\partial y_1} \\ \dfrac{\partial x_1}{\partial y_2} & \dfrac{\partial x_2}{\partial y_2} \end{pmatrix} \right| dy_1 \, dy_2. \tag{1.234}$$

The determinant itself is a **jacobian**

$$J = J(x/y) = \frac{\partial(x_1, x_2)}{\partial(y_1, y_2)} = \det \begin{pmatrix} \dfrac{\partial x_1}{\partial y_1} & \dfrac{\partial x_2}{\partial y_1} \\ \dfrac{\partial x_1}{\partial y_2} & \dfrac{\partial x_2}{\partial y_2} \end{pmatrix}. \tag{1.235}$$

The two equal integrals are

$$\iint_{R_x} f(x_1, x_2)\, dx_1 dx_2 = \iint_{R_y} f \circ x(y_1, y_2)) \left| \frac{\partial(x_1, x_2)}{\partial(y_1, y_2)} \right| dy_1 dy_2 \tag{1.236}$$

in which $f \circ x(y_1, y_2) = f(x_1(y_1, y_2), x_2(y_1, y_2))$ and R_x and R_y are the same region in the two coordinate systems.

In 3 dimensions, with $j = 1, 2$, and 3, the 3 vectors

$$\boldsymbol{dy}^j = \left(\frac{\partial x_1}{\partial y_j} \hat{\boldsymbol{x}}_1 + \frac{\partial x_2}{\partial y_j} \hat{\boldsymbol{x}}_2 + \frac{\partial x_3}{\partial y_j} \hat{\boldsymbol{x}}_3 \right) dy_j \tag{1.237}$$

define a parallelepiped whose volume (1.201) is the absolute value of the determinant

$$\text{volume}(\boldsymbol{dy}^1, \boldsymbol{dy}^2, \boldsymbol{dy}^3) = \left| \det \begin{pmatrix} \dfrac{\partial x_1}{\partial y_1} & \dfrac{\partial x_2}{\partial y_1} & \dfrac{\partial x_3}{\partial y_1} \\ \dfrac{\partial x_1}{\partial y_2} & \dfrac{\partial x_2}{\partial y_2} & \dfrac{\partial x_3}{\partial y_2} \\ \dfrac{\partial x_1}{\partial y_3} & \dfrac{\partial x_2}{\partial y_3} & \dfrac{\partial x_3}{\partial y_2} \end{pmatrix} \right| dy_1\, dy_2\, dy_3. \tag{1.238}$$

The equal integrals are

$$\iiint_{R_x} f(\vec{x})\, d^3x = \iiint_{R_y} f \circ x(\vec{y})) \left| \frac{\partial(x_1, x_2, x_3)}{\partial(y_1, y_2, y_3)} \right| d^3y \tag{1.239}$$

in which $d^3x = dx_1 dx_2 dx_3$, $d^3y = dy_1 dy_2 dy_3$, $f \circ x(\vec{y}) = f(x_1(\vec{y}), x_2(\vec{y}), x_3(\vec{y}))$, and R_x and R_y are the same region in the two coordinate systems.

For n-dimensional integrals over $x = (x_1, \ldots, x_n)$ and $y = (y_1, \ldots, y_n)$, the rule is similar

$$\int_{R_x} f(x)\, d^n x = \int_{R_y} f \circ x(y) \left| \frac{\partial(x_1, \ldots, x_n)}{\partial(y_1, \ldots, y_n)} \right| d^n y \tag{1.240}$$

and uses the absolute value of the n-dimensional jacobian

$$J = J(x/y) = \frac{\partial(x_1, \ldots, x_n)}{\partial(y_1, \ldots, y_n)} = \det \begin{pmatrix} \frac{\partial x_1}{\partial y_1} & \cdots & \frac{\partial x_n}{\partial y_1} \\ \vdots & \ddots & \vdots \\ \frac{\partial x_1}{\partial y_n} & \cdots & \frac{\partial x_n}{\partial y_n} \end{pmatrix}. \tag{1.241}$$

Since the determinant of the transpose of a matrix is the same (1.205) as the determinant of the matrix, some people write jacobians with their rows and columns interchanged.

1.22 Systems of Linear Equations

Suppose we wish to solve the system of n linear equations

$$\sum_{k=1}^{n} A_{ik} x_k = y_i \tag{1.242}$$

for n unknowns x_k. In matrix notation, with A an $n \times n$ matrix and x and y n-vectors, this system of equations is $A\,x = y$. If the matrix A is **nonsingular**, that is, if $\det(A) \neq 0$, then it has an inverse A^{-1} given by (1.215), and we may multiply both sides of $A\,x = y$ by A^{-1} and so find $x = A^{-1}\,y$. When A is nonsingular, this is the unique solution to (1.242).

When A is singular, its determinant vanishes, $\det(A) = 0$, and so its columns are linearly dependent (section 1.20). In this case, the linear dependence of the columns of A implies that $A\,z = 0$ for some nonzero vector z. Thus if x satisfies $A\,x = y$, then so does $x + cz$ for any constant c because $A(x + cz) = A\,x + c\,A\,z = y$. So if $\det(A) = 0$, then the equation $A\,x = y$ may have solutions, but they will not be unique. Whether equation (1.242) has any solutions when $\det(A) = 0$ depends on whether the vector y can be expressed as a linear combination of the columns of A. Since these columns are linearly dependent, they span a subspace of fewer than n dimensions, and so (1.242) has solutions only when the n-vector y lies in that subspace.

A system of $m < n$ equations

$$\sum_{k=1}^{n} A_{ik} x_k = y_i \quad \text{for} \quad i = 1, 2, \ldots, m \tag{1.243}$$

in n unknowns is **under determined**. As long as at least m of the n columns A_{ik} of the matrix A are linearly independent, such a system always has solutions, but they may not be unique.

1.23 Linear Least Squares

Suppose we have a system of $m > n$ equations in n unknowns x_k

$$\sum_{k=1}^{n} A_{ik} x_k = y_i \quad \text{for} \quad i = 1, 2, \ldots, m. \tag{1.244}$$

This problem is **over determined** and, in general, has no solution, but it does have an approximate solution due to Carl Gauss (1777–1855).

If the matrix A and the vector y are real, then Gauss's solution is the n values x_k that minimize the sum E of the squares of the errors

$$E = \sum_{i=1}^{m} \left(y_i - \sum_{k=1}^{n} A_{ik} x_k \right)^2 . \tag{1.245}$$

The minimizing values x_k make the n derivatives of E vanish

$$\frac{\partial E}{\partial x_\ell} = 0 = \sum_{i=1}^{m} 2 \left(y_i - \sum_{k=1}^{n} A_{ik} x_k \right) (-A_{i\ell}) \tag{1.246}$$

or in matrix notation $A^{\mathsf{T}} y = A^{\mathsf{T}} A x$. Since A is real, the matrix $A^{\mathsf{T}} A$ is nonnegative (1.39); if it also is positive (1.40), then it has an inverse, and our **least-squares solution** is

$$x = \left(A^{\mathsf{T}} A \right)^{-1} A^{\mathsf{T}} y. \tag{1.247}$$

If the matrix A and the vector y are complex, and if the matrix $A^\dagger A$ is positive, then one may derive (Exercise 1.25) Gauss's solution

$$x = \left(A^\dagger A \right)^{-1} A^\dagger \, y. \tag{1.248}$$

The operators $\left(A^{\mathsf{T}} A \right)^{-1} A^{\mathsf{T}}$ and $\left(A^\dagger A \right)^{-1} A^\dagger$ are **pseudoinverses** (Section 1.33).

1.24 Lagrange Multipliers

The maxima and minima of a function $f(x)$ of $x = (x_1, x_2, \ldots, x_n)$ are among the points at which its gradient vanishes $\nabla f(x) = 0$, that is,

$$\frac{\partial f(x)}{\partial x_j} = 0 \tag{1.249}$$

for $j = 1, \ldots, n$. These are **stationary** points of f.

Example 1.36 (Minimum) For instance, if $f(x) = x_1^2 + 2x_2^2 + 3x_3^2$, then its minimum is at

$$\nabla f(x) = (2x_1, 4x_2, 6x_3) = 0 \tag{1.250}$$

that is, at $x_1 = x_2 = x_3 = 0$.

How do we find the extrema of $f(x)$ if x also must satisfy a constraint? We use a Lagrange multiplier (Joseph-Louis Lagrange 1736–1813).

In the case of one constraint $c(x) = 0$, we expect the gradient $\nabla f(x)$ to vanish in those directions dx that preserve the constraint. So $dx \cdot \nabla f(x) = 0$ for all dx that make the dot product $dx \cdot \nabla c(x)$ vanish. That is, $\nabla f(x)$ and $\nabla c(x)$ must be parallel. So the extrema of $f(x)$ subject to the constraint $c(x) = 0$ satisfy the equations

$$\nabla f(x) = \lambda \, \nabla c(x) \quad \text{and} \quad c(x) = 0. \tag{1.251}$$

These $n + 1$ equations define the extrema of the unconstrained function

$$L(x, \lambda) = f(x) - \lambda \, c(x) \tag{1.252}$$

of the $n + 1$ variables $x_1, \ldots, x_n, \lambda$

$$\frac{\partial L(x, \lambda)}{\partial x_j} = \frac{\partial \left(f(x) - \lambda \, c(x) \right)}{\partial x_j} = 0 \quad \text{and} \quad \frac{\partial L(x, \lambda)}{\partial \lambda} = -\, c(x) = 0. \tag{1.253}$$

The variable λ is a **Lagrange multiplier**.

In the case of k constraints $c_1(x) = 0, \ldots, c_k(x) = 0$, the projection of ∇f must vanish in those directions dx that preserve all the constraints. So $dx \cdot \nabla f(x) = 0$ for all dx that make all $dx \cdot \nabla c_j(x) = 0$ for $j = 1, \ldots, k$. The gradient ∇f will satisfy this requirement if it's a linear combination

$$\nabla f = \lambda_1 \, \nabla c_1 + \cdots + \lambda_k \, \nabla c_k \tag{1.254}$$

of the k gradients because then $dx \cdot \nabla f$ will vanish if $dx \cdot \nabla c_j = 0$ for $j = 1, \ldots, k$. The extrema also must satisfy the constraints

$$c_1(x) = 0, \ldots, c_k(x) = 0. \tag{1.255}$$

The $n + k$ equations (1.254 & 1.255) define the extrema of the unconstrained function

$$L(x, \lambda) = f(x) - \lambda_1 \, c_1(x) \cdots - \lambda_k \, c_k(x) \tag{1.256}$$

of the $n + k$ variables x and λ

$$\nabla L(x, \lambda) = \nabla f(x) - \lambda_1 \, \nabla c_1(x) \cdots - \lambda_k \, \nabla c_k(x) = 0 \tag{1.257}$$

and

$$\frac{\partial L(x,\lambda)}{\partial \lambda_j} = -\,c_j(x) = 0 \quad \text{for } j = 1, \ldots, k. \tag{1.258}$$

Example 1.37 (Constrained extrema and eigenvectors) Suppose we want to find the extrema of a real, symmetric quadratic form

$$f(x) = x^{\mathsf{T}} A\, x = \sum_{i,j=1}^{n} x_i\, A_{ij}\, x_j \tag{1.259}$$

subject to the constraint $c(x) = x \cdot x - 1$ which says that the n-vector x is of unit length. We form the function

$$L(x,\lambda) = x^{\mathsf{T}} A\, x - \lambda\,(x \cdot x - 1) \tag{1.260}$$

and since the matrix A is real and symmetric, we find its unconstrained extrema as

$$\nabla L(x,\lambda) = 2A\,x - 2\lambda\,x = 0 \quad \text{and} \quad x \cdot x = 1. \tag{1.261}$$

The extrema of $f(x) = x^{\mathsf{T}} A\, x$ subject to the constraint $c(x) = x \cdot x - 1$ are the **normalized eigenvectors**

$$A\,x = \lambda\,x \quad \text{and} \quad x \cdot x = 1 \tag{1.262}$$

of the real, symmetric matrix A.

1.25 Eigenvectors and Eigenvalues

If a linear operator A maps a nonzero vector $|u\rangle$ into a multiple of itself

$$A|u\rangle = \lambda|u\rangle \tag{1.263}$$

then the vector $|u\rangle$ is an **eigenvector** of A with **eigenvalue** λ. (The German adjective *eigen* means *own, special,* or *proper.*)

If the vectors $|k\rangle$ for $k = 1, \ldots, n$ form an orthonormal basis for the vector space in which A acts, then we can write the identity operator for the space as $I = |1\rangle\langle 1| + \cdots + |n\rangle\langle n|$. By inserting this formula for I into the eigenvector equation (1.263), we get

$$\sum_{\ell=1}^{n} \langle k|A|\ell\rangle\langle \ell|u\rangle = \lambda\,\langle k|u\rangle. \tag{1.264}$$

In matrix notation, with $A_{k\ell} = \langle k|A|\ell\rangle$ and $u_\ell = \langle \ell|u\rangle$, this is $A\,u = \lambda\,u$.

A subspace $c_\ell|u_\ell\rangle + \cdots + c_r|u_r\rangle$ spanned by any set of eigenvectors $|u_k\rangle$ of a matrix A is left invariant by its action, that is

$$A\left(\sum_{k\in S} c_k|u_k\rangle\right) = \sum_{k\in S} c_k\, A|u_k\rangle = \sum_{k\in S} c_k\, \lambda_k|u_k\rangle = \sum_{k\in S} c'_k|u_k\rangle \tag{1.265}$$

with $c'_k = c_k\lambda_k$. Eigenvectors span **invariant subspaces.**

Example 1.38 (Eigenvalues of an orthogonal matrix) The matrix equation

$$\begin{pmatrix} \cos\theta & \sin\theta \\ -\sin\theta & \cos\theta \end{pmatrix}\begin{pmatrix} 1 \\ \pm i \end{pmatrix} = e^{\pm i\theta}\begin{pmatrix} 1 \\ \pm i \end{pmatrix} \tag{1.266}$$

tells us that the eigenvectors of this 2×2 orthogonal matrix are $(1, \pm i)$ with eigenvalues $e^{\pm i\theta}$. The eigenvalues λ of a unitary (and of an orthogonal) matrix are unimodular, $|\lambda| = 1$, (Exercise 1.26).

Example 1.39 (Eigenvalues of an antisymmetric matrix) Let us consider an eigenvector equation for a matrix A that is antisymmetric

$$\sum_{k=1}^{n} A_{ik}\, u_k = \lambda\, u_i. \tag{1.267}$$

The antisymmetry $A_{ik} = -A_{ki}$ of A implies that

$$\sum_{i,k=1}^{n} u_i\, A_{ik}\, u_k = 0. \tag{1.268}$$

Thus the last two relations imply that

$$0 = \sum_{i,k=1}^{n} u_i\, A_{ik}\, u_k = \lambda \sum_{i=1}^{n} u_i^2 = 0. \tag{1.269}$$

Thus either the eigenvalue λ or the dot product of the eigenvector with itself vanishes.

1.26 Eigenvectors of a Square Matrix

Let A be an $n \times n$ matrix with complex entries A_{ik}. A vector V with n entries V_k (not all zero) is an **eigenvector** of A with **eigenvalue** λ if

$$\sum_{k=1}^{n} A_{ik} V_k = \lambda\, V_i \quad \text{or} \quad AV = \lambda\, V \tag{1.270}$$

in matrix notation. Every $n\times n$ matrix A has n eigenvectors $V^{(\ell)}$ and eigenvalues λ_ℓ

$$AV^{(\ell)} = \lambda_\ell V^{(\ell)} \tag{1.271}$$

for $\ell = 1, \ldots, n$. To see why, we write the top equation (1.270) as

$$\sum_{k=1}^{n} (A_{ik} - \lambda\, \delta_{ik})\, V_k = 0 \tag{1.272}$$

or in matrix notation as $(A - \lambda\, I)\, V = 0$ in which I is the $n \times n$ matrix with entries $I_{ik} = \delta_{ik}$. This equation and (1.272) say that the columns of the matrix $A - \lambda I$, considered as vectors, are linearly dependent (Section 1.8). The columns of a matrix $A - \lambda I$ are linearly dependent if and only if the determinant $|A - \lambda I|$ vanishes (Section 1.20). Thus a solution of the eigenvalue equation (1.270) exists if and only if the determinant of $A - \lambda I$ vanishes

$$\det(A - \lambda I) = |A - \lambda I| = 0. \tag{1.273}$$

This vanishing of the determinant of $A - \lambda I$ is the **characteristic equation** of the matrix A. For an $n \times n$ matrix A, it is a polynomial equation of the nth degree in the unknown eigenvalue λ

$$\begin{aligned} 0 = |A - \lambda I| &= |A| + \cdots + (-1)^{n-1}\lambda^{n-1}\,\mathrm{Tr}A + (-1)^n\lambda^n \\ &= P(\lambda, A) = \sum_{k=0}^{n} p_k\, \lambda^k \end{aligned} \tag{1.274}$$

in which $p_0 = |A|$, $p_{n-1} = (-1)^{n-1}\mathrm{Tr}A$, and $p_n = (-1)^n$.

All the p_k's are basis independent. For if S is any nonsingular matrix, then multiplication rules (1.222 and 1.226) for determinants imply that the determinant $|A - \lambda I|$ is invariant when A undergoes a similarity transformation (1.67 & (1.278) 1.284) $A \to A' = S^{-1} A S$

$$\begin{aligned} P(\lambda, A') = P(\lambda, S^{-1}AS) &= |S^{-1}AS - \lambda I| = |S^{-1}(A - \lambda I)S| \\ &= |S^{-1}||A - \lambda I||S| = |A - \lambda I| = P(\lambda, A). \end{aligned} \tag{1.275}$$

By the fundamental theorem of algebra (Section 6.9), the characteristic equation (1.274) always has n roots or solutions λ_ℓ lying somewhere in the complex plane. Thus the **characteristic polynomial** $P(\lambda, A)$ has the factored form

$$P(\lambda, A) = (\lambda_1 - \lambda)(\lambda_2 - \lambda)\cdots(\lambda_n - \lambda). \tag{1.276}$$

For every root λ_ℓ, there is a nonzero eigenvector $V^{(\ell)}$ whose components $V_k^{(\ell)}$ are the coefficients that make the n vectors $A_{ik} - \lambda_\ell\, \delta_{ik}$ that are the columns of the matrix $A - \lambda_\ell I$ sum to zero in (1.272). Thus **every $n \times n$ matrix has n eigenvalues λ_ℓ and n eigenvectors $V^{(\ell)}$**.

The $n \times n$ diagonal matrix $A^{(d)}_{k\ell} = \delta_{k\ell}\, \lambda_\ell$ is the **canonical form** of the matrix A; the matrix $V_{k\ell} = V_k^{(\ell)}$ whose columns are the eigenvectors $V^{(\ell)}$ of A is the **modal matrix**; and $A\, V = V\, A_d$ or more explicitly

$$\sum_{k=1}^{n} A_{ik} V_{k\ell} = \sum_{k=1}^{n} A_{ik} V_k^{(\ell)} = \lambda_\ell V_i^{(\ell)} = \sum_{k=1}^{n} V_{ik} \delta_{k\ell} \lambda_\ell = \sum_{k=1}^{n} V_{ik} A_{k\ell}^{(d)}. \quad (1.277)$$

If the eigenvectors $V_{k\ell}$ are linearly independent, then the matrix V, of which they are the columns, is nonsingular and has an inverse V^{-1}. The similarity transformation

$$V^{-1} A V = A^{(d)} \quad (1.278)$$

diagonalizes the matrix A.

Example 1.40 (The Canonical Form of a 3 × 3 Matrix) If in Matlab we set $A = [0\ 1\ 2;\ 3\ 4\ 5;\ 6\ 7\ 8]$ and enter $[V, D] = \mathrm{eig}(A)$, then we get

$$V = \begin{pmatrix} 0.1648 & 0.7997 & 0.4082 \\ 0.5058 & 0.1042 & -0.8165 \\ 0.8468 & -0.5913 & 0.4082 \end{pmatrix} \quad \text{and} \quad A_d = \begin{pmatrix} 13.3485 & 0 & 0 \\ 0 & -1.3485 & 0 \\ 0 & 0 & 0 \end{pmatrix}$$

and one may check that $A V = V A_d$ and that $V^{-1} A V = A_d$.

Setting $\lambda = 0$ in the factored form (1.276) of $P(\lambda, A)$ and in the characteristic equation (1.274), we see that **the determinant of every $n \times n$ matrix is the product of its n eigenvalues**

$$P(0, A) = |A| = p_0 = \lambda_1 \lambda_2 \cdots \lambda_n. \quad (1.279)$$

These n roots usually are all different, and when they are, the eigenvectors $V^{(\ell)}$ are linearly independent. The first eigenvector is trivially linearly independent. Let's assume that the first $k < n$ eigenvectors are linearly independent; we'll show that the first $k+1$ eigenvectors are linearly independent. If they were linearly dependent, then there would be $k+1$ numbers c_ℓ, not all zero, such that

$$\sum_{\ell=1}^{k+1} c_\ell V^{(\ell)} = 0. \quad (1.280)$$

First we multiply this equation from the left by the linear operator A and use the eigenvalue equation (1.271)

$$A \sum_{\ell=1}^{k+1} c_\ell V^{(\ell)} = \sum_{\ell=1}^{k+1} c_\ell A V^{(\ell)} = \sum_{\ell=1}^{k+1} c_\ell \lambda_\ell V^{(\ell)} = 0. \quad (1.281)$$

Now we multiply the same equation (1.280) by λ_{k+1}

$$\sum_{\ell=1}^{k+1} c_\ell \lambda_{k+1} V^{(\ell)} = 0 \quad (1.282)$$

and subtract the product (1.282) from (1.281). The terms with $\ell = k+1$ cancel leaving

$$\sum_{\ell=1}^{k} c_\ell \left(\lambda_\ell - \lambda_{k+1}\right) V^{(\ell)} = 0 \tag{1.283}$$

in which all the factors $(\lambda_\ell - \lambda_{k+1})$ are different from zero since by assumption all the eigenvalues are different. But this last equation says that the first k eigenvectors are linearly dependent, which contradicts our assumption that they were linearly independent. This contradiction tells us that **if all n eigenvectors of an $n \times n$ square matrix have different eigenvalues, then they are linearly independent**. Similarly, if any $k < n$ eigenvectors of an $n \times n$ square matrix have different eigenvalues, then they are linearly independent.

An eigenvalue λ that is a single root of the characteristic equation (1.274) is associated with a single eigenvector; it is called a **simple eigenvalue**. An eigenvalue λ that is a root of multiplicity n of the characteristic equation is associated with n eigenvectors; it is said to be an ***n*-fold degenerate eigenvalue** or to have **algebraic multiplicity** n. Its **geometric multiplicity** is the number $n' \le n$ of linearly independent eigenvectors with eigenvalue λ. A matrix with $n' < n$ for any eigenvalue λ is **defective**. Thus an $n \times n$ matrix with fewer than n linearly independent eigenvectors is defective. **Thus every nondefective square matrix A can be diagonalized by a similarity transformation**

$$\boldsymbol{V^{-1} A V = A^{(d)}} \tag{1.284}$$

(1.278). The elements of the main diagonal of the matrix $A^{(d)}$ are the eigenvalues of the matrix A. Thus the trace of every nondefective matrix A is the sum of its eigenvalues, $\mathrm{Tr} A = \mathrm{Tr} A^{(d)} = \lambda_a + \cdots + \lambda_n$. The columns of the matrix V are the eigenvectors of the matrix A.

Since the determinant of every matrix A is the product (1.279) of its eigenvalues, $\det A = |A| = \lambda_1 \lambda_2 \cdots \lambda_n$, the determinant of every nondefective matrix $A = e^L$ is the exponential of the trace of its logarithm

$$\det A = \exp\left[\mathrm{Tr}\,(\log A)\right] \quad \text{and} \quad \det A = \det(e^L) = \exp[\mathrm{Tr}(L)]. \tag{1.285}$$

Example 1.41 (A defective 2 × 2 matrix) Each of the 2 × 2 matrices

$$\begin{pmatrix} 0 & 1 \\ 0 & 0 \end{pmatrix} \quad \text{and} \quad \begin{pmatrix} 0 & 0 \\ 1 & 0 \end{pmatrix} \tag{1.286}$$

has only one linearly independent eigenvector and so is defective.

1.27 A Matrix Obeys Its Characteristic Equation

Every square matrix obeys its characteristic equation (1.274). That is, the characteristic equation

$$P(\lambda, A) = |A - \lambda I| = \sum_{k=0}^{n} p_k \, \lambda^k = 0 \tag{1.287}$$

remains true when the matrix A replaces the variable λ

$$P(A, A) = \sum_{k=0}^{n} p_k \, A^k = 0. \tag{1.288}$$

To see why, we use the formula (1.215) for the inverse of the matrix $A - \lambda I$

$$(A - \lambda I)^{-1} = \frac{C(\lambda, A)^{\mathsf{T}}}{|A - \lambda I|} \tag{1.289}$$

in which $C(\lambda, A)^{\mathsf{T}}$ is the transpose of the matrix of cofactors of the matrix $A - \lambda I$. Since $|A - \lambda I| = P(\lambda, A)$, we have, rearranging,

$$(A - \lambda I) \, C(\lambda, A)^{\mathsf{T}} = |A - \lambda I| \, I = P(\lambda, A) \, I. \tag{1.290}$$

The transpose of the matrix of cofactors of the matrix $A - \lambda I$ is a polynomial in λ with matrix coefficients

$$C(\lambda, A)^{\mathsf{T}} = C_0 + C_1 \lambda + \cdots + C_{n-1} \lambda^{n-1}. \tag{1.291}$$

Combining these last two equations (1.290 and 1.291) with the characteristic equation (1.287), we have

$$\begin{aligned} (A - \lambda I) C(\lambda, A)^{\mathsf{T}} &= A C_0 + (A C_1 - C_0)\lambda + (A C_2 - C_1)\lambda^2 + \cdots \\ &\quad + (A C_{n-1} - C_{n-2})\lambda^{n-1} - C_{n-1}\lambda^n \\ &= \sum_{k=0}^{n} p_k \, \lambda^k. \end{aligned} \tag{1.292}$$

Equating equal powers of λ on both sides of this equation, we find

$$\begin{aligned} A C_0 &= p_0 I \\ A C_1 - C_0 &= p_1 I \\ A C_2 - C_1 &= p_2 I \\ \cdots &= \cdots \\ A C_{n-1} - C_{n-2} &= p_{n-1} I \\ -C_{n-1} &= p_n I. \end{aligned} \tag{1.293}$$

We now multiply from the left the first of these equations by I, the second by A, the third by A^2, ..., and the last by A^n and then add the resulting equations. All the terms on the left-hand sides cancel, while the sum of those on the right gives $P(A, A)$. Thus a square matrix A obeys its characteristic equation $0 = P(A, A)$ or

$$0 = \sum_{k=0}^{n} p_k A^k = |A|\, I + p_1 A + \cdots + (-1)^{n-1}(\text{Tr}A)\, A^{n-1} + (-1)^n A^n \quad (1.294)$$

a result known as the **Cayley–Hamilton theorem** (Arthur Cayley, 1821–1895, and William Hamilton, 1805–1865). This derivation is due to Israel Gelfand (1913–2009) (Gelfand, 1961, pp. 89–90).

Because every $n \times n$ matrix A obeys its characteristic equation, its nth power A^n can be expressed as a linear combination of its lesser powers

$$A^n = (-1)^{n-1}\big(|A|\, I + p_1 A + p_2 A^2 + \cdots + (-1)^{n-1}(\text{Tr}A)\, A^{n-1}\big). \quad (1.295)$$

For instance, the square A^2 of every 2×2 matrix is given by

$$A^2 = -|A| I + (\text{Tr}A) A. \quad (1.296)$$

Example 1.42 (Spin-one-half rotation matrix) If $\boldsymbol{\theta}$ is a real 3-vector and $\boldsymbol{\sigma}$ is the 3-vector of Pauli matrices (1.32), then the square of the traceless 2×2 matrix $A = \boldsymbol{\theta} \cdot \boldsymbol{\sigma}$ is

$$(\boldsymbol{\theta} \cdot \boldsymbol{\sigma})^2 = -\,|\boldsymbol{\theta} \cdot \boldsymbol{\sigma}|\, I = -\begin{vmatrix} \theta_3 & \theta_1 - i\theta_2 \\ \theta_1 + i\theta_2 & -\theta_3 \end{vmatrix} I = \theta^2\, I \quad (1.297)$$

in which $\theta^2 = \boldsymbol{\theta} \cdot \boldsymbol{\theta}$. One may use this identity to show (Exercise 1.28) that

$$\exp\left(-i\boldsymbol{\theta} \cdot \boldsymbol{\sigma}/2\right) = \cos(\theta/2)\, I - i\hat{\boldsymbol{\theta}} \cdot \boldsymbol{\sigma}\ \sin(\theta/2) \quad (1.298)$$

in which $\hat{\boldsymbol{\theta}}$ is a unit 3-vector. For a spin-one-half object, this matrix represents an active right-handed rotation of θ radians about the axis $\hat{\boldsymbol{\theta}}$.

1.28 Functions of Matrices

What sense can we make of a function f of an $n \times n$ matrix A? And how would we compute it? One way is to use the characteristic equation to express (1.295) every power of A in terms of I, A, ..., A^{n-1} and the coefficients $p_0 = |A|$, $p_1, p_2, \ldots, p_{n-2}$, and $p_{n-1} = (-1)^{n-1}\text{Tr}A$. Then if $f(x)$ is a polynomial or a function with a convergent power series

$$f(x) = \sum_{k=0}^{\infty} c_k\, x^k \quad (1.299)$$

in principle we may express $f(A)$ in terms of n functions $f_k(\boldsymbol{p})$ of the coefficients $\boldsymbol{p} \equiv (p_0, \ldots, p_{n-1})$ as

$$f(A) = \sum_{k=0}^{n-1} f_k(\boldsymbol{p})\, A^k. \tag{1.300}$$

The identity (1.298) for $\exp(-i\boldsymbol{\theta} \cdot \boldsymbol{\sigma}/2)$ is an $n = 2$ example of this technique which can become challenging when $n > 3$.

Example 1.43 (The 3×3 rotation matrix) In Exercise 1.29, one finds the characteristic equation (1.294) for the 3×3 matrix $-i\boldsymbol{\theta} \cdot \boldsymbol{J}$ in which $(J_k)_{ij} = i\epsilon_{ikj}$, and ϵ_{ijk} is totally antisymmetric with $\epsilon_{123} = 1$. The generators J_k satisfy the commutation relations $[J_i, J_j] = i\epsilon_{ijk} J_k$ in which sums over repeated indices from 1 to 3 are understood. In Exercise 1.30, one uses the characteristic equation for $-i\boldsymbol{\theta} \cdot \boldsymbol{J}$ to show that the 3×3 real orthogonal matrix $\exp(-i\boldsymbol{\theta} \cdot \boldsymbol{J})$, which represents a right-handed rotation by θ radians about the axis $\hat{\boldsymbol{\theta}}$, is

$$\exp(-i\boldsymbol{\theta} \cdot \boldsymbol{J}) = \cos\theta\, I - i\hat{\boldsymbol{\theta}} \cdot \boldsymbol{J}\, \sin\theta + (1 - \cos\theta)\, \hat{\boldsymbol{\theta}} (\hat{\boldsymbol{\theta}})^{\mathsf{T}} \tag{1.301}$$

or

$$\exp(-i\boldsymbol{\theta} \cdot \boldsymbol{J})_{ij} = \delta_{ij} \cos\theta - \sin\theta\, \epsilon_{ijk}\hat{\theta}_k + (1 - \cos\theta)\, \hat{\theta}_i \hat{\theta}_j \tag{1.302}$$

in terms of indices.

Direct use of the characteristic equation can become unwieldy for larger values of n. Fortunately, another trick is available if A is a nondefective square matrix, and if the power series (1.299) for $f(x)$ converges. For then A is related to its diagonal form $A^{(d)}$ by a similarity transformation (1.278), and we may define $f(A)$ as

$$f(A) = S f(A^{(d)}) S^{-1} \tag{1.303}$$

in which $f(A^{(d)})$ is the diagonal matrix with entries $f(a_\ell)$

$$f(A^{(d)}) = \begin{pmatrix} f(a_1) & 0 & 0 & \ldots \\ 0 & f(a_2) & 0 & \ldots \\ \vdots & \vdots & \vdots & \vdots \\ 0 & 0 & \ldots & f(a_n) \end{pmatrix} \tag{1.304}$$

and $a_1, a_1, \ldots, a_n$ are the eigenvalues of the matrix A. This definition makes sense if $f(A)$ is a series in powers of A because then

$$f(A) = \sum_{k=0}^{\infty} c_k A^k = \sum_{k=0}^{\infty} c_k \left(S A^{(d)} S^{-1}\right)^k. \tag{1.305}$$

So since $S^{-1}S = I$, we have $\left(SA^{(d)}S^{-1}\right)^k = S\left(A^{(d)}\right)^k S^{-1}$ and thus

$$f(A) = S\left[\sum_{k=0}^{\infty} c_k \left(A^{(d)}\right)^k\right] S^{-1} = Sf(A^{(d)})S^{-1} \tag{1.306}$$

which is (1.303).

Example 1.44 (Momentum operators generate spatial translations) The position operator x and the momentum operator p obey the commutation relation $[x, p] = xp - px = i\hbar$. Thus the a-derivative $\dot{x}(a)$ of the operator $x(a) = e^{iap/\hbar}\, x\, e^{-iap/\hbar}$ is unity

$$\dot{x}(a) = e^{iap/\hbar}\,(i/\hbar)[p, x]\, e^{-iap/\hbar} = e^{iap/\hbar}\, e^{-iap/\hbar} = 1. \tag{1.307}$$

Since $x(0) = x$, we see that the unitary transformation $U(a) = e^{iap/\hbar}$ moves x to $x + a$

$$e^{iap/\hbar}\, x\, e^{-iap/\hbar} = x(a) = x(0) + \int_0^a \dot{x}(a')\, da' = x + a. \tag{1.308}$$

Example 1.45 (Glauber's identity) The commutator of the annihilation operator a and the creation operator $a^\dagger$ for a given mode is the number 1

$$[a, a^\dagger] = a\, a^\dagger - a^\dagger\, a = 1. \tag{1.309}$$

Thus a and $a^\dagger$ commute with their commutator $[a, a^\dagger] = 1$ just as x and p commute with their commutator $[x, p] = i\hbar$.

Suppose that A and B are any two operators that commute with their commutator $[A, B] = AB - BA$

$$[A, [A, B]] = [B, [A, B]] = 0. \tag{1.310}$$

As in the $[x, p]$ example (1.44), we define $A_B(t) = e^{-tB}\, A\, e^{tB}$ and note that because $[B, [A, B]] = 0$, its t-derivative is simply

$$\dot{A}_B(t) = e^{-tB}\,[A, B]\, e^{tB} = [A, B]. \tag{1.311}$$

Since $A_B(0) = A$, an integration gives

$$A_B(t) = A + \int_0^t \dot{A}(t)\, dt = A + \int_0^t [A, B]\, dt = A + t\,[A, B]. \tag{1.312}$$

Multiplication from the left by e^{tB} now gives $e^{tB}\, A_B(t)$ as

$$e^{tB}\, A_B(t) = A\, e^{tB} = e^{tB}\,(A + t\,[A, B])\,. \tag{1.313}$$

Now we define

$$G(t) = e^{tA}\, e^{tB}\, e^{-t(A+B)} \tag{1.314}$$

and use our formula (1.313) to compute its t-derivative as

$$\begin{aligned}\dot{G}(t) &= e^{tA}\left(A\, e^{tB} + e^{tB}\, B - e^{tB}(A+B)\right) e^{-t(A+B)} \\ &= e^{tA}\left(e^{tB}\,(A + t\,[A,B]) + e^{tB}\, B - e^{tB}(A+B)\right) e^{-t(A+B)} \\ &= e^{tA}\, e^{tB}\, t\,[A,B]\, e^{t(A+B)} = t\,[A,B]\, G(t) = t\, G(t)\,[A,B].\end{aligned} \tag{1.315}$$

Since $\dot{G}(t)$, $G(t)$, and $[A, B]$ all commute with each other, we can integrate this operator equation

$$\frac{d}{dt}\log G(t) = \frac{\dot{G}(t)}{G(t)} = t\,[A,B] \tag{1.316}$$

from 0 to 1 and get since $G(0) = 1$

$$\log G(1) - \log G(0) = \log G(1) = \frac{1}{2}\,[A,B]. \tag{1.317}$$

Thus $G(1) = e^{[A,B]/2}$ and so

$$e^A\, e^B\, e^{-(A+B)} = e^{\frac{1}{2}[A,B]} \quad \text{or} \quad e^A\, e^B = e^{A+B+\frac{1}{2}[A,B]} \tag{1.318}$$

which is Glauber's identity.

Example 1.46 (Chemical reactions) The chemical reactions $[\mathrm{A}] \xrightarrow{a} [\mathrm{B}]$, $[\mathrm{B}] \xrightarrow{b} [\mathrm{A}]$, and $[\mathrm{B}] \xrightarrow{c} [\mathrm{C}]$ make the concentrations $[A] \equiv A$, $[B] \equiv B$, and $[C] \equiv C$ of three kinds of molecules vary with time as

$$\dot{A} = -aA + bB, \quad \dot{B} = aA - (b+c)B, \quad \text{and} \quad \dot{C} = cB. \tag{1.319}$$

We can group these concentrations into a 3-vector $V = (A, B, C)$ and write the three equations (1.319) as $\dot{V} = K\,V$ in which K is the matrix

$$K = \begin{pmatrix} -a & b & 0 \\ a & -b-c & 0 \\ 0 & c & 0 \end{pmatrix}. \tag{1.320}$$

The solution to the differential equation $\dot{V} = K\,V$ is $V(t) = e^{Kt}\,V(0)$.

The eigenvalues of the matrix K are the roots of the cubic equation $\det(K - \lambda I) = 0$. One root vanishes, and the other two are the roots of the quadratic equation $\lambda^2 + (a + b + c)\lambda + ac = 0$. Their sum is the trace $\mathrm{Tr}K = -(a + b + c)$. They are real when a, b, and c are positive but are complex when $4ac > (a + b + c)^2$. The eigenvectors are complete unless $4ac = (a + b + c)^2$, but are not orthogonal unless $c = 0$.

The time evolution of the concentrations $[A]$ (dashdot), $[B]$ (solid), and $[C]$ (dashes) are plotted in Fig. 1.1 for the initial conditions $[A] = 1$ and $[B] = [C] = 0$ and rates $a = 0.15$, $b = 0.1$, and $c = 0.1$. The Matlab code is in the repository Linear_algebra at `github.com/kevinecahill`.

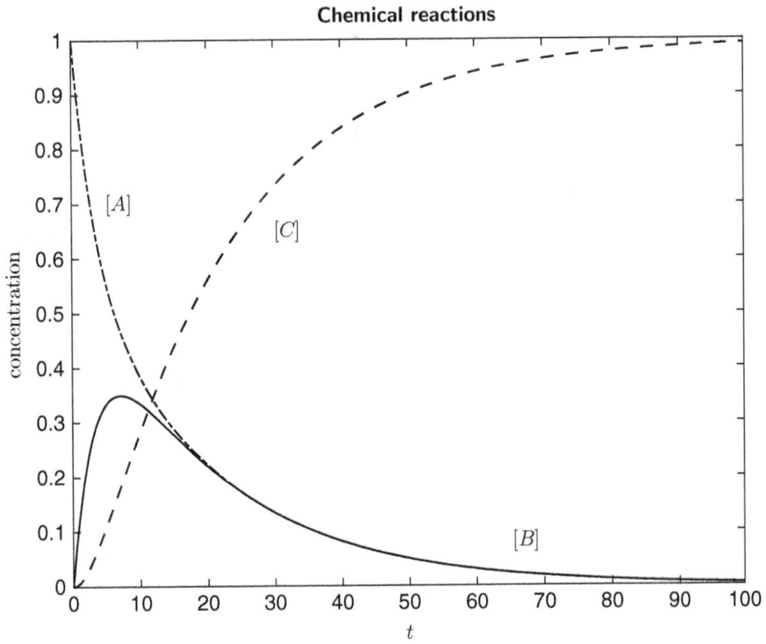

Figure 1.1 The concentrations $[A]$ (dashdot), $[B]$ (solid), and $[C]$ (dashes) as given by the matrix equation $V(t) = e^{Kt}\, V(0)$ for the initial conditions $[A] = 1$ and $[B] = [C] = 0$ and rates $a = 0.15$, $b = 0.1$, and $c = 0.1$.

Example 1.47 (Time-evolution operator) In quantum mechanics, the time-evolution operator is the exponential $\exp(-iHt/\hbar)$ where $H = H^\dagger$ is a hermitian linear operator, the hamiltonian (William Rowan Hamilton 1805–1865), and $\hbar = h/(2\pi) = 1.054 \times 10^{-34}$ Js where h is Planck's constant (Max Planck 1858–1947). As we'll see in the next section, hermitian operators are never defective, so H can be diagonalized by a similarity transformation

$$H = S H^{(d)} S^{-1}. \tag{1.321}$$

The diagonal elements of the diagonal matrix $H^{(d)}$ are the **energies** E_ℓ of the states of the system described by the hamiltonian H. The time-evolution operator $U(t)$ then is

$$U(t) = S\, \exp(-iH^{(d)}t/\hbar)\, S^{-1}. \tag{1.322}$$

For a three-state system with angular frequencies $\omega_i = E_i/\hbar$, it is

$$U(t) = S \begin{pmatrix} e^{-i\omega_1 t} & 0 & 0 \\ 0 & e^{-i\omega_2 t} & \\ 0 & 0 & e^{-i\omega_3 t} \end{pmatrix} S^{-1}. \tag{1.323}$$

Example 1.48 (Entropy) The **entropy** S of a system described by a density operator ρ is the trace $S = -k\,\mathrm{Tr}\,(\rho \ln \rho)$ in which $k = 1.38 \times 10^{-23}$ J/K is the constant named after Ludwig Boltzmann (1844–1906). The density operator ρ is hermitian, nonnegative, and of unit trace. Since ρ is hermitian, the matrix that represents it is never defective (section 1.29), and so it can be diagonalized by a similarity transformation $\rho = S\,\rho^{(d)}\,S^{-1}$. By (1.24), $\mathrm{Tr}ABC = \mathrm{Tr}BCA$, so we can write S as

$$S = -k\mathrm{Tr}\left(S\,\rho^{(d)}\,S^{-1}\,S\,\ln(\rho^{(d)})\,S^{-1}\right) = -k\mathrm{Tr}\left(\rho^{(d)}\,\ln(\rho^{(d)})\right). \tag{1.324}$$

A vanishing eigenvalue $\rho_k^{(d)} = 0$ contributes nothing to this trace since $\lim_{x\to 0} x \ln x = 0$. If the system has three states, populated with probabilities ρ_i, the elements of $\rho^{(d)}$, then the sum

$$\begin{aligned} S &= -k\,(\rho_1 \ln \rho_1 + \rho_2 \ln \rho_2 + \rho_3 \ln \rho_3) \\ &= k\left[\rho_1 \ln(1/\rho_1) + \rho_2 \ln(1/\rho_2) + \rho_3 \ln(1/\rho_3)\right] \end{aligned} \tag{1.325}$$

is its entropy.

Example 1.49 (Logarithm of a determinant) Since every nondefective $n \times n$ matrix A may be diagonalized by a similarity transformation, its determinant is the product of its eigenvalues and its trace is the sum of them, and so the logarithm of its determinant is the trace of its logarithm

$$\ln \det A = \ln(\lambda_1 \dots \lambda_n) = \ln(\lambda_1) + \dots + \ln(\lambda_n) = \mathrm{Tr}(\ln A). \tag{1.326}$$

When none of A's eigenvalues vanishes, this relation implies the earlier result (1.228) that the variation of A's determinant is

$$\delta \det A = \det A\ \mathrm{Tr}(A^{-1}\delta A). \tag{1.327}$$

1.29 Hermitian Matrices

Hermitian matrices have very nice properties. By definition (1.30), a hermitian matrix A is square and unchanged by hermitian conjugation $A^\dagger = A$. Since it is square, the results of section 1.26 ensure that an $n \times n$ hermitian matrix A has n eigenvectors $|k\rangle$ with eigenvalues a_k

$$A|k\rangle = a_k|k\rangle. \tag{1.328}$$

In fact, all its eigenvalues are real. To see why, we take the adjoint

$$\langle k|A^\dagger = a_k^*\langle k| \tag{1.329}$$

and use the property $A^\dagger = A$ to find

$$\langle k|A^\dagger = \langle k|A = a_k^*\langle k|. \tag{1.330}$$

We now form the inner product of both sides of this equation with the ket $|k\rangle$ and use the eigenvalue equation (1.328) to get

$$\langle k|A|k\rangle = a_k\langle k|k\rangle = a_k^*\langle k|k\rangle \tag{1.331}$$

which (since $\langle k|k\rangle > 0$) tells us that the eigenvalues are real

$$a_k^* = a_k. \tag{1.332}$$

Since $A^\dagger = A$, the matrix elements of A between two of its eigenvectors satisfy

$$a_m^* \langle m|k\rangle = (a_m \langle k|m\rangle)^* = \langle k|A|m\rangle^* = \langle m|A^\dagger|k\rangle = \langle m|A|k\rangle = a_k \langle m|k\rangle \tag{1.333}$$

which implies that

$$\left(a_m^* - a_k\right) \langle m|k\rangle = 0. \tag{1.334}$$

But by (1.332), the eigenvalues a_m are real, and so we have

$$(a_m - a_k) \langle m|k\rangle = 0 \tag{1.335}$$

which tells us that when the eigenvalues are different, the eigenvectors are orthogonal. In the absence of a symmetry, all n eigenvalues usually are different, and so the eigenvectors usually are mutually orthogonal.

When two or more eigenvectors $|k_\alpha\rangle$ of a hermitian matrix have the same eigenvalue a_k, their eigenvalues are said to be **degenerate**. In this case, any linear combination of the degenerate eigenvectors also will be an eigenvector with the same eigenvalue a_k

$$A\left(\sum_{\alpha\in D} c_\alpha |k_\alpha\rangle\right) = a_k \left(\sum_{\alpha\in D} c_\alpha |k_\alpha\rangle\right) \tag{1.336}$$

where D is the set of labels α of the eigenvectors with the same eigenvalue. If the degenerate eigenvectors $|k_\alpha\rangle$ are linearly independent, then we may use the Gram–Schmidt procedure (1.113–1.123) to choose the coefficients c_α so as to construct degenerate eigenvectors that are orthogonal to each other and to the nondegenerate eigenvectors. We then may normalize these mutually orthogonal eigenvectors.

But two related questions arise: Are the degenerate eigenvectors $|k_\alpha\rangle$ linearly independent? And if so, what orthonormal linear combinations of them should we choose for a given physical problem? Let's consider the second question first.

We know that unitary transformations preserve the orthonormality of a basis (section 1.16). Any unitary transformation that commutes with the matrix A

$$[A, U] = 0 \tag{1.337}$$

represents a symmetry of A and maps each set of orthonormal degenerate eigenvectors of A into another set of orthonormal degenerate eigenvectors of A with the same eigenvalue because

$$AU|k_\alpha\rangle = UA|k_\alpha\rangle = a_k\, U|k_\alpha\rangle. \tag{1.338}$$

So there's a huge spectrum of choices for the orthonormal degenerate eigenvectors of A with the same eigenvalue. What is the right set for a given physical problem?

A sensible way to proceed is to add to the matrix A a second hermitian matrix B multiplied by a tiny, real scale factor ϵ

$$A(\epsilon) = A + \epsilon B. \tag{1.339}$$

The matrix B must completely break whatever symmetry led to the degeneracy in the eigenvalues of A. Ideally, the matrix B should be one that represents a modification of A that is physically plausible and relevant to the problem at hand. The hermitian matrix $A(\epsilon)$ then will have n different eigenvalues $a_k(\epsilon)$ and n orthonormal nondegenerate eigenvectors

$$A(\epsilon)|k_\beta, \epsilon\rangle = a_{k_\beta}(\epsilon)|k_\beta, \epsilon\rangle. \tag{1.340}$$

These eigenvectors $|k_\beta, \epsilon\rangle$ of $A(\epsilon)$ are orthogonal to each other

$$\langle k_\beta, \epsilon|k_{\beta'}, \epsilon\rangle = \delta_{\beta,\beta'} \tag{1.341}$$

and to the eigenvectors of $A(\epsilon)$ with other eigenvalues, and they remain so as we take the limit

$$|k_\beta\rangle = \lim_{\epsilon\to 0} |k_\beta, \epsilon\rangle. \tag{1.342}$$

We may choose them as the orthogonal degenerate eigenvectors of A. Since one can always find a crooked hermitian matrix B that breaks any particular symmetry, it follows that every $n \times n$ hermitian matrix A possesses n orthonormal eigenvectors, which are complete in the vector space in which A acts. (Any n linearly independent vectors span their n-dimensional vector space, as explained in section 1.9.)

Now let's return to the first question and show by a different argument that an $n \times n$ hermitian matrix has n orthogonal eigenvectors. To do this, we first note that the space $S_{\perp,k}$ of vectors $|y\rangle$ orthogonal to an eigenvector $|k\rangle$ of a hermitian operator A

$$A|k\rangle = a_k|k\rangle \tag{1.343}$$

is **invariant** under the action of A, that is, $\langle k|y\rangle = 0$ implies

$$\langle k|A|y\rangle = a_k\langle k|y\rangle = 0. \tag{1.344}$$

Thus if the vector $|y\rangle$ is in the space $S_{\perp,k}$ of vectors orthogonal to an eigenvector $|k\rangle$ of a hermitian operator A, then the vector $A|y\rangle$ also is in the space $S_{\perp,k}$. This space is invariant under the action of A.

Now a hermitian operator A acting on an n-dimensional vector space S is represented by an $n \times n$ hermitian matrix, and so it has at least one eigenvector $|1\rangle$. The subspace $S_{\perp,1}$ of S consisting of all vectors orthogonal to $|1\rangle$ is an

$(n-1)$-dimensional vector space S_{n-1} that is invariant under the action of A. On this space S_{n-1}, the operator A is represented by an $(n-1) \times (n-1)$ hermitian matrix A_{n-1}. This matrix has at least one eigenvector $|2\rangle$. The subspace $S_{\perp,2}$ of S_{n-1} consisting of all vectors orthogonal to $|2\rangle$ is an $(n-2)$-dimensional vector space S_{n-2} that is invariant under the action of A. On S_{n-2}, the operator A is represented by an $(n-2) \times (n-2)$ hermitian matrix A_{n-2} which has at least one eigenvector $|3\rangle$. By construction, the vectors $|1\rangle$, $|2\rangle$, and $|3\rangle$ are mutually orthogonal. Continuing in this way, we see that A has n orthogonal eigenvectors $|k\rangle$ for $k = 1, 2, \dots, n$. Thus **hermitian matrices are nondefective**.

The n orthogonal eigenvectors $|k\rangle$ of an $n \times n$ matrix A can be normalized and used to write the $n \times n$ identity operator I as

$$I = \sum_{k=1}^{n} |k\rangle\langle k|. \tag{1.345}$$

On multiplying from the left by the matrix A, we find

$$A = AI = A\sum_{k=1}^{n} |k\rangle\langle k| = \sum_{k=1}^{n} a_k |k\rangle\langle k| \tag{1.346}$$

which is the diagonal form of the hermitian matrix A. This expansion of A as a sum over outer products of its eigenstates multiplied by their eigenvalues exhibits the possible values a_k of the physical quantity represented by the matrix A when selective, nondestructive measurements $|k\rangle\langle k|$ of the quantity A are made.

The hermitian matrix A is diagonal in the basis of its eigenstates $|k\rangle$

$$A_{kj} = \langle k|A|j\rangle = a_k \delta_{kj}. \tag{1.347}$$

But in any other basis $|\alpha_k\rangle$, the matrix A appears as

$$A_{k\ell} = \langle \alpha_k|A|\alpha_\ell\rangle = \sum_{n=1}^{n} \langle \alpha_k|n\rangle a_n \langle n|\alpha_\ell\rangle. \tag{1.348}$$

The unitary matrix $U_{kn} = \langle \alpha_k|n\rangle$ relates the matrix $A_{k\ell}$ in an arbitrary basis to its diagonal form $A = UA^{(d)}U^\dagger$ in which $A^{(d)}$ is the diagonal matrix $A^{(d)}_{nm} = a_n\, \delta_{nm}$. An arbitrary $n \times n$ hermitian matrix A can be diagonalized by a unitary transformation.

A matrix that is **real and symmetric** is hermitian; so is one that is **imaginary and antisymmetric**. A real, symmetric matrix R can be diagonalized by an **orthogonal transformation**

$$R = O\, R^{(d)} O^{\mathsf{T}} \tag{1.349}$$

in which the matrix O is a real unitary matrix, that is, an orthogonal matrix (1.184).

Example 1.50 (The seesaw mechanism) Suppose we wish to find the eigenvalues of the real, symmetric mass matrix

$$\mathcal{M} = \begin{pmatrix} 0 & m \\ m & M \end{pmatrix} \tag{1.350}$$

in which m is an ordinary mass and M is a huge mass. The eigenvalues μ of this hermitian mass matrix satisfy $\det(\mathcal{M} - \mu I) = \mu(\mu - M) - m^2 = 0$ with solutions $\mu_\pm = \left(M \pm \sqrt{M^2 + 4m^2}\right)/2$. The larger mass $\mu_+ \approx M + m^2/M$ is approximately the huge mass M and the smaller mass $\mu_- \approx -m^2/M$ is tiny. The physical mass of a fermion is the absolute value of its mass parameter, here m^2/M.

The product of the two eigenvalues is the constant $\mu_+\mu_- = \det\mathcal{M} = -m^2$ so as μ_- goes down, μ_+ must go up. Minkowski, Yanagida, and Gell-Mann, Ramond, and Slansky invented this "**seesaw**" mechanism as an explanation of why neutrinos have such small masses, less than 1 eV/c^2. If $mc^2 = 10$ MeV, and $\mu_- c^2 \approx 0.01$ eV, which is a plausible light-neutrino mass, then the rest energy of the huge mass would be $Mc^2 = 10^7$ GeV suggesting new physics at that scale. But if we set $mc^2 = 0.28$ MeV and use $m_\nu = 0.45$ eV as an average neutrino mass, then the big mass is only $Mc^2 =$ 173 GeV, the mass of the top. Also, the small masses of the neutrinos may be related to the weakness of their interactions.

If we return to the orthogonal transformation (1.349) and multiply column ℓ of the matrix O and row ℓ of the matrix O^{T} by $\sqrt{|R_\ell^{(d)}|}$, then we arrive at the **congruency transformation** of Sylvester's theorem

$$R = C\,\hat{R}^{(d)} C^{\mathsf{T}} \tag{1.351}$$

in which the diagonal entries $\hat{R}_\ell^{(d)}$ are either ± 1 or 0 because the matrices $C_{k\ell} = \sqrt{|R_\ell^{(d)}|}\, O_{k\ell}$ and C^{T} have absorbed the factors $|R_\ell^{(d)}|$.

Example 1.51 (Principle of equivalence) If G is a real, symmetric 4×4 matrix then there's a real 4×4 matrix $D = C^{\mathsf{T}-1}$ such that

$$G_d = D^{\mathsf{T}} G D = \begin{pmatrix} g_1 & 0 & 0 & 0 \\ 0 & g_2 & 0 & 0 \\ 0 & 0 & g_3 & 0 \\ 0 & 0 & 0 & g_4 \end{pmatrix} \tag{1.352}$$

in which the diagonal entries g_i are ± 1 or 0. Thus there's a real 4×4 matrix D that casts any real symmetric metric g_{ik} of spacetime with three positive and one negative eigenvalues into the diagonal metric $\eta_{j\ell}$ of flat spacetime by the congruence

$$g_d = D^{\mathsf{T}} g D = \begin{pmatrix} -1 & 0 & 0 & 0 \\ 0 & 1 & 0 & 0 \\ 0 & 0 & 1 & 0 \\ 0 & 0 & 0 & 1 \end{pmatrix} = \eta \tag{1.353}$$

at any given point x of spacetime. Usually one needs different Ds at different points. The principle of equivalence (section 13.25) says that in the new free-fall coordinates, all physical laws take the same form as in special relativity without acceleration or gravitation in a suitably small region of spacetime about the point x.

1.30 Normal Matrices

The largest set of matrices that can be diagonalized by a unitary transformation is the set of **normal** matrices. These are square matrices that commute with their adjoints

$$[V, V^\dagger] = VV^\dagger - V^\dagger V = 0. \tag{1.354}$$

This broad class of matrices includes not only hermitian matrices but also unitary matrices since

$$[U, U^\dagger] = UU^\dagger - U^\dagger U = I - I = 0. \tag{1.355}$$

A matrix $V = U^\dagger V^{(d)} U$ that can be diagonalized by a unitary transformation U commutes with its adjoint $V^\dagger = U^\dagger V^{(d)*} U$ and so is normal because the commutator of any two diagonal matrices vanishes

$$[V, V^\dagger] = [U^\dagger V^{(d)} U, U^\dagger V^{(d)*} U] = U^\dagger [V, V^{(d)*}] U = 0. \tag{1.356}$$

To see why a normal matrix can be diagonalized by a unitary transformation, we consider an $n \times n$ normal matrix V which since it is square has n eigenvectors $|k\rangle$ with eigenvalues v_k

$$(V - v_k I)\,|k\rangle = 0 \tag{1.357}$$

(section 1.26). The square of the norm (1.85) of this vector must vanish

$$\|\, (V - v_k I)\,|k\rangle \,\|^2 = \langle k|\,(V - v_k I)^\dagger\,(V - v_k I)\,|k\rangle = 0. \tag{1.358}$$

But since V is normal, we also have

$$\langle k|\,(V - v_k I)^\dagger\,(V - v_k I)\,|k\rangle = \langle k|\,(V - v_k I)\,(V - v_k I)^\dagger\,|k\rangle. \tag{1.359}$$

So the square of the norm of the vector $\left(V^\dagger - v_k^* I\right)|k\rangle = (V - v_k I)^\dagger\,|k\rangle$ also vanishes $\|\,\left(V^\dagger - v_k^* I\right)|k\rangle\,\|^2 = 0$ which tells us that $|k\rangle$ also is an eigenvector of $V^\dagger$ with eigenvalue v_k^*

$$V^\dagger |k\rangle = v_k^* |k\rangle \quad \text{and so} \quad \langle k| V = v_k \langle k|. \tag{1.360}$$

If now $|m\rangle$ is an eigenvector of V with eigenvalue v_m

$$V|m\rangle = v_m|m\rangle \tag{1.361}$$

then

$$\langle k|V|m\rangle = v_m\langle k|m\rangle \tag{1.362}$$

and also by (1.360)

$$\langle k|V|m\rangle = v_k\langle k|m\rangle. \tag{1.363}$$

Subtracting (1.362) from (1.363), we get

$$(v_k - v_m)\,\langle k|m\rangle = 0 \tag{1.364}$$

which shows that **any two eigenvectors of a normal matrix V with different eigenvalues are orthogonal.**

To see that a normal $n \times n$ matrix V has n orthogonal eigenvectors, we first note that if $|y\rangle$ is any vector that is orthogonal to any eigenvector $|k\rangle$ of the matrix V, that is both $\langle k|y\rangle = 0$ and $V|k\rangle = v_k|k\rangle$, then the property (1.360) implies that

$$\langle k|V|y\rangle = v_k\langle k|y\rangle = 0. \tag{1.365}$$

Thus the space of vectors orthogonal to an eigenvector of a normal matrix V is invariant under the action of V. The argument following the analogous equation (1.344) applies also to normal matrices and shows that **every $n \times n$ normal matrix has n orthonormal eigenvectors**. It follows then from the argument of equations (1.345–1.348) that every $n \times n$ normal matrix V can be diagonalized by an $n \times n$ unitary matrix U

$$V = UV^{(d)}U^\dagger \tag{1.366}$$

whose kth column $U_{\ell k} = \langle\alpha_\ell|k\rangle$ is the eigenvector $|k\rangle$ in the arbitrary basis $|\alpha_\ell\rangle$ of the matrix $V_{m\ell} = \langle\alpha_m|V|\alpha_\ell\rangle$ as in (1.348).

Since the eigenstates $|k\rangle$ of a normal matrix V

$$V|k\rangle = v_k|k\rangle \tag{1.367}$$

are complete and orthonormal, we can write the identity operator I as

$$I = \sum_{k=1}^{n} |k\rangle\langle k|. \tag{1.368}$$

The product VI is V itself, so

$$V = V\,I = V\sum_{k=1}^{n} |k\rangle\langle k| = \sum_{k=1}^{n} v_k\,|k\rangle\langle k|. \tag{1.369}$$

It follows therefore that if f is a function, then $f(V)$ is

$$f(V) = \sum_{k=1}^{n} f(v_k)\,|k\rangle\langle k| \tag{1.370}$$

which is simpler than the corresponding formula (1.303) for an arbitrary nondefective matrix. This is a good way to think about functions of normal matrices.

Example 1.52 (Time-evolution operator) How do we handle the operator $\exp(-iHt/\hbar)$ that translates states in time by t? The hamiltonian H is hermitian and so is normal. Its orthonormal eigenstates $|k\rangle$ have energy E_k

$$H|k\rangle = E_k|k\rangle. \tag{1.371}$$

So we apply (1.370) with $V \to H$ and get

$$e^{-iHt/\hbar} = \sum_{k=1}^{n} e^{-iE_kt/\hbar}\,|k\rangle\langle k| \tag{1.372}$$

which lets us compute the time evolution of any state $|\psi\rangle$ as

$$e^{-iHt/\hbar}|\psi\rangle = \sum_{k=1}^{n} e^{-iE_kt/\hbar}\,|k\rangle\langle k|\psi\rangle \tag{1.373}$$

if we know the eigenstates $|k\rangle$ and eigenvalues E_k of the hamiltonian H.

The determinant $|V|$ of a normal matrix V satisfies the identities

$$|V| = \exp\left[\mathrm{Tr}(\ln V)\right], \quad \ln|V| = \mathrm{Tr}(\ln V), \quad \text{and} \quad \delta\ln|V| = \mathrm{Tr}\left(V^{-1}\delta V\right). \tag{1.374}$$

1.31 Compatible Normal Matrices

Two normal matrices A and B that **commute**

$$[A, B] \equiv AB - BA = 0 \tag{1.375}$$

are said to be **compatible**. Since these operators are normal, they have complete sets of orthonormal eigenvectors. If $|u\rangle$ is an eigenvector of A with eigenvalue z, then so is $B|u\rangle$ since

$$AB|u\rangle = BA|u\rangle = Bz|u\rangle = z\,B|u\rangle. \tag{1.376}$$

We have seen that any normal matrix A can be written as a sum (1.30) of outer products

$$A = \sum_{k=1}^{n} |a_k\rangle a_k\langle a_k| \tag{1.377}$$

of its orthonormal eigenvectors $|a_k\rangle$ which are complete in the n-dimensional vector space $\mathcal{S}$ on which A acts. Suppose now that the eigenvalues a_k of A are nondegenerate, and that B is another normal matrix acting on $\mathcal{S}$ and that the matrices A and B are compatible. Then in the basis provided by the eigenvectors (or eigenstates) $|a_k\rangle$ of the matrix A, the matrix B must satisfy

$$0 = \langle a_\ell|AB - BA|a_k\rangle = (a_\ell - a_k)\,\langle a_\ell|B|a_k\rangle \tag{1.378}$$

which says that $\langle a_\ell|B|a_k\rangle$ is zero unless $a_\ell = a_k$. Thus if the eigenvalues a_ℓ of the operator A are nondegenerate, then the operator B is diagonal

$$B = IBI = \sum_{\ell=1}^{n} |a_\ell\rangle\langle a_\ell|B \sum_{k=1}^{n} |a_k\rangle\langle a_k| = \sum_{\ell=1}^{n} |a_\ell\rangle\langle a_\ell|B|a_\ell\rangle\langle a_\ell| \tag{1.379}$$

in the $|a_\ell\rangle$ basis. Moreover B maps each eigenket $|a_k\rangle$ of A into

$$B|a_k\rangle = \sum_{\ell=1}^{n} |a_\ell\rangle\langle a_\ell|B|a_\ell\rangle\langle a_\ell|a_k\rangle = \sum_{\ell=1}^{n} |a_\ell\rangle\langle a_\ell|B|a_\ell\rangle\delta_{\ell k} = \langle a_k|B|a_k\rangle|a_k\rangle \tag{1.380}$$

which says that each eigenvector $|a_k\rangle$ of the matrix A also is an eigenvector of the matrix B with eigenvalue $\langle a_k|B|a_k\rangle$. Thus **two compatible normal matrices can be simultaneously diagonalized** if one of them has nondegenerate eigenvalues.

If A's eigenvalues a_ℓ are degenerate, each eigenvalue a_ℓ may have d_ℓ orthonormal eigenvectors $|a_\ell, k\rangle$ for $k = 1, \ldots, d_\ell$. In this case, the matrix elements $\langle a_\ell, k|B|a_m, k'\rangle$ of B are zero unless the eigenvalues are the same, $a_\ell = a_m$. The matrix representing the operator B in this basis consists of square, $d_\ell \times d_\ell$, normal submatrices $\langle a_\ell, k|B|a_\ell, k'\rangle$ arranged along its main diagonal; it is said to be in **block-diagonal form**. Since each submatrix is a $d_\ell \times d_\ell$, normal matrix, we may find linear combinations $|a_\ell, b_k\rangle$ of the degenerate eigenvectors $|a_\ell, k\rangle$ that are orthonormal eigenvectors of both compatible operators

$$A|a_\ell, b_k\rangle = a_\ell|a_\ell, b_k\rangle \quad \text{and} \quad B|a_\ell, b_k\rangle = b_k|a_\ell, b_k\rangle. \tag{1.381}$$

Thus one can simultaneously diagonalize any two compatible operators.

The converse also is true: If the operators A and B can be simultaneously diagonalized as in (1.381), then they commute

$$AB|a_\ell, b_k\rangle = Ab_k|a_\ell, b_k\rangle = a_\ell b_k|a_\ell, b_k\rangle = a_\ell B|a_\ell, b_k\rangle = BA|a_\ell, b_k\rangle \tag{1.382}$$

and so are compatible. Normal matrices can be simultaneously diagonalized if and only if they are compatible, that is, if and only if they commute.

In quantum mechanics, compatible hermitian operators represent physical observables that can be measured simultaneously to arbitrary precision (in principle). A set of compatible hermitian operators $\{A, B, C, \ldots\}$ is said to be **complete** if to every set of eigenvalues $\{a_j, b_k, c_\ell, \ldots\}$ there is only a single eigenvector $|a_j, b_k, c_\ell, \ldots\rangle$.

Example 1.53 (Compatible photon observables) For example, the state of a photon is completely characterized by its momentum and its angular momentum about its direction of motion. For a photon, the momentum operator $\boldsymbol{P}$ and the dot product $\boldsymbol{J} \cdot \boldsymbol{P}$ of the angular momentum $\boldsymbol{J}$ with the momentum form a complete set of compatible hermitian observables. Incidentally, because its mass is zero, the angular momentum $\boldsymbol{J}$ of a photon about its direction of motion can have only two values $\pm\hbar$, which correspond to its two possible states of circular polarization.

Example 1.54 (Thermal density operator) A **density operator** ρ is the most general description of a quantum-mechanical system. It is hermitian, positive definite, and of unit trace. Since it is hermitian, it can be diagonalized (section 1.29)

$$\rho = \sum_n |n\rangle\langle n|\rho|n\rangle\langle n| \tag{1.383}$$

and its eigenvalues $\rho_n = \langle n|\rho|n\rangle$ are real. Each ρ_n is the probability that the system is in the state $|n\rangle$ and so is nonnegative. The unit-trace rule

$$\sum_n \rho_n = 1. \tag{1.384}$$

ensures that these probabilities add up to one – the system is in some state.

The mean value of an operator F is the trace, $\langle F\rangle = \mathrm{Tr}(\rho F)$. So the average energy E is the trace, $E = \langle H\rangle = \mathrm{Tr}(\rho H)$. The **entropy operator** S is the negative logarithm of the density operator multiplied by Boltzmann's constant, $S = -k \ln \rho$, and the mean entropy S is $S = \langle S\rangle = -k\mathrm{Tr}(\rho \ln \rho)$.

A density operator that describes a system in thermal equilibrium at a constant temperature T is time independent and so commutes with the hamiltonian, $[\rho, H] = 0$. Since ρ and H commute, they are compatible operators (1.375), and so they can be simultaneously diagonalized. Each eigenstate $|n\rangle$ of ρ is an eigenstate of H; its energy E_n is its eigenvalue, $H|n\rangle = E_n|n\rangle$.

If we have no information about the state of the system other than its mean energy E, then we take ρ to be the density operator that maximizes the mean entropy S while respecting the constraints $c_1 = \sum_n \rho_n - 1 = 0$ and $c_2 = \mathrm{Tr}(\rho H) - E = 0$. We introduce two Lagrange multipliers (section 1.24) and maximize the unconstrained function

$$\begin{aligned} L(\rho, \lambda_1, \lambda_2) &= S - \lambda_1 c_1 - \lambda_2 c_2 \\ &= -k\sum_n \rho_n \ln \rho_n - \lambda_1\left(\sum_n \rho_n - 1\right) - \lambda_2\left(\sum_n \rho_n E_n - E\right) \end{aligned} \tag{1.385}$$

by setting its derivatives with respect to ρ_n, λ_1, and λ_2 equal to zero

$$\frac{\partial L}{\partial \rho_n} = -k\,(\ln \rho_n + 1) - \lambda_1 - \lambda_2 E_n = 0 \tag{1.386}$$

$$\frac{\partial L}{\partial \lambda_1} = \sum_n \rho_n - 1 = 0 \tag{1.387}$$

$$\frac{\partial L}{\partial \lambda_2} = \sum_n \rho_n E_n - E = 0. \tag{1.388}$$

The first (1.386) of these conditions implies that

$$\rho_n = \exp\left[-(\lambda_1 + \lambda_2 E_n + k)/k\right]. \tag{1.389}$$

We satisfy the second condition (1.387) by choosing λ_1 so that

$$\rho_n = \frac{\exp(-\lambda_2 E_n/k)}{\sum_n \exp(-\lambda_2 E_n/k)}. \tag{1.390}$$

Setting $\lambda_2 = 1/T$, we define the temperature T so that ρ satisfies the third condition (1.388). Its eigenvalue ρ_n then is

$$\rho_n = \frac{\exp(-E_n/kT)}{\sum_n \exp(-E_n/kT)}. \tag{1.391}$$

In terms of the inverse temperature $\beta \equiv 1/(kT)$, the density operator is

$$\rho = \frac{e^{-\beta H}}{\mathrm{Tr}\left(e^{-\beta H}\right)} \tag{1.392}$$

which is the **Boltzmann distribution**, also called the **canonical ensemble**.

Example 1.55 (Grand canonical ensemble) Lagrange's function for the density operator of a system of maximum entropy $S = -k\mathrm{Tr}(\rho \ln \rho)$ given a fixed mean energy $E = \mathrm{Tr}(\rho H)$ and a fixed mean number of particles $\langle N \rangle = \mathrm{Tr}(\rho N)$, in which N is the number operator $N|n\rangle = N_n|n\rangle$, is

$$\begin{aligned} L(\rho, \lambda_1, \lambda_2, \lambda_3) = & -k \sum_n \rho_n \ln \rho_n - \lambda_1 \left(\sum_n \rho_n - 1\right) \\ & - \lambda_2 \left(\sum_n \rho_n E_n - E\right) - \lambda_3 \left(\sum_n \rho_n N_n - \langle N \rangle\right). \end{aligned} \tag{1.393}$$

Setting the partial derivative of L with respect to ρ_n

$$\frac{\partial L}{\partial \rho_n} = -k\,(\ln \rho_n + 1) - \lambda_1 - \lambda_2 E_n - \lambda_3 N_n = 0 \tag{1.394}$$

as well as the partial derivatives of L with respect to the three Lagrange multipliers λ_i equal to zero, we get

$$\rho = \frac{e^{-\beta(H - \mu N)}}{\mathrm{Tr}(e^{-\beta(H - \mu N)})} \tag{1.395}$$

in which μ is the **chemical potential**.

1.32 Singular-Value Decompositions

Every complex $m \times n$ rectangular matrix A is the product of an $m \times m$ unitary matrix U, an $m \times n$ rectangular matrix Σ that is zero except on its main diagonal which consists of A's nonnegative singular values S_k, and an $n \times n$ unitary matrix $V^\dagger$

$$A = U\,\Sigma\,V^\dagger \quad \text{or} \quad A_{ik} = \sum_{\ell=1}^{\min(m,n)} U_{i\ell}\,S_\ell\,V^\dagger_{\ell k}. \tag{1.396}$$

This singular-value decomposition is a key theorem of matrix algebra.

Suppose A is a linear operator that maps vectors in an n-dimensional vector space V_n into vectors in an m-dimensional vector space V_m. The spaces V_n and V_m will have infinitely many orthonormal bases $\left\{|a_j\rangle \in V_n\right\}$ and $\{|b_k\rangle \in V_m\}$ labeled by parameters a and b. Each pair of bases provides a resolution of the identity operator I_n for V_n and I_m for V_m

$$I_n = \sum_{j=1}^{n} |a_j\rangle\langle a_j| \quad \text{and} \quad I_m = \sum_{k=1}^{m} |b_k\rangle\langle b_k| \tag{1.397}$$

and lets us write linear operator A as

$$A = I_m A I_n = \sum_{k=1}^{m}\sum_{j=1}^{n} |b_k\rangle\langle b_k|A|a_j\rangle\langle a_j| \tag{1.398}$$

in which the $\langle b_k|A|a_j\rangle$ are the elements of a complex $m \times n$ matrix.

The singular-value decomposition of the linear operator A is a choice of two special bases $\{|a_j\rangle\}$ and $\{|b_j\rangle\}$ that make $\langle b_k|A|a_j\rangle = S_j\delta_{kj}$ and so express A as

$$A = \sum_{j} |b_j\rangle S_j \langle a_j| \tag{1.399}$$

in which the sum is over the nonzero singular values S_j, which will turn out to be positive.

The kets of the special basis $\{|a_j\rangle\}$ are the eigenstates of the hermitian operator $A^\dagger A$

$$A^\dagger A|a_j\rangle = e_j|a_j\rangle. \tag{1.400}$$

These states $\{|a_j\rangle\}$ are orthogonal because $A^\dagger A$ is hermitian, and we may choose them to be normalized. The eigenvalue e_j is the squared length of the ket $A|a_j\rangle$ and so is positive or zero

$$\langle a_j|A^\dagger A|a_j\rangle = e_j\,\langle a_j|a_j\rangle = e_j \geq 0. \tag{1.401}$$

The singular values are the square roots of these eigenvalues

$$S_j = \sqrt{e_j} = \sqrt{\langle a_j|A^\dagger A|a_j\rangle}. \tag{1.402}$$

For $S_j > 0$, the special ket $|b_j\rangle$ is the suitably normalized image $A|a_j\rangle$ of the special ket $|a_j\rangle$

$$|b_j\rangle = \frac{A|a_j\rangle}{S_j}; \tag{1.403}$$

for $S_j = 0$, the ket $|b_j\rangle$ vanishes. The nonzero special kets $|b_j\rangle$ are orthonormal

$$\langle b_k|b_j\rangle = \frac{1}{S_k S_j}\langle a_k|A^\dagger A|a_j\rangle = \frac{e_j}{S_k S_j}\langle a_k|a_j\rangle = \frac{e_j}{S_k S_j}\delta_{kj} = \delta_{kj}. \tag{1.404}$$

The number of positive singular values, $S_j > 0$, is at most n. It also is at most m because each nonzero ket $|b_j\rangle$ is an orthonormal vector in the space V_m which has only m dimensions. So the number of positive singular values, $S_j > 0$, is at most $\min(m, n)$, the smaller of m and n.

The singular-value decomposition of the linear operator A then is the sum

$$A = A I_n = A \sum_{j=1}^{n} |a_j\rangle\langle a_j| = \sum_{j=1}^{n} A\,|a_j\rangle\langle a_j| = \sum_{j=1}^{n} |b_j\rangle\, S_j\, \langle a_j| \tag{1.405}$$

in which at most $\min(m, n)$ of the singular values are positive.

In terms of any two bases, $|k\rangle$ for $k = 1, \ldots, m$ for the space V_m and $|\ell\rangle$ for $\ell = 1, \ldots, n$ for the space V_n, and their identity operators

$$I_m = \sum_{k=1}^{m} |k\rangle\langle k| \quad \text{and} \quad I_n = \sum_{\ell=1}^{n} |\ell\rangle\langle\ell| \tag{1.406}$$

the singular-value decomposition of the linear operator A is

$$\begin{aligned} A &= \sum_{k=1}^{m} |k\rangle\langle k|A \sum_{\ell=1}^{n} |\ell\rangle\langle\ell| = \sum_{k=1}^{m}\sum_{j=1}^{n}\sum_{\ell=1}^{n} |k\rangle\langle k|b_j\rangle\, S_j\, \langle a_j|\ell\rangle\langle\ell| \\ &= \sum_{k=1}^{m}\sum_{i=1}^{m}\sum_{j=1}^{n}\sum_{\ell=1}^{n} |k\rangle\langle k|b_i\rangle\, S_i\, \delta_{ij}\langle a_j|\ell\rangle\langle\ell| = U\,\Sigma\,V^\dagger. \end{aligned} \tag{1.407}$$

In this expansion, the k, i matrix element of the $m \times m$ unitary matrix U is $U_{ki} = \langle k|b_i\rangle$, the i, j element of the $m \times n$ matrix Σ is $\Sigma_{ij} = S_j\,\delta_{ij}$, and the j, ℓ matrix element of the $n \times n$ unitary matrix $V^\dagger$ is $V^\dagger_{j\ell} = \langle a_j|\ell\rangle$. Thus $V^*_{\ell j} = \langle a_j|\ell\rangle$, and so $V_{\ell j} = \langle a_j|\ell\rangle^* = \langle\ell|a_j\rangle$.

The vectors $|b_j\rangle$ and $|a_j\rangle$ respectively are the left and right singular vectors. Incidentally, the singular-value decomposition (1.405) shows that the left singular vectors $|b_j\rangle$ are the eigenvectors of $A\,A^\dagger$

$$
\begin{aligned}
A\,A^\dagger &= \sum_{j=1}^{n} |b_j\rangle\, S_j\, \langle a_j| \sum_{k=1}^{n} |a_k\rangle\, S_k\, \langle b_k| = \sum_{j,k=1}^{n} |b_j\rangle\, S_j\, \langle a_j|a_k\rangle\, S_k\, \langle b_k| \\
&= \sum_{j,k=1}^{n} |b_j\rangle\, S_j\, \delta_{jk}\, S_k\, \langle b_k| = \sum_{j=1}^{n} |b_j\rangle\, S_j^2\, \langle b_j|
\end{aligned}
\tag{1.408}
$$

just as (1.400) the right singular vectors $|a_j\rangle$ are the eigenvectors of $A^\dagger A$.

The kets $|a_j\rangle$ whose singular values vanish, $S_j = 0$, span the **null space** or **kernel** of the linear operator A.

Example 1.56 (Singular-value decomposition of a 2 × 3 matrix) If A is

$$
A = \begin{pmatrix} 0 & 1 & 0 \\ 1 & 0 & 1 \end{pmatrix} \tag{1.409}
$$

then the positive hermitian matrix $A^\dagger A$ is

$$
A^\dagger A = \begin{pmatrix} 1 & 0 & 1 \\ 0 & 1 & 0 \\ 1 & 0 & 1 \end{pmatrix}. \tag{1.410}
$$

The normalized eigenvectors and eigenvalues of $A^\dagger A$ are

$$
|a_1\rangle = \frac{1}{\sqrt{2}} \begin{pmatrix} 1 \\ 0 \\ 1 \end{pmatrix}, \quad |a_2\rangle = \begin{pmatrix} 0 \\ 1 \\ 0 \end{pmatrix}, \quad |a_3\rangle = \frac{1}{\sqrt{2}} \begin{pmatrix} -1 \\ 0 \\ 1 \end{pmatrix} \tag{1.411}
$$

and their eigenvalues are $e_1 = 2$, $e_2 = 1$, and $e_3 = 0$. The third eigenvalue e_3 had to vanish because A is a 3 × 2 matrix.

The vector $A|a_1\rangle$ (as a row vector) is $(0, \sqrt{2})$, and its norm is $\sqrt{2}$, so the normalized vector is $|b_1\rangle = (0, 1)$. Similarly, the vector $|b_2\rangle$ is $A|a_2\rangle = (1, 0)$. The singular-value decomposition (SVD) of A then is

$$
A = \sum_{n=1}^{2} |b_j\rangle S_j \langle a_j| = U \Sigma V^\dagger \tag{1.412}
$$

where $S_n = \sqrt{e_n}$. The unitary matrices are $U_{k,n} = \langle k|b_n\rangle$ and $V_{\ell,j} = \langle \ell|a_j\rangle$ are

$$
U = \begin{pmatrix} 0 & 1 \\ 1 & 0 \end{pmatrix} \quad \text{and} \quad V = \frac{1}{\sqrt{2}} \begin{pmatrix} 1 & 0 & -1 \\ 0 & \sqrt{2} & 0 \\ 1 & 0 & 1 \end{pmatrix} \tag{1.413}
$$

and the diagonal matrix Σ is

$$
\Sigma = \begin{pmatrix} \sqrt{2} & 0 & 0 \\ 0 & 1 & 0 \end{pmatrix}. \tag{1.414}
$$

So finally the SVD of $A = U\Sigma V^\dagger$ is

$$A = \begin{pmatrix} 0 & 1 \\ 1 & 0 \end{pmatrix} \begin{pmatrix} \sqrt{2} & 0 & 0 \\ 0 & 1 & 0 \end{pmatrix} \frac{1}{\sqrt{2}} \begin{pmatrix} 1 & 0 & 1 \\ 0 & \sqrt{2} & 0 \\ -1 & 0 & 1 \end{pmatrix}. \tag{1.415}$$

The null space or kernel of A is the set of vectors that are real multiples $c|a_3\rangle$ of the eigenvector $|a_3\rangle$ which has a zero eigenvalue, $e_3 = 0$. It is the third column of the matrix V displayed in (1.413).

Example 1.57 (Matlab's singular value decomposition) Matlab's command [U,S,V] = svd(X) performs the singular-value decomposition (SVD) of the matrix X. For instance

```
>> X = rand(3,3) + i*rand(3,3)

     0.6551 + 0.2551i   0.4984 + 0.8909i   0.5853 + 0.1386i
X =  0.1626 + 0.5060i   0.9597 + 0.9593i   0.2238 + 0.1493i
     0.1190 + 0.6991i   0.3404 + 0.5472i   0.7513 + 0.2575i
>> [U,S,V] = svd(X)

    -0.3689 - 0.4587i   0.4056 - 0.2075i   0.4362 - 0.5055i
U = -0.3766 - 0.5002i  -0.5792 - 0.2810i   0.0646 + 0.4351i
    -0.2178 - 0.4626i   0.1142 + 0.6041i  -0.5938 - 0.0901i

     2.2335         0         0
S =       0    0.7172         0
          0         0    0.3742

    -0.4577             0.5749             0.6783
V = -0.7885 - 0.0255i  -0.6118 - 0.0497i  -0.0135 + 0.0249i
    -0.3229 - 0.2527i   0.3881 + 0.3769i  -0.5469 - 0.4900i .
```

The singular values are 2.2335, 0.7172, and 0.3742.

We may use the SVD to solve, when possible, the matrix equation

$$A\,|x\rangle = |y\rangle \tag{1.416}$$

for the n-dimensional vector $|x\rangle$ in terms of the m-dimensional vector $|y\rangle$ and the $m \times n$ matrix A. Using the SVD expansion (1.405), we have

$$\sum_{j=1}^{\min(m,n)} |b_j\rangle S_j\,\langle a_j|x\rangle = |y\rangle. \tag{1.417}$$

The orthonormality (1.404) of the vectors $|b_j\rangle$ then tells us that

$$S_j\,\langle a_j|x\rangle = \langle b_j|y\rangle. \tag{1.418}$$

If the singular value is positive, $S_j > 0$, then we may divide by it to get $\langle a_j|x\rangle = \langle b_j|y\rangle/S_j$ and so find the solution

$$|x\rangle = \sum_{j=1}^{\min(m,n)} \frac{\langle b_j|y\rangle}{S_j} |a_j\rangle. \tag{1.419}$$

But this solution is not always available or unique.

For instance, if for some ℓ the inner product $\langle b_\ell|y\rangle \neq 0$ while the singular value $S_\ell = 0$, then there is no solution to equation (1.416). This problem occurs when $m > n$ because there are at most $n < m$ nonzero singular values.

Example 1.58 Suppose A is the 3×2 matrix

$$A = \begin{pmatrix} r_1 & p_1 \\ r_2 & p_2 \\ r_3 & p_3 \end{pmatrix} \tag{1.420}$$

and the vector $|y\rangle$ is the cross-product $|y\rangle = \boldsymbol{L} = \boldsymbol{r} \times \boldsymbol{p}$. Then no solution $|x\rangle$ exists to the equation $A|x\rangle = |y\rangle$ (unless $\boldsymbol{r}$ and $\boldsymbol{p}$ are parallel) because $A|x\rangle$ is a linear combination of the vectors $\boldsymbol{r}$ and $\boldsymbol{p}$ while $|y\rangle = \boldsymbol{L}$ is perpendicular to both $\boldsymbol{r}$ and $\boldsymbol{p}$.

Even when the matrix A is square, the equation (1.416) sometimes has no solutions. For instance, if A is a square defective matrix (section 1.26), then $A|x\rangle = |y\rangle$ will fail to have a solution when the vector $|y\rangle$ lies outside the space spanned by the linearly dependent eigenvectors of the matrix A.

And when $n > m$, as in for instance

$$\begin{pmatrix} a & b & c \\ d & e & f \end{pmatrix} \begin{pmatrix} x_1 \\ x_2 \\ x_3 \end{pmatrix} = \begin{pmatrix} y_1 \\ y_2 \end{pmatrix} \tag{1.421}$$

the solution (1.419) is never unique, for we may add to it any linear combination of the vectors $|a_j\rangle$ that have zero as their singular values

$$|x\rangle = \sum_{j=1}^{\min(m,n)} \frac{\langle b_j|y\rangle}{S_n} |a_j\rangle + \sum_{j,S_j=0} x_j |a_j\rangle \tag{1.422}$$

of which there are at least $n - m$.

Example 1.59 (CKM matrix) In the standard model, the mass matrices of the u, c, t and d, s, b quarks are 3×3 complex matrices M_u and M_d with singular-value decompositions $M_u = U_u \Sigma_u V_u^\dagger$ and $M_d = U_d \Sigma_d V_d^\dagger$ whose singular-values are the quark masses. The unitary CKM matrix $U_u^\dagger U_d$ (Cabibbo, Kobayashi, Maskawa) describes transitions among the quarks mediated by the $W^\pm$ gauge bosons. By redefining the quark fields, one may make the CKM matrix real, apart from a phase that violates charge-conjugation-parity (CP) symmetry.

The adjoint of a complex symmetric matrix M is its complex conjugate, $M^\dagger = M^*$. So by (1.400), its right singular vectors $|n\rangle$ are the eigenstates of $M^* M$

$$M^* M|n\rangle = S_n^2|n\rangle \tag{1.423}$$

and by (1.408) its left singular vectors $|m_n\rangle$ are the eigenstates of $M M^*$

$$M M^*|m_n\rangle = \left(M^* M\right)^* |m_n\rangle = S_n^2|m_n\rangle. \tag{1.424}$$

Thus its left singular vectors are the complex conjugates of its right singular vectors, $|m_n\rangle = |n\rangle^*$. So the unitary matrix V is the complex conjugate of the unitary matrix U, and the SVD of M is (Autonne, 1915)

$$M = U\Sigma U^{\mathsf{T}}. \tag{1.425}$$

1.33 Moore–Penrose Pseudoinverses

Although a matrix A has an inverse A^{-1} if and only if it is square and has a nonzero determinant, one may use the singular-value decomposition to make a pseudoinverse A^+ for an arbitrary $m \times n$ matrix A. If the singular-value decomposition of the matrix A is

$$A = U\,\Sigma\,V^\dagger \tag{1.426}$$

then the Moore–Penrose pseudoinverse (Eliakim H. Moore 1862–1932, Roger Penrose 1931–) is

$$A^+ = V\,\Sigma^+\,U^\dagger \tag{1.427}$$

in which Σ^+ is the transpose of the matrix Σ with every nonzero entry replaced by its inverse (and the zeros left as they are). One may show that the pseudoinverse A^+ satisfies the four relations

$$\begin{aligned} A\,A^+\,A &= A, & A^+A\,A^+ &= A^+, \\ \left(A\,A^+\right)^\dagger &= A\,A^+, \quad \text{and} & \left(A^+A\right)^\dagger &= A^+\,A, \end{aligned} \tag{1.428}$$

and that it is the only matrix that does so.

Suppose that all the singular values of the $m \times n$ matrix A are positive. In this case, if A has more rows than columns, so that $m > n$, then the product A^+A is the $n \times n$ identity matrix I_n

$$A^+A = V^\dagger\Sigma^+\Sigma V = V^\dagger I_n V = I_n \tag{1.429}$$

and AA^+ is an $m \times m$ matrix that is not the identity matrix I_m. If instead A has more columns than rows, so that $n > m$, then AA^+ is the $m \times m$ identity matrix I_m

$$AA^+ = U\Sigma\Sigma^+U^\dagger = UI_mU^\dagger = I_m \tag{1.430}$$

and A^+A is an $n \times n$ matrix that is not the identity matrix I_n. If the matrix A is square with positive singular values, then it has a true inverse A^{-1} which is equal to its pseudoinverse

$$A^{-1} = A^+. \tag{1.431}$$

If the columns of A are linearly independent, then the matrix $A^\dagger A$ has an inverse, and the pseudoinverse is

$$A^+ = \left(A^\dagger A\right)^{-1} A^\dagger. \tag{1.432}$$

The solution (1.248) to the complex least-squares method used this pseudoinverse.

If the rows of A are linearly independent, then the matrix $AA^\dagger$ has an inverse, and the pseudoinverse is

$$A^+ = A^\dagger \left(AA^\dagger\right)^{-1}. \tag{1.433}$$

If both the rows and the columns of A are linearly independent, then the matrix A has an inverse A^{-1} which is its pseudoinverse

$$A^{-1} = A^+. \tag{1.434}$$

Example 1.60 (The pseudoinverse of a 2 × 3 matrix) The pseudoinverse A^+ of the matrix A

$$A = \begin{pmatrix} 0 & 1 & 0 \\ 1 & 0 & 1 \end{pmatrix} \tag{1.435}$$

with singular-value decomposition (1.415) is

$$\begin{aligned} A^+ &= V\,\Sigma^+\,U^\dagger \\ &= \frac{1}{\sqrt{2}} \begin{pmatrix} 1 & 0 & -1 \\ 0 & \sqrt{2} & 0 \\ 1 & 0 & 1 \end{pmatrix} \begin{pmatrix} 1/\sqrt{2} & 0 \\ 0 & 1 \\ 0 & 0 \end{pmatrix} \begin{pmatrix} 0 & 1 \\ 1 & 0 \end{pmatrix} = \begin{pmatrix} 0 & 1/2 \\ 1 & 0 \\ 0 & 1/2 \end{pmatrix} \end{aligned} \tag{1.436}$$

which satisfies the four conditions (1.428). The product $A\,A^+$ gives the 2 × 2 identity matrix

$$A\,A^+ = \begin{pmatrix} 0 & 1 & 0 \\ 1 & 0 & 1 \end{pmatrix} \begin{pmatrix} 0 & 1/2 \\ 1 & 0 \\ 0 & 1/2 \end{pmatrix} = \begin{pmatrix} 1 & 0 \\ 0 & 1 \end{pmatrix} \tag{1.437}$$

which is an instance of (1.430). Moreover, the rows of A are linearly independent, and so the simple rule (1.433) works:

$$
\begin{aligned}
A^+ &= A^\dagger \left(AA^\dagger\right)^{-1} \\
&= \begin{pmatrix} 1 & 0 \\ 0 & 1 \\ 1 & 0 \end{pmatrix} \left(\begin{pmatrix} 0 & 1 & 0 \\ 1 & 0 & 1 \end{pmatrix} \begin{pmatrix} 1 & 0 \\ 0 & 1 \\ 1 & 0 \end{pmatrix} \right)^{-1} = \begin{pmatrix} 1 & 0 \\ 0 & 1 \\ 1 & 0 \end{pmatrix} \begin{pmatrix} 0 & 1 \\ 2 & 0 \end{pmatrix}^{-1} \\
&= \begin{pmatrix} 1 & 0 \\ 0 & 1 \\ 1 & 0 \end{pmatrix} \begin{pmatrix} 0 & 1/2 \\ 1 & 0 \end{pmatrix} = \begin{pmatrix} 0 & 1/2 \\ 1 & 0 \\ 0 & 1/2 \end{pmatrix}
\end{aligned} \tag{1.438}
$$

which is (1.436).

The columns of the matrix A are not linearly independent, however, and so the simple rule (1.432) fails. Thus the product $A^+ A$

$$
A^+ A = \begin{pmatrix} 0 & 1/2 \\ 1 & 0 \\ 0 & 1/2 \end{pmatrix} \begin{pmatrix} 0 & 1 & 0 \\ 1 & 0 & 1 \end{pmatrix} = \frac{1}{2} \begin{pmatrix} 1 & 0 & 1 \\ 0 & 2 & 0 \\ 1 & 0 & 1 \end{pmatrix} \tag{1.439}
$$

is not the 3×3 identity matrix which it would be if (1.432) held.

1.34 Tensor Products and Entanglement

Tensor products are used to describe composite systems, such as the spins of two electrons. The terms **direct product** and tensor product sometimes are used interchangeably.

If A is an $n \times n$ matrix with elements A_{ij} and B is an $m \times m$ matrix with elements $B_{k\ell}$, then their **tensor product** $C = A \otimes B$ is an $nm \times nm$ matrix with elements $C_{ik,j\ell} = A_{ij} B_{k\ell}$. This tensor-product matrix $A \otimes B$ maps a vector $V_{j\ell}$ into the vector

$$
W_{ik} = \sum_{j=1}^{n} \sum_{\ell=1}^{m} C_{ik,j\ell} V_{j\ell} = \sum_{j=1}^{n} \sum_{\ell=1}^{m} A_{ij} B_{k\ell} V_{j\ell} \tag{1.440}
$$

in which the second double index $j\ell$ of C and the second indices of A and B match the double index $j\ell$ of the vector V.

A tensor-product operator is a product of two operators that act on two different vector spaces. Suppose that an operator A acts on a space S spanned by n kets $|i\rangle$, and that an operator B acts on a space T spanned by m kets $|k\rangle$, and that both operators map vectors into their spaces S and T. Then we may write A as

$$
A = I_S A I_S = \sum_{i,j=1}^{n} |i\rangle\langle i|A|j\rangle\langle j| \tag{1.441}
$$

and B as

$$B = I_T B I_T = \sum_{k,s=1}^{m} |k\rangle\langle k|B|\ell\rangle\langle \ell|. \tag{1.442}$$

Their tensor product $C = A \otimes B$ is

$$C = A \otimes B = \sum_{i,j=1}^{n} \sum_{k,\ell=1}^{m} |i\rangle \otimes |k\rangle \, \langle i|A|j\rangle\langle k|B|\ell\rangle \, \langle j| \otimes \langle \ell| \tag{1.443}$$

and it acts on the tensor-product vector space $S \otimes T$ which is spanned by the tensor-product kets $|i, k\rangle = |i\rangle\,|k\rangle = |i\rangle \otimes |k\rangle$ and has dimension nm.

An arbitrary vector in the space $S \otimes T$ is of the form

$$|\psi\rangle = \sum_{i=1}^{n}\sum_{k=1}^{m} \psi(i,k)\,|i\rangle \otimes |k\rangle = \sum_{i=1}^{n}\sum_{k=1}^{m} |i,k\rangle\langle i,k|\psi\rangle. \tag{1.444}$$

Vectors $|\phi_S, \chi_T\rangle$ that are tensor products $|\phi_S\rangle \otimes |\chi_T\rangle$ of two vectors $|\phi_S\rangle \in S$ and $|\chi_T\rangle \in T$

$$|\phi_S\rangle \otimes |\chi_T\rangle = \left(\sum_{i=1}^{n} \phi_i |i\rangle\right) \otimes \left(\sum_{k=1}^{m} \chi_k |k\rangle\right) = \sum_{i=1}^{n}\sum_{k=1}^{m} \phi_i \chi_k |i,k\rangle \tag{1.445}$$

are **separable**. States represented by vectors that are not separable are said to be **entangled**. Most states in a tensor-product space are entangled.

In the simpler notation $|i, k\rangle$ for $|i\rangle \otimes |k\rangle$, a tensor-product operator $A \otimes B$ maps an arbitrary vector (1.444) to

$$(A \otimes B)\,|\psi\rangle = \sum_{i,j=1}^{n} \sum_{k,\ell=1}^{m} |i,k\rangle \, \langle i|A|j\rangle\langle k|B|\ell\rangle \, \langle j,\ell|\psi\rangle. \tag{1.446}$$

Direct-product operators are special. An arbitrary linear operator on the space $S\otimes T$

$$D = \sum_{i,j=1}^{n} \sum_{k,\ell=1}^{m} |i,k\rangle\langle i,k|D|j,\ell\rangle\langle j,\ell| \tag{1.447}$$

maps an arbitrary vector (1.444) into the vector

$$D\,|\psi\rangle = \sum_{i,j=1}^{n} \sum_{k,\ell=1}^{m} |i,k\rangle\langle i,k|D|j,\ell\rangle\langle j,\ell|\psi\rangle. \tag{1.448}$$

Example 1.61 (States of the hydrogen atom) Suppose the state $|n, \ell, m\rangle$ is an eigenvector of the hamiltonian H, the square $\boldsymbol{L}^2$ of the orbital angular momentum $\boldsymbol{L}$, and the third component of the orbital angular momentum L_3 of a hydrogen atom without spin:

$$
\begin{aligned}
H|n, \ell, m\rangle &= E_n|n, \ell, m\rangle \\
\boldsymbol{L}^2|n, \ell, m\rangle &= \hbar^2\ell(\ell+1)|n, \ell, m\rangle \\
L_3|n, \ell, m\rangle &= \hbar m|n, \ell, m\rangle.
\end{aligned} \tag{1.449}
$$

The state $|n, \ell, m\rangle = |n\rangle \otimes |\ell, m\rangle$ is separable. Suppose the states $|\sigma\rangle$ for $\sigma = \pm$ are eigenstates of the third component S_3 of the operator $\boldsymbol{S}$ that represents the spin of the electron

$$
S_3|\sigma\rangle = \sigma\frac{\hbar}{2}|\sigma\rangle. \tag{1.450}
$$

The separable, tensor-product states

$$
|n, \ell, m, \sigma\rangle \equiv |n, \ell, m\rangle \otimes |\sigma\rangle \equiv |n, \ell, m\rangle|\sigma\rangle \tag{1.451}
$$

represent a hydrogen atom including the spin of its electron. These separable states are eigenvectors of all four operators H, $\boldsymbol{L}^2$, L_3, and S_3:

$$
\begin{aligned}
&H|n, \ell, m, \sigma\rangle = E_n|n, \ell, m, \sigma\rangle \quad \boldsymbol{L}^2|n, \ell, m, \sigma\rangle = \hbar^2\ell(\ell+1)|n, \ell, m, \sigma\rangle \\
&L_3|n, \ell, m, \sigma\rangle = \hbar m|n, \ell, m, \sigma\rangle \quad S_3|n, \ell, m, \sigma\rangle = \sigma\frac{1}{2}\hbar|n, \ell, m, \sigma\rangle.
\end{aligned} \tag{1.452}
$$

Suitable linear combinations of these states are eigenstates of the square $\boldsymbol{J}^2$ of the composite angular momentum $\boldsymbol{J} = \boldsymbol{L} + \boldsymbol{S}$ as well as of J_3, L_3, and S_3. Many of these states are entangled.

Example 1.62 (Adding two spins) The smallest positive value of angular momentum is $\hbar/2$. The spin-one-half angular-momentum operators $\boldsymbol{S}$ are represented by three 2×2 matrices, $S_a = \frac{1}{2}\hbar\,\sigma_a$, the Pauli matrices

$$
\sigma_1 = \begin{pmatrix} 0 & 1 \\ 1 & 0 \end{pmatrix}, \quad \sigma_2 = \begin{pmatrix} 0 & -i \\ i & 0 \end{pmatrix}, \quad \text{and} \quad \sigma_3 = \begin{pmatrix} 1 & 0 \\ 0 & -1 \end{pmatrix} \tag{1.453}
$$

which are both hermitian and unitary. They map the basis vectors

$$
|+\rangle = \begin{pmatrix} 1 \\ 0 \end{pmatrix} \quad \text{and} \quad |-\rangle = \begin{pmatrix} 0 \\ 1 \end{pmatrix} \tag{1.454}
$$

to $\sigma_1|\pm\rangle = |\mp\rangle$, $\sigma_2|\pm\rangle = \pm i|\mp\rangle$, and $\sigma_3|\pm\rangle = \pm|\pm\rangle$.

Suppose two spin operators $\boldsymbol{S}^{(1)}$ and $\boldsymbol{S}^{(2)}$ act on two spin-one-half systems with states $|\pm\rangle_1$ that are eigenstates of $S_3^{(1)}$ and states $|\pm\rangle_2$ that are eigenstates of $S_3^{(2)}$

$$
S_3^{(1)}|\pm\rangle_1 = \pm\frac{1}{2}\hbar|\pm\rangle_1 \quad \text{and} \quad S_3^{(2)}|\pm\rangle_2 = \pm\frac{1}{2}\hbar|\pm\rangle_2. \tag{1.455}
$$

Then the tensor-product states $|\pm,\pm\rangle = |\pm\rangle_1|\pm\rangle_2 = |\pm\rangle_1 \otimes |\pm\rangle_2$ are eigenstates of both $S_3^{(1)}$ and $S_3^{(2)}$

$$
S_3^{(1)}|\pm, s_2\rangle = \pm\frac{1}{2}\hbar\,|+, s_2\rangle \quad \text{and} \quad S_3^{(2)}|s_1, \pm\rangle = \pm\frac{1}{2}\hbar\,|s_1, \pm\rangle. \tag{1.456}
$$

These states also are eigenstates of the third component of the spin operator of the combined system

$$S_3 = S_3^{(1)} + S_3^{(2)} \quad \text{that is} \quad S_3|s_1, s_2\rangle = \frac{1}{2}\hbar\,(s_1 + s_2)\,|s_1, s_2\rangle. \tag{1.457}$$

Thus $S_3|+,+\rangle = \hbar|+,+\rangle$, and $S_3|-,-\rangle = -\hbar|-,-\rangle$, while $S_3|+,-\rangle = 0$ and $S_3|-,+\rangle = 0$.

Using the notation (1.454), we can compute the effect of the operator $\boldsymbol{S}^2$ on the state $|++\rangle$.

We find for S_1^2

$$\begin{aligned} S_1^2|++\rangle &= \left(S_1^{(1)} + S_1^{(2)}\right)^2|++\rangle = \frac{\hbar^2}{4}\left(\sigma_1^{(1)} + \sigma_1^{(2)}\right)^2|++\rangle \\ &= \frac{1}{2}\hbar^2\left(1 + \sigma_1^{(1)}\sigma_1^{(2)}\right)|++\rangle = \frac{1}{2}\hbar^2\left(|++\rangle + \sigma_1^{(1)}|+\rangle\sigma_1^{(2)}|+\rangle\right) \\ &= \frac{1}{2}\hbar^2\left(|++\rangle + |--\rangle\right) \end{aligned} \tag{1.458}$$

and leave S_2^2 and S_3^2 to Exercise 1.36.

Example 1.63 (Entangled states) A neutral pion π^0 has zero angular momentum and negative parity. Its mass is 135 MeV/c^2 and 99% of them decay into two photons with a mean lifetime of 8.5×10^{-17} s. A π^0 at rest decays into two photons moving in opposite directions along the same axis, and the spins of the photons must be either parallel to their momenta $|+,+\rangle$, positive helicity, or antiparallel to their momenta $|-,-\rangle$, negative helicity. Parity reverses helicity, and so the state of negative parity and zero angular momentum is

$$|\gamma,\gamma\rangle = \frac{1}{\sqrt{2}}\Big(|+,+\rangle - |-,-\rangle\Big). \tag{1.459}$$

The two photons have the same helicity. If the helicity of one photon is measured to be positive, then a measurement of the other photon will show it to have positive helicity. The state is entangled.

One π^0 in 17 million will decay into a positron and an electron in a state of zero angular momentum. The spin part of the final state is

$$|e^+, e^-\rangle = \frac{1}{\sqrt{2}}\Big(|+,-\rangle - |-,+\rangle\Big). \tag{1.460}$$

If the spin along any axis of one of the electrons is measured to be positive, then a measurement of the spin of the other electron along the same axis will be negative. The state is entangled.

1.35 Density Operators

A general quantum-mechanical system is represented by a **density operator** ρ that is hermitian $\rho^\dagger = \rho$, of unit trace $\mathrm{Tr}\rho = 1$, and positive $\langle\psi|\rho|\psi\rangle \geq 0$ for all kets $|\psi\rangle$.

If the state $|\psi\rangle$ is normalized, then $\langle\psi|\rho|\psi\rangle$ is the nonnegative probability that the system is in that state. This probability is real because the density matrix is hermitian. If $\{|k\rangle\}$ is any complete set of orthonormal states

$$I = \sum_k |k\rangle\langle k| \tag{1.461}$$

then the probability that the system is in the state $|k\rangle$ is

$$p_k = \langle k|\rho|k\rangle = \mathrm{Tr}\left(\rho|k\rangle\langle k|\right). \tag{1.462}$$

Since $\mathrm{Tr}\rho = 1$, the sum of these probabilities is unity

$$\sum_k p_k = \sum_k \langle k|\rho|k\rangle = \mathrm{Tr}\left(\rho\sum_k |k\rangle\langle k|\right) = \mathrm{Tr}\left(\rho I\right) = \mathrm{Tr}\rho = 1. \tag{1.463}$$

A system that is measured to be in a state $|k\rangle$ cannot simultaneously be measured to be in an orthogonal state $|\ell\rangle$. The probabilities sum to unity because the system must be in some state.

Since the density operator ρ is hermitian and positive, it has a complete, orthonormal set of eigenvectors $|k\rangle$ all of which have nonnegative eigenvalues ρ_k

$$\rho|k\rangle = \rho_k|k\rangle. \tag{1.464}$$

They afford for it an expansion in their outer products

$$\rho = \sum_k \rho_k|k\rangle\langle k| \tag{1.465}$$

each weighted by the probability ρ_k that the system is in the state $|k\rangle$.

A system composed of two systems, one with basis kets $|i\rangle$ and the other with basis kets $|k\rangle$, has basis states $|i,k\rangle = |i\rangle|k\rangle$ and can be described by the density operator

$$\rho = \sum_{ijk\ell} |i,k\rangle\langle i,k|\rho|j,\ell\rangle\langle j,\ell|. \tag{1.466}$$

The density operator for the first system is the trace of ρ over the states $|k\rangle$ of the second system

$$\rho_1 = \sum_k \langle k|\rho|k\rangle = \sum_{ijk} |i\rangle\langle i,k|\rho|j,k\rangle\langle j| \tag{1.467}$$

and similarly the density operator for the second system is the trace of ρ over the states $|i\rangle$ of the first system

$$\rho_2 = \sum_i \langle i|\rho|i\rangle = \sum_{jk\ell} |k\rangle\langle i,k|\rho|i,\ell\rangle\langle\ell|. \tag{1.468}$$

Classical entropy is an **extensive** quantity like volume, mass, and energy. The classical entropy of a composite system is the sum of the classical entropies of its parts. But quantum entropy $S = -k\mathrm{Tr}(\rho \log \rho)$ is not necessarily extensive. The quantum entropy of an entangled system can be less than the sum of the quantum entropies of its parts. The quantum entropy of each of the eigenstates $|\gamma,\gamma\rangle$ and $|e^+,e^-\rangle$ of Example 1.63 is zero, but the sum of the quantum entropies of their parts is in both cases $2k \log 2$.

1.36 Schmidt Decomposition

Suppose $|\psi\rangle$ is an arbitrary vector in the tensor product of the vector spaces B and C

$$|\psi\rangle = \sum_{i=1}^{n}\sum_{k=1}^{m} A_{ik}\,|i\rangle \otimes |k\rangle. \tag{1.469}$$

The arbitrary matrix A has a singular-value decomposition (1.396)

$$A_{ik} = \sum_{\ell=1}^{\min(n,m)} U_{i\ell}\, S_\ell\, V^\dagger_{\ell k}. \tag{1.470}$$

In terms of this SVD, the vector $|\psi\rangle$ is

$$\begin{aligned} |\psi\rangle &= \sum_{i=1}^{n}\sum_{k=1}^{m}\sum_{\ell=1}^{\min(n,m)} U_{i\ell}\, S_\ell\, V^\dagger_{\ell k}\,|i\rangle \otimes |k\rangle \\ &= \sum_{\ell=1}^{\min(n,m)} S_\ell\,|U,\ell\rangle \otimes |V^\dagger,\ell\rangle \end{aligned} \tag{1.471}$$

where the state $|U,\ell\rangle$ is

$$|U,\ell\rangle = \sum_{i=1}^{n} U_{i\ell}\,|i\rangle \tag{1.472}$$

and is in the vector space B, and the state $|V^\dagger,\ell\rangle$ is

$$|V^\dagger,\ell\rangle = \sum_{k=1}^{m} V^\dagger_{\ell k}\,|k\rangle \tag{1.473}$$

and is in the vector space C. The states $|U,\ell\rangle$ and $|V^\dagger,\ell\rangle$ are orthonormal

$$\langle U,\ell|U,\ell'\rangle = \delta_{\ell\ell'} \quad \text{and} \quad \langle V^\dagger,\ell|V^\dagger,\ell'\rangle = \delta_{\ell\ell'} \tag{1.474}$$

because the matrices U and V are unitary.

The outer product of a tensor-product state (1.469) is a pure-state density operator

$$\rho = |\psi\rangle\langle\psi| = \sum_{\ell\ell'} S_\ell S_{\ell'} (|U,\ell\rangle \otimes |V^\dagger,\ell\rangle)(\langle U,\ell'| \otimes \langle V^\dagger,\ell'|). \tag{1.475}$$

Taking the trace over a complete set of orthonormal states in B or C, we get the density operator ρ in the spaces B or C

$$\begin{aligned} \rho_B &= \mathrm{Tr}_C(\rho) = \sum_{\ell''} \langle V^\dagger,\ell''|\rho|V^\dagger,\ell''\rangle = \sum_\ell S_\ell^2\, |U,\ell\rangle\langle U,\ell| \\ \rho_C &= \mathrm{Tr}_B(\rho) = \sum_{\ell''} \langle U,\ell''|\rho|U,\ell''\rangle = \sum_\ell S_\ell^2\, |V^\dagger,\ell\rangle\langle V^\dagger,\ell|. \end{aligned} \tag{1.476}$$

The density operators ρ_B and ρ_C have the same eigenvalues and therefore the same von Neumann entropy

$$\begin{aligned} s(\rho_B) &= -k\,\mathrm{Tr}_B(\rho_B \log \rho_B) = -k \sum_\ell S_\ell^2 \log(S_\ell^2) \\ s(\rho_C) &= -k\,\mathrm{Tr}_C(\rho_C \log \rho_C) = -k \sum_\ell S_\ell^2 \log(S_\ell^2). \end{aligned} \tag{1.477}$$

The nonzero, positive singular values S_ℓ are called Schmidt coefficients. The number of them is the Schmidt rank or Schmidt number of the state $|\psi\rangle$. The state $|\psi\rangle$ is entangled if and only if its Schmidt rank is greater than unity.

1.37 Correlation Functions

We can define two Schwarz inner products for a density matrix ρ. If $|f\rangle$ and $|g\rangle$ are two states, then the inner product

$$(f,g) \equiv \langle f|\rho|g\rangle \tag{1.478}$$

for $g = f$ is nonnegative, $(f,f) = \langle f|\rho|f\rangle \ge 0$, and satisfies the other conditions (1.78, 1.79, and 1.81) for a Schwarz inner product.

The second Schwarz inner product applies to operators A and B and is defined (Titulaer and Glauber, 1965) as

$$(A,B) = \mathrm{Tr}\left(\rho A^\dagger B\right) = \mathrm{Tr}\left(B\rho A^\dagger\right) = \mathrm{Tr}\left(A^\dagger B\rho\right). \tag{1.479}$$

This inner product is nonnegative when $A = B$ and obeys the other rules (1.78, 1.79, and 1.81) for a Schwarz inner product.

These two degenerate inner products are not inner products in the strict sense of (1.78–1.84), but they are Schwarz inner products, and so (1.98–1.99) they satisfy the Schwarz inequality (1.99)

$$(f,f)(g,g) \ge |(f,g)|^2. \tag{1.480}$$

Applied to the first, vector, Schwarz inner product (1.478), the Schwarz inequality gives

$$\langle f|\rho|f\rangle\langle g|\rho|g\rangle \geq |\langle f|\rho|g\rangle|^2 \tag{1.481}$$

which is a useful property of density matrices. Application of the Schwarz inequality to the second, operator, Schwarz inner product (1.479) gives (Titulaer and Glauber, 1965)

$$\mathrm{Tr}\left(\rho A^\dagger A\right)\mathrm{Tr}\left(\rho B^\dagger B\right) \geq \left|\mathrm{Tr}\left(\rho A^\dagger B\right)\right|^2. \tag{1.482}$$

The operator $E_i(x)$ that represents the ith component of the electric field at the point x is the hermitian sum of the "positive-frequency" part $E_i^{(+)}(x)$ and its adjoint $E_i^{(-)}(x) = (E_i^{(+)}(x))^\dagger$

$$E_i(x) = E_i^{(+)}(x) + E_i^{(-)}(x). \tag{1.483}$$

Glauber has defined the first-order correlation function $G_{ij}^{(1)}(x, y)$ as (Glauber, 1963b)

$$G_{ij}^{(1)}(x, y) = \mathrm{Tr}\left(\rho E_i^{(-)}(x) E_j^{(+)}(y)\right) \tag{1.484}$$

or in terms of the operator inner product (1.479) as

$$G_{ij}^{(1)}(x, y) = \left(E_i^{(+)}(x), E_j^{(+)}(y)\right). \tag{1.485}$$

By setting $A = E_i^{(+)}(x)$ and $B = E_j^{(+)}(y)$ in the Schwarz inequality (1.482), we find that the correlation function $G_{ij}^{(1)}(x, y)$ is bounded by (Titulaer and Glauber, 1965)

$$\left|G_{ij}^{(1)}(x, y)\right|^2 \leq G_{ii}^{(1)}(x, x) G_{jj}^{(1)}(y, y). \tag{1.486}$$

Interference fringes are sharpest when this inequality is saturated

$$\left|G_{ij}^{(1)}(x, y)\right|^2 = G_{ii}^{(1)}(x, x) G_{jj}^{(1)}(y, y) \tag{1.487}$$

which can occur only if the correlation function $G_{ij}^{(1)}(x, y)$ factorizes (Titulaer and Glauber, 1965)

$$G_{ij}^{(1)}(x, y) = \mathcal{E}_i^*(x)\mathcal{E}_j(y) \tag{1.488}$$

as it does when the density operator is an outer product of coherent states

$$\rho = |\{\alpha_k\}\rangle\langle\{\alpha_k\}| \tag{1.489}$$

which are eigenstates of $E_i^{(+)}(x)$ with eigenvalue $\mathcal{E}_i(x)$ (Glauber, 1963b,a)

$$E_i^{(+)}(x)|\{\alpha_k\}\rangle = \mathcal{E}_i(x)|\{\alpha_k\}\rangle. \tag{1.490}$$

The higher-order correlation functions

$$G_{i_1 \ldots i_{2n}}^{(n)}(x_1 \cdots x_{2n}) = \mathrm{Tr}\left(\rho E_{i_1}^{(-)}(x_1) \cdots E_{i_n}^{(-)}(x_n) E_{i_{n+1}}^{(+)}(x_{n+1}) \cdots E_{i_{2n}}^{(+)}(x_n)\right) \tag{1.491}$$

satisfy similar inequalities (Glauber, 1963b) which also follow from the Schwarz inequality (1.482).

1.38 Rank of a Matrix

Four equivalent definitions of the **rank** $R(A)$ of an $m \times n$ matrix A are:

1. the number of its linearly independent rows,
2. the number of its linearly independent columns,
3. the number of its nonzero singular values, and
4. the number of rows in its biggest square nonsingular submatrix.

A matrix of rank zero has no nonzero singular values and so is zero.

Example 1.64 (Rank) The 3×4 matrix

$$A = \begin{pmatrix} 1 & 0 & 1 & -2 \\ 2 & 2 & 0 & 2 \\ 4 & 3 & 1 & 1 \end{pmatrix} \tag{1.492}$$

has three rows, so its rank can be at most 3. But twice the first row added to thrice the second row equals twice the third row, $2r_1 + 3r_2 - 2r_3 = 0$, so $R(A) \leq 2$. The first two rows obviously are not parallel, so they are linearly independent. Thus the number of linearly independent rows of A is 2, and so A has rank 2.

1.39 Software

High-quality software for virtually all numerical problems in linear algebra are available in the linear-algebra package Lapack. Lapack codes in Fortran and C++ are available at `netlib.org/lapack/` and at `math.nist.gov/tnt/`. Apple's Xcode command -framework accelerate links this software into gnu executables. The Basic Linear Algebra Subprograms (BLAS) on which Lapack is based are also available in Java at `icl.cs.utk.edu/f2j/` and at `math.nist.gov/javanumerics/`.

Matlab solves a wide variety of numerical problems. A free gnu version is available at `gnu.org/software/octave/`. Maple and Mathematica are good commercial programs for numerical and symbolic problems. Python (`python.org`), Scientific Python (`scipy.org`), and Sage (`sagemath.org`) are websites of free software of broad applicability. Maxima, xMaxima, and wxMaxima (`maxima.sourceforge.net`) are free Lisp programs that excel at computer algebra. Intel gives software to students and teachers (`software.intel.com`).

Exercises

1.1 What is the most general function of three Grassmann numbers $\theta_1, \theta_2, \theta_3$?

1.2 Derive the cyclicity (1.24) of the trace from Eq. (1.23).

1.3 Show that $(AB)^{\mathsf{T}} = B^{\mathsf{T}} A^{\mathsf{T}}$, which is Eq. (1.26).

1.4 Show that a real hermitian matrix is symmetric.

1.5 Show that $(AB)^{\dagger} = B^{\dagger} A^{\dagger}$, which is Eq. (1.29).

1.6 Show that the matrix (1.41) is positive on the space of all real 2-vectors but not on the space of all complex 2-vectors.

1.7 Show that the two 4×4 matrices (1.46) satisfy Grassmann's algebra (1.11) for $n = 2$.

1.8 Show that the operators $a_i = \theta_i$ defined in terms of the Grassmann matrices (1.46) and their adjoints $a_i^{\dagger} = \theta_i^{\dagger}$ satisfy the anticommutation relations (1.47) of the creation and annihilation operators for a system with two fermionic states.

1.9 Derive (1.66) from (1.63–1.65).

1.10 Fill in the steps leading to the formulas (1.74) for the vectors b_1' and b_2' and the formula (1.75) for the matrix a'.

1.11 Show that the antilinearity (1.81) of the inner product follows from its first two properties (1.78 & 1.79).

1.12 Show that the Minkowski product $(x, y) = \boldsymbol{x} \cdot \boldsymbol{y} - x^0 y^0$ of two 4-vectors x and y is an inner product obeying the rules (1.78, 1.79, and 1.84).

1.13 Show that if $f = 0$, then the linearity (1.79) of the inner product implies that (f, f) and (g, f) vanish.

1.14 Show that the condition (1.80) of being positive definite implies nondegeneracy (1.84).

1.15 Show that the nonnegativity (1.82) of the Schwarz inner product implies the condition (1.83). Hint: the inequality $(f - \lambda g, f - \lambda g) \geq 0$ must hold for every complex λ and for all vectors f and g.

1.16 Show that the inequality (1.103) follows from the Schwarz inequality (1.102).

1.17 Show that the inequality (1.105) follows from the Schwarz inequality (1.104).

1.18 Use the Gram-Schmidt method to find orthonormal linear combinations of the three vectors

$$s_1 = \begin{pmatrix} 1 \\ 0 \\ 0 \end{pmatrix}, \quad s_2 = \begin{pmatrix} 1 \\ 1 \\ 0 \end{pmatrix}, \quad s_3 = \begin{pmatrix} 1 \\ 1 \\ 1 \end{pmatrix}. \tag{1.493}$$

1.19 Now use the Gram-Schmidt method to find orthonormal linear combinations of the same three vectors but in a different order

$$s'_1 = \begin{pmatrix} 1 \\ 1 \\ 1 \end{pmatrix}, \quad s'_2 = \begin{pmatrix} 1 \\ 1 \\ 0 \end{pmatrix}, \quad s'_3 = \begin{pmatrix} 1 \\ 0 \\ 0 \end{pmatrix}. \tag{1.494}$$

Did you get the same orthonormal vectors as in the previous exercise?

1.20 Derive the linearity (1.125) of the outer product from its definition (1.124).

1.21 Show that a linear operator A that is represented by a hermitian matrix (1.167) in an orthonormal basis satisfies $(g, A\,f) = (A\,g, f)$.

1.22 Show that a unitary operator maps one orthonormal basis into another.

1.23 Show that the integral (1.186) defines a unitary operator that maps the state $|x'\rangle$ to the state $|x' + a\rangle$.

1.24 For the 2×2 matrices

$$A = \begin{pmatrix} 1 & 2 \\ 3 & -4 \end{pmatrix} \quad \text{and} \quad B = \begin{pmatrix} 2 & -1 \\ 4 & -3 \end{pmatrix} \tag{1.495}$$

verify equations (1.220–1.222).

1.25 Derive the least-squares solution (1.248) for complex A, x, and y when the matrix $A^\dagger A$ is positive.

1.26 Show that the eigenvalues λ of a unitary matrix are unimodular, that is, $|\lambda| = 1$.

1.27 What are the eigenvalues and eigenvectors of the two defective matrices (1.286)?

1.28 Use (1.297) to derive expression (1.298) for the 2×2 rotation matrix $\exp(-i\boldsymbol{\theta} \cdot \boldsymbol{\sigma}/2)$.

1.29 Compute the characteristic equation for the matrix $-i\boldsymbol{\theta} \cdot \boldsymbol{J}$ in which the generators are $(J_k)_{ij} = i\epsilon_{ikj}$ and ϵ_{ijk} is totally antisymmetric with $\epsilon_{123} = 1$.

1.30 Use the characteristic equation of exercise 1.29 to derive identities (1.301) and (1.302) for the 3×3 real orthogonal matrix $\exp(-i\boldsymbol{\theta} \cdot \boldsymbol{J})$.

1.31 Show that the sum of the eigenvalues of a normal antisymmetric matrix vanishes.

1.32 Consider the 2×3 matrix A

$$A = \begin{pmatrix} 1 & 2 & 3 \\ -3 & 0 & 1 \end{pmatrix}. \tag{1.496}$$

Perform the singular value decomposition $A = USV^{\mathsf{T}}$, where V^{T} the transpose of V. Use Matlab or another program to find the singular values and the real orthogonal matrices U and V.

1.33 Consider the 6×9 matrix A with elements $A_{j,k} = x + x^j + i(y - y^k)$ in which $x = 1.1$ and $y = 1.02$. Use Matlab or another program to find the singular values, and the first left and right singular vectors.

1.34 Show that the totally antisymmetric Levi-Civita symbol ϵ_{ijk} where $\epsilon_{123} = 1$ satisfies the useful relation

$$\sum_{i=1}^{3} \epsilon_{ijk}\, \epsilon_{inm} = \delta_{jn}\, \delta_{km} - \delta_{jm}\, \delta_{kn}. \tag{1.497}$$

1.35 Consider the hamiltonian $H = \frac{1}{2}\hbar\omega\sigma_3$ where σ_3 is defined in (1.453). The entropy S of this system at temperature T is $S = -k\mathrm{Tr}\,[\rho \ln(\rho)]$ in which the density operator ρ is

$$\rho = \frac{e^{-H/(kT)}}{\mathrm{Tr}\left[e^{-H/(kT)}\right]}. \tag{1.498}$$

Find expressions for the density operator ρ and its entropy S.

1.36 Use example 1.62 to find the action of the operator $\boldsymbol{S}^2 = \left(\boldsymbol{S}^{(1)} + \boldsymbol{S}^{(2)}\right)^2$ on the four states $| \pm \pm\rangle$ and then find the eigenstates and eigenvalues of $\boldsymbol{S}^2$ in the space spanned by these four states.

1.37 A system that has three fermionic states has three creation operators $a_i^\dagger$ and three annihilation operators a_k which satisfy the anticommutation relations $\{a_i, a_k^\dagger\} = \delta_{ik}$ and $\{a_i, a_k\} = \{a_i^\dagger, a_k^\dagger\} = 0$ for $i, k = 1, 2, 3$. The eight states of the system are $|t, u, v\rangle \equiv (a_1^\dagger)^t (a_2^\dagger)^u (a_3^\dagger)^v |0, 0, 0\rangle$. We can represent them by eight 8-vectors each of which has seven 0's with a 1 in position $4t + 2u + v + 1$. How big should the matrices that represent the creation and annihilation operators be? Write down the three matrices that represent the three creation operators.

1.38 Show that the Schwarz inner product (1.478) is degenerate because it can violate (1.84) for certain density operators and certain pairs of states.

1.39 Show that the Schwarz inner product (1.479) is degenerate because it can violate (1.84) for certain density operators and certain pairs of operators.

1.40 The coherent state $|\{\alpha_k\}\rangle$ is an eigenstate of the annihilation operator a_k with eigenvalue α_k for each mode k of the electromagnetic field, $a_k|\{\alpha_k\}\rangle = \alpha_k|\{\alpha_k\}\rangle$. The positive-frequency part $E_i^{(+)}(x)$ of the electric field is a linear combination of the annihilation operators

$$E_i^{(+)}(x) = \sum_k a_k\, \mathcal{E}_i^{(+)}(k)\, e^{i(kx-\omega t)}. \tag{1.499}$$

Show that $|\{\alpha_k\}\rangle$ is an eigenstate of $E_i^{(+)}(x)$ as in (1.490) and find its eigenvalue $\mathcal{E}_i(x)$.

1.41 Show that if X is a nondefective, nonsingular square matrix, then the variation of the logarithm of its determinant is $\delta \ln(\det X) = \mathrm{Tr}(X^{-1}\delta X)$.

2

Vector Calculus

2.1 Derivatives and Partial Derivatives

The **derivative** of a function $f(x)$ at a point x is the limit of the ratio

$$\frac{df(x)}{dx} = \lim_{x' \to x} \frac{f(x') - f(x)}{x' - x}. \tag{2.1}$$

Example 2.1 (Derivative of a monomial) Setting $x' = x + \epsilon$ and letting $\epsilon \to 0$, we compute the derivative of x^n as

$$\frac{dx^n}{dx} = \lim_{\epsilon \to 0} \frac{(x+\epsilon)^n - x^n}{\epsilon} \approx \frac{x^n + \epsilon\, n\, x^{n-1} - x^n}{\epsilon} = n\, x^{n-1}. \tag{2.2}$$

Similarly, adding fractions, we find

$$\frac{dx^{-n}}{dx} = \lim_{\epsilon \to 0} \frac{(x+\epsilon)^{-n} - x^{-n}}{\epsilon} \approx \frac{x^n - (x^n + \epsilon n x^{n-1})}{\epsilon x^{2n}} = -n\, x^{-n-1}. \tag{2.3}$$

The **partial derivative** of a function with respect to a given variable is the whole derivative of the function with all its other variables held constant. For instance, the partial derivatives of the function $f(x, y, z) = x^\ell\, y^n / z^m$ with respect to x and z are

$$\frac{\partial f(x,y,z)}{\partial x} = \ell\, \frac{x^{\ell-1}\, y^n}{z^m} \quad \text{and} \quad \frac{\partial f(x,y,z)}{\partial z} = -\, m\, \frac{x^\ell\, y^n}{z^{m+1}}. \tag{2.4}$$

One often uses primes or dots to denote derivatives as in

$$f' = \frac{df}{dx}, \quad f'' = \frac{d^2 f}{dx^2} \equiv \frac{d}{dx}\left(\frac{df}{dx}\right), \quad \dot{f} = \frac{df}{dt}, \quad \text{and} \quad \ddot{f} = \frac{d^2 f}{dt^2}. \tag{2.5}$$

For higher or partial derivatives, one sometimes uses superscripts

$$f^{(k)} = \frac{d^k f}{dx^k} \quad \text{and} \quad f^{(k,\ell)} = \frac{\partial^{k+\ell} f}{\partial x^k \partial y^\ell} \tag{2.6}$$

or subscripts, sometimes preceded by commas

$$f_x = f_{,x} = \frac{\partial f}{\partial x} \quad \text{and} \quad f_{xyy} = f_{,xyy} = \frac{\partial^3 f}{\partial x \partial y^2}. \tag{2.7}$$

If variables $x = x_1, \ldots, x_n$ are labeled by indexes, derivatives can be labeled by subscripted indexes, sometimes preceded by commas

$$f_{,k} = \partial_k f = \frac{\partial f}{\partial x_k} \quad \text{and} \quad f_{,k\ell} = \partial_k \partial_\ell f = \frac{\partial^2 f}{\partial x_k \partial x_\ell}. \tag{2.8}$$

2.2 Gradient

The change $\boldsymbol{dp}$ in a point $\boldsymbol{p}$ due to changes du_1, du_2, du_3 in its **orthogonal** coordinates u_1, u_2, u_3 is a linear combination

$$\begin{aligned} \boldsymbol{dp} &= \frac{\partial \boldsymbol{p}}{\partial u_1} du_1 + \frac{\partial \boldsymbol{p}}{\partial u_2} du_2 + \frac{\partial \boldsymbol{p}}{\partial u_3} du_3 \\ &= \boldsymbol{e}_1\, du_1 + \boldsymbol{e}_2\, du_2 + \boldsymbol{e}_3\, du_3 \end{aligned} \tag{2.9}$$

of vectors $\boldsymbol{e}_1, \boldsymbol{e}_2, \boldsymbol{e}_3$ that are orthogonal

$$\boldsymbol{e}_i \cdot \boldsymbol{e}_k = h_i\, h_k\, \delta_{ik}. \tag{2.10}$$

In terms of the orthonormal vectors $\hat{\boldsymbol{e}}_j = \boldsymbol{e}_j/h_j$, the change $\boldsymbol{dp}$ is

$$\boldsymbol{dp} = h_1\, \hat{\boldsymbol{e}}_1\, du_1 + h_2\, \hat{\boldsymbol{e}}_2\, du_2 + h_3\, \hat{\boldsymbol{e}}_3\, du_3. \tag{2.11}$$

The orthonormal vectors $\hat{\boldsymbol{e}}_j$ have cyclic cross products

$$\hat{\boldsymbol{e}}_i \times \hat{\boldsymbol{e}}_j = \sum_{k=1}^{3} \epsilon_{ijk}\, \hat{\boldsymbol{e}}_k \tag{2.12}$$

in which ϵ_{ijk} is the antisymmetric Levi-Civita symbol (1.196) with $\epsilon_{123} = 1$.

In rectangular coordinates, the change $\boldsymbol{dp}$ in a physical point $\boldsymbol{p}$ due to changes dx, dy, and dz in its coordinates is $\boldsymbol{dp} = \hat{\boldsymbol{x}}\, dx + \hat{\boldsymbol{y}}\, dy + \hat{\boldsymbol{z}}\, dz$, and the scale factors are all unity $h_x = h_y = h_z = 1$. In cylindrical coordinates, the change $\boldsymbol{dp}$ in a point $\boldsymbol{p}$ due to changes $d\rho$, $d\phi$, and dz in its coordinates is $\boldsymbol{dp} = \hat{\boldsymbol{\rho}}\, d\rho + \rho\, \hat{\boldsymbol{\phi}}\, d\phi + \hat{\boldsymbol{z}}\, dz$, and the scale factors are $h_\rho = 1$, $h_\phi = \rho$, and $h_z = 1$. In spherical coordinates, the change is $\boldsymbol{dp} = \hat{\boldsymbol{r}}\, dr + r\, \hat{\boldsymbol{\theta}}\, d\theta + r\, \sin\theta\, \hat{\boldsymbol{\phi}}\, d\phi$, and the scale factors are $h_r = 1$, $h_\theta = r$, and $h_\phi = r\, \sin\boldsymbol{\theta}$. In these orthogonal coordinates, the change in a point is

$$
dp = \begin{cases} \hat{\boldsymbol{x}}\, dx + \hat{\boldsymbol{y}}\, dy + \hat{\boldsymbol{z}}\, dz \\ \hat{\boldsymbol{\rho}}\, d\rho + \rho\, \hat{\boldsymbol{\phi}}\, d\phi + \hat{\boldsymbol{z}}\, dz \\ \hat{\boldsymbol{r}}\, dr + r\, \hat{\boldsymbol{\theta}}\, d\theta + r\, \sin\theta\, \hat{\boldsymbol{\phi}}\, d\phi \end{cases} . \tag{2.13}
$$

The **gradient** ∇f of a scalar function f is defined so that its dot product $\nabla f \cdot \boldsymbol{dp}$ with the change $\boldsymbol{dp}$ in the point $\boldsymbol{p}$ is the change df in f

$$
\begin{aligned}
\nabla f \cdot \boldsymbol{dp} &= (\nabla f_1 \hat{\boldsymbol{e}}_1 + \nabla f_2 \hat{\boldsymbol{e}}_2 + \nabla f_3 \hat{\boldsymbol{e}}_3) \cdot (\hat{\boldsymbol{e}}_1 h_1 du_1 + \hat{\boldsymbol{e}}_2 h_2 du_2 + \hat{\boldsymbol{e}}_3 h_3 du_3) \\
&= \nabla f_1 h_1 du_1 + \nabla f_2 h_2 du_2 + \nabla f_3 h_3 du_3 \\
&= df = \frac{\partial f}{\partial u_1} du_1 + \frac{\partial f}{\partial u_2} du_2 + \frac{\partial f}{\partial u_3} du_3.
\end{aligned} \tag{2.14}
$$

Thus the gradient in orthogonal coordinates is

$$
\nabla f = \frac{\hat{\boldsymbol{e}}_1}{h_1} \frac{\partial f}{\partial u_1} + \frac{\hat{\boldsymbol{e}}_2}{h_2} \frac{\partial f}{\partial u_2} + \frac{\hat{\boldsymbol{e}}_3}{h_3} \frac{\partial f}{\partial u_3}, \tag{2.15}
$$

and in rectangular, cylindrical, and spherical coordinates it is

$$
\nabla f = \begin{cases} \hat{\boldsymbol{x}} \dfrac{\partial f}{\partial x} + \hat{\boldsymbol{y}} \dfrac{\partial f}{\partial y} + \hat{\boldsymbol{z}} \dfrac{\partial f}{\partial z} \\[2ex] \hat{\boldsymbol{\rho}} \dfrac{\partial f}{\partial \rho} + \dfrac{\hat{\boldsymbol{\phi}}}{\rho} \dfrac{\partial f}{\partial \phi} + \hat{\boldsymbol{z}} \dfrac{\partial f}{\partial z} \\[2ex] \hat{\boldsymbol{r}} \dfrac{\partial f}{\partial r} + \dfrac{\hat{\boldsymbol{\theta}}}{r} \dfrac{\partial f}{\partial \theta} + \dfrac{\hat{\boldsymbol{\phi}}}{r \sin\theta} \dfrac{\partial f}{\partial \phi} \end{cases} . \tag{2.16}
$$

In particular the gradient of $1/r$ is

$$
\nabla \left(\frac{1}{r} \right) = -\frac{\hat{\boldsymbol{r}}}{r^2} \quad \text{and} \quad \nabla \left(\frac{1}{|\boldsymbol{r} - \boldsymbol{r}'|} \right) = -\frac{\boldsymbol{r} - \boldsymbol{r}'}{|\boldsymbol{r} - \boldsymbol{r}'|^3}. \tag{2.17}
$$

In both of these formulas, the differentiation is with respect to $\boldsymbol{r}$, not $\boldsymbol{r}'$.

2.3 Divergence

The **divergence** of a vector $\boldsymbol{v}$ in an infinitesimal cube C is defined as the integral S of $\boldsymbol{v}$ over the surface of the cube divided by its volume $V = h_1 h_2 h_3\, du_1 du_2 du_3$. The surface integral S is the sum of the integrals of v_1, v_2, and v_3 over the cube's three forward faces $v_1 h_2 du_2 h_3 du_3 + v_2 h_1 du_1 h_3 du_3 + v_3 h_1 du_1 h_2 du_2$ minus the sum of the integrals of v_1, v_2, and v_3 over the cube's three opposite faces. The surface integral is then

$$
S = \left[\frac{\partial(v_1 h_2 h_3)}{\partial u_1} + \frac{\partial(v_2 h_1 h_3)}{\partial u_2} + \frac{\partial(v_3 h_1 h_2)}{\partial u_3} \right] du_1 du_2 du_3. \tag{2.18}
$$

So the divergence $\nabla \cdot \boldsymbol{v}$ is the ratio S/V

$$\nabla \cdot \boldsymbol{v} = \frac{S}{V} = \frac{1}{h_1 h_2 h_3}\left[\frac{\partial(v_1 h_2 h_3)}{\partial u_1} + \frac{\partial(v_2 h_1 h_3)}{\partial u_2} + \frac{\partial(v_3 h_1 h_2)}{\partial u_3}\right]. \tag{2.19}$$

In rectangular coordinates, the divergence is

$$\nabla \cdot \boldsymbol{v} = \frac{\partial v_x}{\partial x} + \frac{\partial v_y}{\partial y} + \frac{\partial v_z}{\partial z}. \tag{2.20}$$

In cylindrical coordinates, it is

$$\nabla \cdot \boldsymbol{v} = \frac{1}{\rho}\left[\frac{\partial(v_\rho \rho)}{\partial \rho} + \frac{\partial v_\phi}{\partial \phi} + \frac{\partial(v_z \rho)}{\partial z}\right] = \frac{1}{\rho}\frac{\partial(\rho v_\rho)}{\partial \rho} + \frac{1}{\rho}\frac{\partial v_\phi}{\partial \phi} + \frac{\partial v_z}{\partial z}, \tag{2.21}$$

and in spherical coordinates it is

$$\nabla \cdot \boldsymbol{v} = \frac{1}{r^2}\frac{\partial(v_r r^2)}{\partial r} + \frac{1}{r\sin\theta}\frac{\partial(v_\theta \sin\theta)}{\partial \theta} + \frac{1}{r\sin\theta}\frac{\partial v_\phi}{\partial \phi}. \tag{2.22}$$

By assembling a suitable number of infinitesimally small cubes, one may create a three-dimensional region of arbitrary shape and volume. The sum of the products of the divergence $\nabla \cdot \boldsymbol{v}$ in each cube times its volume dV is the sum of the surface integrals dS over the faces of these tiny cubes. The integrals over the interior faces cancel leaving just the integral over the surface ∂V of the whole volume V. Thus we arrive at Gauss's theorem

$$\int_V \nabla \cdot \boldsymbol{v}\, dV = \int_{\partial V} \boldsymbol{v} \cdot \boldsymbol{da} \tag{2.23}$$

in which $\boldsymbol{da}$ is an infinitesimal, outward, area element of the surface that is the boundary ∂V of the volume V.

Example 2.2 (Delta function) The integral of the divergence of $\hat{\boldsymbol{r}}/r^2$ over any sphere, however small, centered at the origin is 4π

$$\int \nabla \cdot \left(\frac{\hat{\boldsymbol{r}}}{r^2}\right) dV = \int \frac{\hat{\boldsymbol{r}}}{r^2} \cdot \boldsymbol{da} = \int \frac{\hat{\boldsymbol{r}}}{r^2} \cdot r^2 \hat{\boldsymbol{r}}\, d\Omega = \int d\Omega = 4\pi. \tag{2.24}$$

Similarly, the integral of the divergence of $\boldsymbol{r} - \boldsymbol{r}'/|\boldsymbol{r} - \boldsymbol{r}'|^3$ over any sphere, however small, centered at $\boldsymbol{r}'$ is 4π

$$\int \nabla \cdot \left(\frac{\boldsymbol{r} - \boldsymbol{r}'}{|\boldsymbol{r} - \boldsymbol{r}'|^3}\right) dV = \int \frac{\boldsymbol{r} - \boldsymbol{r}'}{|\boldsymbol{r} - \boldsymbol{r}'|^3} \cdot \boldsymbol{da}$$
$$= \int \frac{\boldsymbol{r} - \boldsymbol{r}'}{|\boldsymbol{r} - \boldsymbol{r}'|^3} \cdot |\boldsymbol{r} - \boldsymbol{r}'|^2 \frac{\boldsymbol{r} - \boldsymbol{r}'}{|\boldsymbol{r} - \boldsymbol{r}'|} d\Omega = \int d\Omega = 4\pi. \tag{2.25}$$

These divergences, vanishing for $\boldsymbol{r} \neq 0$ and for $\boldsymbol{r} - \boldsymbol{r}' \neq 0$, are delta functions

$$\nabla \cdot \left(\frac{\hat{\boldsymbol{r}}}{r^2}\right) = 4\pi\delta^3(\boldsymbol{r}) \quad \text{and} \quad \nabla \cdot \left(\frac{\boldsymbol{r} - \boldsymbol{r}'}{|\boldsymbol{r} - \boldsymbol{r}'|^3}\right) = 4\pi\delta^3(\boldsymbol{r} - \boldsymbol{r}') \tag{2.26}$$

because if $f(\boldsymbol{r})$ is any suitably smooth function, then the integral over any volume that includes the point $\boldsymbol{r}'$ is

$$\int f(\boldsymbol{r})\, \nabla \cdot \left(\frac{\boldsymbol{r} - \boldsymbol{r}'}{|\boldsymbol{r} - \boldsymbol{r}'|^3}\right) d^3r = 4\pi\, f(\boldsymbol{r}'). \tag{2.27}$$

Example 2.3 (Gauss's law) The divergence $\nabla \cdot \boldsymbol{E}$ of the electric field is the charge density ρ divided by the electric constant $\epsilon_0 = 8.854 \times 10^{-12}$ F/m

$$\nabla \cdot \boldsymbol{E} = \frac{\rho}{\epsilon_0}. \tag{2.28}$$

So by Gauss's theorem, the integral of the electric field over a surface ∂V that bounds a volume V is the charge inside divided by ϵ_0

$$\int_{\partial V} \boldsymbol{E} \cdot d\boldsymbol{a} = \int_V \nabla \cdot \boldsymbol{E}\, dV = \int_V \frac{\rho}{\epsilon_0}\, dV = \frac{Q_V}{\epsilon_0}. \tag{2.29}$$

2.4 Laplacian

The laplacian is the divergence (2.19) of the gradient (2.15). So in orthogonal coordinates it is

$$\Delta f \equiv \nabla^2 f \equiv \nabla \cdot \nabla f = \frac{1}{h_1 h_2 h_3}\left[\sum_{k=1}^{3} \frac{\partial}{\partial u_k}\left(\frac{h_1 h_2 h_3}{h_k^2}\frac{\partial f}{\partial u_k}\right)\right]. \tag{2.30}$$

In rectangular coordinates, the laplacian is

$$\Delta f = \frac{\partial^2 f}{\partial x^2} + \frac{\partial^2 f}{\partial y^2} + \frac{\partial^2 f}{\partial z^2}. \tag{2.31}$$

In cylindrical coordinates, it is

$$\Delta f = \frac{1}{\rho}\left[\frac{\partial}{\partial \rho}\left(\rho\frac{\partial f}{\partial \rho}\right) + \frac{1}{\rho}\frac{\partial^2 f}{\partial \phi^2} + \rho\frac{\partial^2 f}{\partial z^2}\right] = \frac{1}{\rho}\frac{\partial}{\partial \rho}\left(\rho\frac{\partial f}{\partial \rho}\right) + \frac{1}{\rho^2}\frac{\partial^2 f}{\partial \phi^2} + \frac{\partial^2 f}{\partial z^2}, \tag{2.32}$$

and in spherical coordinates it is

$$\begin{aligned}
\Delta f &= \frac{1}{r^2 \sin\theta}\left[\frac{\partial}{\partial r}\left(r^2 \sin\theta \frac{\partial f}{\partial r}\right) + \frac{\partial}{\partial \theta}\left(\sin\theta\frac{\partial f}{\partial \theta}\right) + \frac{\partial}{\partial \phi}\left(\frac{1}{\sin\theta}\frac{\partial f}{\partial \phi}\right)\right] \\
&= \frac{1}{r^2}\frac{\partial}{\partial r}\left(r^2\frac{\partial f}{\partial r}\right) + \frac{1}{r^2 \sin\theta}\frac{\partial}{\partial \theta}\left(\sin\theta\frac{\partial f}{\partial \theta}\right) + \frac{1}{r^2 \sin^2\theta}\frac{\partial^2 f}{\partial \phi^2} \\
&= \frac{1}{r}\frac{\partial^2}{\partial r^2}(rf) + \frac{1}{r^2 \sin\theta}\frac{\partial}{\partial \theta}\left(\sin\theta\frac{\partial f}{\partial \theta}\right) + \frac{1}{r^2 \sin^2\theta}\frac{\partial^2 f}{\partial \phi^2}.
\end{aligned} \tag{2.33}$$

Example 2.4 (Delta function as laplacian of $1/r$) By combining the gradient (2.17) of $1/r$ with the representation (2.26) of the delta function as a divergence, we can write delta functions as laplacians (with respect to $\boldsymbol{r}$)

$$-\triangle\left(\frac{1}{r}\right) = 4\pi\delta^3(\boldsymbol{r}) \quad \text{and} \quad -\triangle\left(\frac{1}{|\boldsymbol{r}-\boldsymbol{r}'|}\right) = 4\pi\delta^3(\boldsymbol{r}-\boldsymbol{r}'). \tag{2.34}$$

Example 2.5 (Electric field of a uniformly charged sphere) The electric field $\boldsymbol{E} = -\nabla\phi - \dot{\boldsymbol{A}}$ in static problems is just the gradient $\boldsymbol{E} = -\nabla\phi$ of the scalar potential ϕ. Gauss's law (2.28) then gives us Poisson's equation $\nabla \cdot \boldsymbol{E} = -\Delta\phi = \rho/\epsilon_0$. Writing the laplacian in spherical coordinates (2.33) and using spherical symmetry, we form the differential equation

$$-\frac{1}{r}\frac{d^2}{dr^2}(r\phi) = \frac{\rho}{\epsilon_0} \tag{2.35}$$

in which ρ is the uniform charge density of the sphere. Integrating twice and letting the constant a be $\phi(r)$ at $r = 0$, we find the potential inside the sphere to be $\phi(r) = a - \rho r^2/(6\epsilon_0)$. Outside the sphere, the charge density vanishes and so the second r-derivative of $r\phi$ vanishes, $(r\phi)'' = 0$. Integrating twice, we get for the potential outside the sphere $\phi(r) = b/r$ after dropping a constant term because $\phi(r) \to 0$ as $r \to \infty$. The electric field $\boldsymbol{E} = -\nabla\phi$ is the negative gradient (2.16) of the potential, so on the surface of the sphere where the interior and exterior solutions meet at $r = R$, it is

$$\boldsymbol{E} = -\hat{r}\frac{\rho R}{3\epsilon_0} = \hat{r}\frac{b}{R^2}. \tag{2.36}$$

Thus $b = \rho R^3/(3\epsilon_0)$. Matching the interior potential to the exterior potential on the surface of the sphere at $r = R$ gives $a = \rho R^2/(2\epsilon_0)$. So the potential of a uniformly charged sphere of radius R is

$$\phi(r) = \begin{cases} \rho\left(R^2 - r^2/3\right)/(2\epsilon_0) & r \le R \\ \rho R^3/(3\epsilon_0 r) & r \ge R \end{cases}. \tag{2.37}$$

2.5 Curl

The directed area $d\boldsymbol{S}$ of an infinitesimal rectangle whose sides are the tiny perpendicular vectors $h_i\hat{\boldsymbol{e}}_i du_i$ and $h_j\hat{\boldsymbol{e}}_j du_j$ (fixed i and j) is their cross-product (2.12)

$$d\boldsymbol{S} = h_i\hat{\boldsymbol{e}}_i du_i \times h_j\hat{\boldsymbol{e}}_j du_j = \sum_{k=1}^{3} \hat{\boldsymbol{e}}_k\, h_i h_j\, du_i du_j. \tag{2.38}$$

The line integral of the vector $\boldsymbol{f}$ along the perimeter of this infinitesimal rectangle is

$$\oint \boldsymbol{f} \cdot \boldsymbol{dl} = \left(\frac{\partial\,(h_j f_j)}{\partial u_i} - \frac{\partial\,(h_i f_i)}{\partial u_j} \right) du_i\, du_j. \tag{2.39}$$

The **curl** $\nabla \times \boldsymbol{f}$ of a vector $\boldsymbol{f}$ is defined to be the vector whose dot product with the area (2.38) is the line integral (2.39)

$$(\nabla \times \boldsymbol{f}) \cdot \boldsymbol{dS} = (\nabla \times \boldsymbol{f})_k\, h_i h_j du_i du_j = \left(\frac{\partial\,(h_j f_j)}{\partial u_i} - \frac{\partial\,(h_i f_i)}{\partial u_j} \right) du_i\, du_j \tag{2.40}$$

in which i, j, k are $1, 2, 3$ or a cyclic permutation of $1, 2, 3$. Thus the kth component of the curl is

$$(\nabla \times f)_k = \frac{1}{h_i h_j} \left(\frac{\partial(h_j f_j)}{\partial u_i} - \frac{\partial(h_i f_i)}{\partial u_j} \right) \quad \text{(no sum)}, \tag{2.41}$$

and the curl as a vector field is the sum over i, j, and k from 1 to 3

$$\nabla \times \boldsymbol{f} = \sum_{i,j,k=1}^{3} \epsilon_{ijk} \frac{\hat{\boldsymbol{e}}_k}{h_i h_j} \frac{\partial(h_j f_j)}{\partial u_i}. \tag{2.42}$$

In rectangular coordinates, the scale factors are all unity, and the ith component of the curl $\nabla \times \boldsymbol{f}$ is

$$(\nabla \times \boldsymbol{f})_i = \sum_{j,k=1}^{3} \epsilon_{ijk} \frac{\partial f_k}{\partial x_j} = \sum_{j,k=1}^{3} \epsilon_{ijk}\, \partial_j\, f_k. \tag{2.43}$$

We can write the curl as a determinant

$$\nabla \times \boldsymbol{f} = \frac{1}{h_1 h_2 h_3} \begin{vmatrix} h_1 \hat{\boldsymbol{e}}_1 & h_2 \hat{\boldsymbol{e}}_2 & h_3 \hat{\boldsymbol{e}}_3 \\ \partial_1 & \partial_2 & \partial_3 \\ h_1 f_1 & h_2 f_2 & h_3 f_3 \end{vmatrix}. \tag{2.44}$$

In rectangular coordinates, the curl is

$$\nabla \times \boldsymbol{f} = \begin{vmatrix} \hat{\boldsymbol{x}} & \hat{\boldsymbol{y}} & \hat{\boldsymbol{z}} \\ \partial_x & \partial_y & \partial_z \\ f_x & f_y & f_z \end{vmatrix}. \tag{2.45}$$

In cylindrical coordinates, it is

$$\nabla \times \boldsymbol{f} = \frac{1}{\rho} \begin{vmatrix} \hat{\boldsymbol{\rho}} & \rho\hat{\boldsymbol{\phi}} & \hat{\boldsymbol{z}} \\ \partial_\rho & \partial_\phi & \partial_z \\ f_\rho & \rho f_\phi & f_z \end{vmatrix} \tag{2.46}$$

and in spherical coordinates, it is

$$\nabla \times \boldsymbol{f} = \frac{1}{r^2 \sin\theta} \begin{vmatrix} \hat{\boldsymbol{r}} & r\,\hat{\boldsymbol{\theta}} & r\sin\theta\,\hat{\boldsymbol{\phi}} \\ \partial_r & \partial_\theta & \partial_\phi \\ f_r & r\,f_\theta & r\sin\theta\,f_\phi \end{vmatrix}. \tag{2.47}$$

Sums of products of two Levi-Civita symbols (1.196) yield useful identities

$$\sum_{i=1}^{3} \epsilon_{ijk}\epsilon_{imn} = \delta_{jm}\delta_{kn} - \delta_{jn}\delta_{km} \quad \text{and} \quad \sum_{i,j=1}^{3} \epsilon_{ijk}\epsilon_{ijn} = 2\delta_{kn} \tag{2.48}$$

in which δ_{jm} is Kronecker's delta (1.34). Thus the curl of a curl is

$$\begin{aligned}[\nabla \times (\nabla \times A)]_i &= \sum_{j,k,m,n=1}^{3} \epsilon_{ijk}\partial_j\epsilon_{kmn}\partial_m A_n = \sum_{j,m,n=1}^{3} (\delta_{im}\delta_{jn} - \delta_{in}\delta_{jm})\partial_j\partial_m A_n \\ &= \partial_i \nabla \cdot A - \triangle A_i \quad \text{or} \quad \nabla \times (\nabla \times \boldsymbol{A}) = \nabla(\nabla \cdot \boldsymbol{A}) - \triangle \boldsymbol{A}.\end{aligned} \tag{2.49}$$

By assembling a suitable set of infinitesimal rectangles $\boldsymbol{dS}$, we may create an arbitrary surface S. The surface integral of the dot product $\nabla \times \boldsymbol{f} \cdot \boldsymbol{dS}$ over the tiny rectangles $\boldsymbol{dS}$ that make up the surface S is the sum of the line integrals along the sides of these tiny rectangles. The line integrals over the interior sides cancel leaving just the line integral along the boundary ∂S of the finite surface S. Thus the integral of the curl $\nabla \times \boldsymbol{f}$ of a vector $\boldsymbol{f}$ over a surface is the line integral of the vector $\boldsymbol{f}$ along the boundary of the surface

$$\int_S (\nabla \times \boldsymbol{f}) \cdot \boldsymbol{dS} = \int_{\partial S} \boldsymbol{f} \cdot \boldsymbol{d\ell} \tag{2.50}$$

which is one of Stokes's theorems.

Example 2.6 (Maxwell's equations) In empty space, Maxwell's equations in SI units are $\nabla \cdot \boldsymbol{E} = 0$, $\nabla \cdot \boldsymbol{B} = 0$, $\nabla \times \boldsymbol{E} = -\dot{\boldsymbol{B}}$, and $c^2\,\nabla \times \boldsymbol{B} = \dot{\boldsymbol{E}}$. They imply that the voltage induced in a loop is the negative of the rate of change of the magnetic induction through the loop

$$V = \oint_{\partial S} \boldsymbol{E} \cdot \boldsymbol{dx} = -\,\dot{\Phi}_B = -\int_S \dot{\boldsymbol{B}} \cdot \boldsymbol{da} \tag{2.51}$$

and that the magnetic induction induced in a loop is the rate of change of the electric flux through the loop divided by c^2

$$B = \int_{\partial S} \boldsymbol{B} \cdot \boldsymbol{dx} = \frac{1}{c^2}\,\dot{\Phi}_E = \frac{1}{c^2}\int_S \dot{\boldsymbol{E}} \cdot \boldsymbol{da}. \tag{2.52}$$

Maxwell's equations in empty space and the curl identity (2.49) imply that

$$\nabla \times (\nabla \times \boldsymbol{E}) = \nabla(\nabla \cdot \boldsymbol{E}) - \triangle \boldsymbol{E} = -\triangle \boldsymbol{E} = -\nabla \times \dot{\boldsymbol{B}} = -\ddot{\boldsymbol{E}}/c^2 \tag{2.53}$$

$$\nabla \times (\nabla \times \boldsymbol{B}) = \nabla(\nabla \cdot \boldsymbol{B}) - \triangle \boldsymbol{B} = -\triangle \boldsymbol{B} = \nabla \times \dot{\boldsymbol{E}}/c^2 = -\ddot{\boldsymbol{B}}/c^2 \tag{2.54}$$

or

$$\triangle \boldsymbol{E} = \ddot{\boldsymbol{E}}/c^2 \quad \text{and} \quad \triangle \boldsymbol{B} = \ddot{\boldsymbol{B}}/c^2. \tag{2.55}$$

The exponentials $\boldsymbol{E}(\boldsymbol{k}, \omega) = \boldsymbol{\epsilon}\, e^{i(\boldsymbol{k}\cdot\boldsymbol{r}-\omega t)}$ and $\boldsymbol{B}(\boldsymbol{k}, \omega) = (\hat{\boldsymbol{k}} \times \boldsymbol{\epsilon}/c)\, e^{i(\boldsymbol{k}\cdot\boldsymbol{r}-\omega t)}$ with $\omega = |\boldsymbol{k}|c$ and $\hat{\boldsymbol{k}} \cdot \boldsymbol{\epsilon} = 0$ obey these wave equations.

Exercises

2.1 Derive the Levi-Civita identity

$$\sum_{i=1}^{3} \epsilon_{ijk}\epsilon_{imn} = \delta_{jm}\delta_{kn} - \delta_{jn}\delta_{km}. \tag{2.56}$$

2.2 Derive the Levi-Civita identity

$$\sum_{i,j=1}^{3} \epsilon_{ijk}\epsilon_{ijn} = 2\delta_{kn}. \tag{2.57}$$

2.3 Show that

$$\nabla \times (\boldsymbol{a} \times \boldsymbol{b}) = \boldsymbol{a}\,\nabla \cdot \boldsymbol{b} - \boldsymbol{b}\,\nabla \cdot \boldsymbol{a} + (\boldsymbol{b} \cdot \nabla)\boldsymbol{a} - (\boldsymbol{a} \cdot \nabla)\boldsymbol{b}. \tag{2.58}$$

2.4 Simplify $\nabla \times \nabla\phi$ and $\nabla \cdot (\nabla \times \boldsymbol{a})$ in which ϕ is a scalar field and $\boldsymbol{a}$ is a vector field.

2.5 Simplify $\nabla \cdot (\nabla\phi \times \nabla\psi)$ in which ϕ and ψ are scalar fields.

2.6 Let $\boldsymbol{B} = \nabla \times \boldsymbol{A}$ and $\boldsymbol{E} = -\nabla\phi - \dot{\boldsymbol{A}}$ and show that Maxwell's equations in vacuum (example 2.6) and the Lorentz gauge condition

$$\nabla \cdot \boldsymbol{A} + \dot{\phi}/c^2 = 0 \tag{2.59}$$

imply that $\boldsymbol{A}$ and ϕ obey the wave equations

$$\triangle\phi - \ddot{\phi}/c^2 = 0 \quad \text{and} \quad \triangle \boldsymbol{A} - \ddot{\boldsymbol{A}}/c^2 = 0. \tag{2.60}$$

3
Fourier Series

3.1 Fourier Series

The phases $\exp(inx)/\sqrt{2\pi}$ for integer n are orthonormal on an interval of length 2π

$$\int_0^{2\pi} \frac{e^{-imx}}{\sqrt{2\pi}} \frac{e^{inx}}{\sqrt{2\pi}}\, dx = \int_0^{2\pi} \frac{e^{i(n-m)x}}{2\pi}\, dx = \delta_{m,n} = \begin{cases} 1 & \text{if } m = n \\ 0 & \text{if } m \neq n \end{cases} \tag{3.1}$$

in which $\delta_{n,m}$ is Kronecker's delta (1.34). So if a function $f(x)$ is a sum of these phases, called a **Fourier series**,

$$f(x) = \sum_{n=-\infty}^{\infty} f_n \frac{e^{inx}}{\sqrt{2\pi}}, \tag{3.2}$$

then the orthonormality (3.1) of these phases $\exp(inx)/\sqrt{2\pi}$ gives the nth coefficient f_n as the integral

$$\int_0^{2\pi} \frac{e^{-inx}}{\sqrt{2\pi}} f(x)\, dx = \int_0^{2\pi} \frac{e^{-inx}}{\sqrt{2\pi}} \sum_{m=-\infty}^{\infty} f_m \frac{e^{imx}}{\sqrt{2\pi}}\, dx = \sum_{m=-\infty}^{\infty} \delta_{n,m} f_m = f_n. \tag{3.3}$$

Fourier series can represent functions $f(x)$ that are square integrable on the interval $0 < x < 2\pi$ (Joseph Fourier 1768–1830).

In Dirac's notation, we interpret the phases

$$\langle x|n\rangle = \frac{e^{inx}}{\sqrt{2\pi}} \tag{3.4}$$

as the components of the vector $|n\rangle$ in the $|x\rangle$ basis. These components are inner products $\langle x|n\rangle$ of $|n\rangle$ and $|x\rangle$. The orthonormality integral (3.1) shows that the inner product of $|n\rangle$ and $|m\rangle$ is unity when $n = m$ and zero when $n \neq m$

$$\langle m|n\rangle = \langle m|I|n\rangle = \int_0^{2\pi} \langle m|x\rangle\langle x|n\rangle\, dx = \int_0^{2\pi} \frac{e^{i(n-m)x}}{2\pi}\, dx = \delta_{m,n}. \tag{3.5}$$

Here I is the identity operator of the space spanned by the vectors $|x\rangle$

$$I = \int_0^{2\pi} |x\rangle\langle x|\, dx. \tag{3.6}$$

Since the vectors $|n\rangle$ are orthonormal, a sum of their outer products $|n\rangle\langle n|$ also represents the identity operator

$$I = \sum_{n=-\infty}^{\infty} |n\rangle\langle n| \tag{3.7}$$

of the space they span. This representation of the identity operator, together with the formula (3.4) for $\langle x|n\rangle$, shows that the inner product $f(x) = \langle x|f\rangle$, which is the component of the vector $|f\rangle$ in the $|x\rangle$ basis, is given by the Fourier series (3.2)

$$\begin{aligned} f(x) = \langle x|f\rangle = \langle x|I|f\rangle &= \sum_{n=-\infty}^{\infty} \langle x|n\rangle\langle n|f\rangle \\ &= \sum_{n=-\infty}^{\infty} \frac{e^{inx}}{\sqrt{2\pi}} \langle n|f\rangle = \sum_{n=-\infty}^{\infty} \frac{e^{inx}}{\sqrt{2\pi}} f_n. \end{aligned} \tag{3.8}$$

Similarly, the other representation (3.6) of the identity operator shows that the inner products $f_n = \langle n|f\rangle$, which are the components of the vector $|f\rangle$ in the $|n\rangle$ basis, are the Fourier integrals (3.3)

$$f_n = \langle n|f\rangle = \langle n|I|f\rangle = \int_0^{2\pi} \langle n|x\rangle\langle x|f\rangle\, dx = \int_0^{2\pi} \frac{e^{-inx}}{\sqrt{2\pi}} f(x)\, dx. \tag{3.9}$$

The two representations (3.6 & 3.7) of the identity operator also give two ways of writing the inner product $\langle g|f\rangle$ of two vectors $|f\rangle$ and $|g\rangle$

$$\begin{aligned} \langle g|f\rangle &= \sum_{n=-\infty}^{\infty} \langle g|n\rangle\langle n|f\rangle = \sum_{n=-\infty}^{\infty} g_n^* f_n \\ &= \int_0^{2\pi} \langle g|x\rangle\langle x|f\rangle\, dx = \int_0^{2\pi} g^*(x)\, f(x)\, dx. \end{aligned} \tag{3.10}$$

When the vectors are the same, this identity shows that the sum of the squared absolute values of the Fourier coefficients f_n is equal to the integral of the squared absolute value $|f(x)|^2$

$$\langle f|f\rangle = \sum_{n=-\infty}^{\infty} |\langle n|f\rangle|^2 = \sum_{n=-\infty}^{\infty} |f_n|^2 = \int_0^{2\pi} |\langle x|f\rangle|^2\, dx = \int_0^{2\pi} |f(x)|^2\, dx. \tag{3.11}$$

Fourier series (3.2 and 3.8) are **periodic** with period 2π because the phases $\langle x|n\rangle$ are periodic with period 2π, $\exp(in(x+2\pi)) = \exp(inx)$. Thus even if the function

$f(x)$ which we use in (3.3 and 3.9) to make the Fourier coefficients $f_n = \langle n|f\rangle$ is not periodic, its Fourier series (3.2 and 3.8) will nevertheless be strictly periodic, as illustrated by Figs. 3.2 and 3.4.

The complex conjugate of the Fourier series (3.2 and 3.8) is

$$f^*(x) = \sum_{n=-\infty}^{\infty} f_n^* \frac{e^{-inx}}{\sqrt{2\pi}} = \sum_{n=-\infty}^{\infty} f_{-n}^* \frac{e^{inx}}{\sqrt{2\pi}} \tag{3.12}$$

so the nth Fourier coefficient $f_n(f^*)$ for $f^*(x)$ is the complex conjugate of the $-n$th Fourier coefficient for $f(x)$

$$f_n(f^*) = f_{-n}^*(f). \tag{3.13}$$

Thus if the function $f(x)$ is real, then

$$f_n(f) = f_n(f^*) = f_{-n}^*(f) \quad \text{or} \quad f_n = f_{-n}^*. \tag{3.14}$$

Example 3.1 (Fourier series by inspection) The doubly exponential function $\exp(\exp(ix))$ has the Fourier series

$$\exp\left(e^{ix}\right) = \sum_{n=0}^{\infty} \frac{1}{n!} e^{inx} \tag{3.15}$$

in which $n! = n(n-1)\cdots 1$ is n-factorial with $0! \equiv 1$.

Example 3.2 (Beats) The sum of two sines $f(x) = \sin\omega_1 x + \sin\omega_2 x$ of similar frequencies $\omega_1 \approx \omega_2$ is the product (exercise 3.1)

$$f(x) = 2\cos\frac{1}{2}(\omega_1 - \omega_2)x \;\sin\frac{1}{2}(\omega_1 + \omega_2)x. \tag{3.16}$$

The first factor $\cos\frac{1}{2}(\omega_1 - \omega_2)x$ is the **beat**; it modulates the second factor $\sin\frac{1}{2}(\omega_1 + \omega_2)x$ as illustrated by Fig. 3.1.

Example 3.3 (Laplace's equation) The Fourier series (exercise 3.2)

$$f(\rho,\theta) = \sum_{n=-\infty}^{\infty} \left(\frac{\rho}{a}\right)^{|n|} \left[\int_0^{2\pi} h(\theta') \frac{e^{-in\theta'}}{\sqrt{2\pi}}\, d\theta'\right] \frac{e^{in\theta}}{\sqrt{2\pi}} \tag{3.17}$$

(Ritt, 1970, p. 3) obeys Laplace's equation (7.23)

$$\frac{1}{\rho}\frac{d}{d\rho}\left(\rho\frac{df}{d\rho}\right) + \frac{1}{\rho^2}\frac{\partial^2 f}{\partial\theta^2} = 0 \tag{3.18}$$

for $\rho < a$ and respects the boundary condition $f(a,\theta) = h(\theta)$.

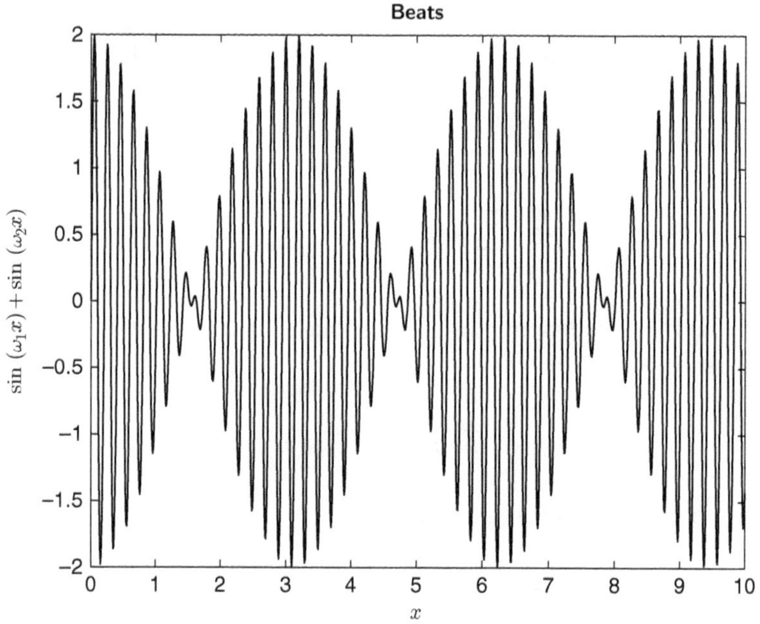

Figure 3.1 The curve $\sin\omega_1 x + \sin\omega_2 x$ for $\omega_1 = 30$ and $\omega_2 = 32$. Matlab scripts for this chapter's figures are in Fourier_series at `github.com/kevinecahill`.

3.2 The Interval

In Section 3.1, we singled out the interval $[0, 2\pi]$, but to represent a periodic function $f(x)$ of period 2π, we could have used any interval of length 2π, such as the interval $[-\pi, \pi]$ or $[r, r + 2\pi]$

$$f_n = \int_r^{r+2\pi} e^{-inx}\, f(x)\, \frac{dx}{\sqrt{2\pi}}. \tag{3.19}$$

This integral is independent of its lower limit r when the function $f(x)$ is periodic with period 2π. The choice $r = -\pi$ is often convenient. With this choice of interval, the coefficient f_n is the integral (3.3) shifted by $-\pi$

$$f_n = \int_{-\pi}^{\pi} e^{-inx}\, f(x)\, \frac{dx}{\sqrt{2\pi}}. \tag{3.20}$$

But if the function $f(x)$ is not periodic with period 2π, then the Fourier coefficients (3.19) do depend upon the choice r of interval.

3.3 Where to Put the 2pi's

In Sections 3.1 and 3.2, we used the orthonormal functions $\exp(inx)/\sqrt{2\pi}$, and so we had factors of $1/\sqrt{2\pi}$ in the Fourier equations (3.2, 3.3, 3.8, and 3.9). One can avoid these square roots by setting $d_n = f_n/\sqrt{2\pi}$ and writing the Fourier series (3.2) and the orthonormality relation (3.3) as

$$f(x) = \sum_{n=-\infty}^{\infty} d_n \, e^{inx} \quad \text{and} \quad d_n = \frac{1}{2\pi}\int_0^{2\pi} dx \, e^{-inx} \, f(x) \tag{3.21}$$

or by setting $c_n = \sqrt{2\pi}\, f_n$ and using the rules

$$f(x) = \frac{1}{2\pi}\sum_{n=-\infty}^{\infty} c_n e^{inx} \quad \text{and} \quad c_n = \int_{-\pi}^{\pi} f(x) e^{-inx}\, dx. \tag{3.22}$$

The cost of these asymmetrical notations is that factors of 2π pop up (exercise 3.3) in equations (3.10 & 3.11) for the inner products $\langle g|f\rangle$ and $\langle f|f\rangle$.

Example 3.4 (Fourier series for $\exp(-m|x|)$) Let's compute the Fourier series for the real function $f(x) = \exp(-m|x|)$ on the interval $(-\pi, \pi)$. Using the shifted interval (3.20) and the 2π-placement convention (3.21), we find that the coefficient d_n is the integral

$$d_n = \int_{-\pi}^{\pi} \frac{dx}{2\pi}\, e^{-inx}\, e^{-m|x|} \tag{3.23}$$

which we may do as two simpler integrals

$$\begin{aligned} d_n &= \int_{-\pi}^{0} \frac{dx}{2\pi}\, e^{(m-in)x} + \int_{0}^{\pi} \frac{dx}{2\pi}\, e^{-(m+in)x} \\ &= \frac{1}{\pi}\frac{m}{m^2+n^2}\left[1-(-1)^n\, e^{-\pi m}\right] \end{aligned} \tag{3.24}$$

which shows that $d_n = d_{-n}$. Since m is real, the coefficients d_n also are real, $d_n = d_n^*$. They therefore satisfy the condition (3.14) that holds for real functions, $d_n = d_{-n}^*$, and give the Fourier series for $\exp(-m|x|)$ as

$$\begin{aligned} e^{-m|x|} &= \sum_{n=-\infty}^{\infty} d_n e^{inx} = \sum_{n=-\infty}^{\infty} \frac{1}{\pi}\frac{m}{m^2+n^2}\left[1-(-1)^n\, e^{-\pi m}\right] e^{inx} \\ &= \frac{\left(1-e^{-\pi m}\right)}{m\pi} + \sum_{n=1}^{\infty} \frac{2}{\pi}\frac{m}{m^2+n^2}\left[1-(-1)^n\, e^{-\pi m}\right]\cos(nx). \end{aligned} \tag{3.25}$$

In Fig. 3.2, the 10-term (dashes) Fourier series for $m = 2$ is plotted from $x = -2\pi$ to $x = 2\pi$. The function $\exp(-2|x|)$ itself is represented by a solid line.

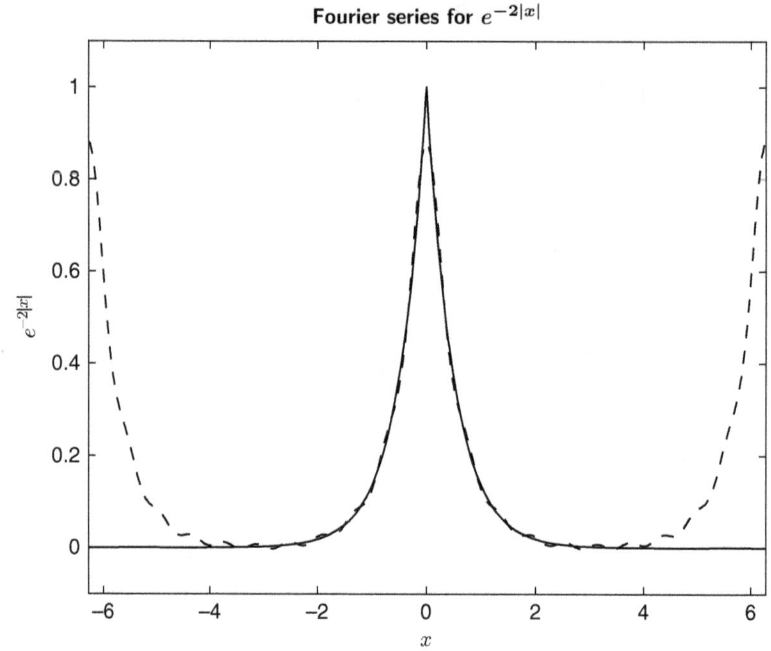

Figure 3.2 The 10-term (dashes) Fourier series (3.25) for the function $\exp(-2|x|)$ on the interval $(-\pi, \pi)$ is plotted from -2π to 2π. All Fourier series are periodic, but the function $\exp(-2|x|)$ (solid) is not.

Although $\exp(-2|x|)$ is not periodic, its Fourier series is periodic with period 2π. The 10-term Fourier series represents the function $\exp(-2|x|)$ quite well within the interval $[-\pi, \pi]$.

In what follows, we usually won't bother to use different letters to distinguish between the symmetric (3.2 and 3.3) and asymmetric conventions (3.21 or 3.22) on the placement of the 2π's.

3.4 Real Fourier Series for Real Functions

The rules (3.1–3.3 and 3.19–3.22) for Fourier series are simple and apply to functions that are continuous and periodic whether complex or real. If a function $f(x)$ is real, then its Fourier coefficients obey the rule (3.14) that holds for real functions, $d_{-n} = d_n^*$. Thus d_0 is real, $d_0 = d_0^*$, and we may write the Fourier series (3.21) for a real function $f(x)$ as

$$f(x) = d_0 + \sum_{n=1}^{\infty} d_n\, e^{inx} + \sum_{n=-\infty}^{-1} d_n\, e^{inx}$$

$$= d_0 + \sum_{n=1}^{\infty} \left[d_n\, e^{inx} + d_{-n}\, e^{-inx} \right] = d_0 + \sum_{n=1}^{\infty} \left[d_n\, e^{inx} + d_n^*\, e^{-inx} \right]$$

$$= d_0 + \sum_{n=1}^{\infty} d_n\,(\cos nx + i\,\sin nx) + d_n^*\,(\cos nx - i\,\sin nx)$$

$$= d_0 + \sum_{n=1}^{\infty} (d_n + d_n^*)\,\cos nx + i(d_n - d_n^*)\,\sin nx. \tag{3.26}$$

In terms of the real coefficients

$$a_n = d_n + d_n^* \quad \text{and} \quad b_n = i(d_n - d_n^*), \tag{3.27}$$

the Fourier series (3.26) of a real function $f(x)$ is

$$f(x) = \frac{a_0}{2} + \sum_{n=1}^{\infty} a_n\,\cos nx + b_n\,\sin nx. \tag{3.28}$$

Using the formulas (3.27) for a_n and (3.21) for d_n as well as the reality of the function $f(x)$, we find that a_n is

$$a_n = \int_0^{2\pi} \left[e^{-inx}\,f(x) + e^{inx}\,f^*(x)\right] \frac{dx}{2\pi} = \int_0^{2\pi} \frac{\left(e^{-inx} + e^{inx}\right)}{2}\,f(x)\,\frac{dx}{\pi}. \tag{3.29}$$

So the coefficient a_n of $\cos nx$ in (3.28) is the cosine integral of $f(x)$

$$a_n = \int_0^{2\pi} \cos nx\;f(x)\,\frac{dx}{\pi}. \tag{3.30}$$

Similarly, equations (3.27 and 3.21) and the reality of $f(x)$ imply that the coefficient b_n is the sine integral of $f(x)$

$$b_n = \int_0^{2\pi} i\,\frac{\left(e^{-inx} - e^{inx}\right)}{2}\,f(x)\,\frac{dx}{\pi} = \int_0^{2\pi} \sin nx\;f(x)\,\frac{dx}{\pi}. \tag{3.31}$$

The real Fourier series (3.28) and the cosine (3.30) and sine (3.31) integrals for the coefficients a_n and b_n also follow from the orthogonality relations

$$\int_0^{2\pi} \sin mx\;\sin nx\,dx = \begin{cases} \pi & \text{if } n = m \neq 0 \\ 0 & \text{otherwise,} \end{cases} \tag{3.32}$$

$$\int_0^{2\pi} \cos mx\;\cos nx\,dx = \begin{cases} \pi & \text{if } n = m \neq 0 \\ 2\pi & \text{if } n = m = 0 \\ 0 & \text{otherwise, and} \end{cases} \tag{3.33}$$

$$\int_0^{2\pi} \sin mx\;\cos nx\,dx = 0, \tag{3.34}$$

which hold for integer values of n and m.

If the function $f(x)$ is periodic with period 2π, then instead of the interval $[0, 2\pi]$, one may choose any interval of length 2π such as $[-\pi, \pi]$.

What if a function $f(x)$ is not periodic? The Fourier series for an aperiodic function is itself strictly periodic, is sensitive to its interval $(r, r+2\pi)$ of definition, may differ somewhat from the function near the ends of the interval, and usually differs markedly from it outside the interval.

Example 3.5 (Fourier series for x^2) The function x^2 is even and so the integrals (3.31) for its sine Fourier coefficients b_n all vanish. Its cosine coefficients a_n are given by (3.30)

$$a_n = \int_{-\pi}^{\pi} \cos nx \, f(x) \, \frac{dx}{\pi} = \int_{-\pi}^{\pi} \cos nx \, x^2 \, \frac{dx}{\pi}. \tag{3.35}$$

Integrating twice by parts, we find for $n \neq 0$

$$a_n = -\frac{2}{n} \int_{-\pi}^{\pi} x \sin nx \, \frac{dx}{\pi} = \left[\frac{2x \cos nx}{\pi n^2} \right]_{-\pi}^{\pi} = (-1)^n \frac{4}{n^2} \tag{3.36}$$

and

$$a_0 = \int_{-\pi}^{\pi} x^2 \, \frac{dx}{\pi} = \frac{2\pi^2}{3}. \tag{3.37}$$

Equation (3.28) now gives for x^2 the cosine Fourier series

$$x^2 = \frac{a_0}{2} + \sum_{n=1}^{\infty} a_n \cos nx = \frac{\pi^2}{3} + 4 \sum_{n=1}^{\infty} (-1)^n \frac{\cos nx}{n^2}. \tag{3.38}$$

This series rapidly converges within the interval $[-1, 1]$ as shown in Fig. 3.3, but not near the endpoints $\pm\pi$.

Example 3.6 (Gibbs overshoot) The function $f(x) = x$ on the interval $[-\pi, \pi]$ is not periodic. So we expect trouble if we represent it as a Fourier series. Since x is an odd function, equation (3.30) tells us that the coefficients a_n all vanish. By (3.31), the b_n's are

$$b_n = \int_{-\pi}^{\pi} \frac{dx}{\pi} \, x \sin nx = 2\,(-1)^{n+1} \, \frac{1}{n}. \tag{3.39}$$

As shown in Fig. 3.4, the series

$$\sum_{n=1}^{\infty} 2\,(-1)^{n+1} \, \frac{1}{n} \, \sin nx \tag{3.40}$$

differs by about 2π from the function $f(x) = x$ for $-3\pi < x < -\pi$ and for $\pi < x < 3\pi$ because the series is periodic while the function x isn't.

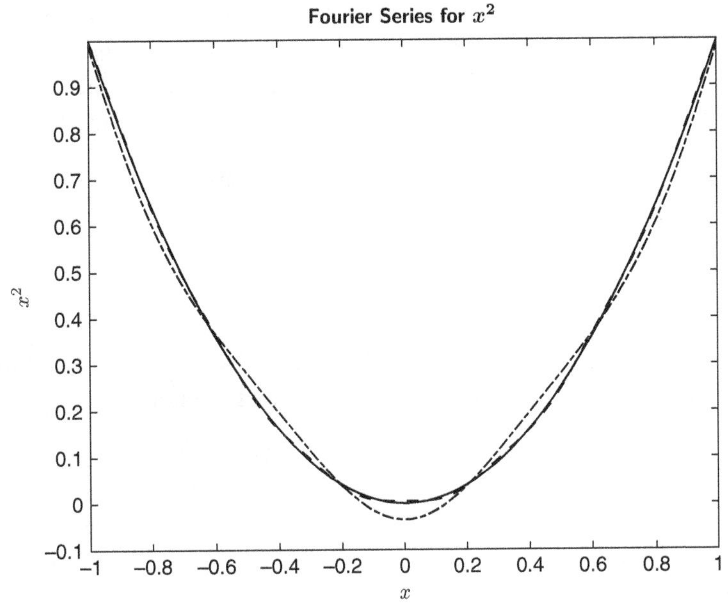

Figure 3.3 The function x^2 (solid) and its Fourier series of 7 terms (dot dash) and 20 terms (dashes). The Fourier series (3.38) for x^2 quickly converges well inside the interval $(-\pi, \pi)$.

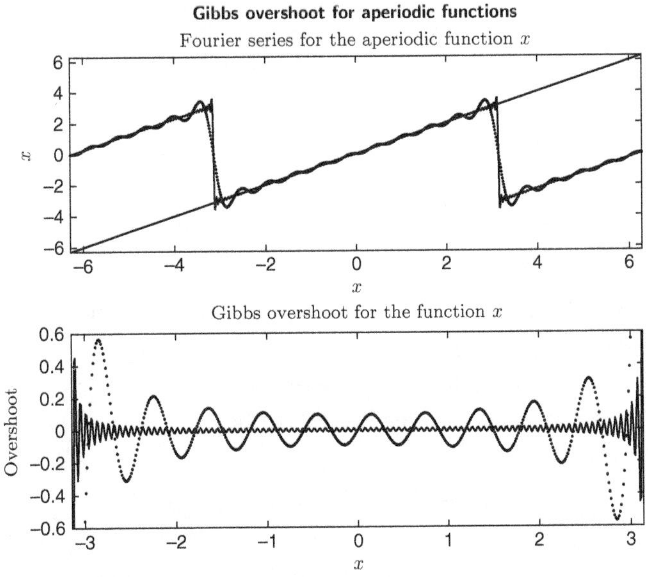

Figure 3.4 (top) The Fourier series (3.40) for the function x (solid line) with 10 terms (dots) and 100 terms (solid curve) for $-2\pi < x < 2\pi$. The Fourier series is periodic, but the function x is not. (bottom) The differences between x and the 10-term (dots) and the 100-term (solid curve) series on $(-\pi, \pi)$ exhibit a Gibbs overshoot of about 9% at $x \gtrsim -\pi$ and at $x \lesssim \pi$.

Within the interval $(-\pi, \pi)$, the series with 100 terms is very accurate except for $x \gtrsim -\pi$ and $x \lesssim \pi$, where it overshoots by about 9% of the 2π discontinuity, a defect called a **Gibbs overshoot** (J. Willard Gibbs 1839–1903. Incidentally, Gibbs's father helped defend the Africans of the schooner *La Amistad*). Any time we use a Fourier series to represent an aperiodic function, a Gibbs overshoot will occur near the endpoints of the interval.

3.5 Stretched Intervals

If the interval of periodicity is of length L instead of 2π, then we may use the phases $\exp(i2\pi nx/\sqrt{L})$ which are orthonormal on the interval $[0, L]$

$$\int_0^L dx \left(\frac{e^{i2\pi nx/L}}{\sqrt{L}}\right)^* \frac{e^{i2\pi mx/L}}{\sqrt{L}} = \delta_{nm}. \tag{3.41}$$

The Fourier series

$$f(x) = \sum_{n=-\infty}^{\infty} f_n \frac{e^{i2\pi nx/L}}{\sqrt{L}} \tag{3.42}$$

is periodic with period L. The coefficient f_n is the integral

$$f_n = \int_0^L \frac{e^{-i2\pi nx/L}}{\sqrt{L}} f(x)\, dx, \tag{3.43}$$

and the sum of their squares $|f_n|^2$ is the integral of $|f(x)|^2$

$$\sum_{n=-\infty}^{\infty} |f_n|^2 = \int_0^L |f(x)|^2\, dx. \tag{3.44}$$

These relations (3.41–3.44) generalize to the interval $[0, L]$ our earlier formulas of Section 3.1 for the interval $[0, 2\pi]$.

If the function $f(x)$ is periodic with period L, that is if $f(x + nL) = f(x)$ for any integer n, then we may shift the domain of integration by any real number r

$$f_n = \int_r^{L+r} \frac{e^{-i2\pi nx/L}}{\sqrt{L}} f(x)\, dx \tag{3.45}$$

without changing the coefficients f_n. An obvious choice is $r = -L/2$ for which (3.42) and (3.43) give

$$f(x) = \sum_{n=-\infty}^{\infty} f_n \frac{e^{i2\pi nx/L}}{\sqrt{L}} \quad \text{and} \quad f_n = \int_{-L/2}^{L/2} \frac{e^{-i2\pi nx/L}}{\sqrt{L}} f(x)\, dx. \tag{3.46}$$

If the function $f(x)$ is real, then on the interval $[0, L]$ in place of Eqs. (3.28), (3.30), and (3.31), one has

$$f(x) = \frac{a_0}{2} + \sum_{n=1}^{\infty} a_n \cos\left(\frac{2\pi n x}{L}\right) + b_n \sin\left(\frac{2\pi n x}{L}\right), \tag{3.47}$$

$$a_n = \frac{2}{L}\int_0^L dx \cos\left(\frac{2\pi n x}{L}\right) f(x), \tag{3.48}$$

and

$$b_n = \frac{2}{L}\int_0^L dx \sin\left(\frac{2\pi n x}{L}\right) f(x). \tag{3.49}$$

The corresponding orthogonality relations, which follow from Eqs. (3.32), (3.33), and (3.34), are:

$$\int_0^L dx \sin\left(\frac{2\pi m x}{L}\right) \sin\left(\frac{2\pi n x}{L}\right) = \begin{cases} L/2 & \text{if } n = m \neq 0 \\ 0 & \text{otherwise,} \end{cases} \tag{3.50}$$

$$\int_0^L dx \cos\left(\frac{2\pi m x}{L}\right) \cos\left(\frac{2\pi n x}{L}\right) = \begin{cases} L/2 & \text{if } n = m \neq 0 \\ L & \text{if } n = m = 0 \\ 0 & \text{otherwise, and} \end{cases} \tag{3.51}$$

$$\int_0^L dx \sin\left(\frac{2\pi m x}{L}\right) \cos\left(\frac{2\pi n x}{L}\right) = 0. \tag{3.52}$$

They hold for integer values of n and m, and they imply Eqs. (3.47)–(3.49).

3.6 Fourier Series of Functions of Several Variables

On an interval $[0, L]$, the Fourier-series formulas (3.42 and 3.43) are

$$f(x) = \sum_{n=-\infty}^{\infty} f_n \frac{e^{i2\pi n x/L}}{\sqrt{L}} \tag{3.53}$$

$$f_n = \int_0^L \frac{e^{-2i\pi n x/L}}{\sqrt{L}} f(x)\, dx. \tag{3.54}$$

We may generalize these equations from a single variable to m variables $x = (x_1, \ldots, x_m)$ with $n \cdot x = n_1 x_1 + \cdots + n_m x_m$

$$f(x) = \sum_{n_1=-\infty}^{\infty} \cdots \sum_{n_m=-\infty}^{\infty} f_n \frac{e^{i2\pi n \cdot x/L}}{L^{m/2}} \tag{3.55}$$

$$f_n = \int_0^L dx_1 \cdots \int_0^L dx_m \frac{e^{-2i\pi n \cdot x/L}}{L^{m/2}} f(x). \tag{3.56}$$

3.7 Integration and Differentiation of Fourier Series

What happens to the convergence of a Fourier series if we integrate or differentiate term by term? If we integrate the series

$$f(x) = \sum_{n=-\infty}^{\infty} f_n \frac{e^{i2\pi nx/L}}{\sqrt{L}} \tag{3.57}$$

then we get a series

$$F(x) = \int_0^x dx' \, f(x') = \frac{f_0}{\sqrt{L}} x - i\frac{\sqrt{L}}{2\pi} \sum_{n=-\infty}^{\infty} \frac{f_n}{n} \left(e^{i2\pi nx/L} - 1\right) \tag{3.58}$$

that converges **better** because of the extra factor of $1/n$. An integrated function $f(x)$ is smoother, and so its Fourier series converges better.

But if we differentiate the same series, then we get a series

$$f'(x) = i\frac{2\pi}{L^{3/2}} \sum_{n=-\infty}^{\infty} n \, f_n \, e^{i2\pi nx/L} \tag{3.59}$$

that converges **less well** because of the extra factor of n. A differentiated function is rougher, and so its Fourier series converges less well.

3.8 How Fourier Series Converge

A Fourier series represents a function $f(x)$ as the limit of a sequence of functions $f_N(x)$ given by

$$f_N(x) = \sum_{n=-N}^{N} f_n \frac{e^{i2\pi nx/L}}{\sqrt{L}} \quad \text{in which} \quad f_n = \int_0^L f(x) \, e^{-i2\pi nx/L} \frac{dx}{\sqrt{L}}. \tag{3.60}$$

Since the exponentials are periodic with period L, a Fourier series always is periodic. So if the function $f(x)$ is not periodic, then its Fourier series will represent the **periodic extension** f_p of f defined by

$$f_p(x + nL) = f(x) \tag{3.61}$$

for all integers n and for $0 \le x \le L$.

A sequence of functions $f_N(x)$ **converges** to a function $f(x)$ on a **closed** interval $[a, b]$ if for every $\epsilon > 0$ and each point $a \le x \le b$, there exists an integer $N(\epsilon, x)$ such that

$$|f(x) - f_N(x)| < \epsilon \quad \text{for all} \quad N > N(\epsilon, x). \tag{3.62}$$

If this holds for an $N(\epsilon, x) = N(\epsilon)$ that is independent of $x \in [a, b]$, then the sequence of functions $f_N(x)$ **converges uniformly** to $f(x)$ on the interval $[a, b]$.

A function $f(x)$ is **continuous** on an **open** interval (a, b) if for every point $a < x < b$ the two limits

$$f(x-0) \equiv \lim_{0<\epsilon\to 0} f(x-\epsilon) \quad \text{and} \quad f(x+0) \equiv \lim_{0<\epsilon\to 0} f(x+\epsilon) \tag{3.63}$$

agree. If $f(x)$ also has the limits $f(a+0) = f(a)$ and $f(b-0) = f(b)$, then f is continuous on the **closed** interval $[a, b]$. A function continuous on a closed interval $[a, b]$ is bounded and integrable on that interval.

If a sequence of continuous functions $f_N(x)$ converges uniformly to a function $f(x)$ on a closed interval $a \le x \le b$, then we know that $|f_N(x) - f(x)| < \epsilon$ for $N > N(\epsilon)$, and so

$$\left| \int_a^b f_N(x)\, dx - \int_a^b f(x)\, dx \right| \le \int_a^b |f_N(x) - f(x)|\, dx < (b-a)\,\epsilon. \tag{3.64}$$

Thus one may integrate a uniformly convergent sequence of continuous functions on a closed interval $[a, b]$ term by term

$$\lim_{N\to\infty} \int_a^b f_N(x)\, dx = \int_a^b \lim_{N\to\infty} f_N(x)\, dx = \int_a^b f(x)\, dx. \tag{3.65}$$

So if a Fourier series (7.370) converges uniformly, then the term-by-term integration implicit in the formula (3.3) for f_n is permitted.

A function is **piecewise continuous** on $[a, b]$ if it is continuous there except for finite jumps from $f(x-0)$ to $f(x+0)$ at a finite number of points x. At such jumps, we *define* the periodically extended function f_p to be the mean $f_p(x) = [f(x-0) + f(x+0)]/2$.

Fourier's convergence theorem (Courant, 1937, p. 439): The Fourier series of a function $f(x)$ that is piecewise continuous with a piecewise continuous first derivative converges to its periodic extension $f_p(x)$. This convergence is uniform on every closed interval on which the function $f(x)$ is continuous (and absolute if the function $f(x)$ has no discontinuities). Examples 3.12 and 3.13 illustrate this result.

A function whose kth derivative is continuous is in **class C^k**. On the interval $[-\pi, \pi]$, its Fourier coefficients (3.22) are

$$f_n = \int_{-\pi}^{\pi} f(x)\, e^{-inx}\, dx. \tag{3.66}$$

If f is both periodic and in C^k, then one integration by parts gives

$$f_n = \int_{-\pi}^{\pi} \left\{ \frac{d}{dx}\left[f(x)\, \frac{e^{-inx}}{-in} \right] - f'(x)\, \frac{e^{-inx}}{-in} \right\} dx = \int_{-\pi}^{\pi} f'(x)\, \frac{e^{-inx}}{in}\, dx$$

and k integrations by parts give

$$f_n = \int_{-\pi}^{\pi} f^{(k)}(x) \, \frac{e^{-inx}}{(in)^k} \, dx \tag{3.67}$$

since the derivatives $f^{(\ell)}(x)$ of a C^k periodic function also are periodic. Moreover if $f^{(k+1)}$ is piecewise continuous, then

$$\begin{aligned} f_n &= \int_{-\pi}^{\pi} \left\{ \frac{d}{dx} \left[f^{(k)}(x) \, \frac{e^{-inx}}{-(in)^{k+1}} \right] - f^{(k+1)}(x) \, \frac{e^{-inx}}{-(in)^{k+1}} \right\} dx \\ &= \int_{-\pi}^{\pi} f^{(k+1)}(x) \, \frac{e^{-inx}}{(in)^{k+1}} \, dx. \end{aligned} \tag{3.68}$$

Since $f^{(k+1)}(x)$ is piecewise continuous on the closed interval $[-\pi, \pi]$, it is bounded there in absolute value by, let us say, M. So the Fourier coefficients of a C^k periodic function with $f^{(k+1)}$ piecewise continuous are bounded by

$$|f_n| \le \frac{1}{n^{k+1}} \int_{-\pi}^{\pi} |f^{(k+1)}(x)| \, dx \le \frac{2\pi M}{n^{k+1}}. \tag{3.69}$$

We often can carry this derivation one step further. In most simple examples, the piecewise continuous periodic function $f^{(k+1)}(x)$ actually is piecewise continuously differentiable between its successive jumps at x_j. In this case, the derivative $f^{(k+2)}(x)$ is a piecewise continuous function plus a sum of a finite number of delta functions with finite coefficients. Thus we can integrate once more by parts. If for instance the function $f^{(k+1)}(x)$ jumps J times between $-\pi$ and π by $\Delta f_j^{(k+1)}$, then its Fourier coefficients are

$$\begin{aligned} f_n &= \int_{-\pi}^{\pi} f^{(k+2)}(x) \, \frac{e^{-inx}}{(in)^{k+2}} \, dx \\ &= \sum_{j=1}^{J} \int_{x_j}^{x_{j+1}} f_{pc}^{(k+2)}(x) \, \frac{e^{-inx}}{(in)^{k+2}} \, dx + \sum_{j=1}^{J} \Delta f_j^{(k+1)} \, \frac{e^{-inx_j}}{(in)^{k+2}} \end{aligned} \tag{3.70}$$

in which the subscript pc means piecewise continuous. The Fourier coefficients (3.70) then are bounded by

$$|f_n| \le \frac{2\pi M}{n^{k+2}} \tag{3.71}$$

in which M is related to the maximum absolute values of $f_{pc}^{(k+2)}(x)$ and of the $\Delta f_j^{(k+1)}$. The Fourier series of periodic C^k functions converge rapidly if k is big.

Example 3.7 (Fourier series of a C^0 function) The function defined by

$$f(x) = \begin{cases} 0 & -\pi \le x < 0 \\ x & 0 \le x < \pi/2 \\ \pi - x & \pi/2 \le x \le \pi \end{cases} \tag{3.72}$$

is continuous on the interval $[-\pi, \pi]$ and its first derivative is piecewise continuous on that interval. By (3.69), its Fourier coefficients f_n should be bounded by M/n. In fact they are (exercise 3.10) bounded by $2\sqrt{2/\pi}/n^2$

$$f_n = \int_{-\pi}^{\pi} f(x) e^{-inx} \frac{dx}{\sqrt{2\pi}} = \frac{(-1)^{n+1}}{\sqrt{2\pi}} \frac{(i^n - 1)^2}{n^2} \tag{3.73}$$

in agreement with the stronger inequality (3.71).

Example 3.8 (Fourier series for a C^1 function) The function defined by $f(x) = 1 + \cos 2x$ for $|x| \le \pi/2$ and $f(x) = 0$ for $|x| \ge \pi/2$ has a periodic extension f_p that is continuous with a continuous first derivative and a piecewise continuous second derivative. Its Fourier coefficients (3.66)

$$f_n = \int_{-\pi/2}^{\pi/2} (1 + \cos 2x)\, e^{-inx} \frac{dx}{\sqrt{2\pi}} = \frac{8 \sin n\pi/2}{\sqrt{2\pi}(4n - n^3)}$$

satisfy the inequalities (3.69) and (3.71) for $k = 1$.

Example 3.9 (Fourier series for $\cos \mu x$) The Fourier series for the even function $f(x) = \cos \mu x$ has only cosines with coefficients (3.30)

$$\begin{aligned} a_n &= \int_{-\pi}^{\pi} \cos nx \, \cos \mu x \, \frac{dx}{\pi} = \int_0^{\pi} [\cos(\mu + n)x + \cos(\mu - n)x] \frac{dx}{\pi} \\ &= \frac{1}{\pi} \left[\frac{\sin(\mu + n)\pi}{\mu + n} + \frac{\sin(\mu - n)\pi}{\mu - n} \right] = \frac{2}{\pi} \frac{\mu(-1)^n}{\mu^2 - n^2} \sin \mu\pi. \end{aligned} \tag{3.74}$$

Thus whether or not μ is an integer, the series (3.28) gives us

$$\cos \mu x = \frac{2\mu \sin \mu\pi}{\pi} \left(\frac{1}{2\mu^2} - \frac{\cos x}{\mu^2 - 1^2} + \frac{\cos 2x}{\mu^2 - 2^2} - \frac{\cos 3x}{\mu^2 - 3^2} + \cdots \right) \tag{3.75}$$

which is continuous at $x = \pm\pi$ (Courant, 1937, chap. IX).

Example 3.10 (The sine as an infinite product) In our series (3.75) for $\cos \mu x$, we set $x = \pi$, divide by $\sin \mu\pi$, replace μ with x, and so find for the cotangent the expansion

$$\cot \pi x = \frac{2x}{\pi} \left(\frac{1}{2x^2} + \frac{1}{x^2 - 1^2} + \frac{1}{x^2 - 2^2} + \frac{1}{x^2 - 3^2} + \cdots \right) \tag{3.76}$$

or equivalently

$$\cot \pi x - \frac{1}{\pi x} = -\frac{2x}{\pi} \left(\frac{1}{1^2 - x^2} + \frac{1}{2^2 - x^2} + \frac{1}{3^2 - x^2} + \cdots \right). \tag{3.77}$$

For $0 \le x \le q < 1$, the absolute value of the nth term on the right is less than $2q/(\pi(n^2 - q^2))$. Thus this series converges uniformly on $[0, x]$, and so we may integrate it term by term. We find (exercise 3.13)

$$\pi \int_0^x \left(\cot \pi t - \frac{1}{\pi t} \right) dt = \ln \frac{\sin \pi x}{\pi x} = \sum_{n=1}^{\infty} \int_0^x \frac{-2t\, dt}{n^2 - t^2} = \sum_{n=1}^{\infty} \ln \left[1 - \frac{x^2}{n^2} \right]. \quad (3.78)$$

Exponentiating, we get the infinite-product formula

$$\frac{\sin \pi x}{\pi x} = \exp \left[\sum_{n=1}^{\infty} \ln \left(1 - \frac{x^2}{n^2} \right) \right] = \prod_{n=1}^{\infty} \left(1 - \frac{x^2}{n^2} \right) \quad (3.79)$$

for the sine from which one can derive the infinite product (exercise 3.14)

$$\cos \pi x = \prod_{n=1}^{\infty} \left(1 - \frac{x^2}{(n - \frac{1}{2})^2} \right) \quad (3.80)$$

for the cosine (Courant, 1937, chap. IX).

3.9 Measure and Lebesgue Integration

Suppose S is a set of points x that lie in an interval $a \le x \le b$ of length $b - a$. All the points of S may also lie inside several subintervals $[a_i, b_i]$, $i = 1, 2, \ldots$, the sum of whose lengths is $b_1 - a_1 + b_2 - a_2 + \cdots$. Now consider all possible such sets of subintervals $[a_i, b_i]$ that contain all the points of S and let m be the greatest lower bound of the sum of their lengths. We may do the same for the complementary set S' consisting of all points of $[a, b]$ that do not lie in the set S. That is, we may let m' be the greatest lower bound of the sum of the lengths of all possible sets of subintervals $[c_i, d_i]$ that contain all the points of S'. If $m + m' = b - a$, then the set S is **measurable** and m is its measure. Every countable set x_i, $i = 1, 2, \ldots$, has measure zero.

Suppose now that for $a \le x \le b$, all the values $f(x)$ of a function f lie in some finite interval J. We partition this interval J into disjoint subintervals J_k and let S_k be the set of points of $[a, b]$ that f maps into each subinterval J_k. If for every subinterval J_k, the set S_k is measurable, then the function $f(x)$ is **measurable** or **summable** on $[a, b]$. Suppose that f is measurable on this interval and let $m(S_k)$ be the measure of the set S_k. Then for each subinterval J_k, we may pick any point $x_k \in S_k$ and approximate the integral of f over the interval $[a, b]$ by the sum $f(x_1)\, m(S_1) + f(x_2)\, m(S_2) + \cdots$. This sum converges (Courant and Hilbert, 1955, pp. 108–111) to the Lebesgue integral as we refine the partition of the interval J into subintervals J_k such that the length L of the longest subinterval goes to zero

$$\lim_{L\to 0} \sum_{k=1}^{\infty} f(x_k)\, m(S_k) = \int_a^b f(x)\, dx \tag{3.81}$$

(Henri Lebesgue, 1875–1941).

Lebesgue integration generalizes Riemann integration and provides a more natural basis for discussions of convergence. One important theorem resulting from measure theory is that of Riesz and Fischer (Hardy and Rogosinski, 1944, p. 16): If a sum

$$\sum_{n=-\infty}^{\infty} |f_n|^2 < \infty \tag{3.82}$$

converges, then (1) there is a function f that is square integrable in the sense of Lebesgue (f is L^2 on $[a, b]$)

$$\int_a^b |f(x)|^2\, dx < \infty, \tag{3.83}$$

whose Fourier coefficients are

$$f_n = \int_a^b \frac{e^{-2\pi i n x/(b-a)}}{\sqrt{b-a}}\, f(x)\, dx, \tag{3.84}$$

(2) the series

$$f_N(x) = \sum_{n=-N}^{N} f_n \, \frac{e^{2\pi i n x/(b-a)}}{\sqrt{b-a}} \tag{3.85}$$

converges to $f(x)$ **in the mean**, that is, as $N \to \infty$

$$\int_a^b |f_N(x) - f(x)|^2\, dx \to 0, \tag{3.86}$$

and (3)

$$\int_a^b |f(x)|^2\, dx = \sum_{n=-\infty}^{\infty} |f_n|^2 \tag{3.87}$$

which is (3.44) for $L = b - a$.

Fourier series can represent a much wider class of functions than those that are continuous. If a function $f(x)$ is square integrable on an interval $[a, b]$, then its N-term Fourier series $f_N(x)$ will converge to $f(x)$ **in the mean**, that is

$$\lim_{N\to\infty} \int_a^b dx\, |f(x) - f_N(x)|^2 = 0. \tag{3.88}$$

3.10 Quantum-Mechanical Examples

Suppose a particle of mass m is trapped in an infinitely deep 1-dimensional square well of potential energy

$$V(x) = \begin{cases} 0 & \text{if } 0 < x < L \\ \infty & \text{otherwise.} \end{cases} \tag{3.89}$$

The hamiltonian operator is

$$H = -\frac{\hbar^2}{2m}\frac{d^2}{dx^2} + V(x), \tag{3.90}$$

in which $\hbar$ is Planck's constant divided by 2π. This tiny bit of action, $\hbar = 1.055 \times 10^{-34}$ J s, sets the scale at which quantum mechanics becomes important. Quantum-mechanical corrections to classical predictions can be big in processes whose action is less than $\hbar$.

An eigenfunction $\psi(x)$ of the hamiltonian H with energy E satisfies the equation $H\psi(x) = E\psi(x)$ which breaks into two simple equations:

$$-\frac{\hbar^2}{2m}\frac{d^2\psi(x)}{dx^2} = E\psi(x) \quad \text{for} \quad 0 < x < L \tag{3.91}$$

and

$$-\frac{\hbar^2}{2m}\frac{d^2\psi(x)}{dx^2} + \infty\,\psi(x) = E\psi(x) \quad \text{for} \quad x < 0 \quad \text{and for} \quad x > L. \tag{3.92}$$

Every solution of these equations with finite energy E must vanish outside the interval $0 < x < L$. So we must find solutions of the first equation (3.91) that satisfy the boundary conditions

$$\psi(x) = 0 \quad \text{for} \quad x \le 0 \text{ and } x \ge L. \tag{3.93}$$

For any integer $n \neq 0$, the function

$$\psi_n(x) = \sqrt{\frac{2}{L}}\,\sin\left(\frac{\pi n x}{L}\right) \quad \text{for} \quad x \in [0, L] \tag{3.94}$$

and $\psi_n(x) = 0$ for $x \notin (0, L)$ satisfies the boundary conditions (3.93). When inserted into equation (3.91)

$$-\frac{\hbar^2}{2m}\frac{d^2}{dx^2}\psi_n(x) = \frac{\hbar^2}{2m}\left(\frac{n\pi}{L}\right)^2 \psi_n(x) = E_n\psi_n(x) \tag{3.95}$$

it reveals its energy to be $E_n = (n\pi\hbar/L)^2/2m$.

These eigenfunctions $\psi_n(x)$ are complete in the sense that they span the space of all functions $f(x)$ that are square-integrable on the interval $(0, L)$ and vanish at its endpoints. They provide for such functions the **sine Fourier series**

$$f(x) = \sum_{n=1}^{\infty} f_n \sqrt{\frac{2}{L}} \sin\left(\frac{\pi n x}{L}\right) \tag{3.96}$$

which is periodic with period $2L$ and is the Fourier series for a function that is odd $f(-x) = -f(x)$ on the interval $(-L, L)$ and zero at both ends.

Example 3.11 (Time evolution of an initially piecewise continuous wave function) Suppose now that at time $t = 0$ the particle is confined to the middle half of the well with the square wave function

$$\psi(x, 0) = \sqrt{\frac{2}{L}} \quad \text{for} \quad \frac{L}{4} < x < \frac{3L}{4} \tag{3.97}$$

and zero otherwise. This piecewise continuous C^{-1} wave function is discontinuous at $x = L/4$ and at $x = 3L/4$. Since the functions $\langle x|n\rangle = \psi_n(x)$ are orthonormal on $[0, L]$

$$\int_0^L dx \sqrt{\frac{2}{L}} \sin\left(\frac{\pi n x}{L}\right) \sqrt{\frac{2}{L}} \sin\left(\frac{\pi m x}{L}\right) = \delta_{nm} \tag{3.98}$$

the coefficients f_n in the Fourier series

$$\psi(x, 0) = \sum_{n=1}^{\infty} f_n \sqrt{\frac{2}{L}} \sin\left(\frac{\pi n x}{L}\right) \tag{3.99}$$

are the inner products

$$f_n = \langle n|\psi, 0\rangle = \int_0^L dx \sqrt{\frac{2}{L}} \sin\left(\frac{\pi n x}{L}\right) \psi(x, 0). \tag{3.100}$$

They are proportional to $1/n$ in accord with (3.71)

$$f_n = \frac{2}{L} \int_{L/4}^{3L/4} dx \sin\left(\frac{\pi n x}{L}\right) = \frac{2}{\pi n} \left[\cos\left(\frac{\pi n}{4}\right) - \cos\left(\frac{3\pi n}{4}\right)\right]. \tag{3.101}$$

Figure 3.5 plots the square wave function $\psi(x, 0)$ (3.97, straight solid lines) and its 10-term (solid curve) and 100-term (dashes) Fourier series (3.99) for an interval of length $L = 2$. Gibbs's overshoot reaches 1.093 at $x = 0.52$ for 100 terms and 1.0898 at $x = 0.502$ for 1000 terms (not shown), amounting to about 9% of the unit discontinuity at $x = 1/2$. A similar overshoot occurs at $x = 3/2$.

How does $\psi(x, 0)$ evolve with time? Since $\psi_n(x)$, the Fourier component (3.94), is an eigenfunction of H with energy E_n, the time-evolution operator $U(t) = \exp(-iHt/\hbar)$ takes $\psi(x, 0)$ into

$$\psi(x, t) = e^{-iHt/\hbar}\, \psi(x, 0) = \sum_{n=1}^{\infty} f_n \sqrt{\frac{2}{L}} \sin\left(\frac{\pi n x}{L}\right) e^{-iE_n t/\hbar}. \tag{3.102}$$

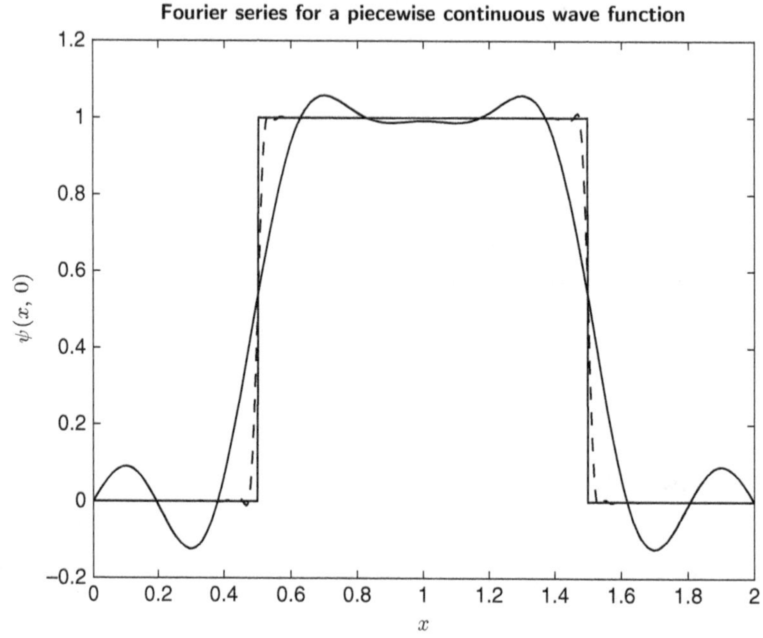

Figure 3.5 The piecewise continuous wave function $\psi(x,0)$ for $L = 2$ (3.97, straight solid lines) and its Fourier series (3.99) with 10 terms (solid curve) and 100 terms (dashes). Gibbs overshoots occur near the discontinuities at $x = 1/2$ and $x = 3/2$.

Because $E_n = (n\pi\hbar/L)^2/2m$, the wave function at time t is

$$\psi(x,t) = \sum_{n=1}^{\infty} f_n \sqrt{\frac{2}{L}} \sin\left(\frac{\pi n x}{L}\right) e^{-i\hbar(n\pi)^2 t/(2mL^2)}. \tag{3.103}$$

It is awkward to plot complex functions, so Fig. 3.6 displays the probability distributions $P(x,t) = |\psi(x,t)|^2$ of the 1000-term Fourier series (3.103) for the wave function $\psi(x,t)$ at $t = 0$ (thick curve), $t = 10^{-3}\,\tau$ (medium curve), and $\tau = 2mL^2/\hbar$ (thin curve). The discontinuities in the initial wave function $\psi(x,0)$ cause both the Gibbs overshoots at $x = 1/2$ and $x = 3/2$ seen in the series for $\psi(x,0)$ plotted in Fig. 3.5 and the choppiness of the probability distribution $P(x,t)$ exhibited in Fig. 3.6.

Example 3.12 (Time evolution of a continuous function) What does the Fourier series of a continuous function look like? How does it evolve with time? Let us take as the wave function at $t = 0$ the C^0 function

$$\psi(x,0) = \frac{2}{\sqrt{L}} \sin\left(\frac{2\pi(x - L/4)}{L}\right) \quad \text{for} \quad \frac{L}{4} < x < \frac{3L}{4} \tag{3.104}$$

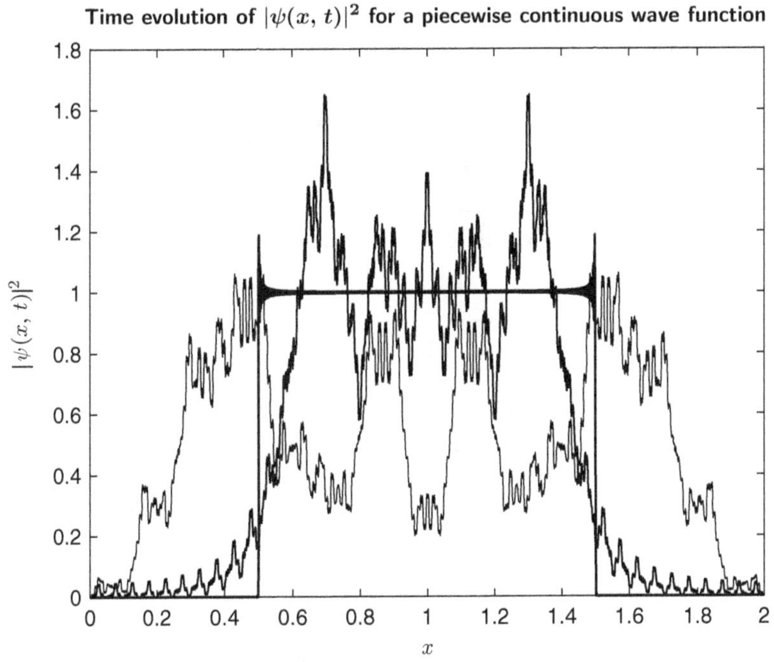

Figure 3.6 For length $L = 2$, the probability distributions $P(x,t) = |\psi(x,t)|^2$ of the 1000-term Fourier series (3.103) for the wave function $\psi(x,t)$ at $t = 0$ (thick curve), $t = 10^{-3}\,\tau$ (medium curve), and $\tau = 2mL^2/\hbar$ (thin curve). The jaggedness of $P(x,t)$ arises from the two discontinuities in the initial wave function $\psi(x,0)$ (3.104) at $x = L/4$ and $x = 3L/4$. A Matlab script for the figure is in Fourier_series at `github.com/kevinecahill`.

and zero otherwise. This initial wave function is a continuous function with a piecewise continuous first derivative on the interval $[0, L]$, and it satisfies the periodic boundary condition $\psi(0,0) = \psi(L,0)$. It therefore satisfies the conditions of Fourier's convergence theorem (Courant, 1937, p. 439), and so its Fourier series converges uniformly (and absolutely) to $\psi(x,0)$ on $[0, L]$.

As in Eq.(3.100), the Fourier coefficients f_n are given by the integrals

$$f_n = \int_0^L dx \sqrt{\frac{2}{L}} \sin\left(\frac{\pi n x}{L}\right) \psi(x,0), \tag{3.105}$$

which now take the form

$$f_n = \frac{2\sqrt{2}}{L} \int_{L/4}^{3L/4} dx \, \sin\left(\frac{\pi n x}{L}\right) \sin\left(\frac{2\pi(x - L/4)}{L}\right). \tag{3.106}$$

Doing the integral, one finds for f_n that for $n \neq 2$

$$f_n = -\frac{\sqrt{2}}{\pi} \frac{4}{n^2 - 4} \left[\sin(3n\pi/4) + \sin(n\pi/4)\right] \tag{3.107}$$

while $c_2 = 0$. These Fourier coefficients satisfy the inequalities (3.69) and (3.71) for $k = 0$. The factor of $1/n^2$ in f_n guarantees the absolute convergence of the series

$$\psi(x,0) = \sum_{n=1}^{\infty} f_n \sqrt{\frac{2}{L}} \sin\left(\frac{\pi n x}{L}\right) \tag{3.108}$$

because asymptotically the coefficient f_n is bounded by $|f_n| \le A/n^2$ where A is a constant ($A = 144/(5\pi\sqrt{L})$ will do) and the sum of $1/n^2$ converges to the Riemann zeta function (5.105)

$$\sum_{n=1}^{\infty} \frac{1}{n^2} = \zeta(2) = \frac{\pi^2}{6}. \tag{3.109}$$

Figure 3.7 plots the 10-term Fourier series (3.108) for $\psi(x,0)$ for $L = 2$. Because this series converges absolutely and uniformly on [0, 2], the 100-term and 1000-term series were too close to $\psi(x,0)$ to be seen clearly in the figure and so were omitted.

As time goes by, the wave function $\psi(x,t)$ evolves from $\psi(x,0)$ to

$$\psi(x,t) = \sum_{n=1}^{\infty} f_n \sqrt{\frac{2}{L}} \sin\left(\frac{\pi n x}{L}\right) e^{-i\hbar(n\pi)^2 t/(2mL^2)} \tag{3.110}$$

in which the Fourier coefficients are given by (3.107). Because $\psi(x,0)$ is continuous and periodic with a piecewise continuous first derivative, its evolution in time is much calmer than that of the piecewise continuous square wave (3.97). Figure 3.8 shows this

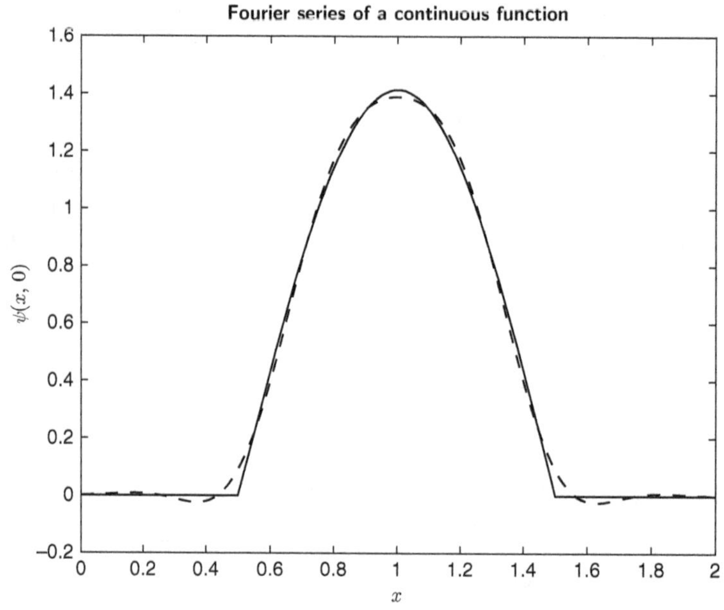

Figure 3.7 The continuous wave function $\psi(x,0)$ (3.104, solid) and its 10-term Fourier series (3.107–3.108, dashes) are plotted for the interval [0, 2].

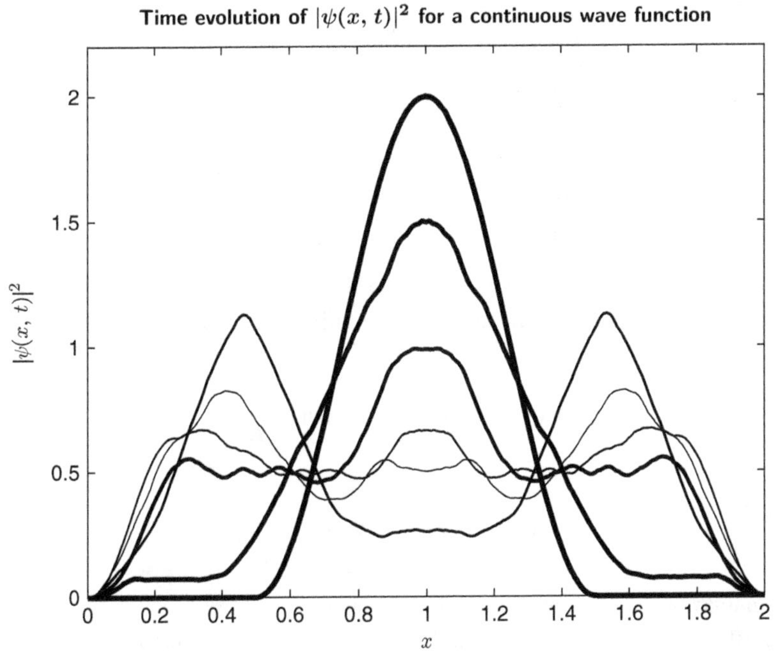

Figure 3.8 For the interval [0, 2], the probability distributions $P(x,t) = |\psi(x,t)|^2$ of the 1000-term Fourier series (3.110) for the continuous wave function $\psi(x,t)$ (3.104) at $t = 0, 10^{-2}\,\tau, 10^{-1}\,\tau, \tau = 2mL^2/\hbar, 10\tau$, and 100τ are plotted as successively thinner curves. A Matlab script for the figure is in Fourier_series at `github.com/kevinecahill`.

evolution in successively thinner curves at times $t = 0, 10^{-2}\,\tau, 10^{-1}\,\tau, \tau = 2mL^2/\hbar$, 10τ, and 100τ. The curves at $t = 0$ and $t = 10^{-2}\,\tau$ are smooth, but some wobbles appear at $t = 10^{-1}\,\tau$ and at $t = \tau$ due to the discontinuities in the first derivative of $\psi(x,0)$ at $x = 0.5$ and at $x = 1.5$.

Example 3.13 (Time evolution of a smooth wave function) Finally, let's try a wave function $\psi(x,0)$ that is periodic and infinitely differentiable on $[0,L]$. An infinitely differentiable function is said to be **smooth** or C^∞. The infinite square-well potential $V(x)$ of Equation (3.89) imposes the periodic boundary conditions $\psi(0,0) = \psi(L,0) = 0$, so we try

$$\psi(x,0) = \sqrt{\frac{2}{3L}}\left[1 - \cos\left(\frac{2\pi x}{L}\right)\right]. \tag{3.111}$$

Its Fourier series

$$\psi(x,0) = \sqrt{\frac{1}{6L}}\left(2 - e^{2\pi i x/L} - e^{-2\pi i x/L}\right) \tag{3.112}$$

has coefficients that satisfy the upper bounds (3.69) by vanishing for $|n| > 1$.

The coefficients of the Fourier sine series for the wave function $\psi(x,0)$ are given by the integrals (3.100)

$$
\begin{aligned}
f_n &= \int_0^L dx \sqrt{\frac{2}{L}} \sin\left(\frac{\pi n x}{L}\right) \psi(x,0) \\
&= \frac{2}{\sqrt{3}\,L} \int_0^L dx \sin\left(\frac{\pi n x}{L}\right) \left[1 - \cos\left(\frac{2\pi x}{L}\right)\right] \\
&= \frac{8\left[(-1)^n - 1\right]}{\pi\sqrt{3}\,n(n^2-4)}
\end{aligned} \tag{3.113}
$$

with all the even coefficients zero, $c_{2n} = 0$. The f_n's are proportional to $1/n^3$ which is more than enough to ensure the absolute and uniform convergence of its Fourier sine series

$$
\psi(x,0) = \sum_{n=1}^{\infty} f_n \sqrt{\frac{2}{L}} \sin\left(\frac{\pi n x}{L}\right). \tag{3.114}
$$

As time goes by, it evolves to

$$
\psi(x,t) = \sum_{n=1}^{\infty} f_n \sqrt{\frac{2}{L}} \sin\left(\frac{\pi n x}{L}\right) e^{-i\hbar(n\pi)^2 t/(2mL^2)} \tag{3.115}
$$

and remains absolutely convergent for all times t.

The effects of the absolute and uniform convergence with $f_n \propto 1/n^3$ are obvious in the graphs. Figure 3.9 shows (for $L = 2$) that only 10 terms are required to nearly overlap the initial wave function $\psi(x,0)$. Figure 3.10 shows that the evolution of the

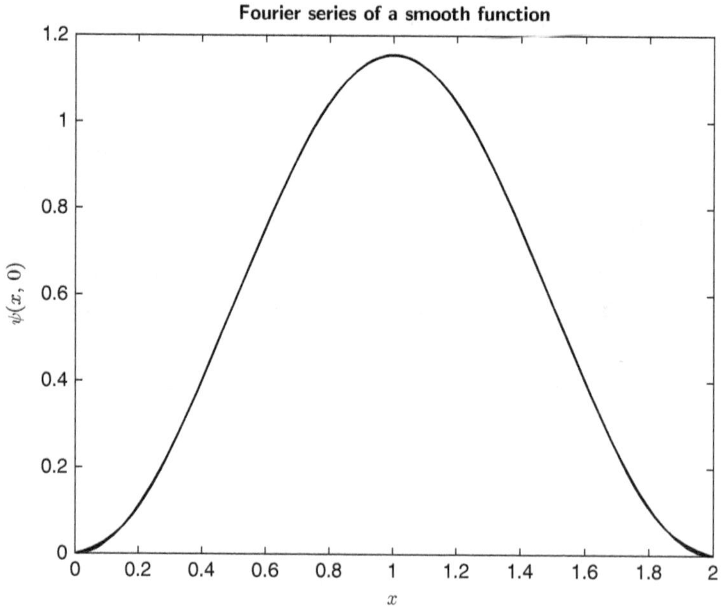

Figure 3.9 The wave function $\psi(x,0)$ (3.111) is infinitely differentiable, and so the first 10 terms of its uniformly convergent Fourier series (3.114) offer a very good approximation to it.

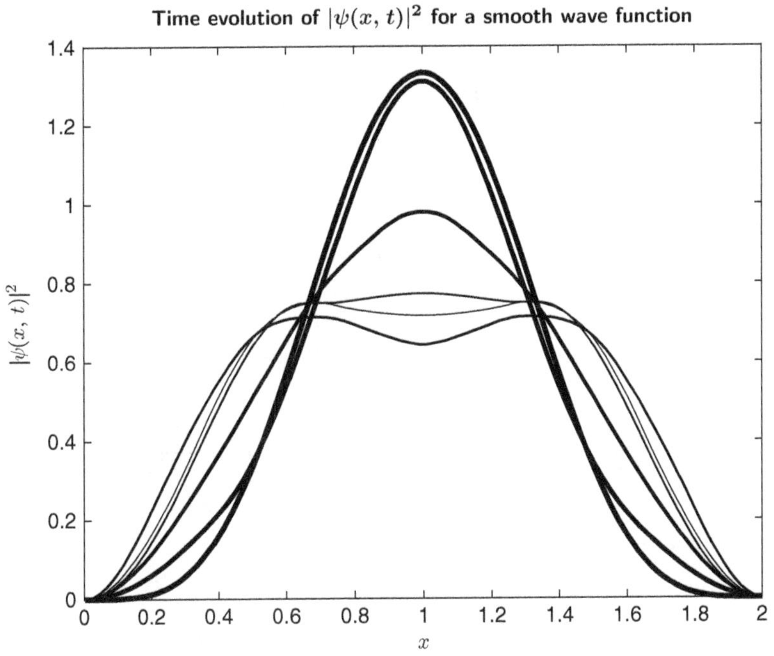

Figure 3.10 The probability distributions $P(x,t) = |\psi(x,t)|^2$ of the 1000-term Fourier series (3.115) for the wave function $\psi(x,t)$ at $t = 0, 10^{-2}\tau, 10^{-1}\tau, \tau = 2mL^2/\hbar, 10\tau$, and 100τ are plotted as successively thinner curves. The time evolution is calm because the wave function $\psi(x,0)$ is smooth.

probability distribution $|\psi(x,t)|^2$ with time is smooth, with no sign of the jaggedness of Fig. 3.6 or the wobbles of Fig. 3.8. Because $\psi(x,0)$ is smooth and periodic, it evolves calmly as time passes.

3.11 Dirac's Delta Function

A Dirac delta function is a (continuous, linear) map from a space of suitably well-behaved functions into the real or complex numbers. It is a **functional** that associates a number with each function in the function space. Thus $\delta(x-y)$ associates the number $f(y)$ with the function $f(x)$. We may write this association as

$$f(y) = \int f(x)\,\delta(x-y)\,dx. \tag{3.116}$$

Delta functions pop up all over physics. Multiplying the identity operator (3.6)

$$I = \int_0^{2\pi} |x\rangle\langle x|\,dx \tag{3.117}$$

from the right by $|f\rangle$ and from the left by $\langle y|$, we get

$$f(y) = \langle y|f\rangle = \langle y|I|f\rangle = \int_0^{2\pi} \langle y|x\rangle\langle x|f\rangle\, dx = \int_0^{2\pi} \langle y|x\rangle f(x)\, dx \quad (3.118)$$

which says that the inner product $\langle y|x\rangle$ is a **delta function**

$$\langle y|x\rangle = \langle x|y\rangle = \delta(x-y). \quad (3.119)$$

Using both Fourier-series formulas (3.2) and (3.3), we get

$$f(x) = \sum_{n=-\infty}^{\infty} f_n \frac{e^{inx}}{\sqrt{2\pi}} = \sum_{n=-\infty}^{\infty} \int_0^{2\pi} \frac{e^{-iny}}{\sqrt{2\pi}} f(y) \frac{e^{inx}}{\sqrt{2\pi}}\, dy. \quad (3.120)$$

Interchanging and rearranging, we have

$$f(x) = \int_0^{2\pi} \left(\sum_{n=-\infty}^{\infty} \frac{e^{in(x-y)}}{2\pi} \right) f(y)\, dy. \quad (3.121)$$

But the phases e^{inx} are **periodic** with period 2π, so we also have

$$f(x+2\pi\ell) = \int_0^{2\pi} \left(\sum_{n=-\infty}^{\infty} \frac{e^{in(x-y)}}{2\pi} \right) f(y)\, dy \quad (3.122)$$

in which the function f of the left-hand side of this equation is the periodic extension (3.61) f_p of f if f is not itself periodic with period 2π. Thus we arrive at the **Dirac comb**

$$\sum_{n=-\infty}^{\infty} \frac{e^{in(x-y)}}{2\pi} = \sum_{\ell=-\infty}^{\infty} \delta(x-y-2\pi\ell) \quad (3.123)$$

or more simply

$$\sum_{n=-\infty}^{\infty} \frac{e^{inx}}{2\pi} = \frac{1}{2\pi} + \frac{1}{\pi} \sum_{n=1}^{\infty} \cos(nx) = \sum_{\ell=-\infty}^{\infty} \delta(x-2\pi\ell). \quad (3.124)$$

Example 3.14 (Dirac's comb) The sum of the first 100,000 terms of this cosine series (3.124) for the Dirac comb is plotted for the interval $(-15, 15)$ in Fig. 3.11. Gibbs overshoots appear at the discontinuities. The integral of the first 100,000 terms from -15 to 15 is 5.0000.

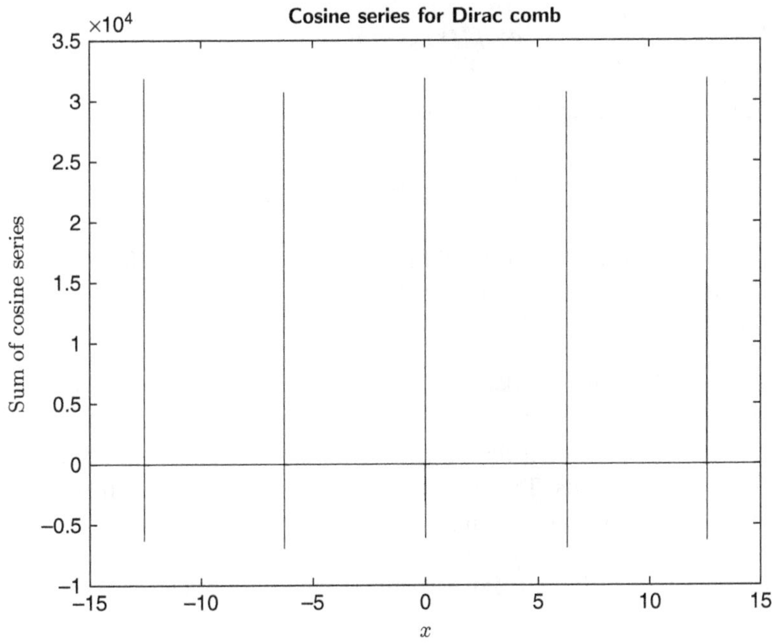

Figure 3.11 The sum of the first 100,000 terms of the series (3.124) for the Dirac comb is plotted for $-15 \le x \le 15$. Both Dirac spikes and Gibbs overshoots are visible.

The stretched Dirac comb is

$$\sum_{n=-\infty}^{\infty} \frac{e^{2\pi i n(x-y)/L}}{L} = \sum_{\ell=-\infty}^{\infty} \delta(x - y - \ell L). \tag{3.125}$$

Example 3.15 (Parseval's identity) Using our formula (3.43) for the Fourier coefficients of a stretched interval, we can relate a sum of products $f_n^* \, g_n$ of the Fourier coefficients of the functions $f(x)$ and $g(x)$ to an integral of the product $f^*(x)\, g(x)$

$$\sum_{n=-\infty}^{\infty} f_n^* \, g_n = \sum_{n=-\infty}^{\infty} \int_0^L dx \, \frac{e^{i2\pi nx/L}}{\sqrt{L}} \, f^*(x) \int_0^L dy \, \frac{e^{-i2\pi ny/L}}{\sqrt{L}} \, g(y). \tag{3.126}$$

This sum contains Dirac's comb (3.125) and so

$$\begin{aligned} \sum_{n=-\infty}^{\infty} f_n^* \, g_n &= \int_0^L dx \int_0^L dy \, f^*(x)\, g(y) \, \frac{1}{L} \sum_{n=-\infty}^{\infty} e^{i2\pi n(x-y)/L} \\ &= \int_0^L dx \int_0^L dy \, f^*(x)\, g(y) \sum_{\ell=-\infty}^{\infty} \delta(x - y - \ell L). \end{aligned} \tag{3.127}$$

But because only the $\ell = 0$ tooth of the comb lies in the interval $[0, L]$, we have more simply

$$\sum_{n=-\infty}^{\infty} f_n^* g_n = \int_0^L dx \int_0^L dy\, f^*(x)\, g(y)\, \delta(x-y) = \int_0^L dx\, f^*(x)\, g(x). \quad (3.128)$$

In particular, if the two functions are the same, then

$$\sum_{n=-\infty}^{\infty} |f_n|^2 = \int_0^L dx\, |f(x)|^2 \quad (3.129)$$

which is **Parseval's identity**. Thus if a function is **square integrable** on an interval, then the sum of the squares of the absolute values of its Fourier coefficients is the integral of the square of its absolute value.

Example 3.16 (Derivatives of delta functions) Delta functions and other generalized functions or distributions map smooth functions that vanish at infinity into numbers in ways that are linear and continuous. Derivatives of delta functions are defined so as to allow integrations by parts. Thus the nth derivative of the delta function $\delta^{(n)}(x-y)$ maps the function $f(x)$ to $(-1)^n$ times its nth derivative $f^{(n)}(y)$ at y

$$\int \delta^{(n)}(x-y)\, f(x)\, dx = \int \delta(x-y)\, (-1)^n\, f^{(n)}(x)\, dx = (-1)^n\, f^{(n)}(y) \quad (3.130)$$

with no surface term.

Example 3.17 (The equation $xf(x) = a$) Dirac's delta function sometimes appears unexpectedly. For instance, the general solution to the equation $x\, f(x) = a(x)$ is $f(x) = a(x)/x + b(x)\, \delta(x)$ in which $b(x)$ is a constant (Dirac, 1967, sec. 15), (Waxman and Peck, 1998) or $x\, b(x) = 0$ at $x = 0$. Similarly, the general solution to the equation $x^2\, f(x) = a(x)$ is $f(x) = a(x)/x^2 + b(x)\, \delta(x)/x + c(x)\, \delta'(x)$ in which $\delta'(x)$ is the derivative of the delta function, $b(x)$ is continuous, $c(x)$ has a continuous first derivative, and $x\, b(x) = x\, c(x) = x^2\, c'(x) = 0$ at $x = 0$.

3.12 Harmonic Oscillators

The hamiltonian for the harmonic oscillator is

$$H = \frac{p^2}{2m} + \frac{1}{2} m\omega^2 q^2. \quad (3.131)$$

The commutation relation $[q, p] \equiv qp - pq = i\hbar$ implies that the **lowering** and **raising** operators

$$a = \sqrt{\frac{m\omega}{2\hbar}} \left(q + \frac{ip}{m\omega} \right) \quad \text{and} \quad a^\dagger = \sqrt{\frac{m\omega}{2\hbar}} \left(q - \frac{ip}{m\omega} \right) \quad (3.132)$$

obey the commutation relation $[a, a^\dagger] = 1$. In terms of a and $a^\dagger$, which also are called the **annihilation** and **creation** operators, the hamiltonian H has the simple form

$$H = \hbar\omega\left(a^\dagger a + \frac{1}{2}\right). \tag{3.133}$$

There is a unique state $|0\rangle$ that is annihilated by the operator a, as may be seen by solving the differential equation

$$\langle q'|a|0\rangle = \sqrt{\frac{m\omega}{2\hbar}}\,\langle q'|\left(q + \frac{ip}{m\omega}\right)|0\rangle = 0. \tag{3.134}$$

Since $\langle q'|q = q'\langle q'|$ and

$$\langle q'|p|0\rangle = \frac{\hbar}{i}\frac{d\langle q'|0\rangle}{dq'} \tag{3.135}$$

the resulting differential equation is

$$\frac{d\langle q'|0\rangle}{dq'} = -\frac{m\omega}{\hbar}q'\langle q'|0\rangle. \tag{3.136}$$

Its suitably normalized solution is the wave function for the ground state of the harmonic oscillator

$$\langle q'|0\rangle = \left(\frac{m\omega}{\pi\hbar}\right)^{1/4}\exp\left(-\frac{m\omega q'^2}{2\hbar}\right). \tag{3.137}$$

For $n = 0, 1, 2, \ldots$, the nth eigenstate of the hamiltonian H is

$$|n\rangle = \frac{1}{\sqrt{n!}}\left(a^\dagger\right)^n|0\rangle \tag{3.138}$$

where $n! \equiv n(n-1)\cdots 1$ is n-**factorial** and $0! = 1$. Its energy is

$$H|n\rangle = \hbar\omega\left(n + \frac{1}{2}\right)|n\rangle. \tag{3.139}$$

The identity operator is

$$I = \sum_{n=0}^{\infty}|n\rangle\langle n|. \tag{3.140}$$

An arbitrary state $|\psi\rangle$ has an expansion in terms of the eigenstates $|n\rangle$

$$|\psi\rangle = I|\psi\rangle = \sum_{n=0}^{\infty}|n\rangle\langle n|\psi\rangle \tag{3.141}$$

and evolves in time like a Fourier series

$$|\psi, t\rangle = e^{-iHt/\hbar}|\psi\rangle = e^{-iHt/\hbar}\sum_{n=0}^{\infty}|n\rangle\langle n|\psi\rangle = e^{-i\omega t/2}\sum_{n=0}^{\infty}e^{-in\omega t}|n\rangle\langle n|\psi\rangle \tag{3.142}$$

with wave function

$$\psi(q,t) = \langle q|\psi,t\rangle = e^{-i\omega t/2}\sum_{n=0}^{\infty} e^{-in\omega t}\langle q|n\rangle\langle n|\psi\rangle. \tag{3.143}$$

The wave functions $\langle q|n\rangle$ of the energy eigenstates are related to the Hermite polynomials (example 9.6)

$$H_n(x) = (-1)^n e^{x^2}\frac{d^n}{dx^n}e^{-x^2} \tag{3.144}$$

by a change of variables $x = \sqrt{m\omega/\hbar}\,q \equiv sq$ and a normalization factor

$$\langle q|n\rangle = \frac{\sqrt{s}\,e^{-(sq)^2/2}}{\sqrt{2^n n!\sqrt{\pi}}}H_n(sq) = \left(\frac{m\omega}{\pi\hbar}\right)^{1/4}\frac{e^{-m\omega q^2/2\hbar}}{\sqrt{2^n n!}}H_n\left(\left(\frac{m\omega}{\hbar}\right)^{1/2}q\right). \tag{3.145}$$

For every complex number α, the **coherent state** $|\alpha\rangle$

$$|\alpha\rangle = e^{-|\alpha|^2/2}e^{\alpha a^\dagger}|0\rangle = e^{-|\alpha|^2/2}\sum_{n=0}^{\infty}\frac{\alpha^n}{\sqrt{n!}}|n\rangle \tag{3.146}$$

is an eigenstate $a|\alpha\rangle = \alpha|\alpha\rangle$ of the lowering (or annihilation) operator a with eigenvalue α. Its time evolution is simply

$$|\alpha,t\rangle = e^{-i\omega t/2}e^{-|\alpha|^2/2}\sum_{n=0}^{\infty}\frac{\left(\alpha e^{-i\omega t}\right)^n}{\sqrt{n!}}|n\rangle = e^{-i\omega t/2}|\alpha e^{-i\omega t}\rangle. \tag{3.147}$$

3.13 Nonrelativistic Strings

If we clamp the ends of a nonrelativistic string at $x = 0$ and $x = L$, then the amplitude $y(x,t)$ will obey the boundary conditions

$$y(0,t) = y(L,t) = 0 \tag{3.148}$$

and the wave equation

$$v^2\frac{\partial^2 y}{\partial x^2} = \frac{\partial^2 y}{\partial t^2} \tag{3.149}$$

as long as $y(x,t)$ remains small. The functions

$$y_n(x,t) = \sin\frac{n\pi x}{L}\left(a_n\cos\frac{n\pi vt}{L} + b_n\sin\frac{n\pi vt}{L}\right) \tag{3.150}$$

satisfy this wave equation (3.149) and the boundary conditions (3.148). They represent waves traveling along the x-axis with speed v.

The space S_L of functions $f(x)$ that satisfy the boundary condition (3.148) is spanned by the functions $\sin(n\pi x/L)$. One may use the integral formula

$$\int_0^L \sin\frac{n\pi x}{L}\,\sin\frac{m\pi x}{L}\,dx = \frac{L}{2}\,\delta_{nm} \tag{3.151}$$

to derive for any function $f \in S_L$ the Fourier series

$$f(x) = \sum_{n=1}^{\infty} f_n \sin\frac{n\pi x}{L} \tag{3.152}$$

with coefficients

$$f_n = \frac{2}{L}\int_0^L \sin\frac{n\pi x}{L} f(x) dx \tag{3.153}$$

and the representation

$$\sum_{m=-\infty}^{\infty} \delta(x - z - 2mL) = \frac{2}{L}\sum_{n=1}^{\infty} \sin\frac{n\pi x}{L}\,\sin\frac{n\pi z}{L} \tag{3.154}$$

for the Dirac comb on S_L.

3.14 Periodic Boundary Conditions

Periodic boundary conditions often are convenient. For instance, rather than study an infinitely long 1-dimensional system, we might study the same system, but of length L. The ends cause effects not present in the infinite system. To avoid them, we imagine that the system forms a circle and impose the periodic boundary condition

$$\psi(x \pm L, t) = \psi(x, t). \tag{3.155}$$

Analogous conditions in 3 dimensions are

$$\psi(x + kL, y + \ell L, z + mL, t) = \psi(x, y, z, t) \tag{3.156}$$

for all integers k, ℓ, and m.

The eigenstates $|\boldsymbol{p}\rangle$ of the free hamiltonian $H = \boldsymbol{p}^2/2m$ have wave functions

$$\psi_{\boldsymbol{p}}(\boldsymbol{x}) = \langle \boldsymbol{x}|\boldsymbol{p}\rangle = e^{i\boldsymbol{x}\cdot\boldsymbol{p}/\hbar}/(2\pi\hbar)^{3/2}. \tag{3.157}$$

The periodic boundary conditions (3.156) require that each component p_i of momentum satisfy $Lp_i/\hbar = 2\pi n_i$ or

$$\boldsymbol{p} = \frac{2\pi\hbar\boldsymbol{n}}{L} = \frac{h\boldsymbol{n}}{L} \tag{3.158}$$

where $\boldsymbol{n}$ is a vector of integers, which may be positive or negative or zero.

Periodic boundary conditions naturally arise in the study of solids. The atoms of a perfect crystal are at the vertices of a **Bravais** lattice

$$\boldsymbol{x}_i = \boldsymbol{x}_0 + \sum_{i=1}^{3} n_i \boldsymbol{a}_i \tag{3.159}$$

in which the three vectors $\boldsymbol{a}_i$ are the **primitive vectors** of the lattice and the n_i are three integers. The hamiltonian of such an infinite crystal is invariant under translations in space by

$$\sum_{i=1}^{3} n_i \boldsymbol{a}_i . \tag{3.160}$$

To keep the notation simple, let's restrict ourselves to a cubic lattice with lattice spacing a. Then since the momentum operator $\boldsymbol{p}$ generates translations in space, the invariance of H under translations by $a\,\boldsymbol{n}$

$$\exp(ia\boldsymbol{n}\cdot\boldsymbol{p})\,H\exp(-ia\boldsymbol{n}\cdot\boldsymbol{p}) = H \tag{3.161}$$

implies that $e^{ia\boldsymbol{n}\cdot\boldsymbol{p}}$ and H are compatible normal operators $[e^{ia\boldsymbol{n}\cdot\boldsymbol{p}}, H] = 0$. As explained in section 1.31, it follows that we may choose the eigenstates of H also to be eigenstates of $e^{ia\boldsymbol{n}\cdot\boldsymbol{p}}$

$$e^{ia\boldsymbol{p}\cdot\boldsymbol{n}/\hbar}|\psi\rangle = e^{ia\boldsymbol{k}\cdot\boldsymbol{n}}\,|\psi\rangle \tag{3.162}$$

which implies that

$$\psi(\boldsymbol{x}+a\boldsymbol{n}) = \langle\boldsymbol{x}+a\boldsymbol{n}|\psi\rangle = \langle\boldsymbol{x}|e^{ia\boldsymbol{p}\cdot\boldsymbol{n}/\hbar}|\psi\rangle = \langle\boldsymbol{x}|e^{ia\boldsymbol{k}\cdot\boldsymbol{n}/\hbar}|\psi\rangle = e^{ia\boldsymbol{k}\cdot\boldsymbol{n}}\,\psi(\boldsymbol{x}). \tag{3.163}$$

Setting

$$\psi(\boldsymbol{x}) = e^{i\boldsymbol{k}\cdot\boldsymbol{x}}\,u(\boldsymbol{x}) \tag{3.164}$$

we see that condition (3.163) implies that $u(\boldsymbol{x})$ is periodic

$$u(\boldsymbol{x}+a\boldsymbol{n}) = u(\boldsymbol{x}). \tag{3.165}$$

For a general Bravais lattice, this **Born–von Karman** periodic boundary condition is

$$u\left(\boldsymbol{x}+\sum_{i=1}^{3} n_i\boldsymbol{a}_i\right) = u(\boldsymbol{x}). \tag{3.166}$$

Equations (3.163) and (3.165) are known as **Bloch's theorem**.

Exercises

3.1 Show that $\sin \omega_1 x + \sin \omega_2 x$ is the same as (3.16).

3.2 Show that the Fourier series (3.17) obeys Laplace's equation (3.18) for $\rho < a$ and respects the boundary condition $f(a, \theta) = h(\theta)$.

3.3 Find the forms that Equations (3.10 and 3.11) for the inner products $\langle g|f\rangle$ and $\langle f|f\rangle$ take when one uses the asymmetrical notations (3.21 and 3.22).

3.4 Find the Fourier series for the function $\exp(ax)$ on the interval $-\pi < x \le \pi$.

3.5 Find the Fourier series for the function $(x^2 - \pi^2)^2$ on the same interval $(-\pi, \pi]$.

3.6 Find the Fourier series for the function $(1 + \cos x) \sin ax$ on the interval $(-\pi, \pi]$.

3.7 Show that the Fourier series for the function $x \cos x$ on the interval $[-\pi, \pi]$ is

$$x \cos x = -\frac{1}{2} \sin x + 2 \sum_{n=2}^{\infty} \frac{(-1)^n \, n}{n^2 - 1} \sin nx. \tag{3.167}$$

3.8 (a) Show that the Fourier series for the function $|x|$ on the interval $[-\pi, \pi]$ is

$$|x| = \frac{\pi}{2} - \frac{4}{\pi} \sum_{n=0}^{\infty} \frac{\cos(2n+1)x}{(2n+1)^2}. \tag{3.168}$$

(b) Use this result to find a neat formula for $\pi^2/8$. Hint: set $x = 0$.

3.9 Show that the Fourier series for the function $|\sin x|$ on the interval $[-\pi, \pi]$ is

$$|\sin x| = \frac{2}{\pi} - \frac{4}{\pi} \sum_{n=1}^{\infty} \frac{\cos 2nx}{4n^2 - 1}. \tag{3.169}$$

3.10 Show that the Fourier coefficients of the C^0 function (3.72) on the interval $[-\pi, \pi]$ are given by (3.73).

3.11 Find by inspection the Fourier series for the function $\exp[\exp(-ix)]$.

3.12 Fill in the steps in the computation (3.36) of the Fourier series for x^2.

3.13 Do the first integral in Equation (3.78). Hint: differentiate $\ln\left(\frac{\sin \pi x}{\pi x}\right)$.

3.14 Use the infinite-product formula (3.79) for the sine and the relation $\cos \pi x = \sin 2\pi x / (2 \sin \pi x)$ to derive the infinite-product formula (3.80) for the cosine. Hint:

$$\prod_{n=1}^{\infty} \left[1 - \frac{x^2}{\frac{1}{4}n^2}\right] = \prod_{n=1}^{\infty} \left[1 - \frac{x^2}{\frac{1}{4}(2n-1)^2}\right] \left[1 - \frac{x^2}{\frac{1}{4}(2n)^2}\right]. \tag{3.170}$$

3.15 What's the general solution to the equation $x^3 f(x) = a(x)$?

3.16 Suppose we wish to approximate the real square-integrable function $f(x)$ by the Fourier series with N terms

$$f_N(x) = \frac{a_0}{2} + \sum_{n=1}^{N} (a_n \cos nx + b_n \sin nx) . \tag{3.171}$$

Then the error

$$E_N = \int_0^{2\pi} [f(x) - f_N(x)]^2 \, dx \tag{3.172}$$

will depend upon the $2N + 1$ coefficients a_n and b_n. The best coefficients minimize this error and satisfy the conditions

$$\frac{\partial E_N}{\partial a_n} = \frac{\partial E_N}{\partial b_n} = 0. \tag{3.173}$$

By using these conditions, find the best coefficients.

3.17 Find the Fourier series for the function $f(x) = \theta(a^2 - x^2)$ on the interval $[-\pi, \pi]$ for the case $a^2 < \pi^2$. The **Heaviside step function** $\theta(x)$ is zero for $x < 0$, one-half for $x = 0$, and unity for $x > 0$ (Oliver Heaviside, 1850–1925). The value assigned to $\theta(0)$ seldom matters, and you need not worry about it in this problem.

3.18 Derive or infer the formula (3.125) for the stretched Dirac comb.

3.19 Use the commutation relation $[q, p] = i\hbar$ to show that the annihilation and creation operators (3.132) satisfy the commutation relation $[a, a^\dagger] = 1$.

3.20 Show that the state $|n\rangle = (a^\dagger)^n |0\rangle / \sqrt{n!}$ is an eigenstate of the hamiltonian (3.133) with energy $\hbar\omega(n + 1/2)$.

3.21 Show that the coherent state $|\alpha\rangle$ (3.146) is an eigenstate of the annihilation operator a with eigenvalue α.

3.22 Derive Equations (3.153 and 3.154) from the expansion (3.152) and the integral formula (3.151).

3.23 Consider a string like the one described in Section 3.13, which satisfies the boundary conditions (3.148) and the wave equation (3.149). The string is at rest $y(x, 0) = 0$ at time $t = 0$ and is struck precisely at $t = 0$ and $x = a$ so that its time derivative at $t = 0$ (denoted by a dot) $\dot{y}(x, 0) = Lv_0\delta(x - a)$. Find $y(x, t)$ and $\dot{y}(x, t)$.

3.24 Same as Exercise 3.23, but now the initial conditions are

$$u(x, 0) = f(x) \quad \text{and} \quad \dot{u}(x, 0) = g(x) \tag{3.174}$$

in which $f(0) = f(L) = 0$ and $g(0) = g(L) = 0$. Find the motion of the amplitude $u(x, t)$ of the string.

3.25 (a) Find the Fourier series for the function $f(x) = x^2$ on the interval $[-\pi, \pi]$.
(b) Use your result at $x = \pi$ to show that

$$\sum_{n=1}^{\infty} \frac{1}{n^2} = \frac{\pi^2}{6} \tag{3.175}$$

which is the value of Riemann's zeta function (5.105) $\zeta(x)$ at $x = 2$.

4

Fourier and Laplace Transforms

The complex exponentials $\exp(i2\pi nx/L)$ are orthonormal and easy to differentiate (and to integrate), but they are periodic with period L. If one wants to represent functions that are not periodic, a better choice is the complex exponentials $\exp(ikx)$, where k is an arbitrary real number. These orthonormal functions are the basis of the Fourier transform. The choice of complex k leads to the transforms of Laplace, Mellin, and Bromwich.

4.1 Fourier Transforms

The interval $[-L/2, L/2]$ is arbitrary in the Fourier series pair (3.46)

$$f(x) = \sum_{n=-\infty}^{\infty} f_n \frac{e^{i2\pi nx/L}}{\sqrt{L}} \quad \text{and} \quad f_n = \int_{-L/2}^{L/2} f(x) \frac{e^{-i2\pi nx/L}}{\sqrt{L}} \, dx. \tag{4.1}$$

What happens when we stretch this interval without limit, letting $L \to \infty$?

We may use the **nearest-integer function** $[y]$ to convert the coefficients f_n into a function of a continuous variable $\hat{f}(y) \equiv f_{[y]}$ such that $\hat{f}(y) = f_n$ when $|y - n| < 1/2$. In terms of this function $\hat{f}(y)$, the Fourier series (4.1) for the function $f(x)$ is

$$f(x) = \sum_{n=-\infty}^{\infty} \int_{n-1/2}^{n+1/2} \hat{f}(y) \frac{e^{i2\pi [y]x/L}}{\sqrt{L}} \, dy = \int_{-\infty}^{\infty} \hat{f}(y) \frac{e^{i2\pi [y]x/L}}{\sqrt{L}} \, dy. \tag{4.2}$$

Since $[y]$ and y differ by no more than $1/2$, the absolute value of the difference between $\exp(i\pi [y]x/L)$ and $\exp(i\pi yx/L)$ for fixed x is

$$\left| e^{i2\pi [y]x/L} - e^{i2\pi yx/L} \right| = \left| e^{i2\pi ([y]-y)x/L} - 1 \right| \approx \frac{\pi |x|}{L} \tag{4.3}$$

which goes to zero as $L \to \infty$. So in this limit, we may replace $[y]$ by y and express $f(x)$ as

$$f(x) = \int_{-\infty}^{\infty} \hat{f}(y) \, \frac{e^{i2\pi yx/L}}{\sqrt{L}} \, dy. \tag{4.4}$$

We let $y = Lk/(2\pi)$ so $k = 2\pi y/L$ and find for $f(x)$ the integral

$$f(x) = \int_{-\infty}^{\infty} \hat{f}\left(\frac{Lk}{2\pi}\right) \frac{e^{ikx}}{\sqrt{L}} \frac{L}{2\pi} \, dk = \int_{-\infty}^{\infty} \sqrt{\frac{L}{2\pi}} \, \hat{f}\left(\frac{Lk}{2\pi}\right) e^{ikx} \, \frac{dk}{\sqrt{2\pi}}. \tag{4.5}$$

Now in terms of the Fourier transform $\tilde{f}(k)$ defined as

$$\tilde{f}(k) = \sqrt{\frac{L}{2\pi}} \, \hat{f}\left(\frac{Lk}{2\pi}\right) \tag{4.6}$$

the integral (4.5) for $f(x)$ is the inverse Fourier transform

$$f(x) = \int_{-\infty}^{\infty} \tilde{f}(k) \, e^{ikx} \, \frac{dk}{\sqrt{2\pi}}. \tag{4.7}$$

To find $\tilde{f}(k)$, we use its definition (4.6), the definition (4.1) of f_n, our formulas $\tilde{f}(k) = \sqrt{Lk/(2\pi)}\,\hat{f}(Lk/(2\pi))$ and $\hat{f}(y) = f_{[y]}$, and the inequality $|2\pi[Lk/2\pi]/L - k| \le \pi/2L$ to write

$$\tilde{f}(k) = \sqrt{\frac{L}{2\pi}} \, f_{\left[\frac{Lk}{2\pi}\right]} = \sqrt{\frac{L}{2\pi}} \int_{-L/2}^{L/2} f(x) \frac{e^{-i2\pi\left[\frac{Lk}{2\pi}\right]\frac{x}{L}}}{\sqrt{L}} dx \approx \int_{-L/2}^{L/2} f(x) e^{-ikx} \frac{dx}{\sqrt{2\pi}}.$$

This formula becomes exact in the limit $L \to \infty$

$$\tilde{f}(k) = \int_{-\infty}^{\infty} f(x) \, e^{-ikx} \, \frac{dx}{\sqrt{2\pi}} \tag{4.8}$$

and so we have the Fourier transformations

$$f(x) = \int_{-\infty}^{\infty} \tilde{f}(k) \, e^{ikx} \, \frac{dk}{\sqrt{2\pi}} \quad \text{and} \quad \tilde{f}(k) = \int_{-\infty}^{\infty} f(x) \, e^{-ikx} \, \frac{dx}{\sqrt{2\pi}}. \tag{4.9}$$

The function $\tilde{f}(k)$ is the **Fourier transform** of $f(x)$, and $f(x)$ is the **inverse Fourier transform** of $\tilde{f}(k)$.

In these symmetrical relations (4.9), the distinction between a Fourier transform and an inverse Fourier transform is entirely a matter of convention. There is no rule for which sign, ikx or $-ikx$, goes with which transform or for where to put the 2π's. Thus one often sees

$$f(x) = \int_{-\infty}^{\infty} \tilde{f}(k) \, e^{\pm ikx} \, dk \quad \text{and} \quad \tilde{f}(k) = \int_{-\infty}^{\infty} f(x) \, e^{\mp ikx} \, \frac{dx}{2\pi} \tag{4.10}$$

as well as

$$f(x) = \int_{-\infty}^{\infty} \tilde{f}(k)\, e^{\pm ikx} \, \frac{dk}{2\pi} \quad \text{and} \quad \tilde{f}(k) = \int_{-\infty}^{\infty} f(x)\, e^{\mp ikx}\, dx. \tag{4.11}$$

One often needs to relate a function's Fourier series to its Fourier transform. So let's compare the Fourier series (4.1) for the function $f(x)$ on the interval $[-L/2, L/2]$ with its Fourier transform (4.9) in the limit of large L setting $k_n = 2\pi n/L$

$$f(x) = \sum_{n=-\infty}^{\infty} f_n \frac{e^{i2\pi nx/L}}{\sqrt{L}} = \sum_{n=-\infty}^{\infty} f_n \frac{e^{ik_n x}}{\sqrt{L}} = \int_{-\infty}^{\infty} \tilde{f}(k)\, e^{ikx} \frac{dk}{\sqrt{2\pi}}. \tag{4.12}$$

Since $f_n = \hat{f}(y) = f_{[y]}$, by using the definition (4.6) of $\tilde{f}(k)$, we have

$$f_n = f_{[n]} = f_{[y]} = \hat{f}(y) = \hat{f}\left(\frac{Lk}{2\pi}\right) = \sqrt{\frac{2\pi}{L}}\, \tilde{f}(k). \tag{4.13}$$

Thus, to get the Fourier series from the Fourier transform, we multiply the series by $2\pi/L$ and use the Fourier transform at k_n divided by $\sqrt{2\pi}$

$$f(x) = \frac{1}{\sqrt{L}} \sum_{n=-\infty}^{\infty} f_n\, e^{ik_n x} = \frac{2\pi}{L} \sum_{n=-\infty}^{\infty} \frac{\tilde{f}(k_n)}{\sqrt{2\pi}}\, e^{ik_n x}. \tag{4.14}$$

Going the other way, we set $\tilde{f}(k) = \sqrt{L/2\pi}\, f_n = \sqrt{L/2\pi}\, f_{[Lk/2\pi]}$ and find

$$f(x) = \int_{-\infty}^{\infty} \tilde{f}(k)\, e^{ikx} \frac{dk}{\sqrt{2\pi}} = \frac{L}{2\pi} \int_{-\infty}^{\infty} \frac{f_{[Lk/2\pi]}}{\sqrt{L}}\, e^{ikx} dk. \tag{4.15}$$

Example 4.1 (The Fourier transform of a gaussian is a gaussian) The Fourier transform of the gaussian $f(x) = \exp(-m^2 x^2)$ is

$$\tilde{f}(k) = \int_{-\infty}^{\infty} \frac{dx}{\sqrt{2\pi}}\, e^{-ikx}\, e^{-m^2 x^2}. \tag{4.16}$$

We complete the square in the exponent:

$$\tilde{f}(k) = e^{-k^2/4m^2} \int_{-\infty}^{\infty} \frac{dx}{\sqrt{2\pi}}\, e^{-m^2 (x+ik/2m^2)^2}. \tag{4.17}$$

As we'll see in Section 6.14 when we study analytic functions, we may replace x by $x - ik/2m^2$ without changing the value of this integral. So we can drop the term $ik/2m^2$ in the exponential and get

$$\tilde{f}(k) = e^{-k^2/4m^2} \int_{-\infty}^{\infty} \frac{dx}{\sqrt{2\pi}}\, e^{-m^2 x^2} = \frac{1}{\sqrt{2}\, m}\, e^{-k^2/4m^2}. \tag{4.18}$$

Thus the Fourier transform of a gaussian is another gaussian

$$\tilde{f}(k) = \int_{-\infty}^{\infty} \frac{dx}{\sqrt{2\pi}} e^{-ikx} e^{-m^2 x^2} = \frac{1}{\sqrt{2}m} e^{-k^2/4m^2}. \tag{4.19}$$

But the two gaussians are very different: if the gaussian $f(x) = \exp(-m^2x^2)$ decreases slowly as $x \to \infty$ because m is small (or quickly because m is big), then its gaussian Fourier transform $\tilde{f}(k) = \exp(-k^2/4m^2)/m\sqrt{2}$ decreases quickly as $k \to \infty$ because m is small (or slowly because m is big).

Can we invert $\tilde{f}(k)$ to get $f(x)$? The inverse Fourier transform (4.7) says

$$f(x) = \int_{-\infty}^{\infty} \frac{dk}{\sqrt{2\pi}} \tilde{f}(k) e^{ikx} = \int_{-\infty}^{\infty} \frac{dk}{\sqrt{2\pi}} \frac{1}{m\sqrt{2}} e^{ikx-k^2/4m^2}. \tag{4.20}$$

By again completing the square in the exponent

$$f(x) = e^{-m^2x^2} \int_{-\infty}^{\infty} \frac{dk}{\sqrt{2\pi}} \frac{1}{m\sqrt{2}} e^{-(k-i2m^2x)^2/4m^2} \tag{4.21}$$

and shifting the variable of integration k to $k + i2m^2x$, we find

$$f(x) = e^{-m^2x^2} \int_{-\infty}^{\infty} \frac{dk}{\sqrt{2\pi}} \frac{1}{m\sqrt{2}} e^{-k^2/(4m^2)} = e^{-m^2x^2} \tag{4.22}$$

which is reassuring.

Using (4.18) for $\tilde{f}(k)$ and the connections (4.12–4.15) between Fourier series and transforms, we see that a Fourier series for this gaussian is in the limit of $L \gg x$

$$f(x) = e^{-m^2x^2} = \frac{2\pi}{L} \sum_{n=-\infty}^{\infty} \frac{1}{\sqrt{4\pi}\, m} e^{-k_n^2/(4m^2)} e^{ik_n x} \tag{4.23}$$

in which $k_n = 2\pi n/L$.

4.2 Fourier Transforms of Real Functions

If a function $f(x)$ is real, then the complex conjugate of its Fourier transform (4.8)

$$\tilde{f}(k) = \int_{-\infty}^{\infty} \frac{dx}{\sqrt{2\pi}} f(x) e^{-ikx} \tag{4.24}$$

is its Fourier transform evaluated at $-k$

$$\tilde{f}^*(k) = \int_{-\infty}^{\infty} \frac{dx}{\sqrt{2\pi}} f(x) e^{ikx} = \tilde{f}(-k). \tag{4.25}$$

It follows (Exercise 4.1) that a real function $f(x)$ satisfies the relation

$$f(x) = \frac{1}{\pi} \int_0^{\infty} dk \int_{-\infty}^{\infty} f(y) \cos k(y-x)\, dy. \tag{4.26}$$

If $f(x)$ is both real and even, then

$$f(x) = \frac{2}{\pi} \int_0^\infty \cos kx \, dk \int_0^\infty f(y) \, \cos ky \, dy \tag{4.27}$$

if it is real and odd then (Exercise 4.2)

$$f(x) = \frac{2}{\pi} \int_0^\infty \sin kx \, dk \int_0^\infty f(y) \, \sin ky \, dy. \tag{4.28}$$

Example 4.2 (Dirichlet's discontinuous factor) Using (4.27), one may write the square wave

$$f(x) = \begin{cases} 1 & |x| < 1 \\ \frac{1}{2} & |x| = 1 \\ 0 & |x| > 1 \end{cases} \tag{4.29}$$

as Dirichlet's discontinuous factor

$$f(x) = \frac{2}{\pi} \int_0^\infty \frac{\sin k \cos kx}{k} \, dk \tag{4.30}$$

(Exercise 4.3).

Example 4.3 (Even and odd exponentials) By using the Fourier-transform formulas (4.27 and 4.28), one may show that the Fourier transform of the even exponential $\exp(-\beta|x|)$ is

$$e^{-\beta|x|} = \frac{2}{\pi} \int_0^\infty \frac{\beta \, \cos kx}{\beta^2 + k^2} \, dk \tag{4.31}$$

while that of the odd exponential $x \exp(-\beta|x|)/|x|$ is

$$\frac{x}{|x|} e^{-\beta|x|} = \frac{2}{\pi} \int_0^\infty \frac{k \, \sin kx}{\beta^2 + k^2} \, dk \tag{4.32}$$

(Exercise 4.4).

4.3 Dirac, Parseval, and Poisson

Combining the basic equations (4.9) that define the Fourier transform, we may do something apparently useless: we may write the function $f(x)$ in terms of itself as

$$f(x) = \int_{-\infty}^\infty \frac{dk}{\sqrt{2\pi}} \, \tilde{f}(k) \, e^{ikx} = \int_{-\infty}^\infty \frac{dk}{\sqrt{2\pi}} \, e^{ikx} \int_{-\infty}^\infty \frac{dy}{\sqrt{2\pi}} \, e^{-iky} \, f(y). \tag{4.33}$$

Let's compare this equation

$$f(x) = \int_{-\infty}^\infty dy \left(\int_{-\infty}^\infty \frac{dk}{2\pi} \, \exp[ik(x-y)] \right) f(y) \tag{4.34}$$

with one (3.116) that describes Dirac's delta function

$$f(x) = \int_{-\infty}^{\infty} dy\, \delta(x-y)\, f(y). \tag{4.35}$$

Thus for functions with sensible Fourier transforms, the delta function is

$$\delta(x-y) = \int_{-\infty}^{\infty} \frac{dk}{2\pi} \exp[ik(x-y)]. \tag{4.36}$$

The same integral from $-N$ to N

$$\delta_N(x-y) = \int_{-N}^{N} \frac{dk}{2\pi} \exp[ik(x-y)] = \frac{\sin(N(x-y))}{\pi(x-y)} \tag{4.37}$$

is plotted in Fig. 4.1 for $y = 0$ and $N = 10^6$. The scales of the two axes differ by more than 10^8.

The inner product (f, g) or $\langle f|g\rangle$ of two functions, $f(x)$ with Fourier transform $\tilde{f}(k)$ and $g(x)$ with Fourier transform $\tilde{g}(k)$, is

$$\langle f|g\rangle = (f, g) = \int_{-\infty}^{\infty} dx\, f^*(x)\, g(x). \tag{4.38}$$

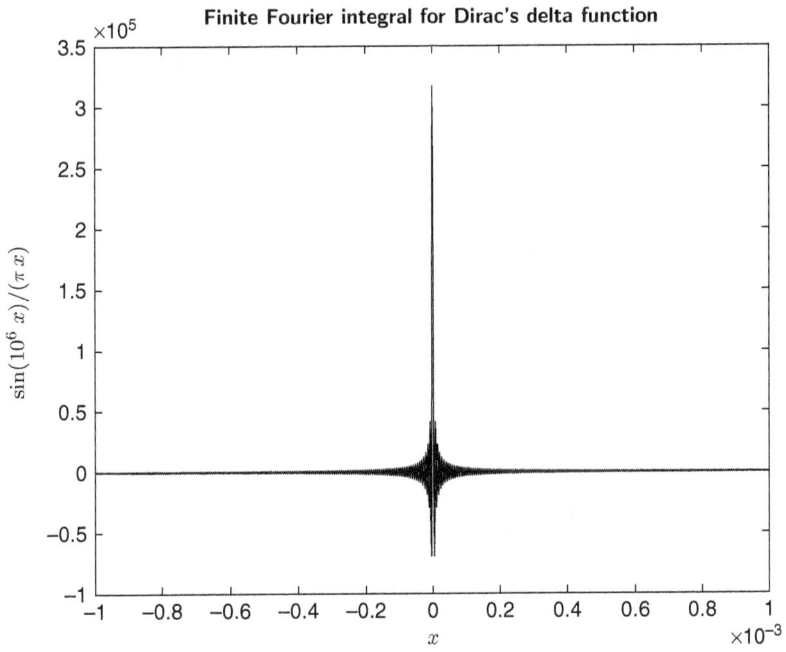

Figure 4.1 The integral (4.37) for $N = 10^6$ is plotted for $|x| \le 0.001$. The scales of the axes differ by more than 10^8. Matlab scripts for this chapter's figures are in Fourier_and_Laplace_transforms at `github.com/kevinecahill`.

Since $f(x)$ and $g(x)$ are related to $\tilde{f}(k)$ and to $\tilde{g}(k)$ by the Fourier transform (4.8), their inner product (f, g) is

$$
\begin{aligned}
(f, g) &= \int_{-\infty}^{\infty} dx \int_{-\infty}^{\infty} \frac{dk}{\sqrt{2\pi}} \left(\tilde{f}(k)\, e^{ikx}\right)^* \int_{-\infty}^{\infty} \frac{dk'}{\sqrt{2\pi}}\, \tilde{g}(k')\, e^{ik'x} \\
&= \int_{-\infty}^{\infty} dk \int_{-\infty}^{\infty} dk' \int_{-\infty}^{\infty} \frac{dx}{2\pi}\, e^{ix(k'-k)}\, \tilde{f}^*(k)\, \tilde{g}(k') \qquad (4.39) \\
&= \int_{-\infty}^{\infty} dk \int_{-\infty}^{\infty} dk'\, \delta(k'-k)\, \tilde{f}^*(k)\, \tilde{g}(k') = \int_{-\infty}^{\infty} dk\, \tilde{f}^*(k)\, \tilde{g}(k).
\end{aligned}
$$

Thus we arrive at **Parseval's relation**

$$
(f, g) = \int_{-\infty}^{\infty} dx\, f^*(x)\, g(x) = \int_{-\infty}^{\infty} dk\, \tilde{f}^*(k)\, \tilde{g}(k) = (\tilde{f}, \tilde{g}) \qquad (4.40)
$$

which says that the inner product of two functions is the same as the inner product of their Fourier transforms. The Fourier transform is a unitary transform. In particular, if $f = g$, then

$$
\langle f|f\rangle = (f, f) = \int_{-\infty}^{\infty} dx\, |f(x)|^2 = \int_{-\infty}^{\infty} dk\, |\tilde{f}(k)|^2 \qquad (4.41)
$$

(Marc-Antoine Parseval des Chênes, 1755–1836).

In fact, one may show that the Fourier transform maps the space of (Lebesgue) square-integrable functions onto itself in a one-to-one manner. Thus the natural space for the Fourier transform is the space of square-integrable functions, and so the representation (4.36) of Dirac's delta function is suitable for continuous square-integrable functions.

This may be a good place to say a few words about how to evaluate integrals involving delta functions of more complicated arguments, such as

$$
J = \int \delta(g(x))\, f(x)\, dx. \qquad (4.42)
$$

To see how this works, let's assume that $g(x)$ vanishes at a single point x_0 at which its derivative $g'(x_0) \neq 0$ isn't zero. Then the integral J involves f only at $f(x_0)$ which we can bring outside as a prefactor

$$
J = f(x_0) \int \delta(g(x))\, dx. \qquad (4.43)
$$

Near x_0 the function $g(x)$ is approximately $g'(x_0)(x - x_0)$, and so the integral is

$$
J = f(x_0) \int \delta(g'(x_0)(x - x_0))\, dx. \qquad (4.44)
$$

Since the delta function is nonnegative, we can write

$$\begin{aligned} J &= \frac{f(x_0)}{|g'(x_0)|} \int \delta(g'(x_0)(x - x_0))|g'(x_0)|\, dx \\ &= \frac{f(x_0)}{|g'(x_0)|} \int \delta(g - g_0)\, dg = \frac{f(x_0)}{|g'(x_0)|}. \end{aligned} \tag{4.45}$$

Thus for a function $g(x)$ that has a single zero at $x = x_0$, we have

$$\int \delta(g(x))\, f(x)\, dx = \frac{f(x_0)}{|g'(x_0)|} \quad \text{or} \quad \delta(g(x)) = \frac{\delta(x - x_0)}{|g'(x_0)|}. \tag{4.46}$$

If $g(x)$ has several zeros x_{0k}, then we must sum over them

$$\int \delta(g(x))\, f(x)\, dx = \sum_k \frac{f(x_{0k})}{|g'(x_{0k})|} \quad \text{or} \quad \delta(g(x)) = \sum_k \frac{\delta(x - x_{0k})}{|g'(x_{0k})|}. \tag{4.47}$$

Example 4.4 (Delta function of a function whose derivative vanishes) The integral (4.42) for J is ill defined when $g(x_0) = g'(x_0) = 0$ unless $f(x_0) = 0$ in which case, with $y = (x - x_0)^2/2$, it is by (4.46)

$$\begin{aligned} J &= \int \delta(g(x))\, f(x)\, dx = \int \delta(\tfrac{1}{2}(x - x_0)^2\, g''(x_0))\, (x - x_0)\, f'(x_0)\, dx \\ &= \int \delta(y\, g''(x_0))\, f'(x_0)\, dy = \frac{f'(x_0)}{|g''(x_0)|}. \end{aligned} \tag{4.48}$$

So if x_0 is the only root of $g(x)$ and $g(x_0) = g'(x_0) = 0$, then

$$\delta(g(x)) = \frac{\delta(x - x_0)}{|g''(x_0)|\, (x - x_0)} \tag{4.49}$$

works in an integral like (4.42 or 4.48) if $f \in C^1$ and $f(x_0) = 0$.

Our Dirac-comb formula (3.124) with $y = 0$ is

$$\sum_{n=-\infty}^{\infty} \frac{e^{-inx}}{2\pi} = \sum_{\ell=-\infty}^{\infty} \delta(x - 2\pi\ell). \tag{4.50}$$

Multiplying both sides of this comb by a function $f(x)$ and integrating over the real line, we have

$$\sum_{n=-\infty}^{\infty} \int_{-\infty}^{\infty} \frac{e^{-inx}}{2\pi} f(x)\, dx = \sum_{\ell=-\infty}^{\infty} \int_{-\infty}^{\infty} \delta(x - 2\pi\ell)\, f(x)\, dx. \tag{4.51}$$

Our formula (4.9) for the Fourier transform $\tilde{f}(n)$ of a function $f(x)$ now gives us the **Poisson summation formula** relating a sum of a function $f(2\pi\ell)$ to a sum of its Fourier transform $\tilde{f}(n)$

$$\frac{1}{\sqrt{2\pi}} \sum_{n=-\infty}^{\infty} \tilde{f}(n) = \sum_{\ell=-\infty}^{\infty} f(2\pi\ell) \tag{4.52}$$

in which n and ℓ are summed over all the integers. The stretched version of the Poisson summation formula is

$$\frac{\sqrt{2\pi}}{L} \sum_{n=-\infty}^{\infty} \tilde{f}(2\pi n/L) = \sum_{\ell=-\infty}^{\infty} f(\ell L). \tag{4.53}$$

Both sides of these formulas make sense for continuous functions that are square integrable on the real line.

Example 4.5 (Poisson summation formula) In Example 4.1, we saw that the gaussian $f(x) = \exp(-m^2x^2)$ has $\tilde{f}(k) = \exp(-k^2/4m^2)/\sqrt{2}\,m$ as its Fourier transform. So in this case, the Poisson summation formula (4.52) gives

$$\frac{1}{2\sqrt{\pi}\,m} \sum_{k=-\infty}^{\infty} e^{-k^2/4m^2} = \sum_{\ell=-\infty}^{\infty} e^{-(2\pi\ell m)^2}. \tag{4.54}$$

For $m \gg 1$, the left-hand sum converges slowly, while the right-hand sum converges quickly. For $m \ll 1$, the right-hand sum converges slowly, while the left-hand sum converges quickly.

A sum that converges slowly in space often converges quickly in momentum space. **Ewald summation** is a technique for summing electrostatic energies, which fall off only with a power of the distance, by summing their Fourier transforms (Darden et al., 1993).

4.4 Derivatives and Integrals of Fourier Transforms

By differentiating the inverse Fourier-transform relation (4.7)

$$f(x) = \int_{-\infty}^{\infty} \frac{dk}{\sqrt{2\pi}}\, \tilde{f}(k)\, e^{ikx} \tag{4.55}$$

we see that the Fourier transform of the derivative $f'(x)$ is $ik\tilde{f}(k)$

$$f'(x) = \int_{-\infty}^{\infty} \frac{dk}{\sqrt{2\pi}}\, ik\, \tilde{f}(k)\, e^{ikx}. \tag{4.56}$$

Differentiation with respect to x corresponds to multiplication by ik.

We may repeat the process and express the second derivative as

$$f''(x) = \int_{-\infty}^{\infty} \frac{dk}{\sqrt{2\pi}}\, (-k^2)\, \tilde{f}(k)\, e^{ikx} \tag{4.57}$$

and the nth derivative as

$$f^{(n)}(x) = \int_{-\infty}^{\infty} \frac{dk}{\sqrt{2\pi}} \, (ik)^n \, \tilde{f}(k) \, e^{ikx}. \tag{4.58}$$

The indefinite integral of the inverse Fourier transform (4.55) is

$$^{(1)}f(x) \equiv \int^x dx_1 \, f(x_1) = \int_{-\infty}^{\infty} \frac{dk}{\sqrt{2\pi}} \, \tilde{f}(k) \, \frac{e^{ikx}}{ik} \tag{4.59}$$

and the nth indefinite integral is

$$^{(n)}f(x) \equiv \int^x dx_1 \cdots \int^{x_{n-1}} dx_n \, f(x_n) = \int_{-\infty}^{\infty} \frac{dk}{\sqrt{2\pi}} \, \tilde{f}(k) \, \frac{e^{ikx}}{(ik)^n}. \tag{4.60}$$

Whether these derivatives and integrals converge better or worse than $f(x)$ depends upon the behavior of $\tilde{f}(k)$ near $k = 0$ and as $|k| \to \infty$.

Example 4.6 (Momentum and momentum space) Let's write the inverse Fourier transform (4.7) with ψ instead of f and with the wave number k replaced by $k = p/\hbar$

$$\psi(x) = \int_{-\infty}^{\infty} \tilde{\psi}(k) \, e^{ikx} \, \frac{dk}{\sqrt{2\pi}} = \int_{-\infty}^{\infty} \frac{\tilde{\psi}(p/\hbar)}{\sqrt{\hbar}} \, e^{ipx/\hbar} \frac{dp}{\sqrt{2\pi\hbar}}. \tag{4.61}$$

For a normalized wave function $\psi(x)$, Parseval's relation (4.41) implies

$$1 = \int_{-\infty}^{\infty} |\psi(x)|^2 \, dx = \int_{-\infty}^{\infty} |\tilde{\psi}(k)|^2 \, dk = \int_{-\infty}^{\infty} \left| \frac{\tilde{\psi}(p/\hbar)}{\sqrt{\hbar}} \right|^2 dp \tag{4.62}$$

or with $\psi(x) = \langle x|\psi\rangle$ and $\varphi(p) = \langle p|\psi\rangle = \tilde{\psi}(p/\hbar)/\sqrt{\hbar}$

$$\begin{aligned} 1 = \langle\psi|\psi\rangle &= \int_{-\infty}^{\infty} |\psi(x)|^2 \, dx = \int_{-\infty}^{\infty} \langle\psi|x\rangle\langle x|\psi\rangle \, dx \\ &= \int_{-\infty}^{\infty} \langle\psi|p\rangle\langle p|\psi\rangle \, dp = \int_{-\infty}^{\infty} |\varphi(p)|^2 \, dp. \end{aligned} \tag{4.63}$$

The inner product of any two states $|\psi\rangle$ and $|\phi\rangle$ is

$$\begin{aligned} \langle\psi|\phi\rangle &= \int_{-\infty}^{\infty} \psi^*(x)\phi(x) \, dx = \int_{-\infty}^{\infty} \langle\psi|x\rangle\langle x|\phi\rangle \, dx \\ &= \int_{-\infty}^{\infty} \psi^*(p)\phi(p) \, dp = \int_{-\infty}^{\infty} \langle\psi|p\rangle\langle p|\phi\rangle \, dp \end{aligned} \tag{4.64}$$

so the outer products $|x\rangle\langle x|$ and $|p\rangle\langle p|$ can represent the identity operator

$$I = \int_{-\infty}^{\infty} dx \, |x\rangle\langle x| = \int_{-\infty}^{\infty} dp \, |p\rangle\langle p|. \tag{4.65}$$

The Fourier transform (4.61) relating the wave function in momentum space to that in position space is

$$\psi(x) = \int_{-\infty}^{\infty} e^{ipx/\hbar}\, \varphi(p)\, \frac{dp}{\sqrt{2\pi\hbar}} \tag{4.66}$$

and the inverse Fourier transform is

$$\varphi(p) = \int_{-\infty}^{\infty} e^{-ipx/\hbar}\, \psi(x)\, \frac{dx}{\sqrt{2\pi\hbar}}. \tag{4.67}$$

In Dirac notation, the first equation (4.66) of this pair is

$$\psi(x) = \langle x|\psi\rangle = \int_{-\infty}^{\infty} \langle x|p\rangle\langle p|\psi\rangle\, dp = \int_{-\infty}^{\infty} \frac{e^{ipx/\hbar}}{\sqrt{2\pi\hbar}}\, \varphi(p)\, dp \tag{4.68}$$

so we identify $\langle x|p\rangle$ with

$$\langle x|p\rangle = \frac{e^{ipx/\hbar}}{\sqrt{2\pi\hbar}} \tag{4.69}$$

which in turn is consistent with the delta-function relation (4.36)

$$\delta(x-y) = \langle x|y\rangle = \int_{-\infty}^{\infty} \langle x|p\rangle\langle p|y\rangle\, dp = \int_{-\infty}^{\infty} \frac{e^{ipx/\hbar}}{\sqrt{2\pi\hbar}}\, \frac{e^{-ipy/\hbar}}{\sqrt{2\pi\hbar}}\, dp$$

$$= \int_{-\infty}^{\infty} \frac{e^{ip(x-y)/\hbar}}{2\pi\hbar}\, dp = \int_{-\infty}^{\infty} e^{ik(x-y)}\, \frac{dk}{2\pi}. \tag{4.70}$$

If we differentiate $\psi(x)$ as given by (4.68), then we find as in (4.56)

$$\frac{\hbar}{i}\frac{d}{dx}\psi(x) = \int_{-\infty}^{\infty} p\, \varphi(p)\, e^{ipx/\hbar}\, \frac{dp}{\sqrt{2\pi\hbar}} \tag{4.71}$$

or

$$\frac{\hbar}{i}\frac{d}{dx}\psi(x) = \langle x|p|\psi\rangle = \int_{-\infty}^{\infty} \langle x|p|p'\rangle\, \langle p'|\psi\rangle\, dp' = \int_{-\infty}^{\infty} p'\, \varphi(p')\, e^{ip'x/\hbar}\, \frac{dp'}{\sqrt{2\pi\hbar}}$$

in Dirac notation.

Example 4.7 (Uncertainty principle) Let's first normalize the gaussian $\psi(x) = N\,\exp(-(x/a)^2)$ to unity over the real axis

$$1 = N^2 \int_{-\infty}^{\infty} e^{-2(x/a)^2}\, dx = \sqrt{\frac{\pi}{2}}\, a\, N^2 \tag{4.72}$$

which gives $N^2 = \sqrt{2/\pi}/a$. So the normalized wave function is

$$\psi(x) \equiv \langle x|\psi\rangle = \left(\frac{2}{\pi}\right)^{1/4} \frac{1}{\sqrt{a}}\, e^{-(x/a)^2}. \tag{4.73}$$

The **mean value** $\langle A\rangle$ of an operator A in a state $|\psi\rangle$ is

$$\langle A\rangle \equiv \langle\psi|A|\psi\rangle. \tag{4.74}$$

More generally, the mean value of an operator A for a system described by a density operator ρ is the trace

$$\langle A \rangle \equiv \mathrm{Tr}\,(\rho A)\,. \tag{4.75}$$

Since the gaussian (4.73) is an even function of x (that is, $\psi(-x) = \psi(x)$), the mean value of the position operator x in the state (4.73) vanishes

$$\langle x \rangle = \langle \psi | x | \psi \rangle = \int_{-\infty}^{\infty} x\, |\psi(x)|^2\, dx = 0. \tag{4.76}$$

The **variance** of an operator A with mean value $\langle A \rangle$ in a state $|\psi\rangle$ is the mean value of the square of the difference $A - \langle A \rangle$

$$(\Delta A)^2 \equiv \langle \psi |\, (A - \langle A \rangle)^2\, |\psi \rangle. \tag{4.77}$$

For a system with density operator ρ, the variance of A is

$$(\Delta A)^2 \equiv \mathrm{Tr}\Big[\rho\,(A - \langle A \rangle)^2\Big]. \tag{4.78}$$

Since $\langle x \rangle = 0$, the variance of the position operator x is

$$\begin{aligned}(\Delta x)^2 &= \langle \psi | (x - \langle x \rangle)^2 | \psi \rangle = \langle \psi | x^2 | \psi \rangle \\ &= \int_{-\infty}^{\infty} x^2\, |\psi(x)|^2\, dx = \frac{a^2}{4}.\end{aligned} \tag{4.79}$$

We can use the Fourier transform to find the variance of the momentum operator. By (4.67), the wave function $\varphi(p)$ in momentum space is

$$\varphi(p) = \langle p | \psi \rangle = \int_{-\infty}^{\infty} \langle p | x \rangle \langle x | \psi \rangle\, dx. \tag{4.80}$$

By (4.69), the inner product $\langle p | x \rangle = \langle x | p \rangle^*$ is $\langle p | x \rangle = e^{-ipx/\hbar}/\sqrt{2\pi\hbar}$, so

$$\varphi(p) = \langle p | \psi \rangle = \int_{-\infty}^{\infty} \frac{dx}{\sqrt{2\pi\hbar}}\, e^{-ipx/\hbar} \langle x | \psi \rangle. \tag{4.81}$$

Thus by (4.72 and 4.73), $\varphi(p)$ is the Fourier transform

$$\varphi(p) = \int_{-\infty}^{\infty} \frac{dx}{\sqrt{2\pi\hbar}}\, e^{-ipx/\hbar} \left(\frac{2}{\pi}\right)^{1/4} \frac{1}{\sqrt{a}}\, e^{-(x/a)^2}. \tag{4.82}$$

Using our formula (4.19) for the Fourier transform of a gaussian, we get

$$\varphi(p) = \sqrt{\frac{a}{2\hbar}} \left(\frac{2}{\pi}\right)^{1/4} e^{-(ap)^2/(2\hbar)^2}. \tag{4.83}$$

Since the gaussian $\varphi(p)$ is an even function of p, the mean value $\langle p \rangle$ of the momentum operator vanishes, like that of the position operator. So the variance of the momentum operator is

$$(\Delta p)^2 = \langle\psi|(p - \langle p\rangle)^2|\psi\rangle = \langle\psi|p^2\,|\psi\rangle = \int_{-\infty}^{\infty} p^2\,|\varphi(p)|^2\,dp$$
$$= \sqrt{\frac{2}{\pi}}\int_{-\infty}^{\infty} p^2\,\frac{a}{2\hbar}\,e^{-(ap)^2/2\hbar^2}\,dp = \frac{\hbar^2}{a^2}. \tag{4.84}$$

Thus in this case, the product of the two variances is

$$(\Delta x)^2\,(\Delta p)^2 = \frac{a^2}{4}\frac{\hbar^2}{a^2} = \frac{\hbar^2}{4} \tag{4.85}$$

which is the minimum value that the product of the variances can assume according to **Heisenberg's uncertainty principle**

$$\Delta x\,\Delta p \ge \frac{\hbar}{2} \tag{4.86}$$

which follows from the Fourier-transform relations between the conjugate variables x and p.

The state $|\psi\rangle$ of a free particle at time $t = 0$

$$|\psi, 0\rangle = \int_{-\infty}^{\infty} |p\rangle\langle p|\psi\rangle\,dp = \int_{-\infty}^{\infty} |p\rangle\varphi(p)\,dp \tag{4.87}$$

evolves under the influence of the hamiltonian $H = p^2/(2m)$ to the state

$$e^{-iHt/\hbar}|\psi, 0\rangle = \int_{-\infty}^{\infty} e^{-iHt/\hbar}|p\rangle\,\varphi(p)\,dp = \int_{-\infty}^{\infty} e^{-ip^2t/(2\hbar m)}|p\rangle\,\varphi(p)\,dp \tag{4.88}$$

at time t.

Example 4.8 (Characteristic function) If $P(x)$ is a **probability distribution** normalized to unity over the range of x

$$\int P(x)\,dx = 1 \tag{4.89}$$

then its Fourier transform is the **characteristic function**

$$\chi(k) = \tilde{P}(k) = \int e^{ikx}\,P(x)\,dx. \tag{4.90}$$

The **expected value** of a function $f(x)$ is the integral

$$E[f(x)] = \int f(x)\,P(x)\,dx. \tag{4.91}$$

So the **characteristic function** $\chi(k) = E[\exp(ikx)]$ is the expected value of the exponential $\exp(ikx)$, and its derivatives at $k = 0$ are the **moments** $E[x^n] \equiv \mu_n$ of the probability distribution

$$E[x^n] = \int x^n\,P(x)\,dx = (-i)^n \left.\frac{d^n\chi(k)}{dk^n}\right|_{k=0}. \tag{4.92}$$

We'll pick up this thread again in Section 15.16.

4.5 Fourier Transforms of Functions of Several Variables

If $f(x_1, x_2)$ is a function of two variables, then its double Fourier transform $\tilde{f}(k_1, k_2)$ is

$$\tilde{f}(k_1, k_2) = \int_{-\infty}^{\infty} \frac{dx_1}{\sqrt{2\pi}} \int_{-\infty}^{\infty} \frac{dx_2}{\sqrt{2\pi}} \, e^{-ik_1x_1 - ik_2x_2} \, f(x_1, x_2). \tag{4.93}$$

By twice using the Fourier representation (4.36) of Dirac's delta function, we may invert this double Fourier transformation

$$\begin{aligned}
&\int_{-\infty}^{\infty}\int_{-\infty}^{\infty} \frac{dk_1 dk_2}{2\pi} \, e^{i(k_1x_1+k_2x_2)} \, \tilde{f}(k_1, k_2) \\
&= \int_{-\infty}^{\infty}\int_{-\infty}^{\infty} \frac{dk_1 dk_2}{2\pi} \int_{-\infty}^{\infty}\int_{-\infty}^{\infty} \frac{dx_1' dx_2'}{2\pi} \, e^{ik_1(x_1-x_1')+ik_2(x_2-x_2')} \, f(x_1', x_2') \\
&= \int_{-\infty}^{\infty} \frac{dk_2}{2\pi} \int_{-\infty}^{\infty}\int_{-\infty}^{\infty} dx_1' dx_2' \, e^{ik_2(x_2-x_2')} \, \delta(x_1 - x_1') \, f(x_1', x_2') \\
&= \int_{-\infty}^{\infty}\int_{-\infty}^{\infty} dx_1' dx_2' \, \delta(x_1 - x_1')\delta(x_2 - x_2') \, f(x_1', x_2') = f(x_1, x_2).
\end{aligned} \tag{4.94}$$

That is

$$f(x_1, x_2) = \int_{-\infty}^{\infty}\int_{-\infty}^{\infty} \frac{dk_1 dk_2}{2\pi} \, e^{i(k_1x_1+k_2x_2)} \, \tilde{f}(k_1, k_2). \tag{4.95}$$

The Fourier transform of a function $f(x_1, \dots, x_n)$ of n variables is

$$\tilde{f}(k_1, \dots, k_n) = \int_{-\infty}^{\infty} \dots \int_{-\infty}^{\infty} \frac{dx_1 \dots dx_n}{(2\pi)^{n/2}} \, e^{-i(k_1x_1+\dots+k_nx_n)} \, f(x_1, \dots, x_n) \tag{4.96}$$

and its inverse is

$$f(x_1, \dots, x_n) = \int_{-\infty}^{\infty} \dots \int_{-\infty}^{\infty} \frac{dk_1 \dots dk_n}{(2\pi)^{n/2}} \, e^{i(k_1x_1+\dots+k_nx_n)} \, \tilde{f}(k_1, \dots, k_n) \tag{4.97}$$

in which all the integrals run from $-\infty$ to ∞.

If we generalize the relations (4.12–4.15) between Fourier series and transforms from one to n dimensions, then we find that the Fourier series corresponding to the Fourier transform (4.97) is

$$f(x_1, \dots, x_n) = \left(\frac{2\pi}{L}\right)^n \sum_{j_1=-\infty}^{\infty} \dots \sum_{j_n=-\infty}^{\infty} e^{i(k_{j_1}x_1+\dots+k_{j_n}x_n)} \, \frac{\tilde{f}(k_{j_1}, \dots, k_{j_n})}{(2\pi)^{n/2}} \tag{4.98}$$

in which $k_{j_\ell} = 2\pi j_\ell / L$. Thus, for $n = 3$ we have

$$f(\boldsymbol{x}) = \frac{(2\pi)^3}{V} \sum_{j_1=-\infty}^{\infty} \sum_{j_2=-\infty}^{\infty} \sum_{j_3=-\infty}^{\infty} e^{i\boldsymbol{k}_j \cdot \boldsymbol{x}} \, \frac{\tilde{f}(\boldsymbol{k}_j)}{(2\pi)^{3/2}} \tag{4.99}$$

in which $\boldsymbol{k}_j = (k_{j_1}, k_{j_2}, k_{j_3})$ and $V = L^3$ is the volume of the box.

Example 4.9 (Feynman propagator) For a spinless quantum field of mass m, Feynman's propagator is the 4-dimensional Fourier transform

$$\Delta_F(x) = \int \frac{\exp(ik \cdot x)}{k^2 + m^2 - i\epsilon} \frac{d^4k}{(2\pi)^4} \tag{4.100}$$

where $k \cdot x = \boldsymbol{k} \cdot \boldsymbol{x} - k^0 x^0$, all physical quantities are in **natural units** ($c = \hbar = 1$), and $x^0 = ct = t$. The tiny imaginary term $-i\epsilon$ makes $\Delta_F(x-y)$ proportional to the mean value in the vacuum state $|0\rangle$ of the **time-ordered product** of the fields $\phi(x)$ and $\phi(y)$ (Section 6.44)

$$\begin{aligned} -i\,\Delta_F(x-y) &= \langle 0|\mathcal{T}\left[\phi(x)\phi(y)\right]|0\rangle \\ &\equiv \theta(x^0 - y^0)\langle 0|\phi(x)\phi(y)|0\rangle + \theta(y^0 - x^0)\langle 0|\phi(y)\phi(x)|0\rangle \end{aligned} \tag{4.101}$$

in which $\theta(a) = (a + |a|)/2|a|$ is the Heaviside function.

4.6 Convolutions

The convolution of $f(x)$ with $g(x)$ is the integral

$$f * g(x) = \int_{-\infty}^{\infty} \frac{dy}{\sqrt{2\pi}} f(x-y)\, g(y). \tag{4.102}$$

The convolution product is symmetric

$$f * g(x) = g * f(x) \tag{4.103}$$

because setting $z = x - y$, we have

$$\begin{aligned} f * g(x) &= \int_{-\infty}^{\infty} \frac{dy}{\sqrt{2\pi}} f(x-y)\, g(y) = -\int_{\infty}^{-\infty} \frac{dz}{\sqrt{2\pi}} f(z)\, g(x-z) \\ &= \int_{-\infty}^{\infty} \frac{dz}{\sqrt{2\pi}} g(x-z)\, f(z) = g * f(x). \end{aligned} \tag{4.104}$$

Convolutions may look strange at first, but they often occur in physics in the 3-dimensional form

$$F(\boldsymbol{x}) = \int G(\boldsymbol{x} - \boldsymbol{x}')\, S(\boldsymbol{x}')\, d^3x' \tag{4.105}$$

in which G is a Green's function and S is a source (George Green, 1793–1841).

Example 4.10 (Gauss's law and the potential for static electric fields) The divergence of the electric field $\boldsymbol{E}$ is the microscopic charge density ρ divided by the electric permittivity of the vacuum $\epsilon_0 = 8.854 \times 10^{-12}$ F/m, that is, $\nabla \cdot \boldsymbol{E} = \rho/\epsilon_0$. This constraint is known as Gauss's law. If the charges and fields are independent of time, then the electric field $\boldsymbol{E}$ is the gradient of a scalar potential $\boldsymbol{E} = -\nabla\phi$. These last two equations imply that ϕ obeys Poisson's equation

$$-\nabla^2\phi = \frac{\rho}{\epsilon_0}. \tag{4.106}$$

We may solve this equation by using Fourier transforms as described in Section 4.13. If $\tilde{\phi}(\boldsymbol{k})$ and $\tilde{\rho}(\boldsymbol{k})$ respectively are the Fourier transforms of $\phi(\boldsymbol{x})$ and $\rho(\boldsymbol{x})$, then Poisson's differential equation (4.106) gives

$$\begin{aligned}-\nabla^2\phi(\boldsymbol{x}) &= -\nabla^2\int e^{i\boldsymbol{k}\cdot\boldsymbol{x}}\,\tilde{\phi}(\boldsymbol{k})\,d^3\boldsymbol{k} = \int \boldsymbol{k}^2\,e^{i\boldsymbol{k}\cdot\boldsymbol{x}}\,\tilde{\phi}(\boldsymbol{k})\,d^3\boldsymbol{k}\\ &= \frac{\rho(\boldsymbol{x})}{\epsilon_0} = \int e^{i\boldsymbol{k}\cdot\boldsymbol{x}}\,\frac{\tilde{\rho}(\boldsymbol{k})}{\epsilon_0}\,d^3\boldsymbol{k}\end{aligned} \tag{4.107}$$

which implies the algebraic equation $\tilde{\phi}(\boldsymbol{k}) = \tilde{\rho}(\boldsymbol{k})/\epsilon_0\boldsymbol{k}^2$ which gives $\tilde{\phi}(\boldsymbol{k})$ as a product of the Fourier transforms $\tilde{\rho}(\boldsymbol{k})$ and $1/\boldsymbol{k}^2$ (and is an instance of (4.166)). The inverse Fourier transform of $\tilde{\phi}(\boldsymbol{k})$ is the scalar potential

$$\begin{aligned}\phi(\boldsymbol{x}) &= \int e^{i\boldsymbol{k}\cdot\boldsymbol{x}}\,\tilde{\phi}(\boldsymbol{k})\,d^3\boldsymbol{k} = \int e^{i\boldsymbol{k}\cdot\boldsymbol{x}}\,\frac{\tilde{\rho}(\boldsymbol{k})}{\epsilon_0\,\boldsymbol{k}^2}\,d^3\boldsymbol{k} \qquad (4.108)\\ &= \int e^{i\boldsymbol{k}\cdot\boldsymbol{x}}\,\frac{1}{\boldsymbol{k}^2}\int e^{-i\boldsymbol{k}\cdot\boldsymbol{x}'}\,\frac{\rho(\boldsymbol{x}')}{\epsilon_0}\,\frac{d^3\boldsymbol{x}'d^3\boldsymbol{k}}{(2\pi)^3} = \int G(\boldsymbol{x}-\boldsymbol{x}')\,\frac{\rho(\boldsymbol{x}')}{\epsilon_0}\,d^3\boldsymbol{x}'\end{aligned}$$

in which

$$G(\boldsymbol{x}-\boldsymbol{x}') = \int \frac{d^3\boldsymbol{k}}{(2\pi)^3}\,\frac{1}{\boldsymbol{k}^2}\,e^{i\boldsymbol{k}\cdot(\boldsymbol{x}-\boldsymbol{x}')}. \tag{4.109}$$

$G(\boldsymbol{x}-\boldsymbol{x}')$ is the Green's function for the differential operator $-\nabla^2$ in the sense that

$$-\nabla^2 G(\boldsymbol{x}-\boldsymbol{x}') = \int \frac{d^3\boldsymbol{k}}{(2\pi)^3}\,e^{i\boldsymbol{k}\cdot(\boldsymbol{x}-\boldsymbol{x}')} = \delta^{(3)}(\boldsymbol{x}-\boldsymbol{x}'). \tag{4.110}$$

We may think of G as the inverse of the operator $-\nabla^2$. The Green's function $G(\boldsymbol{x}-\boldsymbol{x}')$ ensures that $\phi(\boldsymbol{x})$ as given by (4.108) satisfies Poisson's equation (4.106). To integrate (4.109) and compute $G(\boldsymbol{x}-\boldsymbol{x}')$, we use spherical coordinates with the z-axis parallel to the vector $\boldsymbol{x}-\boldsymbol{x}'$

$$\begin{aligned}G(\boldsymbol{x}-\boldsymbol{x}') &= \int \frac{d^3\boldsymbol{k}}{(2\pi)^3}\,\frac{1}{\boldsymbol{k}^2}\,e^{i\boldsymbol{k}\cdot(\boldsymbol{x}-\boldsymbol{x}')} = \int_0^\infty \frac{dk}{(2\pi)^2}\int_{-1}^{1} d\cos\theta\, e^{ik|\boldsymbol{x}-\boldsymbol{x}'|\cos\theta}\\ &= \int_0^\infty \frac{dk}{(2\pi)^2}\,\frac{e^{ik|\boldsymbol{x}-\boldsymbol{x}'|}-e^{-ik|\boldsymbol{x}-\boldsymbol{x}'|}}{ik|\boldsymbol{x}-\boldsymbol{x}'|}\\ &= \frac{1}{2\pi^2|\boldsymbol{x}-\boldsymbol{x}'|}\int_0^\infty \frac{\sin k|\boldsymbol{x}-\boldsymbol{x}'|\,dk}{k} = \frac{1}{2\pi^2|\boldsymbol{x}-\boldsymbol{x}'|}\int_0^\infty \frac{\sin k\,dk}{k}.\end{aligned} \tag{4.111}$$

In Example 6.45 of Section 6.18 on Cauchy's principal value, we'll show that

$$\int_0^\infty \frac{\sin k}{k}\,dk = \frac{\pi}{2}. \tag{4.112}$$

Using this result, we have

$$\int \frac{d^3k}{(2\pi)^3} \frac{1}{k^2} e^{ik\cdot(x-x')} = G(x - x') = \frac{1}{4\pi |x - x'|}. \tag{4.113}$$

Finally by substituting this formula for $G(x - x')$ into Eq. (4.108), we find that the Fourier transform $\phi(x)$ of the product $\tilde{\rho}(k)/k^2$ of the functions $\tilde{\rho}(k)$ and $1/k^2$ is the convolution

$$\phi(x) = \frac{1}{4\pi\epsilon_0} \int \frac{\rho(x')}{|x - x'|} d^3x' \tag{4.114}$$

of their Fourier transforms $1/|x - x'|$ and $\rho(x')$. The Fourier transform of the product of any two functions is the convolution of their Fourier transforms, as we'll see in the next section.

Example 4.11 (Static magnetic vector potential) The magnetic induction B has zero divergence (as long as there are no magnetic monopoles) and so may be written as the curl $B = \nabla \times A$ of a vector potential A. For static currents, Ampère's law is $\nabla \times B = \mu_0 J$ in which $\mu_0 = 1/(\epsilon_0 c^2) = 4\pi \times 10^{-7}$ N A^{-2} is the permeability of the vacuum. It follows that in Coulomb's gauge $\nabla \cdot A = 0$, the magnetostatic vector potential A satisfies the equation

$$\nabla \times B = \nabla \times (\nabla \times A) = \nabla(\nabla \cdot A) - \nabla^2 A = -\nabla^2 A = \mu_0 J. \tag{4.115}$$

Applying the Fourier-transform technique (4.106–4.114), we find that the Fourier transforms of A and J satisfy the algebraic equation

$$\tilde{A}(k) = \mu_0 \frac{\tilde{J}(k)}{k^2} \tag{4.116}$$

which is an instance of (4.166). Performing the inverse Fourier transform, we see that A is the convolution

$$A(x) = \frac{\mu_0}{4\pi} \int d^3x' \frac{J(x')}{|x - x'|}. \tag{4.117}$$

If in the solution (4.114) of Poisson's equation, $\rho(x)$ is translated by a, then so is $\phi(x)$. That is, if $\rho'(x) = \rho(x + a)$ then $\phi'(x) = \phi(x + a)$. Similarly, if the current $J(x)$ in (4.117) is translated by a, then so is the potential $A(x)$. **Convolutions respect translational invariance.** That's one reason why they occur so often in the formulas of physics.

4.7 Fourier Transform of a Convolution

The Fourier transform of the convolution $f * g$ is the product of the Fourier transforms $\tilde{f}$ and $\tilde{g}$:

$$\widetilde{f * g}(k) = \tilde{f}(k)\, \tilde{g}(k). \tag{4.118}$$

To see why, we form the Fourier transform $\widetilde{f * g}(k)$ of the convolution $f * g(x)$

$$\begin{aligned}\widetilde{f * g}(k) &= \int_{-\infty}^{\infty} \frac{dx}{\sqrt{2\pi}} e^{-ikx} \, f * g(x) \\ &= \int_{-\infty}^{\infty} \frac{dx}{\sqrt{2\pi}} e^{-ikx} \int_{-\infty}^{\infty} \frac{dy}{\sqrt{2\pi}} f(x-y) \, g(y).\end{aligned} \tag{4.119}$$

Now we write $f(x-y)$ and $g(y)$ in terms of their Fourier transforms $\tilde{f}(p)$ and $\tilde{g}(q)$

$$\widetilde{f * g}(k) = \int_{-\infty}^{\infty} \frac{dx}{\sqrt{2\pi}} e^{-ikx} \int_{-\infty}^{\infty} \frac{dy}{\sqrt{2\pi}} \int_{-\infty}^{\infty} \frac{dp}{\sqrt{2\pi}} \tilde{f}(p) \, e^{ip(x-y)} \int_{-\infty}^{\infty} \frac{dq}{\sqrt{2\pi}} \tilde{g}(q) \, e^{iqy} \tag{4.120}$$

and use the representation (4.36) of Dirac's delta function twice to get

$$\begin{aligned}\widetilde{f * g}(k) &= \int_{-\infty}^{\infty} \frac{dy}{2\pi} \int_{-\infty}^{\infty} dp \int_{-\infty}^{\infty} dq \, \delta(p-k) \, \tilde{f}(p) \, \tilde{g}(q) \, e^{i(q-p)y} \\ &= \int_{-\infty}^{\infty} dp \int_{-\infty}^{\infty} dq \, \delta(p-k) \, \delta(q-p) \, \tilde{f}(p) \, \tilde{g}(q) \\ &= \int_{-\infty}^{\infty} dp \, \delta(p-k) \, \tilde{f}(p) \, \tilde{g}(p) = \tilde{f}(k) \, \tilde{g}(k)\end{aligned} \tag{4.121}$$

which is (4.118). Examples 4.10 and 4.11 illustrate this result.

4.8 Fourier Transforms and Green's Functions

A Green's function $G(x)$ for a differential operator P turns into a delta function when acted upon by P, that is, $PG(x) = \delta(x)$. If the differential operator is a polynomial $P(\partial) \equiv P(\partial_1, \ldots, \partial_n)$ in the derivatives $\partial_1, \ldots, \partial_n$ with constant coefficients, then a suitable Green's function $G(x) \equiv G(x_1, \ldots, x_n)$ will satisfy

$$P(\partial) G(x) = \delta^{(n)}(x). \tag{4.122}$$

Expressing both $G(x)$ and $\delta^{(n)}(x)$ as Fourier transforms, we get

$$P(\partial) G(x) = \int d^n k \, P(ik) \, e^{ik\cdot x} \, \tilde{G}(k) = \delta^{(n)}(x) = \int \frac{d^n k}{(2\pi)^n} e^{ik\cdot x} \tag{4.123}$$

which gives us the algebraic equation

$$\tilde{G}(k) = \frac{1}{(2\pi)^n \, P(ik)}. \tag{4.124}$$

Thus the Green's function G_P for the differential operator $P(\partial)$ is

$$G_P(x) = \int \frac{d^n k}{(2\pi)^n} \frac{e^{ik\cdot x}}{P(ik)}. \tag{4.125}$$

Example 4.12 (Green and Yukawa) In 1935, Hideki Yukawa (1907–1981) proposed the partial differential equation

$$P_Y(\partial)G_Y(\boldsymbol{x}) \equiv (-\Delta + m^2)G_Y(\boldsymbol{x}) = (-\nabla^2 + m^2)G_Y(\boldsymbol{x}) = \delta(\boldsymbol{x}). \tag{4.126}$$

Here (4.125) gives as the Green's function for $P_Y(\partial)$ the Yukawa potential

$$G_Y(x) = \int \frac{d^3k}{(2\pi)^3} \frac{e^{i\boldsymbol{k}\cdot\boldsymbol{x}}}{P_Y(ik)} = \int \frac{d^3k}{(2\pi)^3} \frac{e^{i\boldsymbol{k}\cdot\boldsymbol{x}}}{\boldsymbol{k}^2 + m^2} = \frac{e^{-mr}}{4\pi r} \tag{4.127}$$

an integration done in example 6.27.

4.9 Laplace Transforms

The Laplace transform $f(s)$ of a function $F(t)$ is the integral

$$f(s) = \int_0^\infty dt\, e^{-st}\, F(t). \tag{4.128}$$

Because the integration is over positive values of t, the exponential $\exp(-st)$ falls off rapidly with the real part of s. As $\operatorname{Re} s$ increases, the Laplace transform $f(s)$ becomes smoother and smaller. For $\operatorname{Re} s > 0$, the exponential $\exp(-st)$ lets many functions $F(t)$ that are not integrable over the half line $[0, \infty)$ have well-behaved Laplace transforms.

For instance, the function $F(t) = 1$ is not integrable over the half line $[0, \infty)$, but its Laplace transform

$$f(s) = \int_0^\infty dt\, e^{-st}\, F(t) = \int_0^\infty dt\, e^{-st} = \frac{1}{s} \tag{4.129}$$

is well defined for $\operatorname{Re} s > 0$ and square integrable for $\operatorname{Re} s > \epsilon$.

The function $F(t) = \exp(kt)$ diverges exponentially for $\operatorname{Re} k > 0$, but its Laplace transform

$$f(s) = \int_0^\infty dt\, e^{-st}\, F(t) = \int_0^\infty dt\, e^{-(s-k)t} = \frac{1}{s-k} \tag{4.130}$$

is well defined for $\operatorname{Re} s > k$ with a simple pole at $s = k$ (section 6.10) and is square integrable for $\operatorname{Re} s > k + \epsilon$.

The Laplace transforms of $\cosh kt$ and $\sinh kt$ are

$$f(s) = \int_0^\infty dt\, e^{-st} \cosh kt = \frac{1}{2}\int_0^\infty dt\, e^{-st}\left(e^{kt} + e^{-kt}\right) = \frac{s}{s^2 - k^2} \tag{4.131}$$

and

$$f(s) = \int_0^\infty dt\, e^{-st} \sinh kt = \frac{1}{2}\int_0^\infty dt\, e^{-st}\left(e^{kt} - e^{-kt}\right) = \frac{k}{s^2 - k^2}. \tag{4.132}$$

The Laplace transform of $\cos \omega t$ is

$$f(s) = \int_0^\infty dt\, e^{-st}\, \cos \omega t = \frac{1}{2} \int_0^\infty dt\, e^{-st}\, \left(e^{i\omega t} + e^{-i\omega t}\right) = \frac{s}{s^2 + \omega^2} \tag{4.133}$$

and that of $\sin \omega t$ is

$$f(s) = \int_0^\infty dt\, e^{-st}\, \sin \omega t = \frac{1}{2i} \int_0^\infty dt\, e^{-st}\, \left(e^{i\omega t} - e^{-i\omega t}\right) = \frac{\omega}{s^2 + \omega^2}. \tag{4.134}$$

Example 4.13 (Lifetime of a fluorophore) Fluorophores are molecules that emit visible light when excited by photons. The probability $P(t, t')$ that a fluorophore with a lifetime τ will emit a photon at time t if excited by a photon at time t' is

$$P(t, t') = \tau\, e^{-(t-t')/\tau}\, \theta(t - t') \tag{4.135}$$

in which $\theta(t - t') = (t - t' + |t - t'|)/2|t - t'|$ is the Heaviside function. One way to measure the lifetime τ of a fluorophore is to modulate the exciting laser beam at a frequency $\nu = 2\pi\omega$ of the order of 60 MHz and to detect the phase-shift ϕ in the light $L(t)$ emitted by the fluorophore. That light is the integral of $P(t, t')$ times the modulated beam $\sin \omega t$ or equivalently the convolution of $e^{-t/\tau}\theta(t)$ with $\sin \omega t$

$$\begin{aligned} L(t) &= \int_{-\infty}^{\infty} P(t, t')\, \sin(\omega t')\, dt' = \int_{-\infty}^{\infty} \tau\, e^{-(t-t')/\tau} \theta(t - t')\, \sin(\omega t')\, dt' \\ &= \int_{-\infty}^{t} \tau\, e^{-(t-t')/\tau}\, \sin(\omega t')\, dt'. \end{aligned} \tag{4.136}$$

Letting $u = t - t'$ and using the trigonometric formula

$$\sin(a - b) = \sin a \cos b - \cos a \sin b \tag{4.137}$$

we may relate this integral to the Laplace transforms of a sine (4.134) and a cosine (4.133)

$$\begin{aligned} L(t) &= -\tau \int_0^\infty e^{-u/\tau}\, \sin \omega(u - t)\, du \\ &= -\tau \int_0^\infty e^{-u/\tau}\, (\sin \omega u \cos \omega t - \cos \omega u \sin \omega t)\, du \\ &= \tau \left(\frac{\sin(\omega t)/\tau}{1/\tau^2 + \omega^2} - \frac{\omega \cos \omega t}{1/\tau^2 + \omega^2} \right). \end{aligned} \tag{4.138}$$

Setting $\cos \phi = (1/\tau)/\sqrt{1/\tau^2 + \omega^2}$ and $\sin \phi = \omega/\sqrt{1/\tau^2 + \omega^2}$, we have

$$L(t) = \frac{\tau}{\sqrt{1/\tau^2 + \omega^2}}\, (\sin \omega t \cos \phi - \cos \omega t \sin \phi) = \frac{\tau}{\sqrt{1/\tau^2 + \omega^2}}\, \sin(\omega t - \phi). \tag{4.139}$$

The phase-shift ϕ then is given by

$$\phi = \arcsin \frac{\omega}{\sqrt{1/\tau^2 + \omega^2}} \le \frac{\pi}{2}. \tag{4.140}$$

So by inverting this formula, we get the lifetime of the fluorophore

$$\tau = (1/\omega)\, \tan \phi \tag{4.141}$$

in terms of the phase-shift ϕ which is much easier to measure.

4.10 Derivatives and Integrals of Laplace Transforms

The derivatives of a Laplace transform $f(s)$ are by its definition (4.128)

$$\frac{d^n f(s)}{ds^n} = \int_0^\infty dt\, (-t)^n\, e^{-st}\, F(t). \tag{4.142}$$

They usually are well defined if $f(s)$ is well defined. For instance, if we differentiate the Laplace transform (4.129) of the function $F(t) = 1$ which is $f(s) = 1/s$, then we get

$$(-1)^n \frac{d^n s^{-1}}{ds^n} = \frac{n!}{s^{n+1}} = \int_0^\infty dt\, e^{-st}\, t^n \tag{4.143}$$

which tells us that the Laplace transform of t^n is $n!/s^{n+1}$.

The result of differentiating the function $F(t)$ also has a simple form. Integrating by parts, we find for the Laplace transform of $F'(t)$

$$\begin{aligned}\int_0^\infty dt\, e^{-st}\, F'(t) &= \int_0^\infty dt \left\{ \frac{d}{dt}\left[e^{-st}\, F(t)\right] - F(t) \frac{d}{dt} e^{-st} \right\} \\ &= -\, F(0) + \int_0^\infty dt\, F(t)\, s\, e^{-st} \\ &= -\, F(0) + s\, f(s)\end{aligned} \tag{4.144}$$

as long as $e^{-st} F(t) \to 0$ as $t \to \infty$.

The indefinite integral of the Laplace transform (4.128) is

$${}^{(1)}f(s) \equiv \int ds_1\, f(s_1) = \int_0^\infty dt\, \frac{e^{-st}}{(-t)}\, F(t) \tag{4.145}$$

and its nth indefinite integral is

$${}^{(n)}f(s) \equiv \int ds_n \cdots \int ds_1\, f(s_1) = \int_0^\infty dt\, \frac{e^{-st}}{(-t)^n}\, F(t). \tag{4.146}$$

If $f(s)$ is a well-behaved function, then these indefinite integrals usually are well defined for $s > 0$ as long as $F(t) \to 0$ suitably as $t \to 0$.

4.11 Laplace Transforms and Differential Equations

Suppose we wish to solve the differential equation

$$P(d/ds)\, f(s) = j(s). \tag{4.147}$$

By writing $f(s)$ and $j(s)$ as Laplace transforms

$$f(s) = \int_0^\infty e^{-st}\, F(t)\, dt \quad \text{and} \quad j(s) = \int_0^\infty e^{-st}\, J(t)\, dt \tag{4.148}$$

and using the formula (4.142) for the nth derivative of a Laplace transform, we see that the differential equation (4.147) amounts to

$$P(d/ds)\, f(s) = \int_0^\infty e^{-st}\, P(-t)\, F(t)\, dt = \int_0^\infty e^{-st}\, J(t)\, dt \tag{4.149}$$

which is equivalent to the algebraic equation

$$F(t) = \frac{J(t)}{P(-t)}. \tag{4.150}$$

A particular solution to the inhomogeneous differential equation (4.147) is then the Laplace transform of this ratio

$$f(s) = \int_0^\infty e^{-st}\, \frac{J(t)}{P(-t)}\, dt. \tag{4.151}$$

A fairly general solution of the associated homogeneous equation

$$P(d/ds)\, f(s) = 0 \tag{4.152}$$

is the Laplace transform

$$f(s) = \int_0^\infty e^{-st}\, \delta(P(-t))\, H(t)\, dt \tag{4.153}$$

because

$$P(d/ds)\, f(s) = \int_0^\infty e^{-st}\, P(-t)\, \delta(P(-t))\, H(t)\, dt = 0 \tag{4.154}$$

as long as the function $H(t)$ is suitably smooth but otherwise arbitrary. Thus our solution of the inhomogeneous equation (4.147) is the sum of the two

$$f(s) = \int_0^\infty e^{-st}\, \frac{J(t)}{P(-t)}\, dt + \int_0^\infty e^{-st}\, \delta(P(-t))\, H(t)\, dt. \tag{4.155}$$

One may generalize this method to differential equations in n variables. But to carry out this procedure, one must be able to find the inverse Laplace transform $J(t)$ of the source function $j(s)$ as outlined in the next section.

4.12 Inversion of Laplace Transforms

How do we invert the Laplace transform

$$f(s) = \int_0^\infty dt\, e^{-st}\, F(t)? \tag{4.156}$$

First we extend the Laplace transform from real s to $s + iu$

$$f(s+iu) = \int_0^\infty dt\, e^{-(s+iu)t}\, F(t) \tag{4.157}$$

and choose s to be sufficiently positive that $f(s + iu)$ is suitably smooth and bounded. Then we apply the delta-function formula (4.36) to the integral

$$\begin{aligned}
\int_{-\infty}^{\infty} \frac{du}{2\pi}\, e^{iut}\, f(s+iu) &= \int_{-\infty}^{\infty} \frac{du}{2\pi} \int_0^\infty dt'\, e^{iut}\, e^{-(s+iu)t'}\, F(t') \\
&= \int_0^\infty dt'\, e^{-st'}\, F(t') \int_{-\infty}^{\infty} \frac{du}{2\pi}\, e^{iu(t-t')} \\
&= \int_0^\infty dt'\, e^{-st'}\, F(t')\, \delta(t-t') = e^{-st}\, F(t).
\end{aligned} \tag{4.158}$$

So our inversion formula is

$$F(t) = e^{st} \int_{-\infty}^{\infty} \frac{du}{2\pi}\, e^{iut}\, f(s+iu) \tag{4.159}$$

for sufficiently large s. Some call this inversion formula a Bromwich integral, others a Fourier–Mellin integral.

4.13 Application to Differential Equations

Let us consider a linear partial differential equation in n variables

$$P(\partial_1, \ldots, \partial_n) f(x_1, \ldots, x_n) = g(x_1, \ldots, x_n) \tag{4.160}$$

in which P is a polynomial in the derivatives

$$\partial_j \equiv \frac{\partial}{\partial x_j} \tag{4.161}$$

with constant coefficients. If $g = 0$, the equation is homogeneous; otherwise it is inhomogeneous. We expand the solution and source as integral transforms

$$\begin{aligned}
f(x_1, \ldots, x_n) &= \int \tilde{f}(k_1, \ldots, k_n)\, e^{i(k_1 x_1 + \cdots + k_n x_n)}\, d^n k \\
g(x_1, \ldots, x_n) &= \int \tilde{g}(k_1, \ldots, k_n)\, e^{i(k_1 x_1 + \cdots + k_n x_n)}\, d^n k
\end{aligned} \tag{4.162}$$

in which the k integrals may run from $-\infty$ to ∞ as in a Fourier transform or up the imaginary axis from 0 to ∞ as in a Laplace transform.

The correspondence (4.58) between differentiation with respect to x_j and multiplication by ik_j tells us that ∂_j^m acting on f gives

$$\partial_j^m f(x_1, \ldots, x_n) = \int \tilde{f}(k_1, \ldots, k_n)\,(ik_j)^m\, e^{i(k_1x_1+\cdots+k_nx_n)}\, d^n k. \tag{4.163}$$

If we abbreviate $f(x_1, \ldots, x_n)$ by $f(x)$ and do the same for g, then we may write our partial differential equation (4.160) as

$$\begin{aligned} P(\partial_1, \ldots, \partial_n) f(x) &= \int \tilde{f}(k)\, P(ik_1, \ldots, ik_n)\, e^{i(k_1x_1+\cdots+k_nx_n)}\, d^n k \\ &= \int \tilde{g}(k)\, e^{i(k_1x_1+\cdots+k_nx_n)}\, d^n k. \end{aligned} \tag{4.164}$$

Thus the inhomogeneous partial differential equation

$$P(\partial_1, \ldots, \partial_n) f_i(x_1, \ldots, x_n) = g(x_1, \ldots, x_n) \tag{4.165}$$

becomes an algebraic equation in k-space

$$P(ik_1, \ldots, ik_n)\, \tilde{f}_i(k_1, \ldots, k_n) = \tilde{g}(k_1, \ldots, k_n) \tag{4.166}$$

where $\tilde{g}(k_1, \ldots, k_n)$ is the mixed Fourier–Laplace transform of $g(x_1, \ldots, x_n)$. So one solution of the inhomogeneous differential equation (4.160) is

$$f_i(x_1, \ldots, x_n) = \int e^{i(k_1x_1+\cdots+k_nx_n)}\, \frac{\tilde{g}(k_1, \ldots, k_n)}{P(ik_1, \ldots, ik_n)}\, d^n k. \tag{4.167}$$

The space of solutions to the **homogeneous** form of equation (4.160)

$$P(\partial_1, \ldots, \partial_n) f_h(x_1, \ldots, x_n) = 0 \tag{4.168}$$

is vast. We will focus on those that satisfy the algebraic equation

$$P(ik_1, \ldots, ik_n)\, \tilde{f}_h(k_1, \ldots, k_n) = 0 \tag{4.169}$$

and that we can write in terms of Dirac's delta function as

$$\tilde{f}_h(k_1, \ldots, k_n) = \delta(P(ik_1, \ldots, ik_n))\, h(k_1, \ldots, k_n) \tag{4.170}$$

in which the function $h(k)$ is arbitrary. That is

$$f_h(x) = \int e^{i(k_1x_1+\cdots+k_nx_n)} \delta(P(ik_1, \ldots, ik_n))\, h(k)\, d^n k. \tag{4.171}$$

Our solution to the differential equation (4.160) then is a sum of a particular solution (4.167) of the inhomogeneous equation (4.166) and our solution (4.171) of the associated homogeneous equation (4.168)

$$\begin{aligned} f(x_1,\ldots,x_n) = \int e^{i(k_1x_1+\cdots+k_nx_n)} \Bigg[& \frac{\tilde g(k_1,\ldots,k_n)}{P(ik_1,\ldots,ik_n)} \\ & + \delta(P(ik_1,\ldots,ik_n))\, h(k_1,\ldots,k_n) \Bigg] d^n k \end{aligned} \tag{4.172}$$

in which $h(k_1,\ldots,k_n)$ is an arbitrary function. The wave equation and the diffusion equation will provide examples of this formula

$$f(x) = \int e^{ik\cdot x} \left[\frac{\tilde g(k)}{P(ik)} + \delta(P(ik))h(k) \right] d^n k. \tag{4.173}$$

Example 4.14 (Wave equation for a scalar field) A free scalar field $\phi(x)$ of mass m in flat spacetime obeys the wave equation

$$\left(\nabla^2 - \partial_t^2 - m^2\right)\phi(x) = 0 \tag{4.174}$$

in natural units ($\hbar = c = 1$). We may use a 4-dimensional Fourier transform to represent the field $\phi(x)$ as

$$\phi(x) = \int e^{ik\cdot x}\, \tilde\phi(k)\, \frac{d^4k}{(2\pi)^2} \tag{4.175}$$

in which $k \cdot x = \boldsymbol{k} \cdot \boldsymbol{x} - k^0 t$ is the Lorentz-invariant inner product.

The homogeneous wave equation (4.174) then says

$$\left(\nabla^2 - \partial_t^2 - m^2\right)\phi(x) = \int \left(-\boldsymbol{k}^2 + (k^0)^2 - m^2\right) e^{ik\cdot x}\, \tilde\phi(k)\, \frac{d^4k}{(2\pi)^2} = 0 \tag{4.176}$$

which implies the algebraic equation

$$\left(-\boldsymbol{k}^2 + (k^0)^2 - m^2\right)\tilde\phi(k) = 0 \tag{4.177}$$

an instance of (4.169). Our solution (4.171) is

$$\phi(x) = \int \delta\left(-\boldsymbol{k}^2 + (k^0)^2 - m^2\right) e^{ik\cdot x}\, h(k)\, \frac{d^4k}{(2\pi)^2} \tag{4.178}$$

in which $h(k)$ is an arbitrary function. The argument of the delta function

$$P(ik) = (k^0)^2 - \boldsymbol{k}^2 - m^2 = \left(k^0 - \sqrt{\boldsymbol{k}^2 + m^2}\right)\left(k^0 + \sqrt{\boldsymbol{k}^2 + m^2}\right) \tag{4.179}$$

has zeros at $k^0 = \pm\sqrt{\boldsymbol{k}^2 + m^2} \equiv \pm\omega_k$ with

$$\left|\frac{dP(\pm\omega_k)}{dk^0}\right| = 2\omega_k. \tag{4.180}$$

So using our formula (4.47) for integrals involving delta functions of functions, we have

$$\phi(x) = \int \left[e^{i(\boldsymbol{k}\cdot\boldsymbol{x}-\omega_k t)}\, h_+(\boldsymbol{k}) + e^{i(\boldsymbol{k}\cdot\boldsymbol{x}+\omega_k t)}\, h_-(\boldsymbol{k}) \right] \frac{d^3k}{(2\pi)^2 2\omega_k} \tag{4.181}$$

where $h_\pm(\boldsymbol{k}) \equiv h(\pm\omega_k, \boldsymbol{k})$. Since ω_k is an even function of $\boldsymbol{k}$, we can write

$$\phi(x) = \int \left[e^{i(\boldsymbol{k}\cdot\boldsymbol{x}-\omega_k t)}\, h_+(\boldsymbol{k}) + e^{-i(\boldsymbol{k}\cdot\boldsymbol{x}-\omega_k t)}\, h_-(-\boldsymbol{k}) \right] \frac{d^3k}{(2\pi)^2 2\omega_k}. \tag{4.182}$$

If $\phi(x) = \phi(\boldsymbol{x}, t)$ is a real-valued classical field, then its Fourier transform $h(\boldsymbol{k})$ must obey the relation (4.25) which says that $h_-(-\boldsymbol{k}) = h_+(\boldsymbol{k})^*$. If ϕ is a hermitian quantum field, then $h_-(-\boldsymbol{k}) = h_+^\dagger(\boldsymbol{k})$. In terms of the **annihilation** operator $a(\boldsymbol{k}) \equiv h_+(\boldsymbol{k})/\sqrt{4\pi\omega_k}$ and its adjoint $a^\dagger(\boldsymbol{k})$, a creation operator, the field $\phi(x)$ is the integral

$$\phi(x) = \int \left[e^{i(\boldsymbol{k}\cdot\boldsymbol{x}-\omega_k t)}\, a(\boldsymbol{k}) + e^{-i(\boldsymbol{k}\cdot\boldsymbol{x}-\omega_k t)}\, a^\dagger(\boldsymbol{k}) \right] \frac{d^3k}{\sqrt{(2\pi)^3 2\omega_k}}. \tag{4.183}$$

The momentum π canonically conjugate to the field is its time derivative

$$\pi(x) = -i \int \left[e^{i(\boldsymbol{k}\cdot\boldsymbol{x}-\omega_k t)}\, a(\boldsymbol{k}) - e^{-i(\boldsymbol{k}\cdot\boldsymbol{x}-\omega_k t)}\, a^\dagger(\boldsymbol{k}) \right] \sqrt{\frac{\omega_k}{2(2\pi)^3}}\, d^3k. \tag{4.184}$$

If the operators a and $a^\dagger$ obey the commutation relations

$$[a(\boldsymbol{k}), a^\dagger(\boldsymbol{k}')] = \delta(\boldsymbol{k} - \boldsymbol{k}') \quad \text{and} \quad [a(\boldsymbol{k}), a(\boldsymbol{k}')] = [a^\dagger(\boldsymbol{k}), a^\dagger(\boldsymbol{k}')] = 0 \tag{4.185}$$

then the field $\phi(\boldsymbol{x}, t)$ and its conjugate momentum $\pi(\boldsymbol{y}, t)$ satisfy (Exercise 4.17) the equal-time commutation relations

$$[\phi(\boldsymbol{x}, t), \pi(\boldsymbol{y}, t)] = i\delta(\boldsymbol{x} - \boldsymbol{y}) \quad \text{and} \quad [\phi(\boldsymbol{x}, t), \phi(\boldsymbol{y}, t)] = [\pi(\boldsymbol{x}, t), \pi(\boldsymbol{y}, t)] = 0 \tag{4.186}$$

which generalize the commutation relations of quantum mechanics

$$[q_j, p_\ell] = i\hbar\delta_{j,\ell} \quad \text{and} \quad [q_j, q_\ell] = [p_j, p_\ell] = 0 \tag{4.187}$$

for a set of coordinates q_j and conjugate momenta p_ℓ.

Example 4.15 (Fourier series for a scalar field) For a field defined in a cube of volume $V = L^3$, one often imposes periodic boundary conditions (Section 3.14) in which a displacement of any spatial coordinate by $\pm L$ does not change the value of the field. A Fourier series can represent a periodic field. Using the relationship (4.99) between Fourier-transform and Fourier-series representations in 3 dimensions, we expect the Fourier series representation for the field (4.183) to be

$$\phi(x) = \frac{(2\pi)^3}{V} \sum_{k} \frac{1}{\sqrt{(2\pi)^3 2\omega_k}} \left[a(\boldsymbol{k}) e^{i(\boldsymbol{k}\cdot\boldsymbol{x}-\omega_k t)} + a^\dagger(\boldsymbol{k}) e^{-i(\boldsymbol{k}\cdot\boldsymbol{x}-\omega_k t)} \right]$$

$$= \sum_{k} \frac{1}{\sqrt{2\omega_k V}} \sqrt{\frac{(2\pi)^3}{V}} \left[a(\boldsymbol{k}) e^{i(\boldsymbol{k}\cdot\boldsymbol{x}-\omega_k t)} + a^\dagger(\boldsymbol{k}) e^{-i(\boldsymbol{k}\cdot\boldsymbol{x}-\omega_k t)} \right] \quad (4.188)$$

in which the sum over $\boldsymbol{k} = (2\pi/L)(\ell, n, m)$ is over all (positive and negative) integers ℓ, n, and m. One can set

$$a_{\boldsymbol{k}} \equiv \sqrt{\frac{(2\pi)^3}{V}}\, a(\boldsymbol{k}) \quad (4.189)$$

and write the field as

$$\phi(x) = \sum_{k} \frac{1}{\sqrt{2\omega_k V}} \left[a_{\boldsymbol{k}}\, e^{i(\boldsymbol{k}\cdot\boldsymbol{x}-\omega_k t)} + a_{\boldsymbol{k}}^\dagger\, e^{-i(\boldsymbol{k}\cdot\boldsymbol{x}-\omega_k t)} \right]. \quad (4.190)$$

The commutator of Fourier-series annihilation and creation operators is by (4.36, 4.185, and 4.189)

$$[a_{\boldsymbol{k}}, a_{\boldsymbol{k}'}^\dagger] = \frac{(2\pi)^3}{V} [a(\boldsymbol{k}), a^\dagger(\boldsymbol{k}')] = \frac{(2\pi)^3}{V} \delta(\boldsymbol{k} - \boldsymbol{k}')$$
$$= \frac{(2\pi)^3}{V} \int e^{i(\boldsymbol{k}-\boldsymbol{k}')\cdot\boldsymbol{x}} \frac{d^3x}{(2\pi)^3} = \frac{(2\pi)^3}{V} \frac{V}{(2\pi)^3} \delta_{\boldsymbol{k},\boldsymbol{k}'} = \delta_{\boldsymbol{k},\boldsymbol{k}'} \quad (4.191)$$

in which the Kronecker delta $\delta_{\boldsymbol{k},\boldsymbol{k}'}$ is $\delta_{\ell,\ell'}\delta_{n,n'}\delta_{m,m'}$.

Example 4.16 (Diffusion) The flow rate $\boldsymbol{J}$ (per unit area, per unit time) of a fixed number of randomly moving particles, such as molecules of a gas or a liquid, is proportional to the negative gradient of their density $\rho(\boldsymbol{x}, t)$

$$\boldsymbol{J}(\boldsymbol{x}, t) = -D\, \nabla \rho(\boldsymbol{x}, t) \quad (4.192)$$

where D is the **diffusion constant**, an equation known as **Fick's law** (Adolf Fick 1829–1901). Since the number of particles is conserved, the 4-vector $J = (\rho, \boldsymbol{J})$ obeys the conservation law

$$\frac{\partial}{\partial t} \int \rho(\boldsymbol{x}, t)\, d^3\boldsymbol{x} = -\oint \boldsymbol{J}(\boldsymbol{x}, t) \cdot d\boldsymbol{a} = -\int \nabla \cdot \boldsymbol{J}(\boldsymbol{x}, t) d^3\boldsymbol{x} \quad (4.193)$$

which with Fick's law (4.192) gives the **diffusion equation**

$$\dot{\rho}(\boldsymbol{x}, t) = -\nabla \cdot \boldsymbol{J}(\boldsymbol{x}, t) = D\, \nabla^2 \rho(\boldsymbol{x}, t) \quad \text{or} \quad \left(D\, \nabla^2 - \partial_t \right) \rho(\boldsymbol{x}, t) = 0. \quad (4.194)$$

Fourier had in mind such equations when he invented his transform.

If we write the density $\rho(\boldsymbol{x}, t)$ as the transform

$$\rho(\boldsymbol{x}, t) = \int e^{i\boldsymbol{k}\cdot\boldsymbol{x}+i\omega t}\, \tilde{\rho}(\boldsymbol{k}, \omega)\, d^3\boldsymbol{k} d\omega \quad (4.195)$$

then the diffusion equation becomes

$$\left(D\,\nabla^2 - \partial_t\right)\rho(\boldsymbol{x},t) = \int e^{i\boldsymbol{k}\cdot\boldsymbol{x}+i\omega t}\left(-D\,\boldsymbol{k}^2 - i\omega\right)\tilde{\rho}(\boldsymbol{k},\omega)\,d^3\boldsymbol{k}\,d\omega = 0 \tag{4.196}$$

which implies the algebraic equation

$$\left(D\,\boldsymbol{k}^2 + i\omega\right)\tilde{\rho}(\boldsymbol{k},\omega) = 0. \tag{4.197}$$

Our solution (4.171) of this homogeneous equation is

$$\rho(\boldsymbol{x},t) = \int e^{i\boldsymbol{k}\cdot\boldsymbol{x}+i\omega t}\,\delta\left(-D\,\boldsymbol{k}^2 - i\omega\right)h(\boldsymbol{k},\omega)\,d^3\boldsymbol{k}\,d\omega \tag{4.198}$$

in which $h(\boldsymbol{k},\omega)$ is an arbitrary function. Dirac's delta function requires ω to be imaginary $\omega = i\,D\boldsymbol{k}^2$, with $D\boldsymbol{k}^2 > 0$. So the ω-integration is up the imaginary axis. It is a Laplace transform, and we have

$$\rho(\boldsymbol{x},t) = \int_{-\infty}^{\infty} e^{i\boldsymbol{k}\cdot\boldsymbol{x}-D\boldsymbol{k}^2 t}\,\tilde{\rho}(\boldsymbol{k})\,d^3\boldsymbol{k} \tag{4.199}$$

in which $\tilde{\rho}(\boldsymbol{k}) \equiv h(\boldsymbol{k}, i\,D\boldsymbol{k}^2)$. Thus the function $\tilde{\rho}(\boldsymbol{k})$ is the Fourier transform of the initial density $\rho(\boldsymbol{x},0)$

$$\rho(\boldsymbol{x},0) = \int_{-\infty}^{\infty} e^{i\boldsymbol{k}\cdot\boldsymbol{x}}\,\tilde{\rho}(\boldsymbol{k})\,d^3\boldsymbol{k}. \tag{4.200}$$

So if the initial density $\rho(\boldsymbol{x},0)$ is concentrated at $\boldsymbol{y}$

$$\rho(\boldsymbol{x},0) = \delta(\boldsymbol{x}-\boldsymbol{y}) = \int_{-\infty}^{\infty} e^{i\boldsymbol{k}\cdot(\boldsymbol{x}-\boldsymbol{y})}\,\frac{d^3\boldsymbol{k}}{(2\pi)^3} \tag{4.201}$$

then its Fourier transform $\tilde{\rho}(\boldsymbol{k})$ is

$$\tilde{\rho}(\boldsymbol{k}) = \frac{e^{-i\boldsymbol{k}\cdot\boldsymbol{y}}}{(2\pi)^3} \tag{4.202}$$

and at later times the density $\rho(\boldsymbol{x},t)$ is given by (4.199) as

$$\rho(\boldsymbol{x},t) = \int_{-\infty}^{\infty} e^{i\boldsymbol{k}\cdot(\boldsymbol{x}-\boldsymbol{y})-D\boldsymbol{k}^2 t}\,\frac{d^3\boldsymbol{k}}{(2\pi)^3}. \tag{4.203}$$

Using our formula (4.19) for the Fourier transform of a gaussian, we find

$$\rho(\boldsymbol{x},t) = \frac{1}{(4\pi D t)^{3/2}}\,e^{-(\boldsymbol{x}-\boldsymbol{y})^2/(4Dt)}. \tag{4.204}$$

Since the diffusion equation is linear, it follows (Exercise 4.18) that an arbitrary initial distribution $\rho(\boldsymbol{y},0)$ evolves to the convolution (Section 4.6)

$$\rho(\boldsymbol{x},t) = \frac{1}{(4\pi D t)^{3/2}}\int e^{-(\boldsymbol{x}-\boldsymbol{y})^2/(4Dt)}\,\rho(\boldsymbol{y},0)\,d^3\boldsymbol{y}. \tag{4.205}$$

Exercises

4.1 Show that the Fourier integral formula (4.26) for real functions follows from (4.9) and (4.25).

4.2 Show that the Fourier integral formula (4.26) for real functions implies (4.27) if f is even and (4.28) if it is odd.

4.3 Derive the formula (4.30) for the square wave (4.29).

4.4 By using the Fourier-transform formulas (4.27 and 4.28), derive the formulas (4.31) and (4.32) for the even and odd extensions of the exponential $\exp(-\beta|x|)$.

4.5 For the state $|\psi, t\rangle$ given by Eqs. (4.83 and 4.88), find the wave function $\psi(x,t) = \langle x|\psi, t\rangle$ at time t. Then find the variance of the position operator at that time. Does it grow as time goes by? How?

4.6 At time $t = 0$, a particle of mass m is in a gaussian superposition of momentum eigenstates centered at $p = \hbar K$

$$\psi(x,0) = N\int_{-\infty}^{\infty} e^{ikx} e^{-L^2(k-K)^2}\, dk. \tag{4.206}$$

(a) Shift k by K and do the integral. Where is the particle most likely to be found? (b) At time t, the wave function $\psi(x,t)$ is $\psi(x,0)$ but with ikx replaced by $ikx - i\hbar k^2 t/2m$. Shift k by K and do the integral. Where is the particle most likely to be found? (c) Does the wave packet spread out like t or like $\sqrt{t}$ as in classical diffusion?

4.7 Express the characteristic function (4.90) of a probability distribution $P(x)$ as its Fourier transform.

4.8 Express the characteristic function (4.90) of a probability distribution as a power series in its moments (4.92).

4.9 Find the characteristic function (4.90) of the gaussian probability distribution

$$P_G(x,\mu,\sigma) = \frac{1}{\sigma\sqrt{2\pi}} \exp\left(-\frac{(x-\mu)^2}{2\sigma^2}\right). \tag{4.207}$$

4.10 Find the moments $\mu_n = E[x^n]$ for $n = 0, \ldots, 3$ of the gaussian probability distribution $P_G(x,\mu,\sigma)$.

4.11 Derive (4.115) from $\boldsymbol{B} = \nabla \times \boldsymbol{A}$ and Ampère's law $\nabla \times \boldsymbol{B} = \mu_0 \boldsymbol{J}$.

4.12 Derive (4.116) from (4.115).

4.13 Derive (4.117) from (4.116).

4.14 Use the Green's function relations (4.110) and (4.111) to show that (4.117) satisfies (4.115).

4.15 Show that the Laplace transform of t^{z-1} is the gamma function (5.58) divided by s^z

$$f(s) = \int_0^\infty e^{-st}\, t^{z-1}\, dt = s^{-z}\, \Gamma(z). \tag{4.208}$$

4.16 Compute the Laplace transform of $1/\sqrt{t}$. Hint: let $t = u^2$.

4.17 Show that the commutation relations (4.185) of the annihilation and creation operators imply the equal-time commutation relations (4.186) for the field ϕ and its conjugate momentum π.

4.18 Use the linearity of the diffusion equation and Equations (4.201–4.204) to derive the general solution (4.205) of the diffusion equation.

5

Infinite Series

5.1 Convergence

A sequence of partial sums

$$S_N = \sum_{n=0}^{N} c_n \tag{5.1}$$

converges to a number S if for every $\epsilon > 0$, there exists an integer $N(\epsilon)$ such that

$$|S - S_N| < \epsilon \quad \text{for all} \quad N > N(\epsilon). \tag{5.2}$$

The number S is then said to be the limit of the **convergent** infinite series

$$S = \sum_{n=0}^{\infty} c_n = \lim_{N\to\infty} S_N = \lim_{N\to\infty} \sum_{n=0}^{N} c_n. \tag{5.3}$$

Some series converge; others wander or **oscillate**; and others **diverge**.

A series whose absolute values converges

$$S = \sum_{n=0}^{\infty} |c_n| \tag{5.4}$$

is said to converge **absolutely**. A convergent series that is not absolutely convergent is said to converge **conditionally**.

Example 5.1 (Two infinite series) The series of inverse factorials converges to the number $e = 2.718281828\ldots$

$$\sum_{n=0}^{\infty} \frac{1}{n!} = e. \tag{5.5}$$

But the harmonic series of inverse integers diverges

$$\sum_{k=1}^{\infty} \frac{1}{k} \to \infty \tag{5.6}$$

as one may see by grouping its terms

$$1+\frac{1}{2}+\left(\frac{1}{3}+\frac{1}{4}\right)+\left(\frac{1}{5}+\frac{1}{6}+\frac{1}{7}+\frac{1}{8}\right)+\cdots \geq 1+\frac{1}{2}+\frac{1}{2}+\frac{1}{2}+\cdots \tag{5.7}$$

to form a series that obviously diverges. This series up to $1/n$ approaches the natural logarithm $\ln n$ to within a constant

$$\gamma = \lim_{n\to\infty}\left(\sum_{k=1}^{n}\frac{1}{k} - \ln n\right) = 0.5772156649\ldots \tag{5.8}$$

known as the Euler–Mascheroni constant (Leonhard Euler 1707–1783, Lorenzo Mascheroni 1750–1800).

Example 5.2 (Geometric series) For any positive integer N, the identity

$$(1-z)(1+z+z^2+\cdots+z^N) = 1-z^{N+1} \tag{5.9}$$

implies that

$$S_N(z) = \sum_{n=0}^{N} z^n = \frac{1-z^{N+1}}{1-z}. \tag{5.10}$$

For $|z| < 1$, the term $|z^{N+1}| \to 0$ as $N \to \infty$, and so the geometric series $S_N(z)$ converges to

$$S(z) = \sum_{n=0}^{\infty} z^n = \frac{1}{1-z} \tag{5.11}$$

as long as the absolute-value of z is less than unity. A useful approximation for $|z| \ll 1$ is

$$\frac{1}{1\pm z} \approx 1 \mp z. \tag{5.12}$$

5.2 Tests of Convergence

The **Cauchy criterion** for the convergence of a sequence S_N is that for every $\epsilon > 0$ there is an integer $N(\epsilon)$ such that for $N > N(\epsilon)$ and $M > N(\epsilon)$ one has

$$|S_N - S_M| < \epsilon. \tag{5.13}$$

Cauchy's criterion is equivalent to the defining condition (5.2).

The **comparison test**: Suppose the convergent series

$$\sum_{n=0}^{\infty} b_n \tag{5.14}$$

has only positive terms $b_n \geq 0$, and that $|c_n| \leq b_n$ for all n. Then the series

$$\sum_{n=0}^{\infty} c_n \tag{5.15}$$

converges absolutely.

Similarly, if for all n, the inequality $0 \leq c_n \leq b_n$ holds and the series of numbers c_n diverges, then so does the series of numbers b_n.

The **Cauchy root test**: If for some N, the terms c_n satisfy

$$|c_n|^{1/n} \leq x < 1 \tag{5.16}$$

for all $n > N$, then for such n we have $|c_n| \leq x^n$ and therefore since the geometric series (5.11) converges for $|z| < 1$, we know that

$$\sum_{n=0}^{\infty} x^n = \frac{1}{1-x}. \tag{5.17}$$

Thus by the comparison test (5.14 and 5.15), the series

$$\sum_{n=0}^{\infty} c_n \tag{5.18}$$

converges absolutely.

The **ratio test of d'Alembert**: The series $\sum_n c_n$ converges if

$$\lim_{n\to\infty} \left| \frac{c_{n+1}}{c_n} \right| = r < 1 \tag{5.19}$$

and diverges if $r > 1$.

The **integral test**: If the terms c_n are positive and monotonically decreasing, $0 \leq c_{n+1} \leq c_n$, and $f(x)$ is a monotonically decreasing function with $f(n) = c_n$, then

$$c_{n+1} \leq \int_n^{n+1} f(x)\, dx \leq c_n, \tag{5.20}$$

and so by the comparison test (5.14 and 5.15), the series converges (diverges) according to whether the integral

$$\int_N^{\infty} f(x)\, dx \tag{5.21}$$

converges (diverges) for suitably large but fixed N.

The **sum** test: One can use Fortran, C, Python, Sage, Matlab, or Mathematica to sum the first N terms of one's series for $N = 100$, $N = 10,000$, $N = 1,000,000$, and so forth as seems appropriate. For simple series, Mathematica's Sum and

NSum commands work, as in sumZeta.nb. For more complicated series, it may be easier to write one's own code. For instance, the Fortran program doSum.f95 sums $\sin(\sin(n))/(n\log(n))$ from 2 to 10^9 and gets 0.4333495. The codes sumZeta.nb and doSum.f95 as well as the Matlab scripts of this chapter are in the GitHub repository Infinite_series at `github.com/kevinecahill`. The sum test has the advantage that one gets partial sums of the series, but if the series converges or diverges slowly, the partial sums may be tricky to interpret.

5.3 Convergent Series of Functions

A sequence of partial sums

$$S_N(z) = \sum_{n=0}^{N} f_n(z) \tag{5.22}$$

of functions $f_n(z)$ **converges** to a function $S(z)$ on a set D if for every $\epsilon > 0$ and every $z \in D$, there exists an integer $N(\epsilon, z)$ such that

$$|S(z) - S_N(z)| < \epsilon \quad \text{for all} \quad N > N(\epsilon, z). \tag{5.23}$$

The numbers z may be real or complex. The function $S(z)$ is said to be the limit on D of the **convergent** infinite series of functions

$$S(z) = \sum_{n=0}^{\infty} f_n(z). \tag{5.24}$$

A sequence of partial sums $S_N(z)$ of functions converges **uniformly** on the set D if the integers $N(\epsilon, z)$ can be chosen independently of the point $z \in D$, that is, if for every $\epsilon > 0$ and every $z \in D$, there exists an integer $N(\epsilon)$ such that

$$|S(z) - S_N(z)| < \epsilon \quad \text{for all} \quad N > N(\epsilon). \tag{5.25}$$

The limit (3.65) of the integral over a closed interval $a \le x \le b$ of a uniformly convergent sequence of partial sums $S_N(x)$ of continuous functions is equal to the integral of the limit

$$\lim_{N\to\infty} \int_a^b S_N(x)\, dx = \int_a^b S(x)\, dx. \tag{5.26}$$

A real or complex-valued function $f(x)$ of a real variable x is **square integrable** on an interval $[a, b]$ if the Riemann or Lebesgue (3.81) integral

$$\int_a^b |f(x)|^2\, dx \tag{5.27}$$

exists and is finite. A sequence of partial sums

$$S_N(x) = \sum_{n=0}^{N} f_n(x) \tag{5.28}$$

of square-integrable functions $f_n(x)$ **converges in the mean** to a function $S(x)$ if

$$\lim_{N\to\infty} \int_a^b |S(x) - S_N(x)|^2 \, dx = 0. \tag{5.29}$$

Convergence in the mean sometimes is defined as

$$\lim_{N\to\infty} \int_a^b \rho(x) \, |S(x) - S_N(x)|^2 \, dx = 0 \tag{5.30}$$

in which $\rho(x) \geq 0$ is a weight function that is positive except at isolated points where it may vanish. If the functions f_n are real, then this definition of convergence in the mean is more simply

$$\lim_{N\to\infty} \int_a^b \rho(x) \, (S(x) - S_N(x))^2 \, dx = 0. \tag{5.31}$$

5.4 Power Series

A power series is a series of functions with $f_n(z) = c_n \, z^n$

$$S(z) = \sum_{n=0}^{\infty} c_n \, z^n. \tag{5.32}$$

By the ratio test (5.19), this power series converges if

$$\lim_{n\to\infty} \left| \frac{c_{n+1} z^{n+1}}{c_n z^n} \right| = |z| \lim_{n\to\infty} \left| \frac{c_{n+1}}{c_n} \right| \equiv \frac{|z|}{R} < 1 \tag{5.33}$$

that is, if z lies within a circle of radius R

$$|z| < R \tag{5.34}$$

given by

$$R = \left(\lim_{n\to\infty} \frac{|c_{n+1}|}{|c_n|} \right)^{-1}. \tag{5.35}$$

Within this circle, the convergence is uniform and absolute.

Example 5.3 (Ratio test) The geometric series (5.11) is a power series with $c_n = 1$ and $R = 1$. Thus, by the ratio test, it converges for $|z| < 1$

$$S(z) = \sum_{n=0}^{\infty} z^n = \frac{1}{1-z}. \tag{5.36}$$

Example 5.4 (Credit) A person who deposits \$100 in a bank has a **credit** of \$100. Suppose banks are required to retain as reserves 10% of their deposits and are free to lend the other 90%. Then the bank getting the \$100 deposit can lend out \$90 to a borrower. That borrower can deposit \$90 in another bank. That bank can then lend \$81 to another borrower. Now three people have credits of \$100 + \$90 + \$81 = \$271. This multiplication of money is the miracle of credit.

If P is the original deposit and r is the fraction of deposits that banks must retain as reserves, then the total credit due to P can be as much as

$$P + P(1-r) + P(1-r)^2 + \cdots = P\sum_{n=0}^{\infty}(1-r)^n = P\left(\frac{1}{1-(1-r)}\right) = \frac{P}{r}. \tag{5.37}$$

An initial deposit of $P = \$100$ with $r = 10\%$ can produce total credits of $P/r = \$1000$. A reserve requirement of $r = 1\%$ can lead to total credits of \$10,000. Since banks charge a higher rate of interest on money they lend than the rate they pay to their depositors, bank profits soar as $r \to 0$. This is why bankers love deregulation.

The funds all the banks hold in reserve due to a deposit P is

$$Pr + (1-r)Pr + (1-r)^2 Pr + \cdots = Pr\sum_{n=0}^{\infty}(1-r)^n = P \tag{5.38}$$

P itself.

5.5 Factorials and the Gamma Function

For any positive integer n, the product

$$n! \equiv n(n-1)(n-2)\cdots 3 \cdot 2 \cdot 1 \tag{5.39}$$

is ***n*-factorial**, with zero-factorial defined as unity

$$0! \equiv 1. \tag{5.40}$$

To estimate $n!$, one can use **Stirling's approximation**

$$n! \approx \sqrt{2\pi\, n}\ (n/e)^n \tag{5.41}$$

derived in example 6.47 or **Ramanujan's correction** to it

$$n! \approx \sqrt{2\pi\, n}\ (n/e)^n \left(1 + 1/2n + 1/8n^2\right)^{1/6} \tag{5.42}$$

or Mermin's first

$$n! \approx \sqrt{2\pi\, n}\, \left(\frac{n}{e}\right)^n \exp\left(\frac{1}{12\, n}\right) \tag{5.43}$$

or second approximation

$$n! \approx \sqrt{2\pi\, n}\, \left(\frac{n}{e}\right)^n \exp\left(\frac{1}{12\, n} - \frac{1}{360\, n^3} + \frac{1}{1260\, n^5}\right) \tag{5.44}$$

which follow from his exact infinite-product formula

$$n! = \sqrt{2\pi\, n}\, \left(\frac{n}{e}\right)^n \prod_{j=1}^{\infty} \frac{(1+1/j)^{j+1/2}}{e}. \tag{5.45}$$

Figure 5.1 plots the relative error of these estimates $E(n!)$ of $n!$

$$10^8 \left(\frac{E(n!) - n!}{n!}\right) \tag{5.46}$$

magnified by 10^8, except for Stirling's formula (5.41), whose relative error is off the chart. Mermin's second approximation (5.43) is the most accurate, followed by

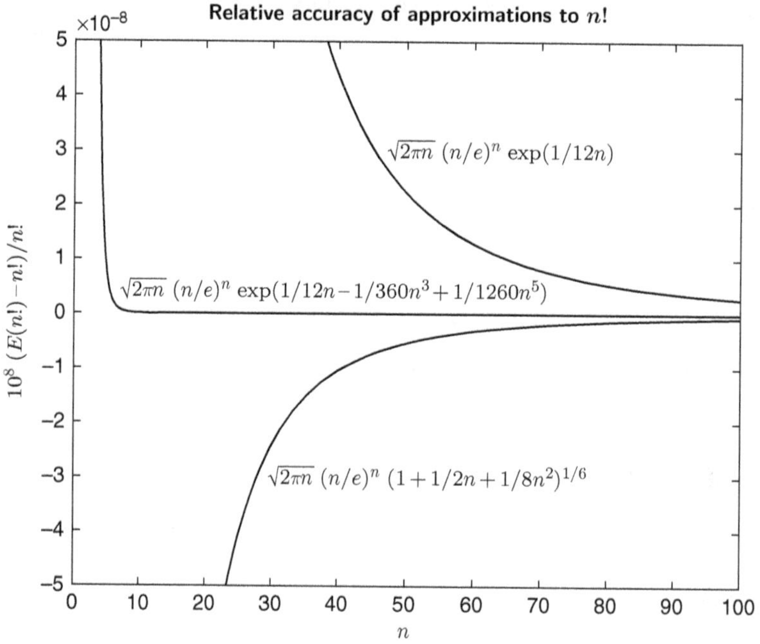

Figure 5.1 The magnified relative error $10^8[E(n!) - n!]/n!$ of Ramanujan's (5.42) and Mermin's (5.43 and 5.44) estimates $E(n!)$ of $n!$ are plotted for $n = 1, 2, \ldots, 100$. This chapter's programs and scripts are in Infinite_series at `github.com/kevinecahill`.

Ramanujan's correction (5.42), and by Mermin's first approximation (5.43) (James Stirling 1692–1770, Srinivasa Ramanujan 1887–1920, N. David Mermin 1935–).

The **binomial coefficient** is a ratio of factorials

$$\binom{n}{k} \equiv \frac{n!}{k!\,(n-k)!}. \tag{5.47}$$

Example 5.5 (Leibniz rule) We can use the notation

$$f^{(n)}(x) \equiv \frac{d^n}{dx^n} f(x) \tag{5.48}$$

to state Leibniz's rule for differentiating the product of two functions:

$$\frac{d^n}{dx^n}\left[f(x)\,g(x)\right] = \sum_{k=0}^{n} \binom{n}{k} f^{(k)}(x)\,g^{(n-k)}(x). \tag{5.49}$$

One may use mathematical induction to prove this rule, which is obviously true for $n = 0$ and $n = 1$ (Exercise 5.4) (Gottfried Leibniz, 1646–1716).

Example 5.6 (Exponential function) The power series with coefficients $c_n = 1/n!$ defines the exponential function

$$e^z = \sum_{n=0}^{\infty} \frac{z^n}{n!}. \tag{5.50}$$

Formula (5.35) shows that the radius of convergence R of this power series is infinite

$$R = \left(\lim_{n\to\infty} \frac{|c_{n+1}|}{|c_n|}\right)^{-1} = \left(\lim_{n\to\infty} \frac{1}{n+1}\right)^{-1} = \infty. \tag{5.51}$$

The series converges uniformly and absolutely inside every circle.

Example 5.7 (Bessel's series) For any integer n, the series

$$\begin{aligned} J_n(\rho) &= \frac{\rho^n}{2^n n!}\left[1 - \frac{\rho^2}{2(2n+2)} + \frac{\rho^4}{2\cdot 4(2n+2)(2n+4)} - \cdots\right] \\ &= \left(\frac{\rho}{2}\right)^n \sum_{m=0}^{\infty} \frac{(-1)^m}{m!(m+n)!}\left(\frac{\rho}{2}\right)^{2m} \end{aligned} \tag{5.52}$$

defines the cylindrical Bessel function of the first kind, which is finite at the origin $\rho = 0$. This series converges even faster (Exercise 5.5) than the one (5.50) for the exponential function.

Double factorials also are useful and are defined as

$$(2n-1)!! \equiv (2n-1)(2n-3)(2n-5)\cdots 1 \tag{5.53}$$

$$(2n)!! \equiv 2n(2n-2)(2n-4)\cdots 2 \tag{5.54}$$

with 0!! and (−1)!! both defined as unity

$$0!! = (-1)!! = 1. \tag{5.55}$$

Thus $5!! = 5 \cdot 3 \cdot 1 = 15$, and $6!! = 6 \cdot 4 \cdot 2 = 48$.

One may extend the definition (5.39) of n-factorial from positive integers to complex numbers by means of the integral formula

$$z! \equiv \int_0^\infty e^{-t}\, t^z\, dt \tag{5.56}$$

for $\mathrm{Re}\, z > -1$. In particular

$$0! = \int_0^\infty e^{-t}\, dt = 1 \tag{5.57}$$

which explains the definition (5.40). The factorial function $(z-1)!$ in turn defines the **gamma function** for $\mathrm{Re}\, z > 0$ as

$$\Gamma(z) = \int_0^\infty e^{-t}\, t^{z-1}\, dt = (z-1)! \tag{5.58}$$

or equivalently as

$$z! = \Gamma(z+1). \tag{5.59}$$

By differentiating the definition (5.58) of the gamma function and integrating by parts, we see that the gamma function satisfies the key identity

$$\begin{aligned}\Gamma(z+1) &= \int_0^\infty \left(-\frac{d}{dt} e^{-t}\right) t^z\, dt = \int_0^\infty e^{-t} \left(\frac{d}{dt} t^z\right) dt = \int_0^\infty e^{-t}\, z\, t^{z-1}\, dt \\ &= z\, \Gamma(z).\end{aligned} \tag{5.60}$$

Iterating, we get

$$\Gamma(z+n) = (z+n-1)\Gamma(z+n-1) = (z+n-1)(z+n-2)\cdots z\,\Gamma(z). \tag{5.61}$$

Thus for any complex z and any integers n and m with $n > m$, ratios of factorials work as one would think

$$\frac{\Gamma(z+n)}{\Gamma(z+m)} = \frac{(z+n-1)!}{(z+m-1)!} = (z+n-1)\cdots(z+m) \tag{5.62}$$

as long as $z+m$ avoids the negative integers.

We can use the identities (5.60 and 5.61) to extend the definition (5.58) of the gamma function in unit steps into the left half-plane

$$\Gamma(z) = \frac{1}{z}\Gamma(z+1) = \frac{1}{z}\frac{1}{z+1}\Gamma(z+2) = \frac{1}{z}\frac{1}{z+1}\frac{1}{z+2}\Gamma(z+3) = \cdots \tag{5.63}$$

as long as we avoid the negative integers and zero. This extension leads to Euler's definition

$$\Gamma(z) = \lim_{n\to\infty} \frac{1\cdot 2\cdot 3\cdots n}{z(z+1)(z+2)\cdots(z+n)}\, n^z \tag{5.64}$$

and to Weierstrass's (Exercise 5.6)

$$\Gamma(z) = \frac{1}{z}\, e^{-\gamma z} \left[\prod_{n=1}^{\infty} \left(1 + \frac{z}{n}\right) e^{-z/n} \right]^{-1} \tag{5.65}$$

(Karl Theodor Wilhelm Weierstrass, 1815–1897), and is an example of analytic continuation (Section 6.12).

One may show (Exercise 5.8) that another formula for $\Gamma(z)$ is

$$\Gamma(z) = 2 \int_0^\infty e^{-t^2} t^{2z-1}\, dt \tag{5.66}$$

for $\mathrm{Re}\, z > 0$ and that

$$\Gamma(n + \tfrac{1}{2}) = \frac{(2n)!}{n!\, 2^{2n}} \sqrt{\pi} \tag{5.67}$$

which implies (Exercise 5.11) that

$$\Gamma\left(n + \frac{1}{2}\right) = \frac{(2n-1)!!}{2^n} \sqrt{\pi}. \tag{5.68}$$

Example 5.8 (Bessel function of nonintegral index) We can use the gamma-function formula (5.58) for $n!$ to extend the definition (5.52) of the Bessel function of the first kind $J_n(\rho)$ to nonintegral values ν of the index n. Replacing n by ν and $(m+n)!$ by $\Gamma(m+\nu+1)$, we get

$$J_\nu(\rho) = \left(\frac{\rho}{2}\right)^\nu \sum_{m=0}^{\infty} \frac{(-1)^m}{m!\,\Gamma(m+\nu+1)} \left(\frac{\rho}{2}\right)^{2m} \tag{5.69}$$

which makes sense even for complex values of ν.

Example 5.9 (Spherical Bessel function) The spherical Bessel function is defined as

$$j_\ell(\rho) \equiv \sqrt{\frac{\pi}{2\rho}}\, J_{\ell+1/2}(\rho). \tag{5.70}$$

For small values of its argument $|\rho| \ll 1$, the first term in the series (5.69) dominates and so (Exercise 5.7)

$$j_\ell(\rho) \approx \frac{\sqrt{\pi}}{2} \left(\frac{\rho}{2}\right)^\ell \frac{1}{\Gamma(\ell+3/2)} = \frac{\ell!\,(2\rho)^\ell}{(2\ell+1)!} = \frac{\rho^\ell}{(2\ell+1)!!} \tag{5.71}$$

as one may show by repeatedly using the key identity $\Gamma(z+1) = z\,\Gamma(z)$.

Example 5.10 (Numerical gamma functions) The function gamma(x) gives $\Gamma(x)$ in Fortran, C, Matlab, and Python. As n increases past 100, $n! = \Gamma(n+1)$ quickly becomes too big for computers to handle, and it becomes essential to work with the logarithm of the gamma function. The Fortran function log_gamma(x), the C function lgamma(x), the Matlab function gammaln(x), and the Python function loggamma(x) all give $\log(\Gamma(x)) = \log((x-1)!)$ for real x.

5.6 Euler's Beta Function

The beta function is the ratio

$$\mathrm{B}(x, y) = \frac{\Gamma(x)\,\Gamma(y)}{\Gamma(x+y)} \tag{5.72}$$

which is symmetric, $\mathrm{B}(x, y) = \mathrm{B}(y, x)$. We can get an explicit formula for it by using the formula (5.66) for the gamma function (5.58) in the product

$$\Gamma(x)\,\Gamma(y) = 4 \int_0^\infty e^{-u^2}\, u^{2x-1}\, du \int_0^\infty e^{-v^2}\, v^{2y-1}\, dv \tag{5.73}$$

and switching to polar coordinates $u = t\cos\theta$, $v = t\sin\theta$, and $t^2 = u^2 + v^2$

$$\begin{aligned} \Gamma(x)\,\Gamma(y) &= 2 \int_0^\infty e^{-t^2} t^{2(x+y)-1}\, dt\, 2 \int_0^{\pi/2} \cos^{2x-1}\theta \sin^{2y-1}\theta d\theta \\ &= \Gamma(x+y)\,\mathrm{B}(x, y). \end{aligned} \tag{5.74}$$

Thus the beta function is the integral

$$\mathrm{B}(x, y) = 2 \int_0^{\pi/2} \cos^{2x-1}\theta \sin^{2y-1}\theta\, d\theta. \tag{5.75}$$

Setting $t = \cos^2\theta$, we find

$$\mathrm{B}(x, y) = \int_0^1 t^{x-1}\,(1-t)^{y-1}\, dt. \tag{5.76}$$

The ratio

$$\mathrm{B}(z_1, z_2, \ldots, z_n) = \frac{\Gamma(z_1)\,\Gamma(z_2)\cdots\Gamma(z_n)}{\Gamma(z_1 + z_2 + \cdots + z_n)} \tag{5.77}$$

naturally generalizes the beta function (5.72).

5.7 Taylor Series

If $f(x)$ is a real-valued function of a real variable x with a continuous Nth derivative, then Taylor's expansion for it is

$$f(x+a) = f(x) + af'(x) + \frac{a^2}{2} f''(x) + \cdots + \frac{a^{N-1}}{(N-1)!} f^{(N-1)} + E_N$$
$$= \sum_{n=0}^{N-1} \frac{a^n}{n!} f^{(n)}(x) + E_N \tag{5.78}$$

in which the error E_N is

$$E_N = \frac{a^N}{N!} f^{(N)}(x+y) \tag{5.79}$$

for some $0 \le y \le a$.

For many functions $f(x)$ the errors go to zero, $E_N \to 0$, as $N \to \infty$; for these functions, the infinite Taylor series converges:

$$f(x+a) = \sum_{n=0}^{\infty} \frac{a^n}{n!} f^{(n)}(x) = \exp\left(a \frac{d}{dx}\right) f(x). \tag{5.80}$$

In quantum mechanics, the momentum operator p acts from the right on $\langle x|$

$$\langle x|p = \frac{\hbar}{i} \frac{d}{dx} \langle x| \tag{5.81}$$

and generates translations in space

$$\psi(x+a) = \langle x+a|\psi\rangle = \langle x|e^{iap/\hbar}|\psi\rangle = \exp\left(a \frac{d}{dx}\right) \langle x|\psi\rangle. \tag{5.82}$$

5.8 Fourier Series as Power Series

The Fourier series (3.46)

$$f(x) = \sum_{n=-\infty}^{\infty} c_n \frac{e^{i2\pi nx/L}}{\sqrt{L}} \tag{5.83}$$

with coefficients (3.54)

$$c_n = \int_{-L/2}^{L/2} \frac{e^{-i2\pi nx/L}}{\sqrt{L}} f(x)\, dx \tag{5.84}$$

is a pair of power series

$$f(x) = \frac{1}{\sqrt{L}} \left(\sum_{n=0}^{\infty} c_n\, z^n + \sum_{n=1}^{\infty} c_{-n}\, (z^{-1})^n \right) \tag{5.85}$$

in the variables

$$z = e^{i2\pi x/L} \quad \text{and} \quad z^{-1} = e^{-i2\pi x/L}. \tag{5.86}$$

Formula (5.35) tells us that the radii of convergence of these two power series are given by

$$R_+^{-1} = \lim_{n\to\infty} \frac{|c_{n+1}|}{|c_n|} \quad \text{and} \quad R_-^{-1} = \lim_{n\to\infty} \frac{|c_{-n-1}|}{|c_{-n}|}. \tag{5.87}$$

Thus the pair of power series (5.85) will converge uniformly and absolutely as long as z satisfies the two inequalities

$$|z| < R_+ \quad \text{and} \quad \frac{1}{|z|} < R_-. \tag{5.88}$$

Since $|z| = 1$, the Fourier series (5.83) converges if $R_-^{-1} < 1 < R_+$.

Example 5.11 (A uniformly and absolutely convergent Fourier series) The Fourier series

$$f(x) = \sum_{n=-\infty}^{\infty} \frac{1}{1+|n|^{|n|}} \frac{e^{i2\pi nx/L}}{\sqrt{L}} \tag{5.89}$$

converges uniformly and absolutely because $R_+ = R_- = \infty$.

5.9 Binomial Series

The Taylor series for the function $f(x) = (1+x)^a$ is

$$\begin{aligned}
(1+x)^a &= \sum_{k=0}^{\infty} \frac{x^k}{k!} \frac{d^k}{dx^k} (1+x)^a \Big|_{x=0} \\
&= 1 + ax + \frac{1}{2}a(a-1)x^2 + \cdots \\
&= 1 + \sum_{k=1}^{\infty} \frac{a(a-1)\cdots(a-k+1)}{k!} x^k.
\end{aligned} \tag{5.90}$$

If a is a positive integer $a = n$, then the kth power of x in this series is multiplied by a binomial coefficient (5.47)

$$(1+x)^n = \sum_{k=0}^{n} \frac{n!}{k!(n-k)!} x^k = \sum_{k=0}^{n} \binom{n}{k} x^k. \tag{5.91}$$

The series (5.90) and (5.91) respectively imply (Exercise 5.13) that

$$(x+y)^a = y^a + \sum_{k=1}^{\infty} \frac{a(a-1)\cdots(a-k+1)}{k!} x^k\, y^{a-k} \tag{5.92}$$

and

$$(x+y)^n = \sum_{k=0}^{n} \binom{n}{k} x^k\, y^{n-k}. \tag{5.93}$$

We can use these versions of the **binomial theorem** to compute approximately or exactly.

Example 5.12 (Neutrino oscillations) The phase of a particle of energy E and momentum p going a distance L in a time t is $\exp(i(pL - Et)/\hbar)$. Neutrinos are nearly massless and travel at nearly the speed of light, hence $cm/p \approx 0$ and $t \approx L/c$. With these excellent approximations, the binomial expansion (5.92) gives

$$pL - Et = pL - \sqrt{c^2p^2 + c^4m^2}\,L/c = pL - pL\sqrt{1 + c^2m^2/p^2} \approx -\frac{c^2m^2L}{2p}. \tag{5.94}$$

Since $E \approx cp$, the phase difference $\Delta\phi$ between two such neutrinos varies with their masses m_1 and m_2 as

$$\Delta\phi = -\frac{c^3(m_1^2 - m_2^2)L}{\hbar E} = -\frac{c^3\,\Delta m^2\,L}{\hbar E}. \tag{5.95}$$

Example 5.13 We can use the binomial expansion (5.93) to compute

$$999^3 = \left(10^3 - 1\right)^3 = 10^9 - 3\times 10^6 + 3\times 10^3 - 1 = 997002999 \tag{5.96}$$

exactly.

When a is not a positive integer, the series (5.90) does not terminate. For instance, the binomial series for $\sqrt{1+x}$ and $1/\sqrt{1+x}$ are (Exercise 5.14)

$$\begin{aligned}(1+x)^{1/2} &= 1 + \sum_{n=1}^{\infty} \frac{\frac{1}{2}\left(\frac{1}{2}-1\right)\cdots\left(\frac{1}{2}-n+1\right)}{n!}\,x^n \\ &= 1 + \sum_{n=1}^{\infty} \frac{(-1)^{n-1}}{2^n}\,\frac{(2n-3)!!}{n!}\,x^n = 1 + \frac{x}{2} - \frac{x^2}{8} + \frac{x^3}{16} + \cdots\end{aligned} \tag{5.97}$$

and

$$\begin{aligned}(1+x)^{-1/2} &= 1 + \sum_{n=1}^{\infty} \frac{-\frac{1}{2}\left(-\frac{3}{2}\right)\cdots\left(-\frac{1}{2}-n+1\right)}{n!}\,x^n \\ &= \sum_{n=0}^{\infty} \frac{(-1)^{n}}{2^n}\,\frac{(2n-1)!!}{n!}\,x^n = 1 - \frac{x}{2} + \frac{3}{8}x^2 - \frac{5}{16}x^3 + \cdots\end{aligned} \tag{5.98}$$

5.10 Logarithmic Series

The Taylor series for the function $f(x) = \ln(1+x)$ is

$$\ln(1+x) = \sum_{n=0}^{\infty} \frac{x^n}{n!} \frac{d^n}{dx^n} \ln(1+x)|_{x=0} \tag{5.99}$$

in which

$$\begin{aligned} f^{(0)}(0) &= \ln(1+x)|_{x=0} = 0 \\ f^{(1)}(0) &= \left.\frac{1}{1+x}\right|_{x=0} = 1 \\ f^{(n)}(0) &= \left.\frac{(-1)^{n-1}\,(n-1)!}{(1+x)^n}\right|_{x=0} = (-1)^{n-1}\,(n-1)!. \end{aligned} \tag{5.100}$$

So the series for $\ln(1+x)$ is

$$\ln(1+x) = \sum_{n=1}^{\infty} \frac{(-1)^{n-1}\,x^n}{n} = x - \frac{1}{2}x^2 + \frac{1}{3}x^3 \pm \cdots \tag{5.101}$$

which converges slowly for $-1 < x \le 1$. Letting $x \to -x$, we see that

$$\ln(1-x) = -\sum_{n=1}^{\infty} \frac{x^n}{n}. \tag{5.102}$$

So the series for the logarithm of the ratio $(1+x)/(1-x)$ is

$$\ln\left(\frac{1+x}{1-x}\right) = 2\sum_{n=0}^{\infty} \frac{x^{2n+1}}{2n+1}. \tag{5.103}$$

5.11 Dirichlet Series and the Zeta Function

A **Dirichlet series** is one in which the nth term is proportional to $1/n^z$

$$f(z) = \sum_{n=1}^{\infty} \frac{c_n}{n^z}. \tag{5.104}$$

An important example is the **Riemann zeta function** $\zeta(z)$

$$\zeta(z) = \sum_{n=1}^{\infty} n^{-z} \tag{5.105}$$

which converges for $\operatorname{Re} z > 1$. Some values are $\zeta(2) = \pi^2/6 \approx 1.645$, $\zeta(3) \approx 1.202$, $\zeta(4) = \pi^4/90 \approx 1.082$, $\zeta(5) \approx 1.037$, and $\zeta(6) = \pi^6/945 \approx 1.017$.

Euler showed that for $\operatorname{Re} z > 1$, the Riemann zeta function is the infinite product

$$\zeta(z) = \prod_p \frac{1}{1 - p^{-z}} \tag{5.106}$$

over all prime numbers $p = 2, 3, 5, 7, 11, \ldots$

Example 5.14 (Planck's distribution) Max Planck (1858–1947) showed that the electromagnetic energy in a closed cavity of volume V at a temperature T in the frequency interval $d\nu$ about ν is

$$dU(\beta, \nu, V) = \frac{8\pi h V}{c^3} \frac{\nu^3}{e^{\beta h \nu} - 1} \, d\nu \tag{5.107}$$

in which $\beta = 1/(kT)$, $k = 1.3806503 \times 10^{-23}$ J/K is **Boltzmann's constant**, and $h = 6.626068 \times 10^{-34}$ Js is **Planck's constant**. The total energy then is the integral

$$U(\beta, V) = \frac{8\pi h V}{c^3} \int_0^\infty \frac{\nu^3}{e^{\beta h \nu} - 1} \, d\nu \tag{5.108}$$

which we may do by letting $x = \beta h \nu$ and using the geometric series (5.11)

$$\begin{aligned} U(\beta, V) &= \frac{8\pi (kT)^4 V}{(hc)^3} \int_0^\infty \frac{x^3}{e^x - 1} \, dx \\ &= \frac{8\pi (kT)^4 V}{(hc)^3} \int_0^\infty \frac{x^3 e^{-x}}{1 - e^{-x}} \, dx \\ &= \frac{8\pi (kT)^4 V}{(hc)^3} \int_0^\infty x^3 e^{-x} \sum_{n=0}^\infty e^{-nx} \, dx. \end{aligned} \tag{5.109}$$

The geometric series is absolutely and uniformly convergent for $x > 0$, and we may interchange the limits of summation and integration. Another change of variables and the definition (5.58) of the gamma function give

$$\begin{aligned} U(\beta, V) &= \frac{8\pi (kT)^4 V}{(hc)^3} \sum_{n=0}^\infty \int_0^\infty x^3 e^{-(n+1)x} \, dx \\ &= \frac{8\pi (kT)^4 V}{(hc)^3} \sum_{n=0}^\infty \frac{1}{(n+1)^4} \int_0^\infty y^3 e^{-y} \, dy \\ &= \frac{8\pi (kT)^4 V}{(hc)^3} 3! \, \zeta(4) = \frac{8\pi^5 (kT)^4 V}{15 (hc)^3}. \end{aligned} \tag{5.110}$$

The power radiated by a "**black body**" is proportional to the fourth power of its temperature and to its area A

$$P = \sigma A T^4 \tag{5.111}$$

in which

$$\sigma = \frac{2\pi^5 k^4}{15 \, h^3 c^2} = 5.670400(40) \times 10^{-8} \, \mathrm{W\,m^{-2}\,K^{-4}} \tag{5.112}$$

is **Stefan's constant**.

The number of photons in the black-body distribution (5.107) at inverse temperature β in the volume V is

$$\begin{aligned} N(\beta, V) &= \frac{8\pi V}{c^3} \int_0^\infty \frac{\nu^2}{e^{\beta h\nu} - 1}\, d\nu = \frac{8\pi V}{(c\beta h)^3} \int_0^\infty \frac{x^2}{e^x - 1}\, dx \\ &= \frac{8\pi V}{(c\beta h)^3} \int_0^\infty \frac{x^2 e^{-x}}{1 - e^{-x}}\, dx = \frac{8\pi V}{(c\beta h)^3} \int_0^\infty x^2 e^{-x} \sum_{n=0}^\infty e^{-nx}\, dx \\ &= \frac{8\pi V}{(c\beta h)^3} \sum_{n=0}^\infty \int_0^\infty x^2 e^{-(n+1)x}\, dx = \frac{8\pi V}{(c\beta h)^3} \sum_{n=0}^\infty \frac{1}{(n+1)^3} \int_0^\infty y^2 e^{-y}\, dy \\ &= \frac{8\pi V}{(c\beta h)^3} \zeta(3) 2! = \frac{8\pi (kT)^3 V}{(ch)^3} \zeta(3) 2!. \end{aligned} \tag{5.113}$$

The mean energy $\langle E \rangle$ of a photon in the black-body distribution (5.107) is the energy $U(\beta, V)$ divided by the number of photons $N(\beta, V)$

$$\langle E \rangle = \langle h\nu \rangle = \frac{3!\,\zeta(4)}{2!\,\zeta(3)}\, kT = \frac{\pi^4}{30\,\zeta(3)}\, kT \tag{5.114}$$

or $\langle E \rangle \approx 2.70118\, kT$ since Apéry's constant $\zeta(3)$ is $1.2020569032\ldots$ (Roger Apéry, 1916–1994).

Example 5.15 (Lerch transcendent) The **Lerch transcendent** is the series

$$\Phi(z, s, \alpha) = \sum_{n=0}^\infty \frac{z^n}{(n+\alpha)^s} \tag{5.115}$$

which converges for $|z| < 1$ and $\operatorname{Re} s > 0$ and reduces to Riemann's zeta function (5.105) when $z = 1$ and $\alpha = 0$, $\Phi(1, s, 0) = \zeta(s)$.

5.12 Bernoulli Numbers and Polynomials

The **Bernoulli numbers** B_n are defined by the infinite series

$$\frac{x}{e^x - 1} = \sum_{n=0}^\infty \frac{x^n}{n!} \left[\frac{d^n}{dx^n} \frac{x}{e^x - 1} \right] \bigg|_{x=0} = \sum_{n=0}^\infty B_n \frac{x^n}{n!} \tag{5.116}$$

for the **generating function** $x/(e^x - 1)$. They are the successive derivatives

$$B_n = \frac{d^n}{dx^n} \frac{x}{e^x - 1} \bigg|_{x=0}. \tag{5.117}$$

So $B_0 = 1$ and $B_1 = -1/2$. The remaining odd Bernoulli numbers vanish

$$B_{2n+1} = 0 \quad \text{for } n > 0 \tag{5.118}$$

and the remaining even ones are given by Euler's zeta function (5.105) as

$$B_{2n} = \frac{(-1)^{n-1}2(2n)!}{(2\pi)^{2n}}\,\zeta(2n) \quad \text{for } n > 0. \tag{5.119}$$

The Bernoulli numbers occur in the power series for many transcendental functions, for instance

$$\coth x = \frac{1}{x} + \sum_{k=1}^{\infty} \frac{2^{2k} B_{2k}}{(2k)!}\, x^{2k-1} \quad \text{for } x^2 < \pi^2. \tag{5.120}$$

Bernoulli's polynomials $B_n(y)$ are defined by the series

$$\frac{xe^{xy}}{e^x - 1} = \sum_{n=0}^{\infty} B_n(y)\,\frac{x^n}{n!} \tag{5.121}$$

for the **generating function** $xe^{xy}/(e^x - 1)$.

Some authors (Whittaker and Watson, 1927, pp. 125–127) define Bernoulli's numbers instead as

$$B_n = \frac{2(2n)!}{(2\pi)^{2n}}\,\zeta(2n) = 4n \int_0^{\infty} \frac{t^{2n-1}\,dt}{e^{2\pi t} - 1} \tag{5.122}$$

a result due to Carda.

5.13 Asymptotic Series

A series

$$s_n(x) = \sum_{k=0}^{n} \frac{a_k}{x^k} \tag{5.123}$$

is an **asymptotic** expansion for a real function $f(x)$ if the **remainder** R_n

$$R_n(x) = f(x) - s_n(x) \tag{5.124}$$

satisfies the condition

$$\lim_{x\to\infty} x^n R_n(x) = 0 \tag{5.125}$$

for fixed n. In this case, one writes

$$f(x) \approx \sum_{k=0}^{\infty} \frac{a_k}{x^k} \tag{5.126}$$

where the wavy equal sign indicates equality in the sense of (5.125). Some authors add the condition:

$$\lim_{n\to\infty} x^n R_n(x) = \infty \tag{5.127}$$

for fixed x.

Example 5.16 (Asymptotic series for E_1) The function

$$E_1(x) = \int_x^\infty e^{-y} \, \frac{dy}{y} \tag{5.128}$$

is related to the exponential-integral function

$$Ei(x) = \int_{-\infty}^x e^{y} \, \frac{dy}{y} \tag{5.129}$$

by the formula $E_1(x) = -Ei(-x)$. Since

$$\frac{e^{-y}}{y} = -\frac{d}{dy}\left(\frac{e^{-y}}{y}\right) - \frac{e^{-y}}{y^2}, \tag{5.130}$$

we may integrate by parts, getting

$$E_1(x) = \frac{e^{-x}}{x} - \int_x^\infty e^{-y} \, \frac{dy}{y^2}. \tag{5.131}$$

Integrating by parts again, we have

$$E_1(x) = \frac{e^{-x}}{x} - \frac{e^{-x}}{x^2} + 2\int_x^\infty e^{-y} \, \frac{dy}{y^3}. \tag{5.132}$$

Eventually, we develop the series

$$E_1(x) = e^{-x}\left(\frac{0!}{x} - \frac{1!}{x^2} + \frac{2!}{x^3} - \frac{3!}{x^4} + \frac{4!}{x^5} - \cdots\right) \tag{5.133}$$

with remainder

$$R_n(x) = (-1)^n \, n! \int_x^\infty e^{-y} \, \frac{dy}{y^{n+1}}. \tag{5.134}$$

Setting $y = u + x$, we find

$$R_n(x) = (-1)^n \, \frac{n! \, e^{-x}}{x^{n+1}} \int_0^\infty e^{-u} \, \frac{du}{\left(1 + \frac{u}{x}\right)^{n+1}} \tag{5.135}$$

which satisfies the condition (5.125) that defines an asymptotic series

$$\begin{aligned} \lim_{x\to\infty} x^n R_n(x) &= \lim_{x\to\infty} (-1)^n \, \frac{n! \, e^{-x}}{x} \int_0^\infty e^{-u} \, \frac{du}{\left(1 + \frac{u}{x}\right)^{n+1}} \\ &= \lim_{x\to\infty} (-1)^n \, \frac{n! \, e^{-x}}{x} \int_0^\infty e^{-u} \, du \\ &= \lim_{x\to\infty} (-1)^n \, \frac{n! \, e^{-x}}{x} = 0 \end{aligned} \tag{5.136}$$

for fixed n as well as (5.127) for fixed x.

Asymptotic series often occur in physics. In such physical problems, a small parameter λ usually plays the role of $1/x$. A perturbative series

$$S_n(\lambda) = \sum_{k=0}^{n} a_k \, \lambda^k \tag{5.137}$$

is an asymptotic expansion of the physical quantity $S(\lambda)$ if the remainder

$$R_n(\lambda) = S(\lambda) - S_n(\lambda) \tag{5.138}$$

satisfies for fixed n

$$\lim_{\lambda \to 0} \lambda^{-n} \, R_n(\lambda) = 0. \tag{5.139}$$

The semiclassical WKB approximation and Dyson's series for quantum electrodynamics are asymptotic expansions in this sense.

5.14 Fractional and Complex Derivatives

For any nonnegative integer k and any complex w, the derivative of the monomial x^w is x^{w-k} times a ratio (5.62) of gamma functions

$$\frac{d^k x^w}{dx^k} = w(w-1)\cdots(w-k+1)x^{w-k} = \frac{w!\, x^{w-k}}{(w-k)!} = \frac{\Gamma(w+1)\, x^{w-k}}{\Gamma(w-k+1)}. \tag{5.140}$$

Since $\Gamma(w-k+1)$ is well defined for complex k as long as $w-k+1$ is not a negative integer, we may use this equation (5.140) to extend the definition of the kth derivative of x^w to fractional and even complex k just by replacing integral k by complex z

$$\frac{d^z x^w}{dx^z} = \frac{\Gamma(w+1)\, x^{w-z}}{\Gamma(w-z+1)}. \tag{5.141}$$

The z derivative of a function $f(x)$ that has a power series expansion (Section 5.4) is then

$$\frac{d^z f(x)}{dx^z} = \sum_{n=0}^{\infty} \frac{\Gamma(n+1)\, x^{n-z}}{\Gamma(n-z+1)}. \tag{5.142}$$

Fractional differential equations are used to describe dispersion, diffusion, and fluid flow in complex media.

Example 5.17 (The half derivative) The definition (5.141) of the z derivative and our formula (5.68) for $\Gamma(n+1/2)$ give for the $\frac{1}{2}$ derivative of x^n

$$\frac{d^{\frac{1}{2}} x^n}{dx^{\frac{1}{2}}} = \frac{\Gamma(n+1)}{\Gamma(n+\frac{1}{2})} x^{n-\frac{1}{2}} = \frac{2^n}{(2n-1)!!\,\sqrt{\pi}}\, x^{n-\frac{1}{2}}. \tag{5.143}$$

Acting twice with the $\frac{1}{2}$ derivative, we get the whole derivative

$$\left(\frac{d^{\frac{1}{2}}}{dx^{\frac{1}{2}}}\right)^2 x^n = \frac{d^{\frac{1}{2}}}{dx^{\frac{1}{2}}}\left(\frac{\Gamma(n+1)\,x^{n-\frac{1}{2}}}{\Gamma(n+\frac{1}{2})}\right) = \frac{\Gamma(n+1)\Gamma(n+\frac{1}{2})\,x^{n-1}}{\Gamma(n+\frac{1}{2})\Gamma(n)} = nx^{n-1}. \tag{5.144}$$

The $\frac{1}{2}$ derivative is the square root of the first derivative.

5.15 Some Electrostatic Problems

Gauss's law $\nabla \cdot \boldsymbol{D} = \rho$ equates the divergence of the **electric displacement** $\boldsymbol{D}$ to the density ρ of **free charges** (charges that are free to move in or out of the dielectric medium – as opposed to those that are part of the medium and bound to it by molecular forces). In electrostatic problems, Maxwell's equations reduce to Gauss's law and the static form $\nabla \times \mathbf{E} = 0$ of Faraday's law which implies that the electric field $\boldsymbol{E}$ is the gradient of an electrostatic potential $\boldsymbol{E} = -\nabla V$ (James Maxwell 1831–1879, Michael Faraday 1791–1867).

Across an interface with normal vector $\hat{\boldsymbol{n}}$ between two dielectrics, the tangential electric field is continuous while the normal electric displacement jumps by the surface density of free charge σ

$$\hat{\boldsymbol{n}} \times (\boldsymbol{E}_2 - \boldsymbol{E}_1) = 0 \quad \text{and} \quad \sigma = \hat{\boldsymbol{n}} \cdot (\boldsymbol{D}_2 - \boldsymbol{D}_1). \tag{5.145}$$

In a **linear isotropic dielectric**, the electric displacement $\boldsymbol{D}$ is proportional to the electric field $\boldsymbol{D} = \epsilon_m \boldsymbol{E}$, where the **permittivity** $\epsilon_m = \epsilon_0 + \chi_m = K_m \epsilon_0$ of the material differs from that of the vacuum ϵ_0 by the **electric susceptibility** χ_m and by the **relative permittivity** K_m. The permittivity of the vacuum is the **electric constant** $\epsilon_0 = 8.85418782 \times 10^{-12}$ F/m.

An electric field $\boldsymbol{E}$ exerts on a charge q a **force** $\boldsymbol{F} = q\boldsymbol{E}$ even in a dielectric medium. The electrostatic energy W of a system of linear dielectrics is the volume integral

$$W = \frac{1}{2}\int \boldsymbol{D} \cdot \boldsymbol{E}\, d^3r. \tag{5.146}$$

Example 5.18 (Field of a charge near an interface) Consider two semi-infinite dielectrics of permittivities ϵ_1 and ϵ_2 separated by an infinite horizontal x–y plane. What is the electrostatic potential due to a charge q in region 1 at a height h above the plane?

The easy way to solve this problem is to put an image charge q' at the same distance from the interface in region 2 so that the potential in region 1 is

$$V_1(\boldsymbol{r}) = \frac{1}{4\pi\epsilon_1}\left(\frac{q}{\sqrt{x^2+y^2+(z-h)^2}} + \frac{q'}{\sqrt{x^2+y^2+(z+h)^2}}\right). \tag{5.147}$$

This potential satisfies Gauss's law $\nabla \cdot \boldsymbol{D} = \rho$ in region 1. In region 2, the potential

$$V_2(\boldsymbol{r}) = \frac{1}{4\pi\epsilon_2}\frac{q''}{\sqrt{x^2+y^2+(z-h)^2}} \tag{5.148}$$

also satisfies Gauss's law. The continuity (5.145) of the tangential component of $\boldsymbol{E}$ tells us that the partial derivatives of V_1 and V_2 in the x (or y) direction must be the same at $z = 0$

$$\frac{\partial V_1(x,y,0)}{\partial x} = \frac{\partial V_2(x,y,0)}{\partial x}. \tag{5.149}$$

The discontinuity equation (5.145) for the electric displacement says that at the interface at $z = 0$ with no surface charge

$$\epsilon_1\frac{\partial V_1(x,y,0)}{\partial z} = \epsilon_2\frac{\partial V_2(x,y,0)}{\partial z}. \tag{5.150}$$

These two equations (5.149 and 5.150) allow one to solve for q' and q''

$$q' = \frac{\epsilon_1-\epsilon_2}{\epsilon_1+\epsilon_2}\,q \quad \text{and} \quad q'' = \frac{2\epsilon_2}{\epsilon_1+\epsilon_2}\,q. \tag{5.151}$$

In the limit $h \to 0$, the potential in region 1 becomes

$$V_1(\boldsymbol{r}) = \frac{1}{4\pi\epsilon_1}\frac{q}{\sqrt{x^2+y^2+z^2}}\left(1+\frac{\epsilon_1-\epsilon_2}{\epsilon_1+\epsilon_2}\right) = \frac{q}{4\pi\bar{\epsilon}r} \tag{5.152}$$

in which $\bar{\epsilon}r$ is the mean permittivity $\bar{\epsilon} = (\epsilon_1+\epsilon_2)/2$. Similarly in region 2, the potential is

$$V_2(\boldsymbol{r}) = \frac{1}{4\pi\epsilon_2}\frac{q}{\sqrt{x^2+y^2+z^2}}\frac{2\epsilon_2}{\epsilon_1+\epsilon_2} = \frac{q}{4\pi\bar{\epsilon}r} \tag{5.153}$$

in the limit $h \to 0$.

Example 5.19 (A charge near a plasma membrane) A eukaryotic cell (the kind with a nucleus) is surrounded by a plasma membrane, which is a phospholipid bilayer about 5 nm thick. Both sides of the plasma membrane are in contact with salty water. The permittivity of the water is $\epsilon_w \approx 80\epsilon_0$ while that of the membrane considered as a simple lipid slab is $\epsilon_\ell \approx 2\epsilon_0$.

Let's think about the potential felt by an ion in the water outside a cell but near its membrane, and let us for simplicity imagine the membrane to be infinitely thick so that we can use the simple formulas we've derived. The potential due to the ion, if its charge is q, is then given by equation (5.147) with $\epsilon_1 = \epsilon_w$ and $\epsilon_2 = \epsilon_\ell$. The image-charge term in $V_1(\boldsymbol{r})$ is the potential due to the polarization of the membrane and the water by the ion. It is the potential felt by the ion. Since the image charge by (5.151) is $q' \approx q$, the potential the ion feels is $V_i(z) \approx q/8\pi e_w z$. The force on the ion then is

$$F = -qV_i'(z) - \frac{q^2}{8\pi e_w z}. \tag{5.154}$$

It always is positive no matter what the sign of the charge is. A lipid slab in water repels ions. Similarly, a charge in a lipid slab is attracted to the water outside the slab.

Now imagine an electric dipole in water near a lipid slab. Now there are two equal and opposite charges and two equal and opposite mirror charges. The net effect is that the slab repels the dipole. So lipids repel water molecules; they are said to be **hydrophobic**. This is one of the reasons why folding proteins move their hydrophobic amino acids inside and their polar or **hydrophilic** ones outside.

With some effort, one may use the method of images to compute the electric potential of a charge in or near a plasma membrane taken to be a lipid slab of finite thickness.

The electric potential in the lipid bilayer $V_\ell(\rho, z)$ of thickness t due to a charge q in the extra-cellular environment at a height h above the bilayer is

$$V_\ell(\rho, z) = \frac{q}{4\pi\epsilon_{w\ell}} \sum_{n=0}^{\infty} (pp')^n \left(\frac{1}{\sqrt{\rho^2 + (z - 2nt - h)^2}} - \frac{p'}{\sqrt{\rho^2 + (z + 2(n+1)t + h)^2}} \right) \tag{5.155}$$

in which $p = (\epsilon_w - \epsilon_\ell)/(\epsilon_w + \epsilon_\ell)$, $p' = (\epsilon_c - \epsilon_\ell)/(\epsilon_c + \epsilon_\ell)$, and $\epsilon_{w\ell} = (\epsilon_w + \epsilon_\ell)/2$. That in the extra-cellular environment is

$$V_w(\rho, z) = \frac{q}{4\pi\epsilon_w} \left(\frac{1}{r} + \frac{p}{\sqrt{\rho^2 + (z + h)^2}} - \frac{\epsilon_w \epsilon_\ell}{\epsilon_{w\ell}^2} \sum_{n=1}^{\infty} \frac{p^{n-1} p'^n}{\sqrt{\rho^2 + (z + 2nt + h)^2}} \right) \tag{5.156}$$

in which r is the distance from the charge q. Finally, the potential in the cytosol is

$$V_c(\rho, z) = \frac{q\,\epsilon_\ell}{4\pi\epsilon_{w\ell}\epsilon_{\ell c}} \sum_{n=0}^{\infty} \frac{(pp')^n}{\sqrt{\rho^2 + (z - 2nt - h)^2}} \tag{5.157}$$

where $\epsilon_{\ell c} = (\epsilon_\ell + \epsilon_c)/2$.

The first 1000 terms of these three series (5.155–5.157) are plotted in Fig. 5.2 for the case of a positive charge $q = |e|$ at $(\rho, z) = (0, 0)$ (top curve), $(0, 1)$ (second curve), $(0, 2)$ nm (third curve), and $(0, 6)$ nm (bottom curve). Although the potential $V(\rho, z)$ is continuous across the two interfaces, its normal derivative isn't due to the different

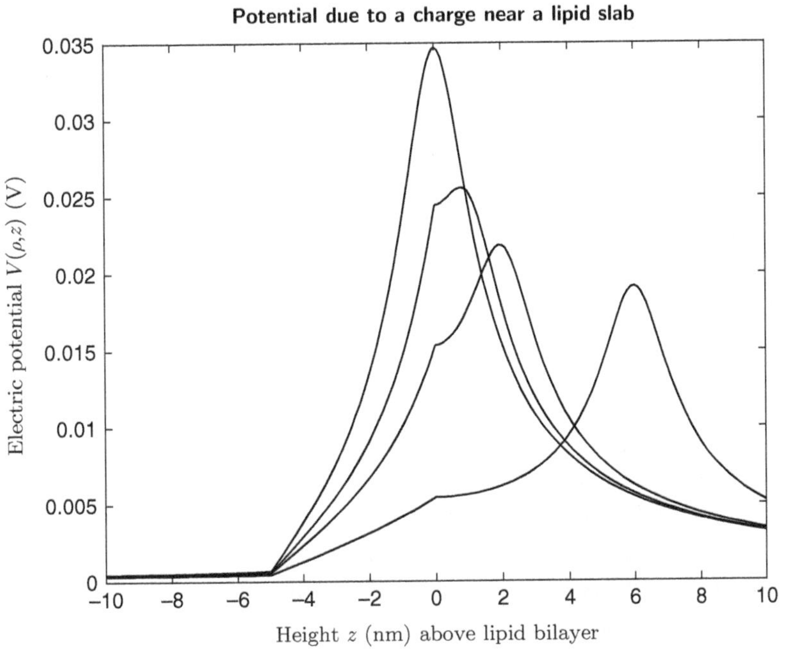

Figure 5.2 The electric potential $V(\rho, z)$ from (5.155–5.157) in volts for $\rho =$ 1 nm as a function of the height z (nm) above (or below) a lipid slab for a unit charge $q = |e|$ at $(\rho, z) = (0, 0)$ (top curve), (0, 1) (second curve), (0, 2) (third curve), and (0, 6) nm (bottom curve). The lipid slab extends from $z = 0$ to $z = -5$ nm, and the cytosol lies below $z = -5$ nm. The relative permittivities were taken to be $\epsilon_w/\epsilon_0 = \epsilon_c/\epsilon_0 = 80$ and $\epsilon_\ell/\epsilon_0 = 2$.

dielectric constants in the three media. Because the potential is small and flat in the cytosol ($z < -5$ nm), charges in the extra-cellular environment ($z > 0$) are nearly decoupled from those in the cytosol.

Real plasma membranes are **phospholipid** bilayers. The lipids avoid the water and so are on the inside. The phosphate groups are dipoles (and phosphatidylserine is negatively charged). So a real membrane is a 4 nm thick lipid layer bounded on each side by dipole layers, each about 0.5 nm thick. The net effect is to weakly *attract* ions that are within 0.5 nm of the membrane.

5.16 Infinite Products

Weierstrass's definition (5.65) of the gamma function, Euler's formula (5.106) for the zeta function, and Mermin's formula (5.45) for $n!$ are useful infinite products. Other examples are the expansions of the trigonometric functions (3.79 and 3.80)

$$\sin z = z \prod_{n=1}^{\infty} \left[1 - \frac{z^2}{\pi^2 n^2} \right] \quad \text{and} \quad \cos z = \prod_{n=1}^{\infty} \left[1 - \frac{z^2}{\pi^2 (n - 1/2)^2} \right] \tag{5.158}$$

which imply these of the hyperbolic functions

$$\sinh z = z \prod_{n=1}^{\infty} \left[1 + \frac{z^2}{\pi^2 n^2} \right] \quad \text{and} \quad \cosh z = \prod_{n=1}^{\infty} \left[1 + \frac{z^2}{\pi^2 (n - 1/2)^2} \right]. \tag{5.159}$$

Exercises

5.1 Test the following series for convergence:

$$\text{(a)} \sum_{n=2}^{\infty} \frac{1}{(\ln n)^2}, \quad \text{(b)} \sum_{n=1}^{\infty} \frac{n!}{20^n}, \quad \text{(c)} \sum_{n=1}^{\infty} \frac{1}{n(n+2)}, \quad \text{(d)} \sum_{n=2}^{\infty} \frac{1}{n \ln n}.$$

In each case, say whether the series converges and how you found out.

5.2 Olber's paradox: Assume a static universe with a uniform density of stars. With you at the origin, divide space into successive shells of thickness t, and assume that the stars in each shell subtend the same solid angle ω (as follows from the first assumption). Take into account the occulting of distant stars by nearer ones and show that the total solid angle subtended by all the stars would be 4π. The sky would be dazzlingly bright at night.

5.3 Use the geometric formula (5.10) to derive the trigonometric summation formula

$$\frac{1}{2} + \cos\alpha + \cos 2\alpha + \cdots + \cos n\alpha = \frac{\sin(n + \frac{1}{2})\alpha}{2 \sin \frac{1}{2}\alpha}. \tag{5.160}$$

Hint: write $\cos n\alpha$ as $[\exp(in\alpha) + \exp(-in\alpha)]/2$.

5.4 Show that

$$\binom{n-1}{k} + \binom{n-1}{k-1} = \binom{n}{k} \tag{5.161}$$

and then use mathematical induction to prove Leibniz's rule (5.49).

5.5 (a) Find the radius of convergence of the series (5.52) for the Bessel function $J_n(\rho)$. (b) Show that this series converges even faster than the one (5.50) for the exponential function.

5.6 Use the formula (5.8) for the Euler–Mascheroni constant to show that Euler's definition (5.64) of the gamma function implies Weierstrass's (5.65).

5.7 Derive the approximation (5.71) for $j_\ell(\rho)$ for $|\rho| \ll 1$.

5.8 Derive formula (5.66) for the gamma function from its definition (5.58).

5.9 Use formula (5.66) to compute $\Gamma(1/2)$.

5.10 Show that $z! = \Gamma(z + 1)$ diverges when z is a negative integer.

5.11 Derive formula (5.68) for $\Gamma(n + \frac{1}{2})$.

5.12 Show that the area of the surface of the unit sphere in d dimensions is

$$A_d = 2\pi^{d/2}/\Gamma(d/2). \tag{5.162}$$

Hint: Compute the integral of the gaussian $\exp(-x^2)$ in d dimensions using both rectangular and spherical coordinates. This formula (5.162) is used in dimensional regularization (Weinberg, 1995, p. 477).

5.13 Derive (5.93) from (5.91) and (5.92) from (5.90).

5.14 Derive the expansions (5.97 and 5.98) for $\sqrt{1+x}$ and $1/\sqrt{1+x}$.

5.15 Find the radii of convergence of the series (5.97) and (5.98).

5.16 Find the first three Bernoulli polynomials $B_n(y)$ by using their generating function (5.121).

5.17 How are the two definitions (5.119) and (5.122) of the Bernoulli numbers related?

5.18 Show that the Lerch transcendent $\Phi(z, s, \alpha)$ defined by the series (5.115) converges when $|z| < 1$ and $\text{Re}\, s > 0$ and $\text{Re}\, \alpha > 0$.

5.19 Langevin's classical formula for the electrical polarization of a gas or liquid of molecules of electric dipole moment p is

$$P(x) = Np\left(\frac{\cosh x}{\sinh x} - \frac{1}{x}\right) \tag{5.163}$$

where $x = pE/(kT)$, E is the applied electric field, and N is the number density of the molecules per unit volume. (a) Expand $P(x)$ for small x as an infinite power series involving the Bernoulli numbers. (b) What are the first three terms expressed in terms of familiar constants? (c) Find the saturation limit of $P(x)$ as $x \to \infty$.

5.20 By using repeatedly the identity (5.60), which is $z\Gamma(z) = \Gamma(z+1)$, show that

$$\frac{\Gamma(-n+\alpha+1)}{\Gamma(-n+\alpha-k+1)} = (\alpha-n)(\alpha-n-1)\cdots(\alpha-n-k+1). \tag{5.164}$$

5.21 Show that the energy of a charge q spread on the surface of a sphere of radius a in an infinite lipid of permittivity ϵ_ℓ is $W = q^2/8\pi\epsilon_\ell a$.

5.22 If the lipid of Exercise 5.21 has finite thickness t and is surrounded on both sides by water of permittivity ϵ_w, then the image charges change the energy W by (Parsegian, 1969)

$$\Delta W = \frac{q^2}{4\pi\epsilon_\ell t}\sum_{n=1}^{\infty}\frac{1}{n}\left(\frac{\epsilon_\ell - \epsilon_w}{\epsilon_\ell + \epsilon_w}\right)^n. \tag{5.165}$$

Sum this series. Hint: read Section 5.10 carefully.

5.23 Consider a stack of three dielectrics of infinite extent in the x–y plane separated by the two infinite x–y planes $z = t/2$ and $z = -t/2$. Suppose the upper region $z > t/2$ is a uniform linear dielectric of permittivity ϵ_1, the central region $-t/2 < z < t/2$ is a uniform linear dielectric of permittivity ϵ_2, and the lower region $z < -t/2$ is a uniform linear dielectric of permittivity ϵ_3. Suppose the lower infinite x–y plane $z = -t/2$ has a uniform surface charge density $-\sigma$, while the upper plane $z = t/2$ has a uniform surface charge density σ. What is the energy per unit area of this system? What is the pressure on the second dielectric? What is the capacitance per unit area of the stack?

6
Complex-Variable Theory

6.1 Analytic Functions

A complex-valued function $f(z)$ of a complex variable z is **differentiable** at z with derivative $f'(z)$ if the limit

$$f'(z) = \lim_{z' \to z} \frac{f(z') - f(z)}{z' - z} \tag{6.1}$$

exists and is unique as z' approaches z from **any direction** in the complex plane. The limit must exist no matter how or from what direction z' approaches z.

If the function $f(z)$ is differentiable in a small disk around a point z_0, then $f(z)$ is said to be **analytic** (or equivalently **holomorphic**) at z_0 (and at all points inside the disk).

Example 6.1 (Polynomials) If $f(z) = z^n$ for some integer n, then for tiny dz and $z' = z + dz$, the difference $f(z') - f(z)$ is

$$f(z') - f(z) = (z + dz)^n - z^n \approx n z^{n-1}\, dz \tag{6.2}$$

and so the limit

$$\lim_{z' \to z} \frac{f(z') - f(z)}{z' - z} = \lim_{dz \to 0} \frac{n z^{n-1}\, dz}{dz} = n z^{n-1} \tag{6.3}$$

exists and is $n z^{n-1}$ *independently* of how z' approaches z. Thus the function z^n is analytic at z for all z with derivative

$$\frac{dz^n}{dz} = n z^{n-1}. \tag{6.4}$$

A function that is analytic everywhere is **entire**. All polynomials

$$P(z) = \sum_{n=0}^{N} c_n\, z^n \tag{6.5}$$

are entire.

Example 6.2 (A function that's not analytic) To see what can go wrong when a function is not analytic, consider the function $f(x, y) = x^2 + y^2 = z\bar{z}$ for $z = x + iy$. If we compute its derivative at $(x, y) = (1, 0)$ by setting $x = 1 + \epsilon$ and $y = 0$, then the limit is

$$\lim_{\epsilon \to 0} \frac{f(1+\epsilon, 0) - f(1, 0)}{\epsilon} = \lim_{\epsilon \to 0} \frac{(1+\epsilon)^2 - 1}{\epsilon} = 2 \tag{6.6}$$

while if we instead set $x = 1$ and $y = \epsilon$, then the limit is

$$\lim_{\epsilon \to 0} \frac{f(1, \epsilon) - f(1, 0)}{i\epsilon} = \lim_{\epsilon \to 0} \frac{1 + \epsilon^2 - 1}{i\epsilon} = -i \lim_{\epsilon \to 0} \epsilon = 0. \tag{6.7}$$

So the derivative depends upon the direction through which $z \to 1$.

6.2 Cauchy–Riemann Conditions

How do we know whether a complex function $f(x, y) = u(x, y) + iv(x, y)$ of two real variables x and y is analytic? We apply the criterion (6.1) of analyticity and require that the change df in the function $f(x, y)$ be proportional to the change $dz = dx + idy$ in the complex variable $z = x + iy$

$$\left(\frac{\partial u}{\partial x} + i\frac{\partial v}{\partial x}\right) dx + \left(\frac{\partial u}{\partial y} + i\frac{\partial v}{\partial y}\right) dy = f'(z)(dx + idy). \tag{6.8}$$

Setting first dy and then dx equal to zero, we have

$$\left(\frac{\partial u}{\partial x} + i\frac{\partial v}{\partial x}\right) = f'(z) = \frac{1}{i}\left(\frac{\partial u}{\partial y} + i\frac{\partial v}{\partial y}\right). \tag{6.9}$$

This complex equation implies the two real equations

$$\frac{\partial u}{\partial x} = \frac{\partial v}{\partial y} \quad \text{and} \quad \frac{\partial v}{\partial x} = -\frac{\partial u}{\partial y} \tag{6.10}$$

which are the Cauchy–Riemann conditions. In a notation in which partial derivatives are labeled by subscripts, the Cauchy–Riemann conditions are $u_x = v_y$ and $v_x = -u_y$.

Example 6.3 (A function analytic except at a point) The real and imaginary parts of the function

$$f(z) = \frac{1}{z - z_0} = \frac{z^* - z_0^*}{|z - z_0|^2} = \frac{x - x_0 + i(y - y_0)}{(x - x_0)^2 + (y - y_0)^2} \tag{6.11}$$

are

$$u(x, y) = \frac{x - x_0}{(x - x_0)^2 + (y - y_0)^2} \quad \text{and} \quad v(x, y) = \frac{y - y_0}{(x - x_0)^2 + (y - y_0)^2}. \tag{6.12}$$

They satisfy the Cauchy–Riemann conditions (6.10)

$$\frac{\partial u(x,y)}{\partial x} = \frac{(y-y_0)^2-(x-x_0)^2}{[(x-x_0)^2+(y-y_0)^2]^2} = \frac{\partial v(x,y)}{\partial y} \tag{6.13}$$

and

$$\frac{\partial v(x,y)}{\partial x} = -\frac{2(x-x_0)(y-y_0)}{[(x-x_0)^2+(y-y_0)^2]^2} = -\frac{\partial u(x,y)}{\partial y} \tag{6.14}$$

except at the point $z = z_0$ where $x = x_0$ and $y = y_0$.

6.3 Cauchy's Integral Theorem

The Cauchy–Riemann conditions imply that the integral of a function along a **closed contour** (one that ends where it starts) vanishes if the function is analytic on the contour and everywhere inside it. To keep the notation simple, let's consider a rectangle R of length ℓ and height h with one corner at the origin and edges running along the x and y axes of the z plane. The integral along the four sides of the rectangle is

$$\begin{aligned}\oint_R f(z)\,dz &= \oint_R (u(x,y)+iv(x,y))\,(dx+idy)\\ &= \int_0^\ell [u(x,0)+iv(x,0)]\,dx + \int_0^h \big[u(\ell,y)+iv(\ell,y)\big]\,idy\\ &\quad + \int_\ell^0 [u(x,h)+iv(x,h)]\,dx + \int_h^0 \big[u(0,y)+iv(0,y)\big]\,idy.\end{aligned} \tag{6.15}$$

The real and imaginary parts of this contour integral are

$$\begin{aligned}\mathrm{Re}\left(\oint_R f(z)dz\right) &= \int_0^\ell [u(x,0)-u(x,h)]\,dx - \int_0^h \big[v(\ell,y)-v(0,y)\big]\,dy\\ \mathrm{Im}\left(\oint_R f(z)dz\right) &= \int_0^\ell [v(x,0)-v(x,h)]\,dx + \int_0^h \big[u(\ell,y)-u(0,y)\big]\,dy.\end{aligned} \tag{6.16}$$

The differences $u(x,0) - u(x,h)$ and $v(\ell,y) - v(0,y)$ in the real part are integrals of the y derivative $u_y(x,y)$ and of the x derivative $v_x(x,y)$

$$\begin{aligned}u(x,0)-u(x,h) &= -\int_0^h u_y(x,y)\,dy\\ v(\ell,y)-v(0,y) &= \int_0^\ell v_x(x,y)\,dx.\end{aligned} \tag{6.17}$$

The real part of the contour integral therefore vanishes due to the second $v_x = -u_y$ of the Cauchy–Riemann conditions (6.10)

$$\begin{aligned} \operatorname{Re}\left(\oint_R f(z)\,dz\right) &= -\int_0^\ell\int_0^h u_y(x,y)\,dy\,dx - \int_0^h\int_0^\ell v_x(x,y)\,dx\,dy \\ &= -\int_0^\ell\int_0^h \left[u_y(x,y)+v_x(x,y)\right]\,dy\,dx = 0. \end{aligned} \tag{6.18}$$

Similarly, differences $v(x,0)-v(x,h)$ and $u(\ell,y)-u(0,y)$ in the imaginary part are integrals of the y derivative $v_y(x,y)$ and of the x derivative $u_x(x,y)$

$$\begin{aligned} v(x,0)-v(x,h) &= -\int_0^h v_y(x,y)\,dy \\ u(\ell,y)-u(0,y) &= \int_0^\ell u_x(x,y)\,dx. \end{aligned} \tag{6.19}$$

Thus the imaginary part of the contour integral vanishes due to the first $u_x = v_y$ of the Cauchy–Riemann conditions (6.10)

$$\begin{aligned} \operatorname{Im}\left(\oint_R f(z)\,dz\right) &= -\int_0^\ell\int_0^h v_y(x,y)\,dy\,dx + \int_0^h\int_0^\ell u_x(x,y)\,dx\,dy \\ &= \int_0^\ell\int_0^h \left[-\,v_y(x,y)+u_x(x,y)\right]\,dy\,dx = 0. \end{aligned} \tag{6.20}$$

A similar argument shows that the contour integral along the four sides of any rectangle vanishes as long as the function $f(z)$ is analytic on and within the rectangle whether or not the rectangle has one corner at the origin $z = 0$.

Suppose a function $f(z)$ is analytic along a closed contour C and also at every point inside it. We can tile the inside area A with a suitable collection of contiguous rectangles some of which might be very small. The integral of $f(z)$ along the perimeter of each rectangle will vanish because each rectangle lies entirely within the region in which $f(z)$ is analytic. Now consider two adjacent rectangles like the two squares in Fig. 6.1. The sum of the two contour integrals around the two adjacent squares is equal to the contour integral around the perimeter of the two squares because the up integral along the right side (dots) of the left square cancels the down integral along the left side of the right square. Thus the sum of the contour integrals around the perimeters of all the rectangles that tile the inside area A amounts to just the integral along the outer contour C. The integral around each rectangle vanishes. So the integral of $f(z)$ along the contour C also must vanish because it is the sum of these vanishing integrals around the rectangles that tile the

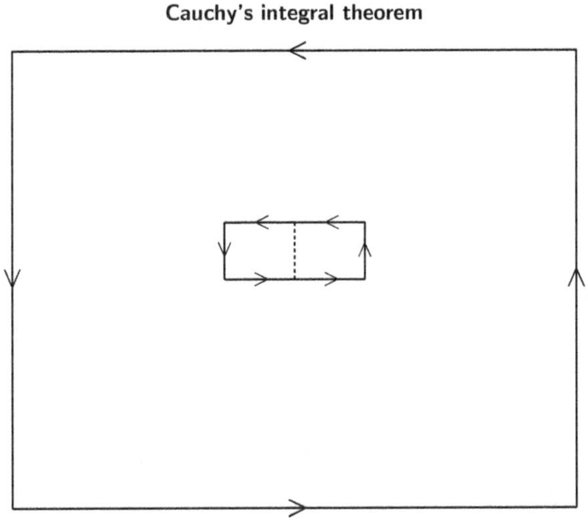

Figure 6.1 The sum of two contour integrals around two adjacent squares is equal to the contour integral around the perimeter of the two squares because the up integral along the right side (dots) of the left square cancels the down integral along the left side (dots) of the right square. A contour integral around a big square is equal to the sum of the contour integrals around the smaller interior squares that tile the big square. Matlab scripts for this chapter's figures of are in Complex-variable_theory at `github.com/kevinecahill`.

inside area A. This is **Cauchy's integral theorem**: The integral of a function $f(z)$ along a closed contour vanishes

$$\oint_C f(z)\, dz = 0 \tag{6.21}$$

if the function $f(z)$ is analytic on the contour and at every point inside it.

What could go wrong? The area A inside the contour might have a hole in it in which the function $f(z)$ is not analytic. To exclude this possibility, we require that the area A inside the contour be **simply connected**, that is, we insist that we be able to shrink every loop in A to a point while keeping the loop inside A. A slice of American cheese is simply connected, a slice of Swiss Emmenthal is not. A dime is simply connected, a washer isn't. The surface of a sphere is simply connected, the surface of a bagel isn't. So another version of Cauchy's integral theorem is that the integral of a function $f(z)$ along a closed contour vanishes if the contour lies within a simply connected region in which $f(z)$ is analytic (Augustin-Louis Cauchy, 1789–1857).

Example 6.4 (Polynomials) Since $dz^{n+1} = (n+1)\, z^n\, dz$, the integral of the entire function z^n along any contour C that ends and starts at the same point z_0 must vanish for any integer $n \geq 0$

$$\oint_C z^n \, dz = \frac{1}{n+1} \oint_C dz^{n+1} = \frac{1}{n+1} \left(z_0^{n+1} - z_0^{n+1} \right) = 0. \tag{6.22}$$

Thus the integral of any polynomial $P(z) = c_0 + c_1 z + c_2 z^2 + \cdots$ along any closed contour C also vanishes

$$\oint_C P(z) \, dz = \oint_C \sum_{n=0}^{m} c_n z^n \, dz = 0. \tag{6.23}$$

Example 6.5 (Tiny circular contour) If $f(z)$ is analytic at z_0, then the definition (6.1) of the derivative $f'(z)$ shows that $f(z) \approx f(z_0) + f'(z_0)\,(z - z_0)$ near z_0 to first order in $z - z_0$. The points of a small circle of radius ϵ and center z_0 are $z = z_0 + \epsilon\, e^{i\theta}$. Since $z - z_0 = \epsilon\, e^{i\theta}$ and $dz = i\epsilon\, e^{i\theta} d\theta$, the closed contour integral around the circle is

$$\begin{aligned} \oint_{\bigcirc} f(z)\, dz &= \int_0^{2\pi} \left[f(z_0) + f'(z_0)\,(z - z_0) \right] i\epsilon\, e^{i\theta} d\theta \\ &= f(z_0) \int_0^{2\pi} i\epsilon\, e^{i\theta} d\theta + f'(z_0) \int_0^{2\pi} \epsilon\, e^{i\theta}\, i\epsilon\, e^{i\theta} d\theta \end{aligned} \tag{6.24}$$

which vanishes because the θ-integrals are zero. Thus the contour integral of an analytic function $f(z)$ around a tiny circle, lying within the region in which $f(z)$ is analytic, vanishes.

Example 6.6 (Tiny square contour) The analyticity of $f(z)$ at $z = z_0$ lets us expand $f(z)$ near z_0 as $f(z) \approx f(z_0) + f'(z_0)\,(z - z_0)$. A tiny square contour consists of four complex segments $dz_1 = \epsilon$, $dz_2 = i\,\epsilon$, $dz_3 = -\epsilon$, and $dz_4 = -i\,\epsilon$. The integral of the constant $f(z_0)$ around the square vanishes

$$\oint_{\square} f(z_0)\, dz = f(z_0) \oint_{\square} dz = f(z_0)\, \left[\epsilon + i\,\epsilon + (-\epsilon) + (-i\,\epsilon)\right] = 0. \tag{6.25}$$

The integral of the second term $f'(z_0)(z - z_0)$ also vanishes. It is the sum of four integrals along the four sides of the tiny square. Like the integral of the constant $f(z_0)$, the integral of the constant $-f'(z_0)\, z_0$ also vanishes. Dropping that term, we are left with the integral of $f'(z_0)\, z$ along the four sides of the tiny square.

The integral from left to right along the bottom of the square where $z = x - i\epsilon/2$ is

$$I_1 = f'(0) \int_{-\epsilon/2}^{\epsilon/2} \left(x - i\frac{\epsilon}{2} \right) dx = -\frac{i\epsilon^2}{2} f'(0). \tag{6.26}$$

The integral up the right side of the square where $z = \epsilon/2 + iy$ is

$$I_2 = f'(0) \int_{-\epsilon/2}^{\epsilon/2} \left(\frac{\epsilon}{2} + iy \right) i dy = \frac{i\epsilon^2}{2} f'(0). \tag{6.27}$$

The integral backwards along the top of the square where $z = x + i\epsilon/2$ is

$$I_3 = f'(0) \int_{\epsilon/2}^{-\epsilon/2} \left(x + i\frac{\epsilon}{2}\right) dx = -\frac{i\epsilon^2}{2} f'(z_0). \tag{6.28}$$

Finally, the integral down the left side where $z = -\epsilon/2 + iy$ is

$$I_4 = f'(0) \int_{\epsilon/2}^{-\epsilon/2} \left(-\frac{\epsilon}{2} + iy\right) i dy = \frac{i\epsilon^2}{2} f'(0). \tag{6.29}$$

These integrals cancel in pairs. Thus the contour integral of an analytic function $f(z)$ around a tiny square of side ϵ is zero to order ϵ^2 as long as the square lies inside the region in which $f(z)$ is analytic.

Suppose a function $f(z)$ is analytic in a simply connected region R and that C and C' are two contours that lie inside R and that both run from z_1 to z_2. The difference of the two contour integrals is an integral along a closed contour C'' that runs from z_1 to z_2 and back to z_1 and that vanishes by Cauchy's theorem

$$\int_{z_1 C}^{z_2} f(z)\, dz - \int_{z_1 C'}^{z_2} f(z)\, dz = \int_{z_1 C}^{z_2} f(z)\, dz + \int_{z_2 C'}^{z_1} f(z)\, dz = \oint_{C''} f(z)\, dz = 0. \tag{6.30}$$

It follows that any two contour integrals that lie within a simply connected region in which $f(z)$ is analytic are equal if they start at the same point z_1 and end at the same point z_2. Thus we may continuously deform the contour of an integral of an analytic function $f(z)$ from C to C' without changing the value of the contour integral as long as as long as these contours and all the intermediate contours lie entirely within the region R and have the same fixed endpoints z_1 and z_2 as in Fig. 6.2

$$\int_{z_1 C}^{z_2} f(z)\, dz = \int_{z_1 C'}^{z_2} f(z)\, dz. \tag{6.31}$$

So a contour integral depends upon its endpoints and upon the function $f(z)$ but not upon the actual contour as long as the deformations of the contour do not push it outside of the region R in which $f(z)$ is analytic.

If the endpoints z_1 and z_2 are the same, then the contour C is closed, and we write the integral as

$$\oint_{z_1 C}^{z_1} f(z)\, dz \equiv \oint_C f(z)\, dz \tag{6.32}$$

with a little circle to denote that the contour is a closed loop. The value of that integral is independent of the contour as long as our deformations of the contour keep it within the domain of analyticity of the function and as long as the contour starts and ends at $z_1 = z_2$. Now suppose that the function $f(z)$ is analytic along the contour and at all points within it. Then we can shrink the

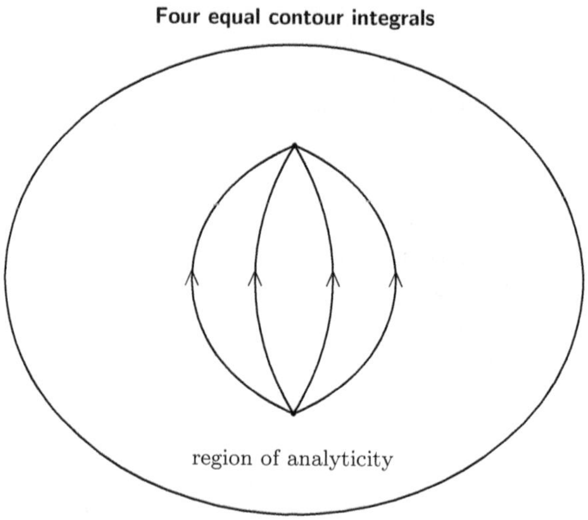

Figure 6.2 As long as the four contours are within the domain of analyticity of $f(z)$ and have the same endpoints, the four contour integrals of that function are all equal.

contour, staying within the domain of analyticity of the function, until the area enclosed is zero and the contour is of zero length – all this without changing the value of the integral. But the value of the integral along such a null contour of zero length is zero. Thus the value of the original contour integral also must be zero

$$\oint_{z_1 C}^{z_1} f(z)\, dz = 0. \tag{6.33}$$

And so we again arrive at **Cauchy's integral theorem**: The contour integral of a function $f(z)$ around a closed contour C lying entirely within the domain of analyticity of the function vanishes

$$\oint_C f(z)\, dz = 0 \tag{6.34}$$

as long as the function $f(z)$ is analytic at all points within the contour.

Example 6.7 (A pole) The function $f(z) = 1/(z - z_0)$ is analytic in a region that is not simply connected because its derivative

$$f'(z) = \lim_{dz \to 0} \left(\frac{1}{z + dz - z_0} - \frac{1}{z - z_0} \right) \frac{1}{dz} = -\frac{1}{(z - z_0)^2} \tag{6.35}$$

exists in the whole complex plane except for the point $z = z_0$.

6.4 Cauchy's Integral Formula

Suppose that $f(z)$ is analytic in a simply connected region R and that z_0 is a point inside this region. We first will integrate the function $f(z)/(z - z_0)$ along a tiny closed counterclockwise contour around the point z_0. The contour is a circle of radius ϵ with center at z_0 with points $z = z_0 + \epsilon\, e^{i\theta}$ for $0 \le \theta \le 2\pi$, and $dz = i\epsilon\, e^{i\theta} d\theta$. Since $z - z_0 = \epsilon\, e^{i\theta}$, the contour integral in the limit $\epsilon \to 0$ is

$$\begin{aligned}\oint_\epsilon \frac{f(z)}{z - z_0}\, dz &= \int_0^{2\pi} \frac{\left[f(z_0) + f'(z_0)\,(z - z_0)\right]}{z - z_0}\, i\epsilon\, e^{i\theta} d\theta \\ &= \int_0^{2\pi} \frac{\left[f(z_0) + f'(z_0)\,\epsilon\, e^{i\theta}\right]}{\epsilon\, e^{i\theta}}\, i\epsilon\, e^{i\theta} d\theta \\ &= \int_0^{2\pi} \left[f(z_0) + f'(z_0)\,\epsilon\, e^{i\theta}\right] i d\theta = 2\pi i\, f(z_0)\end{aligned} \tag{6.36}$$

since the θ-integral involving $f'(z_0)$ vanishes. Thus $f(z_0)$ is the integral

$$f(z_0) = \frac{1}{2\pi i} \oint_\epsilon \frac{f(z)}{z - z_0}\, dz \tag{6.37}$$

which is a miniature version of Cauchy's integral formula.

Now consider the counterclockwise contour C' in Fig. 6.3 which is a big counterclockwise circle, a small clockwise circle, and two parallel straight lines, all within a simply connected region $\mathcal{R}$ in which $f(z)$ is analytic. As we saw in Examples 6.3 and 6.7, the function $1/(z - z_0)$ is analytic except at $z = z_0$. Thus since the product

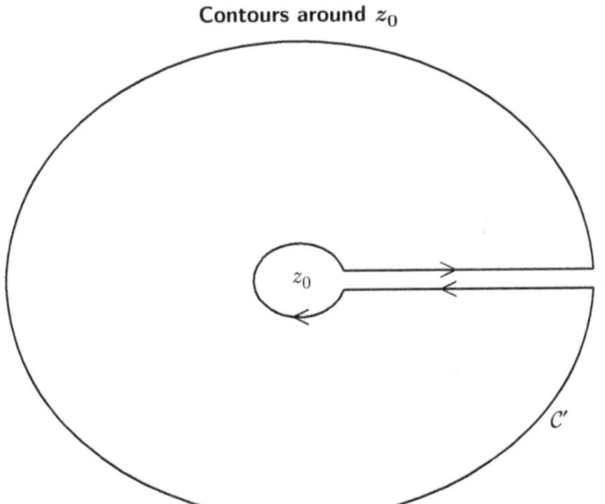

Figure 6.3 The full contour is the sum of a big counterclockwise contour C' and a small clockwise contour, both around z_0, and two straight lines which cancel.

of two analytic functions is analytic (Exercise 6.3), the function $f(z)/(z - z_0)$ is analytic everywhere in $\mathcal{R}$ except at the point z_0. We can withdraw the contour C' to the left of the point z_0 and shrink it to a point without having the contour C' cross z_0. During this process, the integral of $f(z)/(z-z_0)$ does not change. Its final value is zero. So its initial value also is zero

$$0 = \frac{1}{2\pi i} \oint_{C'} \frac{f(z)}{z - z_0}\, dz. \tag{6.38}$$

We let the two straight-line segments approach each other so that they cancel. What remains of contour C' is a big counterclockwise contour C around z_0 and a tiny clockwise circle of radius ϵ around z_0. The tiny clockwise circle integral is the negative of the counterclockwise integral (6.37), so we have

$$0 = \frac{1}{2\pi i} \oint_{C'} \frac{f(z)}{z - z_0}\, dz = \frac{1}{2\pi i} \oint_{C} \frac{f(z)}{z - z_0}\, dz - \frac{1}{2\pi i} \oint_{\epsilon} \frac{f(z)}{z - z_0}\, dz. \tag{6.39}$$

Using the miniature result (6.37), we find

$$f(z_0) = \frac{1}{2\pi i} \oint_{C} \frac{f(z)}{z - z_0}\, dz \tag{6.40}$$

which is **Cauchy's integral formula**.

We can use this formula to compute the first derivative $f'(z)$ of $f(z)$

$$\begin{aligned} f'(z) &= \frac{f(z+dz) - f(z)}{dz} \\ &= \frac{1}{2\pi i} \frac{1}{dz} \oint dz'\, f(z') \left(\frac{1}{z' - z - dz} - \frac{1}{z' - z} \right) \\ &= \frac{1}{2\pi i} \oint dz' \frac{f(z')}{(z' - z - dz)(z' - z)}. \end{aligned} \tag{6.41}$$

So in the limit $dz \to 0$, we get

$$f'(z) = \frac{1}{2\pi i} \oint dz' \frac{f(z')}{(z' - z)^2}. \tag{6.42}$$

The second derivative $f^{(2)}(z)$ of $f(z)$ then is

$$f^{(2)}(z) = \frac{2}{2\pi i} \oint dz' \frac{f(z')}{(z' - z)^3}. \tag{6.43}$$

And its nth derivative $f^{(n)}(z)$ is

$$f^{(n)}(z) = \frac{n!}{2\pi i} \oint dz' \frac{f(z')}{(z' - z)^{n+1}}. \tag{6.44}$$

In these formulas, the contour runs counterclockwise about the point z and lies within the simply connected domain $\mathcal{R}$ in which $f(z)$ is analytic.

Thus a function $f(z)$ that is analytic in a region $\mathcal{R}$ is infinitely differentiable there.

Example 6.8 (Schlaefli's formula for the Legendre polynomials) Rodrigues showed (Section 9.2) that the Legendre polynomial $P_n(x)$ is the nth derivative

$$P_n(x) = \frac{1}{2^n\, n!}\left(\frac{d}{dx}\right)^n (x^2-1)^n. \tag{6.45}$$

Schlaefli used this expression and Cauchy's integral formula (6.44) to represent $P_n(z)$ as the contour integral (Exercise 6.9)

$$P_n(z) = \frac{1}{2^n\, 2\pi i}\oint \frac{(z'^2-1)^n}{(z'-z)^{n+1}}\, dz' \tag{6.46}$$

in which the contour encircles the complex point z counterclockwise. This formula tells us that at $z=1$ the Legendre polynomial is

$$P_n(1) = \frac{1}{2^n\, 2\pi i}\oint \frac{(z'^2-1)^n}{(z'-1)^{n+1}}\, dz' = \frac{1}{2^n\, 2\pi i}\oint \frac{(z'+1)^n}{(z'-1)}\, dz' = 1 \tag{6.47}$$

in which we applied Cauchy's integral formula (6.40) to $f(z) = (z+1)^n$.

Example 6.9 (Bessel functions of the first kind) The counterclockwise integral around the unit circle $z = e^{i\theta}$ of the ratio z^m/z^n in which both m and n are integers is

$$\frac{1}{2\pi i}\oint dz\, \frac{z^m}{z^n} = \frac{1}{2\pi i}\int_0^{2\pi} i e^{i\theta} d\theta\, e^{i(m-n)\theta} = \frac{1}{2\pi}\int_0^{2\pi} d\theta\, e^{i(m+1-n)\theta}. \tag{6.48}$$

If $m+1-n \neq 0$, this integral vanishes because $\exp 2\pi i(m+1-n) = 1$

$$\frac{1}{2\pi}\int_0^{2\pi} d\theta\, e^{i(m+1-n)\theta} = \frac{1}{2\pi}\left[\frac{e^{i(m+1-n)\theta}}{i(m+1-n)}\right]_0^{2\pi} = 0. \tag{6.49}$$

If $m+1-n=0$, the exponential is unity $\exp i(m+1-n)\theta = 1$, and the integral is $2\pi/2\pi = 1$. Thus the original integral is the Kronecker delta

$$\frac{1}{2\pi i}\oint dz\, \frac{z^m}{z^n} = \delta_{m+1,n}. \tag{6.50}$$

The generating function (10.5) for Bessel functions J_m of the first kind is

$$e^{t(z-1/z)/2} = \sum_{m=-\infty}^{\infty} z^m J_m(t). \tag{6.51}$$

Applying our integral formula (6.50) to it, we find

$$\begin{aligned}\frac{1}{2\pi i}\oint dz\, e^{t(z-1/z)/2}\,\frac{1}{z^{n+1}} &= \frac{1}{2\pi i}\oint dz \sum_{m=-\infty}^{\infty} \frac{z^m}{z^{n+1}} J_m(t)\\ &= \sum_{m=-\infty}^{\infty} \delta_{m+1,n+1}\, J_m(t) = J_n(t).\end{aligned} \tag{6.52}$$

Thus letting $z = e^{i\theta}$, we have

$$J_n(t) = \frac{1}{2\pi} \int_0^{2\pi} d\theta \, \exp\left[t \frac{\left(e^{i\theta} - e^{-i\theta}\right)}{2} - in\theta \right] \tag{6.53}$$

or more simply

$$J_n(t) = \frac{1}{2\pi} \int_0^{2\pi} d\theta \, e^{i(t \sin\theta - n\theta)} = \frac{1}{\pi} \int_0^{\pi} d\theta \, \cos(t \sin\theta - n\theta) \tag{6.54}$$

(see Exercise 6.4).

6.5 Harmonic Functions

The Cauchy–Riemann conditions (6.10)

$$u_x = v_y \quad \text{and} \quad u_y = -v_x \tag{6.55}$$

tell us something about the laplacian of the real part u of an analytic function $f = u + iv$. First, the second x-derivative u_{xx} is $u_{xx} = v_{yx} = v_{xy} = -u_{yy}$. So the real part u of an analytic function f is a **harmonic** function

$$u_{xx} + u_{yy} = 0 \tag{6.56}$$

that is, one with a vanishing laplacian. Similarly $v_{xx} = -u_{yx} = -v_{yy}$, so the imaginary part of an analytic function also is a harmonic function

$$v_{xx} + v_{yy} = 0. \tag{6.57}$$

A harmonic function $h(x, y)$ can have saddle points, but not local minima or maxima because at a local minimum both $h_{xx} > 0$ and $h_{yy} > 0$, while at a local maximum both $h_{xx} < 0$ and $h_{yy} < 0$. So in its domain of analyticity, the real and imaginary parts of an analytic function f have neither minima nor maxima.

For static fields, the electrostatic potential $\phi(x, y, z)$ is a harmonic function of the three spatial variables x, y, and z in regions that are free of charge because the electric field is $\mathbf{E} = -\nabla\phi$, and its divergence vanishes $\nabla \cdot \mathbf{E} = 0$ where the charge density is zero. Thus the laplacian of the electrostatic potential $\phi(x, y, z)$ vanishes

$$\nabla \cdot \nabla\phi = \phi_{xx} + \phi_{yy} + \phi_{zz} = 0 \tag{6.58}$$

and $\phi(x, y, z)$ is harmonic where there is no charge. The location of each positive charge is a local maximum of the electrostatic potential $\phi(x, y, z)$ and the location of each negative charge is a local minimum of $\phi(x, y, z)$. But in the absence of charges, the electrostatic potential has neither local maxima nor local minima. Thus one cannot trap charged particles with an electrostatic potential, a result known as Earnshaw's theorem.

The Cauchy–Riemann conditions imply that the real and imaginary parts of an analytic function are harmonic functions with 2-dimensional gradients that are mutually perpendicular

$$(u_x, u_y) \cdot (v_x, v_y) = v_y v_x - v_x v_y = 0. \tag{6.59}$$

And we know that the electrostatic potential is a harmonic function. Thus the real part $u(x, y)$ (or the imaginary part $v(x, y)$) of any analytic function $f(z) = u(x, y) + iv(x, y)$ describes the electrostatic potential $\phi(x, y)$ for some electrostatic problem that does not involve the third spatial coordinate z. The surfaces of constant $u(x, y)$ are the equipotential surfaces, and since the two gradients are orthogonal, the surfaces of constant $v(x, y)$ are the electric field lines.

Example 6.10 (Two-dimensional potentials) The function

$$f(z) = u + iv = E\,z = E\,x + i\,E\,y \tag{6.60}$$

can represent a potential $V(x, y, z) = E\,x$ for which the electric-field lines $\boldsymbol{E} = -E\,\hat{\boldsymbol{x}}$ are lines of constant y. It also can represent a potential $V(x, y, z) = E\,y$ in which $\boldsymbol{E}$ points in the negative y-direction, which is to say along lines of constant x.

Another simple example is the function

$$f(z) = u + iv = z^2 = x^2 - y^2 + 2ixy \tag{6.61}$$

for which $u = x^2 - y^2$ and $v = 2xy$. This function gives us a potential $V(x, y, z)$ whose equipotentials are the hyperbolas $u = x^2 - y^2 = c^2$ and whose electric-field lines are the perpendicular hyperbolas $v = 2xy = d^2$. Equivalently, we may take these last hyperbolas $2xy = d^2$ to be the equipotentials and the other ones $x^2 - y^2 = c^2$ to be the lines of the electric field.

For a third example, we write the variable z as $z = re^{i\theta} = \exp(\ln r + i\theta)$ and use the function

$$f(z) = u(x, y) + iv(x, y) = -\frac{\lambda}{2\pi\epsilon_0}\ln z = -\frac{\lambda}{2\pi\epsilon_0}(\ln r + i\theta) \tag{6.62}$$

which describes the potential $V(x, y, z) = -(\lambda/2\pi\epsilon_0)\ln\sqrt{x^2 + y^2}$ due to a line of charge per unit length $\lambda = q/L$. The electric-field lines are the lines of constant v

$$\boldsymbol{E} = \frac{\lambda}{2\pi\epsilon_0}\frac{(x, y, 0)}{x^2 + y^2} \tag{6.63}$$

or equivalently of constant θ.

6.6 Taylor Series for Analytic Functions

Let's consider the contour integral of the function $f(z')/(z' - z)$ along a circle $\mathcal{C}$ inside a simply connected region $\mathcal{R}$ in which $f(z)$ is analytic. For any point z inside the circle, Cauchy's integral formula (6.40) tells us that

$$f(z) = \frac{1}{2\pi i} \oint_{\mathcal{C}} \frac{f(z')}{z' - z} \, dz'. \tag{6.64}$$

We add and subtract the center z_0 from the denominator $z' - z$

$$f(z) = \frac{1}{2\pi i} \oint_{\mathcal{C}} \frac{f(z')}{z' - z_0 - (z - z_0)} \, dz' \tag{6.65}$$

and then factor the denominator

$$f(z) = \frac{1}{2\pi i} \oint_{\mathcal{C}} \frac{f(z')}{(z' - z_0)\left(1 - \frac{z - z_0}{z' - z_0}\right)} \, dz'. \tag{6.66}$$

From Fig. 6.4, we see that the modulus of the ratio $(z - z_0)/(z' - z_0)$ is less than unity, and so the power series

$$\left(1 - \frac{z - z_0}{z' - z_0}\right)^{-1} = \sum_{n=0}^{\infty} \left(\frac{z - z_0}{z' - z_0}\right)^n \tag{6.67}$$

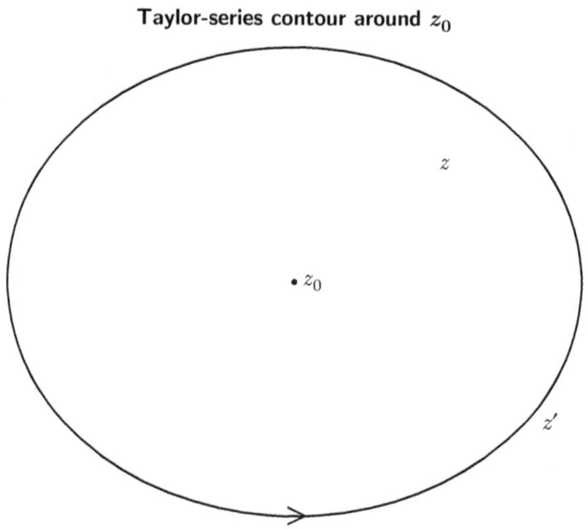

Figure 6.4 Contour of integral for the Taylor series (6.69).

by (5.32–5.35) converges absolutely and uniformly on the circle. We therefore are allowed to integrate the series

$$f(z) = \frac{1}{2\pi i} \oint_C \frac{f(z')}{z' - z_0} \sum_{n=0}^{\infty} \left(\frac{z - z_0}{z' - z_0} \right)^n dz' \tag{6.68}$$

term by term

$$f(z) = \sum_{n=0}^{\infty} (z - z_0)^n \frac{1}{2\pi i} \oint_C \frac{f(z')\, dz'}{(z' - z_0)^{n+1}}. \tag{6.69}$$

Cauchy's integral formula (6.44) tells us that the integral is just the nth derivative $f^{(n)}(z)$ divided by n-factorial. Thus the function $f(z)$ possesses the Taylor series

$$f(z) = \sum_{n=0}^{\infty} \frac{(z - z_0)^n}{n!} f^{(n)}(z_0) \tag{6.70}$$

which converges as long as the point z is inside a circle centered at z_0 that lies within a simply connected region $\mathcal{R}$ in which $f(z)$ is analytic.

6.7 Cauchy's Inequality

Suppose a function $f(z)$ is analytic in a region that includes the disk $|z| \le R$ and that $f(z)$ is bounded by $|f(z)| \le M$ on the circle $|z| = R$ which is the perimeter of the disk. Then by using Cauchy's integral formula (6.44), we may bound the nth derivative $f^{(n)}(0)$ of $f(z)$ at $z = 0$ by

$$\begin{aligned} |f^{(n)}(0)| &\le \frac{n!}{2\pi} \oint \frac{|f(z)||dz|}{|z|^{n+1}} \\ &\le \frac{n!M}{2\pi} \int_0^{2\pi} \frac{R\, d\theta}{R^{n+1}} = \frac{n!M}{R^n} \end{aligned} \tag{6.71}$$

which is Cauchy's inequality. This inequality bounds the terms of the Taylor series (6.70)

$$\sum_{n=0}^{\infty} \frac{|(z - z_0)^n}{n!} |f^{(n)}(z_0)| \le M \sum_{n=0}^{\infty} \frac{|(z - z_0)^n}{R^n} \tag{6.72}$$

showing that it converges (5.35) absolutely and uniformly for $|z - z_0| < R$.

6.8 Liouville's Theorem

Suppose now that $f(z)$ is analytic everywhere (**entire**) and bounded by

$$|f(z)| \le M \quad \text{for all} \quad |z| \ge R_0. \tag{6.73}$$

Then by applying Cauchy's inequality (6.71) at successively larger values of R, we have

$$|f^{(n)}(0)| \le \lim_{R\to\infty} \frac{n!M}{R^n} = 0 \tag{6.74}$$

which shows that for $n \ge 1$ every derivative $f^{(n)}(z)$ vanishes at $z = 0$

$$f^{(n)}(0) = 0. \tag{6.75}$$

But then the Taylor series (5.78) about $z = 0$ for the function $f(z)$ consists of only a single term, and $f(z)$ is a constant

$$f(z) = \sum_{n=0}^{\infty} \frac{z^n}{n!} f^{(n)}(0) = f^{(0)}(0) = f(0). \tag{6.76}$$

So every bounded entire function is a constant (Joseph Liouville, 1809–1882).

6.9 Fundamental Theorem of Algebra

Gauss applied Liouville's theorem to the function

$$f(z) = \frac{1}{P_n(z)} = \frac{1}{c_0 + c_1 z + c_2 z^2 + \cdots + c_n z^n} \tag{6.77}$$

which is the inverse of an arbitrary polynomial of order n. Suppose that the polynomial $P_n(z)$ had no zero, that is, no root anywhere in the complex plane. Then $f(z)$ would be analytic everywhere. Moreover, for sufficiently large $|z|$, the polynomial $P_n(z)$ is approximately $P_n(z) \approx c_n z^n$, and so $f(z)$ would be bounded by something like

$$|f(z)| \le \frac{1}{|c_n| R_0^n} \equiv M \quad \text{for all} \quad |z| \ge R_0. \tag{6.78}$$

So if $P_n(z)$ had no root, then the function $f(z)$ would be a bounded entire function and so would be a constant by Liouville's theorem (6.76). But of course, $f(z) = 1/P_n(z)$ is not a constant unless $n = 0$. Thus any polynomial $P_n(z)$ that is not a constant must have a root, a pole of $f(z)$, so that $f(z)$ is not entire. This is the only exit from the contradiction.

If the root of $P_n(z)$ is at $z = z_1$, then $P_n(z) = (z - z_1)\, P_{n-1}(z)$, in which $P_{n-1}(z)$ is a polynomial of order $n - 1$, and we may repeat the argument for its reciprocal $f_1(z) = 1/P_{n-1}(z)$. In this way, one arrives at the fundamental theorem of algebra: Every polynomial $P_n(z) = c_0 + c_1 z + \cdots + c_n z^n$ has n roots somewhere in the complex plane

$$P_n(z) = c_n\,(z - z_1)(z - z_2)\cdots(z - z_n). \tag{6.79}$$

6.10 Laurent Series

Consider a function $f(z)$ that is analytic in an annulus that contains an outer circle C_1 of radius R_1 and an inner circle C_2 of radius R_2 as in Fig. 6.5. We integrate $f(z)$ along a contour C_{12} within the annulus that encircles the point z in a counterclockwise fashion by following C_1 counterclockwise and C_2 clockwise and a line joining them in both directions. By Cauchy's integral formula (6.40), this contour integral yields $f(z)$

$$f(z) = \frac{1}{2\pi i} \oint_{C_{12}} \frac{f(z')}{z' - z}\, dz'. \tag{6.80}$$

The integrations in opposite directions along the line joining C_1 and C_2 cancel, and we are left with a counterclockwise integral around the outer circle C_1 and a clockwise one around C_2 or *minus* a counterclockwise integral around C_2

$$f(z) = \frac{1}{2\pi i} \oint_{C_1} \frac{f(z')}{z' - z}\, dz' - \frac{1}{2\pi i} \oint_{C_2} \frac{f(z'')}{z'' - z}\, dz''. \tag{6.81}$$

Now from the figure (6.5), the center z_0 of the two concentric circles is closer to the points z'' on the inner circle C_2 than it is to z and also closer to z than to the points z' on C_1

$$\left| \frac{z'' - z_0}{z - z_0} \right| < 1 \quad \text{and} \quad \left| \frac{z - z_0}{z' - z_0} \right| < 1. \tag{6.82}$$

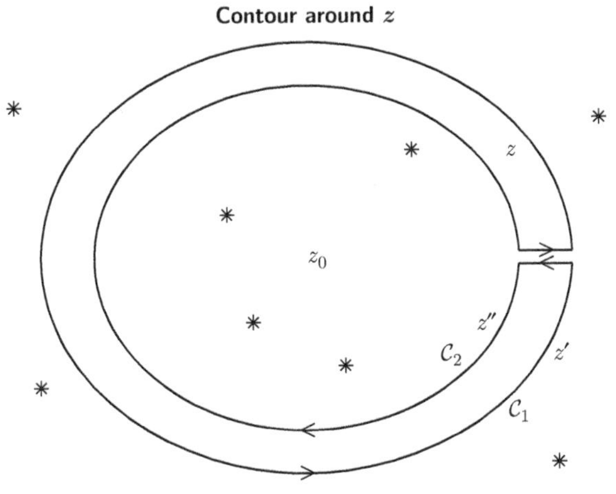

Figure 6.5 A contour consisting of two concentric circles with center at z_0 encircles the point z in a counterclockwise sense. The asterisks are poles or other singularities of the function $f(z)$.

We add and subtract z_0 from each of the denominators (as in (6.65)) and absorb the minus sign before the second integral into its denominator

$$f(z) = \frac{1}{2\pi i}\oint_{C_1} \frac{f(z')}{z' - z_0 - (z - z_0)}\, dz' + \frac{1}{2\pi i}\oint_{C_2} \frac{f(z'')}{z - z_0 - (z'' - z_0)}\, dz''. \tag{6.83}$$

After factoring the two denominators

$$\begin{aligned} f(z) = {} & \frac{1}{2\pi i}\oint_{C_1} \frac{f(z')}{(z' - z_0)\left[1 - (z - z_0)/(z' - z_0)\right]}\, dz' \\ & + \frac{1}{2\pi i}\oint_{C_2} \frac{f(z'')}{(z - z_0)\left[1 - (z'' - z_0)/(z - z_0)\right]}\, dz'' \end{aligned} \tag{6.84}$$

we expand them, as in the series (6.68), in power series that converge absolutely and uniformly on the two contours

$$\begin{aligned} f(z) = {} & \sum_{n=0}^{\infty} (z - z_0)^n \frac{1}{2\pi i}\oint_{C_1} \frac{f(z')}{(z' - z_0)^{n+1}}\, dz' \\ & + \sum_{m=0}^{\infty} \frac{1}{(z - z_0)^{m+1}} \frac{1}{2\pi i}\oint_{C_2} (z'' - z_0)^m\, f(z'')\, dz''. \end{aligned} \tag{6.85}$$

Since the functions being integrated are analytic between the two circles, we may move the contours, without changing the values of the integrals, to a common counterclockwise contour C about any circle of radius $R_2 \le R \le R_1$ between the two circles C_1 and C_2. We then set $m = -n - 1$, or $n = -m - 1$, so as to combine the two sums into one sum on n from $-\infty$ to ∞

$$f(z) = \sum_{n=-\infty}^{\infty} (z - z_0)^n \frac{1}{2\pi i}\oint_{C} \frac{f(z')}{(z' - z_0)^{n+1}}\, dz'. \tag{6.86}$$

This **Laurent series** often is written as

$$f(z) = \sum_{n=-\infty}^{\infty} a_n(z_0)\,(z - z_0)^n \tag{6.87}$$

with

$$a_n(z_0) = \frac{1}{2\pi i}\oint_{C} \frac{f(z)}{(z - z_0)^{n+1}}\, dz \tag{6.88}$$

(Pierre Laurent, 1813–1854).

The coefficient $a_{-1}(z_0)$ is called the **residue** of the function $f(z)$ at z_0. Its significance will be discussed in Section 6.13. Most functions have Laurent series that start at some least integer $-\ell$

$$f(z) = \sum_{n=-\ell}^{\infty} a_n(z_0)\,(z - z_0)^n \tag{6.89}$$

rather than at $-\infty$. For such functions, we can pick off the coefficients a_n one by one without doing the integrals (6.88). The first one $a_{-\ell}$ is the limit

$$a_{-\ell}(z_0) = \lim_{z\to z_0} (z - z_0)^{\ell} f(z). \tag{6.90}$$

The second Laurent coefficient $a_{-\ell+1}(z_0)$ is given by the recipe

$$a_{-\ell+1}(z_0) = \lim_{z\to z_0} (z - z_0)^{\ell-1} \left[f(z) - (z - z_0)^{-\ell} a_{-\ell}(z_0) \right]. \tag{6.91}$$

The third coefficient requires a recipe with two subtractions, and so forth.

6.11 Singularities

A function $f(z)$ that is analytic for all z is **entire**. Entire functions have no singularities except possibly as $|z| \to \infty$, which some call the **point at infinity**.

A function $f(z)$ has an **isolated singularity** at z_0 if it is analytic in a small disk about z_0 but not analytic at that point.

A function $f(z)$ has a **pole** of order $n > 0$ at a point z_0 if $(z-z_0)^n f(z)$ is analytic at z_0 but $(z - z_0)^{n-1} f(z)$ has an isolated singularity at z_0. A pole of order $n = 1$ is called a **simple pole**. Poles are isolated singularities. A function is **meromorphic** if it is analytic for all z except for poles.

Example 6.11 (Poles) The function

$$f(z) = \prod_{j=1}^{n} \frac{1}{(z - j)^j} \tag{6.92}$$

has a pole of order j at $z = j$ for $j = 1, 2, \ldots, n$. It is meromorphic.

An **essential singularity** is a pole of infinite order. If a function $f(z)$ has an essential singularity at z_0, then its Laurent series (6.86) really runs from $n = -\infty$ and not from $n = -\ell$ as in (6.89). Essential singularities are spooky: if a function $f(z)$ has an essential singularity at w, then inside every disk around w, $f(z)$ takes on *every* complex number, with at most one exception, an infinite number of times (Émile Picard, 1856–1941).

Example 6.12 (Essential singularity) The function $f(z) = \exp(1/z)$ has an essential singularity at $z = 0$ because its Laurent series (6.86)

$$f(z) = e^{1/z} = \sum_{m=0}^{\infty} \frac{1}{m!}\frac{1}{z^m} = \sum_{n=-\infty}^{0} \frac{1}{|n|!} z^n \tag{6.93}$$

runs from $n = -\infty$. Near $z = 0$, $f(z) = \exp(1/z)$ takes on every complex number except 0 an infinite number of times.

Example 6.13 (Meromorphic function with two poles) The function $f(z) = 1/z(z+1)$ has poles at $z = 0$ and at $z = -1$ but otherwise is analytic; it is meromorphic. We may expand it in a Laurent series (6.87–6.88)

$$f(z) = \frac{1}{z(z+1)} = \sum_{n=-\infty}^{\infty} a_n z^n \tag{6.94}$$

about $z = 0$ for $|z| < 1$. The coefficient a_n is the integral

$$a_n = \frac{1}{2\pi i} \oint_C \frac{dz}{z^{n+2}(z+1)} \tag{6.95}$$

in which the contour C is a counterclockwise circle of radius $r < 1$. Since $|z| < 1$, we may expand $1/(1+z)$ as the power series

$$\frac{1}{1+z} = \sum_{m=0}^{\infty} (-z)^m. \tag{6.96}$$

Doing the integrals, we find

$$a_n = \sum_{m=0}^{\infty} \frac{1}{2\pi i} \oint_C (-z)^m \frac{dz}{z^{n+2}} = \sum_{m=0}^{\infty} (-1)^m \, r^{m-n-1} \, \delta_{m,n+1} \tag{6.97}$$

for $n \ge -1$ and zero otherwise. Thus the Laurent series about $z = 0$ for $f(z)$ is

$$f(z) = \frac{1}{z(z+1)} = \sum_{n=-1}^{\infty} (-1)^{n+1} \, z^n. \tag{6.98}$$

It starts at $n = -1$, not at $n = -\infty$, because $f(z)$ is meromorphic with only a simple pole at $z = 0$.

Example 6.14 (Argument principle) Consider the counterclockwise integral

$$\frac{1}{2\pi i} \oint_C f(z) \frac{g'(z)}{g(z)} \, dz \tag{6.99}$$

along a contour C that lies inside a simply connected region R in which $f(z)$ is analytic and $g(z)$ meromorphic. If the function $g(z)$ has a zero or a pole of order n at $z = w \in R$ and no other singularity in R so that

$$g(z) = a_n(w)(z - w)^n, \tag{6.100}$$

then the ratio g'/g is

$$\frac{g'(z)}{g(z)} = \frac{n(z-w)^{n-1}}{(z-w)^n} = \frac{n}{z-w} \tag{6.101}$$

and the integral is

$$\frac{1}{2\pi i} \oint_C f(z) \frac{g'(z)}{g(z)} dz = \frac{1}{2\pi i} \oint_C f(z) \frac{n}{z-w} dz = n\, f(w). \tag{6.102}$$

Any function $g(z)$ meromorphic in R will have a Laurent series

$$g(z) = \sum_{k=n}^{\infty} a_k(w)(z-w)^k \tag{6.103}$$

about each point $w \in R$. One may show (Exercise 6.19) that as $z \to w$ the ratio g'/g again approaches (6.101). It follows that the integral (6.99) is a sum of $n_\ell f(w_\ell)$ at the zeros and poles of $g(z)$ that lie within the contour C

$$\frac{1}{2\pi i} \oint_C f(z) \frac{g'(z)}{g(z)} dz = \sum_\ell \frac{1}{2\pi i} \oint_C f(z) \frac{n_\ell}{z-w_\ell} = \sum_\ell n_\ell\, f(w_\ell) \tag{6.104}$$

in which $|n_\ell|$ is the multiplicity of the ℓth zero or pole.

6.12 Analytic Continuation

We saw in Section 6.6 that a function $f(z)$ that is analytic within a circle of radius R about a point z_0 possesses a Taylor series (6.70)

$$f(z) = \sum_{n=0}^{\infty} \frac{(z-z_0)^n}{n!} f^{(n)}(z_0) \tag{6.105}$$

that converges for all z inside the disk $|z - z_0| < R$. Suppose z' is the singularity of $f(z)$ that is closest to z_0. Pick a point z_1 in the disk $|z - z_0| < R$ that is not on the line from z_0 to the nearest singularity z'. The function $f(z)$ is analytic at z_1 because z_1 is within the circle of radius R about the point z_0, and so $f(z)$ has a Taylor series expansion like (6.105) but about the point z_1. Often the circle of convergence of this power series about z_1 will extend beyond the original disk $|z-z_0| < R$. If so, the two power series, one about z_0 and the other about z_1, define the function $f(z)$ and extend its domain of analyticity beyond the original disk $|z - z_0| < R$. Such an extension of the range of an analytic function is an **analytic continuation**.

Example 6.15 (Geometric series) The power series

$$f(z) = \sum_{n=0}^{\infty} z^n \tag{6.106}$$

converges and defines an analytic function for $|z| < 1$. But for such z, we may sum the series to

$$f(z) = \frac{1}{1-z}. \tag{6.107}$$

By summing the series (6.106), we have analytically continued the function $f(z)$ to the whole complex plane apart from its simple pole at $z = 1$.

Example 6.16 (Gamma function) Euler's form of the gamma function is the integral

$$\Gamma(z) = \int_0^{\infty} e^{-t}\, t^{z-1}\, dt = (z-1)! \tag{6.108}$$

which makes $\Gamma(z)$ analytic in the right half-plane $\mathrm{Re}\, z > 0$. But by successively using the relation $\Gamma(z+1) = z\,\Gamma(z)$, we may extend $\Gamma(z)$ into the left half-plane

$$\Gamma(z) = \frac{1}{z}\,\Gamma(z+1) = \frac{1}{z}\,\frac{1}{z+1}\,\Gamma(z+2) = \frac{1}{z}\,\frac{1}{z+1}\,\frac{1}{z+2}\,\Gamma(z+3). \tag{6.109}$$

The last expression defines $\Gamma(z)$ as a function that is analytic for $\mathrm{Re}\, z > -3$ apart from simple poles at $z = 0, -1$, and -2. Proceeding in this way, we may analytically continue the gamma function to the whole complex plane apart from the negative integers and zero. The analytically continued gamma function is represented by Weierstrass's formula

$$\Gamma(z) = \frac{1}{z}\, e^{-\gamma z} \left[\prod_{n=1}^{\infty} \left(1 + \frac{z}{n}\right) e^{-z/n} \right]^{-1}. \tag{6.110}$$

Example 6.17 (Riemann's zeta function) Ser found an analytic continuation

$$\zeta(z) = \sum_{n=1}^{\infty} \frac{1}{n^z} = \frac{1}{z-1} \sum_{n=0}^{\infty} \frac{1}{n+1} \sum_{k=0}^{n} \binom{n}{k} \frac{(-1)^k}{(k+1)^{z-1}} \tag{6.111}$$

of Riemann's zeta function (5.105) to the whole complex plane except for the point $z = 1$ (Joseph Ser, 1875–1954).

Example 6.18 (Dimensional regularization) The loop diagrams of quantum field theory involve badly divergent integrals like

$$I(4) = \int \frac{d^4 q}{(2\pi)^4} \frac{\left(q^2\right)^a}{\left(q^2 + \alpha^2\right)^b} \tag{6.112}$$

where often $a = 0$ and $b = 2$ and $\alpha^2 > 0$. Gerardus 't Hooft (1946–) and Martinus J. G. Veltman (1931–) promoted the number of spacetime dimensions from 4 to a complex number d. The resulting integral has the value (Srednicki, 2007, p. 102)

$$I(d) = \int \frac{d^d q}{(2\pi)^d} \frac{(q^2)^a}{\left(q^2+\alpha^2\right)^b} = \frac{\Gamma(b-a-d/2)\,\Gamma(a+d/2)}{(4\pi)^{d/2}\,\Gamma(b)\,\Gamma(d/2)} \frac{1}{(\alpha^2)^{b-a-d/2}} \tag{6.113}$$

and so defines a function of the complex variable d that is analytic everywhere except for simple poles at $d = 2(n-a+b)$ where $n = 0, 1, 2, \ldots, \infty$. At these poles, the formula

$$\Gamma(-n+z) = \frac{(-1)^n}{n!}\left(\frac{1}{z} - \gamma + \sum_{k=1}^{n} \frac{1}{k} + O(z)\right) \tag{6.114}$$

where $\gamma = 0.5772...$ is the Euler–Mascheroni constant (5.8) can be useful.

6.13 Calculus of Residues

A contour integral of an analytic function $f(z)$ does not change unless the endpoints move or the contour crosses a singularity or leaves the region of analyticity (Section 6.3). Let us consider the integral of a function $f(z)$ along a counterclockwise contour $\mathcal{C}$ that encircles n poles at z_k for $k = 1, \ldots, n$ in a simply connected region $\mathcal{R}$ in which $f(z)$ is meromorphic. We may shrink the area within the contour $\mathcal{C}$ without changing the value of the integral until the area is infinitesimal and the contour is a sum of n tiny counterclockwise circles $\mathcal{C}_k$ around the n poles

$$\oint_{\mathcal{C}} f(z)\,dz = \sum_{k=1}^{n} \oint_{\mathcal{C}_k} f(z)\,dz. \tag{6.115}$$

These tiny counterclockwise integrals around the poles at z_i are $2\pi i$ times the residues $a_{-1}(z_i)$ defined by Laurent's formula (6.88) with $n = -1$. So the whole counterclockwise integral is $2\pi i$ times the sum of the residues of the enclosed poles of the function $f(z)$

$$\oint_{\mathcal{C}} f(z)\,dz = 2\pi i \sum_{k=1}^{n} a_{-1}(z_k) = 2\pi i \sum_{k=1}^{n} \text{Res}(f, z_k) \tag{6.116}$$

a result that is known as the **residue theorem**.

In general, one must do each tiny counterclockwise integral about each pole z_i, but simple poles are an important special case. If w is a simple pole of the function $f(z)$, then near it $f(z)$ is given by its Laurent series (6.87) as

$$f(z) = \frac{a_{-1}(w)}{z-w} + \sum_{n=0}^{\infty} a_n(w)\,(z-w)^n. \tag{6.117}$$

In this case, its residue is by (6.90) with $-\ell = -1$

$$a_{-1}(w) = \lim_{z\to w} (z-w)\,f(z) \tag{6.118}$$

which usually is easier to do than the integral (6.88)

$$a_{-1}(w) = \frac{1}{2\pi i} \oint_C f(z) dz. \tag{6.119}$$

Example 6.19 (A function with simple poles) The integral of the function

$$f(z) = \sum_{n=1}^{\infty} \frac{z}{z - n^{-s}} \tag{6.120}$$

along a circle of radius 2 with center at $z = 0$ is just the sum of its residues

$$\frac{1}{2\pi i} \oint f(z)\, dz = \sum_{n=1}^{\infty} \lim_{z \to n^{-s}} (z - n^{-s}) f(z) = \sum_{n=1}^{\infty} \frac{1}{n^s} = \zeta(s) \tag{6.121}$$

which is the zeta function (5.105).

Example 6.20 (Cauchy's integral formula) Suppose the function $f(z)$ is analytic within a region $\mathcal{R}$ and that C is a counterclockwise contour that encircles a point w in $\mathcal{R}$. Then the counterclockwise contour C encircles the simple pole at w of the function $f(z)/(z - w)$, which is its only singularity in $\mathcal{R}$. By applying the residue theorem and formula (6.118) for the residue $a_{-1}(w)$ of the function $f(z)/(z - w)$, we find

$$\oint_C \frac{f(z)}{z - w} dz = 2\pi i\, a_{-1}(w) = 2\pi i \lim_{z \to w} (z - w) \frac{f(z)}{z - w} = 2\pi i\, f(w). \tag{6.122}$$

So Cauchy's integral formula (6.40) is an example of the calculus of residues.

Example 6.21 (A meromorphic function) By the residue theorem (6.116), the integral of the function

$$f(z) = \frac{1}{z - 1} \frac{1}{(z - 2)^2} \tag{6.123}$$

along the circle $C = 4e^{i\theta}$ for $0 \le \theta \le 2\pi$ is the sum of the residues at $z = 1$ and $z = 2$

$$\oint_C f(z)\, dz = 2\pi i \left[a_{-1}(1) + a_{-1}(2)\right]. \tag{6.124}$$

The function $f(z)$ has a simple pole at $z = 1$, and so we may use the formula (6.118) to evaluate the residue $a_{-1}(1)$ as

$$a_{-1}(1) = \lim_{z \to 1} (z - 1)\, f(z) = \lim_{z \to 1} \frac{1}{(z - 2)^2} = 1 \tag{6.125}$$

instead of using Cauchy's integral formula (6.40) to do the integral of $f(z)$ along a tiny circle about $z = 1$, which gives the same result

$$a_{-1}(1) = \frac{1}{2\pi i} \oint \frac{dz}{z - 1} \frac{1}{(z - 2)^2} = \frac{1}{(1 - 2)^2} = 1. \tag{6.126}$$

The residue $a_{-1}(2)$ is the integral of $f(z)$ along a tiny circle about $z = 2$, which we do by using Cauchy's integral formula (6.42)

$$a_{-1}(2) = \frac{1}{2\pi i} \oint \frac{dz}{(z-2)^2} \frac{1}{z-1} = \frac{d}{dz} \frac{1}{z-1} \bigg|_{z=2} = -\frac{1}{(2-1)^2} = -1 \quad (6.127)$$

getting the same answer as if we had used the recipe (6.90) for a_{-2}

$$a_{-2}(2) = \lim_{z\to 2} (z-2)^2 \frac{1}{(z-1)(z-2)^2} = 1 \quad (6.128)$$

and (6.91) for a_{-1}

$$a_{-1}(2) = \lim_{z\to 2} (z-2) \left[\frac{1}{(z-1)(z-2)^2} - \frac{a_{-2}(2)}{(z-2)^2} \right] = -1. \quad (6.129)$$

The sum of these two residues is zero, and so the integral (6.124) vanishes. Another way of evaluating this integral is to deform it, not into two tiny circles about the two poles, but rather into a huge circle $z = Re^{i\theta}$ and to notice that as $R \to \infty$ the modulus of this integral vanishes

$$\left| \oint f(z)\, dz \right| \approx \frac{2\pi}{R^2} \to 0. \quad (6.130)$$

This contour is an example of a ghost contour.

6.14 Ghost Contours

Often one needs to do an integral that is not a closed counterclockwise contour. Integrals along the real axis occur frequently. One sometimes can convert a line integral into a closed contour by adding a contour along which the integral vanishes, a **ghost contour**. We have just seen an example (6.130) of a ghost contour, and we shall see more of them in what follows.

Example 6.22 (Using ghost contours) Consider the integral

$$I_b = \int_{-\infty}^{\infty} \frac{1}{(x-i)(x-2i)(x-3i)}\, dx. \quad (6.131)$$

We could do the integral by adding a contour $Re^{i\theta}$ from $\theta = 0$ to $\theta = \pi$. In the limit $R \to \infty$, the integral of $1/[(z-i)(z-2i)(z-3i)]$ along this contour vanishes; it is a ghost contour. The original integral I and the ghost contour encircle the three poles, and so we could compute I by evaluating the residues at those poles. But we also could add a ghost contour around the lower half-plane. This contour and the real line encircle no poles. So we get $I = 0$ without doing any work at all.

Example 6.23 (Fourier transform of a gaussian) During our computation of the Fourier transform of a gaussian (4.16–4.19), we promised to justify the shift in the variable of integration from x to $x + ik/2m^2$ in this chapter. So let us consider the contour integral of the entire function $f(z) = \exp(-m^2 z^2)$ over a rectangular closed contour along the real axis from $-R$ to R and then from $z = R$ to $z = R + ic$ and then from there to $z = -R + ic$ and then to $z = -R$. Since $f(z)$ is analytic within the contour, the integral is zero

$$\oint dz\, e^{-m^2z^2} = \int_{-R}^{R} dz\, e^{-m^2z^2} + \int_{R}^{R+ic} dz\, e^{-m^2z^2} + \int_{R+ic}^{-R+ic} dz\, e^{-m^2z^2} + \int_{-R+ic}^{-R} dz\, e^{-m^2z^2} = 0$$

for all finite positive values of R and so also in the limit $R \to \infty$. The two contours in the imaginary direction are of length c and are damped by the factor $\exp(-m^2R^2)$, and so they vanish in the limit $R \to \infty$. They are ghost contours. It follows then from this last equation in the limit $R \to \infty$ that

$$\int_{-\infty}^{\infty} dx\, e^{-m^2(x+ic)^2} = \int_{-\infty}^{\infty} dx\, e^{-m^2x^2} = \frac{\sqrt{\pi}}{m} \tag{6.132}$$

which is the promised result (4.18). Setting $c = k/(2m^2)$ and dividing both sides of (6.132) by $\sqrt{2\pi}$, we see that the Fourier transform of a gaussian is a gaussian (4.19)

$$\tilde{f}(k) = \int_{-\infty}^{\infty} \frac{dx}{\sqrt{2\pi}}\, e^{-ikx}\, e^{-m^2x^2} = \frac{1}{\sqrt{2}\,m}\, e^{-k^2/4m^2}. \tag{6.133}$$

Dividing both sides of this formula by $\sqrt{2\pi}$ and setting $x = p/\hbar$, $k = -\epsilon\dot{q}$, and $m^2 = \epsilon\hbar^2/(2m)$, we get

$$\int_{-\infty}^{\infty} \exp\left(-\epsilon\frac{p^2}{2m} + i\epsilon\frac{\dot{q}\,p}{\hbar}\right)\frac{dp}{2\pi\hbar} = \sqrt{\frac{m}{2\pi\epsilon\hbar^2}}\,\exp\left(-\epsilon\frac{m\dot{q}^2}{2\hbar^2}\right) \tag{6.134}$$

a formula we'll use in Section 20.5 to derive path integrals for partition functions.

The earlier relation (6.132) implies (Exercise 6.22) that

$$\int_{-\infty}^{\infty} dx\, e^{-m^2(x+z)^2} = \int_{-\infty}^{\infty} dx\, e^{-m^2x^2} = \frac{\sqrt{\pi}}{m} \tag{6.135}$$

for $m > 0$ and arbitrary complex z.

Example 6.24 (A cosine integral) To compute the integral

$$I_c = \int_0^{\infty} \frac{\cos x}{q^2 + x^2}\, dx, \qquad q > 0, \tag{6.136}$$

we use the evenness of the integrand to extend the integration

$$I_c = \frac{1}{2}\int_{-\infty}^{\infty} \frac{\cos x}{q^2 + x^2}\, dx, \tag{6.137}$$

write the cosine as $[\exp(ix) + \exp(-ix)]/2$, and factor the denominators

$$I_c = \frac{1}{4}\int_{-\infty}^{\infty} \frac{e^{ix}}{(x - iq)(x + iq)}\, dx + \frac{1}{4}\int_{-\infty}^{\infty} \frac{e^{-ix}}{(x - iq)(x + iq)}\, dx. \tag{6.138}$$

We promote x to a complex variable z and add the contours $z = Re^{i\theta}$ and $z = Re^{-i\theta}$ as θ goes from 0 to π respectively to the first and second integrals. The term $\exp(iz)dz/(q^2 + z^2) = \exp(iR\cos\theta - R\sin\theta)iRe^{i\theta}d\theta/(q^2 + R^2e^{2i\theta})$ vanishes in the limit $R \to \infty$, so the first contour is a counterclockwise ghost contour. A similar argument applies to the second (clockwise) contour, and we have

$$I_c = \frac{1}{4}\oint \frac{e^{iz}}{(z-iq)(z+iq)}\,dz + \frac{1}{4}\oint \frac{e^{-iz}}{(z-iq)(z+iq)}\,dz. \tag{6.139}$$

The first integral picks up the pole at iq and the second the pole at $-iq$

$$I_c = \frac{i\pi}{2}\left(\frac{e^{-q}}{2iq} + \frac{e^{-q}}{2iq}\right) = \frac{\pi e^{-q}}{2q}. \tag{6.140}$$

So the value of the integral is $\pi e^{-q}/2q$.

Example 6.25 (Third-harmonic microscopy) An ultra-short laser pulse intensely focused in a medium generates a third-harmonic electric field E_3 in the forward direction proportional to the integral (Boyd, 2000)

$$E_3 \propto \chi^{(3)}\,E_0^3 \int_{-\infty}^{\infty} e^{i\,\Delta k\, z}\,\frac{dz}{(1+2iz/b)^2} \tag{6.141}$$

along the axis of the beam as in Fig. 6.6. Here $b = 2\pi t_0^2 n/\lambda = kt_0^2$ in which $n = n(\omega)$ is the index of refraction of the medium, λ is the wavelength of the laser light in the medium, and t_0 is the transverse or waist radius of the gaussian beam, defined by $E(r) = E\exp(-r^2/t_0^2)$.

When the dispersion is normal, that is when $dn(\omega)/d\omega > 0$, the shift in the wave vector $\Delta k = 3\omega[n(\omega) - n(3\omega)]/c$ is negative. Since $\Delta k < 0$, the exponential is damped when $z = x + iy$ is in the lower half-plane (LHP)

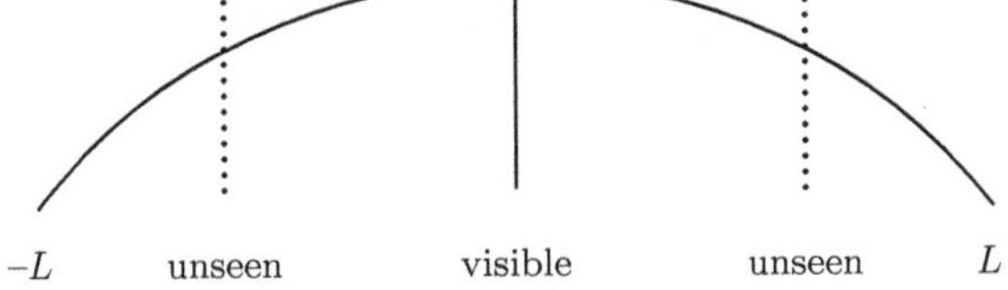

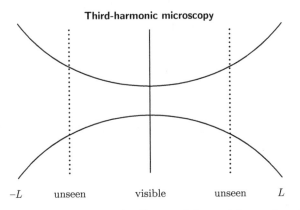

Figure 6.6 In the limit in which the distance L is much larger than the wavelength λ, the integral (6.141) is nonzero when an edge (solid line) lies where the beam is focused but not when a feature (...) lies where the beam is not focused. Only features within the focused region are visible.

$$e^{i\,\Delta k\,z} = e^{i\,\Delta k\,(x+iy)} = e^{i\,\Delta k\,x}\,e^{-\Delta k\,y}. \tag{6.142}$$

So as we did in Example 6.24, we will add a contour around the lower half-plane ($z = R\,e^{i\theta}$, $\pi \le \theta \le 2\pi$, and $dz = i\,Re^{i\theta}d\theta$) because in the limit $R \to \infty$, the integral along it vanishes; it is a ghost contour.

The function $f(z) = \exp(i\,\Delta k\,z)/(1+2iz/b)^2$ has a double pole at $z = ib/2$ which is in the UHP since the length $b > 0$, but no singularity in the LHP $y < 0$. So the integral of $f(z)$ along the closed contour from $z = -R$ to $z = R$ and then along the ghost contour vanishes. But since the integral along the ghost contour vanishes, so does the integral from $-R$ to R. Thus when the dispersion is normal, the third-harmonic signal vanishes, $E_3 = 0$, as long as the medium with constant $\chi^{(3)}(z)$ effectively extends from $-\infty$ to ∞ so that its edges are in the unfocused region like the dotted lines of Fig. 6.6. But an edge with $\Delta k > 0$ in the focused region like the solid line of the figure does make a third-harmonic signal E_3. Third-harmonic microscopy lets us see features instead of background.

Example 6.26 (Green and Bessel) Let us evaluate the Fourier transform

$$I(x) = \int_{-\infty}^{\infty} dk\,\frac{e^{ikx}}{k^2+m^2} \tag{6.143}$$

of the function $1/(k^2+m^2)$. If $x > 0$, then the exponential deceases with $\mathrm{Im}\,k$ in the upper half-plane. So as in Example 6.24, the semicircular contour $k = R\,e^{i\theta}$ for $0 \le \theta \le \pi$ on which $dk = i\,Re^{i\theta}d\theta$ is a ghost contour. So if $x > 0$, then we can add this contour to the integral $I(x)$ without changing it. Thus $I(x)$ is equal to the closed contour integral along the real axis and the semicircular ghost contour

$$I(x) = \oint dk\,\frac{e^{ikx}}{k^2+m^2} = \oint dk\,\frac{e^{ikx}}{(k+im)(k-im)}. \tag{6.144}$$

This closed contour encircles the simple pole at $k = im$ and no other singularity, and so we may shrink the contour into a tiny circle around the pole. Along that tiny circle, the function $e^{ikx}/(k+im)$ is simply $e^{-mx}/2im$, and so

$$I(x) = \frac{e^{-mx}}{2im}\oint \frac{dk}{k-im} = 2\pi i\,\frac{e^{-mx}}{2im} = \frac{\pi\,e^{-mx}}{m} \quad \text{for} \quad x > 0. \tag{6.145}$$

Similarly if $x < 0$, we can add the semicircular ghost contour $k = R\,e^{i\theta}$, $\pi \le \theta \le 2\pi$, $dk = i\,Re^{i\theta}d\theta$ with k running around the perimeter of the lower half-plane. So if $x < 0$, then we can write the integral $I(x)$ as a shrunken closed contour that runs clockwise around the pole at $k = -im$

$$I(x) = \frac{e^{mx}}{-2im}\oint \frac{dk}{k+im} = -2\pi i\,\frac{e^{mx}}{-2im} = \frac{\pi\,e^{mx}}{m} \quad \text{for} \quad x < 0. \tag{6.146}$$

We combine the two cases (6.145) and (6.146) into the result

$$\int_{-\infty}^{\infty} dk \, \frac{e^{ikx}}{k^2 + m^2} = \frac{\pi}{m} \, e^{-m|x|}. \tag{6.147}$$

We can use this formula to develop an expression for the Green's function of the laplacian in cylindical coordinates. Setting $\boldsymbol{x}' = 0$ and $r = |\boldsymbol{x}| = \sqrt{\rho^2 + z^2}$ in the Coulomb Green's function (4.113), we have

$$G(r) = \frac{1}{4\pi r} = \frac{1}{4\pi\sqrt{\rho^2 + z^2}} = \int \frac{d^3\boldsymbol{k}}{(2\pi)^3} \, \frac{1}{\boldsymbol{k}^2} \, e^{i\boldsymbol{k}\cdot\boldsymbol{x}}. \tag{6.148}$$

The integral over the z-component of $\boldsymbol{k}$ is (6.147) with $m^2 = k_x^2 + k_y^2 \equiv k^2$

$$\int_{-\infty}^{\infty} dk_z \, \frac{e^{ik_z z}}{k_z^2 + k^2} = \frac{\pi}{k} \, e^{-k|z|}. \tag{6.149}$$

So with $k_x x + k_y y \equiv k\rho\cos\phi$, the Green's function is

$$\frac{1}{4\pi\sqrt{\rho^2 + z^2}} = \int_0^{\infty} \frac{\pi \, dk}{(2\pi)^3} \int_0^{2\pi} d\phi \, e^{ik\rho\cos\phi} \, e^{-k|z|}. \tag{6.150}$$

The ϕ integral is a representation (6.54 & 10.7) of the Bessel function $J_0(k\rho)$

$$J_0(k\rho) = \int_0^{2\pi} \frac{d\phi}{2\pi} \, e^{ik\rho\cos\phi}. \tag{6.151}$$

Thus we arrive at Bessel's formula for the Coulomb Green's function

$$\frac{1}{4\pi\sqrt{\rho^2 + z^2}} = \int_0^{\infty} \frac{dk}{4\pi} \, J_0(k\rho) \, e^{-k|z|} \tag{6.152}$$

in cylindical coordinates (Schwinger et al., 1998, p. 166).

Example 6.27 (Yukawa and Green) We saw in Example 4.12 that the Green's function for Yukawa's differential operator (4.126) is

$$G_Y(\boldsymbol{x}) = \int \frac{d^3k}{(2\pi)^3} \, \frac{e^{i\boldsymbol{k}\cdot\boldsymbol{x}}}{\boldsymbol{k}^2 + m^2}. \tag{6.153}$$

Letting $\boldsymbol{k}\cdot\boldsymbol{x} = kr\cos\theta$ in which $r = |\boldsymbol{x}|$, we find

$$G_Y(r) = \int_0^{\infty} \frac{k^2 dk}{(2\pi)^2} \int_{-1}^{1} \frac{e^{ikr\cos\theta}}{k^2 + m^2} \, d\cos\theta = \frac{1}{ir} \int_0^{\infty} \frac{dk}{(2\pi)^2} \frac{k}{k^2 + m^2} \left(e^{ikr} - e^{-ikr} \right)$$

$$= \frac{1}{ir} \int_{-\infty}^{\infty} \frac{dk}{(2\pi)^2} \frac{k}{k^2 + m^2} \, e^{ikr} = \frac{1}{ir} \int_{-\infty}^{\infty} \frac{dk}{(2\pi)^2} \frac{k}{(k - im)(k + im)} \, e^{ikr}.$$

We add a ghost contour that loops over the upper-half plane and get

$$G_Y(r) = \frac{2\pi i}{(2\pi)^2 ir} \, \frac{im}{2im} \, e^{-mr} = \frac{e^{-mr}}{4\pi r} \tag{6.154}$$

which Yukawa proposed as the potential between two hadrons due to the exchange of a particle of mass m, the pion. Because the mass of the pion is 140 MeV, the range of the Yukawa potential is $\hbar/mc = 1.4 \times 10^{-15}$ m.

Example 6.28 (Green's function for the laplacian in n dimensions) The Green's function for the laplacian $-\Delta G(x) = \delta^{(n)}(x)$ is

$$G(x) = \int \frac{1}{k^2} e^{ik\cdot x} \frac{d^n k}{(2\pi)^n} \tag{6.155}$$

in n dimensions. We use the formula

$$\frac{1}{k^2} = \int_0^\infty e^{-\lambda k^2} d\lambda \tag{6.156}$$

to write it as a gaussian integral

$$G(x) = \int e^{-\lambda k^2 + ik\cdot x} d\lambda \frac{d^n k}{(2\pi)^n}. \tag{6.157}$$

We now complete the square in the exponent

$$-\lambda k^2 + ik\cdot x = -\lambda\,(k - ix/2\lambda)^2 - x^2/4\lambda, \tag{6.158}$$

use our gaussian formula (6.132), and with $\alpha = x^2/4\lambda$ write the Green's function as

$$\begin{aligned} G(x) &= \int_0^\infty d\lambda \int \frac{d^n k}{(2\pi)^n} e^{-x^2/4\lambda} e^{-\lambda(k-ix/2\lambda)^2} = \int_0^\infty d\lambda \int \frac{d^n k}{(2\pi)^n} e^{-x^2/4\lambda} e^{-\lambda k^2} \\ &= \int_0^\infty e^{-x^2/4\lambda} \frac{d\lambda}{(4\pi\lambda)^{n/2}} = \frac{(x^2)^{1-n/2}}{4\pi^{n/2}} \int_0^\infty e^{-\alpha} \alpha^{n/2-2} d\alpha \\ &= \frac{\Gamma(n/2-1)}{4\pi^{n/2}(x^2)^{(n/2-1)}}. \end{aligned} \tag{6.159}$$

Our formula (5.67) for $\Gamma(n+\frac{1}{2})$ says that $\Gamma(1/2) = \sqrt{\pi}$, and so this formula (6.159) for $n = 3$ gives $G(x) = 1/4\pi|x|$ which is (4.113); since $\Gamma(1) = 1$, it also gives for $n = 4$

$$G(x) = \frac{1}{4\pi^2 x^2}. \tag{6.160}$$

Example 6.29 (Yukawa Green's function in n dimensions) The Yukawa Green's function which satisfies $(-\Delta + m^2)G(x) = \delta^{(n)}(x)$ in n dimensions is the integral (6.155) with k^2 replaced by $k^2 + m^2$

$$G(x) = \int \frac{1}{k^2 + m^2} e^{ik\cdot x} \frac{d^n k}{(2\pi)^n}. \tag{6.161}$$

Using the integral formula (6.156), we write it as a gaussian integral

$$G(x) = \int e^{-\lambda(k^2+m^2)+ik\cdot x} \frac{d\lambda d^n k}{(2\pi)^n}. \tag{6.162}$$

Completing the square as in (6.158), we have

$$G(x) = \int e^{-x^2/4\lambda} e^{-\lambda(k-ix/2\lambda)^2-\lambda m^2} \frac{d\lambda d^n k}{(2\pi)^n} = \int e^{-x^2/4\lambda} e^{-\lambda\,(k^2+m^2)} \frac{d\lambda d^n k}{(2\pi)^n}$$

$$= \int_0^\infty e^{-x^2/4\lambda-\lambda m^2} \frac{d\lambda}{(4\pi\lambda)^{n/2}}. \tag{6.163}$$

We can relate this to a Bessel function by setting $\lambda = (|x|/2m)\exp(-y)$

$$G(x) = \frac{1}{(4\pi)^{n/2}} \left(\frac{2m}{x}\right)^{(n/2-1)} \int_{-\infty}^{\infty} e^{-mx\cosh y+(n/2-1)y}\,dy$$

$$= \frac{2}{(4\pi)^{n/2}} \left(\frac{2m}{x}\right)^{(n/2-1)} \int_0^\infty e^{-mx\cosh y} \cosh(n/2-1)y\,dy$$

$$= \frac{2}{(4\pi)^{n/2}} \left(\frac{2m}{x}\right)^{(n/2-1)} K_{n/2-1}(mx) \tag{6.164}$$

where $x = |x| = \sqrt{x^2}$ and K is a modified Bessel function of the second kind (10.108). If $n = 3$, this is (Exercise 6.29) the Yukawa potential (6.154).

Example 6.30 (A Fourier transform) As another example, let's consider the integral

$$J(x) = \int_{-\infty}^{\infty} \frac{e^{ikx}}{(k^2+m^2)^2}\,dk. \tag{6.165}$$

We may add ghost contours as in the preceding example, but now the integrand has double poles at $k = \pm im$, and so we must use Cauchy's integral formula (6.44) for the case of $n = 1$, which is Eq. (6.42). For $x > 0$, we add a ghost contour in the UHP and find

$$J(x) = \oint \frac{e^{ikx}}{(k+im)^2(k-im)^2}\,dk = 2\pi i \frac{d}{dk} \frac{e^{ikx}}{(k+im)^2}\bigg|_{k=im}$$

$$= \frac{\pi}{2m^2}\left(x + \frac{1}{m}\right) e^{-mx}. \tag{6.166}$$

If $x < 0$, then we add a ghost contour in the LHP and find

$$J(x) = \oint \frac{e^{ikx}}{(k+im)^2(k-im)^2}\,dk = -2\pi i \frac{d}{dk} \frac{e^{ikx}}{(k-im)^2}\bigg|_{k=-im}$$

$$= \frac{\pi}{2m^2}\left(-x + \frac{1}{m}\right) e^{mx}. \tag{6.167}$$

Putting the two together, we get

$$J(x) = \int_{-\infty}^{\infty} \frac{e^{ikx}}{(k^2+m^2)^2}dk = \frac{\pi}{2m^2}\left(|x| + \frac{1}{m}\right) e^{-m|x|} \tag{6.168}$$

as the Fourier transform of $1/(k^2+m^2)^2$.

Example 6.31 (Integral of a complex gaussian) As another example of the use of ghost contours, let us use one to do the integral

$$I = \int_{-\infty}^{\infty} e^{wx^2}\, dx \tag{6.169}$$

in which the real part of the nonzero complex number $w = u + iv = \rho e^{i\phi}$ is negative or zero

$$u \le 0 \quad \Longleftrightarrow \quad \frac{\pi}{2} \le \phi \le \frac{3\pi}{2}. \tag{6.170}$$

We first write the integral I as twice that along half the x-axis

$$I = 2 \int_0^{\infty} e^{wx^2}\, dx. \tag{6.171}$$

If we promote x to a complex variable $z = re^{i\theta}$, then wz^2 will be negative if $\phi + 2\theta = \pi$, that is, if $\theta = (\pi - \phi)/2$ where in view of (6.170) θ lies in the interval $-\pi/4 \le \theta \le \pi/4$.

The closed pie-shaped contour of Fig. 6.7 (down the real axis from $z = 0$ to $z = R$, along the arc $z = R\exp(i\theta')$ as θ' goes from 0 to θ, and then down the line $z = r\exp(i\theta)$ from $z = R\exp(i\theta)$ to $z = 0$) encloses no singularities of the function $f(z) = \exp(wz^2)$. Hence the integral of $\exp(wz^2)$ along that contour vanishes.

To show that the arc is a ghost contour, we bound it by

$$\left| \int_0^{\theta} e^{(u+iv)R^2 e^{2i\theta'}} R\, d\theta' \right| \le \int_0^{\theta} \exp\left[uR^2 \cos 2\theta' - vR^2 \sin 2\theta' \right] R\, d\theta'$$

$$\le \int_0^{\theta} e^{-vR^2 \sin 2\theta'} R\, d\theta'. \tag{6.172}$$

Here $v \sin 2\theta' \ge 0$, and so if v is positive, then so is θ'. Then $0 \le \theta' \le \pi/4$, and so $\sin(2\theta') \ge 4\theta'/\pi$. Thus since $u < 0$, we have the upper bound

$$\left| \int_0^{\theta} e^{(u+iv)R^2 e^{2i\theta'}} R\, d\theta' \right| \le \int_0^{\theta} e^{-4vR^2\theta'/\pi} R\, d\theta' = \frac{\pi(e^{-4vR^2\theta'/\pi} - 1)}{4vR} \tag{6.173}$$

which vanishes in the limit $R \to \infty$. (If v is negative, then so is θ', the pie-shaped contour is in the fourth quadrant, $\sin(2\theta') \le 4\theta'/\pi$, and the inequality (6.173) holds with absolute-value signs around the second integral.)

Since by Cauchy's integral theorem (6.21) the integral along the pie-shaped contour of Fig. 6.7 vanishes, it follows that

$$\frac{1}{2} I + \int_{Re^{i\theta}}^{0} e^{wz^2}\, dz = 0. \tag{6.174}$$

But the choice $\theta = (\pi - \phi)/2$ implies that on the line $z = r\exp(i\theta)$ the quantity wz^2 is negative, $wz^2 = -\rho r^2$. Thus with $dz = \exp(i\theta)dr$, we have

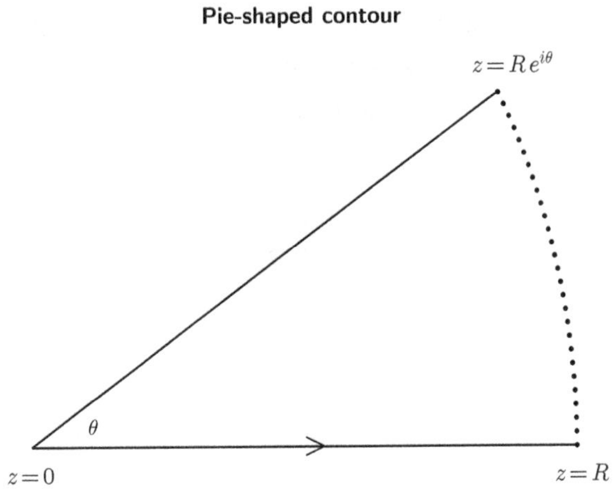

Figure 6.7 The integral of the entire function $\exp(wz^2)$ along the pie-shaped closed contour vanishes by Cauchy's theorem.

$$I = 2\int_0^{Re^{i\theta}} e^{wz^2}\, dz = 2e^{i\theta}\int_0^R e^{-\rho r^2}\, dr \tag{6.175}$$

so that as $R \to \infty$

$$I = 2e^{i\theta}\int_0^\infty e^{-\rho r^2}\, dr = e^{i\theta}\sqrt{\frac{\pi}{\rho}} = \sqrt{\frac{\pi}{\rho e^{-2i\theta}}}. \tag{6.176}$$

Finally from $\theta = (\pi - \phi)/2$ and $w = \rho\exp(i\phi)$, we find that for $\mathrm{Re}\, w \le 0$

$$\int_{-\infty}^{\infty} e^{wx^2}\, dx = \sqrt{\frac{\pi}{-w}} \tag{6.177}$$

as long as $w \neq 0$. Shifting x by a complex number b, we still have

$$\int_{-\infty}^{\infty} e^{w(x-b)^2}\, dx = \sqrt{\frac{\pi}{-w}} \tag{6.178}$$

as long as $\mathrm{Re}\, w < 0$. If $w = ia \neq 0$ and a and b are real, then

$$\int_{-\infty}^{\infty} e^{ia(x-b)^2}\, dx = \sqrt{\frac{i\pi}{a}} \quad \text{or} \quad \int_{-\infty}^{\infty} e^{iax^2-2iabx}\, dx = \sqrt{\frac{i\pi}{a}}\, e^{-iab^2}. \tag{6.179}$$

Setting $x = p$, $a = -\epsilon/(2m\hbar)$, and $b = m\dot{q}$ in the last equation, and dividing both sides by $2\pi\hbar$, we find

$$\int_{-\infty}^{\infty} \exp\left(-i\epsilon\frac{p^2}{2m\hbar} + i\epsilon\frac{\dot{q}\,p}{\hbar}\right)\frac{dp}{2\pi\hbar} = \sqrt{\frac{m}{2\pi i\epsilon\hbar}}\,\exp\left(i\epsilon\frac{m\dot{q}^2}{2\hbar}\right) \tag{6.180}$$

a formula we'll use in Section 20.3 to derive path integrals for probability amplitudes.

Let us try to express the line integral of a not necessarily analytic function $f(x, y) = u(x, y) + iv(x, y)$ along a closed counterclockwise contour C as an integral over the surface enclosed by the contour. The contour integral is

$$\oint_C (u + iv)(dx + idy) = \oint_C (u\,dx - v\,dy) + i\oint_C (v\,dx + u\,dy). \tag{6.181}$$

Now since the contour C is counterclockwise, the differential dx is negative at the top of the curve with coordinates $(x, y_+(x))$ and positive at the bottom $(x, y_-(x))$. So the first line integral is the surface integral

$$\begin{aligned}\oint_C u\,dx &= \int \left[u(x, y_-(x)) - u(x, y_+(x))\right] dx \\ &= -\int \left[\int_{y_-(x)}^{y_+(x)} u_y(x, y)dy\right] dx \\ &= -\int u_y\,|dxdy| = -\int u_y\,da\end{aligned} \tag{6.182}$$

in which $da = |dxdy|$ is a positive element of area. Similarly, we find

$$i\oint_C v\,dx = -i\int v_y\,|dxdy| = -i\int v_y\,da. \tag{6.183}$$

The dy integrals are then:

$$-\oint_C v\,dy = -\int v_x\,|dxdy| = -\int v_x\,da \tag{6.184}$$

$$i\oint_C u\,dy = i\int u_x\,|dxdy| = i\int u_x\,da. \tag{6.185}$$

Combining (6.181–6.185), we find

$$\oint_C (u + iv)(dx + idy) = -\int (u_y + v_x)\,da + i\int (-v_y + u_x)\,da. \tag{6.186}$$

This formula holds whether or not the function $f(x, y)$ is analytic. But if $f(x, y)$ is analytic on and within the contour C, then it satisfies the Cauchy–Riemann conditions (6.10) within the contour, and so both surface integrals vanish. The contour integral then is zero, which is Cauchy's integral theorem (6.34).

The contour integral of the function $f(x, y) = u(x, y) + iv(x, y)$ differs from zero (its value if $f(x, y)$ is analytic in $z = x + iy$) by the surface integrals of $u_y + v_x$ and $u_x - v_y$

$$\left|\oint_C f(z)dz\right|^2 = \left|\oint_C (u + iv)(dx + idy)\right|^2 = \left|\int (u_y + v_x)da\right|^2 + \left|\int (u_x - v_y)da\right|^2 \tag{6.187}$$

which vanish when $f = u + iv$ satisfies the Cauchy–Riemann conditions (6.10).

Example 6.32 (The integral of a nonanalytic function) The integral formula (6.186) can help us evaluate contour integrals of functions that are not analytic. The function

$$f(x, y) = \frac{1}{x + iy + i\epsilon} \, \frac{1}{1 + x^2 + y^2} \tag{6.188}$$

is the product of an analytic function $1/(z + i\epsilon)$, where ϵ is tiny and positive, and a nonanalytic real one $r(x, y) = 1/(1+z^*z)$. The $i\epsilon$ pushes the pole in $u + iv = 1/(z + i\epsilon)$ into the lower half-plane. The real and imaginary parts of f are

$$U(x, y) = u(x, y)\, r(x, y) = \frac{x}{x^2 + (y + \epsilon)^2} \, \frac{1}{1 + x^2 + y^2} \tag{6.189}$$

and

$$V(x, y) = v(x, y)\, r(x, y) = \frac{-y - \epsilon}{x^2 + (y + \epsilon)^2} \, \frac{1}{1 + x^2 + y^2}. \tag{6.190}$$

We will use (6.186) to compute the contour integral I of f along the real axis from $-\infty$ to ∞ and then along the ghost contour $z = x + iy = Re^{i\theta}$ for $0 \le \theta \le \pi$ and $R \to \infty$ around the upper half-plane

$$I = \oint f(x, y)\, dz = \int_{-\infty}^{\infty} dx \int_0^{\infty} dy \left[-U_y - V_x + i\left(-V_y + U_x\right)\right]. \tag{6.191}$$

Since u and v satisfy the Cauchy–Riemann conditions (6.10), the terms in the area integral simplify to $-U_y - V_x = -ur_y - vr_x$ and $-V_y + U_x = -vr_y + ur_x$. So the integral I is

$$I = \int_{-\infty}^{\infty} dx \int_0^{\infty} dy \, \left[-ur_y - vr_x + i(-vr_y + ur_x)\right] \tag{6.192}$$

or explicitly

$$I = \int_{-\infty}^{\infty} dx \int_0^{\infty} dy \frac{-2\epsilon x - 2i(x^2 + y^2 + \epsilon y)}{\left[x^2 + (y + \epsilon)^2\right] \left(1 + x^2 + y^2\right)^2}. \tag{6.193}$$

We let $\epsilon \to 0$ and find

$$I = -2i \int_{-\infty}^{\infty} dx \int_0^{\infty} dy \frac{1}{\left(1 + x^2 + y^2\right)^2}. \tag{6.194}$$

Changing variables to $\rho^2 = x^2 + y^2$, we have

$$I = -4\pi i \int_0^{\infty} d\rho \, \frac{\rho}{(1 + \rho^2)^2} = 2\pi i \int_0^{\infty} d\rho \, \frac{d}{d\rho} \frac{1}{1 + \rho^2} = -2\pi i \tag{6.195}$$

which is simpler than evaluating the integral (6.191) directly.

6.15 Logarithms and Cuts

By definition, a function f is single valued; it maps every number z in its domain into a unique image $f(z)$. A function that maps only one number z in its domain into each $f(z)$ in its range is said to be **one to one**. A one-to-one function $f(z)$ has a well-defined inverse function $f^{-1}(z)$.

The exponential function is one to one when restricted to the real numbers. It maps every real number x into a positive number $\exp(x)$. It has an inverse function $\ln(x)$ that maps every positive number $\exp(x)$ back into x. But the exponential function is not one to one on the complex numbers because $\exp(z + 2\pi ni) = \exp(z)$ for every integer n. Because it is **many to one**, the exponential function has no inverse function on the complex numbers. Its would-be inverse function ln maps it to $\ln(\exp(z))$ or $z + 2\pi ni$ which is not unique. It has in it an arbitrary integer n.

In other words, when exponentiated, the logarithm of a complex number z returns $\exp(\ln z) = z$. So if $z = r\exp(i\theta)$, then a suitable logarithm is $\ln z = \ln r + i\theta$. But what is θ? In the polar representation of z, the argument θ can just as well be $\theta + 2\pi n$ because both give $z = r\exp(i\theta) = r\exp(i\theta + i2\pi n)$. So $\ln r + i\theta + i2\pi n$ is a correct value for $\ln[r\exp(i\theta)]$ for every integer n.

People usually want *one* of the correct values of a logarithm, rather than all of them. Two conventions are common. In the first convention, the angle θ is zero along the positive real axis and increases continuously as the point z moves counterclockwise around the origin, until at points just below the positive real axis, $\theta = 2\pi - \epsilon$ is slightly less than 2π. In this convention, the value of θ drops by 2π as one crosses the positive real axis moving counterclockwise. This discontinuity on the positive real axis is called a **cut**.

The second common convention puts the cut on the negative real axis. Here the value of θ is the same as in the first convention when the point z is in the upper half-plane. But in the lower half-plane, θ decreases from 0 to $-\pi$ as the point z moves clockwise from the positive real axis to just below the negative real axis, where $\theta = -\pi + \epsilon$. As one crosses the negative real axis moving clockwise or up, θ jumps by 2π while crossing the cut. The two conventions agree in the upper half-plane but differ by 2π in the lower half-plane.

Sometimes it is convenient to place the cut on the positive or negative imaginary axis – or along a line that makes an arbitrary angle with the real axis. In any particular calculation, we are at liberty to define the polar angle θ by placing the cut anywhere we like, but we must not change from one convention to another in the same computation.

6.16 Powers and Roots

The logarithm is the key to many other functions to which it passes its arbitrariness. For instance, any power a of $z = r\exp(i\theta)$ is defined as

$$z^a = \exp(a \ln z) = \exp[a(\ln r + i\theta + i2\pi n)] = r^a\, e^{ia\theta}\, e^{i2\pi n a}. \tag{6.196}$$

So z^a is not unique unless a is an integer. The square root, for example, has a sign ambiguity

$$\sqrt{z} = \exp\left[\frac{1}{2}(\ln r + i\theta + i2\pi n)\right] = \sqrt{r}\, e^{i\theta/2}\, e^{in\pi} = (-1)^n \sqrt{r}\, e^{i\theta/2} \tag{6.197}$$

and changes sign when we change θ by 2π as we cross a cut. The mth root

$$\sqrt[m]{z} = z^{1/m} = \exp\left(\frac{\ln z}{m}\right) \tag{6.198}$$

changes by $\exp(\pm 2\pi i/m)$ when we cross a cut and change θ by 2π. And when $a = u + iv$ is a complex number, z^a is

$$z^a = e^{a\ln z} = e^{(u+iv)(\ln r + i\theta + i2\pi n)} = r^{u+iv}\, e^{(-v+iu)(\theta+2\pi n)} \tag{6.199}$$

which changes by $\exp[2\pi(-v+iu)]$ as we cross a cut.

Example 6.33 (i^i) The number $i = \exp(i\pi/2 + i2\pi n)$ for any integer n. So the general value of i^i is $i^i = \exp[i(i\pi/2 + i2\pi n)] = \exp(-\pi/2 - 2\pi n)$.

One can define a sequence of mth-root functions

$$\left(z^{1/m}\right)_n = \exp\left(\frac{\ln r + i(\theta + 2\pi n)}{m}\right) \tag{6.200}$$

one for each integer n. These functions are the **branches** of the mth-root function. One can merge all the branches into one **multivalued** mth-root function. Using a convention for θ, one would extend the $n = 0$ branch to the $n = 1$ branch by winding counterclockwise around the point $z = 0$. One would encounter no discontinuity as one passed from one branch to another. The point $z = 0$, where any cut starts, is called a **branch point** because by winding around it, one passes smoothly from one branch to another. Such branches, introduced by Riemann, can be associated with any multivalued analytic function not just with the mth root.

Example 6.34 (Explicit square roots) If the cut in the square root $\sqrt{z}$ is on the negative real axis, then an explicit formula for the square root of $x + iy$ is

$$\sqrt{x+iy} = \sqrt{\frac{\sqrt{x^2+y^2}+x}{2}} + i\,\mathrm{sign}(y)\sqrt{\frac{\sqrt{x^2+y^2}-x}{2}} \tag{6.201}$$

in which $\mathrm{sign}(y) = \mathrm{sgn}(y) = y/|y|$. On the other hand, if the cut in the square root $\sqrt{z}$ is on the positive real axis, then an explicit formula for the square root of $x + iy$ is

$$\sqrt{x+iy} = \mathrm{sign}(y)\sqrt{\frac{\sqrt{x^2+y^2}+x}{2}} + i\sqrt{\frac{\sqrt{x^2+y^2}-x}{2}} \tag{6.202}$$

(Exercise 6.30).

Example 6.35 (Cuts) Cuts are discontinuities, so people place them where they do the least harm. For the function

$$f(z) = \sqrt{z^2-1} = \sqrt{(z-1)(z+1)} \tag{6.203}$$

two principal conventions work well. We could put the cut in the definition of the angle θ along either the positive or the negative real axis. And we'd get a bonus: the sign discontinuity (a factor of -1) from $\sqrt{z-1}$ would cancel the one from $\sqrt{z+1}$ except for $-1 \le z \le 1$. So the function $f(z)$ would have a discontinuity or a cut only for $-1 \le z \le 1$.

But now suppose we had to work with the function

$$f(z) = \sqrt{z^2+1} = \sqrt{(z-i)(z+i)}. \tag{6.204}$$

If we used one of the usual conventions, we'd have two semi-infinite cuts. So we put the θ-cut on the positive or negative imaginary axis, and the function $f(z)$ now has a cut running along the imaginary axis only from $-i$ to i.

Example 6.36 (Square-root cut) To evaluate the integral

$$I_s = \int_0^\infty \frac{dx}{(x+a)^2\sqrt{x}}, \quad a > 0, \tag{6.205}$$

we put the cut on the positive real axis. The integral backwards along and just below the positive real axis

$$I_- = \int_\infty^0 \frac{dx}{(x+a)^2\sqrt{x-i\epsilon}} = -\int_\infty^0 \frac{dx}{(x+a)^2\sqrt{x+i\epsilon}} = I_s \tag{6.206}$$

is the same as I_s since a minus sign from the square root cancels the minus sign due to the backwards direction.

Since

$$\lim_{|z|\to\infty} \frac{|z|}{|z+a|^2|\sqrt{z}|} = 0, \tag{6.207}$$

the integrals of $f(z) = 1/[\,(z+a)^2\sqrt{z}\,]$ along the contours $z = R\exp(i\theta)$ for $0 < \theta < \pi$ and for $\pi < \theta < 2\pi$ vanish as $R \to \infty$. So these contours are ghost contours. We then add a pair of cancelling integrals along the negative real axis up to the pole at $z = -a$ and then add a clockwise loop C around it. As in Fig. 6.8, the integral along this collection of contours encloses no singularity and therefore vanishes

$$0 = I_s + I_- + I_{\mathcal{G}_+} + I_{\mathcal{G}_-} + I_C. \tag{6.208}$$

Thus $2I_s = -\,I_C$, and so from Cauchy's integral formula (6.44) for $n = 1$, we have

$$I_s = -\frac{1}{2}I_C = -\frac{1}{2}\oint_C \frac{1}{(z+a)^2\sqrt{z}}\,dz = -i\pi \frac{d}{dz}z^{-1/2}\bigg|_{z=-a} = \frac{\pi}{2a^{3/2}} \tag{6.209}$$

which one may check with the Mathematica command Assuming [$a > 0$, Integrate [$1/((x+a)^2$*Sqrt[x]), {x,0,Infinity}].

Example 6.37 (Contour integral with a cut) Let's compute the integral

$$I = \int_0^\infty \frac{x^a}{(x+1)^2}\,dx \tag{6.210}$$

for $-1 < a < 1$. We promote x to a complex variable z and put the cut on the positive real axis. Since

$$\lim_{|z|\to\infty} \frac{|z|^{a+1}}{|z+1|^2} = 0, \tag{6.211}$$

the integrand vanishes faster than $1/|z|$, and we may add two ghost contours, $\mathcal{G}_+$ counterclockwise around the upper half-plane and $\mathcal{G}_-$ counterclockwise around the lower half-plane, as shown in Fig. 6.8.

We add a contour C that runs from $-\infty$ to the double pole at $z = -1$, loops around that pole, and then runs back to $-\infty$; the two long contours along the negative real axis cancel because the cut in θ lies on the positive real axis. So the contour integral along C is just the clockwise integral around the double pole which by Cauchy's integral formula (6.42) is

$$\oint_C \frac{z^a}{(z-(-1))^2}\,dz = -\,2\pi i\,\frac{dz^a}{dz}\bigg|_{z=-1} = 2\pi i\,a\,e^{\pi a i}. \tag{6.212}$$

We also add the integral I_- from ∞ to 0 just below the real axis

$$I_- = \int_\infty^0 \frac{(x-i\epsilon)^a}{(x-i\epsilon+1)^2}\,dx = \int_\infty^0 \frac{\exp(a(\ln(x)+2\pi i))}{(x+1)^2}\,dx \tag{6.213}$$

which is

$$I_- = -\,e^{2\pi a i}\int_0^\infty \frac{x^a}{(x+1)^2}\,dx = -\,e^{2\pi a i}\,I. \tag{6.214}$$

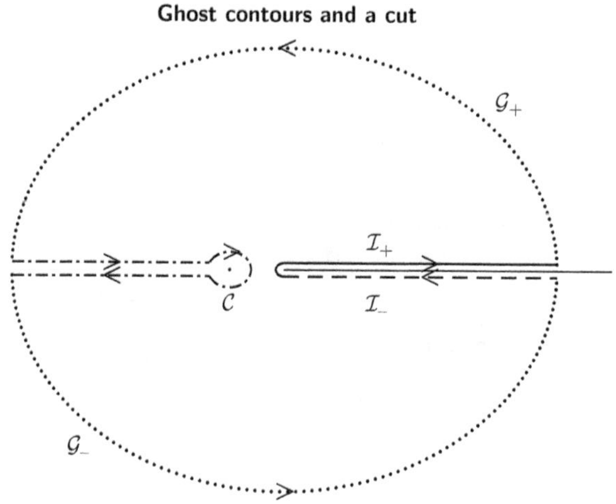

Figure 6.8 The integrals of $f(z) = 1/[\,(x+a)^2\sqrt{z}\,]$ as well as that of $f(z) = z^a/(z+1)^2$ along the ghost contours $\mathcal{G}_+$ and $\mathcal{G}_-$ and the contours $\mathcal{C}$, $\mathcal{I}_-$, and $\mathcal{I}_+$ vanish because the combined contour encircles no poles of either $f(z)$. The cut (solid line) runs from the origin to infinity along the positive real axis.

Now the sum of all these contour integrals is zero because it is a closed contour that encloses no singularity. So we have

$$0 = \left(1 - e^{2\pi a i}\right) I + 2\pi i\, a\, e^{\pi a i} \tag{6.215}$$

or

$$I = \int_0^\infty \frac{x^a}{(x+1)^2}\, dx = \frac{\pi a}{\sin(\pi a)} \tag{6.216}$$

as the value of the integral (6.210).

Example 6.38 (Euler's reflection formula) The beta function (5.76) for $x = z$ and $y = 1 - z$ is the integral

$$\mathrm{B}(z, 1-z) = \Gamma(z)\,\Gamma(1-z) = \int_0^1 t^{z-1}\,(1-t)^{-z}\, dt. \tag{6.217}$$

Setting $t = u/(1+u)$, so that $u = t/(1-t)$ and $dt = 1/(1+u)^2$, we have

$$\mathrm{B}(z, 1-z) = \int_0^\infty \frac{u^{z-1}}{1+u}\, du. \tag{6.218}$$

We integrate $f(u) = u^{z-1}/(1+u)$ along the contour of the preceding example (6.37) which includes the ghost contour $\mathcal{G} = \mathcal{G}_+ \cup \mathcal{G}_-$ and runs down both sides of the cut along the positive real axis. Since $f(u)$ is analytic inside the contour, the integral vanishes

$$0 = \int_{\mathcal{I}_+} f(u)\,du + \int_{\mathcal{G}} f(u)\,du + \int_{\mathcal{C}} f(u)\,du + \int_{\mathcal{I}_-} f(u)\,du. \tag{6.219}$$

The clockwise contour $\mathcal{C}$ is

$$\int_{\mathcal{C}} \frac{u^{z-1}}{1+u}\,du = -2\pi i(-1)^{z-1} = -2\pi i e^{i\pi(z-1)} = 2\pi i e^{i\pi z}. \tag{6.220}$$

The contour $\mathcal{I}_+$ runs just above the positive real axis, and the integral of $f(u)$ along it is the desired integral $\mathrm{B}(z, 1-z)$. The contour $\mathcal{I}_-$ runs backwards and just below the cut where $u = |u| - i\epsilon$

$$\int_{\mathcal{I}_-} f(u)\,du = -\int_0^\infty \frac{(|u|e^{2\pi i - \epsilon})^{z-1}}{1+u}\,du = -e^{2\pi i z}\int_0^\infty \frac{u^{z-1}}{1+u}\,du. \tag{6.221}$$

Thus the vanishing (6.219) of the contour integral

$$0 = \mathrm{B}(z, 1-z) + 2\pi i\, e^{i\pi z} - e^{2\pi i z}\,\mathrm{B}(z, 1-z) \tag{6.222}$$

gives us Euler's reflection formula

$$\mathrm{B}(z, 1-z) = \Gamma(z)\,\Gamma(1-z) = \frac{\pi}{\sin \pi z}. \tag{6.223}$$

Example 6.39 (A Matthews and Walker integral) To do the integral

$$I = \int_0^\infty \frac{dx}{1+x^3} \tag{6.224}$$

we promote x to a complex variable z and consider the function $f(z) = \ln z/(1+z^3)$. If we put the cut in the logarithm on the positive real axis, then $f(z)$ is analytic everywhere except for $z \ge 0$ and at $z^3 = -1$. The integral of $f(z)$ along the ghost contour $z = R\exp(i\theta)$ from $\theta = \epsilon$ to $\theta = 2\pi - \epsilon$ and along both sides of the real axis from $z = i\epsilon$ to $z = R + i\epsilon$ and from $z = R - i\epsilon$ to $z = -i\epsilon$ is by the residue theorem (6.116)

$$\oint f(z)\,dz = \oint \frac{\ln(z)}{(z-1)(z - e^{i\pi/3})(z - e^{2i\pi/3})}\,dz = -\frac{4\pi^2 i}{3\sqrt{3}}. \tag{6.225}$$

Since $\ln(x + i\epsilon) = \ln(x)$ and $\ln(x - i\epsilon) = \ln(x) + 2\pi i$, while $|\epsilon \ln(\epsilon)| \to 0$ as $\epsilon \to 0$, that same integral approaches $-2\pi i I$ as $R \to \infty$. Thus the integral (6.224) is $I = 2\pi/(3\sqrt{3})$.

6.17 Conformal Mapping

An analytic function $f(z)$ maps curves in the z plane into curves in the $f(z)$ plane. In general, this mapping preserves angles. To see why, we consider the angle $d\theta$ between two tiny complex lines $dz = \epsilon\exp(i\theta)$ and $dz' = \epsilon\exp(i\theta')$ that radiate from the same point z. The angle $d\theta = \theta' - \theta$ is the phase of the ratio

$$\frac{dz'}{dz} = \frac{\epsilon e^{i\theta'}}{\epsilon e^{i\theta}} = e^{i(\theta'-\theta)}. \tag{6.226}$$

Let's use $w = \rho e^{i\phi}$ for $f(z)$. Then the analytic function $f(z)$ maps dz into

$$dw = f(z + dz) - f(z) \approx f'(z)\, dz \tag{6.227}$$

and dz' into

$$dw' = f(z + dz') - f(z) \approx f'(z)\, dz'. \tag{6.228}$$

The angle $d\phi = \phi' - \phi$ between dw and dw' is the phase of the ratio

$$\frac{dw'}{dw} = \frac{e^{i\phi'}}{e^{i\phi}} = \frac{f'(z)\, dz'}{f'(z)\, dz} = \frac{dz'}{dz} = \frac{e^{i\theta'}}{e^{i\theta}} = e^{i(\theta'-\theta)}. \tag{6.229}$$

So as long as the derivative $f'(z)$ does not vanish, the angle in the w-plane is the same as the angle in the z-plane

$$d\phi = d\theta. \tag{6.230}$$

Analytic functions preserve angles. They are **conformal** maps.

What if $f'(z) = 0$? In this case, $dw \approx f''(z)\, dz^2/2$ and $dw' \approx f''(z)\, dz'^2/2$, and so the angle $d\phi = d\phi' - d\phi$ between these two tiny complex lines is the phase of the ratio

$$\frac{dw'}{dw} = \frac{e^{i\phi'}}{e^{i\phi}} = \frac{f''(z)\, dz'^2}{f''(z)\, dz^2} = \frac{dz'^2}{dz^2} = e^{2i(\theta'-\theta)}. \tag{6.231}$$

So angles are doubled, $d\phi = 2d\theta$.

In general, if the first nonzero derivative is $f^{(n)}(z)$, then

$$\frac{dw'}{dw} = \frac{e^{i\phi'}}{e^{i\phi}} = \frac{f^{(n)}(z)\, dz'^n}{f^{(n)}(z)\, dz^n} = \frac{dz'^n}{dz^n} = e^{ni(\theta'-\theta)} \tag{6.232}$$

and so $d\phi = nd\theta$. The angles increase by a factor of n.

Example 6.40 (z^n) The function $f(z) = z^n$ has only one nonzero derivative $f^{(k)}(0) = n!\,\delta_{nk}$ at the origin $z = 0$. So at $z = 0$ the map $z \to z^n$ scales angles by n, $d\phi = n\, d\theta$, but at $z \neq 0$ the first derivative $f^{(1)}(z) = nz^{n-1}$ is not equal to zero. So z^n is conformal except at the origin.

Example 6.41 (Möbius transformation) The function

$$f(z) = \frac{az + b}{cz + d} \tag{6.233}$$

maps (straight) lines into lines and circles and maps circles into circles and lines, unless $ad = bc$ in which case it is the constant b/d.

6.18 Cauchy's Principal Value

Suppose that $f(x)$ is differentiable or analytic at and near the point $x = 0$, and that we wish to evaluate the integral

$$K = \lim_{\epsilon \to 0} \int_{-a}^{b} dx \, \frac{f(x)}{x - i\epsilon} \tag{6.234}$$

for $a > 0$ and $b > 0$. First we regularize the pole at $x = 0$ by using a method devised by Cauchy

$$K = \lim_{\delta \to 0} \left[\lim_{\epsilon \to 0} \left(\int_{-a}^{-\delta} dx \, \frac{f(x)}{x - i\epsilon} + \int_{-\delta}^{\delta} dx \, \frac{f(x)}{x - i\epsilon} + \int_{\delta}^{b} dx \, \frac{f(x)}{x - i\epsilon} \right) \right]. \tag{6.235}$$

In the first and third integrals, since $|x| \ge \delta$, we may set $\epsilon = 0$

$$K = \lim_{\delta \to 0} \left(\int_{-a}^{-\delta} dx \, \frac{f(x)}{x} + \int_{\delta}^{b} dx \, \frac{f(x)}{x} \right) + \lim_{\delta \to 0} \lim_{\epsilon \to 0} \int_{-\delta}^{\delta} dx \, \frac{f(x)}{x - i\epsilon}. \tag{6.236}$$

We'll discuss the first two integrals before analyzing the last one.

The limit of the first two integrals is called **Cauchy's principal value**

$$P \int_{-a}^{b} dx \, \frac{f(x)}{x} \equiv \lim_{\delta \to 0} \left(\int_{-a}^{-\delta} dx \, \frac{f(x)}{x} + \int_{\delta}^{b} dx \, \frac{f(x)}{x} \right). \tag{6.237}$$

If the function $f(x)$ is nearly constant near $x = 0$, then the large negative values of $1/x$ for x slightly less than zero cancel the large positive values of $1/x$ for x slightly greater than zero. The point $x = 0$ is not special; Cauchy's principal value about $x = y$ is defined by the limit

$$P \int_{-a}^{b} dx \, \frac{f(x)}{x - y} \equiv \lim_{\delta \to 0} \left(\int_{-a}^{y-\delta} dx \, \frac{f(x)}{x - y} + \int_{y+\delta}^{b} dx \, \frac{f(x)}{x - y} \right). \tag{6.238}$$

Using Cauchy's principal value, we may write the quantity K as

$$K = P \int_{-a}^{b} dx \, \frac{f(x)}{x} + \lim_{\delta \to 0} \lim_{\epsilon \to 0} \int_{-\delta}^{\delta} dx \, \frac{f(x)}{x - i\epsilon}. \tag{6.239}$$

To evaluate the second integral, we use differentiability of $f(x)$ near $x = 0$ to write $f(x) = f(0) + xf'(0)$ and then extract the constants $f(0)$ and $f'(0)$

$$\begin{aligned} \lim_{\delta \to 0} \lim_{\epsilon \to 0} \int_{-\delta}^{\delta} dx \, \frac{f(x)}{x - i\epsilon} &= \lim_{\delta \to 0} \lim_{\epsilon \to 0} \int_{-\delta}^{\delta} dx \, \frac{f(0) + x\, f'(0)}{x - i\epsilon} \\ &= f(0) \lim_{\delta \to 0} \lim_{\epsilon \to 0} \int_{-\delta}^{\delta} \frac{dx}{x - i\epsilon} + f'(0) \lim_{\delta \to 0} \lim_{\epsilon \to 0} \int_{-\delta}^{\delta} \frac{x\, dx}{x - i\epsilon} \end{aligned}$$

$$= f(0) \lim_{\delta\to 0}\lim_{\epsilon\to 0}\int_{-\delta}^{\delta}\frac{dx}{x-i\epsilon} + f'(0)\lim_{\delta\to 0} 2\delta$$
$$= f(0) \lim_{\delta\to 0}\lim_{\epsilon\to 0}\int_{-\delta}^{\delta}\frac{dx}{x-i\epsilon}. \tag{6.240}$$

Since $1/(z - i\epsilon)$ is analytic in the lower half-plane, we may deform the straight contour from $x = -\delta$ to $x = \delta$ into a tiny semicircle that avoids the point $x = 0$ by setting $z = \delta\, e^{i\theta}$ and letting θ run from π to 2π

$$K = P\int_{-a}^{b} dx\,\frac{f(x)}{x} + f(0)\lim_{\delta\to 0}\lim_{\epsilon\to 0}\int_{-\delta}^{\delta} dz\,\frac{1}{z-i\epsilon}. \tag{6.241}$$

We now can set $\epsilon = 0$ and so write K as

$$K = P\int_{-a}^{b} dx\,\frac{f(x)}{x} + f(0)\lim_{\delta\to 0}\int_{\pi}^{2\pi} i\delta e^{i\theta} d\theta\,\frac{1}{\delta e^{i\theta}}$$
$$= P\int_{-a}^{b} dx\,\frac{f(x)}{x} + i\pi f(0). \tag{6.242}$$

Recalling the definition (6.234) of K, we have

$$\lim_{\epsilon\to 0}\int_{-a}^{b} dx\,\frac{f(x)}{x-i\epsilon} = P\int_{-a}^{b} dx\,\frac{f(x)}{x} + i\pi f(0) \tag{6.243}$$

for any function $f(x)$ that is differentiable at $x = 0$. Physicists write this as

$$\frac{1}{x-i\epsilon} = P\,\frac{1}{x} + i\pi\delta(x) \quad\text{and}\quad \frac{1}{x+i\epsilon} = P\,\frac{1}{x} - i\pi\delta(x) \tag{6.244}$$

or as

$$\frac{1}{x-y\pm i\epsilon} = P\,\frac{1}{x-y} \mp i\pi\delta(x-y). \tag{6.245}$$

Example 6.42 (An application of Cauchy's trick) We use (6.244) to evaluate the integral

$$I = \int_{-\infty}^{\infty} dx\,\frac{1}{x+i\epsilon}\,\frac{1}{1+x^2} \tag{6.246}$$

as

$$I = P\int_{-\infty}^{\infty} dx\,\frac{1}{x}\,\frac{1}{1+x^2} - i\pi\int_{-\infty}^{\infty} dx\,\frac{\delta(x)}{1+x^2}. \tag{6.247}$$

Because the function $1/x(1+x^2)$ is odd, the principal part is zero. The integral over the delta function gives unity, so we have $I = -i\pi$.

Example 6.43 (Cubic form of Cauchy's principal value) Cauchy's principal value of the integral

$$P \int_{-a}^{b} \frac{f(x)}{x^3}\, dx \tag{6.248}$$

is finite as long as $f(z)$ is analytic at $z = 0$ with a vanishing first derivative there, $f'(0) = 0$. In this case Cauchy's integral formula (6.43) says that

$$\begin{aligned}\int_{-a}^{b} dx\, \frac{f(x)}{(x - i\epsilon)^3} &= P \int_{-a}^{b} dx\, \frac{f(x)}{x^3} + \lim_{\delta \to 0} \int_{\pi}^{2\pi} i\delta e^{i\theta} d\theta\, \frac{f(\delta e^{i\theta})}{\left(\delta e^{i\theta}\right)^3} \\ &= P \int_{-a}^{b} dx\, \frac{f(x)}{x^3} + i\frac{\pi}{2} f''(0).\end{aligned} \tag{6.249}$$

Example 6.44 (Cauchy's principal value) By explicit use of the formula

$$\int \frac{dx}{x^2 - a^2} = -\frac{1}{2a} \ln \frac{x + a}{x - a} \tag{6.250}$$

one may show (Exercise 6.32) that

$$P \int_{0}^{\infty} \frac{dx}{x^2 - a^2} = \int_{0}^{a-\delta} \frac{dx}{x^2 - a^2} + \int_{a+\delta}^{\infty} \frac{dx}{x^2 - a^2} = 0 \tag{6.251}$$

a result we'll use in Section 6.21.

Example 6.45 ($\sin k/k$) To compute the integral

$$I_s = \int_{0}^{\infty} \frac{dk}{k} \sin k \tag{6.252}$$

which we used to derive the formula (4.113) for the Green's function of the laplacian in 3 dimensions, we first express I_s as an integral along the whole real axis

$$I_s = \int_{0}^{\infty} \frac{dk}{2ik} \left(e^{ik} - e^{-ik}\right) = \int_{-\infty}^{\infty} \frac{dk}{2ik}\, e^{ik} \tag{6.253}$$

by which we actually mean the Cauchy principal part

$$I_s = \lim_{\delta \to 0} \left(\int_{-\infty}^{-\delta} dk\, \frac{e^{ik}}{2ik} + \int_{\delta}^{\infty} dk\, \frac{e^{ik}}{2ik} \right) = P \int_{-\infty}^{\infty} dk\, \frac{e^{ik}}{2ik}. \tag{6.254}$$

Using Cauchy's trick (6.244), we have

$$I_s = P \int_{-\infty}^{\infty} dk\, \frac{e^{ik}}{2ik} = \int_{-\infty}^{\infty} dk\, \frac{e^{ik}}{2i(k + i\epsilon)} + \int_{-\infty}^{\infty} dk\, i\pi\, \delta(k)\, \frac{e^{ik}}{2i}. \tag{6.255}$$

To the first integral, we add a ghost contour around the upper half-plane. For the contour from $k = L$ to $k = L + iH$ and then to $k = -L + iH$ and then down to $k = -L$, one may show (Exercise 6.35) that the integral of $\exp(ik)/k$ vanishes in the double limit $L \to \infty$ and $H \to \infty$. With this ghost contour, the first integral therefore vanishes because the pole at $k = -i\epsilon$ is in the lower half-plane. The delta function in the second integral then gives $\pi/2$, so that

$$I_s = \oint dk \, \frac{e^{ik}}{2i(k + i\epsilon)} + \frac{\pi}{2} = \frac{\pi}{2} \tag{6.256}$$

as stated in (4.112).

Example 6.46 (The Feynman propagator) Adding $\pm i\epsilon$ to the denominator of a pole term of an integral formula for a function $f(x)$ can slightly shift the pole into the upper or lower half-plane, causing the pole to contribute if a ghost contour goes around the upper half-plane or the lower half-plane. Such an $i\epsilon$ can impose a boundary condition on a Green's function.

The Feynman propagator $\Delta_F(x)$ is a Green's function for the Klein–Gordon differential operator (Weinberg, 1995, pp. 274–280)

$$(m^2 - \Box)\Delta_F(x) = \delta^4(x) \tag{6.257}$$

in which $x = (x^0, \boldsymbol{x})$ and

$$\Box = \triangle - \frac{\partial^2}{\partial t^2} = \triangle - \frac{\partial^2}{\partial (x^0)^2} \tag{6.258}$$

is the 4-dimensional version of the laplacian $\triangle \equiv \nabla \cdot \nabla$. Here $\delta^4(x)$ is the four-dimensional Dirac delta function (4.36)

$$\delta^4(x) = \int \frac{d^4q}{(2\pi)^4} \exp[i(\boldsymbol{q} \cdot \boldsymbol{x} - q^0 x^0)] = \int \frac{d^4q}{(2\pi)^4} e^{iqx} \tag{6.259}$$

in which $qx = \boldsymbol{q} \cdot \boldsymbol{x} - q^0 x^0$ is the Lorentz-invariant inner product of the 4-vectors q and x. There are many Green's functions that satisfy Eq. (6.257). Feynman's propagator $\Delta_F(x)$

$$\Delta_F(x) = \int \frac{d^4q}{(2\pi)^4} \frac{\exp(iqx)}{q^2 + m^2 - i\epsilon} = \int \frac{d^3q}{(2\pi)^3} \int_{-\infty}^{\infty} \frac{dq^0}{2\pi} \frac{e^{i\boldsymbol{q}\cdot\boldsymbol{x} - iq^0x^0}}{q^2 + m^2 - i\epsilon} \tag{6.260}$$

is the one that satisfies boundary conditions that will become evident when we analyze the effect of its $i\epsilon$. The quantity $E_{\boldsymbol{q}} = \sqrt{\boldsymbol{q}^2 + m^2}$ is the energy of a particle of mass m and momentum $\boldsymbol{q}$ in natural units with the speed of light $c = 1$. Using this abbreviation and setting $\epsilon' = \epsilon/2E_q$, we may write the denominator as

$$q^2 + m^2 - i\epsilon = \boldsymbol{q} \cdot \boldsymbol{q} - \left(q^0\right)^2 + m^2 - i\epsilon = \left(E_q - i\epsilon' - q^0\right)\left(E_q - i\epsilon' + q^0\right) + \epsilon'^2 \tag{6.261}$$

in which ϵ'^2 is negligible. Dropping the prime on ϵ, we do the q^0 integral

$$I(\boldsymbol{q}) = -\int_{-\infty}^{\infty} \frac{dq^0}{2\pi} e^{-iq^0x^0} \frac{1}{\left[q^0 - (E_{\boldsymbol{q}} - i\epsilon)\right]\left[q^0 - (-E_{\boldsymbol{q}} + i\epsilon)\right]}. \tag{6.262}$$

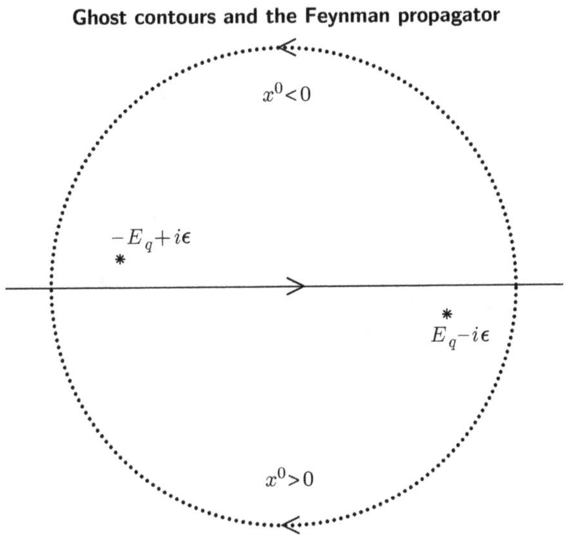

Figure 6.9 In Equation (6.263), the function $f(q^0)$ has poles at $\pm(E_{\boldsymbol{q}} - i\epsilon)$, and the function $\exp(-iq^0x^0)$ is exponentially suppressed in the lower half-plane if $x^0 > 0$ and in the upper half-plane if $x^0 < 0$. So we can add a ghost contour (...) in the LHP if $x^0 > 0$ and in the UHP if $x^0 < 0$.

As shown in Fig. 6.9, the integrand

$$e^{-iq^0x^0}\,\frac{1}{\left[q^0-(E_{\boldsymbol{q}}-i\epsilon)\right]\left[q^0-(-E_{\boldsymbol{q}}+i\epsilon)\right]} \tag{6.263}$$

has poles at $E_{\boldsymbol{q}} - i\epsilon$ and at $-E_{\boldsymbol{q}} + i\epsilon$. When $x^0 > 0$, we can add a ghost contour that goes clockwise around the lower half-plane and get

$$I(\boldsymbol{q}) = ie^{-iE_{\boldsymbol{q}}x^0}\,\frac{1}{2E_{\boldsymbol{q}}} \qquad x^0 > 0. \tag{6.264}$$

When $x^0 < 0$, our ghost contour goes counterclockwise around the upper half-plane, and we get

$$I(\boldsymbol{q}) = ie^{iE_{\boldsymbol{q}}x^0}\,\frac{1}{2E_{\boldsymbol{q}}} \qquad x^0 < 0. \tag{6.265}$$

Using the step function $\theta(x) = (x + |x|)/2$, we combine (6.264) and (6.265)

$$-\,iI(\boldsymbol{q}) = \frac{1}{2E_{\boldsymbol{q}}}\left[\theta(x^0)\,e^{-iE_{\boldsymbol{q}}x^0} + \theta(-x^0)\,e^{iE_{\boldsymbol{q}}x^0}\right]. \tag{6.266}$$

In terms of the Lorentz-invariant function

$$\Delta_+(x) = \frac{1}{(2\pi)^3}\int\frac{d^3q}{2E_{\boldsymbol{q}}}\,\exp[i(\boldsymbol{q}\cdot\boldsymbol{x} - E_{\boldsymbol{q}}x^0)] \tag{6.267}$$

and with a factor of $-i$, Feynman's propagator (6.260) is

$$-i\Delta_F(x) = \theta(x^0)\,\Delta_+(x) + \theta(-x^0)\,\Delta_+(\boldsymbol{x}, -x^0). \tag{6.268}$$

The integral (6.267) defining $\Delta_+(x)$ is insensitive to the sign of $\boldsymbol{q}$, and so

$$\Delta_+(-x) = \frac{1}{(2\pi)^3}\int \frac{d^3q}{2E_{\boldsymbol{q}}}\,\exp[i(-\boldsymbol{q}\cdot\boldsymbol{x} + E_{\boldsymbol{q}}x^0)] \tag{6.269}$$

$$= \frac{1}{(2\pi)^3}\int \frac{d^3q}{2E_{\boldsymbol{q}}}\,\exp[i(\boldsymbol{q}\cdot\boldsymbol{x} + E_{\boldsymbol{q}}x^0)] = \Delta_+(\boldsymbol{x}, -x^0).$$

Thus we arrive at the standard form of the Feynman propagator

$$-i\Delta_F(x) = \theta(x^0)\,\Delta_+(x) + \theta(-x^0)\,\Delta_+(-x). \tag{6.270}$$

The annihilation operators $a(\boldsymbol{q})$ and the creation operators $a^\dagger(\boldsymbol{p})$ of a scalar field $\phi(x)$ satisfy in natural units the commutation relations

$$[a(\boldsymbol{q}), a^\dagger(\boldsymbol{p})] = \delta^3(\boldsymbol{q}-\boldsymbol{p}) \quad \text{and} \quad [a(\boldsymbol{q}), a(\boldsymbol{p})] = [a^\dagger(\boldsymbol{q}), a^\dagger(\boldsymbol{p})] = 0. \tag{6.271}$$

Thus the commutator of the positive-frequency part

$$\phi^+(x) = \int \frac{d^3p}{\sqrt{(2\pi)^3 2p^0}}\,\exp[i(\boldsymbol{p}\cdot\boldsymbol{x} - p^0x^0)]\,a(\boldsymbol{p}) \tag{6.272}$$

of a scalar field $\phi = \phi^+ + \phi^-$ with its negative-frequency part

$$\phi^-(y) = \int \frac{d^3q}{\sqrt{(2\pi)^3 2q^0}}\,\exp[-i(\boldsymbol{q}\cdot\boldsymbol{y} - q^0y^0)]\,a^\dagger(\boldsymbol{q}) \tag{6.273}$$

is the Lorentz-invariant function $\Delta_+(x-y)$

$$[\phi^+(x), \phi^-(y)] = \int \frac{d^3p\,d^3q}{(2\pi)^3 2\sqrt{q^0p^0}}\,e^{ipx-iqy}\,[a(\boldsymbol{p}), a^\dagger(\boldsymbol{q})]$$

$$= \int \frac{d^3p}{(2\pi)^3 2p^0}\,e^{ip(x-y)} = \Delta_+(x-y) \tag{6.274}$$

in which $p(x-y) = \boldsymbol{p}\cdot(\boldsymbol{x}-\boldsymbol{y}) - p^0(x^0-y^0)$.

At points x that are space-like, that is, for which $x^2 = \boldsymbol{x}^2 - (x^0)^2 \equiv r^2 > 0$, the Lorentz-invariant function $\Delta_+(x)$ depends only upon $r = +\sqrt{x^2}$ and has the value (Weinberg, 1995, p. 202)

$$\Delta_+(x) = \frac{m}{4\pi^2 r}\,K_1(mr) \tag{6.275}$$

in which the Hankel function K_1 is

$$K_1(z) = -\frac{\pi}{2}\,[J_1(iz) + iN_1(iz)] = \frac{1}{z} + \frac{z}{2}\left[\ln\left(\frac{z}{2}\right) + \gamma - \frac{1}{2}\right] + \cdots \tag{6.276}$$

where J_1 is the first Bessel function, N_1 is the first Neumann function, and $\gamma = 0.57721\ldots$ is the Euler–Mascheroni constant.

The Feynman propagator arises most simply as the mean value in the vacuum of the **time-ordered product** of the fields $\phi(x)$ and $\phi(y)$

$$\mathcal{T}\{\phi(x)\phi(y)\} \equiv \theta(x^0 - y^0)\phi(x)\phi(y) + \theta(y^0 - x^0)\phi(y)\phi(x). \tag{6.277}$$

The operators $a(\boldsymbol{p})$ and $a^\dagger(\boldsymbol{p})$ respectively annihilate the vacuum ket $a(\boldsymbol{p})|0\rangle = 0$ and bra $\langle 0|a^\dagger(\boldsymbol{p}) = 0$, and so by (6.272 and 6.273) do the positive- and negative-frequency parts of the field $\phi^+(z)|0\rangle = 0$ and $\langle 0|\phi^-(z) = 0$. Thus the mean value in the vacuum of the time-ordered product is

$$\begin{aligned}
\langle 0|\mathcal{T}\{\phi(x)\phi(y)\}|0\rangle &= \langle 0|\theta(x^0 - y^0)\phi(x)\phi(y) + \theta(y^0 - x^0)\phi(y)\phi(x)|0\rangle \\
&= \langle 0|\theta(x^0 - y^0)\phi^+(x)\phi^-(y) + \theta(y^0 - x^0)\phi^+(y)\phi^-(x)|0\rangle \\
&= \langle 0|\theta(x^0 - y^0)[\phi^+(x), \phi^-(y)] \\
&\quad + \theta(y^0 - x^0)[\phi^+(y), \phi^-(x)]|0\rangle.
\end{aligned} \tag{6.278}$$

But by (6.274), these commutators are $\Delta_+(x - y)$ and $\Delta_+(y - x)$. Thus the mean value in the vacuum of the time-ordered product

$$\begin{aligned}
\langle 0|\mathcal{T}\{\phi(x)\phi(y)\}|0\rangle &= \theta(x^0 - y^0)\Delta_+(x - y) + \theta(y^0 - x^0)\Delta_+(y - x) \\
&= -i\Delta_F(x - y)
\end{aligned} \tag{6.279}$$

is the Feynman propagator (6.268) multiplied by $-i$.

6.19 Dispersion Relations

In many physical contexts, functions occur that are analytic in the upper half-plane. Suppose for instance that $\hat{f}(t)$ is a transfer function that determines an effect $e(t)$ due to a cause $c(t)$

$$e(t) = \int_{-\infty}^{\infty} dt'\, \hat{f}(t - t')\, c(t'). \tag{6.280}$$

If the system is **causal**, then the transfer function $\hat{f}(t - t')$ is zero for $t - t' < 0$, and so its Fourier transform

$$f(z) = \int_{-\infty}^{\infty} \frac{dt}{\sqrt{2\pi}}\, \hat{f}(t)\, e^{izt} = \int_{0}^{\infty} \frac{dt}{\sqrt{2\pi}}\, \hat{f}(t)\, e^{izt} \tag{6.281}$$

will be analytic in the upper half-plane and will shrink as the imaginary part of $z = x + iy$ increases.

So let us assume that the function $f(z)$ is analytic in the upper half-plane and on the real axis and further that

$$\lim_{r\to\infty} |f(re^{i\theta})| = 0 \quad \text{for} \quad 0 \le \theta \le \pi. \tag{6.282}$$

By Cauchy's integral formula (6.40), if z_0 lies in the upper half-plane, then $f(z_0)$ is given by the closed counterclockwise contour integral

$$f(z_0) = \frac{1}{2\pi i} \oint \frac{f(z)}{z - z_0}\, dz \tag{6.283}$$

in which the contour runs along the real axis and then loops over the semi-circle

$$\lim_{r\to\infty} r e^{i\theta} \quad \text{for} \quad 0 \le \theta \le \pi. \tag{6.284}$$

Our assumption (6.282) about the behavior of $f(z)$ in the upper half-plane implies that this contour (6.284) is a ghost contour because its modulus is bounded by

$$\lim_{r\to\infty} \frac{1}{2\pi} \int \frac{|f(re^{i\theta})| r}{r}\, d\theta = \lim_{r\to\infty} |f(re^{i\theta})| = 0. \tag{6.285}$$

So we may drop the ghost contour and write $f(z_0)$ as

$$f(z_0) = \frac{1}{2\pi i} \int_{-\infty}^{\infty} \frac{f(x)}{x - z_0}\, dx. \tag{6.286}$$

Letting the imaginary part y_0 of $z_0 = x_0 + i y_0$ shrink to ϵ

$$f(x_0) = \frac{1}{2\pi i} \int_{-\infty}^{\infty} \frac{f(x)}{x - x_0 - i\epsilon}\, dx \tag{6.287}$$

and using Cauchy's trick (6.245), we get

$$f(x_0) = \frac{1}{2\pi i} P \int_{-\infty}^{\infty} \frac{f(x)}{x - x_0}\, dx + \frac{i\pi}{2\pi i} \int_{-\infty}^{\infty} f(x)\, \delta(x - x_0)\, dx \tag{6.288}$$

or

$$f(x_0) = \frac{1}{2\pi i} P \int_{-\infty}^{\infty} \frac{f(x)}{x - x_0}\, dx + \frac{1}{2} f(x_0) \tag{6.289}$$

which is the **dispersion relation**

$$f(x_0) = \frac{1}{\pi i} P \int_{-\infty}^{\infty} \frac{f(x)}{x - x_0}\, dx. \tag{6.290}$$

If we break $f(z) = u(z) + i v(z)$ into its real $u(z)$ and imaginary $v(z)$ parts, then this dispersion relation (6.290)

$$\begin{aligned} u(x_0) + i v(x_0) &= \frac{1}{\pi i} P \int_{-\infty}^{\infty} \frac{u(x) + i v(x)}{x - x_0}\, dx \\ &= \frac{1}{\pi} P \int_{-\infty}^{\infty} \frac{v(x)}{x - x_0}\, dx - \frac{i}{\pi} P \int_{-\infty}^{\infty} \frac{u(x)}{x - x_0}\, dx \end{aligned} \tag{6.291}$$

breaks into its real and imaginary parts

$$u(x_0) = \frac{1}{\pi} P \int_{-\infty}^{\infty} \frac{v(x)}{x - x_0} dx \quad \text{and} \quad v(x_0) = -\frac{1}{\pi} P \int_{-\infty}^{\infty} \frac{u(x)}{x - x_0} dx \tag{6.292}$$

which express u and v as **Hilbert transforms** of each other.

In applications of dispersion relations, the function $f(x)$ for $x < 0$ sometimes is either physically meaningless or experimentally inaccessible. In such cases, there may be a symmetry that relates $f(-x)$ to $f(x)$. For instance, if $f(x)$ is the Fourier transform of a real function $\hat{f}(k)$, then by Eq. (4.25) it obeys the symmetry relation

$$f^*(x) = u(x) - iv(x) = f(-x) = u(-x) + iv(-x), \tag{6.293}$$

which says that u is even, $u(-x) = u(x)$, and v odd, $v(-x) = -v(x)$. Using these symmetries, one may show (Exercise 6.38) that the Hilbert transformations (6.292) become

$$u(x_0) = \frac{2}{\pi} P \int_0^{\infty} \frac{x\, v(x)}{x^2 - x_0^2} dx \quad \text{and} \quad v(x_0) = -\frac{2x_0}{\pi} P \int_0^{\infty} \frac{u(x)}{x^2 - x_0^2} dx \tag{6.294}$$

which do not require input at negative values of x.

6.20 Kramers–Kronig Relations

If we use $\sigma \boldsymbol{E}$ for the current density $\boldsymbol{J}$ and $\boldsymbol{E}(t) = e^{-i\omega t}\boldsymbol{E}$ for the electric field, then Maxwell's equation $\nabla \times \boldsymbol{B} = \mu \boldsymbol{J} + \epsilon\mu \dot{\boldsymbol{E}}$ becomes

$$\nabla \times \boldsymbol{B} = -i\omega\epsilon\mu \left(1 + i\frac{\sigma}{\epsilon\omega}\right) \boldsymbol{E} \equiv -i\omega n^2 \epsilon_0 \mu_0 \boldsymbol{E} \tag{6.295}$$

in which the squared index of refraction is

$$n^2(\omega) = \frac{\epsilon\mu}{\epsilon_0\mu_0} \left(1 + i\frac{\sigma}{\epsilon\omega}\right). \tag{6.296}$$

The imaginary part of n^2 represents the scattering of light mainly by electrons. At high frequencies in nonmagnetic materials $n^2(\omega) \to 1$, and so Kramers and Kronig applied the Hilbert-transform relations (6.294) to the function $n^2(\omega) - 1$ in order to satisfy condition (6.282). Their relations are

$$\mathrm{Re}(n^2(\omega_0)) = 1 + \frac{2}{\pi} P \int_0^{\infty} \frac{\omega\, \mathrm{Im}(n^2(\omega))}{\omega^2 - \omega_0^2} d\omega \tag{6.297}$$

and

$$\mathrm{Im}(n^2(\omega_0)) = -\frac{2\omega_0}{\pi} P \int_0^{\infty} \frac{\mathrm{Re}(n^2(\omega)) - 1}{\omega^2 - \omega_0^2} d\omega. \tag{6.298}$$

What Kramers and Kronig actually wrote was slightly different from these dispersion relations (6.297 and 6.298). H. A. Lorentz had shown that the index of refraction $n(\omega)$ is related to the forward scattering amplitude $f(\omega)$ for the scattering of light by a density N of scatterers (Sakurai, 1982)

$$n(\omega) = 1 + \frac{2\pi c^2}{\omega^2} N f(\omega). \tag{6.299}$$

They used this formula to infer that the real part of the index of refraction approached unity in the limit of infinite frequency and applied the Hilbert transform (6.294)

$$\mathrm{Re}[n(\omega)] = 1 + \frac{2}{\pi} P \int_0^\infty \frac{\omega' \, \mathrm{Im}[n(\omega')]}{\omega'^2 - \omega^2} \, d\omega'. \tag{6.300}$$

The Lorentz relation (6.299) expresses the imaginary part $\mathrm{Im}[n(\omega)]$ of the index of refraction in terms of the imaginary part of the forward scattering amplitude $f(\omega)$

$$\mathrm{Im}[n(\omega)] = 2\pi (c/\omega)^2 N \mathrm{Im}[f(\omega)]. \tag{6.301}$$

And the **optical theorem** relates $\mathrm{Im}[f(\omega)]$ to the **total cross-section**

$$\sigma_{\text{tot}} = \frac{4\pi}{|\boldsymbol{k}|} \mathrm{Im}[f(\omega)] = \frac{4\pi c}{\omega} \mathrm{Im}[f(\omega)]. \tag{6.302}$$

Thus we have $\mathrm{Im}[n(\omega)] = c N \sigma_{\text{tot}}/(2\omega)$, and by the Lorentz relation (6.299) $\mathrm{Re}[n(\omega)] = 1 + 2\pi(c/\omega)^2 N \mathrm{Re}[f(\omega)]$. Insertion of these formulas into the Kramers–Kronig integral (6.300) gives a dispersion relation for the real part of the forward scattering amplitude $f(\omega)$ in terms of the total cross-section

$$\mathrm{Re}[f(\omega)] = \frac{\omega^2}{2\pi^2 c} P \int_0^\infty \frac{\sigma_{\text{tot}}(\omega')}{\omega'^2 - \omega^2} \, d\omega'. \tag{6.303}$$

6.21 Phase and Group Velocities

Suppose $A(\boldsymbol{x}, t)$ is the amplitude

$$A(\boldsymbol{x}, t) = \int e^{i(\boldsymbol{p}\cdot\boldsymbol{x} - Et)/\hbar} A(\boldsymbol{p}) \, d^3p = \int e^{i(\boldsymbol{k}\cdot\boldsymbol{x} - \omega t)} B(\boldsymbol{k}) \, d^3k \tag{6.304}$$

where $B(\boldsymbol{k}) = \hbar^3 A(\hbar \boldsymbol{k})$ varies slowly compared to the phase $\exp[i(\boldsymbol{k}\cdot\boldsymbol{x} - \omega t)]$. The **phase velocity** $\boldsymbol{v}_p$ is the linear relation $\boldsymbol{x} = \boldsymbol{v}_p t$ between $\boldsymbol{x}$ and t that keeps the phase $\phi = \boldsymbol{p}\cdot\boldsymbol{x} - Et$ constant as a function of the time

$$0 = \boldsymbol{p}\cdot d\boldsymbol{x} - E\, dt = (\boldsymbol{p}\cdot\boldsymbol{v}_p - E)\, dt \iff \boldsymbol{v}_p = \frac{E}{p}\hat{\boldsymbol{p}} = \frac{\omega}{k}\hat{\boldsymbol{k}} \tag{6.305}$$

in which $p = |\boldsymbol{p}|$, and $k = |\boldsymbol{k}|$. For light in the vacuum, $\boldsymbol{v}_p = c = (\omega/k)\,\hat{\boldsymbol{k}}$. For a particle of mass $m > 0$, the phase velocity exceeds the speed of light, $v_p = \sqrt{c^2 p^2 + m^2 c^4}/p \geq c$.

The more physical **group velocity** $\boldsymbol{v}_g$ is the linear relation $\boldsymbol{x} = \boldsymbol{v}_g\, t$ between $\boldsymbol{x}$ and t that maximizes the amplitude $A(\boldsymbol{x}, t)$ by keeping the phase $\phi = \boldsymbol{p} \cdot \boldsymbol{x} - Et$ constant as a function of the momentum $\boldsymbol{p}$

$$\nabla_p(\boldsymbol{p} \cdot \boldsymbol{x} - Et) = \boldsymbol{x} - \nabla_p E(\boldsymbol{p})\, t = 0 \tag{6.306}$$

at the maximum of $A(\boldsymbol{p})$. This **condition of stationary phase** gives the group velocity as

$$\boldsymbol{v}_g = \nabla_p E(\boldsymbol{p}) = \nabla_k \omega(\boldsymbol{k}). \tag{6.307}$$

If $E = \boldsymbol{p}^2/(2m)$, then $\boldsymbol{v}_g = \boldsymbol{p}/m$. For a relativistic particle with $E = \sqrt{c^2 p^2 + m^2 c^4}$, the group velocity is $\boldsymbol{v}_g = c^2 \boldsymbol{p}/E$, and $v_g \leq c$.

When light traverses a medium with a complex index of refraction $n(\boldsymbol{k})$, the wave vector $\boldsymbol{k}$ becomes complex, and its (positive) imaginary part represents the scattering of photons in the forward direction, typically by the electrons of the medium. For simplicity, we'll consider the propagation of light through a medium in 1 dimension, that of the forward direction of the beam. Then the (real) frequency $\omega(k)$ and the (complex) wave number k are related by $k = n(k)\,\omega(k)/c$, and the phase velocity of the light is

$$v_p = \frac{\omega}{\mathrm{Re}(k)} = \frac{c}{\mathrm{Re}(n(k))}. \tag{6.308}$$

If we regard the index of refraction as a function of the frequency ω, instead of the wave number k, then by differentiating the real part of the relation $\omega n(\omega) = ck$ with respect to ω, we find

$$n_r(\omega) + \omega \frac{dn_r(\omega)}{d\omega} = c \frac{dk_r}{d\omega} \tag{6.309}$$

in which the subscript r means real part. Thus the group velocity (6.307) of the light is

$$v_g = \frac{d\omega}{dk_r} = \frac{c}{n_r(\omega) + \omega\, dn_r/d\omega}. \tag{6.310}$$

Optical physicists call the denominator the **group index of refraction**

$$n_g(\omega) = n_r(\omega) + \omega \frac{dn_r(\omega)}{d\omega} \tag{6.311}$$

so that as in the expression (6.308) for the phase velocity $v_p = c/n_r(\omega)$, the group velocity is $v_g = c/n_g(\omega)$.

In some media, the derivative $dn_r/d\omega$ is large and positive, and the group velocity v_g of light there can be much less than c (Steinberg et al., 1993; Wang and Zhang, 1995) – as slow as 17 m/s (Hau et al., 1999). This effect is called **slow light**. In certain other media, the derivative $dn/d\omega$ is so negative that the group index of refraction $n_g(\omega)$ is less than unity, and in them the group velocity v_g exceeds c! This effect is called **fast light**. In some media, the derivative $dn_r/d\omega$ is so negative that $dn_r/d\omega < -n_r(\omega)/\omega$, and then $n_g(\omega)$ is not only less than unity but also less than zero. In such a medium, the group velocity v_g of light is negative! This effect is called **backwards light**.

Sommerfeld and Brillouin (Brillouin, 1960, ch. II & III) anticipated fast light and concluded that it would not violate special relativity as long as the **signal velocity**—defined as the speed of the front of a square pulse – remained less than c. Fast light does not violate special relativity (Stenner et al., 2003; Brunner et al., 2004) (Léon Brillouin 1889–1969, Arnold Sommerfeld 1868–1951).

Slow, fast, and backwards light can occur when the frequency ω of the light is near a peak or **resonance** in the total cross-section σ_{tot} for the scattering of light by the atoms of the medium. To see why, recall that the index of refraction $n(\omega)$ is related to the forward scattering amplitude $f(\omega)$ and the density N of scatterers by the formula (6.299)

$$n(\omega) = 1 + \frac{2\pi c^2}{\omega^2} N f(\omega) \tag{6.312}$$

and that the real part of the forward scattering amplitude is given by the Kramers–Kronig integral (6.303) of the total cross-section

$$\text{Re}(f(\omega)) = \frac{\omega^2}{2\pi^2 c} P \int_0^\infty \frac{\sigma_{\text{tot}}(\omega')\, d\omega'}{\omega'^2 - \omega^2}. \tag{6.313}$$

So the real part of the index of refraction is

$$n_r(\omega) = 1 + \frac{cN}{\pi} P \int_0^\infty \frac{\sigma_{\text{tot}}(\omega')\, d\omega'}{\omega'^2 - \omega^2}. \tag{6.314}$$

If the amplitude for forward scattering is of the Breit–Wigner form

$$f(\omega) = f_0 \frac{\Gamma/2}{\omega_0 - \omega - i\Gamma/2} \tag{6.315}$$

then by (6.312) the real part of the index of refraction is

$$n_r(\omega) = 1 + \frac{\pi c^2 N f_0 \Gamma(\omega_0 - \omega)}{\omega^2\left[(\omega - \omega_0)^2 + \Gamma^2/4\right]} \tag{6.316}$$

and by (6.310) the group velocity is

$$v_g = c \left[1 + \frac{\pi c^2 N f_0 \Gamma \, \omega_0}{\omega^2} \frac{\left[(\omega - \omega_0)^2 - \Gamma^2/4\right]}{\left[(\omega - \omega_0)^2 + \Gamma^2/4\right]^2} \right]^{-1} . \tag{6.317}$$

This group velocity v_g is less than c whenever $(\omega - \omega_0)^2 > \Gamma^2/4$. But we get fast light $v_g > c$, if $(\omega - \omega_0)^2 < \Gamma^2/4$, and even backwards light, $v_g < 0$, if $\omega \approx \omega_0$ with $4\pi c^2 N f_0 / \Gamma\omega_0 \gg 1$. Robert W. Boyd's papers explain how to make slow and fast light (Bigelow et al., 2003) and backwards light (Gehring et al., 2006).

We can use the principal-part identity (6.251) to subtract

$$0 = \frac{cN}{\pi} \sigma_{\text{tot}}(\omega) \, P \int_0^\infty \frac{1}{\omega'^2 - \omega^2} \, d\omega' \tag{6.318}$$

from the Kramers–Kronig integral (6.314) so as to write the index of refraction in the regularized form

$$n_r(\omega) = 1 + \frac{cN}{\pi} P \int_0^\infty \frac{\sigma_{\text{tot}}(\omega') - \sigma_{\text{tot}}(\omega)}{\omega'^2 - \omega^2} \, d\omega' \tag{6.319}$$

which we can differentiate and use in the group-velocity formula (6.310)

$$v_g(\omega) = c \left[1 + \frac{cN}{\pi} P \int_0^\infty \frac{\left[\sigma_{\text{tot}}(\omega') - \sigma_{\text{tot}}(\omega)\right] (\omega'^2 + \omega^2)}{(\omega'^2 - \omega^2)^2} \, d\omega' \right]^{-1} . \tag{6.320}$$

6.22 Method of Steepest Descent

Integrals like

$$I(x) = \int_a^b dz \, h(z) \, \exp(x f(z)) \tag{6.321}$$

often are dominated by the exponential. We'll assume that x is real and that the functions $h(z)$ and $f(z)$ are analytic in a simply connected region in which a and b are interior points. Then the value of the integral $I(x)$ is independent of the contour between the endpoints a and b but is sensitive to the real part $u(z)$ of $f(z) = u(z) + iv(z)$. But since $f(z)$ is analytic, its real and imaginary parts $u(z)$ and $v(z)$ are harmonic functions which have no minima or maxima, only saddle points (6.56).

For simplicity, we'll assume that the real part $u(z)$ of $f(z)$ has only one saddle point between the points a and b. (If it has more than one, then we must repeat the computation that follows.) If w is the saddle point, then $u_x = u_y = 0$ which by the Cauchy–Riemann equations (6.10) implies that $v_x = v_y = 0$. Thus the derivative of the function f also vanishes at the saddle point $f'(w) = 0$, and so near w we may approximate $f(z)$ as

$$f(z) \approx f(w) + \frac{1}{2}(z-w)^2 f''(w). \tag{6.322}$$

Let's write the second derivative as $f''(w) = \rho\, e^{i\phi}$ and choose our contour through the saddle point w to be a straight line $z = w + y\, e^{i\theta}$ with θ fixed for z near w. As we vary y along this line, we want

$$(z-w)^2 f''(w) = y^2 \rho\, e^{2i\theta}\, e^{i\phi} < 0 \tag{6.323}$$

so we keep $2\theta + \phi = \pi$ which ensures that near $z = w$

$$f(z) \approx f(w) - \frac{1}{2}\rho\, y^2. \tag{6.324}$$

Since $z = w + y\, e^{i\theta}$, its differential is $dz = e^{i\theta}\, dy$, and the integral $I(x)$ is

$$\begin{aligned} I(x) &\approx \int_{-\infty}^{\infty} h(w) \exp\left\{x\left[f(w) + \frac{1}{2}(z-w)^2 f''(w)\right]\right\} dz \\ &= h(w)\, e^{i\theta}\, e^{xf(w)} \int_{-\infty}^{\infty} \exp\left(-\frac{1}{2}x\rho y^2\right) dy = h(w)\, e^{i\theta}\, e^{xf(w)} \sqrt{\frac{2\pi}{x\rho}}. \end{aligned} \tag{6.325}$$

Moving the phase $e^{i\theta}$ inside the square root

$$I(x) \approx h(w)\, e^{xf(w)} \sqrt{\frac{2\pi}{x\rho\, e^{-2i\theta}}} \tag{6.326}$$

and using $f''(w) = \rho\, e^{i\phi}$ and $2\theta + \phi = \pi$ to show that

$$\rho\, e^{-2i\theta} = \rho\, e^{i\phi - i\pi} = -\rho\, e^{i\phi} = -f''(w) \tag{6.327}$$

we get our formula for the saddle-point integral (6.321)

$$I(x) \approx \left(\frac{2\pi}{-xf''(w)}\right)^{1/2} h(w)\, e^{xf(w)}. \tag{6.328}$$

Example 6.47 (Stirling's formula for $n!$) An exact formula for $n!$

$$n! = (-1)^n \left.\frac{d^n y^{-1}}{dy^n}\right|_{y=1} \tag{6.329}$$

is the integral

$$n! = (-1)^n \frac{d^n}{dy^n} \int_0^{\infty} e^{-yz}\, dz \bigg|_{y=1} = \int_0^{\infty} z^n e^{-z}\, dz = \int_0^{\infty} e^{n \ln z - z}\, dz. \tag{6.330}$$

Comparing it to the integral (6.321) for $I(x)$, we set $f(z) = n \ln z - z$ and $x = h(z) = 1$. The saddle point $z = w$ is where $f'(w) = 0$, or $w = n$. Since $f''(n) = -1/n$, our steepest-descent approximation (6.328) to $n!$ gives us Stirling's formula

$$n! \approx \sqrt{2\pi n}\, n^n\, e^{-n}. \tag{6.331}$$

If there are n saddle points w_j for $j = 1, \ldots, n$, then the steepest-descent approximation to the integral $I(x)$ is the sum

$$I(x) \approx \sum_{j=1}^{N} \left(\frac{2\pi}{-x f''(w_j)} \right)^{1/2} h(w_j)\, e^{x f(w_j)}. \tag{6.332}$$

6.23 Applications to String Theory

This section is optional on a first reading.

String theory may or may not have anything to do with physics, but it does provide many amusing applications of complex-variable theory. The coordinates σ and τ of the world sheet of a string form a complex variable $z = e^{2(\tau - i\sigma)}$. The product of two operators $U(z)$ and $V(w)$ often has poles in $z - w$ as $z \to w$ but is well defined if z and w are radially ordered

$$\mathcal{R}\{U(z)V(w)\} \equiv U(z)\, V(w)\, \theta(|z| - |w|) + V(w)\, U(z)\, \theta(|w| - |z|) \tag{6.333}$$

in which $\theta(x) = (x + |x|)/2|x|$ is the step function. Since the modulus of $z = e^{2(\tau - i\sigma)}$ depends only upon τ, radial order is time order in τ_z and τ_w.

The modes L_n of the principal component of the energy–momentum tensor $T(z)$ are defined by its Laurent series

$$T(z) = \sum_{n=-\infty}^{\infty} \frac{L_n}{z^{n+2}} \tag{6.334}$$

and the inverse relation

$$L_n = \frac{1}{2\pi i} \oint z^{n+1}\, T(z)\, dz. \tag{6.335}$$

Thus the commutator of two modes involves two loop integrals

$$[L_m, L_n] = \left[\frac{1}{2\pi i} \oint z^{m+1}\, T(z)\, dz,\; \frac{1}{2\pi i} \oint w^{n+1}\, T(w)\, dw \right] \tag{6.336}$$

which we may deform as long as we cross no poles. Let's hold w fixed and deform the z loop so as to keep the T's radially ordered when z is near w as in Fig. 6.10. The operator-product expansion of the radially ordered product $\mathcal{R}\{T(z)T(w)\}$ is

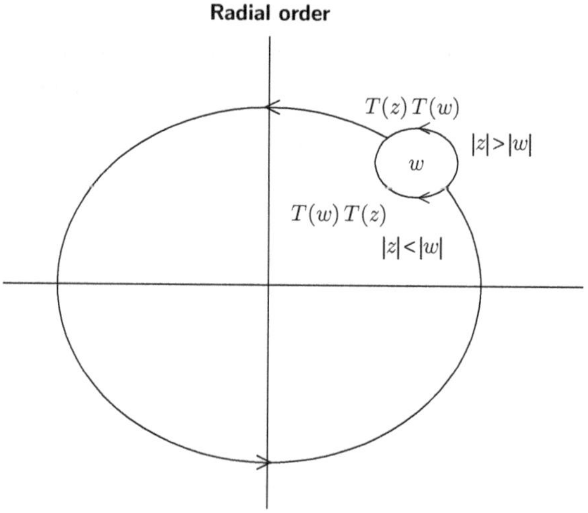

Figure 6.10 The two counterclockwise circles about the origin preserve radial order when z is near w by veering slightly to $|z| > |w|$ for the product $U(z)V(w)$ and to $|z| < |w|$ for the product $V(w)U(z)$.

$$\mathcal{R}\{T(z)T(w)\} = \frac{c/2}{(z-w)^4} + \frac{2}{(z-w)^2}T(w) + \frac{1}{z-w}T'(w) + \cdots \tag{6.337}$$

in which the prime means derivative, c is a constant, and the dots denote terms that are analytic in z and w. The commutator introduces a minus sign that cancels most of the two contour integrals and converts what remains into an integral along a tiny circle C_w about the point w as in Fig. 6.10

$$[L_m, L_n] = \oint \frac{dw}{2\pi i} w^{n+1} \oint_{C_w} \frac{dz}{2\pi i} z^{m+1} \left[\frac{c/2}{(z-w)^4} + \frac{2T(w)}{(z-w)^2} + \frac{T'(w)}{z-w} \right]. \tag{6.338}$$

After doing the z-integral, which is left as a homework exercise (6.41), one may use the Laurent series (6.334) for $T(w)$ to do the w-integral, which one may choose to be along a tiny circle about $w = 0$, and so find the commutator

$$[L_m, L_n] = (m-n)\, L_{m+n} + \frac{c}{12} m(m^2-1)\, \delta_{m+n,0} \tag{6.339}$$

of the Virasoro algebra.

Example 6.48 (Using ghost contours to sum series) Consider the integral

$$I = \oint_C \frac{\csc \pi z}{(z-a)^2}\, dz$$

along the counterclockwise rectangular contour C from $z = N + 1/2 - iY$ to $z = N+1/2+iY$ to $z = -N-1/2+iY$ to $z = -N-1/2-iY$ and back to $z = N+1/2-iY$

in which N is a positive integer, and a is not an integer. In the twin limits $N \to \infty$ and $Y \to \infty$, the integral vanishes because on the contour $1/|z-a|^2 \approx 1/N^2$ or $1/Y^2$ while $|\csc \pi z| \le 1$. We now shrink the contour down to tiny circles about the poles of $\csc \pi z$ at all the integers, $z = n$, and about the nonintegral value, $z = a$. By Cauchy's integral formula (6.42), the tiny contour integral around $z = a$ is

$$\oint_a \frac{\csc \pi z}{(z-a)^2}\, dz = 2\pi i \frac{d \csc \pi z}{dz}\bigg|_{z=a} = -2\pi^2 i \,\frac{\cos \pi a}{\sin^2 \pi a}.$$

In the twin limits $N \to \infty$ and $Y \to \infty$, the tiny counterclockwise integrals around the poles of $1/\sin \pi z$ at $z = n\pi$ are (Exercise 6.44)

$$\sum_{n=-\infty}^{\infty} \oint_n \frac{\csc \pi z}{(z-a)^2}\, dz = 2i \sum_{n=-\infty}^{\infty} (-1)^n \frac{1}{(n-a)^2}.$$

We thus have the sum rule

$$\sum_{n=-\infty}^{\infty} (-1)^n \frac{1}{(n-a)^2} = \pi^2 \cot \pi a \csc \pi a.$$

Further Reading

For examples of conformal mappings see (Lin, 2011, section 3.5.7).

Exercises

6.1 Compute the two limits (6.6) and (6.7) of Example 6.2 but for the function $f(x, y) = x^2 - y^2 + 2ixy$. Do the limits now agree? Explain.

6.2 Show that if $f(z)$ is analytic in a disk, then the integral of $f(z)$ around a tiny (isosceles) triangle of side $\epsilon \ll 1$ inside the disk is zero to order ϵ^2.

6.3 Show that the product $f(z)\, g(z)$ of two functions is analytic at z if both $f(z)$ and $g(z)$ are analytic at z.

6.4 Derive the two integral representations (6.54) for Bessel's functions $J_n(t)$ of the first kind from the integral formula (6.53). Hint: Think of the integral (6.53) as running from $-\pi$ to π.

6.5 Do the integral

$$\oint_C \frac{dz}{z^2 - 1}$$

in which the contour C is counterclockwise about the circle $|z| = 2$.

6.6 The function $f(z) = 1/z$ is analytic in the region $|z| > 0$. Compute the integral of $f(z)$ counterclockwise along the unit circle $z = e^{i\theta}$ for $0 \le \theta \le 2\pi$.

The contour lies entirely within the domain of analyticity of the function $f(z)$. Did you get zero? Why? If not, why not?

6.7 Let $P(z)$ be the polynomial

$$P(z) = (z - a_1)(z - a_2)(z - a_3) \tag{6.340}$$

with roots a_1, a_2, and a_3. Let R be the maximum of the three moduli $|a_k|$. (a) If the three roots are all different, evaluate the integral

$$I = \oint_C \frac{dz}{P(z)} \tag{6.341}$$

along the counterclockwise contour $z = 2Re^{i\theta}$ for $0 \le \theta \le 2\pi$. (b) Same exercise, but for $a_1 = a_2 \neq a_3$.

6.8 Compute the integral of the function $f(z) = e^{az}/(z^2 - 3z + 2)$ along the counterclockwise contour $C_\square$ that follows the perimeter of a square of side 6 centered at the origin. That is, find

$$I = \oint_{C_\square} \frac{e^{az}}{z^2 - 3z + 2}\, dz. \tag{6.342}$$

6.9 Use Cauchy's integral formula (6.44) and Rodrigues's expression (6.45) for Legendre's polynomial $P_n(x)$ to derive Schlaefli's formula (6.46).

6.10 Use Schlaefli's formula (6.46) for the Legendre polynomials and Cauchy's integral formula (6.40) to compute the value of $P_n(-1)$.

6.11 Evaluate the counterclockwise integral around the unit circle $|z| = 1$

$$\oint \left(3 \sinh^2 2z - 4 \cosh^3 z\right) \frac{dz}{z}. \tag{6.343}$$

6.12 Evaluate the counterclockwise integral around the circle $|z| = 2$

$$\oint \frac{z^3}{z^4 - 1}\, dz. \tag{6.344}$$

6.13 Evaluate the contour integral of the function $f(z) = \sin wz/(z - 5)^3$ along the curve $z = 6 + 4(\cos t + i \sin t)$ for $0 \le t \le 2\pi$.

6.14 Evaluate the contour integral of the function $f(z) = \sin wz/(z - 5)^3$ along the curve $z = -6 + 4(\cos t + i \sin t)$ for $0 \le t \le 2\pi$.

6.15 Is the function $f(x, y) = x^2 + iy^2$ analytic?

6.16 Is the function $f(x, y) = x^3 - 3xy^2 + 3ix^2y - iy^3$ analytic? Is the function $x^3 - 3xy^2$ harmonic? Does it have a minimum or a maximum? If so, what are they?

6.17 Is the function $f(x, y) = x^2 + y^2 + i(x^2 + y^2)$ analytic? Is $x^2 + y^2$ a harmonic function? What is its minimum, if it has one?

6.18 Derive the first three nonzero terms of the Laurent series for $f(z) = 1/(e^z - 1)$ about $z = 0$.

6.19 Assume that a function $g(z)$ is meromorphic in R and has a Laurent series (6.103) about a point $w \in R$. Show that as $z \to w$, the ratio $g'(z)/g(z)$ becomes (6.101).

6.20 Use a contour integral to evaluate the integral

$$I_a = \int_0^\pi \frac{d\theta}{a + \cos\theta}, \qquad a > 1. \tag{6.345}$$

6.21 Find the poles and residues of the functions $1/\sin z$ and $1/\cos z$.

6.22 Derive the integral formula (6.135) from (6.132).

6.23 Show that if $\operatorname{Re} w < 0$, then for arbitrary complex z

$$\int_{-\infty}^{\infty} e^{w(x+z)^2}\, dx = \sqrt{\frac{\pi}{-w}}. \tag{6.346}$$

6.24 Use a ghost contour to evaluate the integral

$$\int_{-\infty}^{\infty} \frac{x \sin x}{x^2 + a^2}\, dx.$$

Show your work; do not just quote the result of a commercial math program.

6.25 For $a > 0$ and $b^2 - 4ac < 0$, use a ghost contour to do the integral

$$\int_{-\infty}^{\infty} \frac{dx}{ax^2 + bx + c}. \tag{6.347}$$

6.26 Show that

$$\int_0^\infty \cos ax \, e^{-x^2}\, dx = \frac{1}{2}\sqrt{\pi}\, e^{-a^2/4}. \tag{6.348}$$

6.27 Show that

$$\int_{-\infty}^{\infty} \frac{dx}{1 + x^4} = \frac{\pi}{\sqrt{2}}. \tag{6.349}$$

6.28 Evaluate the integral

$$\int_0^\infty \frac{\cos x}{1 + x^4}\, dx. \tag{6.350}$$

6.29 Show that the Yukawa Green's function (6.164) reproduces the Yukawa potential (6.154) when $n = 3$. Use $K_{1/2}(x) = \sqrt{\pi/2x}\, e^{-x}$ (10.109).

6.30 Derive the two explicit formulas (6.201) and (6.202) for the square root of a complex number.

6.31 What is $(-i)^i$? What is the most general value of this expression?

6.32 Use the indefinite integral (6.250) to derive the principal-part formula (6.251).

6.33 The Bessel function $J_n(x)$ is given by the integral

$$J_n(x) = \frac{1}{2\pi i}\oint_C e^{(x/2)(z-1/z)}\,\frac{dz}{z^{n+1}} \tag{6.351}$$

along a counterclockwise contour about the origin. Find the generating function for these Bessel functions, that is, the function $G(x,z)$ whose Laurent series has the $J_n(x)$'s as coefficients

$$G(x,z) = \sum_{n=-\infty}^{\infty} J_n(x)\,z^n. \tag{6.352}$$

6.34 Show that the Heaviside function $\theta(y) = (y + |y|)/(2|y|)$ is given by the integral

$$\theta(y) = \frac{1}{2\pi i}\int_{-\infty}^{\infty} e^{iyx}\,\frac{dx}{x - i\epsilon} \tag{6.353}$$

in which ϵ is an infinitesimal positive number.

6.35 Show that the integral of $\exp(ik)/k$ along the contour from $k = L$ to $k = L + iH$ and then to $k = -L + iH$ and then down to $k = -L$ vanishes in the double limit $L \to \infty$ and $H \to \infty$.

6.36 Use a ghost contour and a cut to evaluate the integral

$$I = \int_{-1}^{1}\frac{dx}{(x^2+1)\sqrt{1-x^2}} \tag{6.354}$$

by imitating Example 6.39. Be careful when picking up the poles at $z = \pm i$. If necessary, use the explicit square root formulas (6.201) and (6.202).

6.37 Redo the previous exercise (6.36) by defining the square roots so that the cuts run from $-\infty$ to -1 and from 1 to ∞. Take advantage of the evenness of the integrand and integrate on a contour that is slightly above the whole real axis. Then add a ghost contour around the upper half-plane.

6.38 Show that if u is even and v is odd, then the Hilbert transforms (6.292) imply (6.294).

6.39 Show why the principal-part identity (6.251) lets one write the Kramers–Kronig integral (6.314) for the index of refraction in the regularized form (6.319).

6.40 Use the formula (6.310) for the group velocity and the regularized expression (6.319) for the real part of the index of refraction $n_r(\omega)$ to derive formula (6.320) for the group velocity.

6.41 (a) Perform the z-integral in Eq. (6.338). (b) Use the result of part (a) to find the commutator $[L_m, L_n]$ of the **Virasoro algebra**. Hint: use the Laurent series (6.334).

6.42 Assume that $\epsilon(z)$ is analytic in a disk that contains a tiny circular contour C_w about the point w as in Fig. 6.10. Do the contour integral

$$\oint_{C_w} \epsilon(z) \left[\frac{c/2}{(z-w)^4} + \frac{2T(w)}{(z-w)^2} + \frac{T'(w)}{z-w} \right] \frac{dz}{2\pi i} \tag{6.355}$$

and express your result in terms of $\epsilon(w)$, $T(w)$, and their derivatives.

6.43 Show that if the coefficients a_k of the equation $0 = a_0 + a_1 z + \cdots + a_n z^n$ are real, then its n roots z_k are real or come in pairs that are complex conjugates, z_ℓ and z_ℓ^*, of each other.

6.44 Show that if a is not an integer, then the sum of the tiny counterclockwise integrals about the points $z = n$ of Example 6.48 is

$$\sum_{n=-\infty}^{\infty} \oint_n \frac{\csc \pi z}{(z-a)^2} \, dz = 2i \sum_{n=-\infty}^{\infty} (-1)^n \frac{1}{(n-a)^2}.$$

6.45 Use the trick of Example 6.48 with $\csc \pi z \to \cot \pi z$ to show that

$$\sum_{n=-\infty}^{\infty} \frac{1}{(n-a)^2} = \frac{\pi^2}{\sin^2 \pi a}$$

as long as a is not an integer.

7
Differential Equations

7.1 Ordinary Linear Differential Equations

There are many kinds of differential equations – linear and nonlinear, ordinary and partial, homogeneous and inhomogeneous. Any way of correctly solving any of them is fine. We start our overview with some definitions.

An operator of the form

$$L = \sum_{m=0}^{n} h_m(x) \frac{d^m}{dx^m}. \tag{7.1}$$

is an **nth-order, ordinary, linear differential operator**. It is *nth order* because the highest derivative is d^n/dx^n. It is *ordinary* because all the derivatives are with respect to the same independent variable x. It is *linear* because derivatives are linear operators

$$L\left[a_1 f_1(x) + a_2 f_2(x)\right] = a_1\, L\, f_1(x) + a_2\, L\, f_2(x). \tag{7.2}$$

If all the $h_m(x)$ in the operator L are constants, independent of x, then L is an nth-order, ordinary, linear differential operator **with constant coefficients**.

Example 7.1 (Second-order linear differential operators) The operator

$$L = -\frac{d^2}{dx^2} - k^2 \tag{7.3}$$

is a second-order, linear differential operator with constant coefficients. The second-order linear differential operator

$$L = -\frac{d}{dx}\left(p(x)\frac{d}{dx}\right) + q(x) \tag{7.4}$$

is in **self-adjoint form** (Section 7.30).

The differential equation $L\, f(x) = 0$ is **homogeneous** because each of its terms is linear in f or one of its derivatives $f^{(m)}$ – there is no term that is not proportional to f or one of its derivatives. The equation $Lf(x) = s(x)$ is **inhomogeneous** because of the source term $s(x)$.

If a differential equation is linear and homogeneous, then we can add solutions. If $f_1(x)$ and $f_2(x)$ are two solutions of the same linear homogeneous differential equation $L\, f_1(x) = 0$ and $L\, f_2(x) = 0$, then any linear combination of these solutions $f(x) = a_1 f_1(x) + a_2 f_2(x)$ with constant coefficients a_1 and a_2 also is a solution since

$$L\, f(x) = L\, \left[a_1 f_1(x) + a_2 f_2(x)\right] = a_1\, L\, f_1(x) + a_2\, L\, f_2(x) = 0. \tag{7.5}$$

This additivity of solutions makes it possible to find general solutions of linear homogeneous differential equations.

Example 7.2 (Sines and cosines) Two solutions of the second-order, linear, homogeneous, ordinary differential equation (ODE)

$$\left(\frac{d^2}{dx^2} + k^2\right) f(x) = 0 \tag{7.6}$$

are $\sin kx$ and $\cos kx$, and the most general solution is the linear combination $f(x) = a_1 \sin kx + a_2 \cos kx$.

The functions $y_1(x), \ldots, y_n(x)$ are **linearly independent** if the only numbers $k_1, \ldots, k_n$ for which the linear combination vanishes for all x

$$k_1\, y_1(x) + k_2\, y_2(x) + \cdots + k_n\, y_n(x) = 0 \tag{7.7}$$

are $k_1 = \ldots = k_n = 0$. Otherwise they are **linearly dependent**.

Suppose that an nth-order linear, homogeneous, ordinary differential equation $L\, f(x) = 0$ has n linearly independent solutions $f_j(x)$, and that all other solutions to this ODE are linear combinations of these n solutions. Then these n solutions are **complete** in the space of solutions of this equation and form a basis for this space. The **general solution** to $L\, f(x) = 0$ is then a linear combination of the f_j's with n arbitrary constant coefficients

$$f(x) = \sum_{j=1}^{n} a_j\, f_j(x). \tag{7.8}$$

With a source term $s(x)$, the differential equation $L\, f(x) = 0$ becomes an **inhomogeneous** linear ordinary differential equation

$$L\, f_i(x) = s(x). \tag{7.9}$$

If $f_{i1}(x)$ and $f_{i2}(x)$ are any two solutions of this inhomogeneous differential equation, then their difference $f_{i1}(x) - f_{i2}(x)$ is a solution of the associated homogeneous equation $L\, f(x) = 0$

$$L\left[f_{i1}(x) - f_{i2}(x)\right] = L\, f_{i1}(x) - L\, f_{i2}(x) = s(x) - s(x) = 0. \tag{7.10}$$

Thus this difference must be given by the general solution (7.8) of the homogeneous equation for some constants a_j

$$f_{i1}(x) - f_{i2}(x) = \sum_{j=1}^{N} a_j f_j(x). \tag{7.11}$$

It follows therefore that every solution $f_{i1}(x)$ of the inhomogeneous differential equation (7.9) is the sum of a particular solution $f_{i2}(x)$ of that equation and some solution (7.8) of the associated homogeneous equation $L\, f = 0$

$$f_{i1}(x) = f_{i2}(x) + \sum_{j=1}^{N} a_j f_j(x). \tag{7.12}$$

Thus the general solution of a linear inhomogeneous equation is a particular solution of that inhomogeneous equation plus the general solution of the associated homogeneous equation.

A **nonlinear** differential equation is one in which a power $f^n(x)$ of the unknown function or of one of its derivatives $\left(f^{(k)}(x)\right)^n$ other than $n = 1$ or $n = 0$ appears or in which the unknown function f appears in some other nonlinear way. For instance, the equations

$$-\, f''(x) = f^3(x), \quad \left(f'(x)\right)^2 = f(x), \quad \text{and} \quad f'(x) = e^{-f(x)} \tag{7.13}$$

are nonlinear differential equations. We can't add two solutions of a nonlinear equation and expect to get a third solution. Nonlinear equations are much harder to solve.

7.2 Linear Partial Differential Equations

An equation of the form

$$L\, f(x) = \sum_{m_1,\ldots,m_k=0}^{n_1,\ldots,n_k} g_{m_1,\ldots,m_k}(x)\, \frac{\partial^{m_1+\cdots+m_k}}{\partial x_1^{m_1} \cdots \partial x_k^{m_k}}\, f(x) = 0 \tag{7.14}$$

in which x stands for $x_1, \ldots, x_k$ is a linear **partial** differential equation of order $n = n_1 + \cdots + n_k$ in the k variables $x_1, \ldots, x_k$. (A partial differential equation is a whole differential equation that has partial derivatives.)

Linear combinations of solutions of a linear homogeneous partial differential equation also are solutions of the equation. So if f_1 and f_2 are solutions of $L\,f = 0$, and a_1 and a_2 are constants, then $f = a_1 f_1 + a_2 f_2$ is a solution since $L\,f = a_1 L\,f_1 + a_2 L\,f_2 = 0$. Additivity of solutions is a property of all linear homogeneous differential equations, whether ordinary or partial.

The **general** solution $f(x) = f(x_1, \ldots, x_k)$ of a linear homogeneous partial differential equation (7.14) is a sum $f(x) = \sum_j a_j f_j(x)$ over a complete set of solutions $f_j(x)$ of the equation with arbitrary coefficients a_j. A linear partial differential equation $L\,f_i(x) = s(x)$ with a source term $s(x) = s(x_1, \ldots, x_k)$ is an **inhomogeneous** linear partial differential equations because of the added source term.

Just as with ordinary differential equations, the difference $f_{i1} - f_{i2}$ of two solutions of the inhomogeneous linear partial differential equation $L\,f_i = s$ is a solution of the associated homogeneous equation $L\,f = 0$ (7.14)

$$L\left[f_{i1}(x) - f_{i2}(x)\right] = s(x) - s(x) = 0. \tag{7.15}$$

So we can expand this difference in terms of the complete set of solutions f_j of the homogeneous linear partial differential equation $L\,f = 0$

$$f_{i1}(x) - f_{i2}(x) = \sum_j a_j f_j(x). \tag{7.16}$$

Thus the general solution of the inhomogeneous linear partial differential equation $L\,f = s$ is the sum of a particular solution f_{i2} of $L\,f = s$ and the general solution $\sum_j a_j f_j$ of the associated homogeneous equation $L\,f = 0$

$$f_{i1}(x) = f_{i2}(x) + \sum_j a_j f_j(x). \tag{7.17}$$

7.3 Separable Partial Differential Equations

A homogeneous linear partial differential equation (PDE) in n variables $x_1, \ldots, x_n$ is **separable** if it can be decomposed into n ordinary differential equations (ODEs) in each of the n variables $x_1, \ldots, x_n$, and if suitable products of the solutions $X_i(x_i)$ of these n ODEs is a solution

$$f(x_1, \ldots, x_n) = X_1(x_1) X_2(x_2) \cdots X_n(x_n) \tag{7.18}$$

of the original PDE. The general solution to the PDE is then a sum of all of its linearly independent solutions f with arbitrary coefficients.

In general, the separability of a partial differential equation depends upon the choice of coordinates. Several of the fundamental PDEs of classical and quantum field theory are separable in several coordinate systems because they usually involve the laplacian $-\nabla\cdot\nabla$ owing to their rotational symmetry.

Example 7.3 (Laplace's equation) The equation for the electrostatic potential in empty space is Laplace's equation

$$L\,\phi(x,y,z)=\nabla\cdot\nabla\phi(x,y,z)=\left(\frac{\partial^2}{\partial x^2}+\frac{\partial^2}{\partial y^2}+\frac{\partial^2}{\partial z^2}\right)\phi(x,y,z)=0. \qquad (7.19)$$

It is a second-order linear homogeneous partial differential equation.

Example 7.4 (Poisson's equation) Poisson's equation for the electrostatic potential ϕ is

$$-\,\triangle\,\phi(x,y,z)\equiv-\left(\frac{\partial^2}{\partial x^2}+\frac{\partial^2}{\partial y^2}+\frac{\partial^2}{\partial z^2}\right)\phi(x,y,z)=\frac{\rho(x,y,z)}{\epsilon_0} \qquad (7.20)$$

in which ρ is the charge density and ϵ_0 is the electric constant. It is a second-order linear inhomogeneous partial differential equation.

Example 7.5 (Maxwell's equations) In empty space, Maxwell's equations are $\nabla\cdot\boldsymbol{E}=0$, $\nabla\cdot\boldsymbol{B}=0$, $\nabla\times\boldsymbol{E}=-\dot{\boldsymbol{B}}$, and $c^2\,\nabla\times\boldsymbol{B}=\dot{\boldsymbol{E}}$. They imply (Exercise 7.6) the wave equations

$$\triangle\boldsymbol{E}=\ddot{\boldsymbol{E}}/c^2 \quad\text{and}\quad \triangle\boldsymbol{B}=\ddot{\boldsymbol{B}}/c^2 \qquad (7.21)$$

which are separable in rectangular, cylindrical, and spherical coordinates among others. For instance, the exponentials $\boldsymbol{E}(\boldsymbol{k},\omega)=\boldsymbol{\epsilon}\,e^{i(\boldsymbol{k}\cdot\boldsymbol{r}-\omega t)}$ and $\boldsymbol{B}(\boldsymbol{k},\omega)=(\hat{\boldsymbol{k}}\times\boldsymbol{\epsilon}/c)\,e^{i(\boldsymbol{k}\cdot\boldsymbol{r}-\omega t)}$ with $\omega=|\boldsymbol{k}|c$ and $\hat{\boldsymbol{k}}\cdot\boldsymbol{\epsilon}=0$ are solutions.

Example 7.6 (Helmholtz's equation in 2 dimensions) In 2 dimensions and in **rectangular coordinates** (x,y), Helmholtz's linear homogeneous partial differential equation

$$-\,\nabla\cdot\nabla f(x,y)=-\left(\frac{\partial^2}{\partial x^2}+\frac{\partial^2}{\partial y^2}\right)f(x,y)=k^2 f(x,y) \qquad (7.22)$$

is separable. The function $f(x,y)=X(x)\,Y(y)$ is a solution

$$-\left(\frac{\partial^2}{\partial x^2}+\frac{\partial^2}{\partial y^2}\right)X(x)Y(y)=-Y(y)\frac{\partial^2 X(x)}{\partial x^2}-X(x)\frac{\partial^2 Y(y)}{\partial y^2}=k^2X(x)Y(y)$$

as long as X and Y satisfy $-X''(x)=a^2X(x)$ and $-Y''(y)=b^2Y(y)$ with $a^2+b^2=k^2$. For instance, if $X_a(x)=e^{iax}$ and $Y_b(y)=e^{iby}$, then any linear combination of the products $X_a(x)\,Y_b(y)$ with $a^2+b^2=k^2$ will be a solution of Helmholtz's equation (7.22).

Helmholtz's equation (7.22) also is separable in **polar coordinates** (ρ,ϕ) in 2 dimensions with a laplacian that is the 3-dimensional laplacian (2.32) without the z derivative

$$\nabla \cdot \nabla f = \frac{\partial^2 f}{\partial \rho^2} + \frac{1}{\rho}\frac{\partial f}{\partial \rho} + \frac{1}{\rho^2}\frac{\partial^2 f}{\partial \phi^2} = \frac{1}{\rho}\frac{d}{d\rho}\left(\rho \frac{df}{d\rho}\right) + \frac{1}{\rho^2}\frac{\partial^2 f}{\partial \phi^2}. \tag{7.23}$$

Substituting the product $f(\rho, \phi) = \mathrm{P}(\rho)\,\Phi(\phi)$ into Helmholtz's equation (7.22) and multiplying both sides by $\rho^2/\mathrm{P}\,\Phi$, we get

$$\rho^2 \frac{\mathrm{P}''}{\mathrm{P}} + \rho \frac{\mathrm{P}'}{\mathrm{P}} + \rho^2 k^2 = -\frac{\Phi''}{\Phi} = n^2. \tag{7.24}$$

The first three terms are functions of ρ, the fourth term $-\Phi''/\Phi$ is a function of ϕ, and the last term n^2 is a constant. The constant n must be an integer if $\Phi_n(\phi) = e^{in\phi}$ is to be single valued on the interval $[0, 2\pi]$. The function $\mathrm{P}_{kn}(\rho) = J_n(k\rho)$ satisfies

$$\rho^2 \mathrm{P}''_{kn} + \rho \mathrm{P}'_{kn} + \rho^2 k^2 \mathrm{P}_{kn} = n^2 \mathrm{P}_{kn} \tag{7.25}$$

because the **Bessel function** of the first kind $J_n(x)$ obeys Bessel's equation (10.4)

$$x^2 J_n'' + x J_n' + x^2 J_n = n^2 J_n \tag{7.26}$$

(Friedrich Bessel 1784–1846). Thus the product $f_{kn}(\rho, \phi) = J_n(k\rho)\, e^{in\phi}$ is a solution to Helmholtz's equation (7.22), as is any linear combination of such products for different n's.

Example 7.7 (Helmholtz's equation in 3 dimensions) In 3 dimensions and in **rectangular coordinates** (x, y, z), Helmholtz's equation

$$-\nabla \cdot \nabla f(x, y, z) = -\left(\frac{\partial^2}{\partial x^2} + \frac{\partial^2}{\partial y^2} + \frac{\partial^2}{\partial z^2}\right) f(x, y, z) = k^2 f(x, y, z) \tag{7.27}$$

is separable. The function $f(x, y, z) = X(x)Y(y)Z(z)$ is a solution

$$-\triangle (XYZ) = -X''YZ - X(x)Y''Z - XYZ'' = k^2 XYZ \tag{7.28}$$

as long as X, Y, and Z satisfy $-X'' = a^2 X$, $-Y'' = b^2 Y$, and $-Z'' = c^2 Z$ with $a^2 + b^2 + c^2 = k^2$. Thus if $X_a(x) = e^{iax}$, $Y_b(y) = e^{iby}$, and $Z_c(z) = e^{icz}$, then any linear combination of the products $X_a\, Y_b\, Z_c$ with $a^2 + b^2 + c^2 = k^2$ will be a solution of Helmholtz's equation (7.27).

Helmholtz's equation (7.27) also is separable in **polar coordinates** (ρ, ϕ, z) with laplacian (2.32)

$$\nabla \cdot \nabla f = \triangle f = \frac{1}{\rho}\left[\left(\rho\, f_{,\rho}\right)_{,\rho} + \frac{1}{\rho} f_{,\phi\phi} + \rho\, f_{,zz}\right]. \tag{7.29}$$

Setting $f(\rho, \phi, z) = \mathrm{P}(\rho)\,\Phi(\phi)\,Z(z)$ and multiplying both sides of Helmholtz's equation (7.27) by $-\rho^2/\mathrm{P}\,\Phi\,Z$, we get

$$\frac{\rho^2}{f}\triangle f = \frac{\rho^2 \mathrm{P}'' + \rho \mathrm{P}'}{\mathrm{P}} + \frac{\Phi''}{\Phi} + \rho^2 \frac{Z''}{Z} = -k^2 \rho^2. \tag{7.30}$$

If we set $Z_\alpha(z) = e^{\alpha z}$, then this equation becomes (7.24) with k^2 replaced by $\alpha^2 + k^2$. Its solution is

$$f(\rho, \phi, z) = J_n(\sqrt{k^2 + \alpha^2}\rho)\, e^{in\phi}\, e^{\alpha z} \tag{7.31}$$

in which n must be an integer if $f(\rho, \phi, z)$ is to apply to the full range of ϕ from 0 to 2π. The case in which $k = 0$ corresponds to Laplace's equation with solution $f(\rho, \phi, z) = J_n(k\rho)e^{in\phi}e^{\alpha z}$. The alternative choice $Z'' = -\alpha^2 Z$ leads to the solution

$$f(\rho, \phi, z) = J_n(\sqrt{k^2 - \alpha^2}\,\rho)\, e^{in\phi}\, e^{i\alpha z}. \tag{7.32}$$

But if $k^2 < \alpha^2$, this solution involves the **modified Bessel function** $I_n(x) = i^{-n} J_n(ix)$ (Section 10.3)

$$f(\rho, \phi, z) = I_n(\sqrt{\alpha^2 - k^2}\,\rho)\, e^{in\phi}\, e^{i\alpha z}. \tag{7.33}$$

Helmholtz's equation (7.27) also is separable in **spherical coordinates** with the laplacian (2.33)

$$\triangle f = \frac{1}{r^2}\frac{\partial}{\partial r}\left(r^2 \frac{\partial f}{\partial r}\right) + \frac{1}{r^2 \sin\theta}\frac{\partial}{\partial \theta}\left(\sin\theta \frac{\partial f}{\partial \theta}\right) + \frac{1}{r^2 \sin^2\theta}\frac{\partial^2 f}{\partial \phi^2} \tag{7.34}$$

the first term of which can be written as $r^{-1}(rf)_{,rr}$. Setting $f(r, \theta, \phi) = R(r)\,\Theta(\theta)\,\Phi_m(\phi)$ where $\Phi_m = e^{im\phi}$ and multiplying both sides by $-r^2/R\Theta\Phi$, we get

$$\frac{(r^2 R')'}{R} + \frac{(\sin\theta\,\Theta')'}{\sin\theta\,\Theta} - \frac{m^2}{\sin^2\theta} = -k^2 r^2. \tag{7.35}$$

The first term is a function of r, the next two terms are functions of θ, and the last term is a constant. We set the r-dependent terms equal to a constant $\ell(\ell+1) - k^2$ and the θ-dependent terms equal to $-\ell(\ell+1)$. The **associated Legendre function** $\Theta_{\ell m}(\theta)$ satisfies (9.93)

$$\left(\sin\theta\,\Theta'_{\ell m}\right)'/\sin\theta + \left[\ell(\ell+1) - m^2/\sin^2\theta\right]\Theta_{\ell m} = 0. \tag{7.36}$$

If $\Phi(\phi) = e^{im\phi}$ is to be single valued for $0 \le \phi \le 2\pi$, then the parameter m must be an integer. The constant ℓ also must be an integer with $-\ell \le m \le \ell$ (Example 7.39, Section 9.12) if $\Theta_{\ell m}(\theta)$ is to be single valued and finite for $0 \le \theta \le \pi$. The product $f = R\,\Theta\,\Phi$ then will obey Helmholtz's equation (7.27) if the radial function $R_{k\ell}(r) = j_\ell(kr)$ satisfies

$$\left(r^2 R'_{k\ell}\right)' + \left[k^2 r^2 - \ell(\ell+1)\right] R_{k\ell} = 0 \tag{7.37}$$

which it will because the **spherical Bessel function** $j_\ell(x)$ obeys Bessel's equation (10.67)

$$\left(x^2 j'_\ell\right)' + [x^2 - \ell(\ell+1)]\, j_\ell = 0. \tag{7.38}$$

In 3 dimensions, Helmholtz's equation separates in 11 standard coordinate systems (Morse and Feshbach, 1953, pp. 655–664).

In special relativity, the time and space coordinates ct and $\boldsymbol{x}$ often are written as x^0, x^1, x^2, and x^3 or as the 4-vector $(x^0, \boldsymbol{x})$. To form the invariant inner product $px \equiv \boldsymbol{x} \cdot \boldsymbol{x} - p^0 x^0 = \boldsymbol{x} \cdot \boldsymbol{x} - Et$ as $p_a x^a$ with a summed from 0 to 3, one attaches a minus sign to the time components of 4-vectors with lowered indexes so that $p_0 = -p^0$ and $x_0 = -x^0$. The derivatives $\partial_a f$ and $\partial^a f$ are

$$\partial_a f = \frac{\partial f}{\partial x^a} \quad \text{and} \quad \partial^a f = \frac{\partial f}{\partial x_a} = -\frac{\partial f}{\partial x^a} = -\partial_a f. \tag{7.39}$$

The opposite metric with px set equal to $p^0 x^0 - \boldsymbol{x} \cdot \boldsymbol{x}$ is also in use.

Example 7.8 (Klein–Gordon equation) In Minkowski space, the analog of the laplacian in natural units ($\hbar = c = 1$) is (summing over a from 0 to 3)

$$\Box = \partial_a \partial^a = \Delta - \frac{\partial^2}{\partial x^{02}} = \nabla \cdot \nabla - \frac{\partial^2}{\partial t^2} = \frac{\partial^2}{\partial x^2} + \frac{\partial^2}{\partial y^2} + \frac{\partial^2}{\partial z^2} - \frac{\partial^2}{\partial t^2} \tag{7.40}$$

in which for time derivatives $\partial^0 = -\partial_0$, but for spatial derivatives $\partial^i = \partial_i$. The Klein–Gordon equation (4.174) is

$$\left(\Box - m^2\right) A(x) = \left(\Delta - \frac{\partial^2}{\partial t^2} - m^2\right) A(x) = 0. \tag{7.41}$$

If we set $A(x) = B(px)$ where $px = p_a x^a = \boldsymbol{p} \cdot \boldsymbol{x} - p^0 x^0$, then the kth partial derivative of A is p_k times the first derivative of B

$$\frac{\partial}{\partial x^k} A(x) = \frac{\partial}{\partial x^k} B(px) = p_k B'(px) \tag{7.42}$$

and so the Klein–Gordon equation (7.41) becomes

$$\left(\Box - m^2\right) A = (\boldsymbol{p}^2 - (p^0)^2) B'' - m^2 B = p^2 B'' - m^2 B = 0 \tag{7.43}$$

in which $p^2 = \boldsymbol{p}^2 - (p^0)^2$. Thus if $B(p \cdot x) = \exp(ip \cdot x)$ so that $B'' = -B$, and if the energy–momentum 4-vector $(p^0, \boldsymbol{p})$ satisfies $p^2 + m^2 = 0$, then $A(x)$ will satisfy the Klein-Gordon equation. The condition $p^2 + m^2 = 0$ relates the energy $p^0 = \sqrt{\boldsymbol{p}^2 + m^2}$ to the momentum $\boldsymbol{p}$ for a particle of mass m.

Example 7.9 (Field of a spinless boson) The quantum field (4.183)

$$\phi(x) = \int \frac{d^3 p}{\sqrt{2p^0 (2\pi)^3}} \left[a(\boldsymbol{p}) e^{ipx} + a^\dagger(\boldsymbol{p}) e^{-ipx} \right] \tag{7.44}$$

describes spinless bosons of mass m. It satisfies the Klein–Gordon equation $\left(\Box - m^2\right) \phi(x) = 0$ because $p^0 = \sqrt{\boldsymbol{p}^2 + m^2}$. The operators $a(\boldsymbol{p})$ and $a^\dagger(\boldsymbol{p})$ respectively represent the annihilation and creation of the bosons and obey the commutation relations

$$[a(\boldsymbol{p}), a^\dagger(\boldsymbol{p}')] = \delta^3(\boldsymbol{p} - \boldsymbol{p}') \quad \text{and} \quad [a(\boldsymbol{p}), a(\boldsymbol{p}')] = [a^\dagger(\boldsymbol{p}), a^\dagger(\boldsymbol{p}')] = 0 \tag{7.45}$$

in units with $\hbar = c = 1$. These relations make the field $\phi(x)$ and its time derivative $\dot{\phi}(y)$ satisfy **the canonical equal-time commutation relations** $[\phi(\boldsymbol{x},t), \dot{\phi}(\boldsymbol{y},t)] = i\,\delta^3(\boldsymbol{x}-\boldsymbol{y})$ and $[\phi(\boldsymbol{x},t), \phi(\boldsymbol{y},t)] = [\dot{\phi}(\boldsymbol{x},t), \dot{\phi}(\boldsymbol{y},t)] = 0$ in which a dot means a time derivative.

Example 7.10 (Field of the photon) The electromagnetic field has four components, but in the Coulomb or radiation gauge $\nabla \cdot \boldsymbol{A}(x) = 0$, the component A_0 is a function of the charge density, and the vector potential $\boldsymbol{A}$ in the absence of charges and currents satisfies the wave equation $\Box \boldsymbol{A}(x) = 0$ for a spin-one massless particle. We write it as

$$\boldsymbol{A}(x) = \sum_{s=1}^{2} \int \frac{d^3p}{\sqrt{2p^0(2\pi)^3}} \left[\boldsymbol{e}(\boldsymbol{p},s)\, a(\boldsymbol{p},s)\, e^{ipx} + \boldsymbol{e}^*(\boldsymbol{p},s)\, a^\dagger(\boldsymbol{p},s)\, e^{-ipx} \right] \tag{7.46}$$

in which the sum is over the two possible polarizations s. The energy p^0 is equal to the modulus $|\boldsymbol{p}|$ of the momentum because the photon is massless, $p^2 = 0$. The dot product of the polarization vectors $\boldsymbol{e}(\boldsymbol{p},s)$ with the momentum vanishes $\boldsymbol{p}\cdot\boldsymbol{e}(\boldsymbol{p},s) = 0$ so as to respect the gauge condition $\nabla \cdot \boldsymbol{A}(x) = 0$. The annihilation and creation operators obey the commutation relation

$$[a(\boldsymbol{p},s), a^\dagger(\boldsymbol{p}',s')] = \delta^3(\boldsymbol{p}-\boldsymbol{p}')\,\delta_{s,s'} \tag{7.47}$$

as well as $[a(\boldsymbol{p},s), a(\boldsymbol{p}',s')] = 0$ and $[a^\dagger(\boldsymbol{p},s), a^\dagger(\boldsymbol{p}',s')] = 0$. Because of the Coulomb-gauge condition $\nabla \cdot \boldsymbol{A}(x) = 0$, the commutation relations of the vector potential $\boldsymbol{A}(x)$ involve the transverse delta function

$$\begin{aligned}\left[A_i(t,\boldsymbol{x}), \dot{A}_j(t,\boldsymbol{y})\right] &= i\,\delta_{ij}\delta^{(3)}(\boldsymbol{x}-\boldsymbol{y}) + i\frac{\partial^2}{\partial x^i \partial x^j}\frac{1}{4\pi|\boldsymbol{x}-\boldsymbol{y}|} \\ &= i\int e^{i\boldsymbol{k}\cdot(\boldsymbol{x}-\boldsymbol{y})}\left(\delta_{ij} - \frac{k_i k_j}{\boldsymbol{k}^2}\right)\frac{d^3k}{(2\pi)^3}.\end{aligned} \tag{7.48}$$

Example 7.11 (Dirac's equation) Fields $\chi_b(x)$ that describe particles of spin one-half have four components, $b = 1, \ldots, 4$. In the absence of interactions, they satisfy the Dirac equation

$$\left(\gamma^a_{bc}\partial_a + m\delta_{bc}\right)\chi_c(x) = 0 \tag{7.49}$$

in which repeated indices are summed over – b, c from 1 to 4 and a from 0 to 3. In matrix notation, the **Dirac equation** is

$$\left(\gamma^a\partial_a + m\right)\chi(x) = 0. \tag{7.50}$$

The four Dirac gamma matrices are defined by the 10 rules

$$\{\gamma^a, \gamma^b\} \equiv \gamma^a\gamma^b + \gamma^b\gamma^a = 2\eta^{ab} \tag{7.51}$$

in which η is the 4×4 diagonal matrix $\eta^{00} = \eta_{00} = -1$ and $\eta^{bc} = \eta_{bc} = \delta_{bc}$ for $b, c = 1, 2,$ or 3.

If $\phi(x)$ is a 4-component field that satisfies the Klein–Gordon equation $(\Box - m^2)\phi = 0$, then the field $\chi(x) = (\gamma^b \partial_b - m)\phi(x)$ satisfies (Exercise 7.7) the Dirac equation (7.50)

$$
\begin{aligned}
(\gamma^a \partial_a + m)\, \chi(x) &= (\gamma^a \partial_a + m)\, (\gamma^b \partial_b - m)\phi(x) = \left(\gamma^a \gamma^b \partial_a \partial_b - m^2\right) \phi(x) \\
&= \left[\frac{1}{2}\left(\{\gamma^a, \gamma^b\} + [\gamma^a, \gamma^b]\right) \partial_a \partial_b - m^2\right] \phi(x) \\
&= \left(\eta^{ab} \partial_a \partial_b - m^2\right) \phi(x) = (\Box - m^2)\phi(x) = 0.
\end{aligned}
\tag{7.52}
$$

The simplest Dirac field is the Majorana field

$$
\chi_b(x) = \int \frac{d^3p}{(2\pi)^{3/2}} \sum_s \left[u_b(\boldsymbol{p}, s)\, a(\boldsymbol{p}, s) e^{ipx} + v_b(\boldsymbol{p}, s)\, a^\dagger(\boldsymbol{p}, s) e^{-ipx}\right] \tag{7.53}
$$

in which $p^0 = \sqrt{\boldsymbol{p}^2 + m^2}$, s labels the two spin states, and the operators a and $a^\dagger$ obey **the anticommutation relations**

$$
\begin{aligned}
\{a(\boldsymbol{p}, s), a^\dagger(\boldsymbol{p}', s')\} &\equiv a(\boldsymbol{p}, s)\, a^\dagger(\boldsymbol{p}', s') + a^\dagger(\boldsymbol{p}', s')\, a(\boldsymbol{p}, s) = \delta_{ss'}\, \delta(\boldsymbol{p} - \boldsymbol{p}') \\
\{a(\boldsymbol{p}, s), a(\boldsymbol{p}', s')\} &= \{a^\dagger(\boldsymbol{p}, s), a^\dagger(\boldsymbol{p}', s')\} = 0.
\end{aligned}
\tag{7.54}
$$

It describes a neutral particle of mass m.

If two Majorana fields χ_1 and χ_2 represent particles of the same mass, then one may combine them into one Dirac field

$$
\psi(x) = \frac{1}{\sqrt{2}} [\chi_1(x) + i\chi_2(x)] \tag{7.55}
$$

which describes a charged particle such as a quark or a lepton.

7.4 First-Order Differential Equations

The equation

$$
\frac{dy}{dx} = f(x, y) = -\frac{P(x, y)}{Q(x, y)} \tag{7.56}
$$

or **system**

$$
P(x, y)\, dx + Q(x, y)\, dy = 0 \tag{7.57}
$$

is a **first-order ordinary differential equation**.

7.5 Separable First-Order Differential Equations

If in a first-order ordinary differential equation like (7.57) one can separate the dependent variable y from the independent variable x

$$
F(x)\, dx + G(y)\, dy = 0 \tag{7.58}
$$

then the equation (7.57) is **separable**, and (7.58) is **separated**.

Once the variables are separated, one can integrate and obtain an equation, called the **general integral**, that relates y to x

$$0 = \int_{x_0}^{x} F(x')\,dx' + \int_{y_0}^{y} G(y')\,dy' \tag{7.59}$$

and so provides a solution $y(x)$ of the differential equation.

Example 7.12 (Zipf's law) In 1913, Auerbach noticed that many quantities are distributed as (Gell-Mann, 1994, pp. 92–100)

$$dn = -a\frac{dx}{x^{k+1}} \tag{7.60}$$

an ODE that is separable and separated. For $k \neq 0$, we may integrate this to $n + c = a/kx^k$ or

$$x = \left(\frac{a}{k(n+c)}\right)^{1/k} \tag{7.61}$$

in which c is a constant.

The case $k = 1$ occurs frequently $x = a/(n+c)$ and is called **Zipf's law**. With $c = 0$, it applies approximately to the populations of cities: if the largest city ($n = 1$) has population x, then the populations of the second, third, fourth cities ($n = 2, 3, 4$) will be $x/2$, $x/3$, and $x/4$.

Again with $c = 0$, Zipf's law applies to the occurrence of numbers x in a table of some sort. Since $x = a/n$, the rank n of the number x is approximately $n = a/x$. So the number of numbers that occur with first digit d and, say, 4 trailing digits will be

$$\begin{aligned} n(d0000) - n(d9999) &= a\left(\frac{1}{d0000} - \frac{1}{d9999}\right) = a\left(\frac{9999}{d0000 \times d9999}\right) \\ &\approx a\left(\frac{10^4}{d(d+1)\,10^8}\right) = \frac{a\,10^{-4}}{d(d+1)}. \end{aligned} \tag{7.62}$$

The ratio of the number of numbers with first digit d to the number with first digit d' is then $d'(d'+1)/d(d+1)$. For example, the first digit is more likely to be 1 than 9 by a factor of 45. The German government uses such formulas to catch tax evaders.

Example 7.13 (Logistic equation)

$$\frac{dy}{dt} = ay\left(1 - \frac{y}{Y}\right) \tag{7.63}$$

is separable and separated. It describes a wide range of phenomena whose evolution with time t is sigmoidal such as (Gell-Mann, 2008) the cumulative number of casualties in a war, the cumulative number of deaths in London's great plague, and the cumulative number of papers in an academic's career. It also describes the effect y on an animal of a given dose t of a drug.

With $f = y/Y$, the logistic equation (7.63) is $\dot{f} = af\,(1-f)$ or

$$a\,dt = \frac{df}{f(1-f)} = \frac{df}{f} + \frac{df}{1-f} \tag{7.64}$$

which we may integrate to $a(t - t_h) = \ln\left[f/(1-f)\right]$. Taking the exponential of both sides, we find $\exp[a(t-t_h)] = f/(1-f)$ which we can solve for f

$$f(t) = \frac{e^{a(t-t_h)}}{1 + e^{a(t-t_h)}}. \tag{7.65}$$

The sigmoidal shape of $f(t)$ is like a smoothed Heaviside function. In terms of $y_0 = y(0)$, the value of $y(t)$ is

$$y(t) = \frac{Y y_0\, e^{at}}{Y + y_0\,(e^{at} - 1)}. \tag{7.66}$$

Example 7.14 (Lattice QCD) In lattice field theory, the beta-function

$$\beta(g) \equiv -\frac{dg}{d\ln a} \tag{7.67}$$

tells us how we must adjust the coupling constant g in order to keep the physical predictions of the theory constant as we vary the lattice spacing a. In quantum chromodynamics $\beta(g) = -\beta_0\, g^3 - \beta_1\, g^5 + \cdots$ where

$$\beta_0 = \frac{1}{(4\pi)^2}\left(11 - \frac{2}{3}n_f\right) \quad \text{and} \quad \beta_1 = \frac{1}{(4\pi)^4}\left(102 - 10\,n_f - \frac{8}{3}n_f\right) \tag{7.68}$$

in which n_f is the number of light quark flavors. Combining the definition (7.67) of the β-function with the first term of its expansion $\beta(g) = -\beta_0\, g^3$ for small g, one arrives at the differential equation

$$\frac{dg}{d\ln a} = \beta_0\, g^3 \tag{7.69}$$

which one may integrate

$$\int d\ln a = \ln a + c = \int \frac{dg}{\beta_0 g^3} = -\frac{1}{2\beta_0 g^2} \tag{7.70}$$

to find

$$\Lambda\, a(g) = e^{-1/2\beta_0 g^2} \tag{7.71}$$

in which Λ is a constant of integration. As g approaches 0, which is an essential singularity (Section 6.11), the lattice spacing $a(g)$ goes to zero *very fast* (as long as $n_f \le 16$). The inverse of this relation $g(a) \approx 1/\sqrt{\beta_0 \ln(1/a^2\Lambda^2)}$ shows that the coupling constant $g(a)$ slowly goes to zero as the lattice spacing (or shortest wave length) a goes to zero. The strength of the interaction shrinks logarithmically as the energy $1/a$ increases in this lattice version of **asymptotic freedom**.

7.6 Hidden Separability

As long as each of the functions $P(x, y)$ and $Q(x, y)$ in the ODE

$$P(x, y)dx + Q(x, y)dy = U(x)V(y)dx + R(x)S(y)dy = 0 \tag{7.72}$$

can be factored $P(x, y) = U(x)V(y)$ and $Q(x, y) = R(x)S(y)$ into the product of a function of x times a function of y, then the ODE is separable. Following (Ince, 1956), we divide the ODE by $R(x)V(y)$, separate the variables

$$\frac{U(x)}{R(x)}\,dx + \frac{S(y)}{V(y)}\,dy = 0 \tag{7.73}$$

and integrate

$$\int \frac{U(x)}{R(x)}\,dx + \int \frac{S(y)}{V(y)}\,dy = C \tag{7.74}$$

in which C is a constant of integration.

Example 7.15 (Hidden separability) We separate the variables in

$$x(y^2 - 1)\,dx - y(x^2 - 1)\,dy = 0 \tag{7.75}$$

by dividing by $(y^2 - 1)(x^2 - 1)$ so as to get

$$\frac{x}{x^2 - 1}\,dx - \frac{y}{y^2 - 1}\,dy = 0. \tag{7.76}$$

Integrating, we find $\ln(x^2 - 1) - \ln(y^2 - 1) = -\ln C$ or $C\,(x^2 - 1) = y^2 - 1$ which we solve for $y(x) = \sqrt{1 + C(x^2 - 1)}$.

7.7 Exact First-Order Differential Equations

The differential equation

$$P(x, y)\,dx + Q(x, y)\,dy = 0 \tag{7.77}$$

is **exact** if its left-hand side is the differential of some function $\phi(x, y)$

$$P\,dx + Q\,dy = d\phi = \phi_x\,dx + \phi_y\,dy. \tag{7.78}$$

We'll have more to say about the **exterior derivative** d in section 12.6.

The **criteria of exactness** are

$$P(x, y) = \frac{\partial \phi(x, y)}{\partial x} \equiv \phi_x(x, y) \quad \text{and} \quad Q(x, y) = \frac{\partial \phi(x, y)}{\partial y} \equiv \phi_y(x, y). \tag{7.79}$$

Thus, if the ODE (7.77) is exact, then

$$P_y(x, y) = \phi_{yx}(x, y) = \phi_{xy}(x, y) = Q_x(x, y) \tag{7.80}$$

which is called the **condition of integrability**. This condition implies that the ODE (7.77) is exact and integrable, as we'll see in Section 7.8.

A first-order ODE that is separable and separated

$$P(x)dx + Q(y)dy = 0 \tag{7.81}$$

is exact because

$$P_y = 0 = Q_x. \tag{7.82}$$

But a first-order ODE may be exact without being separable.

Example 7.16 (Boyle's law) At a fixed temperature T, changes in the pressure P and volume V of an ideal gas are related by $PdV + VdP = 0$. This ODE is exact because $PdV + VdP = d(PV)$. Its integrated form is the **ideal-gas law** $PV = NkT$ in which N is the number of molecules in the gas and k is **Boltzmann's constant**, $k = 1.38066 \times 10^{-23}$ J/K $= 8.617385 \times 10^{-5}$ eV/K.

A more accurate formula, due to van der Waals (1837–1923) is

$$\left[P + \left(\frac{N}{V}\right)^2 a'\right]\left(V - Nb'\right) = NkT \tag{7.83}$$

in which $a' > 0$ represents the mutual attraction of the molecules and $b' > 0$ is the effective volume of a single molecule. This equation was a sign that molecules were real particles, a fact finally accepted after Einstein in 1905 related the viscous-friction coefficient ζ and the diffusion constant D to the energy kT of a thermal fluctuation by the equation $\zeta\, D = kT$, as explained in section 15.12 (Albert Einstein 1879–1955).

Example 7.17 (Human population growth) If the number of people rises as the square of the population, then $\dot{N} = N^2/b$. The separated and hence exact form of this differential equation is $dN/N^2 = dt/b$ which we integrate to $N(t) = b/(T - t)$ where T is the time at which the population becomes infinite. With $T = 2025$ years and $b = 2 \times 10^{11}$ years, this formula is a fair model of the world's population between the years 1 and 1970. For a more accurate account, see (von Foerster et al., 1960).

7.8 Meaning of Exactness

We can integrate the differentials of a first-order ODE

$$P(x, y)\, dx + Q(x, y)\, dy = 0 \tag{7.84}$$

along any contour C in the x–y-plane, but in general we'd get a functional

$$\phi(x, y, C, x_0, y_0) = \int_{(x_0, y_0)C}^{(x,y)} P(x', y')\, dx' + Q(x', y')\, dy' \tag{7.85}$$

that depends upon the contour $\mathcal{C}$ of integration as well as upon the endpoints (x_0, y_0) and (x, y).

But if the differential $P dx + Q dy$ is exact, then it's the differential or exterior derivative $d\phi = P(x, y)\, dx + Q(x, y)\, dy$ of a function $\phi(x, y)$ that depends upon the variables x and y without any reference to a contour of integration. Thus if $P dx + Q dy = d\phi$, then the contour integral (7.85) is

$$\int_{(x_0,y_0)\mathcal{C}}^{(x,y)} P(x', y')\, dx' + Q(x', y')\, dy' = \int_{(x_0,y_0)}^{(x,y)} d\phi = \phi(x, y) - \phi(x_0, y_0). \quad (7.86)$$

This integral defines a function $\phi(x, y; x_0, y_0) \equiv \phi(x, y) - \phi(x_0, y_0)$ whose differential vanishes $d\phi = P dx + Q dy = 0$ according to the original differential equation (7.84). Thus the ODE and its exactness leads to an equation

$$\phi(x, y; x_0, y_0) = B \quad (7.87)$$

that we can solve for y, our solution of the ODE (7.84)

$$d\phi(x, y; x_0, y_0) = P(x, y)\, dx + Q(x, y)\, dy = 0. \quad (7.88)$$

Example 7.18 (Explicit use of exactness) We'll now explicitly use the criteria of exactness

$$P(x, y) = \frac{\partial \phi(x, y)}{\partial x} \equiv \phi_x(x, y) \ \text{ and } \ Q(x, y) = \frac{\partial \phi(x, y)}{\partial y} \equiv \phi_y(x, y) \quad (7.89)$$

to integrate the general exact differential equation

$$P(x, y)\, dx + Q(x, y)\, dy = 0. \quad (7.90)$$

We use the first criterion $P = \phi_x$ to integrate the condition $\phi_x = P$ in the x-direction getting a known integral $R(x, y)$ and an unknown function $C(y)$

$$\phi(x, y) = \int P(x, y)\, dx + C(y) = R(x, y) + C(y). \quad (7.91)$$

The second criterion $Q = \phi_y$ says that

$$Q(x, y) = \phi_y(x, y) = R_y(x, y) + C_y(y). \quad (7.92)$$

We get $C(y)$ by integrating its known derivative $C_y = Q - R_y$

$$C(y) = \int Q(x, y) - R_y(x, y)\, dy + D. \quad (7.93)$$

We now put C into the formula $\phi = R + C$ which is (7.91). Setting $\phi = E$ a constant, we find an equation that we can solve for y

$$\phi(x, y) = R(x, y) + C(y) = R(x, y) + \int Q(x, y) - R_y(x, y)\, dy + D = E.$$

Example 7.19 (Using exactness) The functions P and Q in the differential equation

$$P(x, y)\,dx + Q(x, y)\,dy = \ln(y^2 + 1)\,dx + \frac{2y(x-1)}{y^2+1}\,dy = 0 \tag{7.94}$$

are factorized, so the ODE is separable. It's also exact since

$$P_y = \frac{2y}{y^2+1} = Q_x \tag{7.95}$$

and so we can apply the method just outlined. First as in (7.91), we integrate $\phi_x = P$ in the x-direction

$$\phi(x, y) = \int \ln(y^2 + 1)\,dx + C(y) = x\,\ln(y^2 + 1) + C(y). \tag{7.96}$$

Then as in (7.92), we use $\phi_y = Q$

$$\phi(x, y)_y = \frac{2xy}{y^2+1} + C_y(y) = Q(x, y) = \frac{2y(x-1)}{y^2+1} \tag{7.97}$$

to find that $C_y = -2y/(y^2+1)$. We integrate C_y in the y-direction as in (7.93) getting $C(y) = -\ln(y^2 + 1) + D$. We now put $C(y)$ into our formula (7.96) for $\phi(x, y)$

$$\phi(x, y) = (x - 1)\,\ln(y^2 + 1) + D$$

which we set equal to a constant

$$\phi(x, y) = (x - 1)\,\ln(y^2 + 1) + D = E \tag{7.98}$$

or more simply $(x - 1)\,\ln(y^2 + 1) = F$. Unraveling this equation we find

$$y(x) = \left(e^{F/(x-1)} - 1\right)^{1/2} \tag{7.99}$$

as our solution to the differential equation (7.94).

7.9 Integrating Factors

With great luck, one might invent an **integrating factor** $\alpha(x, y)$ that makes an ordinary differential equation $P\,dx + Q\,dy = 0$ exact

$$\alpha\,P\,dx + \alpha\,Q\,dy = d\phi \tag{7.100}$$

and therefore integrable. Such an integrating factor α must satisfy both

$$\alpha\,P = \phi_x \quad \text{and} \quad \alpha\,Q = \phi_y \tag{7.101}$$

so that

$$(\alpha\,P)_y = \phi_{xy} = (\alpha\,Q)_x. \tag{7.102}$$

Example 7.20 (Two simple integrating factors) The ODE $y dx - x dy = 0$ is not exact, but $\alpha(x, y) = 1/x^2$ is an integrating factor. For after multiplying by α, we have

$$-\frac{y}{x^2} dx + \frac{1}{x} dy = 0 \tag{7.103}$$

so that $P = -y/x^2$, $Q = 1/x$, and

$$P_y = -\frac{1}{x^2} = Q_x \tag{7.104}$$

which shows that (7.103) is exact.

Another integrating factor is $\alpha(x, y) = 1/xy$ which separates the variables

$$\frac{dx}{x} = \frac{dy}{y} \tag{7.105}$$

so that we can integrate and get $\ln(y/y_0) = \ln(x/x_0)$ or $\ln(y x_0/x y_0) = 0$ which implies that $y = (y_0/x_0)x$.

7.10 Homogeneous Functions

A function $f(x) = f(x_1, \ldots, x_k)$ of k variables x_i is **homogeneous** of degree n if

$$f(tx) = f(tx_1, \ldots, tx_k) = t^n f(x). \tag{7.106}$$

For instance, $z^2 \ln(x/y)$ is homogeneous of degree 2 because

$$(tz)^2 \ln(tx/ty) = t^2 \left(z^2 \ln(x/y)\right). \tag{7.107}$$

By differentiating (7.106) with respect to t, we find

$$\frac{d}{dt} f(tx) = \sum_{i=1}^{k} \frac{dtx_i}{dt} \frac{\partial f(tx)}{\partial tx_i} = \sum_{i=1}^{k} x_i \frac{\partial f(tx)}{\partial tx_i} = n t^{n-1} f(x). \tag{7.108}$$

Setting $t = 1$, we see that a function that is homogeneous of degree n satisfies

$$\sum_{i=1}^{k} x_i \frac{\partial f(x)}{\partial x_i} = n f(x) \tag{7.109}$$

which is one of Euler's many theorems.

Example 7.21 (Internal energy and entropy) The internal energy U is a first-degree homogeneous function of the entropy S, the volume V, and the numbers N_j of molecules of kind j because $U(tS, tV, tN) = t\,U(S, V, N)$. The temperature T, pressure p, and chemical potential μ_j are defined as

$$T = \left.\frac{\partial U}{\partial S}\right|_{V,N}, \quad p = -\left.\frac{\partial U}{\partial V}\right|_{S,N}, \text{ and } \quad \mu_j = \left.\frac{\partial U}{\partial N_j}\right|_{S,V,N_{i\neq j}} \tag{7.110}$$

so the change dU in the internal energy is

$$dU = T\,dS - p\,dV + \sum_j \mu_j dN_j. \tag{7.111}$$

Euler's theorem (7.109) for $n = 1$ expresses the internal energy as

$$U = T\,S - p\,V + \sum_j \mu_j\,N_j. \tag{7.112}$$

If the energy density of empty space is positive, then the internal energy of the universe rises as it expands, and the definition (7.110) of the pressure implies that the pressure on the universe is negative. Thus **dark energy** accelerates the expansion of the universe.

The formula (7.111) for the change dU implies that the change in the entropy is

$$dS = dU/T + (p/T)\,dV - \sum_j (\mu_j/T)\,dN_j \tag{7.113}$$

which leads to relations

$$\frac{1}{T} = \left.\frac{\partial S}{\partial U}\right|_{V,N}, \quad \frac{p}{T} = \left.\frac{\partial S}{\partial V}\right|_{U,N}, \text{ and } \quad \frac{\mu_j}{T} = -\left.\frac{\partial S}{\partial N_j}\right|_{U,V,N_{i\neq j}} \tag{7.114}$$

analogous to the definitions (7.110). Unlike the internal energy, however, the entropy $S(U, V, N)$ is not a homogeneous function of U, V, and N. The Sackur–Tetrode formula for the entropy of an ideal gas is

$$S \approx kN\left[\ln\left(\frac{V}{N}\left(\frac{4\pi m\,U}{3Nh^2}\right)^{3/2}\right) + \frac{5}{2}\right] \tag{7.115}$$

in which h is Planck's constant and m is the mass of a molecule of the gas.

7.11 Virial Theorem

Consider N particles moving nonrelativistically in a potential $V(x)$ of $3N$ variables that is homogeneous of degree n. Their **virial** is the sum of the products of the coordinates x_i multiplied by the momenta p_i

$$G = \sum_{i=1}^{3N} x_i\,p_i. \tag{7.116}$$

In terms of the kinetic energy $T = (v_1 p_1 + \cdots + v_{3N} p_{3N})/2$, the time derivative of the virial is

$$\frac{dG}{dt} = \sum_{i=1}^{3N} (v_i\, p_i + x_i\, F_i) = 2T + \sum_{i=1}^{3N} x_i\, F_i \tag{7.117}$$

in which the time derivative of a momentum $\dot{p}_i = F_i$ is a component of the force. We now form the infinite-time average of both sides of this equation

$$\lim_{t\to\infty} \frac{G(t) - G(0)}{t} = \left\langle \frac{dG}{dt} \right\rangle = 2\,\langle T \rangle + \left\langle \sum_{i=1}^{3N} x_i\, F_i \right\rangle. \tag{7.118}$$

If the particles are bound by a potential V, then it is reasonable to assume that the positions and momenta of the particles and their virial $G(t)$ are bounded for all times, and we will make this assumption. It follows that as $t \to \infty$, the time average of the time derivative $\dot{G}$ of the virial must vanish

$$0 = 2\,\langle T \rangle + \left\langle \sum_{i=1}^{3N} x_i\, F_i \right\rangle. \tag{7.119}$$

Newton's law

$$F_i = -\frac{\partial V(x)}{\partial x_i} \tag{7.120}$$

now implies that

$$2\,\langle T \rangle = \left\langle \sum_{i=1}^{3N} x_i\, \frac{\partial V(x)}{x_i} \right\rangle. \tag{7.121}$$

If, further, the potential $V(x)$ is a *homogeneous function of degree n*, then Euler's theorem (7.109) gives us $x_i\, \partial_i V = nV$ and the **virial theorem**

$$\langle T \rangle = \frac{n}{2}\, \langle V \rangle\,. \tag{7.122}$$

The long-term time average of the kinetic energy of particles trapped in a homogeneous potential of degree n is $n/2$ times the long-term time average of their potential energy.

Example 7.22 (Coulomb forces) A $1/r$ gravitational or electrostatic potential is homogeneous of degree -1, and so the virial theorem asserts that particles bound in such wells must have long-term time averages that satisfy

$$\langle T \rangle = -\frac{1}{2}\, \langle V \rangle. \tag{7.123}$$

In natural units ($\hbar = c = 1$), the energy of an electron of momentum p a distance r from a proton is $E = p^2/2m - e^2/r$ in which e is the charge of the electron. The uncertainty principle (Example 4.7) gives us an approximate lower bound on the product $r\, p \gtrsim 1$ which we will use in the form $r\, p = 1$ to estimate the energy E of the ground state of the hydrogen atom. Using $1/r = p$, we have $E = p^2/2m - e^2 p$. Differentiating, we find the minimum of E is at $0 = p/m - e^2$. Thus the kinetic energy of the ground state is $T = p^2/2m = me^4/2$ while its potential energy is $V = -e^2 p = -me^4$. Since $T = -V/2$, these values satisfy the virial theorem. They give the ground-state energy as $E = -me^4/2 = -mc^2(e^2/\hbar c)^2 = 13.6$ eV.

Example 7.23 (Dark matter) In 1933, Zwicky applied the gravitational version of the virial theorem (7.123) to the galaxies of the Coma cluster. He used his observations of their luminosities to estimate the gravitational term $-\frac{1}{2}\langle V\rangle$ and found it to be much less than their mean kinetic energy $\langle T\rangle$. He called the transparent mass **dark matter** (Fritz Zwicky, 1898–1974).

Example 7.24 (Harmonic forces) Particles confined in a harmonic potential $V(r) = \sum_k m_k \omega_k^2 \boldsymbol{r}_k^2$ which is homogeneous of degree 2 must have long-term time averages that satisfy $\langle T\rangle = \langle V(x)\rangle$.

7.12 Legendre's Transform

The change in a function $A(x, y)$ of two independent variables x and y is

$$dA = \frac{\partial A}{\partial x}\,dx + \frac{\partial A}{\partial y}\,dy. \tag{7.124}$$

Suppose we define a variable v as

$$v \equiv \frac{\partial A}{\partial y} \tag{7.125}$$

and a new function B as

$$B \equiv A(x, y) - v\,y. \tag{7.126}$$

Then the change in B is

$$dB = \frac{\partial A}{\partial x}\,dx + \frac{\partial A}{\partial y}\,dy - dv\,y - \frac{\partial A}{\partial y}\,dy = \frac{\partial A}{\partial x}\,dx - y\,dv \tag{7.127}$$

which says that B is properly a function $B(x, v)$ of x and v. To really make B a function of x and v, however, we must invert the definition (7.125) of v and express y as a function of x and v so that we can write

$$B(x, v) = A(x, y(x, v)) - v\,y(x, v). \tag{7.128}$$

We also could define a new variable u as

$$u \equiv \frac{\partial A}{\partial x} \tag{7.129}$$

and a new function C as

$$C \equiv A(x, y) - u\,x. \tag{7.130}$$

The change in C is

$$dC = \frac{\partial A}{\partial x}\,dx + \frac{\partial A}{\partial y}\,dy - x\,du - \frac{\partial A}{\partial x}\,dx = \frac{\partial A}{\partial y}\,dy - x\,du \tag{7.131}$$

which says that C is properly a function $C(u, y)$ of u and y. To really make C a function of u and y, however, we must invert the definition (7.129) of u and express x as a function of u and y so that we can write

$$C(u, y) = A(x(u, y), y) - u\,x(u, y). \tag{7.132}$$

We also could combine the two transformations by defining two variables

$$u \equiv \frac{\partial A}{\partial x} \quad \text{and} \quad v \equiv \frac{\partial A}{\partial y} \tag{7.133}$$

and the new function

$$D \equiv A(x, y) - u\,x - v\,y. \tag{7.134}$$

The change in D is

$$dD = \frac{\partial A}{\partial x}\,dx + \frac{\partial A}{\partial y}\,dy - \frac{\partial A}{\partial x}\,dx - x\,du - \frac{\partial A}{\partial y}\,dy - y\,dv = -x\,du - y\,dv \tag{7.135}$$

which says that D is properly a function $D(u, v)$ of u and v. To really make D a function of u and v, however, we must invert the definitions (7.133) and express x and y in terms of u and v so that

$$D(u, v) = A(x(u, v), y(u, v)) - u\,x(u, v) - v\,y(u, v). \tag{7.136}$$

Example 7.25 (The functions of Lagrange and Hamilton) The lagrangian $L(q, \dot{q})$ of a time-independent system is a function of the two independent variables q and $\dot{q}$, and the change in L is

$$dL = \frac{\partial L}{\partial q}\,dq + \frac{\partial L}{\partial \dot{q}}\,d\dot{q}. \tag{7.137}$$

One defines a new variable called the momentum p as

$$p \equiv \frac{\partial L}{\partial \dot{q}} \tag{7.138}$$

and the hamiltonian H as

$$H \equiv p\,\dot{q} - L(q, \dot{q}) \tag{7.139}$$

in which an overall minus sign makes H positive in most cases. The change in H is

$$dH = p\,d\dot{q} + \dot{q}\,dp - dL = \frac{\partial L}{\partial \dot{q}}\,d\dot{q} + \dot{q}\,dp - \frac{\partial L}{\partial q}\,dq - \frac{\partial L}{\partial \dot{q}}\,d\dot{q} = \dot{q}\,dp - \frac{\partial L}{\partial q}\,dq \tag{7.140}$$

which says that H is properly a function $H(q, p)$ of q and p. When L is quadratic in the time derivative $\dot{q}$, it's easy to invert the definition (7.138) of the momentum and express the hamiltonian H as a function $H(q, p)$.

For the harmonic oscillator of mass m and angular frequency ω, the lagrangian is

$$L = \frac{m\,\dot{q}^2}{2} - \frac{m\,\omega^2\,q^2}{2}, \tag{7.141}$$

and the momentum is

$$p \equiv \frac{\partial L}{\partial \dot{q}} = m\,\dot{q}. \tag{7.142}$$

We easily invert this definition and get $\dot{q} = p/m$. Inserting this formula for $\dot{q}$ into the definition

$$H \equiv p\,\dot{q} - L = p\,\dot{q} - \frac{m\,\dot{q}^2}{2} + \frac{m\,\omega^2\,q^2}{2}, \tag{7.143}$$

we have as the hamiltonian

$$H = \frac{p^2}{m} - \frac{p^2}{2m} + \frac{m\,\omega^2\,q^2}{2} = \frac{p^2}{2m} + \frac{m\,\omega^2\,q^2}{2}. \tag{7.144}$$

One also can go backwards, from H to L. Define $\dot{q}$ as

$$\dot{q} \equiv \frac{\partial H}{\partial p} \tag{7.145}$$

and set

$$L = p\,\dot{q} - H. \tag{7.146}$$

Then the change in L is

$$dL = \dot{q}dp + pd\dot{q} - dH = \frac{\partial H}{\partial p}dp + pd\dot{q} - \frac{\partial H}{\partial q}dq - \frac{\partial H}{\partial p}dp = pd\dot{q} - \frac{\partial H}{\partial q}dq \tag{7.147}$$

which shows that L is a function of q and $\dot{q}$.

Example 7.26 (Thermodynamic potentials) The internal energy U varies with the entropy S, the volume V, and the numbers of molecules $N_1, N_2, \ldots$ as (7.111)

$$dU = T\,dS - p\,dV + \sum_j \mu_j\,dN_j \tag{7.148}$$

in which T is the temperature, p is the pressure, and μ_j is a chemical potential. To change the independent variables from S, V, and $\boldsymbol{N}$ to S, p, and $\boldsymbol{N}$, we add $p\,V$ and get the enthalpy

$$H \equiv U + p\,V. \tag{7.149}$$

Its differential dH is

$$dH = T\,dS - p\,dV + \sum_j \mu_j\,dN_j + p\,dV + V\,dp = T\,dS + V\,dp + \sum_j \mu_j\,dN_j \tag{7.150}$$

which shows that the enthalpy depends upon the entropy S, the pressure p, and the number N_j of each kind of molecule.

Subtracting $T\,S$ from the enthalpy H, we get the Gibbs free energy

$$G \equiv U + p\,V - T\,S = H - T\,S \tag{7.151}$$

whose differential is

$$\begin{aligned} dG &= dH - T\,dS - S\,dT \\ &= T\,dS + V\,dp + \sum_j \mu_j\,dN_j - T\,dS - S\,dT \\ &= V\,dp - S\,dT + \sum_j \mu_j\,dN_j \end{aligned} \tag{7.152}$$

which shows that the Gibbs free energy is a function $G(p, T, \boldsymbol{N})$ of the pressure p, the temperature T, and the numbers $N_1, N_2, \ldots$ of molecules.

Finally, Helmholtz's free energy F is

$$F \equiv U - T\,S \tag{7.153}$$

so its differential is

$$\begin{aligned} dF = dU - T\,dS - S\,dT &= T\,dS - p\,dV + \sum_j \mu_j\,dN_j - T\,dS - S\,dT \\ &= -\,p\,dV - S\,dT + \sum_j \mu_j\,dN_j \end{aligned} \tag{7.154}$$

which shows that the Helmholtz free energy is a function $F(V, T, \boldsymbol{N})$ of the volume V, the temperature T, and the numbers $\boldsymbol{N}$ of molecules.

7.13 Principle of Stationary Action in Mechanics

In classical mechanics, the motion of n particles in 3 dimensions is described by an action density or lagrangian $L(q, \dot{q}, t)$ in which q stands for the $3n$ generalized coordinates $q_1, q_2, \ldots, q_{3n}$ and $\dot{q}$ for their time derivatives. The action of a motion $q(t)$ is the time integral

$$S = \int_{t_1}^{t_2} L(q, \dot{q}, t)\,dt. \tag{7.155}$$

If $q(t)$ changes slightly by $\delta q(t)$, then the first-order change in the action is

$$\delta S = \int_{t_1}^{t_2} \sum_{i=1}^{3n} \left[\frac{\partial L(q, \dot{q}, t)}{\partial q_i} \, \delta q_i(t) + \frac{\partial L(q, \dot{q}, t)}{\partial \dot{q}_i} \, \delta \dot{q}_i(t) \right] dt. \tag{7.156}$$

The change in $\dot{q}_i$ is the time derivative of the change δq_i

$$\delta \, \frac{dq_i}{dt} = \frac{d(q_i + \delta q_i)}{dt} - \frac{dq_i}{dt} = \frac{d \, \delta q_i}{dt}, \tag{7.157}$$

so we have

$$\delta S = \int_{t_1}^{t_2} \sum_i \left[\frac{\partial L(q, \dot{q}, t)}{\partial q_i} \, \delta q_i(t) + \frac{\partial L(q, \dot{q}, t)}{\partial \dot{q}_i} \, \frac{d \, \delta q_i(t)}{dt} \right] dt. \tag{7.158}$$

Integrating by parts, we find

$$\delta S = \int_{t_1}^{t_2} \sum_i \left[\left(\frac{\partial L}{\partial q_i} - \frac{d}{dt} \frac{\partial L}{\partial \dot{q}_i} \right) \delta q_i(t) \right] dt + \left[\sum_i \frac{\partial L}{\partial \dot{q}_i} \delta q_i(t) \right]_{t_1}^{t_2}. \tag{7.159}$$

According to the **principle of stationary action**, a classical process is one that makes the action **stationary** to first order in $\delta q(t)$ for changes that vanish at the endpoints $\delta q(t_1) = 0 = \delta q(t_2)$. Thus a classical process satisfies Lagrange's equations

$$\frac{d}{dt} \frac{\partial L}{\partial \dot{q}_i} - \frac{\partial L}{\partial q_i} = 0 \quad \text{for} \quad i = 1, \ldots, 3n. \tag{7.160}$$

Moreover, if the lagrangian L does not depend explicitly on the time t, as in **autonomous** systems, then the **energy**

$$E = \sum_i \frac{\partial L}{\partial \dot{q}_i} \dot{q}_i \; - \; L \tag{7.161}$$

does not change with time because its time derivative is the vanishing explicit time dependence of the lagrangian $- \; \partial L / \partial t \; = \; 0$. That is, **the energy is conserved**

$$\dot{E} = \sum_i \left(\frac{d}{dt} \frac{\partial L}{\partial \dot{q}_i} \right) \dot{q}_i + \frac{\partial L}{\partial \dot{q}_i} \ddot{q}_i - \dot{L} = \sum_i \frac{\partial L}{\partial q_i} \dot{q}_i + \frac{\partial L}{\partial \dot{q}_i} \ddot{q}_i - \dot{L} = - \frac{\partial L}{\partial t} = 0. \tag{7.162}$$

The **momentum** p_i **canonically conjugate** to the coordinate q_i is

$$p_i = \frac{\partial L}{\partial \dot{q}_i}. \tag{7.163}$$

If we can write the time derivatives $\dot{q}_i$ of the coordinates in terms of the q_k's and p_ks, that is, $\dot{q}_i = \dot{q}_i(q, p)$, then the **hamiltonian** is a Legendre transform of the lagrangian (Example 7.25)

$$H(q, p) = \sum_{i=1}^{3n} p_i \, \dot{q}_i(q, p) - L(q, p). \tag{7.164}$$

This rewriting of the velocities $\dot{q}_i$ in terms of the q's and p's is easy to do when the lagrangian is quadratic in the $\dot{q}_i$'s but not so easy in most other cases.

The change (7.159) in the action due to a tiny detour $\delta q(t)$ that differs from zero only at t_2 is proportional to the momenta (7.163)

$$\delta S = \sum_i \frac{\partial L}{\partial \dot{q}_i} \delta q_i(t_2) = \sum_i p_i \, \delta q_i(t_2) \tag{7.165}$$

whence

$$\frac{\partial S}{\partial q_i} = p_i. \tag{7.166}$$

We can write the total time derivative of the action S, which by construction (7.155) is the lagrangian L, in terms of the $3n$ momenta (7.166) as

$$\frac{dS}{dt} = L = \frac{\partial S}{\partial t} + \sum_i \frac{\partial S}{\partial q_i} \dot{q}_i = \frac{\partial S}{\partial t} + \sum_i p_i \dot{q}_i. \tag{7.167}$$

Thus apart from a minus sign, the partial time derivative of the action S is the energy function (7.161) or the hamiltonian (7.164) (if we can find it)

$$\frac{\partial S}{\partial t} = L - \sum_i p_i \dot{q}_i = -E = -H. \tag{7.168}$$

7.14 Symmetries and Conserved Quantities in Mechanics

A transformation $q_i'(t) = q_i(t) + \delta q_i(t)$ and its time derivative

$$\dot{q}_i'(t) = \frac{dq_i'(t)}{dt} = \frac{dq_i(t)}{dt} + \frac{d\,\delta q_i(t)}{dt} = \dot{q}_i(t) + \delta \dot{q}_i(t) \tag{7.169}$$

is a **symmetry** of a lagrangian L if the resulting change δL vanishes

$$\delta L = \sum_i \frac{\partial L}{\partial q_i(t)} \delta q_i(t) + \frac{\partial L}{\partial \dot{q}_i(t)} \delta \dot{q}_i(t) = 0. \tag{7.170}$$

This symmetry and Lagrange's equations (7.160) imply that the quantity

$$Q = \sum_i \frac{\partial L}{\partial \dot{q}_i} \delta q_i \tag{7.171}$$

is **conserved**. That is, the time derivative of Q vanishes

$$\frac{d}{dt}\left(\sum_i \frac{\partial L}{\partial \dot{q}_i}\delta q_i\right) = \sum_i \left(\frac{d}{dt}\frac{\partial L}{\partial \dot{q}_i}\right)\delta q_i + \frac{\partial L}{\partial \dot{q}_i}\frac{d\,\delta q_i}{dt} = \sum_i \frac{\partial L}{\partial q_i}\delta q_i + \frac{\partial L}{\partial \dot{q}_i}\delta \dot{q}_i = 0. \tag{7.172}$$

Example 7.27 (Conservation of momentum and angular momentum) Suppose the coordinates q_i are the spatial coordinates $\boldsymbol{r}_i = (x_i, y_i, z_i)$ of a system of particles with time derivatives $\boldsymbol{v}_i = (\dot{x}_i, \dot{y}_i, \dot{z}_i)$. If the lagrangian is unchanged $\delta L = 0$ by **spatial displacement** or **spatial translation** by a constant vector $\boldsymbol{d} = (a, b, c)$, that is, by $\delta x_i = a$, $\delta y_i = b$, $\delta z_i = c$, then the momentum in the direction $\boldsymbol{d}$

$$\boldsymbol{P}\cdot\boldsymbol{d} = \sum_i \frac{\partial L}{\partial \boldsymbol{v}_i}\cdot\boldsymbol{d} = \sum_i \boldsymbol{p}_i\cdot\boldsymbol{d} \tag{7.173}$$

is conserved.

If the lagrangian is unchanged $\delta L = 0$ when the system is rotated by an angle $\boldsymbol{\theta}$, that is, if $\delta\boldsymbol{r}_i = \boldsymbol{\theta}\times\boldsymbol{r}_i$ is a symmetry of the lagrangian, then the angular momentum $\boldsymbol{J}$ about the axis $\boldsymbol{\theta}$

$$\boldsymbol{J}\cdot\boldsymbol{\theta} = \sum_i \frac{\partial L}{\partial \boldsymbol{v}_i}\cdot(\boldsymbol{\theta}\times\boldsymbol{r}_i) = \sum_i \boldsymbol{p}_i\cdot(\boldsymbol{\theta}\times\boldsymbol{r}_i) = \left(\sum_i \boldsymbol{r}_i\times\boldsymbol{p}_i\right)\cdot\boldsymbol{\theta} \tag{7.174}$$

is conserved.

7.15 Homogeneous First-Order Ordinary Differential Equations

Suppose the functions $P(x, y)$ and $Q(x, y)$ in the first-order ODE

$$P(x, y)\,dx + Q(x, y)\,dy = 0 \tag{7.175}$$

are homogeneous of degree n (Ince, 1956). We change variables from x and y to x and $y(x) = xv(x)$ so that $dy = xdv + vdx$, and

$$P(x, xv)dx + Q(x, xv)(xdv + vdx) = 0. \tag{7.176}$$

The homogeneity of $P(x, y)$ and $Q(x, y)$ imply that

$$x^n P(1, v)dx + x^n Q(1, v)(xdv + vdx) = 0. \tag{7.177}$$

Rearranging this equation, we are able to separate the variables

$$\frac{dx}{x} + \frac{Q(1, v)}{P(1, v) + vQ(1, v)}dv = 0. \tag{7.178}$$

We integrate this equation

$$\ln x + \int^{v} \frac{Q(1, v')}{P(1, v') + v'Q(1, v')} dv' = C \tag{7.179}$$

and find $v(x)$ and so too the solution $y(x) = xv(x)$.

Example 7.28 (Using homogeneity) In the differential equation

$$(x^2 - y^2)\, dx + 2xy\, dy = 0 \tag{7.180}$$

the coefficients of the differentials $P(x, y) = x^2 - y^2$ and $Q(x, y) = 2xy$ are homogeneous functions of degree $n = 2$, so the above method applies. With $y(x) = xv(x)$, we have

$$x^2(1 - v^2)dx + 2x^2 v(vdx + xdv) = 0 \tag{7.181}$$

in which x^2 cancels out, leaving $(1 + v^2)dx + 2vxdv = 0$. Separating variables and integrating, we find

$$\int \frac{dx}{x} + \int \frac{2v\, dv}{1 + v^2} = \ln C \tag{7.182}$$

or $\ln x + \ln(1 + v^2) = \ln C$. So $(1 + v^2)x = C$ which leads to the general integral $x^2 + y^2 = Cx$ and so to $y(x) = \sqrt{Cx - x^2}$ as the solution of the ODE (7.180).

7.16 Linear First-Order Ordinary Differential Equations

The general form of a linear first-order ODE is

$$\frac{dy}{dx} + r(x)\, y = s(x). \tag{7.183}$$

We always can find an integrating factor $\alpha(x)$ that makes

$$0 = \alpha(ry - s)dx + \alpha dy \tag{7.184}$$

exact. With $P \equiv \alpha(ry - s)$ and $Q \equiv \alpha$, the condition (7.80) for this equation to be exact is $P_y = \alpha r = Q_x = \alpha_x$ or $\alpha_x/\alpha = r$. So

$$\frac{d \ln \alpha}{dx} = r \tag{7.185}$$

which we integrate and exponentiate

$$\alpha(x) = \alpha(x_0) \exp\left(\int_{x_0}^{x} r(x')dx'\right). \tag{7.186}$$

Now since $\alpha r = \alpha_x$, the original equation (7.183) multiplied by this integrating factor is

$$\alpha y_x + \alpha r y = \alpha y_x + \alpha_x y = (\alpha y)_x = \alpha s. \tag{7.187}$$

Integrating we find

$$\alpha(x) y(x) = \alpha(x_0) y(x_0) + \int_{x_0}^{x} \alpha(x') s(x') dx' \tag{7.188}$$

so that

$$y(x) = \frac{\alpha(x_0) y(x_0)}{\alpha(x)} + \frac{1}{\alpha(x)} \int_{x_0}^{x} \alpha(x') s(x') dx' \tag{7.189}$$

in which $\alpha(x)$ is the exponential (7.186). More explicitly, $y(x)$ is

$$y(x) = \exp\left(-\int_{x_0}^{x} r(x')dx'\right)\left[y(x_0) + \int_{x_0}^{x} \exp\left(\int_{x_0}^{x'} r(x'')dx''\right) s(x')dx'\right]. \tag{7.190}$$

The first term in the square brackets multiplied by the prefactor $\alpha(x_0)/\alpha(x)$ is the general solution of the homogeneous equation $y_x + ry = 0$. The second term in the square brackets multiplied by the prefactor $\alpha(x_0)/\alpha(x)$ is a particular solution of the inhomogeneous equation $y_x + ry = s$. Thus Equation (7.190) expresses the general solution of the inhomogeneous equation (7.183) as the sum of a particular solution of the inhomogeneous equation and the general solution of the associated homogeneous equation as in Section 7.1.

We were able to find an integrating factor α because the original equation (7.183) was *linear* in y. So we could set $P = \alpha(ry - s)$ and $Q = \alpha$. When P and Q are more complicated, integrating factors are harder to find or nonexistent.

Example 7.29 (Bodies falling in air) The downward speed v of a mass m in a gravitational field of constant acceleration g is described by the inhomogeneous first-order ODE $m v_t = mg - bv$ in which b represents air resistance. This equation is like (7.183) but with t instead of x as the independent variable, $r = b/m$, and $s = g$. Thus by (7.190), its solution is

$$v(t) = \frac{mg}{b} + \left(v(0) - \frac{mg}{b}\right) e^{-bt/m}. \tag{7.191}$$

The terminal speed mg/b is nearly 200 km/h for a falling man. A diving Peregrine falcon can exceed 320 km/h; so can a falling bullet. But mice can fall down mine shafts and run off unhurt, and insects and birds can fly.

If the falling bodies are microscopic, a statistical model is appropriate. The potential energy of a mass m at height h is $V = mgh$. The heights of particles at temperature T K follow Boltzmann's distribution (1.392)

$$P(h) = P(0)e^{-mgh/kT} \tag{7.192}$$

in which $k = 1.380\,6504 \times 10^{-23}$ J/K $= 8.617\,343 \times 10^{-5}$eV/K is his constant. The probability depends exponentially upon the mass m and drops by a factor of e with the **scale height** $S = kT/mg$, which can be a few kilometers for a small molecule.

Example 7.30 (R-C circuit) The **capacitance** C of a capacitor is the charge Q it holds (on each plate) divided by the applied voltage V, that is, $C = Q/V$. The current I through the capacitor is the time derivative of the charge $I = \dot{Q} = C\dot{V}$. The voltage across a **resistor** of R Ω (Ohms) through which a current I flows is $V = IR$ by Ohm's law. So if a time-dependent voltage $V(t)$ is applied to a capacitor in series with a resistor, then $V(t) = Q/C + IR$. The current I therefore obeys the first-order differential equation

$$\dot{I} + I/RC = \dot{V}/R \tag{7.193}$$

or (7.183) with $x \to t$, $y \to I$, $r \to 1/RC$, and $s \to \dot{V}/R$. Since r is a constant, the integrating factor $\alpha(x) \to \alpha(t)$ is

$$\alpha(t) = \alpha(t_0)\, e^{(t-t_0)/RC}. \tag{7.194}$$

Our general solution (7.190) of linear first-order ODEs gives us the expression

$$I(t) = e^{-(t-t_0)/(RC)} \left[I(t_0) + \int_{t_0}^{t} e^{(t'-t_0)/RC} \frac{\dot{V}(t')}{R}\, dt' \right] \tag{7.195}$$

for the current $I(t)$.

Example 7.31 (Emission rate from fluorophores) A fluorophore is a molecule that emits light when illuminated. The frequency of the emitted photon usually is less than that of the incident one. Consider a population of N fluorophores of which N_+ are excited and can emit light and $N_- = N - N_+$ are unexcited. If the fluorophores are exposed to an illuminating photon flux I, and the cross-section for the excitation of an unexcited fluorophore is σ, then the rate at which unexcited fluorophores become excited is $I\sigma N_-$. The time derivative of the number of excited fluorophores is then

$$\dot{N}_+ = I\sigma N_- - \frac{1}{\tau} N_+ = -\frac{1}{\tau} N_+ + I\sigma\,(N - N_+) \tag{7.196}$$

in which $1/\tau$ is the decay rate (also the emission rate) of the excited fluorophores. Using the shorthand $a = I\sigma + 1/\tau$, we have $\dot{N}_+ = -aN_+ + I\sigma N$ which we solve using the general formula (7.190) with $r = a$ and $s = I\sigma N$

$$N_+(t) = e^{-at} \left[N_+(0) + \int_0^t e^{at'} I(t')\sigma N\, dt' \right]. \tag{7.197}$$

If the illumination $I(t)$ is constant, then by doing the integral we find

$$N_+(t) = \frac{I\sigma N}{a}\left(1 - e^{-at}\right) + N_+(0)e^{-at}. \tag{7.198}$$

The emission rate $E = N_+(t)/\tau$ of photons from the $N_+(t)$ excited fluorophores then is

$$E = \frac{I\sigma N}{a\tau}\left(1 - e^{-at}\right) + \frac{N_+(0)}{\tau}e^{-at} \tag{7.199}$$

which with $a = I\sigma + 1/1\tau$ gives for the emission rate per fluorophore

$$\frac{E}{N} = \frac{I\sigma}{1 + I\sigma\tau}\left(1 - e^{-(I\sigma+1/\tau)t}\right) \tag{7.200}$$

if no fluorophores were excited at $t = 0$, so that $N_+(0) = 0$.

7.17 Small Oscillations

Actual physical problems often involve **systems of differential equations**. The motion of n particles in 3 dimensions is described by a system (7.160) of $3n$ differential equations. Electrodynamics involves the four Maxwell equations (12.34 and 12.35). Thousands of coupled differential equations describe the chemical reactions among the many molecular species in a living cell. Despite this bewildering complexity, many systems merely execute **small oscillations** about the minimum of their potential energy.

The lagrangian

$$L = \sum_{i=1}^{3n} \frac{m_i}{2}\dot{x}_i^2 - U(x) \tag{7.201}$$

describes n particles of mass m_i interacting through a potential $U(x)$ that has no explicit time dependence. By letting $q_i = \sqrt{m_i/m}\, x_i$ we may scale the masses to the same value m and set $V(q) = U(x)$, so that we have

$$L = \frac{m}{2}\sum_{i=1}^{3n} \dot{q}_i^2 - V(q) = \frac{m}{2}\dot{q} \cdot \dot{q} - V(q) \tag{7.202}$$

which describes n particles of mass m interacting through a potential $V(q)$. Since $U(x)$ and $V(q)$ depend upon time only because the variables x_i and q_i vary with the time, the energy or equivalently the hamiltonian

$$H = \sum_{i=1}^{3n} \frac{p_i^2}{2m} + V(q) \tag{7.203}$$

is conserved. It has a minimum energy E_0 at q_0, and so its first derivatives there vanish. So near q_0, the potential $V(q)$ to lowest order is a quadratic form in the displacements $r_i \equiv q_i - q_{i0}$ from their minima q_{i0}, and the lagrangian, apart from the constant $V(q_0)$, is

$$L \approx \frac{m}{2} \sum_{i=1}^{3n} \dot{r}_i^2 - \frac{1}{2} \sum_{j,k=1}^{3n} r_j \, r_k \, \frac{\partial^2 V(q_0)}{\partial q_j \partial q_k}. \tag{7.204}$$

The matrix V'' of second derivatives is real and symmetric, and so we may diagonalize it $V'' = O^{\mathsf{T}} \, V_d'' \, O$ by an orthogonal transformation O. The lagrangian is diagonal in the new coordinates $s = O\, r$

$$L \approx \frac{1}{2} \sum_{i=1}^{3n} \left(m \, \dot{s}_i^2 - V_{di}'' \, s_i^2 \right) \tag{7.205}$$

and Lagrange's equations are $m \, \ddot{s}_i = -V_{di}'' \, s_i$. These **normal modes** are uncoupled harmonic oscillators $s_i(t) = a_i \cos \sqrt{V_{di}''/m}\, t + b_i \sin \sqrt{V_{di}''/m}\, t$ with frequencies that are real because q_0 is the minimum of the potential.

7.18 Systems of Ordinary Differential Equations

An **autonomous** system of n first-order ordinary differential equations

$$\begin{aligned} \dot{x}_1 &= F_1(x_1, x_2, \ldots, x_n) \\ \dot{x}_2 &= F_2(x_1, x_2, \ldots, x_n) \\ &\vdots \\ \dot{x}_n &= F_n(x_1, x_2, \ldots, x_n) \end{aligned} \tag{7.206}$$

is one in which the functions $F_i(x_1, \ldots, x_n)$ do not depend upon the time t. First-order autonomous systems are very general. They can even represent systems that do depend upon time and systems of higher-order differential equations.

To make a time-dependent system of n equations with time derivatives $F_i(x_1, \ldots, x_n, t)$ autonomous, one sets $t = x_{n+1}$ so that $F_i(x_1, \ldots, x_n, t)$ becomes $F_i(x_1, \ldots, x_n, x_{n+1})$, and $F_{n+1}(x_1, \ldots, x_{n+1}) = \dot{x}_{n+1} = \dot{t} = 1$. For example, one can turn the explicitly time-dependent differential equation $\dot{x} = -x^3 + \sin^3(t)$ into the autonomous system $\dot{x}_1 = -x_1^3 + \sin^3(x_2)$ and $\dot{x}_2 = 1$ by setting $x_1 = x$ and $x_2 = t$.

A similar trick turns a system of n higher-order ordinary differential equations into a system of more than n first-order differential equations. For instance, we can replace the third-order differential equation $\dddot{x} = x^3 + \dot{x}^2$ by the first-order system

$$\dot{x}_1 = x_3^3 + x_2^2, \quad \dot{x}_2 = x_1, \quad \text{and} \quad \dot{x}_3 = x_2 \tag{7.207}$$

in which $x_1 = \dot{x}_2$, $x_2 = \dot{x}_1$, and $x_3 = x$.

Example 7.32 (First-order form of a second-order system) One can turn the second-order nonlinear differential equations

$$\ddot{x} = -rx(x^2 - a^2) - gxy^2 \quad \text{and} \quad \ddot{y} = -ry(y^2 - b^2) - gyx^2 \tag{7.208}$$

into the first-order system

$$\begin{aligned}
\dot{x}_1 &= -rx_3(x_3^2 - a^2) - gx_3x_4^2 \\
\dot{x}_2 &= -rx_4(x_4^2 - b^2) - gx_4x_3^2 \\
\dot{x}_3 &= x_1 \\
\dot{x}_4 &= x_2
\end{aligned} \tag{7.209}$$

by setting $x_1 = \dot{x}$, $x_2 = \dot{y}$, $x_3 = x$, and $x_4 = y$. A representative trajectory $(x(t), y(t))$ is plotted in Fig. 7.1.

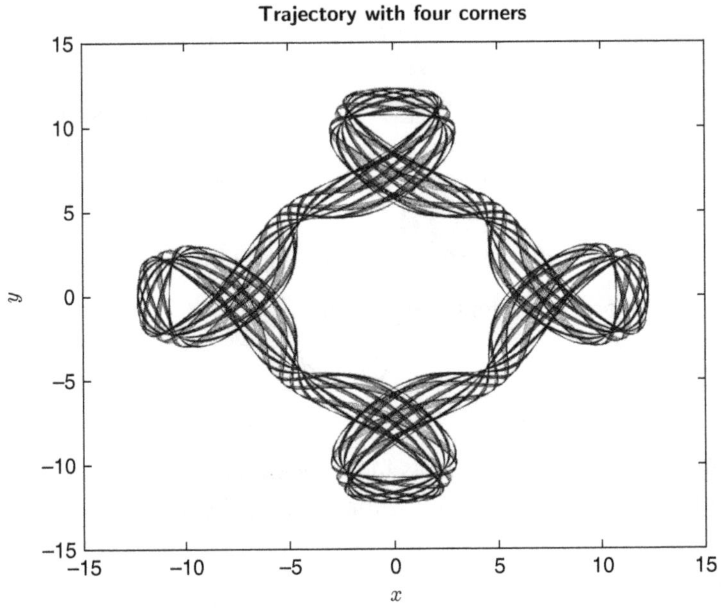

Figure 7.1 A trajectory $(x(t), y(t))$ of the autonomous system (7.209) for $0 \le t \le 400$ with $r = .2, a = 10, b = 10, g = 1$ and initial conditions $x(0) = 5, y(0) = 5, \dot{x}(0) = 10, \dot{y}(0) = 0$. Matlab scripts for this chapter's figures are in Differential_equations at `github.com/kevinecahill`.

7.19 Exact Higher-Order Differential Equations

An nth-order ordinary differential equation of the form

$$0 = \frac{d}{dx}\left(P_{n-1}(x)\,\frac{d^{n-1}y(x)}{dx^{n-1}} + P_{n-2}(x)\,\frac{d^{n-2}y(x)}{dx^{n-2}} + \cdots + P_0(x)\,y(x)\right) \quad (7.210)$$

is **exact** (Bender and Orszag, 1978, p. 13).

Example 7.33 (Exact second-order equation) The differential equation

$$0 = y'' + x^2\,y' + 2x\,y = \frac{d}{dx}\left(y' + x^2\,y\right) \quad (7.211)$$

is exact. We can solve it by integrating the first-order equation $y' + x^2\,y = c$ where c is a constant. Using the method of Section 7.16 with $r(x) = x^2$ and $s(x) = c$, we find

$$y(x) = e^{-x^3/3}\left[c' + c\int_0^x e^{x'^3/3}\,dx'\right] \quad (7.212)$$

where c' is another constant.

Example 7.34 (Integrating factor) The exponential e^x is an integrating factor for the differential equation $y'' + (1 + 1/x)\,y' + (1/x - 1/x^2)\,y = 0$ because

$$0 = e^x\,y'' + \frac{x+1}{x}\,e^x\,y' + \frac{x-1}{x^2}\,e^x\,y = \frac{d}{dx}\left(e^x\,y' + \frac{e^x}{x}\,y\right). \quad (7.213)$$

Solving the equation $y' + y/x = ce^{-x}$ by using the method of Section 7.16 with $r(x) = 1/x$ and $s(x) = e^{-x}$, we find $y(x) = -c_1(1 + 1/x)\,e^{-x} + c_2/x$.

7.20 Constant-Coefficient Equations

The exact, higher-order differential equations

$$0 = \sum_{k=0}^{n} c_k\,\frac{d^k y}{dt^k} \quad (7.214)$$

with coefficients c_k that are constants independent of t are soluble. We try $y(t) = e^{zt}$ and find that the complex number z must satisfy the equation

$$0 = \sum_{k=0}^{n} c_k\,z^k\,e^{zt} \quad \text{or} \quad 0 = \sum_{k=0}^{n} c_k\,z^k. \quad (7.215)$$

The fundamental theorem of algebra (Section 6.9) tells us that this equation always has n solutions $z_i = z_1, z_2, \ldots, z_n$. For each root z_i of the polynomial $P(z) = c_0 + c_1 z + \cdots + c_n z^n$, the function $y(t) = e^{z_i t}$ is a solution of the differential equation (7.214).

The roots need not all be different. But since partial derivatives commute, the jth z-derivative of the differential equation (7.214) is by Leibniz's rule (5.49)

$$\sum_{k=0}^{n} c_k \frac{\partial^k}{\partial t^k} (t^j e^{zt}) = \frac{\partial^j}{\partial z^j} \sum_{k=0}^{n} c_k \frac{\partial^k}{\partial t^k} e^{zt} = \frac{\partial^j}{\partial z^j} \sum_{k=0}^{n} c_k z^k e^{zt}$$
$$= \frac{\partial^j}{\partial z^j} \left(P(z)\, e^{zt} \right) = \sum_{\ell=0}^{j} \binom{j}{\ell} \left(\frac{\partial^\ell}{\partial z^\ell} P(z) \right) t^{j-\ell} e^{zt}. \tag{7.216}$$

Thus if $z = z'$ is a multiple root of order m, then the ℓth z derivative of $P(z)$ will vanish at $z = z'$ for $\ell < m$, and so for $j < m$, the functions $e^{z't}, te^{z't}, \ldots, t^j e^{z't}$ will be solutions of the differential equation (7.214).

Example 7.35 (A fourth-order constant-coefficient differential equation) The differential equation $\ddddot{y} - 4\dddot{y} + 6\ddot{y} - 4\dot{y} + y = 0$ has four solutions e^t, te^t, t^2e^t, and t^3e^t.

Example 7.36 (Euler equations) An equation of the form

$$0 = \sum_{k=0}^{n} \frac{c_k}{t^{n-k}} \frac{d^k y}{dt^k} \tag{7.217}$$

is called an Euler equation or an equidimensional equation. It is an exact, constant-coefficient differential equation in the variable $t = e^x$

$$0 = \sum_{k=0}^{n} c'_k \frac{d^k y}{dx^k} \tag{7.218}$$

whose solutions $y(x) = e^{z_i x}$ or $y(t) = t^{z_i}$ are given in terms of the roots z_i of the polynomial $P'(z) = c'_0 + c'_1 z + \cdots + c'_n z^n$. If z' is a root order m of this polynomial, then the functions $(\ln t)\, t^{z'}, \ldots, (\ln t)^{m-1}\, t^{z'}$ also are solutions.

7.21 Singular Points of Second-Order Ordinary Differential Equations

If in the ODE $y'' = f(x, y, y')$, the acceleration $y'' = f(x_0, y, y')$ is finite for all finite y and y', then x_0 is **a regular point** of the ODE. If $y'' = f(x_0, y, y')$ is infinite for any finite y and y', then x_0 is **a singular point** of the ODE.

If a second-order ODE $y'' + P(x)y' + Q(x)y = 0$ is linear and homogeneous and both $P(x_0)$ and $Q(x_0)$ are finite, then x_0 is a regular point of the ODE. But if $P(x_0)$ or $Q(x_0)$ or both are infinite, then x_0 is a singular point.

Some singular points are regular. If $P(x)$ or $Q(x)$ diverges as $x \to x_0$, but both $(x - x_0)P(x)$ and $(x - x_0)^2 Q(x)$ remain finite as $x \to x_0$, then x_0 is **a regular singular point** or equivalently a **nonessential singular point**. But if either

$(x-x_0)P(x)$ or $(x-x_0)^2Q(x)$ diverges as $x \to x_0$, then x_0 is **an irregular singular point** or equivalently an **essential singularity**.

To treat the point at infinity, one sets $z = 1/x$. Then if $(2z - P(1/z))/z^2$ and $Q(1/z)/z^4$ remain finite as $z \to 0$, the point $x_0 = \infty$ is **a regular point** of the ODE. If they don't remain finite, but $(2z - P(1/z))/z$ and $Q(1/z)/z^2$ do remain finite as $z \to 0$, then $x_0 = \infty$ is **a regular singular point**. Otherwise the point at infinity is **an irregular singular point** or an **essential singularity**.

Example 7.37 (Legendre's equation) Its self-adjoint form is

$$\left[\left(1-x^2\right)y'\right]' + \ell(\ell+1)y = 0 \tag{7.219}$$

which is $(1-x^2)y'' - 2xy' + \ell(\ell+1)y = 0$ or

$$y'' - \frac{2x}{1-x^2}y' + \frac{\ell(\ell+1)}{1-x^2}y = 0. \tag{7.220}$$

It has regular singular points at $x = \pm 1$ and $x = \infty$ (Exercise 7.15).

7.22 Frobenius's Series Solutions

Frobenius showed how to find a power-series solution of a second-order linear homogeneous ordinary differential equation $y'' + P(x)\,y' + Q(x)\,y = 0$ at any of its regular or regular singular points. We set $p(x) = x\,P(x)$ and $q(x) = x^2 Q(x)$ and assume that p and q are polynomials or analytic functions, and that $x = 0$ is a regular or regular singular point of the ODE so that $p(0)$ and $q(0)$ are both finite. Then writing the differential equation as

$$x^2 y'' + x\,p(x)\,y' + q(x)\,y = 0, \tag{7.221}$$

we expand y as a power series in x about $x = 0$

$$y(x) = x^r \sum_{n=0}^{\infty} a_n\,x^n \tag{7.222}$$

in which $a_0 \neq 0$ is the coefficient of the lowest power of x in $y(x)$. Differentiating, we have

$$y'(x) = \sum_{n=0}^{\infty} (r+n)\,a_n\,x^{r+n-1} \tag{7.223}$$

and

$$y''(x) = \sum_{n=0}^{\infty} (r+n)(r+n-1)\,a_n\,x^{r+n-2}. \tag{7.224}$$

When we substitute the three series (7.222–7.224) into our differential equation (7.221), we find

$$\sum_{n=0}^{\infty} \left[(n+r)(n+r-1) + (n+r)p(x) + q(x)\right] a_n x^{n+r} = 0. \tag{7.225}$$

If this equation is to be satisfied for all x, then the coefficient of every power of x must vanish. The lowest power of x is x^r, and it occurs when $n = 0$ with coefficient $\left[r(r-1+p(0)) + q(0)\right] a_0$. Thus since $a_0 \neq 0$, we have

$$r(r-1+p(0)) + q(0) = 0. \tag{7.226}$$

This quadratic **indicial equation** has two roots r_1 and r_2.

To analyze higher powers of x, we introduce the notation

$$p(x) = \sum_{j=0}^{\infty} p_j x^j \quad \text{and} \quad q(x) = \sum_{j=0}^{\infty} q_j x^j \tag{7.227}$$

in which $p_0 = p(0)$ and $q_0 = q(0)$. The requirement (exercise 7.16) that the coefficient of x^{r+k} vanish gives us a **recurrence relation**

$$a_k = -\left[\frac{1}{(r+k)(r+k-1+p_0)+q_0}\right] \sum_{j=0}^{k-1} \left[(j+r)p_{k-j} + q_{k-j}\right] a_j \tag{7.228}$$

that expresses a_k in terms of $a_0, a_1, \ldots, a_{k-1}$. When $p(x)$ and $q(x)$ are polynomials of low degree, these equations become much simpler.

When the roots r_1 and r_2 are complex, the coefficients a_n also are complex, and the real and imaginary parts of the complex solution $y(x)$ are two solutions of the differential equation.

Example 7.38 (Sines and cosines) To apply Frobenius's method the ODE $y'' + \omega^2 y = 0$, we first write it in the form $x^2 y'' + x p(x) y' + q(x) y = 0$ in which $p(x) = 0$ and $q(x) = \omega^2 x^2$. So both $p(0) = p_0 = 0$ and $q(0) = q_0 = 0$, and the indicial equation (7.226) is $r(r-1) = 0$ with roots $r_1 = 0$ and $r_2 = 1$.

We first set $r = r_1 = 0$. Since the p's and q's vanish except for $q_2 = \omega^2$, the recurrence relation (7.228) is $a_k = -q_2\, a_{k-2}/k(k-1) = -\omega^2 a_{k-2}/k(k-1)$. Thus $a_2 = -\omega^2 a_0/2$, and $a_{2n} = (-1)^n \omega^{2n} a_0/(2n)!$. The recurrence relation (7.228) gives no information about a_1, so to find the simplest solution, we set $a_1 = 0$. The recurrence relation $a_k = -\omega^2 a_{k-2}/k(k-1)$ then makes all the terms a_{2n+1} of odd index vanish. Our solution for the first root $r_1 = 0$ then is

$$y(x) = \sum_{n=0}^{\infty} a_n\, x^n = a_0 \sum_{n=0}^{\infty} (-1)^n \frac{(\omega x)^{2n}}{(2n)!} = a_0 \cos \omega x. \tag{7.229}$$

Similarly, the recurrence relation (7.228) for the second root $r_2 = 1$ is $a_k = -\omega^2 a_{k-2}/k(k+1)$, so that $a_{2n} = (-1)^n \omega^{2n} a_0/(2n+1)!$, and we again set all the terms of odd index equal to zero. Thus we have

$$y(x) = x \sum_{n=0}^{\infty} a_n \, x^n = \frac{a_0}{\omega} \sum_{n=0}^{\infty} (-1)^n \frac{(\omega x)^{2n+1}}{(2n+1)!} = \frac{a_0}{\omega} \sin \omega x \tag{7.230}$$

as our solution for the second root $r_2 = 1$.

Frobenius's method sometimes shows that solutions exist only when a parameter in the ODE assumes a special value called an **eigenvalue**.

Example 7.39 (Legendre's equation) If one rewrites Legendre's equation $(1 - x^2)y'' - 2xy' + \lambda y = 0$ as $x^2 y'' + xpy' + qy = 0$, then one finds $p(x) = -2x^2/(1 - x^2)$ and $q(x) = x^2\lambda/(1 - x^2)$, which are analytic but not polynomials. In this case, it is simpler to substitute the expansions (7.222–7.224) directly into Legendre's equation $(1 - x^2)y'' - 2xy' + \lambda y = 0$. We then find

$$\sum_{n=0}^{\infty} \left[(n+r)(n+r-1)(1-x^2)x^{n+r-2} - 2(n+r)x^{n+r} + \lambda x^{n+r} \right] a_n = 0.$$

The coefficient of the lowest power of x is $r(r-1)a_0$, and so the indicial equation is $r(r-1) = 0$. For $r = 0$, we shift the index n on the term $n(n-1)x^{n-2}a_n$ to $n = j+2$ and replace n by j in the other terms:

$$\sum_{j=0}^{\infty} \left\{ (j+2)(j+1)\, a_{j+2} - \left[j(j-1) + 2j - \lambda \right] a_j \right\} x^j = 0. \tag{7.231}$$

Since the coefficient of x^j must vanish, we get the recursion relation

$$a_{j+2} = \frac{j(j+1) - \lambda}{(j+2)(j+1)} a_j \tag{7.232}$$

which for big j says that $a_{j+2} \approx a_j$. Thus the series (7.222) does not converge for $|x| \geq 1$ unless $\lambda = j(j+1)$ for some integer j in which case the series (7.222) is a Legendre polynomial (Chapter 9).

Frobenius's method also allows one to expand solutions about $x_0 \neq 0$

$$y(x) = (x - x_0)^r \sum_{n=0}^{\infty} a_n \, (x - x_0)^n. \tag{7.233}$$

7.23 Fuch's Theorem

The method of Frobenius can run amok, especially if one expands about a singular point x_0. One can get only one solution or none at all. But Fuch has shown that if one applies Frobenius's method to a linear homogeneous second-order ODE and expands about a regular point or a regular singular point, then one always gets at least one power-series solution:

1. If the two roots of the indicial equation are equal, one gets only one solution.
2. If the two roots differ by a noninteger, one gets two solutions.
3. If the two roots differ by an integer, then the bigger root yields a solution.

Example 7.40 (Roots that differ by an integer) If one applies the method of Frobenius to Legendre's equation as in Example 7.39, then one finds (Exercise 7.18) that the $k = 0$ and $k = 1$ roots lead to the same solution.

7.24 Even and Odd Differential Operators

Under the parity transformation $x \to -x$, a typical term transforms as

$$x^n \left(\frac{d}{dx}\right)^p x^k = \frac{k!}{(k-p)!} x^{n+k-p} \to (-1)^{n+k-p} \frac{k!}{(k-p)!} x^{n+k-p} \tag{7.234}$$

and so the corresponding differential operator transforms as

$$x^n \left(\frac{d}{dx}\right)^p \to (-1)^{n-p} x^n \left(\frac{d}{dx}\right)^p . \tag{7.235}$$

The reflected form of the second-order linear differential operator

$$L(x) = h_0(x) + h_1(x)\frac{d}{dx} + h_2(x)\frac{d^2}{dx^2} \tag{7.236}$$

therefore is

$$L(-x) = h_0(-x) - h_1(-x)\frac{d}{dx} + h_2(-x)\frac{d^2}{dx^2}. \tag{7.237}$$

The operator $L(x)$ is **even** if it is unchanged by reflection, that is, if $h_0(-x) = h_0(x)$, $h_1(-x) = -h_1(x)$, and $h_2(-x) = h_2(x)$, so that

$$L(-x) = L(x). \tag{7.238}$$

It is **odd** if it changes sign under reflection, that is, if $h_0(-x) = -h_0(x)$, $h_1(-x) = h_1(x)$, and $h_2(-x) = -h_2(x)$, so that

$$L(-x) = -L(x). \tag{7.239}$$

Not every differential operator $L(x)$ is even or odd. But just as we can write every function $f(x)$ whose reflected form $f(-x)$ is well defined as the sum of $[f(x)+f(-x)]/2$ which is even and $[f(x)-f(-x)]/2$ which is odd

$$f(x) = \frac{1}{2}\big[f(x) + f(-x)\big] + \frac{1}{2}\big[f(x) - f(-x)\big] \tag{7.240}$$

so too we can write every differential operator $L(x)$ whose reflected form $L(-x)$ is well defined as the sum of one that is even and one that is odd

$$L(x) = \frac{1}{2}[L(x) + L(-x)] + \frac{1}{2}[L(x) - L(-x)]. \tag{7.241}$$

Many of the standard differential operators have $h_0 = 1$ and are even.

If $y(x)$ is a solution of the ODE $L(x)\,y(x) = 0$ and $L(-x)$ and $y(-x)$ are well defined, then we have $L(-x)\,y(-x) = 0$. If further $L(-x) = \pm L(x)$, then $y(-x)$ also is a solution $L(x)\,y(-x) = 0$. Thus if a differential operator $L(x)$ has **a definite parity**, that is, if $L(x)$ is either even or odd, then $y(-x)$ is a solution if $y(x)$ is, and solutions come in pairs $y(x) \pm y(-x)$, one even, one odd.

7.25 Wronski's Determinant

If the N functions $y_1(x), \dots, y_N(x)$ are linearly dependent, then by (7.7) there is a set of coefficients $k_1, \dots, k_N$, **not all zero**, such that the sum

$$0 = k_1\, y_1(x) + \cdots + k_N\, y_N(x) \tag{7.242}$$

vanishes for all x. Differentiating i times, we get

$$0 = k_1\, y_1^{(i)}(x) + \cdots + k_N\, y_N^{(i)}(x) \tag{7.243}$$

for all x. So if we use the y_j and their derivatives to define the matrix

$$Y_{ij}(x) \equiv y_j^{(i-1)}(x) \tag{7.244}$$

then we may express the linear dependence (7.242) and (7.243) of the functions $y_1, \dots, y_N$ in matrix notation as $0 = Y(x)\,k$ for some nonzero vector $k = (k_1, k_2, \dots, k_N)$. Since the matrix $Y(x)$ maps the nonzero vector k to zero, its determinant must vanish: $\det(Y(x)) \equiv |Y(x)| = 0$. This determinant

$$W(x) = |Y(x)| = \left| y_j^{(i-1)}(x) \right| \tag{7.245}$$

is called **Wronski's determinant** or **the wronskian**. It vanishes on an interval if and only if the functions $y_j(x)$ or their derivatives are linearly dependent on the interval.

7.26 Second Solutions

If we have one solution to a second-order linear homogeneous ODE, then we may use the wronskian to find a second solution. Here's how: If y_1 and y_2 are two linearly independent solutions of the second-order linear homogeneous ordinary differential equation

$$y''(x) + P(x)\,y'(x) + Q(x)\,y(x) = 0 \tag{7.246}$$

then their wronskian does not vanish

$$W(x) = \begin{vmatrix} y_1(x) & y_2(x) \\ y_1'(x) & y_2'(x) \end{vmatrix} = y_1(x)\,y_2'(x) - y_2(x)\,y_1'(x) \neq 0 \tag{7.247}$$

except perhaps at isolated points. Its derivative

$$\begin{aligned} W' &= y_1'\,y_2' + y_1\,y_2'' - y_2'\,y_1' - y_2\,y_1'' \\ &= y_1\,y_2'' - y_2\,y_1'' \end{aligned} \tag{7.248}$$

must obey

$$\begin{aligned} W' &= -y_1\,\left(P\,y_2' + Q\,y_2\right) + y_2\,\left(P\,y_1' + Q\,y_1\right) \\ &= -P\,\left(y_1\,y_2' - y_2\,y_1'\right) \end{aligned} \tag{7.249}$$

or $W'(x) = -P(x)\,W(x)$ which we integrate to

$$W(x) = W(x_0)\,\exp\left[-\int_{x_0}^{x} P(x')dx'\right]. \tag{7.250}$$

This is Abel's formula for the wronskian (Niels Abel, 1802–1829).

Having expressed the wronskian in terms of the known function $P(x)$, we now use it to find $y_2(x)$ from $y_1(x)$. We note that

$$W = y_1\,y_2' - y_2\,y_1' = y_1^2\,\frac{d}{dx}\left(\frac{y_2}{y_1}\right). \tag{7.251}$$

So

$$\frac{d}{dx}\left(\frac{y_2}{y_1}\right) = \frac{W}{y_1^2} \tag{7.252}$$

which we integrate to

$$y_2(x) = y_1(x)\left[\int^{x} \frac{W(x')}{y_1^2(x')}dx' + c\right]. \tag{7.253}$$

Using our formula (7.250) for the wronskian, we find as the second solution

$$y_2(x) = y_1(x) \int^{x} \frac{1}{y_1^2(x')}\,\exp\left[-\int^{x'} P(x'')dx''\right]\,dx' \tag{7.254}$$

apart from additive and multiplicative constants.

In the important special case in which $P(x) = 0$, the wronskian is a constant, $W'(x) = 0$, and the second solution is simply

$$y_2(x) = y_1(x) \int^x \frac{dx'}{y_1^2(x')}. \tag{7.255}$$

By Fuchs's theorem, Frobenius's expansion about a regular point or a regular singular point yields at least one solution. From this solution, we can use Wronski's trick to find a (linearly independent) second solution. So we always get two linearly independent solutions if we expand a second-order linear homogeneous ODE about a regular point or a regular singular point.

7.27 Why Not Three Solutions?

We have seen that a second-order linear homogeneous ODE has two linearly independent solutions. Why not three?

If y_1, y_2, and y_3 were three linearly independent solutions of the second-order linear homogeneous ODE

$$0 = y_j'' + P\, y_j' + Q\, y_j, \tag{7.256}$$

then their third-order wronskian

$$W = \begin{vmatrix} y_1 & y_2 & y_3 \\ y_1' & y_2' & y_3' \\ y_1'' & y_2'' & y_3'' \end{vmatrix} \tag{7.257}$$

would not vanish except at isolated points.

But the ODE (7.256) relates the second derivatives $y_j'' = -(P\, y_j' + Q\, y_j)$ to the y_j' and the y_j, and so the third row of this third-order wronskian is a linear combination of the first two rows. Thus it vanishes identically

$$W = \begin{vmatrix} y_1 & y_2 & y_3 \\ y_1' & y_2' & y_3' \\ -Py_1' - Qy_1 & -Py_2' - Qy_2 & -Py_3' - Qy_3 \end{vmatrix} = 0 \tag{7.258}$$

and so any three solutions of a second-order ODE (7.256) are linearly dependent.

One may extend this argument to show that an nth-order linear homogeneous ODE can have at most n linearly independent solutions. To do so, we'll use superscript notation (2.6) in which $y^{(n)}$ denotes the nth derivative of $y(x)$ with respect to x

$$y^{(n)} \equiv \frac{d^n y}{dx^n}. \tag{7.259}$$

Suppose there were $n + 1$ linearly independent solutions y_j of the ODE

$$y^{(n)} + P_1 \, y^{(n-1)} + P_2 \, y^{(n-2)} + \cdots + P_{n-1} \, y' + P_n \, y = 0 \tag{7.260}$$

in which the P_k's are functions of x. Then we could form a Wronskian of order $(n + 1)$ in which row 1 would be $y_1, \ldots, y_{n+1}$, row 2 would be the first derivatives $y'_1, \ldots, y'_{n+1}$, and row $n + 1$ would be the nth derivatives $y_1^{(n)}, \ldots, y_{n+1}^{(n)}$. We could then replace each term $y_k^{(n)}$ in the last row by

$$y_k^{(n)} = -P_1 \, y_k^{(n-1)} - P_2 \, y_k^{(n-2)} - \cdots - P_{n-1} \, y'_k - P_n \, y_k. \tag{7.261}$$

But then the last row would be a linear combination of the first n rows, the determinant would vanish, and the $n+1$ solutions would be linearly dependent. This is why an nth-order linear homogeneous ODE can have at most n linearly independent solutions.

7.28 Boundary Conditions

Since an nth-order linear homogeneous ordinary differential equation can have at most n linearly independent solutions, it follows that we can make a solution unique by requiring it to satisfy n boundary conditions. We'll see that the n arbitrary coefficients c_k of the general solution

$$y(x) = \sum_{k=1}^{n} c_k \, y_k(x) \tag{7.262}$$

of the differential equation (7.260) are fixed by the n boundary conditions

$$y(x_1) = b_1, \quad y(x_2) = b_2, \quad \ldots, \quad y(x_n) = b_n \tag{7.263}$$

as long as the functions $y_k(x)$ are linearly independent, and as long as the matrix Y with entries $Y_{jk} = y_k(x_j)$ is nonsingular, that is, $\det Y \neq 0$. In matrix notation, with B a vector with components b_j and C a vector with components c_k, the n boundary conditions (7.263) are

$$y(x_j) = \sum_{k=1}^{n} c_k \, y_k(x_j) = b_j \quad \text{or} \quad Y \, C = B. \tag{7.264}$$

Thus since $\det Y \neq 0$, the coefficients are uniquely given by $C = Y^{-1} \, B$.

The boundary conditions can involve the derivatives $y_k^{(\ell_j)}(x_j)$. We then define the matrix $Y_{jk} = y_k^{(\ell_j)}(x_j)$, write the n boundary conditions as

$$y^{(\ell_j)}(x_j) = \sum_{k=1}^{n} c_k \, y_k^{(\ell_j)}(x_j) = b_j \tag{7.265}$$

or as $Y\,C = B$, and show (Exercise 7.20) that as long as the matrix Y is nonsingular, the n coefficients are uniquely $C = Y^{-1}\,B$.

But what if all the b_j are zero? If all the boundary conditions are homogeneous $Y\,C = 0$, and $\det Y \neq 0$, then $Y^{-1}\,Y\,C = C = 0$, and $y_k(x) \equiv 0$ is the only solution. So there is no solution if $B = 0$ and the matrix Y is nonsingular. But if the $n \times n$ matrix Y has rank $n-1$, then (Section 1.38) it maps a unique vector C to zero (apart from an overall factor). So if all the boundary conditions are homogeneous, and the matrix Y has rank $n - 1$, then the solution $y = c_k y_k$ is unique. But if the rank of Y is less than $n - 1$, the solution is not unique.

Since a matrix of rank zero vanishes identically, any nonzero 2×2 matrix Y must be of rank 1 or 2. Thus a second-order ODE with two homogeneous boundary conditions either has a unique solution or none at all.

Example 7.41 (Boundary conditions and eigenvalues) The solutions y_k of the differential equation $-\,y'' = k^2\,y$ are $y_1(x) = \sin kx$ and $y_2(x) = \cos kx$. If we impose the boundary conditions $y(-a) = 0$ and $y(a) = 0$, then the matrix $Y_{jk} = y_k(x_j)$ is

$$Y = \begin{pmatrix} -\sin ka & \cos ka \\ \sin ka & \cos ka \end{pmatrix} \tag{7.266}$$

with determinant $\det Y = -\,2 \sin ka \cos ka = -\sin 2ka$. This determinant vanishes only if $ka = n\pi/2$ for some integer n, so if $ka \neq n\pi/2$, then no solution y of the differential equation $-y'' = k^2\,y$ satisfies the boundary conditions $y(-a) = 0 = y(a)$. But if $ka = n\pi/2$, then there is a solution, and it is unique because for even (odd) n, the first (second) column of Y vanishes, but not the second (first), which implies that Y has rank 1. One may regard the condition $ka = n\pi/2$ either as determining the eigenvalue k^2 or as telling us what interval to use.

7.29 A Variational Problem

For what functions $u(x)$ is the "energy" functional

$$E[u] \equiv \int_a^b \left[p(x)u'^2(x) + q(x)u^2(x)\right] dx \tag{7.267}$$

stationary? That is, for what functions u is $E[u + \delta u]$ unchanged to first order in δu when $u(x)$ is changed by an arbitrary but tiny function $\delta u(x)$ to $u(x) + \delta u(x)$? Our equations will be less cluttered if we drop explicit mention of the x-dependence of p, q, and u which we assume to be real functions of x.

The first-order change in E is

$$\delta E[u] \equiv \int_a^b \left(p\,2u'\,\delta u' + q\,2u\,\delta u\right) dx \tag{7.268}$$

in which the change in the derivative of u is $\delta u' = u' + (\delta u)' - u' = (\delta u)'$. Setting $\delta E = 0$ and integrating by parts, we have

$$
\begin{aligned}
0 = \delta E &= \int_a^b \left[p\,u'(\delta u)' + q\,u\,\delta u \right] dx \\
&= \int_a^b \left[\left(p\,u'\delta u\right)' - \left(p\,u'\right)'\delta u + q\,u\,\delta u \right] dx \\
&= \int_a^b \left[-\left(p\,u'\right)' + q\,u \right] \delta u\, dx + \left[p\,u'\delta u \right]_a^b .
\end{aligned} \tag{7.269}
$$

So if E is to be stationary with respect to all tiny changes δu that vanish at the endpoints a and b, then u must satisfy the differential equation

$$
L\,u = -\left(p\,u'\right)' + q\,u = 0. \tag{7.270}
$$

If instead E is to be stationary with respect to ***all*** tiny changes δu, then u must satisfy the differential equation (7.270) as well as the **natural** boundary conditions

$$
0 = p(b)\,u'(b) \quad \text{and} \quad 0 = p(a)\,u'(a). \tag{7.271}
$$

If $p(a) \neq 0 \neq p(b)$, then these natural boundary conditions imply **Neumann's** boundary conditions

$$
u'(a) = 0 \quad \text{and} \quad u'(b) = 0 \tag{7.272}
$$

(Carl Neumann, 1832–1925).

7.30 Self-Adjoint Differential Operators

If $p(x)$ and $q(x)$ are real, then the differential operator

$$
L = -\frac{d}{dx}\left(p(x)\frac{d}{dx} \right) + q(x) \tag{7.273}
$$

is formally **self adjoint**. Such operators are interesting because if we take any two functions u and v that are twice differentiable on an interval $[a, b]$ and integrate $v\,L\,u$ twice by parts over the interval, we get

$$
\begin{aligned}
(v, L\,u) &= \int_a^b v\,L\,u\,dx = \int_a^b v\left[-\left(pu'\right)' + qu \right] dx \\
&= \int_a^b \left[pu'v' + uqv \right] dx - \left[vpu' \right]_a^b \\
&= \int_a^b \left[-(pv')' + qv \right] u\,dx + \left[puv' - vpu' \right]_a^b \\
&= \int_a^b (L\,v)\,u\,dx + \left[p(uv' - vu') \right]_a^b
\end{aligned} \tag{7.274}
$$

which is **Green's formula**

$$\int_a^b (vLu - uLv)\, dx = \left[p(uv' - vu')\right]_a^b = \left[pW(u,v)\right]_a^b \tag{7.275}$$

(George Green, 1793–1841). Its differential form is **Lagrange's identity**

$$v\,L\,u - u\,L\,v = \left[p\,W(u,v)\right]' \tag{7.276}$$

(Joseph-Louis Lagrange, 1736–1813). Thus if the twice-differentiable functions u and v satisfy boundary conditions at $x = a$ and $x = b$ that make the boundary term (7.275) vanish

$$\left[p(uv' - vu')\right]_a^b = \left[pW(u,v)\right]_a^b = 0 \tag{7.277}$$

then the real differential operator L is **symmetric**

$$(v, L\,u) = \int_a^b v\,L\,u\,dx = \int_a^b u\,L\,v\,dx = (u, L\,v). \tag{7.278}$$

A real linear operator A that acts in a real vector space and satisfies the analogous relation (1.176)

$$(g, A\,f) = (f, A\,g) = (A\,g, f) \tag{7.279}$$

for all vectors in the space is said to be symmetric and **self adjoint**. In this sense, the differential operator (7.273) is self adjoint on the space of functions that satisfy the boundary condition (7.277).

In quantum mechanics, we often deal with wave functions that are complex. So keeping L real, let's replace u and v by twice-differentiable, complex-valued functions $\psi = u_1 + iu_2$ and $\chi = v_1 + iv_2$. If u_1, u_2, v_1, and v_2 satisfy boundary conditions at $x = a$ and $x = b$ that make the boundary terms (7.277) vanish

$$\left[p(u_i v_j' - v_j u_i')\right]_a^b = \left[pW(u_i, v_j)\right]_a^b = 0 \quad \text{for} \quad i, j = 1, 2 \tag{7.280}$$

then (7.278) implies that

$$\int_a^b v_j\,L\,u_i\,dx = \int_a^b \left(L\,v_j\right) u_i\,dx \quad \text{for} \quad i, j = 1, 2. \tag{7.281}$$

Under these assumptions, one may show (Exercise 7.21) that the boundary condition (7.280) makes the complex boundary term vanish

$$\left[p\,W(\psi, \chi^*)\right]_a^b = \left[p\,\left(\psi\,\chi^{*\prime} - \psi'\,\chi^*\right)\right]_a^b = 0 \tag{7.282}$$

and (Exercise 7.22) that since L is real, the identity (7.281) holds for complex functions

$$(\chi, L\,\psi) = \int_a^b \chi^* L\,\psi\,dx = \int_a^b (L\,\chi)^*\,\psi\,dx = (L\,\chi, \psi). \tag{7.283}$$

A linear operator A that satisfies the analogous relation (1.169)

$$(g, A\, f) = (A\, g,\, f) \tag{7.284}$$

is said to be self adjoint or **hermitian**. In this sense, the differential operator (7.273) is self adjoint on the space of functions that satisfy the boundary condition (7.282).

The formally self-adjoint differential operator (7.273) will satisfy the inner-product integral equations (7.278 or 7.283) only when the function p and the twice-differentiable functions u and v or ψ and χ conspire to make the boundary terms (7.277 or 7.282) vanish. This requirement leads us to define a self-adjoint differential system.

7.31 Self-Adjoint Differential Systems

A self-adjoint **differential system** consists of a real formally self-adjoint differential operator, a differential equation on an interval, boundary conditions, and a set of twice differentiable functions that obey them.

A second-order differential equation needs two boundary conditions to make a solution unique (Section 7.28). In a **self-adjoint differential system**, the two boundary conditions are **linear and homogeneous** so that the set of all twice differentiable functions u that satisfy them is a vector space. This space D is the **domain** of the system. For an interval $[a, b]$, **Dirichlet's boundary conditions** (Johann Dirichlet 1805–1859) are

$$u(a) = 0 \quad \text{and} \quad u(b) = 0 \tag{7.285}$$

and **Neumann's** (7.272) are

$$u'(a) = 0 \quad \text{and} \quad u'(b) = 0. \tag{7.286}$$

We will require that all the functions in the domain D either obey Dirichlet boundary conditions or obey Neumann boundary conditions.

The **adjoint domain** D^* of a differential system is the set of all twice-differentiable functions v that make the boundary term (7.277) vanish

$$\left[p(uv' - vu')\right]_a^b = \left[pW(u, v)\right]_a^b = 0 \tag{7.287}$$

for all functions u that are in the domain D, that is, that satisfy either Dirichlet or Neumann boundary conditions.

A differential system is **regular** and **self adjoint** if the differential operator $Lu = -(pu')' + qu$ is formally self adjoint, if the interval $[a, b]$ is finite, if p, p', and q are continuous real functions of x on the interval, if $p(x) > 0$ on $[a, b]$, and if the two domains D and D^* coincide, $D = D^*$.

One may show (Exercises 7.23 and 7.24) that if D is the set of all twice-differentiable functions $u(x)$ on $[a, b]$ that satisfy either Dirichlet's boundary conditions (7.285) or Neumann's boundary conditions (7.286), and if the function $p(x)$ is continuous and positive on $[a, b]$, then the adjoint set D^* is the same as D. A real formally self-adjoint differential operator $Lu = -(pu')' + qu$ together with Dirichlet (7.285) or Neumann (7.286) boundary conditions therefore forms a regular and self-adjoint system if p, p', and q are real and continuous on a finite interval $[a, b]$, and p is positive on $[a, b]$.

Since any two functions u and v in the domain D of a regular and self-adjoint differential system make the boundary term (7.287) vanish, a real formally self-adjoint differential operator L is symmetric and self adjoint (7.278) on all functions in its domain

$$(v, L\,u) = \int_a^b v\,L\,u\,dx = \int_a^b u\,L\,v\,dx = (u, L\,v). \tag{7.288}$$

If functions in the domain are complex, then by (7.282 and 7.283) the operator L is self adjoint or hermitian

$$(\chi, L\,\psi) = \int_a^b \chi^* L\,\psi\,dx = \int_a^b (L\,\chi)^*\,\psi\,dx = (L\,\chi, \psi) \tag{7.289}$$

on all complex functions ψ and χ in its domain.

Example 7.42 (Sines and cosines) The differential system with the formally self-adjoint differential operator

$$L = -\frac{d^2}{dx^2} \tag{7.290}$$

on an interval $[a, b]$ and the differential equation $L\,u = -u'' = \lambda\,u$ has the function $p(x) = 1$. If we choose the interval to be $[-\pi, \pi]$ and the domain D to be the set of all functions that are twice differentiable on this interval and satisfy Dirichlet boundary conditions (7.285), then we get a self-adjoint differential system in which the domain includes linear combinations of $u_n(x) = \sin nx$. If instead, we impose Neumann boundary conditions (7.286), then the domain D contains linear combinations of $u_n(x) = \cos nx$. In both cases, the system is regular and self adjoint.

Some important differential systems are self adjoint but **singular** because the function $p(x)$ vanishes at one or both of the endpoints of the interval $[a, b]$ or because the interval is infinite, for instance $[0, \infty)$ or $(-\infty, \infty)$. In these singular, self-adjoint differential systems, the boundary term (7.287) vanishes if u and v are in the domain $D = D^*$.

Example 7.43 (Legendre's system) Legendre's formally self-adjoint differential operator is

$$L = -\frac{d}{dx}\left[(1-x^2)\frac{d}{dx}\right] \tag{7.291}$$

and his differential equation is

$$L\,u = -\left[(1-x^2)u'\right]' = \ell(\ell+1)\,u \tag{7.292}$$

on the interval $[-1, 1]$. The function $p(x) = 1-x^2$ vanishes at both endpoints $x = \pm 1$, and so this self-adjoint system is singular. Because $p(\pm 1) = 0$, the boundary term (7.287) is zero as long as the functions u and v are differentiable on the interval. The domain D is the set of all functions that are twice differentiable on the interval $[-1, 1]$.

Example 7.44 (Hermite's system) Hermite's formally self-adjoint differential operator is

$$L = -\frac{d^2}{dx^2} + x^2 \tag{7.293}$$

and his differential equation is

$$L\,u = -u'' + x^2\,u = (2n+1)\,u \tag{7.294}$$

on the interval $(-\infty, \infty)$. This system has $p(x) = 1$ and $q(x) = x^2$. It is self adjoint but singular because the interval is infinite. The domain D consists of all functions that are twice-differentiable and that go to zero as $x \to \pm\infty$ faster than $1/x^{3/2}$, which ensures that the relevant integrals converge and that the boundary term (7.287) vanishes.

7.32 Making Operators Formally Self-Adjoint

We can make a generic real second-order linear homogeneous differential operator

$$L_0 = h_2\,\frac{d^2}{dx^2} + h_1\,\frac{d}{dx} + h_0 \tag{7.295}$$

formally self adjoint

$$L = -\frac{d}{dx}\left[p(x)\frac{d}{dx}\right] + q(x) = -p(x)\,\frac{d^2}{dx^2} - p'(x)\,\frac{d}{dx} + q(x) \tag{7.296}$$

by first dividing through by $-h_2(x)$

$$L_1 = -\frac{1}{h_2}L_0 = -\frac{d^2}{dx^2} - \frac{h_1}{h_2}\frac{d}{dx} - \frac{h_0}{h_2} \tag{7.297}$$

and then by multiplying L_1 by the positive prefactor

$$p(x) = \exp\left(\int^x \frac{h_1(y)}{h_2(y)}\,dy\right) > 0. \tag{7.298}$$

The product $p\,L_1$ then is formally self adjoint

$$\begin{aligned}
L = p(x)\,L_1 &= -\exp\left(\int^x \frac{h_1(y)}{h_2(y)}\,dy\right)\left[\frac{d^2}{dx^2} + \frac{h_1(x)}{h_2(x)}\frac{d}{dx} + \frac{h_0(x)}{h_2(x)}\right] \\
&= -\frac{d}{dx}\left[\exp\left(\int^x \frac{h_1(y)}{h_2(y)}\,dy\right)\frac{d}{dx}\right] - \exp\left(\int^x \frac{h_1(y)}{h_2(y)}\,dy\right)\frac{h_0(x)}{h_2(x)} \\
&= -\frac{d}{dx}\left(p\frac{d}{dx}\right) + q
\end{aligned} \tag{7.299}$$

with $q(x) = -p(x)\,h_0(x)/h_2(x)$. So we may turn any second-order linear homogeneous differential operator L_0 (7.295) into a formally self-adjoint operator L by multiplying it by

$$\rho(x) = -\frac{\exp\left(\int^x h_1(y)/h_2(y)dy\right)}{h_2(x)} = -\frac{p(x)}{h_2(x)}. \tag{7.300}$$

The two differential equations $L_0 u = 0$ and $Lu = \rho L_0 u = 0$ have the same solutions, but under the transformation (7.300), an eigenvalue equation $L_0 u = \lambda\, u$ becomes $Lu = \rho L_0 u = \rho\lambda u$ which is an eigenvalue equation

$$Lu = -(pu')' + qu = \lambda\,\rho\,u \tag{7.301}$$

with a **weight function** $\rho(x)$. Such an eigenvalue problem is known as a **Sturm–Liouville** problem (Jacques Sturm, 1803–1855; Joseph Liouville, 1809–1882). If $h_2(x)$ is negative (as for many positive operators), then the weight function $\rho(x) = -p(x)/h_2(x)$ is positive.

7.33 Wronskians of Self-Adjoint Operators

We saw in (7.246–7.250) that if $y_1(x)$ and $y_2(x)$ are two linearly independent solutions of the ODE

$$y''(x) + P(x)\,y'(x) + Q(x)\,y(x) = 0 \tag{7.302}$$

then their Wronskian $W(x) = y_1(x)\,y_2'(x) - y_2(x)\,y_1'(x)$ is

$$W(x) = W(x_0)\,\exp\left[-\int_{x_0}^{x} P(x')dx'\right]. \tag{7.303}$$

Thus if we convert the ODE (7.302) to its formally self-adjoint form

$$-\left[p(x)y'(x)\right]' + q(x)y(x) = -p(x)\frac{d^2y(x)}{dx^2} - p'(x)\frac{dy(x)}{dx} + q(x)y(x) = 0 \tag{7.304}$$

then $P(x) = p'(x)/p(x)$, and so the Wronskian (7.303) is

$$W(x) = W(x_0)\,\exp\left[-\int_{x_0}^{x} p'(x')/p(x')dx'\right] \tag{7.305}$$

which we may integrate directly to

$$W(x) = W(x_0)\,\exp\left[-\ln\left[p(x)/p(x_0)\right]\right] = W(x_0)\,\frac{p(x_0)}{p(x)}. \tag{7.306}$$

We learned in (7.246–7.254) that if we had one solution $y_1(x)$ of the ODE (7.302 or 7.304), then we could find another solution $y_2(x)$ that is linearly independent of $y_1(x)$ as

$$y_2(x) = y_1(x) \int^{x} \frac{W(x')}{y_1^2(x')} dx'. \tag{7.307}$$

In view of (7.303), this is an iterated integral. But if the ODE is formally self adjoint, then the formula (7.306) reduces it to

$$y_2(x) = y_1(x) \int^{x} \frac{1}{p(x')\,y_1^2(x')} dx' \tag{7.308}$$

apart from a constant factor.

Example 7.45 (Legendre functions of the second kind) Legendre's self-adjoint differential equation (9.4) is

$$-\left[(1-x^2)\,y'\right]' = \ell(\ell+1)\,y \tag{7.309}$$

and an obvious solution for $\ell = 0$ is $y(x) \equiv P_0(x) = 1$. Since $p(x) = 1 - x^2$, the integral formula (7.308) gives us as a second solution

$$Q_0(x) = P_0(x) \int^{x} \frac{1}{p(x')\,P_0^2(x')} dx' = \int^{x} \frac{1}{(1-x^2)} dx' = \frac{1}{2}\ln\left(\frac{1+x}{1-x}\right). \tag{7.310}$$

This second solution $Q_0(x)$ is singular at both ends of the interval $[-1, 1]$ and so does not satisfy the Dirichlet (7.285) or Neumann (7.286) boundary conditions that make the system self adjoint or hermitian.

7.34 First-Order Self-Adjoint Differential Operators

The first-order differential operator

$$L = u\,\frac{d}{dx} + v \tag{7.311}$$

will be self adjoint if

$$\int_a^b \chi^* L\psi \, dx = \int_a^b \left(L^\dagger \chi\right)^* \psi \, dx = \int_a^b (L\chi)^* \, \psi \, dx. \tag{7.312}$$

Starting from the first term, we find

$$\begin{aligned}
\int_a^b \chi^* L\psi \, dx &= \int_a^b \chi^* \left(u\,\psi' + v\psi\right) dx \\
&= \int_a^b \left[(-\chi^* u)' + \chi^* v\right] \psi \, dx + \left[\chi^* u \psi\right]_a^b \\
&= \int_a^b \left[(-\chi u^*)' + \chi v^*\right]^* \psi \, dx + \left[\chi^* u \psi\right]_a^b \\
&= \int_a^b \left[-u^* \chi' + (v^* - u^{*\prime})\chi\right]^* \psi \, dx + \left[\chi^* u \psi\right]_a^b .
\end{aligned} \tag{7.313}$$

So if the boundary terms vanish

$$\left[\chi^* u \psi\right]_a^b = 0 \tag{7.314}$$

and if both $u^* = -u$ and $v^* - u^{*\prime} = v$, then

$$\int_a^b \chi^* L\psi \, dx = \int_a^b \left[u\chi' + v\chi\right]^* \, \psi \, dx = \int_a^b (L\chi)^* \, \psi \, dx \tag{7.315}$$

and so L will be self adjoint or hermitian, $L^\dagger = L$. The general form of a first-order self-adjoint linear operator is then

$$L = ir(x)\,\frac{d}{dx} + s(x) + \frac{i}{2} r'(x) \tag{7.316}$$

in which r and s are arbitrary real functions of x.

Example 7.46 (Momentum and angular momentum) The momentum operator

$$p = \frac{\hbar}{i}\frac{d}{dx} \tag{7.317}$$

has $r = -\hbar$, which is real, and $s = 0$ and so is formally self adjoint. The boundary terms (7.314) are zero if the functions ψ and χ vanish at a and b, which often are $\pm\infty$.

The angular-momentum operators $L_i = \epsilon_{ijk}\, x_j\, p_k$, where $p_k = -i\hbar\,\partial_k$, also are formally self adjoint because the total antisymmetry of ϵ_{ijk} ensures that j and k are different as they are summed from 1 to 3.

Example 7.47 (Momentum in a magnetic field) In a magnetic field $\boldsymbol{B} = \nabla \times \boldsymbol{A}$, the differential operator

$$\frac{\hbar}{i}\nabla - e\,\boldsymbol{A} \tag{7.318}$$

that (in SI units) represents the kinetic momentum $m\boldsymbol{v}$ is formally self adjoint as is its Yang–Mills analog (13.308) when divided by i.

7.35 A Constrained Variational Problem

In quantum mechanics, we usually deal with normalizable wave functions. So let's find the function $u(x)$ that minimizes the energy functional

$$E[u] = \int_a^b \left[p(x)\, u'^2(x) + q(x)\, u^2(x)\right] dx \tag{7.319}$$

subject to the constraint that $u(x)$ be normalized on $[a, b]$ with respect to a positive weight function $\rho(x)$

$$N[u] = \|u\|^2 = \int_a^b \rho(x)\, u^2(x)\, dx \; = \; 1. \tag{7.320}$$

Introducing λ as a Lagrange multiplier (Section 1.24) and suppressing explicit mention of the x-dependence of the real functions p, q, ρ, and u, we minimize the unconstrained functional

$$\mathcal{E}[u, \lambda] = \int_a^b \left(p\, u'^2 + q\, u^2\right) dx - \lambda \left(\int_a^b \rho\, u^2\, dx \; - \; 1\right) \tag{7.321}$$

which will be stationary at the function u that minimizes it. The first-order change in $\mathcal{E}[u, \lambda]$ is

$$\delta\mathcal{E}[u, \lambda] = \int_a^b \left(p\, 2u'\, \delta u' + q\, 2u\, \delta u - \lambda\, \rho\, 2u\, \delta u\right) dx \tag{7.322}$$

in which the change in the derivative of u is $\delta u' = u' + (\delta u)' - u' = (\delta u)'$. Setting $\delta\mathcal{E} = 0$ and integrating by parts, we have

$$\begin{aligned} 0 = \frac{1}{2}\,\delta\mathcal{E} &= \int_a^b \left[p\, u'(\delta u)' + (q - \lambda\, \rho)\, u\, \delta u\right] dx \\ &= \int_a^b \left[\left(p\, u'\delta u\right)' - \left(p\, u'\right)' \delta u + (q - \lambda\, \rho)\, u\, \delta u\right] dx \\ &= \int_a^b \left[-\left(p\, u'\right)' + (q - \lambda\, \rho)\, u\right] \delta u\, dx + \left[p\, u'\delta u\right]_a^b. \end{aligned} \tag{7.323}$$

So if $\mathcal{E}$ is to be stationary with respect to all tiny changes δu, then u must satisfy both the self-adjoint differential equation

$$0 = \; - \left(p\, u'\right)' + (q - \lambda\, \rho)\, u \tag{7.324}$$

and the **natural** boundary conditions

$$0 = p(b)\, u'(b) \quad \text{and} \quad 0 = p(a)\, u'(a). \tag{7.325}$$

If instead, we require $\mathcal{E}[u, \lambda]$ to be stationary with respect to all variations δu that vanish at the endpoints, $\delta u(a) = \delta u(b) = 0$, then u must satisfy the differential equation (7.324) but need not satisfy the natural boundary conditions (7.325).

In both cases, the function $u(x)$ that minimizes the energy $E[u]$ subject to the normalization condition $N[u] = 1$ is an **eigenfunction** of the formally self-adjoint differential operator

$$L = -\frac{d}{dx}\left(p(x)\frac{d}{dx}\right) + q(x) \tag{7.326}$$

with **eigenvalue** λ

$$Lu = -\left(p\, u'\right)' + q\, u = \lambda\, \rho\, u. \tag{7.327}$$

The Lagrange multiplier λ has become an eigenvalue of a Sturm–Liouville equation (7.301).

Is the eigenvalue λ related to $E[u]$ and $N[u]$? To keep things simple, we restrict ourselves to a regular and self-adjoint differential system (Section 7.31) consisting of the self-adjoint differential operator (7.326), the differential equation (7.327), and a domain $D = D^*$ of functions $u(x)$ that are twice differentiable on $[a, b]$ and that satisfy two homogeneous Dirichlet (7.285) or Neumann (7.286) boundary conditions on $[a, b]$. All functions u in the domain D therefore satisfy

$$\left[upu'\right]_a^b = 0. \tag{7.328}$$

We multiply the Sturm–Liouville equation (7.327) from the left by u and integrate by parts from a to b. Noting the vanishing of the boundary terms (7.328), we find

$$\begin{aligned}
\lambda \int_a^b \rho\, u^2\, dx &= \int_a^b u\, Lu\, dx = \int_a^b u\left[-\left(p\, u'\right)' + q\, u\right] dx \\
&= \int_a^b \left[p\, u'^2 + q\, u^2\right] dx - \left[upu'\right]_a^b \\
&= \int_a^b \left[p\, u'^2 + q\, u^2\right] dx = E[u].
\end{aligned} \tag{7.329}$$

Thus in view of the normalization constraint (7.320), we see that the eigenvalue λ is the ratio of the energy $E[u]$ to the norm $N[u]$

$$\lambda = \frac{\displaystyle\int_a^b \left[p\, u'^2 + q\, u^2\right] dx}{\displaystyle\int_a^b \rho\, u^2\, dx} = \frac{E[u]}{N[u]}. \tag{7.330}$$

But is the function that minimizes the ratio

$$R[u] \equiv \frac{E[u]}{N[u]} \tag{7.331}$$

an eigenfunction u of the Sturm–Liouville equation (7.327)? And is the minimum of $R[u]$ the least eigenvalue λ of the Sturm–Liouville equation (7.327)? To see that the answers are *yes* and *yes*, we require $\delta R[u]$ to vanish

$$\delta R[u] = \frac{\delta E[u]}{N[u]} - \frac{E[u]\,\delta N[u]}{N^2[u]} = 0 \tag{7.332}$$

to first order in tiny changes $\delta u(x)$ that are zero at the endpoints of the interval, $\delta u(a) = \delta u(b) = 0$. Multiplying both sides by $N[u]$, we have

$$\delta E[u] = R[u]\,\delta N[u]. \tag{7.333}$$

Referring back to our derivation (7.321–7.323) of the Sturm–Liouville equation, we see that since $\delta u(a) = \delta u(b) = 0$, the change δE is

$$\begin{aligned} \delta E[u] &= 2 \int_a^b \left[- (p\,u')' + q\,u \right] \delta u \, dx + 2\left[p\,u' \delta u \right]_a^b \\ &= 2 \int_a^b \left[- (p\,u')' + q\,u \right] \delta u \, dx \end{aligned} \tag{7.334}$$

while δN is

$$\delta N[u] = 2 \int_a^b \rho\, u\, \delta u \, dx. \tag{7.335}$$

Substituting these changes (7.334) and (7.335) into the condition (7.333) that $R[u]$ be stationary, we find that the integral

$$\int_a^b \left[- (p\,u')' + (q - R[u]\,\rho\,)\,u \right] \delta u \, dx = 0 \tag{7.336}$$

must vanish for all tiny changes $\delta u(x)$ that are zero at the endpoints of the interval. Thus on $[a, b]$, the function u that minimizes the ratio $R[u]$ must satisfy the Sturm–Liouville equation (7.327)

$$- \left(p\,u'\right)' + q\,u = R[u]\,\rho\,u \tag{7.337}$$

with an eigenvalue $\lambda \equiv R[u]$ that is the minimum value of the ratio $R[u]$.

So the eigenfunction u_1 with the smallest eigenvalue λ_1 is the one that minimizes the ratio $R[u]$, and $\lambda_1 = R[u_1]$. What about other eigenfunctions with larger eigenvalues? How do we find the eigenfunction u_2 with the next smallest eigenvalue λ_2? Simple: we minimize $R[u]$ with respect to all functions u that are in the domain D and that are orthogonal to u_1.

Example 7.48 (Infinite square well) Let us consider a particle of mass m trapped in an interval $[a, b]$ by a potential that is V for $a < x < b$ but infinite for $x < a$ and for $x > b$. Because the potential is infinite outside the interval, the wave function $u(x)$ will satisfy the boundary conditions

$$u(a) = u(b) = 0. \tag{7.338}$$

The mean value of the hamiltonian is then the energy functional

$$\langle u|H|u\rangle = E[u] = \int_a^b \left[p(x)\, u'^2(x) + q(x)\, u^2(x)\right] dx \tag{7.339}$$

in which $p(x) = \hbar^2/2m$ and $q(x) = V$ a constant independent of x. Wave functions in quantum mechanics are normalized when possible. So we need to minimize the functional

$$E[u] = \int_a^b \left[\frac{\hbar^2}{2m}\, u'^2(x) + V\, u^2(x)\right] dx \tag{7.340}$$

subject to the constraint

$$c = \int_a^b u^2(x)\, dx - 1 = 0 \tag{7.341}$$

for all tiny variations δu that vanish at the endpoints of the interval. The weight function $\rho(x) = 1$, and the eigenvalue equation (7.327) is

$$-\frac{\hbar^2}{2m}\, u'' + V\, u = \lambda\, u. \tag{7.342}$$

For any positive integer n, the normalized function

$$u_n(x) = \left(\frac{2}{b-a}\right)^{1/2} \sin\left(n\pi\,\frac{x-a}{b-a}\right) \tag{7.343}$$

satisfies the boundary conditions (7.338) and the eigenvalue equation (7.342) with energy eigenvalue

$$\lambda_n = E[u_n] = \frac{1}{2m}\left(\frac{n\pi\hbar}{b-a}\right)^2 + V. \tag{7.344}$$

The second eigenfunction u_2 minimizes the energy functional $E[u]$ over the space of normalized functions that satisfy the boundary conditions (7.338) and are orthogonal to the first eigenfunction u_1. The eigenvalue λ_2 is higher than λ_1 (4 times higher). As the quantum number n increases, the energy $\lambda_n = E[u_n]$ goes to infinity as n^2. That $\lambda_n \to \infty$ as $n \to \infty$ is related (Section 7.38) to the completeness of the eigenfunctions u_n.

Example 7.49 (Harmonic oscillator) We'll minimize the energy

$$E[u] = \int_{-\infty}^{\infty} \left[\frac{\hbar^2}{2m}\, u'^2(x) + \frac{1}{2} m\, \omega^2\, x^2\, u^2(x)\right] dx \tag{7.345}$$

subject to the normalization condition

$$N[u] = \|u\|^2 = \int_{-\infty}^{\infty} u^2(x)\,dx = 1. \tag{7.346}$$

We introduce λ as a Lagrange multiplier and find the minimum of the unconstrained function $E[u]-\lambda\,(N[u]-1)$. Following Equations (7.319–7.327), we find that u must satisfy Schrödinger's equation

$$-\frac{\hbar^2}{2m}u'' + \frac{1}{2}m\,\omega^2\,x^2\,u = \lambda u \tag{7.347}$$

which we write as

$$\hbar\omega\left[\frac{m\omega}{2\hbar}\left(x - \frac{\hbar}{m\omega}\frac{d}{dx}\right)\left(x + \frac{\hbar}{m\omega}\frac{d}{dx}\right) + \frac{1}{2}\right]u = \lambda u. \tag{7.348}$$

The lowest eigenfunction u_0 is mapped to zero by the second factor

$$\left(x + \frac{\hbar}{m\omega}\frac{d}{dx}\right)u_0(x) = 0 \tag{7.349}$$

so its eigenvalue λ_0 is $\hbar\omega/2$. Integrating this differential equation, we get

$$u_0(x) = \left(\frac{m\omega}{\pi\hbar}\right)^{1/4}\exp\left(-\frac{m\omega x^2}{2\hbar}\right) \tag{7.350}$$

in which the prefactor is a normalization constant. As in Section 3.12, one may get the higher eigenfunctions by acting on u_0 with powers of the first factor inside the square brackets (7.348)

$$u_n(x) = \frac{1}{\sqrt{n!}}\left(\frac{m\omega}{2\hbar}\right)^{n/2}\left(x - \frac{\hbar}{m\omega}\frac{d}{dx}\right)^n u_0(x). \tag{7.351}$$

The eigenvalue of u_n is $\lambda_n = \hbar\omega(n+1/2)$. Because the eigenfunctions are complete, the eigenvalues increase without limit $\lambda_n \to \infty$ as $n \to \infty$.

Example 7.50 (Bessel's system) Bessel's energy functional is

$$E[u] = \int_0^1 \left[x\,u'^2(x) + \frac{n^2}{x}u^2(x)\right]dx \tag{7.352}$$

in which $n \ge 0$ is an integer. We seek the minimum of this functional over the set of twice differentiable functions $u(x)$ on $[0, 1]$ that are normalized

$$N[u] = \|u\|^2 = \int_0^1 x\,u^2(x)\,dx = 1 \tag{7.353}$$

and that satisfy the boundary conditions $u(0) = 0$ for $n > 0$ and $u(1) = 0$. We'll use a Lagrange multiplier λ (Section 1.24) and minimize the unconstrained functional $E[u] - \lambda\,(N[u] - 1)$. Proceeding as in (7.319–7.327), we find that u must obey the formally self-adjoint differential equation

$$L\,u = -\,(x\,u')' + \frac{n^2}{x}\,u = \lambda\,x\,u. \tag{7.354}$$

The ratio formula (7.330) and the positivity of Bessel's energy functional (7.352) tell us that the eigenvalues $\lambda = E[u]/N[u]$ are positive (Exercise 7.25). As we'll see in a moment, the boundary conditions largely determine these eigenvalues $\lambda_{n,m} \equiv k_{n,m}^2$. By changing variables to $\rho = k_{n,m}x$ and letting $u_n(x) = J_n(\rho)$, we arrive (Exercise 7.26) at

$$\frac{d^2 J_n}{d\rho^2} + \frac{1}{\rho}\frac{dJ_n}{d\rho} + \left(1 - \frac{n^2}{\rho^2}\right) J_n = 0 \tag{7.355}$$

which is Bessel's equation. The eigenvalues are determined by the condition $u_n(1) = J_n(k_{n,m}) = 0$; they are the squares of the zeros of $J_n(\rho)$. The eigenfunction of the self-adjoint differential equation (7.354) with eigenvalue $\lambda_{n,m} = k_{n,m}^2$ (and $u_n(0) = 0$ for $n > 0$) is $u_m(x) = J_n(k_{n,m}x)$. The parameter n labels the differential system; it is not an eigenvalue. Asymptotically as $m \to \infty$, one has (Courant and Hilbert, 1955, p. 416)

$$\lim_{m\to\infty} \frac{\lambda_{n,m}}{m^2\pi^2} = 1 \tag{7.356}$$

which shows that the eigenvalues $\lambda_{n,m}$ rise like m^2 as $m \to \infty$.

7.36 Eigenfunctions and Eigenvalues of Self-Adjoint Systems

A **regular Sturm–Liouville system** is a set of regular and self-adjoint differential systems (Section 7.31) that have the same differential operator, interval $[a, b]$, boundary conditions, and domain, and whose differential equations are of Sturm–Liouville (7.327) type

$$L\,\psi = -\,(p\,\psi')' + q\,\psi = \lambda\,\rho\,\psi \tag{7.357}$$

each distinguished by an **eigenvalue** λ. The functions p, q, and ρ are real and continuous, p and ρ are positive on $[a, b]$, but the weight function ρ may vanish at isolated points of the interval.

Since the differential systems are self adjoint, the real or complex functions in the common domain D are twice differentiable on the interval $[a, b]$ and satisfy two homogeneous boundary conditions that make the boundary terms (7.287) vanish

$$p\,W(\psi', \psi^*)\Big|_a^b = 0 \tag{7.358}$$

and so the differential operator L obeys the condition (7.289)

$$(\chi, L\,\psi) = \int_a^b \chi^* L\,\psi\,dx = \int_a^b (L\,\chi)^*\,\psi\,dx = (L\,\chi, \psi) \tag{7.359}$$

of being self adjoint or hermitian.

Let ψ_i and ψ_j be eigenfunctions of L with eigenvalues λ_i and λ_j

$$L\,\psi_i = \lambda_i\,\rho\,\psi_i \quad \text{and} \quad L\,\psi_j = \lambda_j\,\rho\,\psi_j \tag{7.360}$$

in a regular Sturm–Liouville system. Multiplying the first of these eigenvalue equations by ψ_j^* and the complex conjugate of the second by ψ_i, we get

$$\psi_j^* L\,\psi_i = \psi_j^* \lambda_i\,\rho\,\psi_i \quad \text{and} \quad \psi_i (L\,\psi_j)^* = \psi_i \lambda_j^*\,\rho\,\psi_j^*. \tag{7.361}$$

Integrating the difference of these equations over the interval $[a, b]$ and using the hermiticity (7.359) of L in the form $\int_a^b \psi_j^* L\,\psi_i\,dx = \int_a^b (L\,\psi_j)^*\,\psi_i\,dx$, we have

$$0 = \int_a^b \left[\psi_j^* L\,\psi_i - (L\,\psi_j)^*\,\psi_i\right] dx = \left(\lambda_i - \lambda_j^*\right) \int_a^b \psi_j^*\,\psi_i\,\rho\,dx. \tag{7.362}$$

Setting $i = j$, we find

$$0 = \left(\lambda_i^* - \lambda_i\right) \int_a^b \rho\,|\psi_i|^2\,dx \tag{7.363}$$

which since the integral is positive, shows that the eigenvalue λ_i must be **real**. All the eigenvalues of a regular Sturm–Liouville system are real. Using $\lambda_j^* = \lambda_j$ in (7.362), we see that eigenfunctions that have different eigenvalues are **orthogonal** on the interval $[a, b]$ with weight function $\rho(x)$

$$0 = \left(\lambda_i - \lambda_j\right) \int_a^b \psi_j^*\,\rho\,\psi_i\,dx. \tag{7.364}$$

Since the differential operator L, the eigenvalues λ_i, and the weight function ρ are all real, we may write the first of the eigenvalue equations in (7.360) both as $L\,\psi_i = \lambda_i\,\rho\,\psi_i$ and as $L\,\psi_i^* = \lambda_i\,\rho\,\psi_i^*$. By adding these two equations, we see that the real part of ψ_i satisfies them, and by subtracting them, we see that the imaginary part of ψ_i also satisfies them. So it might seem that $\psi_i = u_i + i v_i$ is made of two real eigenfunctions with the same eigenvalue.

But each eigenfunction u_i in the domain D satisfies two homogeneous boundary conditions as well as its second-order differential equation

$$-\,(p\,u_i')' + q\,u_i = \lambda_i\,\rho\,u_i \tag{7.365}$$

and so u_i is the unique solution in D to this equation. There can be no other eigenfunction in D with the same eigenvalue. In a regular Sturm–Liouville system, **there is no degeneracy**. All the eigenfunctions u_i are orthogonal and can be normalized on the interval $[a, b]$ with weight function $\rho(x)$

$$\int_a^b u_j^*\,\rho\,u_i\,dx = \delta_{ij}. \tag{7.366}$$

They may be taken to be **real**.

It is true that the eigenfunctions of a second-order differential equation come in pairs because one can use Wronski's formula (7.308)

$$y_2(x) = y_1(x) \int^x \frac{dx'}{p(x')\, y_1^2(x')} \tag{7.367}$$

to find a linearly independent second solution with the same eigenvalue. But the second solutions don't obey the boundary conditions of the domain. Bessel functions of the second kind, for example, are infinite at the origin.

A set of eigenfunctions u_i

$$-\,(p\, u_i')' + q\, u_i = \lambda_i\, \rho\, u_i \tag{7.368}$$

is **complete in the mean** in the space $L^2(a, b)$ of functions $f(x)$ that are square integrable with weight function ρ on the interval $[a, b]$

$$\int_a^b |f(x)|^2\, \rho(x)\, dx < \infty \tag{7.369}$$

(in the sense of Lebesgue, Section 3.9) if every such function $f(x)$ can be represented by a Fourier series

$$f(x) = \sum_{i=1}^{\infty} a_i\, u_i(x) \tag{7.370}$$

that converges **in the mean**

$$\lim_{n\to\infty} \int_a^b \left| f(x) - \sum_{i=1}^{n} a_i\, u_i(x) \right|^2 \rho(x)\, dx = 0. \tag{7.371}$$

The orthonormality (7.366) of the eigenfunctions $u_i(x)$ implies that the coefficients a_i are the integrals

$$a_i = \int_a^b f(x)\, u_i(x)\, \rho(x)\, dx. \tag{7.372}$$

Putting this formula into the Fourier series (7.370) and interchanging the orders of summation and integration, we find that

$$f(x) = \sum_{i=1}^{\infty} a_i\, u_i(x) = \int_a^b f(y) \left(\sum_{i=1}^{n} u_i(y)\, u_i(x)\, \rho(y) \right) dy \tag{7.373}$$

which gives us a representation for Dirac's delta function

$$\delta(x - y) = \sum_{i=1}^{\infty} u_i(y)\, u_i(x)\, \rho(y). \tag{7.374}$$

The orthonormal eigenfunctions of every regular Sturm–Liouville system on an interval $[a, b]$ are complete in the mean in $L^2(a, b)$. The completeness of these eigenfunctions follows (Section 7.38) from the fact that the eigenvalues λ_n of a regular Sturm–Liouville system are **unbounded**: when arranged in ascending order $\lambda_n < \lambda_{n+1}$ they go to infinity with the index n

$$\lim_{n\to\infty} \lambda_n = \infty \tag{7.375}$$

as we'll see in the next section.

7.37 Unboundedness of Eigenvalues

We have seen (Section 7.35) that the function $u(x)$ that minimizes the ratio

$$R[u] = \frac{E[u]}{N[u]} = \frac{\int_a^b \left[p\,u'^2 + q\,u^2\right] dx}{\int_a^b \rho\,u^2\,dx} \tag{7.376}$$

is a solution of the Sturm–Liouville equation

$$Lu = -\left(p\,u'\right)' + q\,u = \lambda\,\rho\,u \tag{7.377}$$

with eigenvalue

$$\lambda = \frac{E[u]}{N[u]}. \tag{7.378}$$

Let us call this least value of the ratio (7.376) λ_1; it also is the smallest eigenvalue of the differential equation (7.377). The second smallest eigenvalue λ_2 is the minimum of the same ratio (7.376) but for functions that are orthogonal to u_1

$$\int_a^b \rho\,u_1\,u_2\,dx = 0. \tag{7.379}$$

And λ_3 is the minimum of the ratio $R[u]$ but for functions that are orthogonal to both u_1 and u_2. Continuing in this way, we make a sequence of orthogonal eigenfunctions $u_n(x)$ (which we can normalize, $N[u_n] = 1$) with eigenvalues $\lambda_1 \le \lambda_2 \le \lambda_3 \le \cdots \lambda_n$. How do the eigenvalues λ_n behave as $n \to \infty$?

Since the function $p(x)$ is positive for $a < x < b$, it is clear that the energy functional (7.319)

$$E[u] = \int_a^b \left[p\,u'^2 + q\,u^2\right] dx \tag{7.380}$$

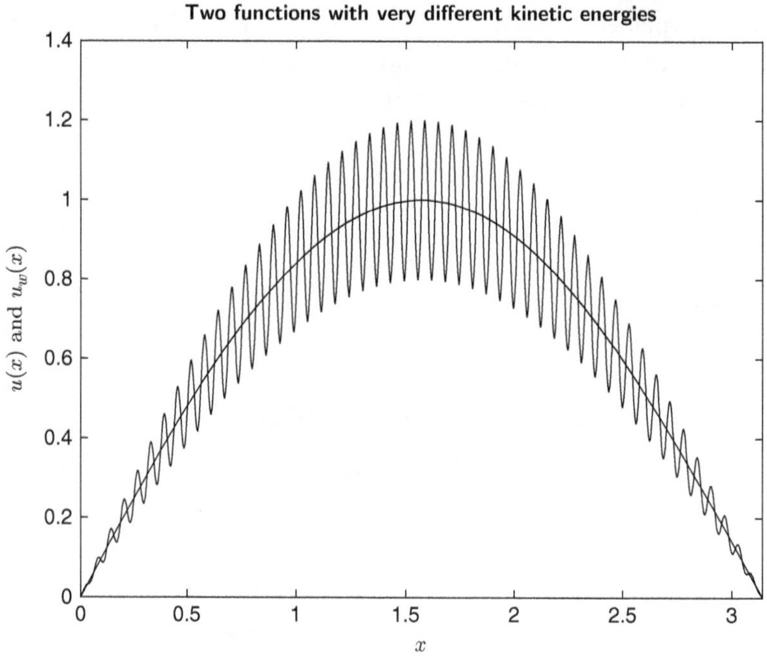

Figure 7.2 The energy functional $E[u]$ of Equation (7.319) assigns a much higher energy to the function $u_\omega(x) = u(x)(1 + 0.2\sin(\omega x))$ (zigzag curve with $\omega = 100$) than to the function $u(x) = \sin(x)$ (smooth curve). As the frequency $\omega \to \infty$, the energy $E[u_2] \to \infty$.

gets bigger as u'^2 increases. In fact, if we let the function $u(x)$ zigzag up and down about a given curve $\bar{u}$, then the kinetic energy $\int pu'^2 dx$ will rise but the potential energy $\int qu^2 dx$ will remain approximately constant. Thus by increasing the frequency of the zigzags, we can drive the energy $E[u]$ to infinity. For instance, if $u(x) = \sin x$, then its zigzag version $u_\omega(x) = u(x)(1 + 0.2 \sin \omega x)$ will have higher energy. The case of $\omega = 100$ is illustrated in Fig. 7.2. As $\omega \to \infty$, its energy $E[u_\omega] \to \infty$.

It is therefore intuitively clear (or at least plausible) that if the real functions $p(x)$, $q(x)$, and $\rho(x)$ are continuous on $[a, b]$ and if $p(x) > 0$ and $\rho(x) > 0$ on (a, b), then there are infinitely many energy eigenvalues λ_n, and that they increase without limit as $n \to \infty$

$$\lim_{n\to\infty} \lambda_n = \infty. \tag{7.381}$$

Courant and Hilbert (Richard Courant 1888–1972 and David Hilbert 1862–1943) provide several proofs of this result (Courant and Hilbert, 1955, pp. 397–429). One of their proofs involves the change of variables $f = (p\rho)^{1/4}$ and $v = fu$, after which the eigenvalue equation

$$L\,u = -\left(p\,u'\right)' + q\,u = \lambda\rho u \tag{7.382}$$

becomes $L_f\,v = -v'' + r v = \lambda_v v$ with $r = f''/f + q/\rho$. Were this $r(x)$ a constant, the eigenfunctions of L_f would be $v_n(x) = \sin(n\pi/(b-a))$ with eigenvalues

$$\lambda_{v_n} = \left(\frac{n\,\pi}{b-a}\right)^2 + r \tag{7.383}$$

rising as n^2. Courant and Hilbert show that as long as $r(x)$ is bounded for $a \le x \le b$, the actual eigenvalues of L_f are $\lambda_{v,n} = c\,n^2 + d_n$ in which d_n is bounded and that the eigenvalues λ_n of L differ from the $\lambda_{v,n}$ by a scale factor, so that they too diverge as $n \to \infty$

$$\lim_{n\to\infty} \frac{n^2}{\lambda_n} = g \tag{7.384}$$

where g is a constant.

7.38 Completeness of Eigenfunctions

We have seen in Section 7.37 that the eigenvalues of every regular Sturm–Liouville system when arranged in ascending order tend to infinity with the index n

$$\lim_{n\to\infty} \lambda_n = \infty. \tag{7.385}$$

We'll now use this property to show that the corresponding eigenfunctions $u_n(x)$ are complete in the mean (7.371) in the domain D of the system.

To do so, we follow Courant and Hilbert (Courant and Hilbert, 1955, pp. 397–428) and extend the energy E and norm N functionals to inner products on the domain of the system

$$E[f,g] \equiv \int_a^b \left[p(x)\,f'(x)\,g'(x) + q(x)\,f(x)\,g(x)\right]\,dx \tag{7.386}$$

$$N[f,g] \equiv \int_a^b \rho(x)\,f(x)\,g(x)\,dx \tag{7.387}$$

for any f and g in D. Integrating $E[f,g]$ by parts, we have

$$\begin{aligned} E[f,g] &= \int_a^b \left[\left(p\,f\,g'\right)' - f\left(pg'\right)' + q\,f\,g\right]\,dx \\ &= \int_a^b \left[-f\left(pg'\right)' + f\,q\,g\right]\,dx + \left[p\,f\,g'\right]_a^b \end{aligned} \tag{7.388}$$

or in terms of the self-adjoint differential operator L of the system

$$E[f,g] = \int_a^b f\,L\,g\,dx + \left[p\,f\,g'\right]_a^b. \tag{7.389}$$

Since the boundary term vanishes (7.328) when the functions f and g are in the domain D of the system, it follows that for f and g in D

$$E[f, g] = \int_a^b f\, L\, g\, dx. \tag{7.390}$$

We can use the first n orthonormal eigenfunctions u_k of the system

$$L\, u_k = \lambda_k\, \rho\, u_k \tag{7.391}$$

to approximate an arbitrary function in $f \in D$ as the linear combination

$$f(x) \sim \sum_{k=1}^{n} c_k\, u_k(x) \tag{7.392}$$

with coefficients c_k given by

$$c_k = N[f, u_k] = \int_a^b \rho\, f\, u_k\, dx. \tag{7.393}$$

We'll show that this series converges in the mean to the function f.

By construction (7.393), the remainder or error of the nth sum

$$r_n(x) = f(x) - \sum_{k=1}^{n} c_k\, u_k(x) \tag{7.394}$$

is orthogonal to the first n eigenfunctions

$$N[r_n, u_k] = 0 \quad \text{for} \quad k = 1, \ldots, n. \tag{7.395}$$

The next eigenfunction u_{n+1} minimizes the ratio

$$R[\phi] = \frac{E[\phi, \phi]}{N[\phi, \phi]} \tag{7.396}$$

over all ϕ that are orthogonal to the first n eigenfunctions u_k in the sense that $N[\phi, u_k] = 0$ for $k = 1, \ldots n$. That minimum is the eigenvalue λ_{n+1}

$$R[u_{n+1}] = \lambda_{n+1} \tag{7.397}$$

which therefore must be less than the ratio $R[r_n]$ of the remainder r_n

$$\lambda_{n+1} \le R[r_n] = \frac{E[r_n, r_n]}{N[r_n, r_n]}. \tag{7.398}$$

Thus the square of the norm of the remainder is bounded by the ratio

$$\|r_n\|^2 \equiv N[r_n, r_n] \le \frac{E[r_n, r_n]}{\lambda_{n+1}}. \tag{7.399}$$

So since $\lambda_{n+1} \to \infty$ as $n \to \infty$, we're done if we can show that the energy $E[r_n, r_n]$ of the remainder is bounded.

This energy is

$$
\begin{aligned}
E[r_n, r_n] &= E\left[f - \sum_{k=1}^{n} c_k u_k,\, f - \sum_{k=1}^{n} c_k u_k\right] \\
&= E[f, f] - \sum_{k=1}^{n} c_k \left(E[f, u_k] + E[u_k, f]\right) + \sum_{k=1}^{n}\sum_{\ell=1}^{n} c_k c_\ell \, E[u_k, u_\ell] \\
&= E[f, f] - 2\sum_{k=1}^{n} c_k \, E[f, u_k] + \sum_{k=1}^{n}\sum_{\ell=1}^{n} c_k c_\ell \, E[u_k, u_\ell]. \qquad (7.400)
\end{aligned}
$$

Since f and all the u_k are in the domain of the system, they satisfy the boundary condition (7.287 or 7.358), and so (7.389, 7.391, and 7.366) imply that

$$E[f, u_k] = \int_a^b f \, L u_k \, dx = \lambda_k \int_a^b \rho \, f \, u_k \, dx = \lambda_k \, c_k \qquad (7.401)$$

and that

$$E[u_k, u_\ell] = \int_a^b u_k \, L u_\ell \, dx = \lambda_\ell \int_a^b \rho \, u_k \, u_\ell \, dx = \lambda_k \, \delta_{k,\ell}. \qquad (7.402)$$

Using these relations to simplify our formula (7.400) for $E[r_n, r_n]$ we find

$$E[r_n, r_n] = E[f, f] - \sum_{k=1}^{n} \lambda_k c_k^2. \qquad (7.403)$$

Since $\lambda_n \to \infty$ as $n \to \infty$, we can be sure that for high enough n, the sum

$$\sum_{k=1}^{n} \lambda_k c_k^2 > 0 \quad \text{for} \quad n > N \qquad (7.404)$$

is positive. It follows from (7.403) that the energy of the remainder r_n is bounded by that of the function f

$$E[r_n, r_n] = E[f, f] - \sum_{k=1}^{n} \lambda_k c_k^2 \le E[f, f]. \qquad (7.405)$$

By substituting this upper bound $E[f, f]$ on $E[r_n, r_n]$ into our upper bound (7.399) on the squared norm $\|r_n\|^2$ of the remainder, we find

$$\|r_n\|^2 \le \frac{E[f, f]}{\lambda_{n+1}}. \qquad (7.406)$$

Thus since $\lambda_n \to \infty$ as $n \to \infty$, we see that the series (7.392) converges in the mean (Section 5.3) to f

$$\lim_{n\to\infty} \|r_n\|^2 = \lim_{n\to\infty} \|f - \sum_{k=1}^{n} c_k u_k\|^2 \le \lim_{n\to\infty} \frac{E[f,f]}{\lambda_{n+1}} = 0. \tag{7.407}$$

The eigenfunctions u_k of a regular Sturm–Liouville system are therefore complete in the mean in the domain D of the system. They span D.

It is a short step from spanning D to spanning the space $L^2(a,b)$ of functions that are square integrable on the interval $[a,b]$ of the system. To take this step, we assume that the domain D is dense in (a,b), that is, that for every function $g \in L^2(a,b)$ there is a sequence of functions $f_n \in D$ that converges to it in the mean so that for any $\epsilon > 0$ there is an integer N_1 such that

$$\|g - f_n\|^2 \equiv \int_a^b |g(x) - f_n(x)|^2 \, \rho(x) \, dx < \epsilon \quad \text{for} \quad n > N_1. \tag{7.408}$$

Since $f_n \in D$, we can find a series of eigenfunctions u_k of the system that converges in the mean to f_n so that for any $\epsilon > 0$ there is an integer N_2 such that

$$\|f_n - \sum_{k=1}^{N} c_{n,k} \, u_k\|^2 \equiv \int_a^b \left| f_n(x) - \sum_{k=1}^{N} c_{n,k} \, u_k(x) \right|^2 \rho(x) \, dx < \epsilon \quad \text{for} \quad N > N_2. \tag{7.409}$$

The Schwarz inequality (1.105) applies to these inner products, and so

$$\|g - \sum_{k=1}^{N} c_{n,k} \, u_k\| \le \|g - f_n\| + \|f_n(x) - \sum_{k=1}^{N} c_{n,k} \, u_k\|. \tag{7.410}$$

Combining the last three inequalities, we have for $n > N_1$ and $N > N_2$

$$\|g - \sum_{k=1}^{N} c_{n,k} \, u_k\| < 2\sqrt{\epsilon}. \tag{7.411}$$

So the eigenfunctions u_k of a regular Sturm–Liouville system span the space of functions that are square integrable on its interval $L^2(a,b)$.

One may further show (Courant and Hilbert 1955, p. 360; Stakgold 1967, p. 220) that the eigenfunctions $u_k(x)$ of any regular Sturm–Liouville system form a complete orthonormal set in the sense that every function $f(x)$ that satisfies Dirichlet (7.285) or Neumann (7.286) boundary conditions and has a continuous first and a piecewise continuous second derivative may be expanded in a series

$$f(x) = \sum_{k=1}^{\infty} a_k \, u_k(x) \tag{7.412}$$

that converges absolutely and uniformly on the interval $[a,b]$ of the system.

Our discussion (7.385–7.407) of the completeness of the eigenfunctions of a regular Sturm–Liouville system was insensitive to the finite length of the interval $[a, b]$ and to the positivity of $p(x)$ on $[a, b]$. What was essential was the vanishing of the boundary terms (7.287) which can happen if p vanishes at the endpoints of a finite interval or if the functions u and v tend to zero as $|x| \to \infty$ on an infinite one. This is why the results of this section have been extended to singular Sturm–Liouville systems made of self-adjoint differential systems that are singular because the interval is infinite or has p vanishing at one or both of its ends.

The eigenfunctions $u_i(x)$ of a Sturm–Liouville system provide a representation (7.374) of Dirac's delta function $\delta(x-y)$ as a sum of the terms $\rho(y)u_i(x)u_i(y)$. Since this series is nonzero only for $x = y$, the weight function $\rho(y)$ is just a scale factor, and we can write for $0 \le \alpha \le 1$

$$\delta(x-y) = \rho^{\alpha}(x)\,\rho^{1-\alpha}(y) \sum_{i=1}^{\infty} u_i(x)\,u_i(y). \tag{7.413}$$

These representations of the delta functional are suitable for functions f in the domain D of the regular Sturm–Liouville system.

Example 7.51 (A Bessel representation of the delta function) Bessel's nth system $L\,u = -(x\,u')' + n^2\,u/x = \lambda\,x\,u$ has eigenvalues $\lambda = z_{n,k}^2$ that are the squares of the zeros of the Bessel function $J_n(x)$. The eigenfunctions (Section 6.9) that are orthonormal with weight function $\rho(x) = x$ are $u_k^{(n)}(x) = \sqrt{2}\,J_n(z_{n,k}x)/J_{n+1}(z_{n,k})$. Thus by (7.413), we can represent Dirac's delta functional for functions in the domain D of Bessel's system as

$$\delta(x-y) = x^{\alpha}\,y^{1-\alpha} \sum_{k=1}^{\infty} u_k^{(n)}(x)\,u_k^{(n)}(y). \tag{7.414}$$

For $n = 0$, this Bessel representation is

$$\delta(x-y) = 2\,x^{\alpha}\,y^{1-\alpha} \sum_{k=1}^{\infty} \frac{J_0(z_{0,k}x)\,J_0(z_{0,k}y)}{J_1^2(z_{0,k})}. \tag{7.415}$$

Figure 7.3 plots the sum of the first 10,000 terms of this series for $\alpha = 0$ and $y = 0.47$, for $\alpha = 1/2$ and $y = 1/2$, and for $\alpha = 1$ and $y = 0.53$. This plot illustrates the Sturm–Liouville representation (7.413) of the delta function and its validity for $0 \le \alpha \le 1$.

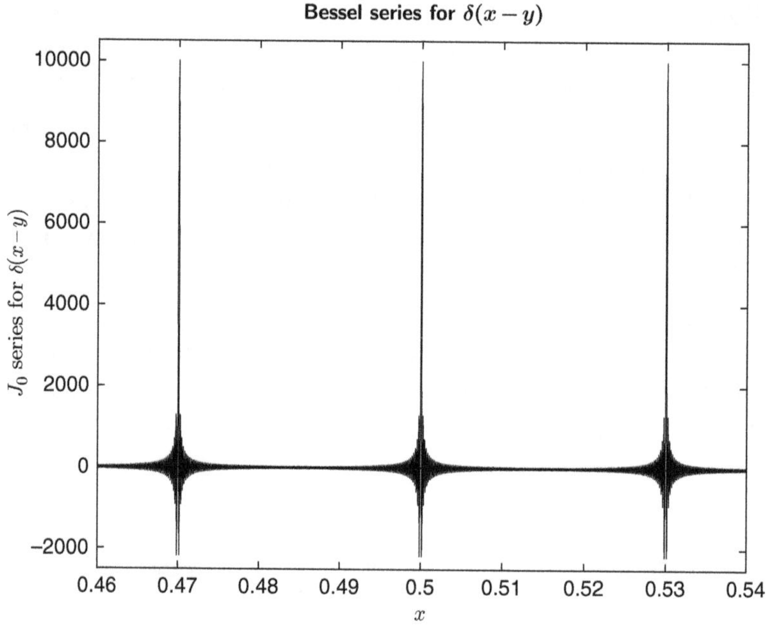

Figure 7.3 The sum of the first 10,000 terms of the Bessel representation (7.415) for the Dirac delta function $\delta(x - y)$ is plotted for $y = 0.47$ and $\alpha = 0$, for $y = 1/2$ and $\alpha = 1/2$, and for $y = 0.53$ and $\alpha = 1$.

7.39 Inequalities of Bessel and Schwarz

The inequality

$$\int_a^b \rho(x) \left| f(x) - \sum_{k=1}^{n} a_k u_k(x) \right|^2 dx \geq 0 \tag{7.416}$$

and the formula (7.372) for a_k lead (exercise 7.27) to Bessel's inequality

$$\int_a^b \rho(x)\, |f(x)|^2 \, dx \geq \sum_{k=1}^{\infty} |a_k|^2. \tag{7.417}$$

The argument we used to derive the Schwarz inequality (1.100) for vectors applies also to functions and leads to the Schwarz inequality

$$\int_a^b \rho(x)|f(x)|^2 \, dx \int_a^b \rho(x)|g(x)|^2 \, dx \geq \left| \int_a^b \rho(x) g^*(x) f(x) \, dx \right|^2. \tag{7.418}$$

7.40 Green's Functions

Physics is full of equations of the form

$$L\,G(x) = \delta^{(n)}(x) \tag{7.419}$$

in which L is a differential operator in n variables. The solution $G(x)$ is a Green's function (Section 4.8) for the operator L.

Example 7.52 (Poisson's Green's function) Probably the most important Green's function arises when the interaction is of long range as in gravity and electrodynamics. The divergence of the electric field is related to the charge density ρ by Gauss's law $\nabla \cdot \boldsymbol{E} = \rho/\epsilon_0$ where $\epsilon_0 = 8.854 \times 10^{-12}$ F/m is the electric constant. The electric field is $\boldsymbol{E} = -\nabla\phi - \dot{\boldsymbol{A}}$ in which ϕ is the scalar potential. In the Coulomb or radiation gauge, the divergence of $\boldsymbol{A}$ vanishes, $\nabla \cdot \boldsymbol{A} = 0$, and so $-\triangle\phi = -\nabla\cdot\nabla\phi = \rho/\epsilon_0$. The needed Green's function satisfies

$$-\triangle G(\boldsymbol{x}) = -\nabla \cdot \nabla G(\boldsymbol{x}) = \delta^{(3)}(\boldsymbol{x}) \tag{7.420}$$

and expresses the scalar potential ϕ as the integral

$$\phi(t,\boldsymbol{x}) = \int G(\boldsymbol{x}-\boldsymbol{x}')\,\frac{\rho(t,\boldsymbol{x}')}{\epsilon_0}\,d^3x'. \tag{7.421}$$

For when we apply (minus) the laplacian to it, we get

$$\begin{aligned} -\triangle\phi(t,\boldsymbol{x}) &= -\int \triangle\, G(\boldsymbol{x}-\boldsymbol{x}')\,\frac{\rho(t,\boldsymbol{x}')}{\epsilon_0}\,d^3x' \\ &= \int \delta^{(3)}(\boldsymbol{x}-\boldsymbol{x}')\,\frac{\rho(t,\boldsymbol{x}')}{\epsilon_0}\,d^3x' = \frac{\rho(t,\boldsymbol{x})}{\epsilon_0} \end{aligned} \tag{7.422}$$

which is **Poisson's equation**.

The reader might wonder how the potential $\phi(t,\boldsymbol{x})$ can depend upon the charge density $\rho(t,\boldsymbol{x}')$ at different points *at the same time*. The scalar potential is instantaneous because the Coulomb gauge condition $\nabla \cdot \boldsymbol{A} = 0$ is not Lorentz invariant. The gauge-invariant physical fields $\boldsymbol{E}$ and $\boldsymbol{B}$ are not instantaneous and do describe Lorentz-invariant electrodynamics.

It is easy to find the Green's function $G(\boldsymbol{x})$ by expressing it as a Fourier transform

$$G(\boldsymbol{x}) = \int e^{i\boldsymbol{k}\cdot\boldsymbol{x}}\,g(\boldsymbol{k})\,d^3k \tag{7.423}$$

and by using the 3-dimensional version

$$\delta^{(3)}(\boldsymbol{x}) = \int \frac{d^3k}{(2\pi)^3}\,e^{i\boldsymbol{k}\cdot\boldsymbol{x}} \tag{7.424}$$

of Dirac's delta function (4.36). If we insert these Fourier transforms into the equation (7.420) that defines the Green's function $G(\boldsymbol{x})$, then we find

$$
\begin{aligned}
-\triangle G(\boldsymbol{x}) &= -\triangle \int e^{i\boldsymbol{k}\cdot\boldsymbol{x}}\, g(\boldsymbol{k})\, d^3k \\
&= \int e^{i\boldsymbol{k}\cdot\boldsymbol{x}}\, \boldsymbol{k}^2\, g(\boldsymbol{k})\, d^3k = \delta^{(3)}(\boldsymbol{x}) = \int e^{i\boldsymbol{k}\cdot\boldsymbol{x}} \frac{d^3k}{(2\pi)^3}.
\end{aligned} \tag{7.425}
$$

Thus the Green's function $G(\boldsymbol{x})$ is the Fourier transform (4.109)

$$
G(\boldsymbol{x}) = \int \frac{e^{i\boldsymbol{k}\cdot\boldsymbol{x}}}{\boldsymbol{k}^2} \frac{d^3k}{(2\pi)^3} \tag{7.426}
$$

which we may integrate to (4.113)

$$
G(\boldsymbol{x}) = \frac{1}{4\pi|\boldsymbol{x}|} = \frac{1}{4\pi r} \tag{7.427}
$$

where $r = |\boldsymbol{x}|$ is the length of the vector $\boldsymbol{x}$. This formula is generalized to n dimensions in Example 6.28.

Example 7.53 (Helmholtz's Green's functions) The Green's function for the Helmholtz equation $(-\triangle - m^2)V(\boldsymbol{x}) = \rho(\boldsymbol{x})$ must satisfy

$$
(-\triangle - m^2)\, G_H(\boldsymbol{x}) = \delta^{(3)}(\boldsymbol{x}). \tag{7.428}
$$

By using the Fourier-transform method of the previous example, one may show that G_H is

$$
G_H(\boldsymbol{x}) = \frac{e^{imr}}{4\pi r} \tag{7.429}
$$

in which $r = |\boldsymbol{x}|$ and m has units of inverse length.

Similarly, the Green's function G_{mH} for the modified Helmholtz equation

$$
(-\triangle + m^2)\, G_{mH}(\boldsymbol{x}) = \delta^{(3)}(\boldsymbol{x}) \tag{7.430}
$$

is (Example 6.27)

$$
G_{mH}(\boldsymbol{x}) = \frac{e^{-mr}}{4\pi r} \tag{7.431}
$$

which is a Yukawa potential.

Of these Green's functions, probably the most important is $G(\boldsymbol{x}) = 1/4\pi r$ which has the expansion

$$
G(\boldsymbol{x} - \boldsymbol{x}') = \frac{1}{4\pi|\boldsymbol{x} - \boldsymbol{x}'|} = \sum_{\ell=0}^{\infty} \sum_{m=-\ell}^{\ell} \frac{1}{2\ell+1} \frac{r_<^\ell}{r_>^{\ell+1}} Y_{\ell,m}(\theta, \phi) Y_{\ell,m}^*(\theta', \phi') \tag{7.432}
$$

in terms of the spherical harmonics $Y_{\ell,m}(\theta, \phi)$. Here r, θ, and ϕ are the spherical coordinates of the point $\boldsymbol{x}$, and r', θ', and ϕ' are those of the point $\boldsymbol{x}'$; $r_>$ is the larger of r and r', and $r_<$ is the smaller of r and r'. If we substitute this expansion

(7.432) into the formula (7.421) for the potential ϕ, then we arrive at the multipole expansion

$$\begin{aligned}\phi(t,\boldsymbol{x}) &= \int G(\boldsymbol{x}-\boldsymbol{x}')\,\frac{\rho(t,\boldsymbol{x}')}{\epsilon_0}\,d^3x' \\ &= \sum_{\ell=0}^{\infty}\sum_{m=-\ell}^{\ell}\frac{1}{2\ell+1}\int \frac{r_<^\ell}{r_>^{\ell+1}}\,Y_{\ell,m}(\theta,\phi)Y^*_{\ell,m}(\theta',\phi')\,\frac{\rho(t,\boldsymbol{x}')}{\epsilon_0}\,d^3x'.\end{aligned} \tag{7.433}$$

Physicists often use this expansion to compute the potential at $\boldsymbol{x}$ due to a localized, remote distribution of charge $\rho(t,\boldsymbol{x}')$. In this case, the integration is only over the restricted region where $\rho(t,\boldsymbol{x}') \neq 0$, and so $r_< = r'$ and $r_> = r$, and the multipole expansion is

$$\phi(t,\boldsymbol{x}) = \sum_{\ell=0}^{\infty}\frac{1}{2\ell+1}\sum_{m=-\ell}^{\ell}\frac{Y_{\ell,m}(\theta,\phi)}{r^{\ell+1}}\int r'^\ell\, Y^*_{\ell m}(\theta',\phi')\,\frac{\rho(t,\boldsymbol{x}')}{\epsilon_0}\,d^3x'. \tag{7.434}$$

In terms of the multipoles

$$Q_\ell^m = \int r'^\ell\, Y^*_{\ell,m}(\theta',\phi')\,\frac{\rho(t,\boldsymbol{x}')}{\epsilon_0}\,d^3x' \tag{7.435}$$

the potential is

$$\phi(t,\boldsymbol{x}) = \sum_{\ell=0}^{\infty}\frac{1}{2\ell+1}\frac{1}{r^{\ell+1}}\sum_{m=-\ell}^{\ell} Q_\ell^m\, Y_{\ell,m}(\theta,\phi). \tag{7.436}$$

The spherical harmonics provide for the Legendre polynomial the expansion

$$P_\ell(\hat{\boldsymbol{x}}\cdot\hat{\boldsymbol{x}}') = \frac{4\pi}{2\ell+1}\sum_{m=-\ell}^{\ell} Y_{\ell m}(\theta,\phi)Y^*_{\ell m}(\theta',\phi') \tag{7.437}$$

which abbreviates the Green's function formula (7.432) to

$$G(\boldsymbol{x}-\boldsymbol{x}') = \frac{1}{4\pi|\boldsymbol{x}-\boldsymbol{x}'|} = \frac{1}{4\pi}\sum_{\ell=0}^{\infty}\frac{r_<^\ell}{r_>^{\ell+1}}\,P_\ell(\hat{\boldsymbol{x}}\cdot\hat{\boldsymbol{x}}'). \tag{7.438}$$

Example 7.54 (Feynman's propagator) The Feynman propagator

$$\Delta_F(x) = \int \frac{d^4q}{(2\pi)^4}\,\frac{\exp(iqx)}{q^2+m^2-i\epsilon} \tag{7.439}$$

is a Green's function (6.257) for the operator $L = m^2 - \Box$

$$(m^2 - \Box)\Delta_F(x) = \delta^4(x). \tag{7.440}$$

By integrating over q^0 while respecting the $i\epsilon$ (Example 4.100), one may write the propagator in terms of the Lorentz-invariant function

$$\Delta_+(x) = \frac{1}{(2\pi)^3} \int \frac{d^3q}{2E_{\boldsymbol{q}}} \exp[i(\boldsymbol{q} \cdot \boldsymbol{x} - E_{\boldsymbol{q}} x^0)] \tag{7.441}$$

as (6.268)

$$\Delta_F(x) = i\theta(x^0)\, \Delta_+(x) + i\theta(-x^0)\, \Delta_+(\boldsymbol{x}, -x^0) \tag{7.442}$$

which for space-like x, that is, for $x^2 = \boldsymbol{x}^2 - (x^0)^2 \equiv r^2 > 0$, depends only upon $r = +\sqrt{x^2}$ and has the value $\Delta_+(x) = (m/(4\pi^2 r))\, K_1(mr)$ in which K_1 is the Hankel function (6.276) (Weinberg, 1995, p. 202).

7.41 Eigenfunctions and Green's Functions

The Green's function

$$G(x-y) = \int \frac{dk}{2\pi} \frac{1}{k^2 + m^2} e^{ik\,(x-y)} \tag{7.443}$$

is based on the resolution of the delta function

$$\delta(x-y) = \int \frac{dk}{2\pi} e^{ik\,(x-y)} \tag{7.444}$$

in terms of the eigenfunctions $\exp(ik\,x)$ of the differential operator $-\partial_x^2 + m^2$ with eigenvalues $k^2 + m^2$.

We may generalize this way of making Green's functions to a regular Sturm–Liouville system (Section 7.36) with a differential operator L, eigenvalues λ_n

$$L\, u_n(x) = \lambda_n\, \rho(x)\, u_n(x) \tag{7.445}$$

and eigenfunctions $u_n(x)$ that are orthonormal with respect to a positive weight function $\rho(x)$

$$\delta_{n\ell} = (u_n, u_k) = \int \rho(x)\, u_n(x) u_k(x)\, dx \tag{7.446}$$

and that span in the mean the domain D of the system.

To make a Green's function $G(x-y)$ that satisfies

$$L\, G(x-y) = \delta(x-y) \tag{7.447}$$

we write $G(x-y)$ in terms of the complete set of eigenfunctions u_k as

$$G(x-y) = \sum_{k=1}^{\infty} \frac{u_k(x) u_k(y)}{\lambda_k} \tag{7.448}$$

so that the action $L u_k = \lambda_k \rho u_k$ turns G into

$$L\, G(x-y) = \sum_{k=1}^{\infty} \frac{L\, u_k(x) u_k(y)}{\lambda_k} = \sum_{k=1}^{\infty} \rho(x)\, u_k(x)\, u_k(y) = \delta(x-y) \quad (7.449)$$

an $\alpha = 1$ series expansion (7.413) of the delta function.

7.42 Green's Functions in One Dimension

In 1 dimension, we can explicitly solve the inhomogeneous ordinary differential equation $L\, f(x) = g(x)$ in which

$$L = -\frac{d}{dx}\left(p(x)\frac{d}{dx}\right) + q(x) \quad (7.450)$$

is formally self adjoint. We'll build a Green's function from two solutions u and v of the homogeneous equation $L\, u(x) = L\, v(x) = 0$ as

$$G(x,y) = \frac{1}{A}\left[\theta(x-y)u(y)v(x) + \theta(y-x)u(x)v(y)\right] \quad (7.451)$$

in which $\theta(x) = (x+|x|)/(2|x|)$ is **the Heaviside step function** (Oliver Heaviside 1850–1925), and A is a constant which we'll presently identify. We'll show that the expression

$$f(x) = \int_a^b G(x,y)\, g(y)\, dy = \frac{v(x)}{A}\int_a^x u(y)\, g(y)\, dy + \frac{u(x)}{A}\int_x^b v(y)\, g(y)\, dy$$

solves $L\, f(x) = g(x)$. Differentiating $f(x)$, we find after a cancellation

$$f'(x) = \frac{v'(x)}{A}\int_a^x u(y)\, g(y)\, dy + \frac{u'(x)}{A}\int_x^b v(y)\, g(y)\, dy. \quad (7.452)$$

Differentiating again, we have

$$\begin{aligned} f''(x) &= \frac{v''(x)}{A}\int_a^x u(y)\, g(y)\, dy + \frac{u''(x)}{A}\int_x^b v(y)\, g(y)\, dy \\ &\quad + \frac{v'(x)u(x)g(x)}{A} - \frac{u'(x)v(x)g(x)}{A} \\ &= \frac{v''(x)}{A}\int_a^x u(y)\, g(y)\, dy + \frac{u''(x)}{A}\int_x^b v(y)\, g(y)\, dy + \frac{W(x)}{A}\, g(x) \end{aligned} \quad (7.453)$$

in which $W(x)$ is the wronskian $W(x) = u(x)v'(x) - u'(x)v(x)$. The result (7.306) for the wronskian of two linearly independent solutions of a self-adjoint homogeneous ODE gives us $W(x) = W(x_0)\,p(x_0)/p(x)$. We set the constant $A = -W(x_0)\,p(x_0)$ so that the last term in (7.453) is $-g(x)/p(x)$. It follows that

$$Lf(x) = \frac{[Lv(x)]}{A}\int_a^x u(y)\,g(y)\,dy + \frac{[Lu(x)]}{A}\int_x^b v(y)\,g(y)\,dy + g(x) = g(x). \tag{7.454}$$

But $Lu(x) = Lv(x) = 0$, so we see that f satisfies our inhomogeneous equation $Lf(x) = g(x)$.

Example 7.55 (Green's functions with boundary conditions) To use the Green's function (7.451) with $A = W(x_0)p(x_0)$ to solve the ODE (7.450) subject to the Dirichelet boundary conditions $f(a) = 0 = f(b)$, we choose solutions $u(x)$ and $v(x)$ of the homogeneous equations $Lu(x) = 0 = Lv(x)$ that obey these boundary conditions, $u(a) = 0 = v(b)$. For then our formula $f(x) = \int_a^b G(x,y)g(y)dy$ gives

$$f(a) = \frac{u(a)}{A}\int_a^b v(y)\,g(y)\,dy = 0 = f(b) = \frac{v(b)}{A}\int_a^b u(y)\,g(y)\,dy. \tag{7.455}$$

Similarly, to impose the Neumann boundary conditions $f'(a) = 0 = f'(b)$, we choose solutions $u(x)$ and $v(x)$ of the homogeneous equations $Lu(x) = 0 = Lv(x)$ that obey these boundary conditions, $u'(a) = 0 = v'(b)$, so that our formula (7.452) for $f'(x)$ gives

$$f'(a) = \frac{u'(a)}{A}\int_a^b v(y)\,g(y)\,dy = 0 = f'(b) = \frac{v'(b)}{A}\int_a^b u(y)\,g(y)\,dy. \tag{7.456}$$

For instance, to solve the equation $-f''(x) - f(x) = \exp x$, with the mixed boundary conditions $f(-\pi) = 0$ and $f'(\pi) = 0$, we choose from among the solutions $\alpha\cos x + \beta\sin x$ of the homogeneous equation $-f'' - f = 0$, the functions $u(x) = \sin x$ and $v(x) = \cos x$. Substituting them into the formula (7.451) and setting $p(x) = 1$ and $A = -W(x_0) = \sin^2(x_0) + \cos^2(x_0) = 1$, we find as the Green's function

$$G(x,y) = \theta(x-y)\sin y\cos x + \theta(y-x)\sin x\cos y. \tag{7.457}$$

The solution $f(x) = \int_{-\pi}^{\pi} G(x,y)\,e^y\,dy$ then is

$$\begin{aligned} f(x) &= \int_{-\pi}^{\pi}\left[\theta(x-y)\sin y\cos x + \theta(y-x)\sin x\cos y\right]e^y\,dy \\ &= -\frac{1}{2}\left(e^{-\pi}\cos x + e^{\pi}\sin x + e^x\right). \end{aligned} \tag{7.458}$$

7.43 Principle of Stationary Action in Field Theory

If $\phi(x)$ is a scalar field, and $L(\phi)$ is its action density, then its action $S[\phi]$ is the integral over all of spacetime

$$S[\phi] = \int L(\phi(x))\, d^4x. \tag{7.459}$$

The principle of least (or stationary) action says that the field $\phi(x)$ that satisfies the classical equation of motion is the one for which the first-order change in the action due to any tiny variation $\delta\phi(x)$ in the field vanishes, $\delta S[\phi] = 0$. To keep things simple, we'll assume that the action (or Lagrange) density $L(\phi)$ is a function only of the field ϕ and its first derivatives $\partial_a\phi = \partial\phi/\partial x^a$. The first-order change in the action then is

$$\delta S[\phi] = \int \left[\frac{\partial L}{\partial \phi}\delta\phi + \frac{\partial L}{\partial(\partial_a\phi)}\delta(\partial_a\phi) \right] d^4x \tag{7.460}$$

in which we sum over the repeated index a from 0 to 3. Now $\delta(\partial_a\phi) = \partial_a(\phi+\delta\phi) - \partial_a\phi = \partial_a\delta\phi$. So we may integrate by parts and drop the surface terms because we set $\delta\phi = 0$ on the surface at infinity

$$\delta S[\phi] = \int \left[\frac{\partial L}{\partial \phi}\delta\phi + \frac{\partial L}{\partial(\partial_a\phi)}\partial_a(\delta\phi) \right] d^4x = \int \left[\frac{\partial L}{\partial \phi} - \partial_a \frac{\partial L}{\partial(\partial_a\phi)} \right] \delta\phi\, d^4x.$$

This first-order variation is zero for arbitrary $\delta\phi$ only if the field $\phi(x)$ satisfies Lagrange's equation

$$\partial_a \left(\frac{\partial L}{\partial(\partial_a\phi)} \right) \equiv \frac{\partial}{\partial x^a}\left[\frac{\partial L}{\partial(\partial\phi/\partial x^a)} \right] = \frac{\partial L}{\partial \phi} \tag{7.461}$$

which is the classical equation of motion.

Example 7.56 (Theory of a scalar field) The action density of a single scalar field ϕ of mass m is $L = \frac{1}{2}\,(\dot{\phi})^2 - \frac{1}{2}\,(\nabla\phi)^2 - \frac{1}{2}\,m^2\phi^2$ or equivalently $L = -\frac{1}{2}\,\partial_a\phi\,\partial^a\phi - \frac{1}{2}\,m^2\phi^2$. Lagrange's equation (7.461) is then

$$\nabla^2\phi - \ddot{\phi} = \partial_a\,\partial^a\phi = m^2\phi \tag{7.462}$$

which is the Klein–Gordon equation (7.41).

In a theory of several fields $\phi_1, \ldots, \phi_n$ with action density $L(\phi_k, \partial_a\phi_k)$, the fields obey n copies of Lagrange's equation one for each field ϕ_k

$$\frac{\partial}{\partial x^a}\left(\frac{\partial L}{\partial(\partial_a\phi_k)} \right) = \frac{\partial L}{\partial \phi_k}. \tag{7.463}$$

7.44 Symmetries and Conserved Quantities in Field Theory

A transformation of the coordinates x^a or of the fields ϕ_i and their derivatives $\partial_a \phi_i$ that leaves the action density $L(\phi_i, \partial_a \phi_i)$ invariant is a **symmetry** of the theory. Such a symmetry implies that something is conserved or time independent.

Suppose that an action density $L(\phi_i, \partial_a \phi_i)$ is unchanged when the fields ϕ_i and their derivatives $\partial_a \phi_i$ change by $\delta\phi_i$ and by $\delta(\partial_a \phi_i) = \partial_a(\delta\phi_i)$

$$0 = \delta L = \sum_i \frac{\partial L}{\partial \phi_i} \delta\phi_i + \frac{\partial L}{\partial \partial_a \phi_i} \partial_a \delta\phi_i . \tag{7.464}$$

Then using Lagrange's equations (7.463) to rewrite $\partial L / \partial \phi_i$, we find

$$0 = \sum_i \left(\partial_a \frac{\partial L}{\partial \partial_a \phi_i} \right) \delta\phi_i + \frac{\partial L}{\partial \partial_a \phi_i} \partial_a \delta\phi_i = \partial_a \sum_i \frac{\partial L}{\partial \partial_a \phi_i} \delta\phi_i \tag{7.465}$$

which says that the **current**

$$J^a = \sum_i \frac{\partial L}{\partial \partial_a \phi_i} \delta\phi_i \tag{7.466}$$

has zero divergence, $\partial_a J^a = 0$. Thus the time derivative of the volume integral of the charge density J^0

$$Q_V = \int_V J^0 \, d^3x \tag{7.467}$$

is the flux of current $\vec{J}$ entering through the boundary S of the volume V

$$\dot{Q}_V = \int_V \partial_0 J^0 \, d^3x = - \int_V \partial_k J^k \, d^3x = - \int_S J^k d^2 S_k . \tag{7.468}$$

If no current enters V, then the charge Q inside V is conserved. When the volume V is the whole universe, the charge is the integral over all of space

$$Q = \int J^0 \, d^3x = \int \sum_i \frac{\partial L}{\partial \dot{\phi}_i} \delta\phi_i \, d^3x = \int \sum_i \pi_i \, \delta\phi_i \, d^3x \tag{7.469}$$

in which π_i is the momentum conjugate to the field ϕ_i

$$\pi_i = \frac{\partial L}{\partial \dot{\phi}_i} . \tag{7.470}$$

Example 7.57 ($O(n)$ symmetry and its charge) Suppose the action density L is the sum of n copies of the quadratic action density

$$L = \sum_{i=1}^{n} \frac{1}{2} (\dot{\phi}_i)^2 - \frac{1}{2} (\nabla \phi_i)^2 - \frac{1}{2} m^2 \phi_i^2 = -\frac{1}{2} \partial_a \phi \, \partial^a \phi - \frac{1}{2} m^2 \phi^2 , \tag{7.471}$$

and A_{ij} is any constant antisymmetric matrix, $A_{ij} = -A_{ji}$. Then if the fields change by $\delta\phi_i = \epsilon \sum_j A_{ij}\,\phi_j$, the change (7.464) in the action density

$$\delta L = -\epsilon \sum_{i,j=1}^{n} \left[m^2 \phi_i A_{ij} \phi_j + \partial^a \phi_i A_{ij} \partial_a \phi_j \right] = 0 \tag{7.472}$$

vanishes. Thus the charge (7.469) associated with the matrix A

$$Q_A = \int \sum_i \pi_i \, \delta\phi_i \, d^3x = \epsilon \int \sum_i \pi_i \, A_{ij} \, \phi_j \, d^3x \tag{7.473}$$

is conserved. There are $n(n-1)/2$ antisymmetric $n \times n$ imaginary matrices; they generate the group $O(n)$ of $n \times n$ orthogonal matrices (Example 11.3).

An action density $L(\phi_i, \partial_a \phi_i)$ that is invariant under a spacetime translation, $x'^a = x^a + \delta x^a$, depends upon x^a only through the fields ϕ_i and their derivatives $\partial_a \phi_i$

$$\frac{\partial L}{\partial x^a} = \sum_i \left(\frac{\partial L}{\partial \phi_i} \frac{\partial \phi_i}{\partial x^a} + \frac{\partial L}{\partial \partial_b \phi_i} \frac{\partial^2 \phi_i}{\partial x^b \partial x^a} \right). \tag{7.474}$$

Using Lagrange's equations (7.463) to rewrite $\partial L / \partial \phi_i$, we find

$$\begin{aligned} 0 &= \left(\sum_i \partial_b \left(\frac{\partial L}{\partial \partial_b \phi_i} \right) \partial_a \phi_i + \frac{\partial L}{\partial \partial_b \phi_i} \frac{\partial^2 \phi_i}{\partial x^b \partial x^a} \right) - \frac{\partial L}{\partial x^a} \\ 0 &= \partial_b \left[\left(\sum_i \frac{\partial L}{\partial \partial_b \phi_i} \frac{\partial \phi_i}{\partial x^a} \right) - \delta^b_a L \right] \end{aligned} \tag{7.475}$$

that the **energy–momentum tensor**

$$T^b_{\ a} = \sum_i \frac{\partial L}{\partial \partial_b \phi_i} \frac{\partial \phi_i}{\partial x^a} - \delta^b_a L \tag{7.476}$$

has zero divergence, $\partial_b T^b_{\ a} = 0$.

Thus the time derivative of the 4-momentum P_{aV} inside a volume V

$$P_{aV} = \int_V \left(\sum_i \frac{\partial L}{\partial \partial_0 \phi_i} \frac{\partial \phi_i}{\partial x^a} - \delta^0_a L \right) d^3x = \int_V T^0_{\ a} \, d^3x \tag{7.477}$$

is equal to the flux entering through V's boundary S

$$\partial_0 P_{aV} = \int_V \partial_0 T^0_{\ a} \, d^3x = -\int_V \partial_k T^k_{\ a} \, d^3x = -\int_S T^k_{\ a} \, d^2 S_k. \tag{7.478}$$

The invariance of the action density L under spacetime translations implies the **conservation of energy** P_0 and **momentum** $\vec{P}$.

The momentum $\pi_i(x)$ that is canonically conjugate to the field $\phi_i(x)$ is the derivative of the action density L with respect to the time derivative of the field

$$\pi_i = \frac{\partial L}{\partial \dot{\phi}_i}. \tag{7.479}$$

If one can express the time derivatives $\dot{\phi}_i$ of the fields in terms of the fields ϕ_i and their momenta π_i, then the hamiltonian of the theory is the spatial integral of

$$H = P_0 = T^0_{\ 0} = \left(\sum_{i=1}^{n} \pi_i \, \dot{\phi}_i \right) - L \tag{7.480}$$

in which $\dot{\phi}_i = \dot{\phi}_i(\phi, \pi)$.

Example 7.58 (Hamiltonian of a scalar field) For the lagrangian of example 7.56, the hamiltonian density (7.480) is $H = \frac{1}{2}\pi^2 + \frac{1}{2}(\nabla\phi)^2 + \frac{1}{2}m^2\phi^2$.

Example 7.59 (Euler's theorem and the Nambu–Goto string) When the action density is a first-degree homogeneous function (Section 7.10) of the time derivatives of the fields, as is that of the Nambu-Goto string

$$L = -\frac{T_0}{c}\sqrt{\left(\dot{X} \cdot X'\right)^2 - \left(\dot{X}\right)^2 (X')^2}, \tag{7.481}$$

Euler's theorem (7.109) implies that the energy density (7.480) vanishes identically, independently of the equations of motion,

$$E^0 = \frac{\partial L}{\partial \dot{X}^\mu}\,\dot{X}^\mu - L = 0. \tag{7.482}$$

7.45 Nonlinear Differential Equations

Nonlinear differential equations are very interesting. Because linear combinations of the solutions to a given nonlinear differential equation are not also solutions, the actual solutions tend to have intrinsic sizes and shapes. But for the same reason, finding the general solution to a nonlinear differential equation is difficult. Fortunately, modern computers give us numerical solutions in most cases. And sometimes we get lucky.

Example 7.60 (Riccati's equation) The nonlinear differential equation $y' = a^2 x^{-4} - y^2$ has the solution $y(x) = x^{-1} + a x^{-2}(c - e^{2a/x})/(c + e^{2a/x})$.

Example 7.61 (Equidimensional equation) The second-order nonlinear ordinary differential equation $y'' = y\,y'/x$ has a general solution given by $y(x) = 2c$

$\tan(c\ln(x)+c')-1$ and two special solutions $y(x)=c''$ and $y(x)=-2/(c'''+\ln x)-1$ (Bender and Orszag, 1978, p. 30).

Example 7.62 (Fluid mechanics) The continuity equation (4.193) for a fluid of mass density ρ, velocity $\boldsymbol{v}$, and current density $\boldsymbol{j}=\rho\,\boldsymbol{v}$ is

$$\begin{aligned}\frac{\partial}{\partial t}\int \rho(\boldsymbol{x},t)\,d^3\boldsymbol{x} &= -\oint \boldsymbol{j}(\boldsymbol{x},t)\cdot d\boldsymbol{a} = -\int \nabla\cdot\boldsymbol{j}(\boldsymbol{x},t)\,d^3\boldsymbol{x}\\ &= -\int \nabla\cdot(\rho\,\boldsymbol{v})\,d^3\boldsymbol{x}.\end{aligned} \tag{7.483}$$

Its differential form is $\dot\rho=-\nabla\cdot(\rho\,\boldsymbol{v})=-\rho\,\nabla\cdot\boldsymbol{v}-\boldsymbol{v}\cdot\nabla\rho$ so that the total time derivative of the density is

$$\frac{d\rho}{dt}\equiv\frac{\partial\rho}{\partial t}+\boldsymbol{v}\cdot\nabla\rho=-\rho\,\nabla\cdot\boldsymbol{v}. \tag{7.484}$$

An **incompressible fluid** is one for which both sides of this continuity equation vanish

$$\frac{d\rho}{dt}=0\quad\text{and}\quad\nabla\cdot\boldsymbol{v}=0. \tag{7.485}$$

In the absence of viscosity, the force $\boldsymbol{F}$ acting on a tiny volume dV of a fluid is the integral of the pressure p over the surface $d\boldsymbol{A}$ of dV

$$-\oint p\,d\boldsymbol{A}=-\int\nabla p\,dV \tag{7.486}$$

in which $d\boldsymbol{A}$ is the outward normal to the surface. Equating this force per unit volume to the density times the acceleration, we find

$$\rho\,\frac{d\boldsymbol{v}}{dt}=\rho\left(\frac{\partial\boldsymbol{v}}{\partial t}+(\boldsymbol{v}\cdot\nabla)\,\boldsymbol{v}\right)=-\nabla p \tag{7.487}$$

which is Euler's equation for a fluid without viscosity (Leonhard Euler 1707–1783).

An incompressible fluid with a constant viscosity η obeys the **Navier–Stokes equation**

$$\rho\left(\frac{\partial\boldsymbol{v}}{\partial t}+(\boldsymbol{v}\cdot\nabla)\,\boldsymbol{v}\right)=-\nabla p+\eta\,\nabla^2\boldsymbol{v} \tag{7.488}$$

(Claude-Louis Navier 1785–1836, George Stokes 1819–1903).

7.46 Nonlinear Differential Equations in Cosmology

On large scales our universe is homogeneous and isotropic in space but not in time. The invariant squared distance between nearby points is

$$ds^2=-c^2dt^2+a^2(t)\left(\frac{dr^2}{1-kr^2/L^2}+r^2d\theta^2+r^2\sin^2\theta\,d\phi^2\right) \tag{7.489}$$

in which the magnitude of the **scale factor** $a(t)$ describes the expansion of space, and $k = 0$ or ± 1 corresponding to flat space ($k = 0$), a closed universe ($k = 1$), or an open universe ($k = -1$). The Friedmann equations of general relativity (13.273 and 13.288) for the dimensionless scale factor $a(t)$ are

$$\frac{\ddot{a}}{a} = -\frac{4\pi G}{3}\left(\rho + \frac{3p}{c^2}\right) \quad \text{and} \quad \left(\frac{\dot{a}}{a}\right)^2 = \frac{8\pi G}{3}\rho - \frac{c^2 k}{L^2 a^2} \tag{7.490}$$

in which $L > 2 \times 10^{13}$ ly, and $k = \pm 1$ or 0.

These equations are simpler when the pressure p is related to the mass density ρ by an equation of state $p = c^2 w \rho$. This happens when the mass density ρ is due to a single constituent – radiation ($w = 1/3$), matter ($w = 0$), or dark energy ($w = -1$). Conservation of energy $\dot{\rho} = -3\,(\rho + p/c^2)\,\dot{a}/a$ (13.279–13.284) then gives (Exercise 7.32) the mass density as $\rho = \rho_0\, a^{-3(1+w)}$ in which ρ_0 is its present value, and the present value a_0 of the scale factor has been set equal to unity, $a_0 = 1$. Friedmann's equations then are

$$\frac{\ddot{a}}{a} = -\frac{4\pi G(1+3w)\rho_0}{3}\,a^{-3(1+w)} \quad \text{and} \quad \left(\frac{\dot{a}}{a}\right)^2 = \frac{8\pi G\rho_0}{3}\,a^{-3(1+w)} - \frac{c^2 k}{L^2\, a^2}. \tag{7.491}$$

Example 7.63 (de Sitter spacetimes) For a universe dominated by a cosmological constant $\Lambda = 8\pi G\rho_0$, the equation-of-state parameter w is -1, and Friedmann's equations (7.491) are

$$\frac{\ddot{a}}{a} = \frac{8\pi G\,\rho_0}{3} \quad \text{and} \quad \dot{a}^2 = \frac{8\pi G\rho_0}{3}\,a^2 - \frac{c^2 k}{L^2}. \tag{7.492}$$

If $\Lambda = 8\pi G\rho_0 < 0$ and $k = -1$, then the scale factor $a(t)$ satisfies the harmonic equation $\ddot{a} = -\lambda^2 a$ where $\lambda^2 = 8\pi G\,|\rho_0|/3$. So the scale factor is a periodic function, $a(t) = \alpha e^{i\lambda t} + \bar{\alpha} e^{-i\lambda t}$. The nonlinear first-order equation

$$-\lambda^2(\alpha^2 e^{2i\lambda t} - 2|\alpha|^2 + \bar{\alpha}^2 e^{-2i\lambda t}) = -\lambda^2(\alpha^2 e^{2i\lambda t} + 2|\alpha|^2 + \bar{\alpha}^2 e^{-2i\lambda t}) + c^2/L^2 \tag{7.493}$$

fixes the magnitude of α to be $|\alpha| = c/(2\lambda L)$. With the initial condition $a(0) = 0$, we get $a(t) = (c/(L\,\lambda))\,\sin(\lambda\, t)$ in which the sign of $a(t)$ is irrelevant because only a^2 appears in the invariant distance (7.489). This maximally symmetric (section 13.24) spacetime is called **anti-de Sitter space**. There are no $k = 0$ or $k = 1$ solutions for $\rho_0 < 0$.

If $\Lambda = 8\pi G\rho_0 > 0$, the scale factor $a(t)$ obeys the equation $\ddot{a} = \lambda^2\, a$ with $\lambda^2 = 8\pi G\,\rho_0/3 > 0$. So $a(t) = \alpha\, e^{\lambda t} + \beta\, e^{-\lambda t}$ with α, β real. The nonlinear first-order equation $\dot{a}^2 = \lambda^2\, a^2 - c^2 k/L^2$ implies that $\alpha\beta = c^2 k/(4\lambda^2 L^2)$. The solutions

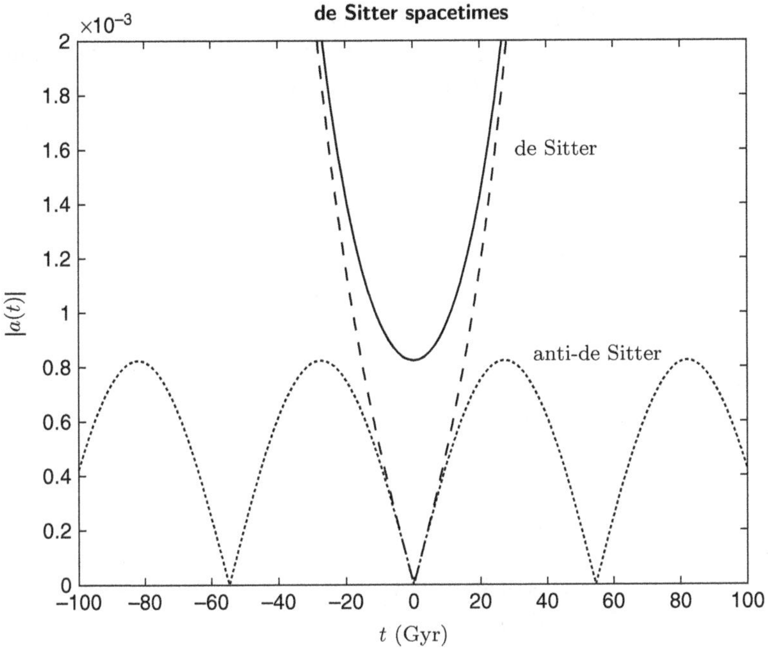

Figure 7.4 The absolute value $|a(t)|$ of the scale factor is plotted for de Sitter universes (7.494) with $\Lambda > 0$ and $k = 1$ (solid) and $k = -1$ (dashed) as well as for an anti-de Sitter universe $a(t) = (c/(L\,\lambda))\,\sin(\lambda\,t)$ with $\Lambda < 0$ and $k = -1$ (dotted).

$$a(t) = \begin{cases} \frac{c}{L\,\lambda}\,\cosh(\lambda\,t) & \text{if } k = 1 \\ a(0)\,e^{\pm\lambda t} & \text{if } k = 0 \\ \frac{c}{L\,\lambda}\,\sinh(\lambda\,t) & \text{if } k = -1 \end{cases} \tag{7.494}$$

remarkably represent in different coordinates the same maximally symmetric spacetime known as **de Sitter space**.

Figure 7.4 plots the absolute value $|a(t)|$ of the scale factor for de Sitter universes (7.494) with $\Lambda > 0$ and $k = 1$ (solid) and $k = -1$ (dashed) as well as for an anti-de Sitter universe $a(t) = (c/(L\,\lambda))\,\sin(\lambda\,t)$ with $\Lambda < 0$ and $k = -1$ (dotted). The present value of the dark-energy mass density $\rho_0 = 5.924 \times 10^{-27}$ kg/m^3 is used with $L = 2.1 \times 10^{13}$ ly.

Example 7.64 (Universe of radiation) For a universe in which the mass density ρ is entirely due to radiation, the parameter $w = 1/3$, and the Friedmann equations (7.491) are

$$a^3\,\ddot{a} = -\frac{8\pi G\rho_0}{3} \quad \text{and} \quad \dot{a}^2 = \frac{8\pi G\rho_0}{3a^2} - \frac{c^2 k}{L^2}. \tag{7.495}$$

In terms of $r^4 \equiv 32\pi\, G\,\rho_0/3 > 0$, the first-order $k = 0$ Friedmann equation (7.495) is $\dot{a} = r^2/(2a)$ or $2a\,da = r^2\,dt$ so that $a^2 = r\,t$. Thus the $k = 0$ solution for a universe of radiation is $a(t) = r\,\sqrt{t}$. In our universe the term $-\,c^2k/L^2$ is so small as to be consistent with zero, so this solution is a good approximation during the first 50,000 years, which is the era of radiation. The $k = \pm 1$ solutions are

$$\begin{aligned} a(t) &= \frac{Lr^2}{2c}\left[1 - \left(1 - \frac{2c^2t}{L^2r^2}\right)^2\right]^{1/2} \quad \text{for} \quad k = 1 \\ a(t) &= \frac{Lr^2}{2c}\left[\left(1 + \frac{2c^2t}{L^2r^2}\right)^2 - 1\right]^{1/2} \quad \text{for} \quad k = -1. \end{aligned} \tag{7.496}$$

The scale factor of a $k = 1$ universe reaches a maximum size

$$a_{\max} = \frac{Lr^2}{2c} = \frac{L}{c}\sqrt{\frac{8\pi G\rho_0}{3}} \tag{7.497}$$

and collapses back to zero at $t = L^2r^2/c^2 > 5.6$ Gyr.

Example 7.65 (Universe of matter) A universe made entirely of matter has no pressure, so $w = 0$. Its Friedmann equations (7.491) are

$$a^2\,\ddot{a} = -\frac{4\pi G\rho_0}{3} \quad \text{and} \quad \dot{a}^2 = \frac{8\pi G\rho_0}{3}\frac{1}{a} - \frac{c^2k}{L^2}. \tag{7.498}$$

In terms of $m^3 = 6\pi\, G\,\rho_0 > 0$, the first-order $k = 0$ Friedmann equation (7.498) is $(3/2)\sqrt{a}\,\dot{a} = m^{3/2}$ or $(3/2)\sqrt{a}\,da = m^{3/2}dt$ which integrates to $a^{3/2} = m^{3/2}t$. So the $k = 0$ scale factor for a universe of matter rises as the two-thirds power of the time $a(t) = m\,t^{2/3}$ which is a reasonable approximation during the first few Gyr of the era of matter because the curvature term $-\,c^2k/L^2$ is tiny.

The scale factor of a $k = 1$ universe reaches a maximum size

$$a_{\max} = \frac{8\pi\, G\,\rho_0\,L^2}{3\,c^2} \tag{7.499}$$

and then collapses to zero.

Figure 7.5 plots the $k = 0$ scale factors $a(t) = r\,\sqrt{t}$ for a universe of radiation (dash-dot) and $a(t) = m\,t^{2/3}$ for a universe of matter (dashed) as well as the scale factor for a multi-component universe (13.296, solid) based on the parameters of the Planck collaboration (Aghanim et al., 2018).

7.47 Nonlinear Differential Equations in Particle Physics

The equations of particle physics are nonlinear. Physicists usually use perturbation theory to cope with the nonlinearities. But occasionally they focus on the nonlinearities and treat the fields classically or semi-classically. To keep things relatively simple, we'll work in a spacetime of only 2 dimensions and consider a model field theory described by the action density

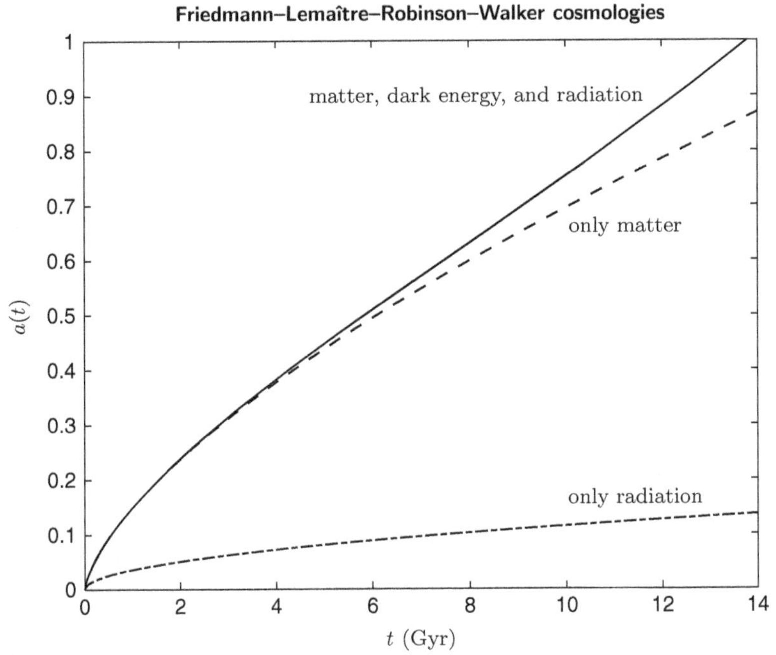

Figure 7.5 The $k = 0$ scale factors $a(t) = m\,t^{2/3}$ for a universe of matter (dashes) and $a(t) = r\,\sqrt{t}$ for a universe of radiation (dashdot) are compared with a multi-component scale factor (13.296, solid) based on the parameters of the Planck collaboration (Aghanim et al., 2018).

$$\mathcal{L} = \frac{1}{2}\left(\dot{\phi}^2 - \phi'^2\right) - V(\phi) \tag{7.500}$$

in which a dot means a time derivative, a prime means a spatial derivative, and V is a function of the field ϕ. Lagrange's equation for this theory is

$$\ddot{\phi} - \phi'' = -\frac{dV}{d\phi}. \tag{7.501}$$

We can convert this partial differential equation to an ordinary one by making the field ϕ depend only upon the combination $u = x - v\,t$ rather than upon both x and t. We then have $\dot{\phi} = -v\,\phi_u$. With this restriction to traveling-wave solutions, Lagrange's equation reduces to

$$(1 - v^2)\,\phi_{uu} = \frac{dV}{d\phi} = V_{,\phi}. \tag{7.502}$$

We multiply both sides of this equation by ϕ_u $(1-v^2)\,\phi_u\,\phi_{uu} = \phi_u\,V_{,\phi}$ and integrate to get $(1 - v^2)\,\frac{1}{2}\,\phi_u^2 = V + E$ in which E is a constant of integration $E = \frac{1}{2}\,(1 - v^2)\,\phi_u^2 - V(\phi)$. We can convert (Exercise 7.36) this equation into a problem of integration

$$u - u_0 = \int \frac{\sqrt{1-v^2}}{\sqrt{2(E + V(\phi))}}\, d\phi. \tag{7.503}$$

By inverting the resulting equation relating u to ϕ, we may find the **soliton** solution $\phi(u - u_0)$, which is a lump of energy traveling with speed v.

Example 7.66 (Soliton of the ϕ^4 theory) To simplify the integration (7.503), we take as the action density

$$\mathcal{L} = \frac{1}{2}\left(\dot{\phi}^2 - \phi'^2\right) - \left[\frac{\lambda^2}{2}\left(\phi^2 - \phi_0^2\right)^2 - E\right]. \tag{7.504}$$

Our formal solution (7.503) gives

$$u - u_0 = \pm \int \frac{\sqrt{1-v^2}}{\lambda\left(\phi^2 - \phi_0^2\right)}\, d\phi = \mp \frac{\sqrt{1-v^2}}{\lambda\phi_0} \tanh^{-1}(\phi/\phi_0) \tag{7.505}$$

or

$$\phi(x - vt) = \mp\phi_0 \tanh\left[\lambda\phi_0 \frac{x - x_0 - v(t - t_0)}{\sqrt{1-v^2}}\right] \tag{7.506}$$

which is a soliton (or an antisoliton) at $x_0 + v(t - t_0)$. A unit soliton at rest is plotted in Fig. 7.6. Its energy is concentrated at $x = 0$ where $|\phi^2 - \phi_0^2|$ is maximal.

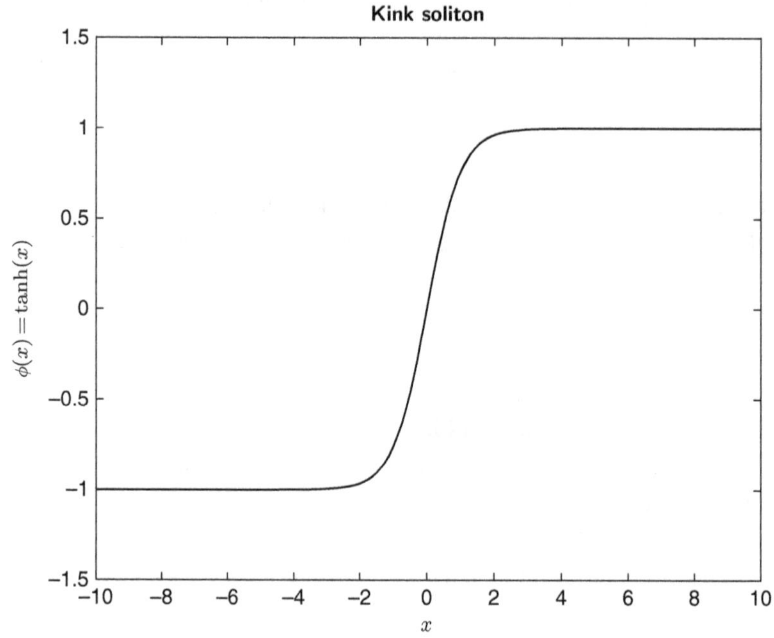

Figure 7.6 The field $\phi(x)$ of the soliton (7.506) at rest ($v = 0$) at position $x_0 = 0$ for $\lambda = 1 = \phi_0$. The energy density of the field vanishes when $\phi = \pm\phi_0 = \pm 1$. The energy of this soliton is concentrated at $x = 0$.

Further Reading

One can learn more about differential equations in *Advanced Mathematical Methods for Scientists and Engineers* (Bender and Orszag, 1978).

Exercises

7.1 In rectangular coordinates, the curl of a curl is by definition (2.43)

$$(\nabla \times (\nabla \times \boldsymbol{E}))_i = \sum_{j,k=1}^{3} \epsilon_{ijk} \partial_j (\nabla \times \boldsymbol{E})_k = \sum_{j,k,\ell,m=1}^{3} \epsilon_{ijk} \partial_j \epsilon_{k\ell m} \partial_\ell E_m. \tag{7.507}$$

Use Levi-Civita's identity (1.497) to show that

$$\nabla \times (\nabla \times \boldsymbol{E}) = \nabla(\nabla \cdot \boldsymbol{E}) - \Delta \boldsymbol{E}. \tag{7.508}$$

This formula defines $\Delta \boldsymbol{E}$ in any system of orthogonal coordinates.

7.2 Show that since the Bessel function $J_n(x)$ satisfies Bessel's equation (7.26), the function $P_n(\rho) = J_n(k\rho)$ satisfies (7.25).

7.3 Show that (7.38) implies that $R_{k,\ell}(r) = j_\ell(kr)$ satisfies (7.37).

7.4 Use (7.36, 7.37), and $\Phi''_m = -m^2 \Phi_m$ to show in detail that the product $f(r, \theta, \phi) = R_{k\ell}(r)\, \Theta_{\ell m}(\theta)\, \Phi_m(\phi)$ satisfies $-\triangle f = k^2 f$.

7.5 Replacing Helmholtz's k^2 by $2m(E - V(r))/\hbar^2$, we get Schrödinger's equation $-(\hbar^2/2m)\triangle\psi(r,\theta,\phi) + V(r)\psi(r,\theta,\phi) = E\psi(r,\theta,\phi)$. Now let $\psi(r,\theta,\phi) = R_{n\ell}(r)\Theta_{\ell m}(\theta)e^{im\phi}$ in which $\Theta_{\ell m}$ satisfies (7.36) and show that the radial function $R_{n\ell}$ must obey

$$-\frac{\left(r^2 R'_{n,\ell}\right)'}{r^2} + \left[\frac{\ell(\ell+1)}{r^2} + \frac{2mV}{\hbar^2}\right] R_{n,\ell} = \frac{2mE_{n,\ell}}{\hbar^2} R_{n,\ell}. \tag{7.509}$$

7.6 Use the empty-space Maxwell's equations $\nabla \cdot \boldsymbol{B} = 0$, $\nabla \times \boldsymbol{E} + \dot{\boldsymbol{B}} = 0$, $\nabla \cdot \boldsymbol{E} = 0$, and $\nabla \times \boldsymbol{B} - \dot{\boldsymbol{E}}/c^2 = 0$ and the formula (7.508) to show that in vacuum $\Delta \boldsymbol{E} = \ddot{\boldsymbol{E}}/c^2$ and $\Delta \boldsymbol{B} = \ddot{\boldsymbol{B}}/c^2$.

7.7 Argue from symmetry and antisymmetry that $[\gamma^a, \gamma^b]\partial_a \partial_b = 0$ in which the sums over a and b run from 0 to 3.

7.8 Suppose a voltage $V(t) = V \sin(\omega t)$ is applied to a resistor of R (Ω) in series with a capacitor of capacitance C (F). If the current through the circuit at time $t = 0$ is zero, what is the current at time t?

7.9 (a) Is $\left(1 + x^2 + y^2\right)^{-3/2} \left[(1 + y^2) y\, dx + (1 + x^2) x\, dy\right] = 0$ exact? (b) Find its general integral and solution $y(x)$. Use Section 7.8.

7.10 (a) Separate the variables of the ODE $(1 + y^2) y\, dx + (1 + x^2) x\, dy = 0$.
(b) Find its general integral and solution $y(x)$.

7.11 Find the general solution to the differential equation $y' + y/x = c/x$.

7.12 Find the general solution to the differential equation $y' + xy = ce^{-x^2/2}$.

7.13 James Bernoulli studied ODEs of the form $y' + p\, y = q\, y^n$ in which p and q are functions of x. Division by y^n and the substitution $v = y^{1-n}$ gives us the equation $v' + (1-n) p\, v = (1-n)\, q$ which is soluble as shown in Section 7.16. Use this method to solve the ODE $y' - y/2x = 5x^2 y^5$.

7.14 Integrate the ODE $(xy+1)\, dx + 2x^2(2xy-1)\, dy = 0$. Hint: Use the variable $v(x) = xy(x)$ instead of $y(x)$.

7.15 Show that the points $x = \pm 1$ and ∞ are regular singular points of Legendre's equation (7.220).

7.16 Use the vanishing of the coefficient of every power of x in (7.225) and the notation (7.227) to derive the recurrence relation (7.228).

7.17 In Example 7.39, derive the recursion relation for $r = 1$ and discuss the resulting eigenvalue equation.

7.18 In Example 7.39, show that the solutions associated with the roots $r = 0$ and $r = 1$ are the same.

7.19 For a hydrogen atom with $V(r) = -e^2/4\pi\epsilon_0 r \equiv -q^2/r$, Equation (7.509) is $(r^2 R_{n\ell}')' + \left[(2m/\hbar^2)\left(E_{n\ell} + Zq^2/r\right) r^2 - \ell(\ell+1)\right] R_{n\ell} = 0$. So at big r, $R_{n\ell}'' \approx -2mE_{n\ell} R_{n\ell}/\hbar^2$ and $R_{n\ell} \sim \exp(-\sqrt{-2mE_{n\ell}}\, r/\hbar)$. At tiny r, $(r^2 R_{n\ell}')' \approx \ell(\ell+1) R_{n\ell}$ and $R_{n\ell}(r) \sim r^\ell$. Set $R_{n\ell}(r) = r^\ell \exp(-\sqrt{-2mE_{n\ell}}\, r/\hbar) P_{n\ell}(r)$ and apply the method of Frobenius to find the values of $E_{n\ell}$ for which $R_{n\ell}$ is suitably normalizable.

7.20 Show that as long as the matrix $\mathcal{Y}_{kj} = y_k^{(\ell_j)}(x_j)$ is nonsingular, the n boundary conditions $b_j = y^{(\ell_j)}(x_j) = c_1\, y_1^{(\ell_j)}(x_j) + \cdots c_n\, y_n^{(\ell_j)}(x_j)$ determine the n coefficients c_k of the expansion (7.262) to be

$$C^{\mathsf{T}} = B^{\mathsf{T}} \mathcal{Y}^{-1} \quad \text{or} \quad C_k = \sum_{j=1}^{n} b_j \mathcal{Y}_{jk}^{-1}. \tag{7.510}$$

7.21 Show that if the real and imaginary parts u_1, u_2, v_1, and v_2 of ψ and χ satisfy boundary conditions at $x = a$ and $x = b$ that make the boundary term (7.280) vanish, then its complex analog (7.282) also vanishes.

7.22 Show that if the real and imaginary parts u_1, u_2, v_1, and v_2 of ψ and χ satisfy boundary conditions at $x = a$ and $x = b$ that make the boundary term (7.280) vanish, and if the differential operator L is real and self adjoint, then (7.278) implies (7.283).

7.23 Show that if D is the set of all twice-differentiable functions $u(x)$ on $[a, b]$ that satisfy Dirichlet's boundary conditions (7.285) and if the function $p(x)$ is continuous and positive on $[a, b]$, then the adjoint set D^* defined as the set of all twice-differentiable functions $v(x)$ that make the boundary term (7.287) vanish for all functions $u \in D$ is D itself.

7.24 Same as Exercise 7.23 but for Neumann boundary conditions (7.286).

7.25 Use Bessel's equation (7.354) and the boundary conditions $u(0) = 0$ for $n > 0$ and $u(1) = 0$ to show that the eigenvalues λ are all positive.

7.26 Show that after the change of variables $u(x) = J_n(kx) = J_n(\rho)$, the self-adjoint differential equation (7.354) becomes Bessel's equation (7.355).

7.27 Derive Bessel's inequality (7.417) from the inequality (7.416).

7.28 Repeat Example 7.51 using J_1's instead of J_0's. Hint: the *Mathematica* command Do[Print[N[BesselJZero[1, k], 10]], {k, 1, 100, 1}] gives the first 100 zeros $z_{1,k}$ of the Bessel function $J_1(x)$ to 10 significant figures.

7.29 Derive the Yukawa potential (7.431) as the Green's function for the modified Helmholtz equation (7.430).

7.30 Use Lagrange's equation to derive the Klein–Gordon equation (7.462).

7.31 Derive the formula for the hamiltonian density of the theory of Example 7.58.

7.32 Derive the relation $\rho = \overline{\rho}(\overline{a}/a)^{3(1+w)}$ between the energy density ρ and the scale factor $a(t)$ from the conservation law $d\rho/da = -3(\rho + p)/a$ and the equation of state $p = w\rho$.

7.33 Derive the three de Sitter solutions (7.494).

7.34 Derive the two solutions (7.496) for a universe of radiation.

7.35 How do we know that the maximum value of the scale factor of a closed universe of matter is (7.499).

7.36 Use $E = \frac{1}{2}\,(1 - v^2)\, \phi_u^2 - V(\phi)$ to derive the soliton solution (7.503).

7.37 Find the solution of the differential equation $-f''(x) - f(x) = 1$ that satisfies the boundary conditions $f(-\pi) = 0 = f'(\pi)$. Hint: Use Example 7.55.

7.38 Show that the sum of exponentials $f(x) = c_1 e^{z_1 x} + c_2 e^{z_2 x} + \cdots + c_n e^{z_n x}$ in which the z_k's are the n roots of the algebraic equation

$$0 = a_0 + a_1 z + a_2 z^2 + \cdots + a_n z^n$$

and the c_k's are any constants is a solution of the homogeneous ordinary differential equation $0 = a_0 f(x) + a_1 f'(x) + a_2 f''(x) + \cdots + a_n f^{(n)}(x)$ with constant coefficients a_k. When the roots are all different, $f(x)$ is the most general solution.

7.39 Find two linearly independent solutions of the ODE $f'' - 2f' + f = 0$.

8

Integral Equations

8.1 Differential Equations as Integral Equations

Differential equations when integrated become integral equations with built-in boundary conditions. Thus if we integrate the first-order ODE

$$\frac{du(x)}{dx} \equiv u_x(x) = p(x)\,u(x) + q(x) \tag{8.1}$$

then we get the integral equation

$$u(x) = \int_a^x p(y)\,u(y)\,dy + \int_a^x q(y)\,dy + u(a). \tag{8.2}$$

To transform a second-order differential equation into an integral equation, one uses Cauchy's identity (Exercise 8.1)

$$\int_a^x dz \int_a^z dy\, f(y) = \int_a^x (x-y)\, f(y)\,dy, \tag{8.3}$$

which is a special case of his formula for repeated integration

$$\int_a^x \int_a^{x_1} \cdots \int_a^{x_{n-1}} f(x_n)\,dx_n \cdots dx_2\,dx_1 = \frac{1}{(n-1)!} \int_a^x (x-y)^{n-1} f(y)\,dy. \tag{8.4}$$

Using the special case (8.3), one may integrate (Exercise 8.2) the second-order ODE $u'' = pu' + qu + r$ to

$$u(x) = f(x) + \int_a^x k(x,y)\,u(y)\,dy \tag{8.5}$$

with $k(x,y) = p(y) + (x-y)\left[q(y) - p'(y)\right]$ and

$$f(x) = u(a) + (x-a)\left[u'(a) - p(a)\,u(a)\right] + \int_a^x (x-y)r(y)\,dy. \tag{8.6}$$

In some physical problems, integral equations arise independently of differential equations. Whatever their origin, integral equations tend to have properties more suitable to mathematical analysis because derivatives are unbounded operators.

8.2 Fredholm Integral Equations

An equation of the form

$$\int_a^b k(x, y)\, u(y)\, dy = \lambda\, u(x) + f(x) \tag{8.7}$$

for $a \le x \le b$ with a given **kernel** $k(x, y)$ and a specified function $f(x)$ is an **inhomogeneous Fredholm equation of the second kind** for the function $u(x)$ and the parameter λ (Erik Ivar Fredholm, 1866–1927).

If $f(x) = 0$, then it is a **homogeneous Fredholm equation of the second kind**

$$\int_a^b k(x, y)\, u(y)\, dy = \lambda\, u(x), \qquad a \le x \le b. \tag{8.8}$$

Such an equation typically has nontrivial solutions only for certain **eigenvalues** λ. Each solution $u(x)$ is an **eigenfunction.**

If $\lambda = 0$ but $f(x) \neq 0$, then Equation (8.7) is an **inhomogeneous Fredholm equation of the first kind**

$$\int_a^b k(x, y)\, u(y)\, dy = f(x), \qquad a \le x \le b. \tag{8.9}$$

Finally, if both $\lambda = 0$ and $f(x) = 0$, then (8.7) is a **homogeneous Fredholm equation of the first kind**

$$\int_a^b k(x, y)\, u(y)\, dy = 0, \qquad a \le x \le b. \tag{8.10}$$

These Fredholm equations (8.7–8.10) are **linear** because they involve only the first (and zeroth) power of the unknown function $u(x)$.

8.3 Volterra Integral Equations

If the kernel $k(x, y)$ in the Equations (8.7– 8.10) that define the Fredholm integral equations is **causal,** that is, if $k(x, y) = k(x, y)\, \theta(x - y)$, in which $\theta(x) = (x + |x|)/2|x|$ is the Heaviside function, then the corresponding equations bear the name **Volterra** (Vito Volterra, 1860–1941). Thus an equation of the form

$$\int_a^x k(x, y)\, u(y)\, dy = \lambda\, u(x) + f(x) \tag{8.11}$$

in which the **kernel** $k(x, y)$ and the function $f(x)$ are given, is an **inhomogeneous Volterra equation of the second kind** for the function $u(x)$ and the parameter λ.

If $f(x) = 0$, then it is a **homogeneous Volterra equation of the second kind**

$$\int_a^x k(x, y)\, u(y)\, dy = \lambda\, u(x). \tag{8.12}$$

Such an equation typically has nontrivial solutions only for certain **eigenvalues** λ. The solutions $u(x)$ are the **eigenfunctions**.

If $\lambda = 0$ but $f(x) \neq 0$, then Equation (8.11) is an **inhomogeneous Volterra equation of the first kind**

$$\int_a^x k(x, y)\, u(y)\, dy = f(x). \tag{8.13}$$

Finally, if both $\lambda = 0$ and $f(x) = 0$, then it is a **homogeneous Volterra equation of the first kind**

$$\int_a^x k(x, y)\, u(y)\, dy = 0. \tag{8.14}$$

These Volterra equations (8.11–8.14) are **linear** because they involve only the first (and zeroth power) of the unknown function $u(x)$.

In what follows, we'll mainly discuss Fredholm integral equations, since those of the Volterra type are a special case of the Fredholm type.

8.4 Implications of Linearity

Because the Fredholm and Volterra integral equations are linear, one may add solutions of the homogeneous equations (8.8, 8.10, 8.12, and 8.14) and get new solutions. Thus if $u_1, u_2, \ldots$ are eigenfunctions

$$\int_a^b k(x, y)\, u_j(y)\, dy = \lambda\, u_j(x), \qquad a \le x \le b \tag{8.15}$$

with the same eigenvalue λ, then the sum $\sum_j a_j u_j(x)$ also is an eigenfunction with the same eigenvalue

$$\begin{aligned}\int_a^b k(x, y) \left(\sum_j a_j u_j(y) \right) dy &= \sum_j a_j \int_a^b k(x, y)\, u_j(y)\, dy \\ &= \sum_j a_j \lambda\, u_j(x) = \lambda \left(\sum_j a_j u_j(x) \right).\end{aligned} \tag{8.16}$$

It also is true that the difference between any two solutions $u_1^i(x)$ and $u_2^i(x)$ of one of the inhomogeneous Fredholm (8.7, 8.9) or Volterra (8.11, 8.13) equations is a solution of the associated homogeneous equation (8.8, 8.10, 8.12, or 8.14). Thus if $u_1^i(x)$ and $u_2^i(x)$ satisfy the inhomogeneous Fredholm equation of the second kind

$$\int_a^b k(x, y)\, u_j^i(y)\, dy = \lambda\, u_j^i(x) + f(x), \qquad j = 1, 2 \tag{8.17}$$

then their difference $u_1^i(x) - u_2^i(x)$ satisfies the homogeneous Fredholm equation of the second kind

$$\int_a^b k(x, y) \left[u_1^i(y) - u_2^i(y)\right] dy = \lambda \left[u_1^i(x) - u_2^i(x)\right]. \tag{8.18}$$

Thus the most general solution $u^i(x)$ of the inhomogeneous Fredholm equation of the second kind (8.17) is a particular solution $u_p^i(x)$ of that equation plus the general solution of the homogeneous Fredholm equation of the second kind (8.15)

$$u^i(x) = u_p^i(x) + \sum_j a_j u_j(x). \tag{8.19}$$

Linear integral equations are much easier to solve than nonlinear ones.

8.5 Numerical Solutions

Let us break the real interval $[a, b]$ into N segments $[y_k, y_{k+1}]$ of equal length $\Delta y = (b - a)/N$ with $y_0 = a$, $y_k = a + k\,\Delta y$, and $y_N = b$. Let's also set $x_k = y_k$ and define U as the vector with entries $U_k = u(y_k)$ and K as the $(N+1) \times (N+1)$ square matrix with elements $K_{k\ell} = k(x_k, y_\ell)\,\Delta y$. Then we may approximate the homogeneous Fredholm equation of the second kind (8.8)

$$\int_a^b k(x, y)\, u(y)\, dy = \lambda\, u(x), \qquad a \le x \le b \tag{8.20}$$

as the algebraic equation

$$\sum_{\ell=0}^{N} K_{k,\ell}\, U_\ell = \lambda\, U_k \tag{8.21}$$

or in matrix notation $K\,U = \lambda\,U$.

We saw in Section 1.26 that every such equation has $N+1$ eigenvectors $U^{(\alpha)}$ and eigenvalues $\lambda^{(\alpha)}$, and that the eigenvalues $\lambda^{(\alpha)}$ are the solutions of the characteristic equation (1.273)

$$\det(K - \lambda^{(\alpha)} I) = \left|K - \lambda^{(\alpha)} I\right| = 0. \tag{8.22}$$

In general, as $N \to \infty$ and $\Delta y \to 0$, the number $N + 1$ of eigenvalues $\lambda^{(\alpha)}$ and eigenvectors $U^{(\alpha)}$ becomes infinite.

We may apply the same technique to the inhomogeneous Fredholm equation of the first kind

$$\int_a^b k(x, y)\, u(y)\, dy = f(x) \quad \text{for} \quad a \le x \le b. \tag{8.23}$$

The resulting matrix equation is $K\, U = F$ in which the kth entry in the vector F is $F_k = f(x_k)$. This equation has the solution $U = K^{-1}\, F$ as long as the matrix K is nonsingular, that is, as long as $\det K \ne 0$.

This technique applied to the inhomogeneous Fredholm equation of the second kind

$$\int_a^b k(x, y)\, u(y)\, dy = \lambda\, u(x) + f(x) \tag{8.24}$$

leads to the matrix equation $K\, U = \lambda\, U + F$. The associated homogeneous matrix equation $K\, U = \lambda\, U$ has $N + 1$ eigenvalues $\lambda^{(\alpha)}$ and eigenvectors $U^{(\alpha)} \equiv |\alpha\rangle$. For any value of λ that is *not* one of the eigenvalues $\lambda^{(\alpha)}$, the matrix $K - \lambda I$ has a nonzero determinant and hence an inverse, and so the vector $U^i = (K - \lambda\, I)^{-1}\, F$ is a solution of the inhomogeneous matrix equation $K\, U = \lambda\, U + F$.

If $\lambda = \lambda^{(\beta)}$ is one of the eigenvalues $\lambda^{(\alpha)}$ of the homogeneous matrix equation $K\, U = \lambda\, U$ then the matrix $K - \lambda^{(\beta)} I$ will not have an inverse, but it will have a pseudoinverse (Section 1.33). If its singular-value decomposition (1.396) is

$$K - \lambda^{(\beta)} I = \sum_{n=1}^{N+1} |m_n\rangle S_n \langle n| \tag{8.25}$$

then its pseudoinverse (1.427) is

$$\left(K - \lambda^{(\beta)} I\right)^+ = \sum_{\substack{n=1 \\ S_n \ne 0}}^{N+1} |n\rangle S_n^{-1} \langle m_n| \tag{8.26}$$

in which the sum is over the positive singular values. So if the vector F is a linear combination of the left singular vectors $|m_n\rangle$ whose singular values are positive

$$F = \sum_{\substack{n=1 \\ S_n \ne 0}}^{N+1} f_n |m_n\rangle \tag{8.27}$$

then the vector

$$U^i = \left(K - \lambda^{(\beta)} I\right)^+ \, F \tag{8.28}$$

will be a solution of $K\,U = \lambda\,U + F$. For in this case

$$\left(K - \lambda^{(\beta)} I\right) U^i = \left(K - \lambda^{(\beta)} I\right) \left(K - \lambda^{(\beta)} I\right)^+ F$$

$$= \sum_{n''=1}^{N+1} |m_{n''}\rangle S_{n''} \langle n''| \sum_{\substack{n'=1 \\ S_{n'}\neq 0}}^{N+1} |n'\rangle S_{n'}^{-1} \langle m_{n'}| \sum_{\substack{n=1 \\ S_n \neq 0}}^{N+1} f_n |m_n\rangle$$

$$= \sum_{\substack{n=1 \\ S_n \neq 0}}^{N+1} f_n |m_n\rangle = F. \tag{8.29}$$

The most general solution will be the sum of this particular solution of the inhomogeneous equation $K\,U = \lambda\,U + F$ and the most general solution of the homogeneous equation $K\,U = \lambda\,U$

$$U = U^i + \sum_k f_{\beta,k}\, U^{(\beta,k)} = \left(K - \lambda^{(\beta)} I\right)^+ F + \sum_k f_{\beta,k}\, U^{(\beta,k)}. \tag{8.30}$$

Open-source programs are available in C++ (math.nist.gov/tnt/) and in FORTRAN (www.netlib.org/lapack/) that can solve such equations for the $N+1$ eigenvalues $\lambda^{(\alpha)}$ and eigenvectors $U^{(\alpha)}$ and for the inverse K^{-1} for $N = 100$, 1000, 10,000, and so forth in milliseconds on a PC.

8.6 Integral Transformations

Integral transformations (Courant and Hilbert, 1955, chap. VII) help us solve linear homogeneous differential equations like

$$L\,u + c\,u = 0 \tag{8.31}$$

in which L is a linear operator involving derivatives of $u(z)$ with respect to its complex argument $z = x + iy$ and c is a constant. We choose a **kernel** $K(z, w)$ analytic in both variables and write $u(z)$ as an integral along a contour in the complex w-plane weighted by an unknown function $v(w)$

$$u(z) = \int_C K(z, w)\, v(w)\, dw. \tag{8.32}$$

If the differential operator L commutes with the contour integration as it usually would, then our differential equation (8.31) is

$$\int_C [L\,K(z, w) + c\,K(z, w)]\; v(w)\, dw = 0. \tag{8.33}$$

The next step is to find a linear operator M that acting on $K(z, w)$ with w-derivatives (but no z-derivatives) gives L acting on $K(z, w)$

$$M\,K(z,w) = L\,K(z,w). \tag{8.34}$$

We then get an integral equation

$$\int_C [M\,K(z,w) + c\,K(z,w)]\; v(w)\,dw = 0 \tag{8.35}$$

involving w-derivatives which we can integrate by parts. We choose the contour C so that the resulting boundary terms vanish. By using our freedom to pick the kernel and the contour, we often can make the resulting differential equation for v simpler than the one (8.31) we started with.

Example 8.1 (Fourier, Laplace, and Euler kernels) The kernel $K(z,w) = \exp(izw)$ leads to the Fourier transform (Chapter 4)

$$u(z) = \int_{-\infty}^{\infty} e^{izw}\; v(w)\,dw. \tag{8.36}$$

The kernel $K(z,w) = \exp(-zw)$ gives us the Laplace transform (Section 4.9).

$$u(z) = \int_0^{\infty} e^{-zw}\; v(w)\,dw. \tag{8.37}$$

Euler's kernel $K(z,w) = (z-w)^a$ occurs in applications of Cauchy's integral theorem (6.34) and integral formula (6.44).

Example 8.2 (Bessel functions) The differential operator L for Bessel's equation (7.355)

$$z^2\,u'' + z\,u' + z^2\,u - \lambda^2\,u = 0 \tag{8.38}$$

is

$$L = z^2\,\frac{d^2}{dz^2} + z\,\frac{d}{dz} + z^2 \tag{8.39}$$

and the constant c is $-\lambda^2$. We choose $M = -d^2/dw^2$ and seek a suitable (8.34) kernel K that satisfies $M\,K = L\,K$

$$-K_{ww} = z^2\,K_{zz} + z\,K_z + z^2\,K \tag{8.40}$$

in which subscripts indicate differentiation as in (2.7). The kernel

$$K(z,w) = e^{\pm iz\sin w} \tag{8.41}$$

is a solution of (8.40) that is entire in both variables (Exercise 8.3). In terms of it, our integral equation (8.35) is

$$\int_C \left[K_{ww}(z,w) + \lambda^2\,K(z,w)\right]\, v(w)\,dw = 0. \tag{8.42}$$

We now integrate by parts once

$$\int_C \left[-K_w\, v' + \lambda^2\, K\, v + \frac{dK_w\, v}{dw} \right] dw \tag{8.43}$$

and then again

$$\int_C \left[K\, \left(v'' + \lambda^2\, v \right) + \frac{d(K_w\, v - K\, v')}{dw} \right] dw. \tag{8.44}$$

If we choose the contour so that $K_w\, v - K\, v'$ vanishes at both ends, then the unknown function v need only satisfy the differential equation

$$v'' + \lambda^2\, v = 0 \tag{8.45}$$

which is much simpler than Bessel's equation (7.355). The solution $v(w) = \exp(i\lambda w)$ is an entire function of w for every complex λ. The contour integral (8.32) now gives us Bessel's function as the integral transform

$$u(z) = \int_C K(z, w)\, v(w)\, dw = \int_C e^{\pm iz \sin w}\, e^{i\lambda w}\, dw. \tag{8.46}$$

For $\text{Re}(z) > 0$ and any complex λ, the contour C_1 that runs from $-i\infty$ to the origin $w = 0$, then to $w = -\pi$, and finally up to $-\pi + i\infty$ has $K_w\, v - K\, v' = 0$ at its ends (Exercise 8.4) provided we use the minus sign in the exponential. The function defined by this choice

$$H_\lambda^{(1)}(z) = -\frac{1}{\pi} \int_{C_1} e^{-iz \sin w + i\lambda w}\, dw \tag{8.47}$$

is the **first Hankel function** (Hermann Hankel, 1839–1873). The **second Hankel function** is defined for $\text{Re}(z) > 0$ and any complex λ by a contour C_2 that runs from $\pi + i\infty$ to $w = \pi$, then to $w = 0$, and lastly to $-i\infty$

$$H_\lambda^{(2)}(z) = -\frac{1}{\pi} \int_{C_2} e^{-iz \sin w + i\lambda w}\, dw. \tag{8.48}$$

Because the integrand $\exp(-iz \sin w + i\lambda w)$ is an entire function of z and w, one may deform the contours C_1 and C_2 and analytically continue the Hankel functions beyond the right half-plane (Courant and Hilbert, 1955, chap. VII). One may verify (Exercise 8.5) that the Hankel functions are related by complex conjugation

$$H_\lambda^{(1)}(z) = H_\lambda^{(2)*}(z) \tag{8.49}$$

when both $z > 0$ and λ are real.

Exercises

8.1 Show that

$$\int_a^x dz \int_a^z dy\, f(y) = \int_a^x (x - y)\, f(y)\, dy. \tag{8.50}$$

Hint: differentiate both sides with respect to x.

8.2 Use this identity (8.50) to integrate $u'' = pu' + qu + r$ and derive Equations (8.5), $k(x, y) = p(y) + (x - y)\left[q(y) - p'(y)\right]$, and (8.6).

8.3 Show that the kernel $K(z, w) = \exp(\pm iz \sin w)$ satisfies the differential equation (8.40).

8.4 Show that for $\operatorname{Re} z > 0$ and arbitrary complex λ, the boundary terms in the integral (8.44) vanish for the two contours C_1 and C_2 that define the two Hankel functions.

8.5 Show that the Hankel functions are related by complex conjugation (8.49) when both $z > 0$ and λ are real.

9

Legendre Polynomials and Spherical Harmonics

9.1 Legendre's Polynomials

The monomials x^n span the space of functions $f(x)$ that have power-series expansions on an interval about the origin

$$f(x) = \sum_{n=0}^{\infty} c_n\, x^n = \sum_{n=0}^{\infty} \frac{f^{(n)}(0)}{n!}\, x^n. \tag{9.1}$$

They are complete but not orthogonal or normalized. We can make them into real, **orthogonal** polynomials $P_n(x)$ of degree n on the interval $[-1, 1]$

$$(P_n, P_m) = \int_{-1}^{1} P_n(x)\, P_m(x)\, dx = 0 \qquad n \neq m \tag{9.2}$$

by requiring that each $P_n(x)$ be orthogonal to all monomials x^m for $m < n$

$$\int_{-1}^{1} P_n(x)\, x^m\, dx = 0 \qquad m < n. \tag{9.3}$$

If we also impose the **normalization** condition

$$P_n(1) = 1 \tag{9.4}$$

then they are unique and are the **Legendre polynomials** as in Fig. 9.1 (Adrien-Marie Legendre, 1752–1833).

The coefficients a_k of the nth Legendre polynomial

$$P_n(x) = a_0 + a_1 x + \cdots + a_n x^n \tag{9.5}$$

must satisfy (Exercise 9.3) the n conditions (9.3) of orthogonality

$$\int_{-1}^{1} P_n(x)\, x^m\, dx = \sum_{k=0}^{n} \frac{1-(-1)^{m+k+1}}{m+k+1}\, a_k = 0 \quad \text{for} \quad 0 \le m < n \tag{9.6}$$

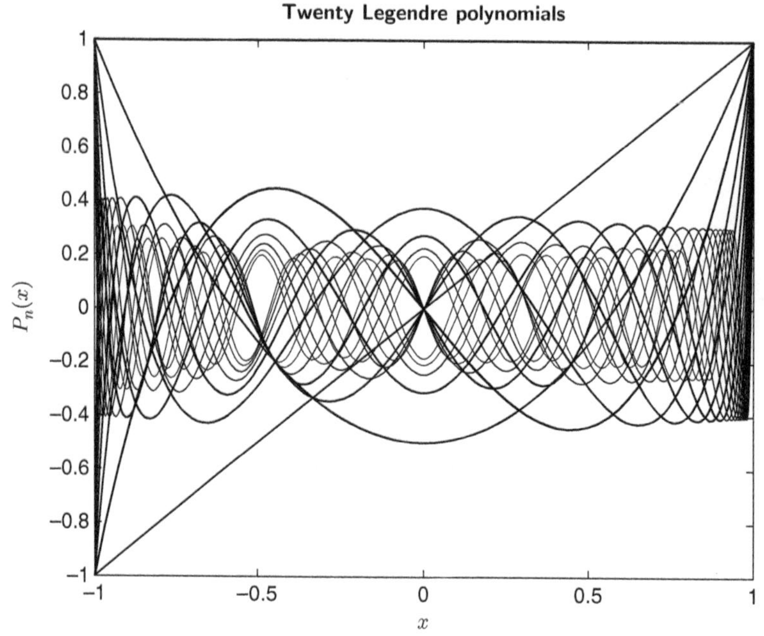

Figure 9.1 The first 20 Legendre polynomials in successively finer linewidths. The straight lines are $P_0(x) = 1$ and $P_1(x) = x$. Matlab scripts for this chapter's figures of are in Legendre_polynomials_and_spherical_harmonics at `github.com/kevinecahill`.

and the normalization condition (9.4)

$$P_n(1) = a_0 + a_1 + \cdots + a_n = 1. \tag{9.7}$$

Example 9.1 (Building the Legendre polynomials) Conditions (9.6) and (9.7) give $P_0(x) = 1$ and $P_1(x) = x$. To make $P_2(x)$, we set $n = 2$ in the orthogonality condition (9.6) and find $2a_0 + 2a_2/3 = 0$ for $m = 0$, and $2a_1/3 = 0$ for $m = 1$. The normalization condition (9.7) then says that $a_0 + a_1 + a_2 = 1$. These three equations give $P_2(x) = (3x^2 - 1)/2$. Similarly, one finds $P_3(x) = (5x^3 - 3x)/2$ and $P_4(x) = (35x^4 - 30x^2 + 3)/8$.

9.2 The Rodrigues Formula

Perhaps the easiest way to compute the Legendre polynomials is to apply Leibniz's rule (5.49) to the **Rodrigues formula**

$$P_n(x) = \frac{1}{2^n n!} \frac{d^n (x^2 - 1)^n}{dx^n} \tag{9.8}$$

which leads to (Exercise 9.5)

$$P_n(x) = \frac{1}{2^n} \sum_{k=0}^{n} \binom{n}{k}^2 (x-1)^{n-k}(x+1)^k. \tag{9.9}$$

This formula at $x = 1$ is

$$P_n(1) = \frac{1}{2^n} \sum_{k=0}^{n} \binom{n}{k}^2 0^{n-k} 2^k = \frac{1}{2^n} \binom{n}{n}^2 2^n = 1 \tag{9.10}$$

which shows that Rodrigues got the normalization right (Benjamin Rodrigues, 1795–1851).

Example 9.2 (Using Rodrigues's formula) By (9.8) or (9.9) and with more effort, one finds

$$P_5(x) = \frac{1}{2^5 5!} \frac{d^5(x^2-1)^5}{dx^5} = \frac{1}{8}\left(63x^5 - 70x^3 + 15x\right) \tag{9.11}$$

$$P_6(x) = \frac{1}{2^6} \sum_{k=0}^{6} \binom{6}{k}^2 (x-1)^{6-k}(x+1)^k = \frac{1}{16}\left(231x^6 - 315x^4 + 105x^2 - 5\right)$$

$$P_7(x) = \frac{(x-1)^7}{2^7} \sum_{k=0}^{7} \binom{7}{k}^2 \left(\frac{x+1}{x-1}\right)^k = \frac{1}{16}\left(429x^7 - 693x^5 + 315x^3 - 35x\right)$$

In Matlab, mfun('P',n,x) returns the numerical value of $P_n(x)$.

To check that the polynomial $P_n(x)$ generated by Rodrigues's formula (9.8) is orthogonal to x^m for $m < n$, we integrate $x^m\, P_n(x)$ by parts n times and drop all the surface terms (which vanish because $x^2 - 1$ is zero at $x = \pm 1$)

$$\begin{aligned} \int_{-1}^{1} x^m\, P_n(x)\, dx &= \frac{1}{2^n n!} \int_{-1}^{1} x^m \frac{d^n}{dx^n}(x^2-1)^n\, dx \\ &= \frac{(-1)^n}{2^n n!} \int_{-1}^{1} (x^2-1)^n \frac{d^n x^m}{dx^n}\, dx = 0 \quad \text{for} \quad n > m. \end{aligned} \tag{9.12}$$

Thus the polynomial $P_n(x)$ generated by Rodrigues's formula (9.8) satisfies the orthogonality condition (9.3). It also satisfies the normalization condition (9.4) as shown by (9.10). The Rodrigues formula does generate Legendre's polynomials.

One may show (Exercises 9.9, 9.10, and 9.11) that the inner product of two Legendre polynomials is

$$\int_{-1}^{1} P_n(x)\, P_m(x)\, dx = \frac{2}{2n+1}\, \delta_{nm}. \tag{9.13}$$

9.3 Generating Function for Legendre Polynomials

In the expansion

$$g(t,x) = \left(1 - 2xt + t^2\right)^{-1/2} = \sum_{n=0}^{\infty} p_n(x)\, t^n \tag{9.14}$$

the coefficient $p_n(x)$ is the nth partial derivative of $g(t,x)$

$$p_n(x) = \frac{\partial^n}{\partial t^n} \left(1 - 2xt + t^2\right)^{-1/2} \Big|_{t=0} \tag{9.15}$$

and is a function of x alone. Explicit calculation shows that it is a polynomial of degree n.

To identify these polynomials $p_n(x)$, we apply the logarithmic series (5.103) to the integral formula

$$\begin{aligned} \int_{-1}^{1} g(t,x)\, g(v,x)\, dx &= \int_{-1}^{1} \frac{dx}{\sqrt{1-2xt+t^2}\sqrt{1-2xv+v^2}} \\ &= \frac{1}{\sqrt{tv}} \ln \frac{1+\sqrt{tv}}{1-\sqrt{tv}} = \sum_{k=0}^{\infty} \frac{2}{2k+1} (tv)^k. \end{aligned} \tag{9.16}$$

The same integral of $g(t,x)\, g(v,x)$ over the interval $-1 \le x \le 1$ also is

$$\int_{-1}^{1} g(t,x)\, g(v,x)\, dx = \int_{-1}^{1} \sum_{n,m=0}^{\infty} p_n(x)\, p_m(x)\, t^n\, v^m\, dx = \sum_{k=0}^{\infty} \frac{2}{2k+1} (tv)^k. \tag{9.17}$$

Equating the coefficients of $t^n v^m$ in the second and third terms of this equation, we see that the polynomials $p_n(x)$ satisfy the inner product rule (9.13) obeyed by the Legendre polynomials

$$\int_{-1}^{1} p_n(x)\, p_m(x)\, dx = \frac{2}{2n+1}\, \delta_{n,m} \tag{9.18}$$

which ensures their orthogonality. Next, setting $x = 1$ in the definition (9.14) of $g(t,x)$, we get from (5.11)

$$\frac{1}{1-t} = \sum_{n=0}^{\infty} t^n = \sum_{n=0}^{\infty} p_n(1)\, t^n \tag{9.19}$$

which says that $p_n(1) = 1$ for all nonnegative integers $n = 0, 1, 2$, and so forth. The polynomials $p_n(x)$ are therefore the Legendre polynomials $P_n(x)$, and the function $g(t,x)$ is their **generating function**

$$\frac{1}{\sqrt{1-2xt+t^2}} = \sum_{n=0}^{\infty} t^n\, P_n(x). \tag{9.20}$$

Example 9.3 (Green's function for Poisson's equation) The Green's function (4.113) for the laplacian is

$$G(\boldsymbol{R}-\boldsymbol{r}) = \frac{1}{4\pi|\boldsymbol{R}-\boldsymbol{r}|} = \frac{1}{4\pi\sqrt{R^2 - 2\boldsymbol{R}\cdot\boldsymbol{r} + r^2}} \tag{9.21}$$

in which $R = |\boldsymbol{R}|$ and $r = |\boldsymbol{r}|$. It occurs throughout physics and satisfies

$$-\nabla^2 G(\boldsymbol{R}-\boldsymbol{r}) = \delta^{(3)}(\boldsymbol{R}-\boldsymbol{r}) \tag{9.22}$$

where the derivatives can act on $\boldsymbol{R}$ or on $\boldsymbol{r}$.

We set $x = \cos\theta = \boldsymbol{R}\cdot\boldsymbol{r}/rR$ and $t = r/R$, and then factor out $1/R$

$$\begin{aligned}\frac{1}{|\boldsymbol{R}-\boldsymbol{r}|} &= \frac{1}{\sqrt{R^2 - 2Rr\cos\theta + r^2}} \\ &= \frac{1}{R}\frac{1}{\sqrt{1-2(r/R)x + (r/R)^2}} \\ &= \frac{1}{R}\frac{1}{\sqrt{1-2xt+t^2}} = \frac{1}{R}\,g(t,x)\end{aligned} \tag{9.23}$$

With $t = r/R$ and $x = \cos\theta$, the series (9.20) is the well-known expansion

$$\frac{1}{|\boldsymbol{R}-\boldsymbol{r}|} = \frac{1}{R}\sum_{n=0}^{\infty}\left(\frac{r}{R}\right)^n P_n(\cos\theta) \tag{9.24}$$

of the Green's function $G(\boldsymbol{R}-\boldsymbol{r}) = 1/4\pi|\boldsymbol{R}-\boldsymbol{r}| = g(t,x)/4\pi R$.

9.4 Legendre's Differential Equation

Apart from the prefactor $1/(2^n n!)$, Rodrigues's formula (9.8) expresses the Legendre polynomial $P_n(x)$ as the nth derivative $u^{(n)}$ of $u = (x^2-1)^n$. Since $u' = 2nx(x^2-1)^{n-1}$, the function u satisfies $(x^2-1)u' = 2nxu$. Using Leibniz's rule (5.49) to differentiate $(n+1)$ times both sides of this equation $2nxu = (x^2-1)u'$, we find

$$(2nxu)^{(n+1)} = 2n\sum_{k=0}^{n+1}\binom{n+1}{k}x^{(k)}\,u^{(n+1-k)} = 2n\left(x\,u^{(n+1)} + (n+1)\,u^{(n)}\right) \tag{9.25}$$

and

$$\begin{aligned}\left((x^2-1)u'\right)^{(n+1)} &= \sum_{k=0}^{n+1}\binom{n+1}{k}(x^2-1)^{(k)}\,u^{(n+2-k)} \\ &= (x^2-1)u^{(n+2)} + 2(n+1)xu^{(n+1)} + n(n+1)u^{(n)}.\end{aligned} \tag{9.26}$$

Equating the two and setting $u^{(n)} = 2^n n! P_n$, we get

$$- \left[(1 - x^2)\, P_n' \right]' = n(n+1)\, P_n \tag{9.27}$$

which is Legendre's equation in self-adjoint form.

The differential operator

$$L = -\frac{d}{dx}\, p(x)\, \frac{d}{dx} = -\frac{d}{dx}\, (1 - x^2)\, \frac{d}{dx} \tag{9.28}$$

is formally self adjoint and the real function $p(x) = 1 - x^2$ is positive on the open interval $(-1, 1)$ and vanishes at $x = \pm 1$, so Legendre's differential operator L, his differential equation (9.27), and the domain D of functions that are twice differentiable on the interval $[-1, 1]$ form a singular self-adjoint system (Example 7.43). The Legendre polynomial $P_n(x)$ is an eigenfunction of L with eigenvalue $n(n+1)$ and weight function $w(x) = 1$. The orthogonality relation (7.364) tells us that eigenfunctions of a self-adjoint differential operator that have different eigenvalues are orthogonal on the interval $[-1, 1]$ with respect to the weight function $w(x)$. Thus $P_n(x)$ and $P_m(x)$ are orthogonal for $n \neq m$; their normalization

$$\int_{-1}^{1} P_n(x)\, P_m(x)\, dx = \frac{2}{2n+1}\, \delta_{nm} \tag{9.29}$$

follows from the Rodrigues formula (9.13).

The eigenvalues $n(n+1)$ increase without limit, and so the argument of Section 7.38 shows that the eigenfunctions $P_n(x)$ are complete. Since the weight function of the Legendre polynomials is unity $w(x) = 1$, the expansion (7.413) of Dirac's delta function in terms of the Legendre polynomials is

$$\delta(x - x') = \sum_{n=0}^{\infty} \frac{2n+1}{2}\, P_n(x)\, P_n(x'). \tag{9.30}$$

By multiplying both sides of this equation by any twice-differentiable function $f(x')$ and integrating over x' from -1 to 1, we get its Fourier–Legendre expansion

$$f(x) = \int_{-1}^{1} \delta(x - x')\, f(x')\, dx' = \sum_{n=0}^{\infty} \frac{2n+1}{2}\, P_n(x) \int_{-1}^{1} P_n(x')\, f(x')\, dx'. \tag{9.31}$$

Figure 9.2 displays for $x' = 0$ the sum of the first 10,001 terms of the series (9.30); the sum approximates a delta function $\delta(x)$ suitable for functions that vary little for $\Delta x \lesssim 0.01$.

Changing variables to $\cos\theta = x$, we have $(1 - x^2) = \sin^2\theta$ and

$$\frac{d}{d\theta} = \frac{d\cos\theta}{d\theta}\, \frac{d}{dx} = -\sin\theta\, \frac{d}{dx} \tag{9.32}$$

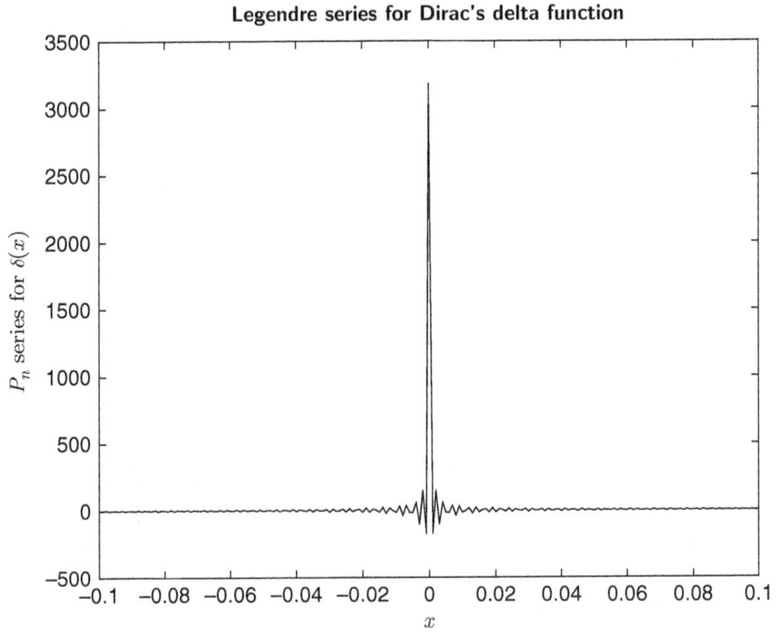

Figure 9.2 The sum of the first 10,001 terms of the expansion (9.30) of Dirac's delta function $\delta(x)$ as sums of products of Legendre polynomials.

so that

$$\frac{d}{dx} = -\frac{1}{\sin\theta}\frac{d}{d\theta}. \tag{9.33}$$

In these variables, Legendre's equation (9.27) is

$$\frac{1}{\sin\theta}\frac{d}{d\theta}\left[\sin\theta\,\frac{d}{d\theta}P_n(\cos\theta)\right] + n(n+1)\,P_n(\cos\theta) = 0. \tag{9.34}$$

9.5 Recurrence Relations

The t-derivative of the generating function $g(t,x) = 1/\sqrt{1-2xt+t^2}$ is

$$\frac{\partial g(t,x)}{\partial t} = \frac{x-t}{(1-2xt+t^2)^{3/2}} = \sum_{n=1}^{\infty} n\,P_n(x)\,t^{n-1} \tag{9.35}$$

which we can rewrite as

$$(1-2xt+t^2)\sum_{n=1}^{\infty} n\,P_n(x)\,t^{n-1} = (x-t)\,g(t,x) = (x-t)\sum_{n=0}^{\infty} P_n(x)\,t^n. \tag{9.36}$$

By equating the coefficients of t^n in the final sums of these equations (9.35 and 9.36), we get the **recurrence relation**

$$P_{n+1}(x) = \frac{1}{n+1}\left[(2n+1)\,x\,P_n(x) - n\,P_{n-1}(x)\right]. \tag{9.37}$$

Example 9.4 (Building the Legendre polynomials) Since $P_1(x) = x$ and $P_0(x) = 1$, this recurrence relation for $n = 1$ gives

$$P_2(x) = \frac{1}{2}\left[3\,x\,P_1(x) - P_0(x)\right] = \frac{1}{2}\left(3x^2 - 1\right). \tag{9.38}$$

Similarly for $n = 2$ it gives

$$P_3(x) = \tfrac{1}{3}\left[5\,x\,P_2(x) - 2\,P_1(x)\right] = \frac{1}{2}(5x^2 - 3x). \tag{9.39}$$

It builds Legendre polynomials faster than Rodrigues's formula (9.8).

The x-derivative of the generating function is

$$\frac{\partial g(t,x)}{\partial x} = \frac{t}{(1-2xt+t^2)^{3/2}} = \sum_{n=1}^{\infty} P_n'(x)\,t^n \tag{9.40}$$

which we can rewrite as

$$(1-2xt+t^2)\sum_{n=1}^{\infty} P_n'(x)\,t^n = t\,g(t,x) = \sum_{n=0}^{\infty} P_n(x)\,t^{n+1}. \tag{9.41}$$

Equating coefficients of t^n, we have

$$P_{n+1}'(x) + P_{n-1}'(x) = 2x\,P_n'(x) + P_n(x). \tag{9.42}$$

By differentiating the recurrence relation (9.37) and combining it with this last equation, we get

$$P_{n+1}'(x) - P_{n-1}'(x) = (2n+1)\,P_n(x). \tag{9.43}$$

The last two recurrence relations (9.42 and 9.43) lead to several more:

$$P_{n+1}'(x) = (n+1)\,P_n(x) + x\,P_n'(x) \tag{9.44}$$
$$P_{n-1}'(x) = -n\,P_n(x) + x\,P_n'(x) \tag{9.45}$$
$$(1-x^2)\,P_n'(x) = n\,P_{n-1}(x) - nx\,P_n(x) \tag{9.46}$$
$$(1-x^2)\,P_n'(x) = (n+1)x\,P_n(x) - (n+1)\,P_{n+1}(x). \tag{9.47}$$

By differentiating (9.47) and using (9.44) for P_{n+1}', we recover Legendre's equation $-\left[(1-x^2)P_n'\right]' = n(n+1)P_n$.

9.6 Special Values of Legendre Polynomials

At $x = -1$, the generating function is

$$g(t,-1) = \left(1 + t^2 + 2t\right)^{-1/2} = \frac{1}{1+t} = \sum_{n=0}^{\infty}(-t)^n = \sum_{n=0}^{\infty} P_n(-1)\, t^n \tag{9.48}$$

which implies that

$$P_n(-1) = (-1)^n. \tag{9.49}$$

Similarly, $g(t, 1)$ leads to $P_n(1) = 1$ which is the condition (9.4).

The generating function $g(t, x)$ is even under the reflection of both independent variables, so

$$g(t,x) = \sum_{n=0}^{\infty} t^n\, P_n(x) = \sum_{n=0}^{\infty}(-t)^n\, P_n(-x) = g(-t,-x) \tag{9.50}$$

which implies that

$$P_n(-x) = (-1)^n\, P_n(x) \quad \text{whence} \quad P_{2n+1}(0) = 0. \tag{9.51}$$

With more effort, one can show that

$$P_{2n}(0) = (-1)^n\, \frac{(2n-1)!!}{(2n)!!} \quad \text{and that} \quad |P_n(x)| \le 1. \tag{9.52}$$

Our formulas (9.8 and 9.9) for the nth Legendre polynomial $P_n(x)$ involve huge factorials when n is big. NIST's webpage (dlmf.nist.gov/14.15#iii) offers the useful approximation for $|x| < 1$ as $n \to \infty$

$$P_n(x) \approx \frac{\sqrt{\arccos(x)}}{(1-x^2)^{1/4}}\, J_0\left((n+1/2)\arccos(x)\right) \tag{9.53}$$

plus terms of order $1/n$ in which J_0 is a Bessel function of the first kind (Section 10.1). Applying to this approximation the further approximation (10.3), we get for $|x| < 1$ as $n \to \infty$

$$P_n(x) \approx \left(\frac{2}{\pi(n+\frac{1}{2})\sqrt{1-x^2}}\right)^{1/2} \cos\left((n+\frac{1}{2})\arccos(x) - \frac{1}{2}n\pi - \tfrac{1}{4}\pi\right). \tag{9.54}$$

9.7 Schlaefli's Integral

Schlaefli used Cauchy's integral formula (6.44) and Rodrigues's formula

$$P_n(x) = \frac{1}{2^n\, n!}\left(\frac{d}{dx}\right)^n (x^2-1)^n \tag{9.55}$$

to express $P_n(x)$ as a counterclockwise contour integral around the point x

$$P_n(x) = \frac{1}{2^n \, 2\pi i} \oint \frac{(z'^2 - 1)^n}{(z' - x)^{n+1}} \, dz'. \tag{9.56}$$

9.8 Orthogonal Polynomials

Rodrigues's formula (9.8) generates other families of orthogonal polynomials. The nth order polynomials R_n in which the e_n are constants

$$R_n(x) = \frac{1}{e_n w(x)} \frac{d^n}{dx^n} \left[w(x) \, q^n(x) \right] \tag{9.57}$$

are orthogonal on the interval from a to b with weight function $w(x)$

$$\int_a^b R_n(x) \, R_k(x) \, w(x) \, dx = N_n \, \delta_{nk} \tag{9.58}$$

as long as $q(x)$ vanishes at a and b (Exercise 9.8)

$$q(a) = q(b) = 0. \tag{9.59}$$

Example 9.5 (Jacobi's polynomials) The choice $q(x) = (x^2 - 1)$ with weight function $w(x) = (1 - x)^\alpha (1 + x)^\beta$ and normalization $e_n = 2^n n!$ leads for $\alpha > -1$ and $\beta > -1$ to the Jacobi polynomials

$$P_n^{(\alpha,\beta)}(x) = \frac{1}{2^n n!} (1 - x)^{-\alpha} (1 + x)^{-\beta} \frac{d^n}{dx^n} \left[(1 - x)^\alpha (1 + x)^\beta \, (x^2 - 1)^n \right] \tag{9.60}$$

which are orthogonal on $[-1, 1]$

$$\int_{-1}^1 P_n^{(\alpha,\beta)}(x) \, P_m^{(\alpha,\beta)}(x) \, w(x) \, dx = \frac{2^{\alpha+\beta+1} \Gamma(n + \alpha + 1) \, \Gamma(n + \beta + 1)}{(2n + \alpha + \beta + 1) \, \Gamma(n + \alpha + \beta + 1)} \, \delta_{nm} \tag{9.61}$$

and satisfy the normalization condition

$$P_n^{(\alpha,\beta)}(1) = \binom{n + \alpha}{n} \tag{9.62}$$

and the differential equation

$$(1 - x^2) \, y'' + (\beta - \alpha - (\alpha + \beta + 2)x) \, y' + n(n + \alpha + \beta + 1) \, y = 0. \tag{9.63}$$

In terms of $R(x, y) = \sqrt{1 - 2xy + y^2}$, their generating function is

$$2^{\alpha+\beta} (1 - y + R(x, y))^{-\alpha} (1 + w + R(x, y))^{-\beta} / R(x, y) = \sum_{n=0}^{\infty} P_n^{(\alpha,\beta)}(x) y^n. \tag{9.64}$$

When $\alpha = \beta$, they are the Gegenbauer polynomials, which for $\alpha = \beta = \pm 1/2$ are the Chebyshev polynomials (of the second and first kind, respectively). For $\alpha = \beta = 0$, they are Legendre's polynomials. The recursion relations

$$\begin{aligned} T_0(x) &= 1 & U_0(x) &= 1 \\ T_1(x) &= x & U_1(x) &= 2x \\ T_{n+1}(x) &= 2xT_n(x) - T_{n-1}(x) & U_{n+1}(x) &= 2xU_n(x) - U_{n-1}(x) \end{aligned} \tag{9.65}$$

define the Chebyshev polynomials of the first $T_n(x)$ and second $U_n(x)$ kinds.

Example 9.6 (Hermite's polynomials) The choice $q(x) = 1$ with weight function $w(x) = \exp(-x^2)$ leads to the Hermite polynomials

$$H_n(x) = (-1)^n e^{x^2} \frac{d^n}{dx^n} e^{-x^2} = e^{x^2/2} \left(x - \frac{d}{dx} \right)^n e^{-x^2/2} = 2^n \, e^{-D^2/4} \, x^n \tag{9.66}$$

where $D = d/dx$ is the x-derivative. They are orthogonal on the real line

$$\int_{-\infty}^{\infty} H_n(x) \, H_m(x) \, e^{-x^2} \, dx = \sqrt{\pi} \; 2^n \; n! \, \delta_{nm} \tag{9.67}$$

and satisfy the differential equation

$$y'' - 2 \, x \, y' + 2 \, n \, y = 0. \tag{9.68}$$

Their generating function is

$$e^{2xy - y^2} = \sum_{n=0}^{\infty} H_n(x) \, \frac{y^n}{n!}. \tag{9.69}$$

The nth excited state of the harmonic oscillator of mass m and angular frequency ω is proportional to $H_n(x)$ in which $x = \sqrt{m\omega/\hbar} \; q$ is the dimensionless position of the oscillator (Section 3.12).

Example 9.7 (Laguerre's polynomials) The choices $q(x) = x$ and weight function $w(x) = x^\alpha e^{-x}$ lead to the generalized Laguerre polynomials

$$L_n^{(\alpha)}(x) = \frac{e^x}{n! \, x^\alpha} \frac{d^n}{dx^n} \left(e^{-x} x^{n+\alpha} \right). \tag{9.70}$$

They are orthogonal on the interval $[0, \infty)$

$$\int_0^{\infty} L_n^{(\alpha)}(x) \, L_m^{(\alpha)}(x) \, x^\alpha e^{-x} \, dx = \frac{\Gamma(n + \alpha + 1)}{n!} \, \delta_{n,m} \tag{9.71}$$

and satisfy the differential equation

$$x \, y'' + (\alpha + 1 - x) \, y' + n \, y = 0. \tag{9.72}$$

Their generating function is

$$(1-y)^{-\alpha-1}\exp\left(\frac{x\,y}{y-1}\right)=\sum_{n=0}^{\infty}L_n^{(\alpha)}(x)\,y^n. \tag{9.73}$$

The radial wave function for the state of the nonrelativistic hydrogen atom with quantum numbers n and ℓ is $\rho^\ell\,L_{n-\ell-1}^{2\ell+1}(\rho)\,e^{-\rho/2}$ in which $\rho = 2r/na_0$ and $a_0 = 4\pi\epsilon_0\hbar^2/m_e e^2$ is the Bohr radius.

9.9 Azimuthally Symmetric Laplacians

We saw in Section 7.3 that the laplacian $\triangle = \nabla\cdot\nabla$ separates in spherical coordinates r, θ, ϕ. A system with no dependence on the angle ϕ is said to have **azimuthal symmetry**. An azimuthally symmetric function

$$f(r,\theta,\phi) = R_{k,\ell}(r)\,\Theta_\ell(\theta) \tag{9.74}$$

will be a solution of Helmholtz's equation

$$-\triangle f = k^2 f \tag{9.75}$$

if the functions $R_{k,\ell}(r)$ and $\Theta_\ell(\theta)$ satisfy

$$\frac{1}{r^2}\frac{d}{dr}\left(r^2\frac{dR_{k,\ell}}{dr}\right)+\left[k^2-\frac{\ell(\ell+1)}{r^2}\right]R_{k,\ell}=0 \tag{9.76}$$

for a nonnegative integer ℓ and Legendre's equation (9.34)

$$\frac{1}{\sin\theta}\frac{d}{d\theta}\left(\sin\theta\frac{d\Theta_\ell}{d\theta}\right)+\ell(\ell+1)\Theta_\ell=0 \tag{9.77}$$

so that we may set $\Theta_\ell(\theta) = P_\ell(\cos\theta)$. For $k > 0$, the solutions of the radial equation (9.76) that are finite at $r = 0$ are the spherical Bessel functions

$$R_{k,\ell}(r) = j_\ell(kr) \tag{9.78}$$

which are given by Rayleigh's formula (10.71)

$$j_\ell(x) = (-1)^\ell\,x^\ell\left(\frac{d}{xdx}\right)^\ell\left(\frac{\sin x}{x}\right). \tag{9.79}$$

So the general azimuthally symmetric solution of the Helmholtz equation (9.75) that is finite at $r = 0$ is

$$f(r,\theta) = \sum_{\ell=0}^{\infty} a_{k,\ell}\,j_\ell(kr)\,P_\ell(\cos\theta) \tag{9.80}$$

in which the $a_{k,\ell}$ are constants. If the solution can be infinite at the origin, then the Neumann functions

$$n_\ell(x) = -(-1)^\ell \, x^\ell \left(\frac{d}{x dx}\right)^\ell \left(\frac{\cos x}{x}\right) \tag{9.81}$$

must be included, and the general solution then is

$$f(r,\theta) = \sum_{\ell=0}^{\infty} \left[a_{k,\ell}\, j_\ell(kr) + b_{k,\ell}\, n_\ell(kr)\right] P_\ell(\cos\theta) \tag{9.82}$$

in which the $a_{k,\ell}$ and $b_{k,\ell}$ are constants.

When $k = 0$, Helmholtz's equation reduces to Laplace's equation

$$\triangle f = 0 \tag{9.83}$$

which describes the Coulomb-gauge electrostatic potential in the absence of charge and the Newtonian gravitational potential in the absence of mass. Now the radial equation is simply

$$\frac{d}{dr}\left(r^2 \frac{dR_\ell}{dr}\right) = \ell(\ell+1) R_\ell \tag{9.84}$$

since $k = 0$. Setting

$$R_\ell(r) = r^n \tag{9.85}$$

we get $n(n+1) = \ell(\ell+1)$ so that $n = \ell$ or $n = -(\ell+1)$. Thus the general solution to (9.83) is

$$f(r,\theta) = \sum_{\ell=0}^{\infty} \left[a_\ell\, r^\ell + b_\ell\, r^{-\ell-1}\right] P_\ell(\cos\theta). \tag{9.86}$$

If the solution must be finite at $r = 0$, then all the b_ℓ's must vanish.

9.10 Laplace's Equation in Two Dimensions

In Section 7.3, we saw that Helmholtz's equation separates in cylindrical coordinates, and that the equation for $\mathrm{P}(\rho)$ is Bessel's equation (7.25). But if $\alpha = 0$, Helmholtz's equation (7.30) reduces to Laplace's equation $\triangle f = 0$, and if the potential f also is independent of z, then simpler solutions exist. For now $\alpha = 0 = k$, and so if $\Phi_m'' = -m^2 \Phi_m$, then Equation (7.30) becomes

$$\rho \frac{d}{d\rho}\left(\rho \frac{d\mathrm{P}_m}{d\rho}\right) = m^2\, \mathrm{P}_m. \tag{9.87}$$

The function $\Phi(\phi)$ may be taken to be $\Phi(\phi) = \exp(im\phi)$ or a linear combination of $\cos(m\phi)$ and $\sin(m\phi)$. If the whole range of ϕ from 0 to 2π is physically relevant, then $\Phi(\phi)$ must be periodic, and so m must be an integer. To solve the equation (9.87) for P_m, we set $P_m = \rho^n$ and get

$$n^2 \rho^n = m^2 \rho^n \tag{9.88}$$

which says that $n = \pm m$. The general z-independent solution of Laplace's equation in cylindrical coordinates then is

$$f(\rho, \phi) = \sum_{m=0}^{\infty} \left(c_m \rho^m + d_m \rho^{-m}\right) e^{im\phi}. \tag{9.89}$$

9.11 Helmholtz's Equation in Spherical Coordinates

The laplacian $\triangle$ separates in spherical coordinates, as we saw in Section 7.3. Thus a function

$$f(r, \theta, \phi) = R_{k,\ell}(r)\, \Theta_{\ell,m}(\theta)\, \Phi_m(\phi) \tag{9.90}$$

will be a solution of the Helmholtz equation $-\triangle f = k^2 f$ if $R_{k,\ell}$ is a linear combination of the spherical Bessel functions j_ℓ (9.79) and n_ℓ (9.81)

$$R_{k,\ell}(r) = a_{k,\ell}\, j_\ell(kr) + b_{k,\ell}\, n_\ell(kr) \tag{9.91}$$

if $\Phi_m = e^{im\phi}$, and if $\Theta_{\ell,m}$ satisfies the associated Legendre equation

$$\frac{1}{\sin\theta}\frac{d}{d\theta}\left(\sin\theta \frac{d\Theta_{\ell,m}}{d\theta}\right) + \left[\ell(\ell+1) - \frac{m^2}{\sin^2\theta}\right]\Theta_{\ell,m} = 0. \tag{9.92}$$

9.12 Associated Legendre Polynomials

The associated Legendre functions $P_\ell^m(x) \equiv P_{\ell,m}(x)$ are polynomials in $\sin\theta$ and $\cos\theta$. They arise as solutions of the separated θ equation (9.92)

$$\frac{1}{\sin\theta}\frac{d}{d\theta}\left(\sin\theta \frac{dP_{\ell,m}}{d\theta}\right) + \left[\ell(\ell+1) - \frac{m^2}{\sin^2\theta}\right]P_{\ell,m} = 0 \tag{9.93}$$

of the laplacian in spherical coordinates. In terms of $x = \cos\theta$, this self-adjoint ordinary differential equation is

$$\left[(1-x^2)P'_{\ell,m}(x)\right]' + \left[\ell(\ell+1) - \frac{m^2}{1-x^2}\right]P_{\ell,m}(x) = 0. \tag{9.94}$$

The associated Legendre function $P_{\ell,m}(x)$, for $m = 0, 1, 2, \ldots$, is simply related to the mth derivative $P_\ell^{(m)}(x)$

$$P_{\ell,m}(x) \equiv (1-x^2)^{m/2}\, P_\ell^{(m)}(x). \tag{9.95}$$

To see why this function satisfies the differential equation (9.94), we differentiate

$$P_\ell^{(m)}(x) = (1-x^2)^{-m/2} P_{\ell,m}(x) \tag{9.96}$$

twice getting

$$P_\ell^{(m+1)} = (1-x^2)^{-m/2}\left(P'_{\ell,m} + \frac{mx\,P_{\ell,m}}{1-x^2}\right) \tag{9.97}$$

and

$$P_\ell^{(m+2)} = (1-x^2)^{-m/2}\left[P''_{\ell,m} + \frac{2mx\,P'_{\ell,m}}{1-x^2} + \frac{m\,P_{\ell,m}}{1-x^2} + \frac{m(m+2)x^2 P_{\ell,m}}{(1-x^2)^2}\right]. \tag{9.98}$$

Next we use Leibniz's rule (5.49) to differentiate Legendre's equation (9.27)

$$\left[(1-x^2)\,P'_\ell\right]' + \ell(\ell+1)\,P_\ell = 0 \tag{9.99}$$

m times, obtaining

$$(1-x^2)P_\ell^{(m+2)} - 2x(m+1)P_\ell^{(m+1)} + (\ell-m)(\ell+m+1)P_\ell^{(m)} = 0. \tag{9.100}$$

Now we put the formulas for the three derivatives (9.96–9.98) into this equation (9.100) and find that the $P_{\ell,m}(x)$ as defined (9.95) obey the desired differential equation (9.94).

Thus the associated Legendre functions are

$$P_{\ell,m}(x) = (1-x^2)^{m/2}\,P_\ell^{(m)}(x) = (1-x^2)^{m/2}\,\frac{d^m}{dx^m}P_\ell(x). \tag{9.101}$$

They are simple polynomials in $x = \cos\theta$ and $\sqrt{1-x^2} = \sin\theta$

$$P_{\ell,m}(\cos\theta) = \sin^m\theta\,\frac{d^m}{d(\cos\theta)^m}\,P_\ell(\cos\theta). \tag{9.102}$$

One can use Rodrigues's formula (9.8) for the Legendre polynomial $P_\ell(x)$ to write the definition (9.95) of $P_{\ell,m}(x)$ as

$$P_{\ell,m}(x) = \frac{(1-x^2)^{m/2}}{2^\ell \ell!}\,\frac{d^{\ell+m}}{dx^{\ell+m}}\,(x^2-1)^\ell. \tag{9.103}$$

This formula extends the associated Legendre polynomial $P_{\ell,m}(x)$ to negative values of the integer m and also tells us that under parity $P_\ell^m(x)$ changes by $(-1)^{\ell+m}$

$$P_{\ell,m}(-x) = (-1)^{\ell+m}\,P_{\ell,m}(x). \tag{9.104}$$

Rodrigues's formula (9.103) for the associated Legendre function makes sense as long as $\ell+m \geq 0$. This last condition is the requirement in quantum mechanics that m not be less than $-\ell$. And if m exceeds ℓ, then $P_{\ell,m}(x)$ is given by more than 2ℓ derivatives of a polynomial of degree 2ℓ; so $P_{\ell,m}(x) = 0$ if $m > \ell$. This last condition is the requirement in quantum mechanics that m not be greater than ℓ. So we have

$$-\ell \le m \le \ell. \tag{9.105}$$

One may show that

$$P_{\ell,-m}(x) = (-1)^m \frac{(\ell-m)!}{(\ell+m)!} P_{\ell,m}(x). \tag{9.106}$$

In fact, since m occurs only as m^2 in the ordinary differential equation (9.94), $P_{\ell,-m}(x)$ must be proportional to $P_{\ell,m}(x)$.

Under reflections, the parity of $P_{\ell,m}$ is $(-1)^{\ell+m}$, that is,

$$P_{\ell,m}(-x) = (-1)^{\ell+m} P_{\ell,m}(x). \tag{9.107}$$

If $m \ne 0$, then $P_{\ell,m}(x)$ has a power of $\sqrt{1-x^2}$ in it, so

$$P_{\ell,m}(\pm 1) = 0 \quad \text{for} \quad m \ne 0. \tag{9.108}$$

We may consider either $\ell(\ell+1)$ or m^2 as the eigenvalue in the ODE (9.94)

$$\left[(1-x^2)P'_{\ell,m}(x)\right]' + \left[\ell(\ell+1) - \frac{m^2}{1-x^2}\right] P_{\ell,m}(x) = 0. \tag{9.109}$$

If $\ell(\ell+1)$ is the eigenvalue, then the weight function is unity, and since this ODE is self adjoint on the interval $[-1, 1]$ (at the ends of which $p(x) = (1-x^2) = 0$), the eigenfunctions $P_{\ell,m}(x)$ and $P_{\ell',m}(x)$ must be orthogonal on that interval when $\ell \ne \ell'$. The full integral formula is

$$\int_{-1}^{1} P_{\ell,m}(x)\, P_{\ell',m}(x)\, dx = \frac{2}{2\ell+1} \frac{(\ell+m)!}{(\ell-m)!} \delta_{\ell,\ell'}. \tag{9.110}$$

If m^2 for fixed ℓ is the eigenvalue, then the weight function is $1/(1-x^2)$, and the eigenfunctions $P_{\ell,m}(x)$ and $P_{\ell,m'}(x)$ must be orthogonal on $[-1, 1]$ when $m \ne m'$. The full formula is

$$\int_{-1}^{1} P_{\ell,m}(x)\, P_{\ell,m'}(x) \frac{dx}{1-x^2} = \frac{(\ell+m)!}{|m|(\ell-m)!} \delta_{m,m'}. \tag{9.111}$$

9.13 Spherical Harmonics

The spherical harmonic $Y_\ell^m(\theta,\phi) \equiv Y_{\ell,m}(\theta,\phi)$ is the product

$$Y_{\ell,m}(\theta,\phi) = \Theta_{\ell,m}(\theta)\, \Phi_m(\phi) \tag{9.112}$$

in which $\Theta_{\ell,m}(\theta)$ is proportional to the associated Legendre function $P_{\ell,m}$

$$\Theta_{\ell,m}(\theta) = (-1)^m \sqrt{\frac{2\ell+1}{2}\frac{(\ell-m)!}{(\ell+m)!}}\, P_{\ell,m}(\cos\theta) \tag{9.113}$$

and

$$\Phi_m(\phi) = \frac{e^{im\phi}}{\sqrt{2\pi}}. \tag{9.114}$$

The big square root in the definition (9.113) ensures that

$$\int_0^{2\pi} d\phi \int_0^{\pi} \sin\theta\, d\theta \;\, Y^*_{\ell,m}(\theta,\phi)\, Y_{\ell',m'}(\theta,\phi) = \delta_{\ell\ell'}\,\delta_{mm'}. \tag{9.115}$$

In spherical coordinates, the parity transformation

$$\boldsymbol{x}' = -\boldsymbol{x} \tag{9.116}$$

is $r' = r$, $\theta' = \pi - \theta$, and $\phi' = \phi \pm \pi$. So under parity, $\cos\theta' = -\cos\theta$ and $\exp(im\phi') = (-1)^m \exp(im\phi)$. This factor of $(-1)^m$ cancels the m-dependence (9.104) of $P_{\ell,m}(\theta)$ under parity, so that under parity

$$Y_{\ell,m}(\theta',\phi') = Y_{\ell,m}(\pi-\theta,\phi\pm\pi) = (-1)^\ell Y_{\ell,m}(\theta,\phi). \tag{9.117}$$

Thus the parity of the state $|n,\ell,m\rangle$ is $(-1)^\ell$.

The spherical harmonics are complete on the unit sphere. They may be used to expand any smooth function $f(\theta,\phi)$ as

$$f(\theta,\phi) = \sum_{\ell=0}^{\infty}\sum_{m=-\ell}^{\ell} a_{\ell m} Y_{\ell,m}(\theta,\phi). \tag{9.118}$$

The orthonormality relation (9.115) says that the coefficients $a_{\ell m}$ are

$$a_{\ell m} = \int_0^{2\pi} d\phi \int_0^{\pi} \sin\theta\, d\theta \;\, Y^*_{\ell,m}(\theta,\phi)\, f(\theta,\phi). \tag{9.119}$$

Putting the last two equations together, we find

$$f(\theta,\phi) = \int_0^{2\pi} d\phi' \int_0^{\pi} \sin\theta'\, d\theta' \left[\sum_{\ell=0}^{\infty}\sum_{m=-\ell}^{\ell} Y^*_{\ell,m}(\theta',\phi')\, Y_{\ell,m}(\theta,\phi)\right] f(\theta',\phi') \tag{9.120}$$

and so, we may identify the sum within the brackets as an angular delta function

$$\sum_{\ell=0}^{\infty}\sum_{m=-\ell}^{\ell} Y^*_{\ell,m}(\theta',\phi')\, Y_{\ell,m}(\theta,\phi) = \frac{1}{\sin\theta}\,\delta(\theta-\theta')\,\delta(\phi-\phi') \tag{9.121}$$

which sometimes is abbreviated as

$$\sum_{\ell=0}^{\infty} \sum_{m=-\ell}^{\ell} Y^*_{\ell,m}(\Omega')\, Y_{\ell,m}(\Omega) = \delta^{(2)}(\Omega - \Omega'). \qquad (9.122)$$

The spherical-harmonic expansion (9.118) of the Legendre polynomial $P_\ell(\hat{\boldsymbol{n}} \cdot \hat{\boldsymbol{n}}')$ of the cosine $\hat{\boldsymbol{n}} \cdot \hat{\boldsymbol{n}}'$ in which the polar angles of the unit vectors respectively are θ, ϕ and θ', ϕ' is the **addition theorem** (Example 11.21)

$$P_\ell(\hat{\boldsymbol{n}} \cdot \hat{\boldsymbol{n}}') = \frac{4\pi}{2\ell + 1} \sum_{m=-\ell}^{\ell} Y_{\ell,m}(\theta, \phi) Y^*_{\ell,m}(\theta', \phi'). \qquad (9.123)$$

9.14 Cosmic Microwave Background Radiation

Instruments on the Wilkinson Microwave Anisotropy Probe (WMAP) and on the Planck satellite in orbit at the Lagrange point L_2 (in the Earth's shadow, 1.5×10^6 km farther from the Sun) have measured the temperature $T(\theta, \phi)$ of the cosmic microwave background (CMB) radiation as a function of the polar angles θ and ϕ in the sky as shown in Fig. 9.3. This radiation is photons last scattered when the visible universe became transparent at an age of 380,000 years and a temperature (3000 K) cool enough for hydrogen atoms to be stable. This **initial transparency** is usually (and inexplicably) called **recombination**.

Temperature fluctuations of the cosmic microwave background radiation

Figure 9.3 CMB temperature fluctuations over the celestial sphere as measured by the Planck satellite. The average temperature is 2.7255 K. White regions are cooler and black ones warmer by about 300 μK. © ESA and the Planck Collaboration, 2018.

Since the spherical harmonics $Y_{\ell,m}(\theta, \phi)$ are complete on the sphere, we can expand the temperature as

$$T(\theta, \phi) = \sum_{\ell=0}^{\infty} \sum_{m=-\ell}^{\ell} a_{\ell,m} Y_{\ell,m}(\theta, \phi) \tag{9.124}$$

in which the coefficients are by (9.119)

$$a_{\ell,m} = \int_0^{2\pi} d\phi \int_0^{\pi} \sin\theta \, d\theta \;\; Y^*_{\ell,m}(\theta, \phi) \, T(\theta, \phi). \tag{9.125}$$

The average temperature $\overline{T}$ contributes only to $a_{0,0} = \overline{T} = 2.7255$ K. The other coefficients describe the difference $\Delta T(\theta, \phi) = T(\theta, \phi) - \overline{T}$. The angular power spectrum is

$$C_\ell = \frac{1}{2\ell + 1} \sum_{m=-\ell}^{\ell} |a_{\ell,m}|^2. \tag{9.126}$$

If we let the unit vector $\hat{\boldsymbol{n}}$ point in the direction θ, ϕ and use the addition theorem (9.123), then we can write the angular power spectrum as

$$C_\ell = \frac{1}{4\pi} \int d^2\hat{\boldsymbol{n}} \int d^2\hat{\boldsymbol{n}}' \;\; P_\ell(\hat{\boldsymbol{n}} \cdot \hat{\boldsymbol{n}}') \, T(\hat{\boldsymbol{n}}) \, T(\hat{\boldsymbol{n}}'). \tag{9.127}$$

In Fig. 9.4, the measured values (arXiv:1807.06205) of the power spectrum $\mathcal{D}_\ell = \ell(\ell + 1) C_\ell / 2\pi$ are plotted against ℓ for $1 < \ell < 2500$ with the angles *decreasing* with ℓ as $\theta \sim 180°/\ell$. The power spectrum is a snapshot at the moment of initial transparency of the temperature distribution of the rapidly expanding plasma of photons, electrons, and nuclei undergoing tiny (2×10^{-4})

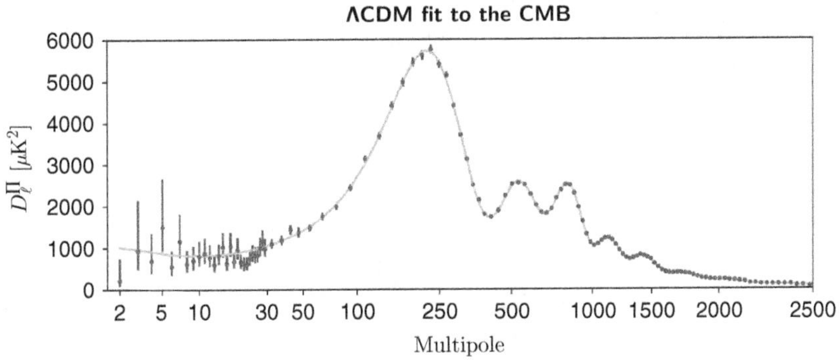

Figure 9.4 The power spectrum $\mathcal{D}_\ell^{TT} = \ell(\ell + 1)C_\ell/2\pi$ of the CMB temperature fluctuations in μK^2 is plotted against the multipole moment ℓ. The solid curve is the ΛCDM prediction. (Source: Planck Collaboration, arXiv:1807.06205, https://arxiv.org/pdf/1807.06205.pdf)

acoustic oscillations. In these oscillations, gravity opposes radiation pressure, and $|\Delta T(\theta, \phi)|$ is maximal both when the oscillations are most compressed and when they are most rarefied. Regions that gravity has squeezed to maximum compression at transparency form the first and highest peak. Regions that have bounced off their first maximal compression and that radiation pressure has expanded to minimum density at transparency form the second peak. Those at their second maximum compression at transparency form the third peak, and so forth.

The solid curve is the prediction of a model with inflation, cold dark matter, ordinary matter, and a cosmological constant Λ. In this model, the age of the visible universe is 13.8 Gyr; the Hubble constant is $H_0 = 67.7$ km/(s Mpc); the energy density of the universe is enough to make the universe flat; and the fractions of the energy density due to ordinary matter, dark matter, and dark energy are 5%, 26%, and 69% (Edwin Hubble 1889–1953).

We can learn a lot from the data in the CMB Figure 9.4. The radius of the maximum causally connected region at transparency is 380,000 light-years. The radius of the maximum compressed region is smaller by a factor of $\sqrt{3}$ because the speed of "sound" in the plasma is nearly $c/\sqrt{3}$. The expansion of the universe since transparency has stretched the wavelength of light and reduced its frequency from that of 3000 K to 2.7255 K, an expansion factor $z \approx 3000/2.7255 \approx 1100$. The diameter of the maximum compressed region is now bigger by 1100, which (too simply) suggests an angle of about

$$2\,\frac{3.8 \times 10^5 \times 1100 \times 180^\circ}{\sqrt{3} \times 13.8 \times 10^9 \times \pi} = 2^\circ. \tag{9.128}$$

A more accurate estimate of that angle and of the location of the first peak in Figure 9.4 is about one degree. This result tells us that space is flat. For if the universe were closed ($k = 1$), then the angle would appear bigger than 1°; and if it were open ($k = -1$), the angle would appear smaller than 1°. The heights and locations of the peaks in Figure 9.4 also tell us about the density of dark matter and the density of dark energy (Aghanim et al., 2018).

Further Reading

Much is known about Legendre functions. The books *A Course of Modern Analysis* (Whittaker and Watson, 1927, chap. XV) and *Methods of Mathematical Physics* (Courant and Hilbert, 1955) are classics. The NIST Digital Library of Mathematical Functions (dlmf.nist.gov) and the companion *NIST Handbook of Mathematical Functions* (Olver et al., 2010) are outstanding. You can learn more about the CMB in Steven Weinberg's book *Cosmology* (Weinberg, 2010, chap. 7) and at the website `camb.info`.

Exercises

9.1 Use conditions (9.6) and (9.7) to find $P_0(x)$ and $P_1(x)$.

9.2 Using the Gram–Schmidt method (section 1.10) to turn the functions x^n into a set of functions $L_n(x)$ that are orthonormal on the interval $[-1, 1]$ with inner product (9.2), find $L_n(x)$ for $n = 0, 1, 2$, and 3. Isn't Rodrigues's formula (9.8) easier to use?

9.3 Derive the conditions (9.6–9.7) on the coefficients a_k of the Legendre polynomial $P_n(x) = a_0 + a_1 x + \cdots + a_n x^n$.

9.4 Use equations (9.6–9.7) to find $P_3(x)$ and $P_4(x)$.

9.5 In superscript notation (2.6), Leibniz's rule (5.49) for derivatives of products $u\,v$ of functions is

$$(uv)^{(n)} = \sum_{k=0}^{n} \binom{n}{k} u^{(n-k)}\, v^{(k)}. \tag{9.129}$$

Use it and Rodrigues's formula (9.8) to derive the explicit formula (9.9).

9.6 The product rule for derivatives in superscript notation (2.6) is

$$(uv)^{(n)} = \sum_{k=0}^{n} \binom{n}{k} u^{(n-k)}\, v^{(k)}. \tag{9.130}$$

Apply it to Rodrigues's formula (9.8) with $x^2 - 1 = (x-1)(x+1)$ and show that the Legendre polynomials satisfy $P_n(1) = 1$.

9.7 Use Cauchy's integral formula (6.44) and Rodrigues's formula (9.55) to derive Schlaefli's integral formula (9.56).

9.8 Show that the polynomials (9.57) are orthogonal (9.58) as long as they satisfy the endpoint condition (9.59).

9.9 Derive the orthogonality relation (9.2) from Rodrigues's formula (9.8).

9.10 (a) Use the fact that the quantities $w = x^2 - 1$ and $w_n = w^n$ vanish at the endpoints ± 1 to show by repeated integrations by parts that in superscript notation (2.6)

$$\int_{-1}^{1} w_n^{(n)} w_n^{(n)} dx = -\int_{-1}^{1} w_n^{(n-1)} w_n^{(n+1)} dx = (-1)^n \int_{-1}^{1} w_n w_n^{(2n)} dx. \tag{9.131}$$

(b) Show that the final integral is equal to

$$I_n = (2n)! \int_{-1}^{1} (1-x)^n\,(1+x)^n\,dx. \tag{9.132}$$

9.11 (a) Show by integrating by parts that $I_n = (n!)^2\, 2^{2n+1}/(2n+1)$. (b) Prove (9.13).

9.12 Suppose that $P_n(x)$ and $Q_n(x)$ are two solutions of (9.27). Find an expression for their wronskian, apart from an overall constant.

9.13 Use the method of sections (7.26 and 7.33) and the solution $f(r) = r^\ell$ to find a second solution of the ODE (9.84).

9.14 For a uniformly charged circle of radius a, find the resulting scalar potential $\phi(r, \theta)$ for $r < a$.

9.15 (a) Find the electrostatic potential $V(r, \theta)$ outside an uncharged perfectly conducting sphere of radius R in a vertical uniform static electric field that tends to $\boldsymbol{E} = E\hat{z}$ as $r \to \infty$. (b) Find the potential if the free charge on the sphere is q_f.

9.16 Derive (9.127) from (9.125) and (9.126).

9.17 Find the electrostatic potential $V(r, \theta)$ inside a hollow sphere of radius R if the potential on the sphere is $V(R, \theta) = V_0 \cos^2 \theta$.

9.18 Find the electrostatic potential $V(r, \theta)$ outside a hollow sphere of radius R if the potential on the sphere is $V(R, \theta) = V_0 \cos^2 \theta$.

10

Bessel Functions

10.1 Cylindrical Bessel Functions of the First Kind

The cylindrical Bessel functions are defined for any integer $n \geq 0$ by the series

$$J_n(z) = \frac{z^n}{2^n n!}\left[1 - \frac{z^2}{2(2n+2)} + \frac{z^4}{2\cdot 4(2n+2)(2n+4)} - \cdots\right]$$
$$= \left(\frac{z}{2}\right)^n \sum_{m=0}^{\infty} \frac{(-1)^m}{m!\,(m+n)!}\left(\frac{z}{2}\right)^{2m} \tag{10.1}$$

(Friedrich Bessel, 1784–1846). The first term of this series tells us that for small $|z| \ll 1$

$$J_n(z) \approx \frac{z^n}{2^n n!}. \tag{10.2}$$

The alternating signs in the series (10.1) make the waves plotted in Fig. 10.1 and for $|z| \gg 1$ give us the approximation (Courant and Hilbert, 1955, chap. VII)

$$J_n(z) \approx \sqrt{\frac{2}{\pi z}}\, \cos\left(z - \frac{n\pi}{2} - \frac{\pi}{4}\right) + O(|z|^{-3/2}). \tag{10.3}$$

The $J_n(z)$ are entire functions. They obey Bessel's equation (7.355)

$$\frac{d^2 J_n}{dz^2} + \frac{1}{z}\frac{dJ_n}{dz} + \left(1 - \frac{n^2}{z^2}\right) J_n = 0 \tag{10.4}$$

as one may show (Exercise 10.1) by substituting the series (10.1) into the differential equation (10.4). Their generating function is

$$\exp\left[\frac{z}{2}(u - 1/u)\right] = \sum_{n=-\infty}^{\infty} u^n\, J_n(z) \tag{10.5}$$

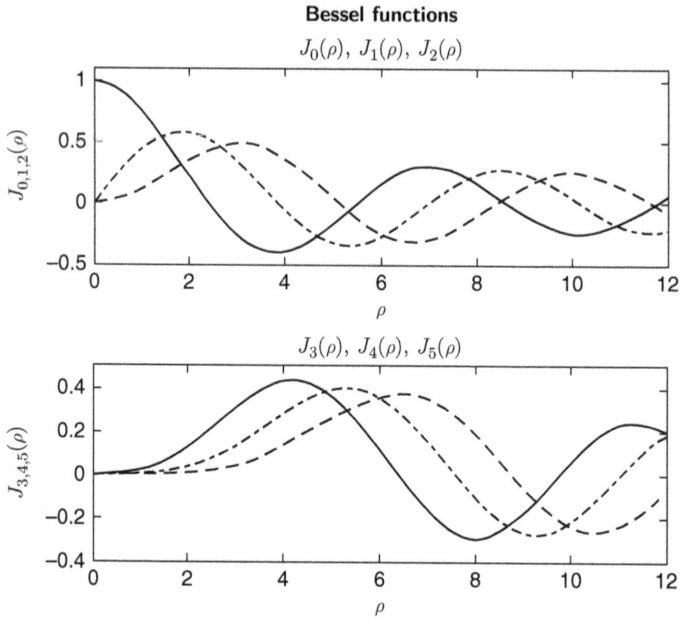

Figure 10.1 Top: Plots of $J_0(\rho)$ (solid curve), $J_1(\rho)$ (dot-dash), and $J_2(\rho)$ (dashed) for real ρ. Bottom: Plots of $J_3(\rho)$ (solid curve), $J_4(\rho)$ (dot-dash), and $J_5(\rho)$(dashed). The points at which Bessel functions cross the ρ-axis are called **zeros** or **roots**; we use them to satisfy boundary conditions. This chapter's Matlab scripts are in Bessel_functions at `github.com/kevinecahill`.

from which one may derive (Exercise 10.5) the series expansion (10.1) and (Exercise 10.6) the integral representation (6.54)

$$J_n(z) = \frac{1}{\pi} \int_0^{\pi} \cos(z \sin\theta - n\theta)\, d\theta = J_{-n}(-z) = (-1)^n J_{-n}(z) \tag{10.6}$$

for all complex z. For $n = 0$, this integral is (Exercise 10.7) more simply

$$J_0(z) = \frac{1}{2\pi} \int_0^{2\pi} e^{iz\cos\theta}\, d\theta = \frac{1}{2\pi} \int_0^{2\pi} e^{iz\sin\theta}\, d\theta. \tag{10.7}$$

These integrals (Exercise 10.8) give $J_n(0) = 0$ for $n \neq 0$, and $J_0(0) = 1$.

By differentiating the generating function (10.5) with respect to u and identifying the coefficients of powers of u, one finds the recursion relation

$$J_{n-1}(z) + J_{n+1}(z) = \frac{2n}{z} J_n(z). \tag{10.8}$$

Similar reasoning after taking the z derivative gives (Exercise 10.10)

$$J_{n-1}(z) - J_{n+1}(z) = 2\, J_n'(z). \tag{10.9}$$

By using the gamma function (Section 6.16), one may extend Bessel's equation (10.4) and its solutions $J_n(z)$ to nonintegral values of n

$$J_\nu(z) = \left(\frac{z}{2}\right)^\nu \sum_{m=0}^{\infty} \frac{(-1)^m}{m!\,\Gamma(m+\nu+1)} \left(\frac{z}{2}\right)^{2m}. \tag{10.10}$$

The self-adjoint form (7.354) of Bessel's equation (10.4) with $z = kx$ is (Exercise 10.11)

$$-\frac{d}{dx}\left(x\frac{d}{dx}J_n(kx)\right) + \frac{n^2}{x}J_n(kx) = k^2 x\, J_n(kx). \tag{10.11}$$

In the notation of Equation (7.327), $p(x) = x$, k^2 is an eigenvalue, and $\rho(x) = x$ is a weight function. To have a self-adjoint system (Section 7.31) on an interval $[0, b]$, we need the boundary condition (7.287)

$$0 = \left[p(J_n v' - J_n' v)\right]_0^b = \left[x(J_n(kx)v'(kx) - J_n'(kx)v(kx))\right]_0^b \tag{10.12}$$

for all functions $v(kx)$ in the domain D of the system. The definition (10.1) of $J_n(z)$ implies that $J_0(0) = 1$, and that $J_n(0) = 0$ for integers $n > 0$. Thus since $p(x) = x$, the terms in this boundary condition vanish at $x = 0$ as long as the domain consists of functions $v(kx)$ that are twice differentiable on the interval $[0, b]$. To make these terms vanish at $x = b$, we require that $J_n(kb) = 0$ and that $v(kb) = 0$. Thus kb must be a zero $z_{n,m}$ of $J_n(z)$, and so $J_n(kb) = J_n(z_{n,m}) = 0$. With $k = z_{n,m}/b$, Bessel's equation (10.11) is

$$-\frac{d}{dx}\left(x\frac{d}{dx}J_n\left(z_{n,m}x/b\right)\right) + \frac{n^2}{x}\,J_n\left(z_{n,m}x/b\right) = \frac{z_{n,m}^2}{b^2}\,x\,J_n\left(z_{n,m}x/b\right). \tag{10.13}$$

For fixed n, the eigenvalue $k^2 = z_{n,m}^2/b^2$ is different for each positive integer m. Moreover as $m \to \infty$, the zeros $z_{n,m}$ of $J_n(x)$ rise as $m\pi$ as one might expect since the leading term of the asymptotic form (10.3) of $J_n(x)$ is proportional to $\cos(x - n\pi/2 - \pi/4)$ which has zeros at $m\pi + (n+1)\pi/2 + \pi/4$. It follows that the eigenvalues $k^2 \approx (m\pi)^2/b^2$ increase without limit as $m \to \infty$ in accordance with the general result of Section 7.37. It follows then from the argument of Section 7.38 and from the orthogonality relation (7.366) that for every fixed n, the eigenfunctions $J_n(z_{n,m}x/b)$, one for each zero, are complete in the mean, orthogonal, and normalizable on the interval $[0, b]$ with weight function $\rho(x) = x$

$$\int_0^b x\, J_n\left(\frac{z_{n,m}x}{b}\right) J_n\left(\frac{z_{n,m'}x}{b}\right) dx = \delta_{m,m'}\,\frac{b^2}{2}\,J_n'^2(z_{n,m}) = \delta_{m,m'}\,\frac{b^2}{2}\,J_{n+1}^2(z_{n,m}) \tag{10.14}$$

and a normalization constant (Exercise 10.12) that depends upon the first derivative of the Bessel function or the square of the next Bessel function at the zero.

Because they are complete, sums of Bessel functions $J_n(z_{n,k}x/b)$ can represent Dirac's delta function on the interval [0, b] as in the sum (7.413)

$$\delta(x-y) = \left(2\,x^{\alpha}\,y^{1-\alpha}/b^2\right) \sum_{k=1}^{\infty} \frac{J_n(z_{n,k}x/b)\,J_n(z_{n,k}y/b)}{J_{n+1}^2(z_{n,k})}. \tag{10.15}$$

For $n = 1$ and $b = 1$, this sum is

$$\delta(x-y) = 2\,x^{\alpha}\,y^{1-\alpha} \sum_{k=1}^{\infty} \frac{J_1(z_{1,k}x)\,J_1(z_{1,k}y)}{J_2^2(z_{1,k})}. \tag{10.16}$$

Figure 10.2 plots the sum of the first 100,000 terms of this series for $\alpha = 1/2$ and $y = 1/2$. In the figure the scales of the two axes differ by a factor of 10^8. This series adequately represents $\delta(x - \frac{1}{2})$ for functions that vary little on a scale of $\Delta x = 0.001$.

The expansion (10.15) of the delta function expresses the completeness of the eigenfunctions $J_n(z_{n,k}x/b)$ for fixed n and lets us write any twice differentiable

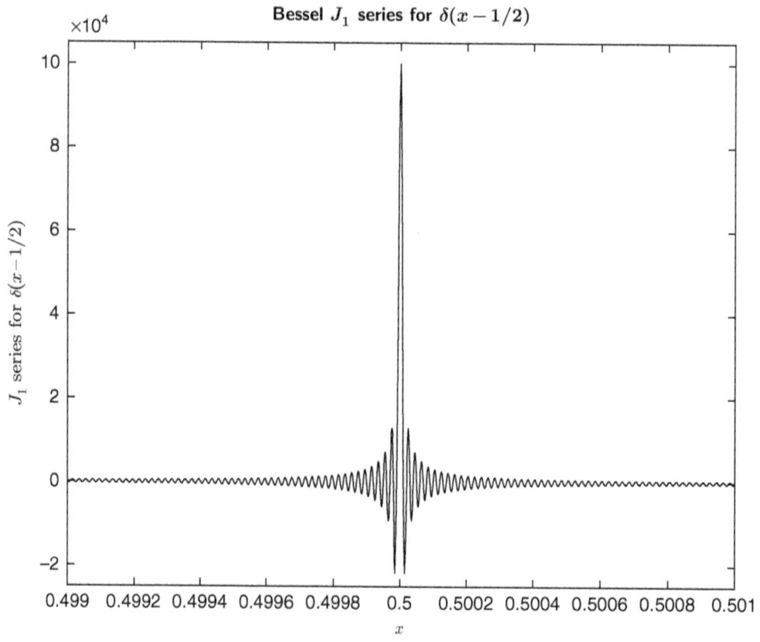

Figure 10.2 The sum of the first 100,000 terms of the $\alpha = 1/2$ Bessel J_1 series (10.16) for the Dirac delta function $\delta(x - 1/2)$ is plotted from $x = 0.499$ to $x = 0.501$. The scales of the axes differ by 10^8.

function $f(x)$ that vanishes at $x = b$ as the Fourier series

$$f(x) = x^{\alpha} \sum_{k=1}^{\infty} a_k \, J_n(z_{n,k}x/b) \tag{10.17}$$

where

$$a_k = \frac{2}{b^2 \, J_{n+1}^2(z_{n,k})} \int_0^b y^{1-\alpha} \, J_n(z_{n,k}y/b) \, f(y) \, dy. \tag{10.18}$$

The orthogonality relation on an infinite interval is

$$\delta(x - y) = x \int_0^{\infty} k \, J_n(kx) \, J_n(ky) \, dk \tag{10.19}$$

for positive values of x and y. One may generalize these relations (10.11–10.19) from integral n to complex ν with $\text{Re}\,\nu > -1$.

Example 10.1 (Bessel's drum) The top of a drum is a circular membrane with a fixed circumference. The membrane's potential energy is approximately proportional to the extra area it has when it's not flat. Let $h(x, y)$ be the displacement of the membrane in the z direction normal to the x–y-plane of the flat membrane, and let h_x and h_y denote its partial derivatives (2.7). The extra length of a line segment dx on the stretched membrane is $\sqrt{1 + h_x^2}\, dx$, and so the extra area of an element $dx\, dy$ is

$$dA \approx \left(\sqrt{1 + h_x^2}\sqrt{1 + h_y^2} - 1\right) dx\, dy \approx \frac{1}{2}\left(h_x^2 + h_y^2\right) dx\, dy. \tag{10.20}$$

The (nonrelativistic) kinetic energy of the area element is proportional to the square of its speed. So if σ is the surface tension and μ the surface mass density of the membrane, then to lowest order in d the action functional (Sections 7.13 and 7.43) is

$$S[h] = \int \left[\frac{\mu}{2} h_t^2 - \frac{\sigma}{2}\left(h_x^2 + h_y^2\right)\right] dx\, dy\, dt. \tag{10.21}$$

We minimize this action for h's that vanish on the boundary $x^2 + y^2 = r_d^2$

$$0 = \delta S[h] = \int \left[\mu \, h_t \, \delta h_t - \sigma \left(h_x \, \delta h_x + h_y \, \delta h_y\right)\right] dx\, dy\, dt. \tag{10.22}$$

Since (7.157) $\delta h_t = (\delta h)_t$, $\delta h_x = (\delta h)_x$, and $\delta h_y = (\delta h)_y$, we can integrate by parts and get

$$0 = \delta S[h] = \int \left[-\mu \, h_{tt} + \sigma \left(h_{xx} + h_{yy}\right)\right] \delta h \, dx\, dy\, dt \tag{10.23}$$

apart from a surface term proportional to δh which vanishes because $\delta h = 0$ on the circumference of the membrane. The membrane therefore obeys the wave equation $\mu \, h_{tt} = \sigma \left(h_{xx} + h_{yy}\right) \equiv \sigma \, \Delta h$.

This equation is separable, and so letting $h(x, y, t) = s(t)\, v(x, y)$, we have

$$\frac{s_{tt}}{s} = \frac{\sigma}{\mu}\frac{\Delta v}{v} = -\omega^2. \tag{10.24}$$

The eigenvalues of the Helmholtz equation $-\Delta v = \lambda\, v$ give the angular frequencies as $\omega = \sqrt{\sigma\lambda/\mu}$. The time dependence then is

$$s(t) = a \sin(\sqrt{\sigma\lambda/\mu}\,(t - t_0)) \tag{10.25}$$

in which a and t_0 are constants.

In polar coordinates, Helmholtz's equation is separable (7.23–7.26)

$$-\Delta v = -v_{rr} - r^{-1}v_r - r^{-2}v_{\theta\theta} = \lambda v. \tag{10.26}$$

We set $v(r, \theta) = u(r)h(\theta)$ and find $-u''h - r^{-1}u'h - r^{-2}uh'' = \lambda uh$. After multiplying both sides by r^2/uh, we get

$$r^2\frac{u''}{u} + r\frac{u'}{u} + \lambda r^2 = -\frac{h''}{h} = n^2. \tag{10.27}$$

The general solution for h then is $h(\theta) = b\sin(n(\theta - \theta_0))$ in which b and θ_0 are constants and n must be an integer so that h is single valued on the circumference of the membrane.

The displacement $u(r)$ must obey the differential equation

$$-(r\,u')' + n^2\,u/r = \lambda\, r\, u \tag{10.28}$$

and vanish at $r = r_d$, the rim of the drum. Therefore it is an eigenfunction $u(r) = J_n(z_{n,m}r/r_d)$ of the self-adjoint differential equation (10.13)

$$-\frac{d}{dr}\left(r\frac{d}{dr}J_n\left(z_{n,m}r/r_d\right)\right) + \frac{n^2}{r}\,J_n\left(z_{n,m}r/r_d\right) = \frac{z_{n,m}^2}{r_d^2}\,x\,J_n\left(z_{n,m}r/r_d\right) \tag{10.29}$$

with an eigenvalue $\lambda = z_{n,m}^2/r_d^2$ in which $z_{n,m}$ is the mth zero of the nth Bessel function, $J_n(z_{n,m}) = 0$. For each integer $n \geq 0$, there are infinitely many zeros $z_{n,m}$ at which the Bessel function vanishes, and for each zero the frequency is $\omega = (z_{n,m}/r_d)\sqrt{\sigma/\mu}$. The general solution to the wave equation of the membrane $\mu\, h_{tt} = \sigma\left(h_{xx} + h_{yy}\right) \equiv \sigma\,\Delta h$ thus is

$$h(r, \theta, t) = \sum_{n=0}^{\infty}\sum_{m=1}^{\infty} c_{n,m}\sin\left[\frac{z_{n,m}}{r_d}\sqrt{\frac{\sigma}{\mu}}\,(t - t_0)\right]\sin\left[n(\theta - \theta_0)\right] J_n\left(z_{n,m}\frac{r}{r_d}\right) \tag{10.30}$$

in which t_0 and θ_0 can depend upon n and m. For each n and big m, the zero $z_{n,m}$ is near $m\pi + (n+1)\pi/2 + \pi/4$.

Helmholtz's equation $-\Delta V = \alpha^2\, V$ separates in cylindrical coordinates in 3 dimensions (Section 7.3). Thus the function $V(\rho, \phi, z) = B(\rho)\Phi(\phi)Z(z)$ satisfies the equation

$$- \triangle V = - \frac{1}{\rho} \left[\left(\rho\, V_{,\rho}\right)_{,\rho} + \frac{1}{\rho}\, V_{,\phi\phi} + \rho\, V_{,zz} \right] = \alpha^2\, V \tag{10.31}$$

both if $B(\rho)$ obeys Bessel's equation

$$\rho \frac{d}{d\rho} \left(\rho \frac{dB}{d\rho} \right) + \left((\alpha^2 + k^2)\rho^2 - n^2 \right) B = 0 \tag{10.32}$$

while Φ and Z satisfy

$$- \frac{d^2\Phi}{d\phi^2} = n^2 \Phi(\phi) \quad \text{and} \quad \frac{d^2 Z}{dz^2} = k^2 Z(z), \tag{10.33}$$

and if $B(\rho)$ obeys the Bessel equation

$$\rho \frac{d}{d\rho} \left(\rho \frac{dB}{d\rho} \right) + \left((\alpha^2 - k^2)\rho^2 - n^2 \right) B = 0 \tag{10.34}$$

while Φ and Z satisfy

$$- \frac{d^2\Phi}{d\phi^2} = n^2 \Phi(\phi) \quad \text{and} \quad \frac{d^2 Z}{dz^2} = -k^2 Z(z). \tag{10.35}$$

In the first case (10.32 and 10.33), the solution is

$$V_{k,n}(\rho, \phi, z) = J_n \left(\sqrt{\alpha^2 + k^2}\, \rho \right) e^{\pm i n \phi} e^{\pm k z}. \tag{10.36}$$

In the second case (10.34 and 10.35), it is

$$V_{k,n}(\rho, \phi, z) = J_n \left(\sqrt{\alpha^2 - k^2}\, \rho \right) e^{\pm i n \phi} e^{\pm i k z}. \tag{10.37}$$

In both cases, n must be an integer if the solution is to be single valued on the full range of ϕ from 0 to 2π.

When $\alpha = 0$, Helmholtz's equation reduces to Laplace's equation $\triangle V = 0$ which is satisfied by the simpler functions

$$V_{k,n}(\rho, \phi, z) = J_n(k\rho) e^{\pm i n \phi} e^{\pm k z} \quad \text{and} \quad V_{k,n}(\rho, \phi, z) = J_n(ik\rho) e^{\pm i n \phi} e^{\pm i k z}. \tag{10.38}$$

The product $i^{-\nu}\, J_\nu(ik\rho)$ is real and is known as the **modified Bessel function**

$$I_\nu(k\rho) \equiv i^{-\nu}\, J_\nu(ik\rho). \tag{10.39}$$

It occurs in various solutions of the **diffusion equation** $D\triangle V = \dot{V}$. The function $V(\rho, \phi, z) = B(\rho)\Phi(\phi)Z(z)$ satisfies

$$\triangle V = \frac{1}{\rho} \left[\left(\rho\, V_{,\rho}\right)_{,\rho} + \frac{1}{\rho}\, V_{,\phi\phi} + \rho\, V_{,zz} \right] = \alpha^2\, V \tag{10.40}$$

both if $B(\rho)$ obeys Bessel's equation

$$\rho \frac{d}{d\rho} \left(\rho \frac{dB}{d\rho} \right) - \left((\alpha^2 - k^2)\rho^2 + n^2 \right) B = 0 \tag{10.41}$$

while Φ and Z satisfy

$$-\frac{d^2\Phi}{d\phi^2} = n^2\Phi(\phi) \quad \text{and} \quad \frac{d^2Z}{dz^2} = k^2 Z(z) \tag{10.42}$$

and if $B(\rho)$ obeys the Bessel equation

$$\rho\frac{d}{d\rho}\left(\rho\frac{dB}{d\rho}\right) - \left((\alpha^2+k^2)\rho^2 + n^2\right) B = 0 \tag{10.43}$$

while Φ and Z satisfy

$$-\frac{d^2\Phi}{d\phi^2} = n^2\Phi(\phi) \quad \text{and} \quad \frac{d^2Z}{dz^2} = -k^2 Z(z). \tag{10.44}$$

In the first case (10.41 and 10.42), the solution is

$$V_{k,n}(\rho,\phi,z) = I_n(\sqrt{\alpha^2-k^2}\,\rho)e^{\pm in\phi}e^{\pm kz}. \tag{10.45}$$

In the second case (10.43 and 10.44), it is

$$V_{k,n}(\rho,\phi,z) = I_n(\sqrt{\alpha^2+k^2}\,\rho)e^{\pm in\phi}e^{\pm ikz}. \tag{10.46}$$

In both cases, n must be an integer if the solution is to be single valued on the full range of ϕ from 0 to 2π.

Example 10.2 (Charge near a membrane) We will use ρ to denote the density of **free charges** – those that are free to move in or out of a dielectric medium, as opposed to those that are part of the medium, bound in it by molecular forces. The time-independent Maxwell equations are Gauss's law $\nabla \cdot \boldsymbol{D} = \rho$ for the divergence of the electric displacement $\boldsymbol{D}$, and the static form $\nabla \times \mathbf{E} = 0$ of Faraday's law which implies that the electric field $\boldsymbol{E}$ is the gradient of an electrostatic potential $\boldsymbol{E} = -\nabla V$.

Across an interface between two dielectrics with normal vector $\hat{\boldsymbol{n}}$, the tangential electric field is continuous, $\hat{\boldsymbol{n}} \times \boldsymbol{E}_2 = \hat{\boldsymbol{n}} \times \boldsymbol{E}_1$, while the normal component of the electric displacement jumps by the surface density σ of free charge, $\hat{\boldsymbol{n}} \cdot (\boldsymbol{D}_2 - \boldsymbol{D}_1) = \sigma$. In a linear dielectric, the electric displacement $\boldsymbol{D}$ is the electric field multiplied by the permittivity ϵ of the material, $\boldsymbol{D} = \epsilon\,\boldsymbol{E}$.

The membrane of a eukaryotic cell is a phospholipid bilayer whose area is some 3×10^8 nm^2, and whose thickness t is about 5 nm. On a scale of nanometers, the membrane is flat. We will take it to be a slab extending to infinity in the x and y directions. If the interface between the lipid bilayer and the extracellular salty water is at $z = 0$, then the cytosol extends thousands of nm down from $z = -t = -5$ nm. We will ignore the phosphate head groups and set the permittivity ϵ_ℓ of the lipid bilayer to twice that of the vacuum $\epsilon_\ell \approx 2\epsilon_0$; the permittivity of the extracellular water and that of the cytosol are $\epsilon_w \approx \epsilon_c \approx 80\epsilon_0$.

We will compute the electrostatic potential V due to a charge q at a point $(0, 0, h)$ on the z-axis above the membrane. This potential is cylindrically symmetric about the z-axis, so $V = V(\rho, z)$. The functions $J_n(k\rho)\, e^{in\phi}\, e^{\pm kz}$ form a complete set of solutions (10.38) of Laplace's equation, but due to the symmetry, we only need the $n = 0$ functions $J_0(k\rho)\, e^{\pm kz}$. Since there are no free charges in the lipid bilayer or in the cytosol, we may express the potential in the lipid bilayer V_ℓ and in the cytosol V_c as

$$\begin{aligned} V_\ell(\rho, z) &= \int_0^\infty dk\, J_0(k\rho) \left[m(k)\, e^{kz} + f(k)\, e^{-kz} \right] \\ V_c(\rho, z) &= \int_0^\infty dk\, J_0(k\rho)\, d(k)\, e^{kz}. \end{aligned} \tag{10.47}$$

The Green's function (4.113) for Poisson's equation $-\triangle G(\boldsymbol{x}) = \delta^{(3)}(\boldsymbol{x})$ in cylindrical coordinates is (6.152)

$$G(\boldsymbol{x}) = \frac{1}{4\pi |\boldsymbol{x}|} = \frac{1}{4\pi \sqrt{\rho^2 + z^2}} = \int_0^\infty \frac{dk}{4\pi} J_0(k\rho)\, e^{-k|z|}. \tag{10.48}$$

Thus we may expand the potential in the salty water as

$$V_w(\rho, z) = \int_0^\infty dk\, J_0(k\rho) \left[\frac{q}{4\pi \epsilon_w} e^{-k|z-h|} + u(k)\, e^{-kz} \right]. \tag{10.49}$$

Using $\hat{\boldsymbol{n}} \times \boldsymbol{E}_2 = \hat{\boldsymbol{n}} \times \boldsymbol{E}_1$ and $\hat{\boldsymbol{n}} \cdot (\boldsymbol{D}_2 - \boldsymbol{D}_1) = \sigma$, suppressing k, and setting $\beta \equiv q e^{-kh}/4\pi\epsilon_w$ and $y = e^{2kt}$, we get four equations

$$\begin{aligned} m + f - u = \beta \quad &\text{and} \quad \epsilon_\ell m - \epsilon_\ell f + \epsilon_w u = \epsilon_w \beta \\ \epsilon_\ell m - \epsilon_\ell y f - \epsilon_c d = 0 \quad &\text{and} \quad m + y f - d = 0. \end{aligned} \tag{10.50}$$

In terms of the abbreviations $\epsilon_{w\ell} = (\epsilon_w + \epsilon_\ell)/2$ and $\epsilon_{c\ell} = (\epsilon_c + \epsilon_\ell)/2$ as well as $p = (\epsilon_w - \epsilon_\ell)/(\epsilon_w + \epsilon_\ell)$ and $p' = (\epsilon_c - \epsilon_\ell)/(\epsilon_c + \epsilon_\ell)$, the solutions are

$$\begin{aligned} u(k) = \beta\, \frac{p - p'/y}{1 - pp'/y} \quad &\text{and} \quad m(k) = \beta\, \frac{\epsilon_w}{\epsilon_{w\ell}} \frac{1}{1 - pp'/y} \\ f(k) = -\beta\, \frac{\epsilon_w}{\epsilon_{w\ell}} \frac{p'/y}{1 - pp'/y} \quad &\text{and} \quad d(k) = \beta\, \frac{\epsilon_w \epsilon_\ell}{\epsilon_{w\ell}\epsilon_{c\ell}} \frac{1}{1 - pp'/y}. \end{aligned} \tag{10.51}$$

Inserting these solutions into the Bessel expansions (10.47) for the potentials, expanding their denominators

$$\frac{1}{1 - pp'/y} = \sum_0^\infty (pp')^n\, e^{-2nkt} \tag{10.52}$$

and using the integral (10.48), we find that the potential V_w in the extracellular water of a charge q at $(0, 0, h)$ in the water is

$$V_w(\rho, z) = \frac{q}{4\pi\epsilon_w}\left[\frac{1}{r} + \frac{p}{\sqrt{\rho^2 + (z+h)^2}} - \sum_{n=1}^{\infty} \frac{p'\left(1-p^2\right)(pp')^{n-1}}{\sqrt{\rho^2 + (z+2nt+h)^2}}\right] \tag{10.53}$$

in which $r = \sqrt{\rho^2 + (z-h)^2}$ is the distance to the charge q. The principal image charge pq is at $(0, 0, -h)$. Similarly, the potential V_ℓ in the lipid bilayer is

$$V_\ell(\rho, z) = \frac{q}{4\pi\epsilon_{w\ell}} \sum_{n=0}^{\infty} \left[\frac{(pp')^n}{\sqrt{\rho^2 + (z-2nt-h)^2}} - \frac{p^n p'^{n+1}}{\sqrt{\rho^2 + (z+2(n+1)t+h)^2}}\right] \tag{10.54}$$

and that in the cytosol is

$$V_c(\rho, z) = \frac{q\,\epsilon_\ell}{4\pi\epsilon_{w\ell}\epsilon_{c\ell}} \sum_{n=0}^{\infty} \frac{(pp')^n}{\sqrt{\rho^2 + (z-2nt-h)^2}}. \tag{10.55}$$

These potentials are the same as those of Example 5.19, but this derivation is much simpler and less error prone than the method of images.

Since $p = (\epsilon_w - \epsilon_\ell)/(\epsilon_w + \epsilon_\ell) > 0$, the principal image charge pq at $(0, 0, -h)$ has the same sign as the charge q and so contributes to the energy a positive term proportional to pq^2. So a lipid slab repels a nearby charge in water no matter what the sign of the charge.

A cell membrane is a phospholipid bilayer. The lipids avoid water and form a 4 nm-thick layer that lies between two 0.5 nm layers of phosphate groups which are electric dipoles. These electric dipoles cause the cell membrane to weakly *attract* ions that are within 0.5 nm of the membrane.

Example 10.3 (Cylindrical waveguides) An electromagnetic wave traveling in the z-direction down a cylindrical waveguide looks like

$$\boldsymbol{E}\, e^{in\phi}\, e^{i(kz-\omega t)} \quad \text{and} \quad \boldsymbol{B}\, e^{in\phi}\, e^{i(kz-\omega t)} \tag{10.56}$$

in which $\boldsymbol{E}$ and $\boldsymbol{B}$ depend upon ρ

$$\boldsymbol{E} = E^\rho \hat{\boldsymbol{\rho}} + E^\phi \hat{\boldsymbol{\phi}} + E^z \hat{\boldsymbol{z}} \quad \text{and} \quad \boldsymbol{B} = B^\rho \hat{\boldsymbol{\rho}} + B^\phi \hat{\boldsymbol{\phi}} + B^z \hat{\boldsymbol{z}} \tag{10.57}$$

in cylindrical coordinates. If the waveguide is an evacuated, perfectly conducting cylinder of radius r, then on the surface of the waveguide the parallel components of $\boldsymbol{E}$ and the normal component of $\boldsymbol{B}$ must vanish which leads to the boundary conditions

$$E^z(r) = 0, \quad E^\phi(r) = 0, \quad \text{and} \quad B^\rho(r) = 0. \tag{10.58}$$

In a notation (2.7) in which subscripts denote differentiation, the vacuum forms $\nabla \times \boldsymbol{E} = -\dot{\boldsymbol{B}}$ and $\nabla \times \boldsymbol{B} = \dot{\boldsymbol{E}}/c^2$ of the Faraday and Maxwell–Ampère laws give us (exercise 10.15) the field equations

$$\begin{aligned} E^z_\phi/\rho - ikE^\phi &= i\omega B^\rho & \qquad inB^z/\rho - ikB^\phi &= -i\omega E^\rho/c^2 \\ ikE^\rho - E^z_\rho &= i\omega B^\phi & \qquad ikB^\rho - B^z_\rho &= -i\omega E^\phi/c^2 \end{aligned} \tag{10.59}$$

$$[(\rho E^\phi)_\rho - inE^\rho]/\rho = i\omega B^z \qquad [(\rho B^\phi)_\rho - inB^\rho]/\rho = -i\omega E^z/c^2.$$

Solving them for the ρ and ϕ components of $\boldsymbol{E}$ and $\boldsymbol{B}$ in terms of their z components (Exercise 10.16), we find

$$E^\rho = \frac{-ikE^z_\rho + n\omega B^z/\rho}{k^2 - \omega^2/c^2} \qquad E^\phi = \frac{nkE^z/\rho + i\omega B^z_\rho}{k^2 - \omega^2/c^2}$$
$$B^\rho = \frac{-ikB^z_\rho - n\omega E^z/c^2\rho}{k^2 - \omega^2/c^2} \qquad B^\phi = \frac{nkB^z/\rho - i\omega E^z_\rho/c^2}{k^2 - \omega^2/c^2}. \tag{10.60}$$

The fields E^z and B^z obey the wave equations (12.43 , Exercise 7.6)

$$-\triangle E^z = -\ddot{E}^z/c^2 = \omega^2 E^z/c^2 \quad \text{and} \quad -\triangle B^z = -\ddot{B}^z/c^2 = \omega^2 B^z/c^2. \tag{10.61}$$

Because their z-dependence (10.56) is periodic, they are (Exercise 10.17) linear combinations of $J_n(\sqrt{\omega^2/c^2 - k^2}\,\rho)e^{in\phi}e^{i(kz-\omega t)}$.

Modes with $B_z = 0$ are **transverse magnetic** or **TM** modes. For them the boundary conditions (10.58) will be satisfied if $\sqrt{\omega^2/c^2 - k^2}\,r$ is a zero $z_{n,m}$ of J_n. So the frequency $\omega_{n,m}(k)$ of the n, m TM mode is

$$\omega_{n,m}(k) = c\sqrt{k^2 + z^2_{n,m}/r^2}. \tag{10.62}$$

Since the first zero of J_0 is $z_{0,1} \approx 2.4048$, the minimum frequency $\omega_{0,1}(0) = c\,z_{0,1}/r \approx 2.4048\,c/r$ occurs for $n = 0$ and $k = 0$. If the radius of the waveguide is $r = 1$ cm, then $\omega_{0,1}(0)/2\pi$ is about 11 GHz, which is a microwave frequency with a wavelength of 2.6 cm. In terms of the frequencies (10.62), the field of a pulse moving in the $+z$-direction is

$$E^z(\rho,\phi,z,t) = \sum_{n=0}^{\infty}\sum_{m=1}^{\infty}\int_0^\infty c_{n,m}(k)\,J_n\left(\frac{z_{n,m}\,\rho}{r}\right)e^{in\phi}\exp i\left[kz - \omega_{n,m}(k)t\right]dk. \tag{10.63}$$

Modes with $E_z = 0$ are **transverse electric** or **TE** modes. Their boundary conditions (10.58) are satisfied (Exercise 10.19) when $\sqrt{\omega^2/c^2 - k^2}\,r$ is a zero $z'_{n,m}$ of J'_n. Their frequencies are $\omega_{n,m}(k) = c\sqrt{k^2 + z'^2_{n,m}/r^2}$. Since the first zero of a first derivative of a Bessel function is $z'_{1,1} \approx 1.8412$, the minimum frequency $\omega_{1,1}(0) = c\,z'_{1,1}/r \approx 1.8412\,c/r$ occurs for $n = 1$ and $k = 0$. If the radius of the waveguide is $r = 1$ cm, then $\omega_{1,1}(0)/2\pi$ is about 8.8 GHz, which is a microwave frequency with a wavelength of 3.4 cm.

Example 10.4 (Cylindrical cavity) The modes of an evacuated, perfectly conducting cylindrical cavity of radius r and height h are like those of a cylindrical waveguide (Example 10.3) but with extra boundary conditions

$$B^z(\rho,\phi,0,t) = E^\rho(\rho,\phi,0,t) = E^\phi(\rho,\phi,0,t) = 0$$
$$B^z(\rho,\phi,h,t) = E^\rho(\rho,\phi,h,t) = E^\phi(\rho,\phi,h,t) = 0 \tag{10.64}$$

at the two ends of the cylinder. If ℓ is an integer and if $\sqrt{\omega^2/c^2 - \pi^2\ell^2/h^2}\, r$ is a zero $z'_{n,m}$ of J'_n, then the TE fields $E^z = 0$ and

$$B^z = J_n(z'_{n,m}\,\rho/r)\, e^{in\phi}\, \sin(\pi \ell z/h)\, e^{-i\omega t} \tag{10.65}$$

satisfy both these (10.64) boundary conditions at $z = 0$ and h and those (10.58) at $\rho = r$ as well as the separable wave equations (10.61). The frequencies of the resonant TE modes then are $\omega_{n,m,\ell} = c\sqrt{z'^2_{n,m}/r^2 + \pi^2\ell^2/h^2}$.

The TM modes are $B^z = 0$ and $E^z = J_n(z_{n,m}\,\rho/r)\, e^{in\phi}\, \sin(\pi\ell z/h)\, e^{-i\omega t}$ with resonant frequencies $\omega_{n,m,\ell} = c\sqrt{z^2_{n,m}/r^2 + \pi^2\ell^2/h^2}$.

10.2 Spherical Bessel Functions of the First Kind

The **spherical Bessel function** $j_\ell(x)$ is proportional to the cylindrical Bessel function $J_{\ell+1/2}(x)$ divided by the square root $\sqrt{x}$

$$j_\ell(x) \equiv \sqrt{\frac{\pi}{2x}}\, J_{\ell+1/2}(x). \tag{10.66}$$

By setting $n = \ell + 1/2$ and $J_{\ell+1/2}(x) = \sqrt{2x/\pi}\, j_\ell(x)$ in Bessel's equation (10.4), one may derive (Exercise 10.22) the equation

$$x^2\, j''_\ell(x) + 2x\, j'_\ell(x) + [x^2 - \ell(\ell+1)]\, j_\ell(x) = 0 \tag{10.67}$$

for the spherical Bessel function j_ℓ.

We saw in Example 7.7 that by setting $V(r,\theta,\phi) = R_{k,\ell}(r)\,\Theta_{\ell,m}(\theta)\,\Phi_m(\phi)$ we could separate the variables of Helmholtz's equation $-\triangle V = k^2 V$ in spherical coordinates

$$\frac{r^2\triangle V}{V} = \frac{(r^2 R'_{k,\ell})'}{R_{k,\ell}} + \frac{(\sin\theta\,\Theta'_{\ell,m})'}{\sin\theta\,\Theta_{\ell,m}} + \frac{\Phi''}{\sin^2\theta\,\Phi} = -k^2 r^2. \tag{10.68}$$

Thus if $\Phi_m(\phi) = e^{im\phi}$ so that $\Phi''_m = -m^2\Phi_m$, and if $\Theta_{\ell,m}$ satisfies the **associated Legendre equation** (9.93)

$$\sin\theta\left(\sin\theta\,\Theta'_{\ell,m}\right)' + [\ell(\ell+1)\sin^2\theta - m^2]\,\Theta_{\ell,m} = 0, \tag{10.69}$$

then the product $V(r,\theta,\phi) = R_{k,\ell}(r)\,\Theta_{\ell,m}(\theta)\,\Phi_m(\phi)$ will obey (10.68) because by Bessel's equation (10.67) the radial function $R_{k,\ell}(r) = j_\ell(kr)$ satisfies

$$(r^2 R'_{k,\ell})' + [k^2r^2 - \ell(\ell+1)]R_{k,\ell} = 0. \tag{10.70}$$

In terms of the spherical harmonic $Y_{\ell,m}(\theta,\phi) = \Theta_{\ell,m}(\theta)\,\Phi_m(\phi)$, the solution is $V(r,\theta,\phi) = j_\ell(kr)\,Y_{\ell,m}(\theta,\phi)$.

Rayleigh's formula gives the spherical Bessel function $j_\ell(x)$ as the ℓth derivative of $\sin x/x$

$$j_\ell(x) = (-1)^\ell \, x^\ell \left(\frac{1}{x}\frac{d}{dx}\right)^\ell \left(\frac{\sin x}{x}\right). \tag{10.71}$$

In particular, $j_0(x) = \sin x/x$ and $j_1(x) = \sin x/x^2 - \cos x/x$ (Lord Rayleigh (John William Strutt), 1842–1919).

His formula (10.71) leads (Exercise 10.23) to the recursion relation

$$j_{\ell+1}(x) = \frac{\ell}{x}\, j_\ell(x) - j_\ell'(x) \tag{10.72}$$

with which one can show (Exercise 10.24) that the spherical Bessel functions $j_\ell(x)$ defined by (10.71) satisfy the differential equation (10.70) with $x = kr$.

The series expansion (10.1) for J_n and the definition (10.66) of j_ℓ give us for small $|\rho| \ll 1$ the approximation

$$j_\ell(\rho) \approx \frac{\ell\,!\,(2\rho)^\ell}{(2\ell+1)!} = \frac{\rho^\ell}{(2\ell+1)!!}. \tag{10.73}$$

To see how $j_\ell(\rho)$ behaves for large $|\rho| \gg 1$, we use Rayleigh's formula (10.71) to compute $j_1(\rho)$ and notice that the derivative $d/d\rho$

$$j_1(\rho) = -\frac{d}{d\rho}\left(\frac{\sin\rho}{\rho}\right) = -\frac{\cos\rho}{\rho} + \frac{\sin\rho}{\rho^2} \tag{10.74}$$

adds a factor of $1/\rho$ when it acts on $1/\rho$ but not when it acts on $\sin\rho$. Thus the dominant term is the one in which all the derivatives act on the sine, and so for large $|\rho| \gg 1$, we have approximately

$$j_\ell(\rho) = (-1)^\ell \rho^\ell \left(\frac{1}{\rho}\frac{d}{d\rho}\right)^\ell \left(\frac{\sin\rho}{\rho}\right) \approx \frac{(-1)^\ell}{\rho}\frac{d^\ell \sin\rho}{d\rho^\ell} = \frac{\sin\left(\rho - \ell\pi/2\right)}{\rho} \tag{10.75}$$

with an error that falls off as $1/\rho^2$. The quality of the approximation, which is exact for $\ell = 0$, is illustrated for $\ell = 1$ and 2 in Fig. 10.3.

The spherical Bessel functions $j_\ell(kr)$ satisfy the self-adjoint Sturm–Liouville (7.377) equation (10.70)

$$-\left(r^2 j_\ell'\right)' + \ell(\ell+1)\, j_\ell = k^2 r^2 j_\ell \tag{10.76}$$

with eigenvalue k^2 and weight function $\rho = r^2$. If $j_\ell(z_{\ell,n}) = 0$, then the functions $j_\ell(kr) = j_\ell(z_{\ell,n} r/a)$ vanish at $r = a$ and form an orthogonal basis

$$\int_0^a j_\ell(z_{\ell,n} r/a)\, j_\ell(z_{\ell,m} r/a)\, r^2\, dr = \frac{a^3}{2}\, j_{\ell+1}^2(z_{\ell,n})\, \delta_{n,m} \tag{10.77}$$

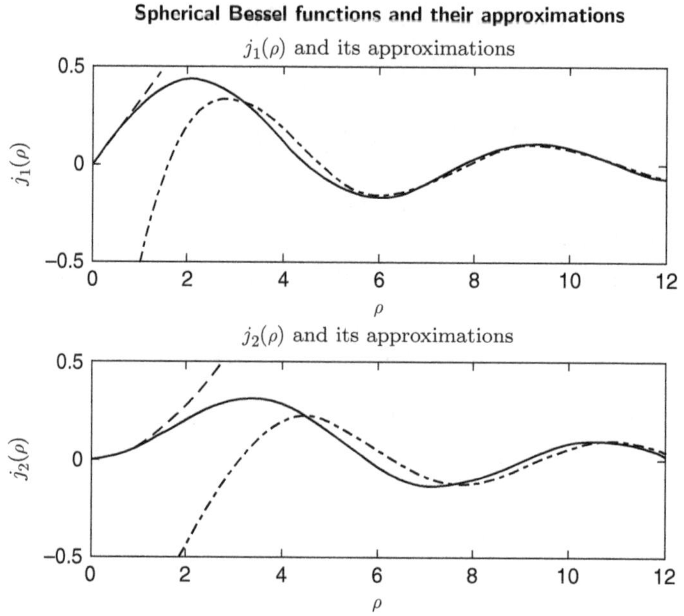

Figure 10.3 Top: Plot of $j_1(\rho)$ (solid curve) and its approximations $\rho/3$ for small ρ (10.73, dashes) and $\sin(\rho - \pi/2)/\rho$ for big ρ (10.75, dot-dash). Bottom: Plot of $j_2(\rho)$ (solid curve) and its approximations $\rho^2/15$ for small ρ (10.73, dashed) and $\sin(\rho - \pi)/\rho$ for big ρ (10.75, dot-dash). The values of ρ at which $j_\ell(\rho) = 0$ are the **zeros** or **roots** of j_ℓ; we use them to fit boundary conditions.

for a self-adjoint system on the interval $[0, a]$. Moreover, since as $n \to \infty$ the eigenvalues $k_{\ell,n}^2 = z_{\ell,n}^2/a^2 \approx [(n + \ell/2)\pi]^2/a^2 \to \infty$, the eigenfunctions $j_\ell(z_{\ell,n} r/a)$ also are complete in the mean (Section 7.38) as shown by the expansion of the delta function

$$\delta(r - r') = \frac{2 r^\alpha r'^{2-\alpha}}{a^3} \sum_{n=1}^{\infty} \frac{j_\ell(z_{\ell,n} r/a)\, j_\ell(z_{\ell,n} r'/a)}{j_{\ell+1}^2(z_{\ell,n})} \tag{10.78}$$

for $0 \le \alpha \le 2$. This formula lets us expand any twice-differentiable function $f(r)$ that vanishes at $r = a$ as

$$f(r) = r^\alpha \sum_{n=1}^{\infty} a_n\, j_\ell(z_{\ell,n} r/a) \tag{10.79}$$

where

$$a_n = \frac{2}{a^3\, j_{\ell+1}^2(z_{\ell,n})} \int_0^a r'^{2-\alpha}\, j_\ell(z_{\ell,n} r'/a)\, f(r')\, dr'. \tag{10.80}$$

On an infinite interval, the delta-function formulas are for $k, k' > 0$

$$\delta(k-k') = \frac{2kk'}{\pi}\int_0^\infty j_\ell(kr)\, j_\ell(k'r)\, r^2\, dr \tag{10.81}$$

and for $r, r' > 0$

$$\delta(r-r') = \frac{2rr'}{\pi}\int_0^\infty j_\ell(kr)\, j_\ell(kr')\, k^2\, dk. \tag{10.82}$$

One may iterate the trick

$$\begin{aligned}\frac{d}{zdz}\int_{-1}^{1} e^{izx}\, dx &= \frac{i}{z}\int_{-1}^{1} x e^{izx}\, dx = \frac{i}{2z}\int_{-1}^{1} e^{izx}\, d(x^2-1)\\ &= -\frac{i}{2z}\int_{-1}^{1}(x^2-1)de^{izx} = \frac{1}{2}\int_{-1}^{1}(x^2-1)e^{izx}dx\end{aligned} \tag{10.83}$$

to show (Exercise 10.25) that (Schwinger et al., 1998, p. 227)

$$\left(\frac{d}{zdz}\right)^{\ell}\int_{-1}^{1} e^{izx}\, dx = \int_{-1}^{1}\frac{(x^2-1)^\ell}{2^\ell \ell!}\, e^{izx}\, dx. \tag{10.84}$$

Using this relation and the integral

$$j_0(z) = \frac{\sin z}{z} = \frac{1}{2}\int_{-1}^{1} e^{izx}\, dx, \tag{10.85}$$

we can write Rayleigh's formula (10.71) for the spherical Bessel function as

$$\begin{aligned}j_\ell(z) &= (-z)^\ell \left(\frac{1}{z}\frac{d}{dz}\right)^{\ell}\left(\frac{\sin z}{z}\right) = (-z)^\ell \left(\frac{1}{z}\frac{d}{dz}\right)^{\ell}\frac{1}{2}\int_{-1}^{1} e^{izx}\, dx\\ &= \frac{z^\ell}{2}\int_{-1}^{1}\frac{(1-x^2)^\ell}{2^\ell \ell!}\, e^{izx}\, dx = \frac{(-i)^\ell}{2}\int_{-1}^{1}\frac{(1-x^2)^\ell}{2^\ell \ell!}\,\frac{d^\ell}{dx^\ell}e^{izx}\, dx.\end{aligned} \tag{10.86}$$

Now integrating ℓ times by parts and using Rodrigues's formula (9.8) for the Legendre polynomial $P_\ell(x)$, we get

$$j_\ell(z) = \frac{(-i)^\ell}{2}\int_{-1}^{1} e^{izx}\,\frac{d^\ell}{dx^\ell}\frac{(x^2-1)^\ell}{2^\ell \ell!}\, dx = \frac{(-i)^\ell}{2}\int_{-1}^{1} P_\ell(x)\, e^{izx}\, dx. \tag{10.87}$$

This formula with $z = kr$ and $x = \cos\theta$

$$i^\ell\, j_\ell(kr) = \frac{1}{2}\int_{-1}^{1} P_\ell(\cos\theta) e^{ikr\cos\theta}\, d\cos\theta \tag{10.88}$$

turns the Fourier–Legendre expansion (9.31) for $e^{ikr\cos\theta}$ into

$$
\begin{aligned}
e^{ikr\cos\theta} &= \sum_{\ell=0}^{\infty} \frac{2\ell+1}{2} P_\ell(\cos\theta) \int_{-1}^{1} P_\ell(\cos\theta')\, e^{ikr\cos\theta'}\, d\cos\theta' \\
&= \sum_{\ell=0}^{\infty} (2\ell+1)\, P_\ell(\cos\theta)\, i^\ell\, j_\ell(kr).
\end{aligned}
\tag{10.89}
$$

If θ, ϕ and θ', ϕ' are the polar angles of the vectors $\boldsymbol{r}$ and $\boldsymbol{k}$, then by using the addition theorem (9.123) we get the plane-wave expansion

$$
e^{i\boldsymbol{k}\cdot\boldsymbol{r}} = \sum_{\ell=0}^{\infty} \sum_{m=-\ell}^{\ell} 4\pi\, i^\ell\, j_\ell(kr)\, Y_{\ell,m}(\theta,\phi)\, Y^*_{\ell,m}(\theta',\phi'). \tag{10.90}
$$

Example 10.5 (Partial waves) Spherical Bessel functions occur in the wave functions of free particles with well-defined angular momentum.

The hamiltonian $H_0 = \boldsymbol{p}^2/2m$ for a free particle of mass m and the square $\boldsymbol{L}^2$ of the orbital angular-momentum operator are both invariant under rotations; thus they commute with the orbital angular-momentum operator $\boldsymbol{L}$. Since the operators H_0, $\boldsymbol{L}^2$, and L_z commute with each other, simultaneous eigenstates $|k,\ell,m\rangle$ of these **compatible** operators (Section 1.31) exist

$$
\begin{aligned}
H_0\, |k,\ell,m\rangle = \frac{\boldsymbol{p}^2}{2m}\, |k,\ell,m\rangle &= \frac{(\hbar k)^2}{2m}\, |k,\ell,m\rangle \\
\boldsymbol{L}^2\, |k,\ell,m\rangle = \hbar^2\, \ell(\ell+1)\, |k,\ell,m\rangle \quad &\text{and} \quad L_z\, |k,\ell,m\rangle = \hbar m\, |k,\ell,m\rangle.
\end{aligned}
\tag{10.91}
$$

By (10.67–10.70), their wave functions are products of spherical Bessel functions and spherical harmonics (9.112)

$$
\langle \boldsymbol{r}|k,\ell,m\rangle = \langle r,\theta,\phi|k,\ell,m\rangle = \sqrt{\frac{2}{\pi}}\, k\, j_\ell(kr)\, Y_{\ell,m}(\theta,\phi). \tag{10.92}
$$

They satisfy the normalization condition

$$
\begin{aligned}
\langle k,\ell,m|k',\ell',m'\rangle &= \frac{2kk'}{\pi} \int_0^\infty j_\ell(kr)\, j_\ell(k'r)\, r^2\, dr \int Y^*_{\ell,m}(\theta,\phi) Y_{\ell',m'}(\theta,\phi)\, d\Omega \\
&= \delta(k-k')\, \delta_{\ell,\ell'}\, \delta_{m,m'}
\end{aligned}
\tag{10.93}
$$

and the completeness relation

$$
1 = \int_0^\infty dk \sum_{\ell=0}^{\infty} \sum_{m=-\ell}^{\ell} |k,\ell,m\rangle\langle k,\ell,m|. \tag{10.94}
$$

Their inner products with an eigenstate $|\boldsymbol{k}'\rangle$ of a free particle of momentum $\boldsymbol{p}' = \hbar\boldsymbol{k}'$ are

$$\langle k, \ell, m|\boldsymbol{k}'\rangle = \frac{i^\ell}{k}\,\delta(k-k')\,Y^*_{\ell,m}(\theta',\phi') \tag{10.95}$$

in which the polar coordinates of $\boldsymbol{k}'$ are θ', ϕ'.

Using the resolution (10.94) of the identity operator and the inner-product formulas (10.92 and 10.95), we recover the expansion (10.90)

$$\begin{aligned}\frac{e^{i\boldsymbol{k}'\cdot\boldsymbol{r}}}{(2\pi)^{3/2}} = \langle\boldsymbol{r}|\boldsymbol{k}'\rangle &= \int_0^\infty dk \sum_{\ell=0}^{\infty}\sum_{m=-\ell}^{\ell} \langle\boldsymbol{r}|k,\ell,m\rangle\langle k,\ell,m|\boldsymbol{k}'\rangle \\ &= \sum_{\ell=0}^{\infty}\sqrt{\frac{2}{\pi}}\; i^\ell\; j_\ell(kr)\; Y_{\ell,m}(\theta,\phi)\; Y^*_{\ell,m}(\theta',\phi').\end{aligned} \tag{10.96}$$

The small kr approximation (10.73), the definition (10.92), and the normalization (9.115) of the spherical harmonics tell us that the probability that a particle with angular momentum $\hbar\ell$ about the origin has $r = |\boldsymbol{r}| \ll 1/k$ is

$$P(r) = \frac{2k^2}{\pi}\int_0^r j_\ell^2(kr')r'^2dr' \approx \frac{2}{\pi[(2\ell+1)!!]^2}\int_0^r (kr)^{2\ell+2}dr = \frac{(4\ell+6)(kr)^{2\ell+3}}{\pi[(2\ell+3)!!]^2k} \tag{10.97}$$

which is very small for big ℓ and tiny k. So a short-range potential can only affect partial waves of low angular momentum. When physicists found that nuclei scattered low-energy hadrons into s-waves, they knew that the range of the nuclear force was short, about 10^{-15}m.

If the potential $V(r)$ that scatters a particle is of short range, then at big r the radial wave function $u_\ell(r)$ of the scattered wave should look like that of a free particle (10.96) which by the big kr approximation (10.75) is

$$u_\ell^{(0)}(r) = j_\ell(kr) \approx \frac{\sin(kr-\ell\pi/2)}{kr} = \frac{1}{2ikr}\left[e^{i(kr-\ell\pi/2)} - e^{-i(kr-\ell\pi/2)}\right]. \tag{10.98}$$

Thus at big r the radial wave function $u_\ell(r)$ differs from $u_\ell^{(0)}(r)$ only by a **phase shift** δ_ℓ

$$u_\ell(r) \approx \frac{\sin(kr-\ell\pi/2+\delta_\ell)}{kr} = \frac{1}{2ikr}\left[e^{i(kr-\ell\pi/2+\delta_\ell)} - e^{-i(kr-\ell\pi/2+\delta_\ell)}\right]. \tag{10.99}$$

The phase shifts determine the **cross-section** σ to be (Cohen-Tannoudji et al., 1977, chap. VIII)

$$\sigma = \frac{4\pi}{k^2}\sum_{\ell=0}^{\infty}(2\ell+1)\,\sin^2\delta_\ell. \tag{10.100}$$

If the potential $V(r)$ is negligible for $r > r_0$, then for momenta $k \ll 1/r_0$ the cross-section is $\sigma \approx 4\pi\sin^2\delta_0/k^2$.

Example 10.6 (Quantum dots) The active region of some quantum dots is a CdSe sphere whose radius a is less than 2 nm. Photons from a laser excite electron–hole pairs which fluoresce in nanoseconds.

I will model a quantum dot simply as an electron trapped in a sphere of radius a. Its wave function $\psi(r, \theta, \phi)$ satisfies Schrödinger's equation

$$-\frac{\hbar^2}{2m} \triangle \psi + V \psi = E \psi \tag{10.101}$$

with the boundary condition $\psi(a, \theta, \phi) = 0$ and the potential V constant and negative for $r < a$ and infinitely positive for $r > a$. With $k^2 = 2m(E - V)/\hbar^2 = z_{\ell,n}^2/a^2$, the unnormalized eigenfunctions are

$$\psi_{n,\ell,m}(r, \theta, \phi) = j_\ell(z_{\ell,n} r/a)\, Y_{\ell,m}(\theta, \phi)\, \theta(a - r) \tag{10.102}$$

in which the Heaviside function $\theta(a - r)$ makes ψ vanish for $r > a$, and ℓ and m are integers with $-\ell \le m \le \ell$ because ψ must be single valued for all angles θ and ϕ.

The zeros $z_{\ell,n}$ of $j_\ell(x)$ fix the energy levels as $E_{n,\ell,m} = (\hbar z_{\ell,n}/a)^2/2m + V$. For $j_0(x) = \sin x/x$, they are $z_{0,n} = n\pi$. So $E_{n,0,0} = (\hbar n\pi/a)^2/2m + V$. If the coupling to a photon is via a term like $\boldsymbol{p} \cdot \boldsymbol{A}$, then one expects $\Delta\ell = 1$. The energy gap from the $n, \ell = 1$ state to the $n = 1, \ell = 0$ ground state is

$$\Delta E_n = E_{n,1,0} - E_{1,0,0} = (z_{1,n}^2 - \pi^2)\frac{\hbar^2}{2ma^2}. \tag{10.103}$$

Inserting factors of c^2 and using $\hbar c = 197$ eV nm, and $mc^2 = 0.511$ MeV, we find from the zero $z_{1,2} = 7.72525$ that $\Delta E_2 = 1.89\ (\text{nm}/a)^2$ eV, which is red light if $a =$ 1 nm. The next zero $z_{1,3} = 10.90412$ gives $\Delta E_3 = 4.14\ (\text{nm}/a)^2$ eV, which is in the visible if $1.2 < a < 1.5$ nm. The Mathematica command Do[Print[N[BesselJZero[1.5, k]]], {k, 1, 5, 1}] gives the first five zeros of $j_1(x)$ to six significant figures.

10.3 Bessel Functions of the Second Kind

In Section 8.6, we derived integral representations (8.47 and 8.48) for the Hankel functions $H_\lambda^{(1)}(z)$ and $H_\lambda^{(2)}(z)$ for $\text{Re}\, z > 0$. One may analytically continue them (Courant and Hilbert, 1955, chap. VII) to the upper half z-plane

$$H_\lambda^{(1)}(z) = \frac{1}{\pi i} e^{-i\lambda/2} \int_{-\infty}^{\infty} e^{iz\cosh x - \lambda x}\, dx \qquad \text{Im} z \ge 0 \tag{10.104}$$

and to the lower half z-plane

$$H_\lambda^{(2)}(z) = -\frac{1}{\pi i} e^{+i\lambda/2} \int_{-\infty}^{\infty} e^{-iz\cosh x - \lambda x}\, dx \qquad \text{Im} z \le 0. \tag{10.105}$$

When both $z = \rho$ and $\lambda = \nu$ are real, the two Hankel functions are complex conjugates of each other

$$H_\nu^{(1)}(\rho) = H_\nu^{(2)*}(\rho). \tag{10.106}$$

Hankel functions, called **Bessel functions of the third kind**, are linear combinations of Bessel functions of the first $J_\lambda(z)$ and **second** $Y_\lambda(z)$ **kind**

$$\begin{aligned} H_\lambda^{(1)}(z) &= J_\lambda(z) + iY_\lambda(z) \\ H_\lambda^{(2)}(z) &= J_\lambda(z) - iY_\lambda(z). \end{aligned} \tag{10.107}$$

Bessel functions of the second kind also are called **Neumann functions**; the symbols $Y_\lambda(z) = N_\lambda(z)$ refer to the same function. They are infinite at $z = 0$ as illustrated in Fig. 10.4.

When $z = ix$ is imaginary, we get the **modified Bessel functions**

$$\begin{aligned} I_\alpha(x) &= i^{-\alpha} J_\alpha(ix) = \sum_{m=0}^{\infty} \frac{1}{m!\,\Gamma(m+\alpha+1)} \left(\frac{x}{2}\right)^{2m+\alpha} \\ K_\alpha(x) &= \frac{\pi}{2} i^{\alpha+1} H_\alpha^{(1)}(ix) = \int_0^\infty e^{-x\cosh t} \cosh \alpha t \, dt. \end{aligned} \tag{10.108}$$

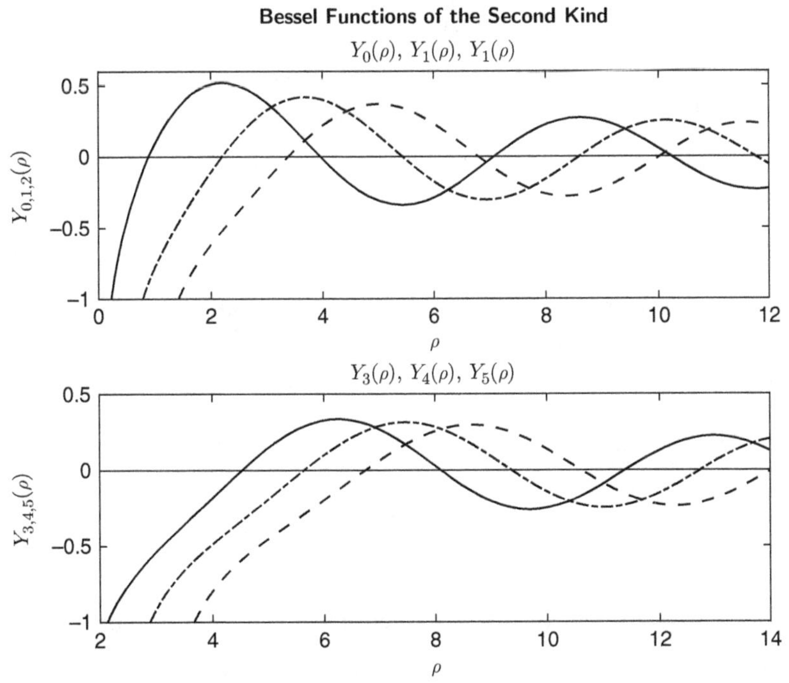

Figure 10.4 Top: $Y_0(\rho)$ (solid curve), $Y_1(\rho)$ (dot-dash), and $Y_2(\rho)$ (dashed) for $0 < \rho < 12$. Bottom: $Y_3(\rho)$ (solid curve), $Y_4(\rho)$ (dot-dash), and $Y_5(\rho)$ (dashed) for $2 < \rho < 14$. Bessel functions cross the ρ-axis at **zeros** or **roots**.

Some simple cases are

$$I_{-1/2}(z) = \sqrt{\frac{2}{\pi z}} \cosh z, \quad I_{1/2}(z) = \sqrt{\frac{2}{\pi z}} \sinh z, \quad \text{and} \quad K_{1/2}(z) = \sqrt{\frac{\pi}{2z}} e^{-z}. \tag{10.109}$$

When do we need to use these functions? If we are representing functions that are finite at the origin $\rho = 0$, then we don't need them. But if the point $\rho = 0$ lies outside the region of interest or if the function we are representing is infinite at that point, then we do need the $Y_\nu(\rho)$'s.

Example 10.7 (Coaxial waveguides) An ideal coaxial waveguide is perfectly conducting for $\rho < r_0$ and $\rho > r$, and the waves occupy the region $r_0 < \rho < r$. Since points with $\rho = 0$ are not in the physical domain of the problem, the electric field $\boldsymbol{E}(\rho, \phi) \exp(i(kz - \omega t))$ is a linear combination of Bessel functions of the first and second kinds with

$$E^z(\rho, \phi) = a\, J_n\left(\sqrt{\omega^2/c^2 - k^2}\, \rho\right) + b\, Y_n\left(\sqrt{\omega^2/c^2 - k^2}\, \rho\right) \tag{10.110}$$

in the notation of Example 10.3. A similar equation represents the magnetic field B^z. The fields $\boldsymbol{E}$ and $\boldsymbol{B}$ obey the equations and boundary conditions of Example 10.3 as well as

$$E^z(r_0, \phi) = 0, \quad E^\phi(r_0, \phi) = 0, \quad \text{and} \quad B^\rho(r_0, \phi) = 0 \tag{10.111}$$

at $\rho = r_0$. In TM modes with $B^z = 0$, one may show (Exercise 10.28) that the boundary conditions $E^z(r_0, \phi) = 0$ and $E^z(r, \phi) = 0$ can be satisfied if

$$J_n(x)\, Y_n(vx) - J_n(vx)\, Y_n(x) = 0 \tag{10.112}$$

in which $v = r/r_0$ and $x = \sqrt{\omega^2/c^2 - k^2}\, r_0$. The Matlab code JYequation.m in Bessel_functions at `github.com/kevinecahill` shows that for $n = 0$ and $v = 10$, the first three solutions are $x_{0,1} = 0.3314$, $x_{0,2} = 0.6858$, and $x_{0,3} = 1.0377$. Setting $n = 1$ and adjusting the guesses in the code, one finds $x_{1,1} = 0.3941$, $x_{1,2} = 0.7331$, and $x_{1,3} = 1.0748$. The corresponding dispersion relations are $\omega_{n,i}(k) = c\sqrt{k^2 + x_{n,i}^2/r_0^2}$.

10.4 Spherical Bessel Functions of the Second Kind

Spherical Bessel functions of the second kind are defined as

$$y_\ell(\rho) = \sqrt{\frac{\pi}{2\rho}}\, Y_{\ell+1/2}(\rho) \tag{10.113}$$

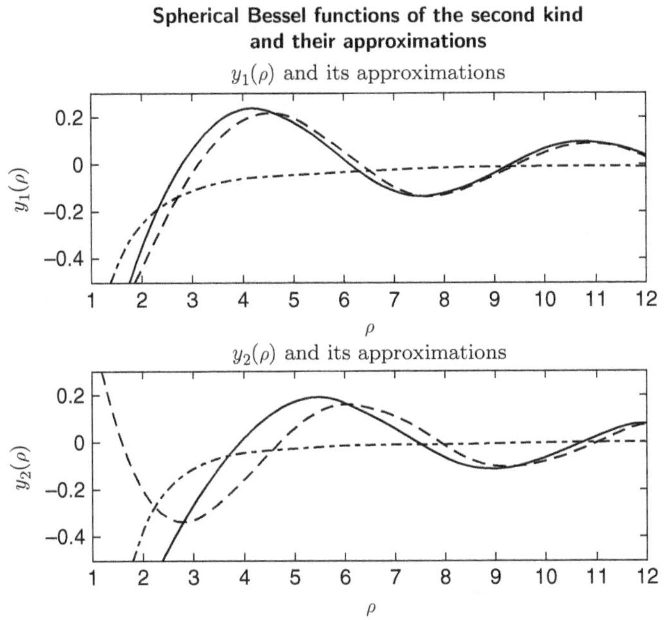

Figure 10.5 Top: Plot of $y_1(\rho)$ (solid curve) and its approximations $-1/\rho^2$ for small ρ (10.116, dot-dash) and $-\cos(\rho - \pi/2)/\rho$ for big ρ (10.115, dashed). Bottom: Plot of $y_2(\rho)$ (solid curve) and its approximations $-3/\rho^3$ for small ρ (10.116, dot-dash) and $-\cos(\rho - \pi)/\rho$ for big ρ (10.115, dashed). The values of ρ at which $y_\ell(\rho) = 0$ are the **zeros** or **roots** of y_ℓ; we use them to fit boundary conditions. All six plots run from $\rho = 1$ to $\rho = 12$.

and Rayleigh formulas express them as

$$y_\ell(\rho) = (-1)^{\ell+1} \rho^\ell \left(\frac{d}{\rho\, d\rho} \right)^\ell \left(\frac{\cos \rho}{\rho} \right). \tag{10.114}$$

The term in which all the derivatives act on the cosine dominates at big ρ

$$y_\ell(\rho) \approx (-1)^{\ell+1} \frac{1}{\rho} \frac{d^\ell \cos \rho}{d\rho^\ell} = -\cos\left(\rho - \ell\pi/2\right)/\rho. \tag{10.115}$$

The second kind of spherical Bessel functions at small ρ are approximately

$$y_\ell(\rho) \approx -(2\ell - 1)!!/\rho^{\ell+1}. \tag{10.116}$$

They all are infinite at $x = 0$ as illustrated in Fig. 10.5.

Example 10.8 (Scattering off a hard sphere) In the notation of Example 10.5, the potential of a hard sphere of radius r_0 is $V(r) = \infty\,\theta(r_0 - r)$ in which $\theta(x) = (x + |x|)/2|x|$ is Heaviside's function. Since the point $r = 0$ is not in the physical region, the scattered wave function is a linear combination of spherical Bessel functions of the first and second kinds

$$u_\ell(r) = c_\ell \, j_\ell(kr) + d_\ell \, y_\ell(kr). \tag{10.117}$$

The boundary condition $u_\ell(kr_0) = 0$ fixes the ratio $v_\ell = d_\ell/c_\ell$ of the constants c_ℓ and d_ℓ. Thus for $\ell = 0$, Rayleigh's formulas (10.71 and 10.114) and the boundary condition say that $kr_0 \, u_0(r_0) = c_0 \, \sin(kr_0) - d_0 \, \cos(kr_0) = 0$ or $d_0/c_0 = \tan kr_0$. The s-wave then is $u_0(kr) = c_0 \, \sin(kr - kr_0)/(kr \cos kr_0)$, which tells us that the tangent of the phase shift is $\tan \, \delta_0(k) = -kr_0$. By (10.100), the cross-section at low energy is $\sigma \approx 4\pi r_0^2$ or four times the classical value.

Similarly, one finds (Exercise 10.29) that the tangent of the p-wave phase shift is

$$\tan \, \delta_1(k) = \frac{kr_0 \cos kr_0 - \sin kr_0}{\cos kr_0 - kr_0 \sin kr_0}. \tag{10.118}$$

For $kr_0 \ll 1$, we have $\delta_1(k) \approx -(kr_0)^3/3$; more generally, the ℓth phase shift is $\delta_\ell(k) \approx -(kr_0)^{2\ell+1}/\left\{(2\ell+1)[(2\ell-1)!!]^2\right\}$ for a potential of range r_0 at low energy $k \ll 1/r_0$.

Further Reading

A great deal is known about Bessel functions. Students may find *Mathematical Methods for Physics and Engineering* (Riley et al., 2006) as well as the classics *A Treatise on the Theory of Bessel Functions* (Watson, 1995), *A Course of Modern Analysis* (Whittaker and Watson, 1927, chap. XVII), and *Methods of Mathematical Physics* (Courant and Hilbert, 1955) of special interest. The NIST Digital Library of Mathematical Functions (dlmf.nist.gov) and the companion *NIST Handbook of Mathematical Functions* (Olver et al., 2010) are superb.

Exercises

10.1 Show that the series (10.1) for $J_n(\rho)$ satisfies Bessel's equation (10.4).

10.2 Show that the generating function $\exp(z(u - 1/u)/2)$ for Bessel functions is invariant under the substitution $u \to -1/u$.

10.3 Use the invariance of $\exp(z(u - 1/u)/2)$ under $u \to -1/u$ to show that $J_{-n}(z) = (-1)^n J_n(z)$.

10.4 By writing the generating function (10.5) as the product of the exponentials $\exp(zu/2)$ and $\exp(-z/2u)$, derive the expansion

$$\exp\left[\frac{z}{2}\left(u - u^{-1}\right)\right] = \sum_{n=0}^{\infty} \sum_{m=-n}^{\infty} \left(\frac{z}{2}\right)^{m+n} \frac{u^{m+n}}{(m+n)!} \left(-\frac{z}{2}\right)^n \frac{u^{-n}}{n!}. \tag{10.119}$$

10.5 From this expansion (10.119) of the generating function (10.5), derive the power-series expansion (10.1) for $J_n(z)$.

10.6 In the formula (10.5) for the generating function $\exp(z(u-1/u)/2)$, replace u by $\exp i\theta$ and then derive the integral representation (10.6) for $J_n(z)$. Start with the interval $[-\pi, \pi]$.

10.7 From the general integral representation (10.6) for $J_n(z)$, derive the two integral formulas (10.7) for $J_0(z)$.

10.8 Show that the integral representations (10.6 and 10.7) imply that for any integer $n \neq 0$, $J_n(0) = 0$, while $J_0(0) = 1$.

10.9 By differentiating the generating function (10.5) with respect to u and identifying the coefficients of powers of u, derive the recursion relation

$$J_{n-1}(z) + J_{n+1}(z) = \frac{2n}{z} J_n(z). \tag{10.120}$$

10.10 By differentiating the generating function (10.5) with respect to z and identifying the coefficients of powers of u, derive the recursion relation

$$J_{n-1}(z) - J_{n+1}(z) = 2\, J_n'(z). \tag{10.121}$$

10.11 Change variables to $z = ax$ and turn Bessel's equation (10.4) into the self-adjoint form (10.11).

10.12 If $y = J_n(ax)$, then Equation (10.11) is $(xy')' + (xa^2 - n^2/x)y = 0$. Multiply this equation by xy', integrate from 0 to b, and so show that if $ab = z_{n,m}$ and $J_n(z_{n,m}) = 0$, then

$$2 \int_0^b x\, J_n^2(ax)\, dx = b^2 J_n'^2(z_{n,m}) \tag{10.122}$$

which is the normalization condition (10.14).

10.13 Use the expansion (10.15) of the delta function to find formulas for the coefficients a_k in the expansion for $0 < \alpha < 1$

$$f(x) = \sum_{k=1}^{\infty} a_k\, x^{\alpha}\, J_n(z_{n,k}x/b) \tag{10.123}$$

of twice-differentiable functions that vanish at at $x = b$.

10.14 Show that with $\lambda \equiv z^2/r_d^2$, the change of variables $\rho = zr/r_d$ and $u(r) = J_n(\rho)$ turns $-\left(r\, u'\right)' + n^2\, u/r = \lambda\, r\, u$ into Bessel's equation (10.29).

10.15 Use the formula (2.46) for the curl in cylindrical coordinates and the vacuum forms $\nabla \times \boldsymbol{E} = -\dot{\boldsymbol{B}}$ and $\nabla \times \boldsymbol{B} = \dot{\boldsymbol{E}}/c^2$ of the laws of Faraday and Maxwell–Ampère to derive the field equations (10.59).

10.16 Derive Equations (10.60) from (10.59).

10.17 Show that $J_n\left(\sqrt{\omega^2/c^2 - k^2}\,\rho\right) e^{in\phi} e^{i(kz-\omega t)}$ is a traveling-wave solution (10.56) of the wave equations (10.61).

10.18 Find expressions for the nonzero TM fields in terms of the formula (10.63) for E_z.

10.19 Show that the TE field $E_z = 0$ and $B_z = J_n\left(\sqrt{\omega^2/c^2 - k^2}\,\rho\right) e^{in\phi} e^{i(kz-\omega t)}$ will satisfy the boundary conditions (10.58) if $\sqrt{\omega^2/c^2 - k^2}\,r$ is a zero $z'_{n,m}$ of J'_n.

10.20 Show that if ℓ is an integer and if $\sqrt{\omega^2/c^2 - \pi^2\ell^2/h^2}\,r$ is a zero $z'_{n,m}$ of J'_n, then the fields $E_z = 0$ and $B_z = J_n(z'_{n,m}\rho/r)\, e^{in\phi} \sin(\ell\pi z/h)\, e^{-i\omega t}$ satisfy both the boundary conditions (10.58) at $\rho = r$ and those (10.64) at $z = 0$ and h as well as the wave equations (10.61). Hint: Use Maxwell's equations $\nabla \times \boldsymbol{E} = -\dot{\boldsymbol{B}}$ and $\nabla \times \boldsymbol{B} = \dot{\boldsymbol{E}}/c^2$ as in (10.59).

10.21 Show that the resonant frequencies of the TM modes of the cavity of Example 10.4 are $\omega_{n,m,\ell} = c\sqrt{z_{n,m}^2/r^2 + \pi^2\ell^2/h^2}$.

10.22 By setting $n = \ell + 1/2$ and $j_\ell = \sqrt{\pi/2x}\, J_{\ell+1/2}$, show that Bessel's equation (10.4) implies that the spherical Bessel function j_ℓ satisfies (10.67).

10.23 Show that Rayleigh's formula (10.71) implies the recursion relation (10.72).

10.24 Use the recursion relation (10.72) to show by induction that the spherical Bessel functions $j_\ell(x)$ as given by Rayleigh's formula (10.71) satisfy their differential equation (10.70) which with $x = kr$ is

$$-x^2 j''_\ell - 2x j'_\ell + \ell(\ell+1)\, j_\ell = x^2 j_\ell. \tag{10.124}$$

Hint: start by showing that $j_0(x) = \sin(x)/x$ satisfies this equation. This problem involves some tedium.

10.25 Iterate the trick (10.83) and derive the identity (10.84).

10.26 Use the expansions (10.89 and 10.90) to show that the inner product of the ket $|\boldsymbol{r}\rangle$ that represents a particle at $\boldsymbol{r}$ with polar angles θ and ϕ and the one $|\boldsymbol{k}\rangle$ that represents a particle with momentum $\boldsymbol{p} = \hbar\boldsymbol{k}$ with polar angles θ' and ϕ' is with $\boldsymbol{k}\cdot\boldsymbol{r} = kr\cos\theta$

$$\begin{aligned}\langle \boldsymbol{r}|\boldsymbol{k}\rangle &= \frac{1}{(2\pi)^{3/2}} e^{ikr\cos\theta} = \frac{1}{(2\pi)^{3/2}} \sum_{\ell=0}^{\infty} (2\ell+1)\, P_\ell(\cos\theta)\, i^\ell\, j_\ell(kr) \\ &= \frac{1}{(2\pi)^{3/2}} e^{i\boldsymbol{k}\cdot\boldsymbol{r}} = \sqrt{\frac{2}{\pi}} \sum_{\ell=0}^{\infty} i^\ell\, j_\ell(kr)\, Y_{\ell,m}(\theta,\phi)\, Y^*_{\ell,m}(\theta',\phi').\end{aligned} \tag{10.125}$$

10.27 Show that $(-1)^\ell d^\ell \sin\rho/d\rho^\ell = \sin(\rho - \pi\ell/2)$ and so complete the derivation of the approximation (10.75) for $j_\ell(\rho)$ for big ρ.

10.28 In the context of Examples 10.3 and 10.7, show that the boundary conditions $E_z(r_0, \phi) = 0$ and $E_z(r, \phi) = 0$ imply (10.112).

10.29 Show that for scattering off a hard sphere of radius r_0 as in Example 10.8, the p-wave phase shift is given by (10.118).

11

Group Theory

11.1 What Is a Group?

A group G is a set of elements $f, g, h, \ldots$, and an operation called multiplication such that for all elements f, g, and h in the group G:

1. The product fg is in the group G (**closure**);
2. $f(gh) = (fg)h$ (**associativity**);
3. there is an **identity** element e in the group G such that $ge = eg = g$; and
4. every g in G has an **inverse** g^{-1} in G such that $gg^{-1} = g^{-1}g = e$.

Physical transformations naturally form groups. The elements of a group might be all physical transformations on a given set of objects that leave invariant a chosen property of the set of objects. For instance, the objects might be the points (x, y) in a plane. The chosen property could be their distances $\sqrt{x^2 + y^2}$ from the origin. The physical transformations that leave unchanged these distances are the rotations about the origin

$$\begin{pmatrix} x' \\ y' \end{pmatrix} = \begin{pmatrix} \cos\theta & \sin\theta \\ -\sin\theta & \cos\theta \end{pmatrix} \begin{pmatrix} x \\ y \end{pmatrix}. \tag{11.1}$$

These rotations form the special orthogonal group in 2 dimensions, $SO(2)$.

More generally, suppose the transformations $T, T', T'', \ldots$ change a set of objects in ways that leave invariant a chosen property of the objects. Suppose the product $T'\,T$ of the transformations T and T' represents the action of T followed by the action of T' on the objects. Since both T and T' leave the chosen property unchanged, so will their product $T'\,T$. Thus the closure condition is satisfied. The triple products $T''\,(T'\,T)$ and $(T''\,T')\,T$ both represent the action of T followed by the action of T' followed by action of T''. Thus the action of $T''\,(T'\,T)$ is the same as the action of $(T''\,T')\,T$, and so the transformations are associative. The identity element e is the null transformation, the one that does nothing. The inverse T^{-1} is the transformation that reverses the action of T. Thus physical transformations that leave a chosen property unchanged naturally form a group.

Example 11.1 (Permutations) A **permutation** of an ordered set of n objects changes the order but leaves the set unchanged.

Example 11.2 (Groups of coordinate transformations) The set of all transformations that leave invariant the distance from the origin of every point in n-dimensional space is the group $O(n)$ of **rotations** and **reflections**. The rotations in n-space form the special orthogonal group $SO(n)$.

Linear transformations $\boldsymbol{x}' = \boldsymbol{x} + \boldsymbol{a}$, for different n-dimensional vectors $\boldsymbol{a}$, leave invariant the spatial difference $\boldsymbol{x} - \boldsymbol{y}$ between every pair of points $\boldsymbol{x}$ and $\boldsymbol{y}$ in n-dimensional space. They form the group of **translations**. Here, group multiplication is vector addition.

The set of all linear transformations that leave invariant the square of the Minkowski distance $x_1^2 + x_2^2 + x_3^2 - x_0^2$ between any 4-vector x and the origin is the **Lorentz group** (Hermann Minkowski 1864–1909, Hendrik Lorentz 1853–1928).

The set of all linear transformations that leave invariant the square of the Minkowski distance $(x_1 - y_1)^2 + (x_2 - y_2)^2 + (x_3 - y_3)^2 - (x^0 - y^0)^2$ between any two 4-vectors x and y is the **Poincaré group**, which includes Lorentz transformations and translations (Henri Poincaré 1854–1912).

In the group of translations, the order of multiplication (which is vector addition) does not matter. A group whose elements all **commute**

$$[g, h] \equiv g\,h - h\,g = 0 \tag{11.2}$$

is said to be **abelian**. Except for the group of translations and the group $SO(2)$, the order of the physical transformations in these examples does matter: the transformation $T'\,T$ is not in general the same as $T\,T'$. Such groups are **nonabelian** (Niels Abel 1802–1829).

Matrices naturally form groups with group multiplication defined as matrix multiplication. Since matrix multiplication is associative, any set of $n \times n$ nonsingular matrices D that includes the inverse D^{-1} of every matrix in the set as well as the identity matrix I automatically satisfies three of the four properties that characterize a group, leaving only the closure property uncertain. Such a set $\{D\}$ of matrices will form a group as long as the product of any two matrices is in the set. As with physical transformations, one way to ensure closure is to have every matrix leave something unchanged.

Example 11.3 (Orthogonal groups) The group of all real matrices that leave unchanged the squared distance $x_1^2 + \cdots + x_n^2$ of a point $x = (x_1, \ldots, x_n)$ from the origin is the group $O(n)$ of all $n \times n$ orthogonal (1.37) matrices (Exercises 11.1 and 11.2).

The group $O(2n)$ leaves unchanged the anticommutation relations (Section 11.19) of the real and imaginary parts of n complex fermionic operators $\psi_1, \ldots, \psi_n$. The $n \times n$ orthogonal matrices of unit determinant form the special orthogonal group $SO(n)$. The group $SO(3)$ describes rotations in 3-space.

Example 11.4 (Unitary groups) The set of all $n \times n$ complex matrices that leave invariant the quadratic form $z_1^* z_1 + z_2^* z_2 + \cdots + z_n^* z_n$ forms the unitary group $U(n)$ of all $n \times n$ unitary (1.36) matrices (Exercises 11.3 and 11.4). Those of unit determinant form the special unitary group $SU(n)$ (Exercise 11.5).

Like $SO(3)$, the group $SU(2)$ represents rotations. The group $SU(3)$ is the symmetry group of the strong interactions, quantum chromodynamics. Physicists have used the groups $SU(5)$ and $SO(10)$ to unify the electroweak and strong interactions; whether nature also does so is unclear.

Example 11.5 (Symplectic groups) The set of all $2n \times 2n$ real matrices R that leave invariant the commutation relations $[q_i, p_k] = i\hbar\delta_{ik}$, $[q_i, q_k] = 0$, and $[p_i, p_k] = 0$ of quantum mechanics is the symplectic group $Sp(2n, \mathbb{R})$.

The number of elements in a group is the **order** of the group. A **finite group** is a group with a finite number of elements, or equivalently a group of finite order.

Example 11.6 (Z_2, Z_n, and $\mathbb{Z}$) The **parity** group whose elements are 1 and -1 under ordinary multiplication is the cyclic group Z_2. It is abelian and of order 2. The cyclic group Z_n for any positive integer n is made of the phases $\exp(i2k\pi/n)$ for $k = 1, 2, \ldots, n$. It is abelian and of order n. The integers $\mathbb{Z}$ form a group $\mathbb{Z}$ with multiplication defined as addition.

A group whose elements $g = g(\{\alpha\})$ depend continuously upon a set of parameters α_k is a **continuous group** or a **Lie group**. Continuous groups are of infinite order.

A group G of matrices D is **compact** if the (squared) norm as given by the trace

$$\mathrm{Tr}\left(D^\dagger D\right) \leq M \tag{11.3}$$

is bounded for all the $D \in G$.

Example 11.7 ($SO(n)$, $O(n)$, $SU(n)$, and $U(n)$) The groups $SO(n)$, $O(n)$, $SU(n)$, and $U(n)$ are continuous Lie groups of infinite order. Since for any matrix D in one of these groups

$$\mathrm{Tr}\left(D^\dagger D\right) = \mathrm{Tr}\, I = n \leq M \tag{11.4}$$

these groups also are compact.

Example 11.8 (Noncompact groups) The set of all real $n \times n$ matrices forms the general linear group $GL(n, \mathbb{R})$; those of unit determinant form the special linear group $SL(n, \mathbb{R})$. The corresponding groups of matrices with complex entries are $GL(n, \mathbb{C})$ and $SL(n, \mathbb{C})$. These four groups and the symplectic groups $Sp(2n, \mathbb{R})$ and $Sp(2n, \mathbb{C})$ have matrix elements that are unbounded; they are noncompact. They are continuous Lie groups of infinite order like the orthogonal and unitary groups. The group $SL(2, \mathbb{C})$ represents Lorentz transformations.

Incidentally, a **semigroup** is a set of elements $G = \{f, g, h, \dots\}$ and an operation called multiplication that is closed, $fg \in G$, and associative, $f(gh) = (fg)h$, but that may lack an identity element and inverses.

11.2 Representations of Groups

If one can associate with every element g of a group G a square matrix $D(g)$ and have matrix multiplication imitate group multiplication

$$D(f)\, D(g) = D(fg) \tag{11.5}$$

for all elements f and g of the group G, then the set of matrices $D(g)$ is said to form a **representation** of the group G. If the matrices of the representation are $n \times n$, then n is the **dimension** of the representation. The dimension of a representation also is the dimension of the vector space on which the matrices act. If the matrices $D(g)$ are unitary $D^\dagger(g) = D^{-1}(g)$, then they form a **unitary representation** of the group.

Example 11.9 (Representations of the groups $SU(2)$ and $SO(3)$) The defining representations of $SU(2)$ and $SO(3)$ are the 2×2 complex matrix

$$D(\boldsymbol{\theta}) = \begin{pmatrix} \cos\frac{1}{2}\theta + i\hat{\theta}_3 \sin\frac{1}{2}\theta & i(\hat{\theta}_1 - i\hat{\theta}_2)\sin\frac{1}{2}\theta \\ i(\hat{\theta}_1 + i\hat{\theta}_2)\sin\frac{1}{2}\theta & \cos\frac{1}{2}\theta - i\hat{\theta}_3 \sin\frac{\theta}{2} \end{pmatrix} \tag{11.6}$$

and the 3×3 real matrix

$$D(\boldsymbol{\theta}) = \begin{pmatrix} c + \hat{\theta}_1^2(1-c) & \hat{\theta}_1\hat{\theta}_2(1-c) - \hat{\theta}_3 s & \hat{\theta}_1\hat{\theta}_3(1-c) + \hat{\theta}_2 s \\ \hat{\theta}_2\hat{\theta}_1(1-c) + \hat{\theta}_3 s & c + \hat{\theta}_2^2(1-c) & \hat{\theta}_2\hat{\theta}_3(1-c) - \hat{\theta}_1 s \\ \hat{\theta}_3\hat{\theta}_1(1-c) - \hat{\theta}_2 s & \hat{\theta}_3\hat{\theta}_2(1-c) + \hat{\theta}_1 s & c + \hat{\theta}_3^2(1-c) \end{pmatrix} \tag{11.7}$$

in which $\theta = |\boldsymbol{\theta}|$, $\hat{\theta}_i = \theta_i/\theta$, $c = \cos\theta$, and $s = \sin\theta$.

Compact groups possess finite-dimensional unitary representations; noncompact groups do not. A group of bounded (11.3) matrices is compact. An abstract group of elements $g(\{\alpha\})$ is compact if its space of parameters $\{\alpha\}$ is

closed and bounded. (A set is **closed** if the limit of every convergent sequence of its points lies in the set. A set is **open** if each of its elements lies in a neighborhood that lies in the set. For example, the interval $[a, b] \equiv \{x | a \le x \le b\}$ is closed, while $(a, b) \equiv \{x | a < x < b\}$ is open.) The group of rotations is compact, but the group of translations and the Lorentz group are noncompact.

Every $n \times n$ matrix S that is nonsingular ($\det S \neq 0$) maps any $n \times n$ representation $D(g)$ of a group G into an **equivalent representation** $D'(g)$ through the **similarity transformation**

$$D'(g) = S^{-1} D(g) S \tag{11.8}$$

which preserves the law of multiplication

$$\begin{aligned} D'(f)\, D'(g) &= S^{-1} D(f) S\; S^{-1} D(g) S \\ &= S^{-1} D(f)\, D(g) S = S^{-1} D(fg) S = D'(fg). \end{aligned} \tag{11.9}$$

A **proper subspace** W of a vector space V is a subspace of lower (but not zero) dimension. A proper subspace W is **invariant** under the action of a representation $D(g)$ if $D(g)$ maps every vector $v \in W$ to a vector $D(g)\, v = v' \in W$. A representation that has a **proper invariant subspace** is **reducible**. A representation that is not reducible is **irreducible**.

A representation $D(g)$ is **completely reducible** if it is equivalent to a representation whose matrices are in **block-diagonal form**

$$S^{-1} D(g)\, S = \begin{pmatrix} D_1(g) & 0 & \dots \\ 0 & D_2(g) & \dots \\ \vdots & \vdots & \vdots \end{pmatrix} \tag{11.10}$$

in which each representation $D_i(g)$ is irreducible. A representation in block-diagonal form is said to be a **direct sum** of its irreducible representations

$$S^{-1} D\, S = D_1 \oplus D_2 \oplus \cdots . \tag{11.11}$$

11.3 Representations Acting in Hilbert Space

A symmetry transformation g is a map (1.190) of states $\psi \to \psi'$ that preserves probabilities

$$|\langle \phi' | \psi' \rangle|^2 = |\langle \phi | \psi \rangle|^2. \tag{11.12}$$

The action of a group G of symmetry transformations g on the Hilbert space of a quantum theory can be represented either by operators $U(g)$ that are linear and unitary (the usual case) or by ones $K(g)$ that are antilinear (1.188) and antiunitary (1.189), as in the case of time reversal. Wigner proved this theorem in the 1930s, and Weinberg improved it in his 1995 classic (Weinberg, 1995, p. 51) (Eugene Wigner, 1902–1995; Steven Weinberg, 1933–).

Two operators F_1 and F_2 that commute $F_1 F_2 = F_2 F_1$ are **compatible** (1.375). A set of compatible operators $F_1, F_2, \ldots$ is **complete** if to every set of eigenvalues there belongs only a single eigenvector (Section 1.31).

Example 11.10 (Rotation operators) Suppose that the hamiltonian H, the square of the angular momentum $\boldsymbol{J}^2$, and its z-component J_z form a complete set of compatible observables, so that the identity operator can be expressed as a sum over the eigenvectors of these operators

$$I = \sum_{E,j,m} |E, j, m\rangle\langle E, j, m|. \tag{11.13}$$

Then the matrix element of a unitary operator $U(g)$ between two states $|\psi\rangle$ and $|\phi\rangle$ is

$$\langle\phi|U(g)|\psi\rangle = \langle\phi| \sum_{E',j',m'} |E', j', m'\rangle\langle E', j', m'|\, U(g) \sum_{E,j,m} |E, j, m\rangle\langle E, j, m|\psi\rangle. \tag{11.14}$$

Let H and $\boldsymbol{J}^2$ be invariant under the action of $U(g)$ so that $U^\dagger(g)HU(g) = H$ and $U^\dagger(g)\boldsymbol{J}^2U(g) = \boldsymbol{J}^2$. Then $HU(g) = U(g)H$ and $\boldsymbol{J}^2U(g) = U(g)\boldsymbol{J}^2$, and so if $H|E, j, m\rangle = E|E, j, m\rangle$ and $\boldsymbol{J}^2|E, j, m\rangle = j(j+1)|E, j, m\rangle$, we have

$$\begin{aligned} HU(g)|E, j, m\rangle &= U(g)H|E, j, m\rangle = EU(g)|E, j, m\rangle \\ \boldsymbol{J}^2U(g)|E, j, m\rangle &= U(g)\boldsymbol{J}^2|E, j, m\rangle = j(j+1)U(g)|E, j, m\rangle. \end{aligned} \tag{11.15}$$

Thus $U(g)$ cannot change E or j, and so

$$\langle E', j', m'|U(g)|E, j, m\rangle = \delta_{E'E}\delta_{j'j}\langle j, m'|U(g)|j, m\rangle = \delta_{E'E}\delta_{j'j}D^{(j)}_{m'm}(g). \tag{11.16}$$

The matrix element (11.14) is a single sum over E and j in which the irreducible representations $D^{(j)}_{m'm}(g)$ of the rotation group $SU(2)$ appear

$$\langle\phi|U(g)|\psi\rangle = \sum_{E,j,m',m} \langle\phi|E, j, m'\rangle D^{(j)}_{m'm}(g)\langle E, j, m|\psi\rangle. \tag{11.17}$$

This is how the block-diagonal form (11.10) usually appears in calculations. The matrices $D^{(j)}_{m'm}(g)$ inherit the unitarity of the operator $U(g)$.

11.4 Subgroups

If all the elements of a group S also are elements of a group G, then S is a **subgroup** of G. Every group G has two **trivial subgroups** – the identity element e and the whole group G itself. Many groups have more interesting subgroups. For example, the rotations about a fixed axis is an abelian subgroup of the group of all rotations in 3-dimensional space.

A subgroup $S \subset G$ is an **invariant** subgroup if every element s of the subgroup S is left inside the subgroup under the **action** of every element g of the whole group G, that is, if

$$g^{-1} s\, g = s' \in S \quad \text{for all} \quad g \in G. \tag{11.18}$$

This condition often is written as $g^{-1} S g = S$ for all $g \in G$ or as

$$S\, g = g\, S \quad \text{for all } g \in G. \tag{11.19}$$

Invariant subgroups also are called **normal subgroups**.

A set $C \subset G$ is called a **conjugacy class** if it's invariant under the action of the whole group G, that is, if $Cg = g\,C$ or

$$g^{-1} C\, g = C \quad \text{for all } g \in G. \tag{11.20}$$

A subgroup that is the union of a set of conjugacy classes is invariant.

The **center** C of a group G is the set of all elements $c \in G$ that **commute** with every element g of the group, that is, their **commutators**

$$[c, g] \equiv cg - gc = 0 \tag{11.21}$$

vanish for all $g \in G$.

Example 11.11 (Centers are abelian subgroups) Does the center C always form an abelian subgroup of its group G? The product $c_1 c_2$ of any two elements c_1 and c_2 of the center commutes with every element g of G since $c_1 c_2 g = c_1 g c_2 = g c_1 c_2$. So the center is closed under multiplication. The identity element e commutes with every $g \in G$, so $e \in C$. If $c' \in C$, then $c' g = g c'$ for all $g \in G$, and so multiplication of this equation from the left and the right by c'^{-1} gives $g c'^{-1} = c'^{-1} g$, which shows that $c'^{-1} \in C$. The subgroup C is abelian because each of its elements commutes with all the elements of G including those of C itself.

So the center of any group always is one of its abelian invariant subgroups. The center may be trivial, however, consisting either of the identity or of the whole group. But a group with a nontrivial center can not be simple or semisimple (Section 11.28).

11.5 Cosets

If H is a subgroup of a group G, then for every element $g \in G$ the set of elements $Hg \equiv \{hg | h \in H\}$ is a **right coset of the subgroup $H \subset G$**. (Here $\subset$ means *is a subset of* or equivalently *is contained in*.)

If H is a subgroup of a group G, then for every element $g \in G$ the set of elements gH is a **left coset of the subgroup $H \subset G$**.

The number of elements in a coset is the same as the number of elements of H, which is the order of H.

An element g of a group G is in one and only one right coset (and in one and only one left coset) of the subgroup $H \subset G$. For suppose instead that g were in two right cosets $g \in Hg_1$ and $g \in Hg_2$, so that $g = h_1 g_1 = h_2 g_2$ for suitable $h_1, h_2 \in H$ and $g_1, g_2 \in G$. Then since H is a (sub)group, we have $g_2 = h_2^{-1} h_1 g_1 = h_3 g_1$, which says that $g_2 \in Hg_1$. But this means that every element $hg_2 \in Hg_2$ is of the form $hg_2 = hh_3 g_1 = h_4 g_1 \in Hg_1$. So every element $hg_2 \in Hg_2$ is in Hg_1: the two right cosets are identical, $Hg_1 = Hg_2$.

The right (or left) cosets are the points of the **quotient coset space G/H**.

If H is an invariant subgroup of G, then by definition (11.19) $Hg = gH$ for all $g \in G$, and so the left cosets are the same sets as the right cosets. In this case, the coset space G/H is itself a group with multiplication defined by

$$
\begin{aligned}
(Hg_1)(Hg_2) &= \left\{ h_i g_1 h_j g_2 | h_i, h_j \in H \right\} \\
&= \left\{ h_i g_1 h_j g_1^{-1} g_1 g_2 | h_i, h_j \in H \right\} \\
&= \{ h_i h_k g_1 g_2 | h_i, h_k \in H \} \\
&= \{ h_\ell g_1 g_2 | h_\ell \in H \} = Hg_1 g_2
\end{aligned} \tag{11.22}
$$

which is the multiplication rule of the group G. This group G/H is called the **quotient** or **factor** group of ***G*** by ***H***

11.6 Morphisms

An **isomorphism** is a one-to-one map between groups that respects their multiplication laws. For example, a similarity transformation (11.8) relates two equivalent representations

$$
D'(g) = S^{-1} D(g) S \tag{11.23}
$$

and is an isomorphism (Exercise 11.8). An **automorphism** is an isomorphism between a group and itself. The map $g_i \to g\, g_i\, g^{-1}$ is one to one because $g\, g_1\, g^{-1} = g\, g_2\, g^{-1}$ implies that $g\, g_1 = g\, g_2$, and so that $g_1 = g_2$. This map also preserves the law of multiplication since $g\,g_1\,g^{-1}\,g\,g_2\,g^{-1} = g\,g_1\,g_2\,g^{-1}$. So the map

$$G \to gGg^{-1} \tag{11.24}$$

is an automorphism. It is called an **inner automorphism** because g is an element of G. An automorphism not of this form (11.24) is an **outer automorphism**.

11.7 Schur's Lemma

Part 1: If D_1 and D_2 are inequivalent, irreducible representations of a group G, and if $D_1(g)A = AD_2(g)$ for some matrix A and for all $g \in G$, then the matrix A must vanish, $A = 0$.

Proof: First suppose that A annihilates some vector $|x\rangle$, that is, $A|x\rangle = 0$. Let P be the projection operator into the subspace that A annihilates, which is of at least 1 dimension. This subspace, incidentally, is called the **null space** $\mathcal{N}(A)$ or the **kernel** of the matrix A. The representation D_2 must leave this null space $\mathcal{N}(A)$ invariant since

$$AD_2(g)P = D_1(g)AP = 0. \tag{11.25}$$

If $\mathcal{N}(A)$ were a proper subspace, then it would be a proper invariant subspace of the representation D_2, and so D_2 would be reducible, which is contrary to our assumption that D_1 and D_2 are irreducible. So the null space $\mathcal{N}(A)$ must be the whole space upon which A acts, that is, $A = 0$.

A similar argument shows that if $\langle y|A = 0$ for some bra $\langle y|$, then $A = 0$.

So either A is zero or it annihilates no ket and no bra. In the latter case, A must be square and invertible, which would imply that $D_2(g) = A^{-1}D_1(g)A$, that is, that D_1 and D_2 are equivalent representations, which is contrary to our assumption that they are inequivalent. The only way out is that A vanishes.

Part 2: If for a finite-dimensional, irreducible representation $D(g)$ of a group G, we have $D(g)A = AD(g)$ for some matrix A and for all $g \in G$, then $A = cI$. That is, any matrix that commutes with every element of a finite-dimensional, irreducible representation must be a multiple of the identity matrix.

Proof: Every square matrix A has at least one eigenvector $|x\rangle$ and eigenvalue c so that $A|x\rangle = c|x\rangle$ because its characteristic equation $\det(A - cI) = 0$ always has at least one root by the fundamental theorem of algebra (6.79). So the null space $\mathcal{N}(A - cI)$ has dimension greater than zero. The assumption $D(g)A = AD(g)$ for all $g \in G$ implies that $D(g)(A - cI) = (A - cI)D(g)$ for all $g \in G$. Let P be the projection operator onto the null space $\mathcal{N}(A-cI)$. Then we have $(A-cI)D(g)P = D(g)(A - cI)P = 0$ for all $g \in G$ which implies that $D(g)P$ maps vectors into the null space $\mathcal{N}(A - cI)$. This null space therefore is a subspace that is invariant under $D(g)$, which means that D is reducible unless the null space $\mathcal{N}(A - cI)$ is the whole space. Since by assumption D is irreducible, it follows that $\mathcal{N}(A - cI)$ is the whole space, that is, that $A = cI$ (Issai Schur, 1875–1941). □

Example 11.12 (Schur, Wigner, and Eckart) Suppose an arbitrary observable O is invariant under the action of the rotation group $SU(2)$ represented by unitary operators $U(g)$ for $g \in SU(2)$

$$U^\dagger(g)\, O\, U(g) = O \quad \text{or} \quad [O, U(g)] = 0. \tag{11.26}$$

These unitary rotation operators commute with the square $\boldsymbol{J}^2$ of the angular momentum $[\boldsymbol{J}^2, U] = 0$. Suppose that they also leave the hamiltonian H unchanged $[H, U] = 0$. Then as shown in Example 11.10, the state $U|E, j, m\rangle$ is a sum of states all with the same values of j and E. It follows that

$$\begin{aligned} &\sum_{m'} \langle E, j, m|O|E', j', m'\rangle\langle E', j', m'|U(g)|E', j', m''\rangle \\ &\quad = \sum_{m'} \langle E, j, m|U(g)|E, j, m'\rangle\langle E, j, m'|O|E', j', m''\rangle \end{aligned} \tag{11.27}$$

or in the notation of (11.16)

$$\sum_{m'} \langle E, j, m|O|E', j', m'\rangle D^{(j')}(g)_{m'm''} = \sum_{m'} D^{(j)}(g)_{mm'}\langle E, j, m'|O|E', j', m''\rangle. \tag{11.28}$$

Now Part 1 of Schur's lemma tells us that the matrix $\langle E, j, m|O|E', j', m'\rangle$ must vanish unless the representations are equivalent, which is to say unless $j = j'$. So we have

$$\sum_{m'} \langle E, j, m|O|E', j, m'\rangle D^{(j)}(g)_{m'm''} = \sum_{m'} D^{(j)}(g)_{mm'}\langle E, j, m'|O|E', j, m''\rangle. \tag{11.29}$$

Now Part 2 of Schur's lemma tells us that the matrix $\langle E, j, m|O|E', j, m'\rangle$ must be a multiple of the identity. Thus the symmetry of O under rotations simplifies the matrix element to

$$\langle E, j, m|O|E', j', m'\rangle = \delta_{jj'}\delta_{mm'} O_j(E, E'). \tag{11.30}$$

This result is a special case of the **Wigner–Eckart theorem** (Eugene Wigner 1902–1995, Carl Eckart 1902–1973).

11.8 Characters

Suppose the $n \times n$ matrices $D_{ij}(g)$ form a representation of a group $G \ni g$. The **character** $\chi_D(g)$ of the matrix $D(g)$ is the trace

$$\chi_D(g) = \mathrm{Tr}D(g) = \sum_{i=1}^{n} D_{ii}(g). \tag{11.31}$$

Traces are cyclic, that is, $\mathrm{Tr}ABC = \mathrm{Tr}BCA = \mathrm{Tr}CAB$. So if two representations D and D' are equivalent, so that $D'(g) = S^{-1}D(g)S$, then they have the same characters because

$$\chi_{D'}(g) = \mathrm{Tr}D'(g) = \mathrm{Tr}\left(S^{-1}D(g)S\right) = \mathrm{Tr}\left(D(g)SS^{-1}\right) = \mathrm{Tr}D(g) = \chi_D(g). \tag{11.32}$$

If two group elements g_1 and g_2 are in the same conjugacy class, that is, if $g_2 = gg_1g^{-1}$ for all $g \in G$, then they have the same character in a given representation $D(g)$ because

$$\begin{aligned}\chi_D(g_2) &= \mathrm{Tr}\,D(g_2) = \mathrm{Tr}\,D(gg_1g^{-1}) = \mathrm{Tr}\left(D(g)D(g_1)D(g^{-1})\right)\\ &= \mathrm{Tr}\left(D(g_1)D^{-1}(g)D(g)\right) = \mathrm{Tr}\,D(g_1) = \chi_D(g_1).\end{aligned} \tag{11.33}$$

11.9 Direct Products

Suppose $D^{(a)}(g)$ is a k-dimensional representation of a group G, and $D^{(b)}(g)$ is an n-dimensional representation of the same group. Then their product

$$D^{(a,b)}_{im,j\ell}(g) = D^{(a)}_{ij}(g)\, D^{(b)}_{m\ell}(g) \tag{11.34}$$

is a (kn)-dimensional **direct-product** representation of the group G. Direct products are also called **tensor products**. They occur in quantum systems that have two or more parts, each described by a different space of vectors.

Suppose the vectors $|i\rangle$ for $i = 1, \ldots, k$ are the basis vectors of the k-dimensional space V_k on which the representation $D^{(a)}(g)$ acts, and that the vectors $|m\rangle$ for $m = 1 \ldots n$ are the basis vectors of the n-dimensional space V_n on which $D^{(b)}(g)$ acts. The kn vectors $|i, m\rangle$ are basis vectors for the kn-dimensional tensor-product space V_{kn}. The matrices $D^{(a,b)}(g)$ defined as

$$\langle i, m|D(g)^{(a,b)}|j, \ell\rangle = \langle i|D^{(a)}(g)|j\rangle\langle m|D^{(b)}(g)|\ell\rangle \tag{11.35}$$

act in this kn-dimensional space V_{kn} and form a representation of the group G; this direct-product representation usually is reducible. Many tricks help one to decompose reducible tensor-product representations into direct sums of irreducible representations (Georgi, 1999; Zee, 2016).

Example 11.13 (Adding angular momenta) The addition of angular momenta illustrates both the direct product and its reduction to a direct sum of irreducible representations. Let $D^{(j_1)}(g)$ and $D^{(j_2)}(g)$ respectively be the $(2j_1+1)\times(2j_1+1)$ and the

$(2j_2+1) \times (2j_2+1)$ representations of the rotation group $SU(2)$. The direct-product representation $D^{(j_1,j_2)}$

$$\langle m_1', m_2' | D^{(j_1,j_2)} | m_1, m_2 \rangle = \langle m_1' | D^{(j_1)}(g) | m_1 \rangle \langle m_2' | D^{(j_2)}(g) | m_2 \rangle \tag{11.36}$$

is reducible into a direct sum of all the irreducible representations of $SU(2)$ from $D^{(j_1+j_2)}(g)$ down to $D^{(|j_1-j_2|)}(g)$ in integer steps:

$$D^{(j_1,j_2)} = D^{(j_1+j_2)} \oplus D^{(j_1+j_2-1)} \oplus \cdots \oplus D^{(|j_1-j_2|+1)} \oplus D^{(|j_1-j_2|)} \tag{11.37}$$

each irreducible representation occurring once in the direct sum.

Example 11.14 (Adding two spins) When one adds $j_1 = 1/2$ to $j_2 = 1/2$, one finds that the tensor-product matrix $D^{(1/2,1/2)}$ is equivalent to the direct sum $D^{(1)} \oplus D^{(0)}$

$$D^{(1/2,1/2)}(\boldsymbol{\theta}) = S^{-1} \begin{pmatrix} D^{(1)}(\boldsymbol{\theta}) & 0 \\ 0 & D^{(0)}(\boldsymbol{\theta}) \end{pmatrix} S \tag{11.38}$$

where the matrices S, $D^{(1)}$, and $D^{(0)}$ are 4×4, 3×3, and 1×1.

11.10 Finite Groups

A **finite group** is one that has a finite number of elements. The number of elements in a group is the **order** of the group.

Example 11.15 (Z_2) The group $\mathbf{Z_2}$ consists of two elements e and p with multiplication rules

$$ee = e, \quad ep = pe = p, \quad \text{and} \quad pp = e. \tag{11.39}$$

Clearly, Z_2 is abelian, and its order is 2. The identification $e \to 1$ and $p \to -1$ gives a 1-dimensional representation of the group Z_2 in terms of 1×1 matrices, which are just numbers.

It is tedious to write the multiplication rules as individual equations. Normally people compress them into a multiplication table like this:

×	e	p
e	e	p
p	p	e

(11.40)

A simple generalization of Z_2 is the group $\mathbf{Z}_n$ whose elements may be represented as $\exp(i2\pi m/n)$ for $m = 1, \ldots, n$. This group is also abelian, and its order is n.

Example 11.16 (Z_3) The multiplication table for $\mathbf{Z_3}$ is

$\times$	e	a	b
e	e	a	b
a	a	b	e
b	b	e	a

(11.41)

which says that $a^2 = b$, $b^2 = a$, and $ab = ba = e$.

11.11 Regular Representations

For any finite group G we can associate an orthonormal vector $|g_i\rangle$ with each element g_i of the group. So $\langle g_i|g_j\rangle = \delta_{ij}$. These orthonormal vectors $|g_i\rangle$ form a basis for a vector space whose dimension is the order of the group. The matrix $D(g_k)$ of the regular representation of G is defined to map any vector $|g_i\rangle$ into the vector $|g_k g_i\rangle$ associated with the product $g_k g_i$

$$D(g_k)|g_i\rangle = |g_k g_i\rangle. \tag{11.42}$$

Since group multiplication is associative, we have

$$D(g_j)D(g_k)|g_i\rangle = D(g_j)|g_k g_i\rangle = |g_j(g_k g_i)\rangle = |(g_j g_k)g_i\rangle = D(g_j g_k)|g_i\rangle. \tag{11.43}$$

Because the vector $|g_i\rangle$ was an arbitrary basis vector, it follows that

$$D(g_j)D(g_k) = D(g_j g_k) \tag{11.44}$$

which means that the matrices $D(g)$ satisfy the criterion (11.5) for their being a representation of the group G. The matrix $D(g)$ has entries

$$[D(g)]_{ij} = \langle g_i|D(g)|g_j\rangle. \tag{11.45}$$

The sum of dyadics $|g_\ell\rangle\langle g_\ell|$ over all the elements g_ℓ of a finite group G is the unit matrix

$$\sum_{g_\ell \in G} |g_\ell\rangle\langle g_\ell| = I_n \tag{11.46}$$

in which n is the order of G, that is, the number of elements in G. The matrix (11.45) respects the product law (11.44) and so is a representation of the group

$$\begin{aligned}[D(g_j g_k)]_{m,n} &= \langle g_m|D(g_j g_k)|g_n\rangle = \langle g_m|D(g_j)D(g_k)|g_n\rangle \qquad (11.47)\\ &= \sum_{g_\ell \in G} \langle g_m|D(g_j)|g_\ell\rangle\langle g_\ell|D(g_k)|g_n\rangle = \sum_{g_\ell \in G} [D(g_j)]_{m,\ell}[D(g_k)]_{\ell,n}.\end{aligned}$$

Example 11.17 (Z_3's regular representation) The regular representation of Z_3 is

$$D(e) = \begin{pmatrix} 1 & 0 & 0 \\ 0 & 1 & 0 \\ 0 & 0 & 1 \end{pmatrix}, \quad D(a) = \begin{pmatrix} 0 & 0 & 1 \\ 1 & 0 & 0 \\ 0 & 1 & 0 \end{pmatrix}, \quad D(b) = \begin{pmatrix} 0 & 1 & 0 \\ 0 & 0 & 1 \\ 1 & 0 & 0 \end{pmatrix} \tag{11.48}$$

so $D(a)^2 = D(b)$, $D(b)^2 = D(a)$, and $D(a)D(b) = D(b)D(a) = D(e)$.

11.12 Properties of Finite Groups

In his book (Georgi, 1999, chap. 1), Georgi proves the following theorems:

1. Every representation of a finite group is equivalent to a unitary representation.
2. Every representation of a finite group is completely reducible.
3. The irreducible representations of a finite abelian group are 1 dimensional.
4. If $D^{(a)}(g)$ and $D^{(b)}(g)$ are two unitary irreducible representations of dimensions n_a and n_b of a group G of N elements $g_1, \dots, g_N$, then the functions

$$\sqrt{\frac{n_a}{N}}\, D^{(a)}_{jk}(g) \tag{11.49}$$

are orthonormal and complete in the sense that

$$\frac{n_a}{N} \sum_{j=1}^{N} D^{(a)*}_{ik}(g_j) D^{(b)}_{\ell m}(g_j) = \delta_{ab}\delta_{i\ell}\delta_{km}. \tag{11.50}$$

5. The order N of a finite group is the sum of the squares of the dimensions of its inequivalent irreducible representations

$$N = \sum_a n_a^2. \tag{11.51}$$

Example 11.18 (Z_N) The abelian cyclic group Z_N with elements

$$g_j = e^{2\pi i j/N} \tag{11.52}$$

has N 1-dimensional irreducible representations

$$D^{(a)}(g_j) = e^{2\pi i a j/N} \tag{11.53}$$

for $a = 1, 2, \dots, N$. Their orthonormality relation (11.50) is the Fourier formula

$$\frac{1}{N} \sum_{j=1}^{N} e^{-2\pi i a j/N} e^{2\pi i b j/N} = \delta_{ab}. \tag{11.54}$$

The n_a are all unity, there are N of them, and the sum of the n_a^2 is N as required by the sum rule (11.51).

11.13 Permutations

The permutation group on n objects is called S_n. Permutations are made of **cycles** that change the order of the n objects. For instance, the permutation $(1\,2) = (2\,1)$ is a 2-cycle that means $x_1 \to x_2 \to x_1$; the unitary operator $U((1\,2))$ that represents it interchanges states like this:

$$U((1\,2))|+,-\rangle = U((1\,2))|+,1\rangle\,|-,2\rangle = |-,1\rangle|+,2\rangle = |-,+\rangle. \tag{11.55}$$

The 2-cycle $(3\,4)$ means $x_3 \to x_4 \to x_3$, it changes (a,b,c,d) into (a,b,d,c). The 3-cycle $(1\,2\,3) = (2\,3\,1) = (3\,1\,2)$ means $x_1 \to x_2 \to x_3 \to x_1$, it changes (a,b,c,d) into (b,c,a,d). The 4-cycle $(1\,3\,2\,4)$ means $x_1 \to x_3 \to x_2 \to x_4 \to x_1$ and changes (a,b,c,d) into (c,d,b,a). The 1-cycle (2) means $x_2 \to x_2$ and leaves everything unchanged.

The identity element of S_n is the product of 1-cycles $e = (1)(2)\cdots(n)$. The inverse of the cycle $(1\,3\,2\,4)$ must invert $x_1 \to x_3 \to x_2 \to x_4 \to x_1$, so it must be $(1\,4\,2\,3)$ which means $x_1 \to x_4 \to x_2 \to x_3 \to x_1$ so that it changes (c,d,b,a) back into (a,b,c,d). Every element of S_n has each integer from 1 to n in one and only one cycle. So an arbitrary element of S_n with ℓ_k k-cycles must satisfy

$$\sum_{k=1}^{n} k\,\ell_k = n. \tag{11.56}$$

11.14 Compact and Noncompact Lie Groups

Imagine rotating an object repeatedly. Notice that the biggest rotation is by an angle of $\pm\pi$ about some axis. The possible angles form a circle; the space of parameters is a circle. The parameter space of a compact group is compact – closed and bounded. The rotations form a **compact group**.

Now consider the translations. Imagine moving a pebble to the Sun, then moving it to the next-nearest star, then moving it to the nearest galaxy. If space is flat, then there is no limit to how far one can move a pebble. The parameter space of a noncompact group is not compact. The translations form a **noncompact group**.

We'll see that compact Lie groups possess unitary representations, with $n \times n$ unitary matrices $D(\alpha)$, while noncompact ones don't. Here α stands for the parameters $\alpha_1, \ldots, \alpha_n$ that label the elements of the group, three for the rotation group. The α's usually are real, but can be complex.

11.15 Generators

To study continuous groups, we will use calculus and algebra, and we will focus on the simplest part of the group – the elements $g(\alpha)$ for $\alpha \approx 0$ which are near the identity $e = g(0)$ for which all $\alpha_a = 0$. Each element $g(\alpha)$ of the group is represented by a matrix $D(\alpha) \equiv D(g(\alpha))$ in the D representation of the group and by another matrix $D'(\alpha) \equiv D'(g(\alpha))$ in any other D' representation of the group. Every representation respects the multiplication law of the group. So if $g(\beta)g(\alpha) = g(\gamma)$, then the matrices of the D representation must satisfy $D(\beta)D(\alpha) = D(\gamma)$, and those of any other representation D' must satisfy $D'(\beta)D'(\alpha) = D'(\gamma)$.

A **generator** t_a of a representation D is the partial derivative of the matrix $D(\alpha)$ with respect to the component α_a of α evaluated at $\alpha = 0$

$$t_a = -i \left. \frac{\partial D(\alpha)}{\partial \alpha_a} \right|_{\alpha=0}. \tag{11.57}$$

When all the parameters α_a are infinitesimal, $|\alpha_a| \ll 1$, the matrix $D(\alpha)$ is very close to the identity matrix I

$$D(\alpha) \simeq I + i \sum_a \alpha_a \, t_a. \tag{11.58}$$

Replacing α by α/n, we get a relation that becomes exact as $n \to \infty$

$$D\left(\frac{\alpha}{n}\right) = I + i \sum_a \frac{\alpha_a}{n} \, t_a. \tag{11.59}$$

The nth power of this equation is the matrix $D(\alpha)$ that represents the group element $g(\alpha)$ in the **exponential parametrization**

$$D(\alpha) = D\left(\frac{\alpha}{n}\right)^n = \lim_{n\to\infty} \left(I + i \sum_a \frac{\alpha_a}{n} \, t_a \right)^n = \exp\left(i \sum_a \alpha_a \, t_a \right). \tag{11.60}$$

The i's appear in these equations so that when the generators t_a are hermitian matrices, $(t_a)^\dagger = t_a$, and the α's are real, the matrices $D(\alpha)$ are unitary

$$D^{-1}(\alpha) = \exp\left(-i \sum_a \alpha_a \, t_a \right) = D^\dagger(\alpha) = \exp\left(-i \sum_a \alpha_a \, t_a \right). \tag{11.61}$$

Compact groups have finite-dimensional, unitary representations with hermitian generators.

11.16 Lie Algebra

If t_a and t_b are any two generators of a representation D, then the matrices

$$D(\alpha) = e^{i\epsilon t_a} \quad \text{and} \quad D(\beta) = e^{i\epsilon t_b} \tag{11.62}$$

represent the group elements $g(\alpha)$ and $g(\beta)$ with infinitesimal exponential parameters $\alpha_i = \epsilon \, \delta_{ia}$ and $\beta_i = \epsilon \, \delta_{ib}$. The inverses of these group elements $g^{-1}(\alpha) =$

$g(-\alpha)$ and $g^{-1}(\beta) = g(-\beta)$ are represented by the matrices $D(-\alpha) = e^{-i\epsilon t_a}$ and $D(-\beta) = e^{-i\epsilon t_b}$. The multiplication law of the group determines the parameters $\gamma(\alpha, \beta)$ of the product

$$g(\beta)\, g(\alpha)\, g(-\beta)\, g(-\alpha) = g(\gamma(\alpha, \beta)). \tag{11.63}$$

The matrices of any two representations D with generators t_a and D' with generators t'_a obey the same multiplication law

$$\begin{aligned} D(\beta)\, D(\alpha)\, D(-\beta)\, D(-\alpha) &= D(\gamma(\alpha, \beta)) \\ D'(\beta)\, D'(\alpha)\, D'(-\beta)\, D'(-\alpha) &= D'(\gamma(\alpha, \beta)) \end{aligned} \tag{11.64}$$

with the same infinitesimal exponential parameters α, β, and $\gamma(\alpha, \beta)$. To order ϵ^2, the product of the four D's is

$$\begin{aligned} e^{i\epsilon t_b}\, e^{i\epsilon t_a}\, e^{-i\epsilon t_b}\, e^{-i\epsilon t_a} &\approx (1 + i\epsilon\, t_b - \frac{\epsilon^2}{2}\, t_b^2)(1 + i\epsilon\, t_a - \frac{\epsilon^2}{2}\, t_a^2) \\ &\quad \times (1 - i\epsilon\, t_b - \frac{\epsilon^2}{2}\, t_b^2)(1 - i\epsilon\, t_a - \frac{\epsilon^2}{2}\, t_a^2) \\ &\approx 1 + \epsilon^2 (t_a\, t_b - t_b\, t_a) = 1 + \epsilon^2 [t_a, t_b]. \end{aligned} \tag{11.65}$$

The other representation gives the same result but with primes

$$e^{i\epsilon t'_b}\, e^{i\epsilon t'_a}\, e^{-i\epsilon t'_b}\, e^{-i\epsilon t'_a} \approx 1 + \epsilon^2 [t'_a, t'_b]. \tag{11.66}$$

The products (11.65 and 11.66) represent the same group element $g(\gamma(\alpha, \beta))$, so they have the same infinitesimal parameters $\gamma(\alpha, \beta)$ and therefore are the same linear combinations of their respective generators t_c and t'_c

$$\begin{aligned} D(\gamma(\alpha, \beta)) &\approx 1 + \epsilon^2 [t_a, t_b] = 1 + i\epsilon^2 \sum_{c=1}^{n} f_{ab}^c\, t_c \\ D'(\gamma(\alpha, \beta)) &\approx 1 + \epsilon^2 [t'_a, t'_b] = 1 + i\epsilon^2 \sum_{c=1}^{n} f_{ab}^c\, t'_c \end{aligned} \tag{11.67}$$

which in turn imply the **Lie algebra** formulas with the *same* f_{ab}^c

$$[t_a, t_b] = \sum_{c=1}^{n} f_{ab}^c\, t_c \quad \text{and} \quad [t'_a, t'_b] = \sum_{c=1}^{n} f_{ab}^c\, t'_c. \tag{11.68}$$

The commutator of any two generators is a linear combination of the generators. The coefficients $f^c{}_{ab}$ are the structure constants of the group. They are the same for all representations of the group.

Unless the parameters α_a are redundant, the generators are linearly independent. They span a vector space, and any linear combination may be called a generator.

By using the Gram–Schmidt procedure (Section 1.10), we may make the generators t_a orthogonal with respect to the inner product (1.91)

$$(t_a, t_b) = \mathrm{Tr}\left(t_a^\dagger t_b\right) = k\,\delta_{ab} \tag{11.69}$$

in which k is a nonnegative normalization constant that depends upon the representation. We can't normalize the generators, making k unity, because the structure constants f_{ab}^c are the same in all representations.

In what follows, I will often omit the summation symbol $\sum$ when an index is repeated. In this notation, the structure-constant formulas (11.68) are

$$[t_a, t_b] = f_{ab}^c\, t_c \quad \text{and} \quad [t_a', t_b'] = f_{ab}^c\, t_c'. \tag{11.70}$$

This summation convention avoids unnecessary summation symbols.

By multiplying both sides of the first of the two Lie algebra formulas (11.68) by $t_d^\dagger$ and using the orthogonality (11.69) of the generators, we find

$$\mathrm{Tr}\left([t_a, t_b]\, t_d^\dagger\right) = i\, f^c{}_{ab}\, \mathrm{Tr}\left(t_c\, t_d^\dagger\right) = i\, f^c{}_{ab}\, k\, \delta_{cd} = ik\, f^d{}_{ab} \tag{11.71}$$

which implies that the structure constant $f^c{}_{ab}$ is the trace

$$f^c{}_{ab} = -\frac{i}{k}\, \mathrm{Tr}\left([t_a, t_b]\, t_c^\dagger\right). \tag{11.72}$$

Because of the antisymmetry of the commutator $[t_a, t_b]$, structure constants are **antisymmetric in their lower indices**

$$f^c{}_{ab} = -f^c{}_{ba}. \tag{11.73}$$

From any $n \times n$ matrix A, one may make a hermitian matrix $A + A^\dagger$ and an antihermitian one $A - A^\dagger$. Thus one may separate the n_G generators into a set that are hermitian $t_a^{(h)}$ and a set that are antihermitian $t_a^{(ah)}$. The exponential of any imaginary linear combination of $n \times n$ hermitian generators $D(\alpha) = \exp\left(i\alpha_a\, t_a^{(h)}\right)$ is an $n \times n$ unitary matrix since

$$D^\dagger(\alpha) = \exp\left(-i\alpha_a\, t_a^{\dagger(h)}\right) = \exp\left(-i\alpha_a\, t_a^{(h)}\right) = D^{-1}(\alpha). \tag{11.74}$$

A group with only hermitian generators is **compact** and has finite-dimensional unitary representations.

On the other hand, the exponential of any imaginary linear combination of antihermitian generators $D(\alpha) = \exp\left(i\alpha_a\, t_a^{(ah)}\right)$ is a real exponential of their hermitian counterparts $i\, t_a^{(ah)}$ whose squared norm

$$\|D(\alpha)\|^2 = \mathrm{Tr}\left[D(\alpha)^\dagger D(\alpha)\right] = \mathrm{Tr}\left[\exp\left(2\alpha_a\, i t_a^{(ah)}\right)\right] \tag{11.75}$$

grows exponentially and without limit as the parameters $\alpha_a \to \pm\infty$. A group with some antihermitian generators is **noncompact** and does not have finite-dimensional unitary representations. (The unitary representations of the translations and of the Lorentz and Poincaré groups are infinite dimensional.)

Compact Lie groups have hermitian generators, and so the structure-constant formula (11.72) reduces in this case to

$$f^c_{\ ab} = (-i/k)\mathrm{Tr}\left([t_a, t_b]\, t_c^\dagger\right) = (-i/k)\mathrm{Tr}\left([t_a, t_b]\, t_c\right). \tag{11.76}$$

Now, since the trace is cyclic, we have

$$\begin{aligned} f^b_{\ ac} &= (-i/k)\mathrm{Tr}\left([t_a, t_c]\, t_b\right) = (-i/k)\mathrm{Tr}\left(t_a t_c t_b - t_c t_a t_b\right) \\ &= (-i/k)\mathrm{Tr}\left(t_b t_a t_c - t_a t_b t_c\right) \\ &= (-i/k)\mathrm{Tr}\left([t_b, t_a]\, t_c\right) = f^c_{\ ba} = -f^c_{\ ab}. \end{aligned} \tag{11.77}$$

Interchanging a and b, we get

$$f^a_{\ bc} = f^c_{\ ab} = -f^c_{\ ba}. \tag{11.78}$$

Finally, interchanging b and c in (11.77) gives

$$f^c_{\ ab} = f^b_{\ ca} = -f^b_{\ ac}. \tag{11.79}$$

Combining (11.77, 11.78, and 11.79), we see that **the structure constants of a compact Lie group are totally antisymmetric**

$$f^b_{\ ac} = -f^b_{\ ca} = f^c_{\ ba} = -f^c_{\ ab} = -f^a_{\ bc} = f^a_{\ cb}. \tag{11.80}$$

Because of this antisymmetry, it is usual to lower the upper index

$$f^c_{\ ab} = f_{cab} = f_{abc} \tag{11.81}$$

and write the antisymmetry of the structure constants of compact Lie groups as

$$f_{acb} = -f_{cab} = f_{bac} = -f_{abc} = -f_{bca} = f_{cba}. \tag{11.82}$$

For compact Lie groups, the generators are hermitian, and so the **structure constants f_{abc} are real**, as we may see by taking the complex conjugate of the formula (11.76) for f_{abc}

$$f^*_{abc} = (i/k)\mathrm{Tr}\left(t_c\, [t_b, t_a]\right) = (-i/k)\mathrm{Tr}\left([t_a, t_b]\, t_c\right) = f_{abc}. \tag{11.83}$$

It follows from (11.68 and 11.81–11.83) that **the commutator of any two generators of a Lie group is a linear combination**

$$[t_a, t_b] = i\, f^c_{ab}\, t_c \tag{11.84}$$

of its generators t_c, and that the structure constants $f_{abc} \equiv f^c_{ab}$ are real and totally antisymmetric if the group is compact.

11.17 Yang and Mills Invent Local Nonabelian Symmetry

The action of a Yang–Mills theory is unchanged when a spacetime-dependent unitary matrix $U(x) = \exp(-it_a\,\theta^a(x))$ maps a vector $\psi(x)$ of matter fields to $\psi'(x) = U(x)\psi(x)$. The symmetry $\psi^\dagger(x)U^\dagger(x)U(x)\psi(x) = \psi^\dagger(x)\psi(x)$ is obvious, but how can kinetic terms like $\partial_i\psi^\dagger\,\partial^i\psi$ be made invariant? Yang and Mills introduced matrices $A_i = -it_a\,A^a_i$ of gauge fields, replaced ordinary derivatives ∂_i by **covariant derivatives** $D_i \equiv \partial_i + A_i$, and required that covariant derivatives of fields transform like fields

$$\left(\partial_i + A'_i\right)U\psi = \left(\partial_i U + U\partial_i + A'_i U\right)\psi = U\left(\partial_i + A_i\right)\psi. \tag{11.85}$$

Their nonabelian gauge transformation is

$$A'_i(x) = U(x)A_i(x)U^\dagger(x) - \left(\partial_i U(x)\right)U^\dagger(x). \tag{11.86}$$

Their Faraday tensor $F_{ik} = [D_i, D_k] = \partial_i A_k - \partial_k A_i + [A_i, A_k]$ transforms as

$$F'_{ik}(x) = U(x)F_{ik}U^{-1}(x) = U(x)[D_i, D_k]U^{-1}(x). \tag{11.87}$$

11.18 Rotation Group

The rotations and reflections in 3-dimensional space form a compact group $O(3)$ whose elements R are 3×3 real matrices that leave invariant the dot product of any two 3-vectors

$$(R\boldsymbol{x})\cdot(R\boldsymbol{y}) = \boldsymbol{x}^{\mathsf{T}}R^{\mathsf{T}}R\,\boldsymbol{y} = \boldsymbol{x}^{\mathsf{T}}I\,\boldsymbol{y} = \boldsymbol{x}\cdot\boldsymbol{y}. \tag{11.88}$$

These matrices therefore are orthogonal (1.184)

$$R^{\mathsf{T}}R = I. \tag{11.89}$$

Taking the determinant of both sides and using the transpose (1.205) and product (1.225) rules, we have

$$(\det R)^2 = 1 \tag{11.90}$$

whence $\det R = \pm 1$. The group $O(3)$ contains reflections as well as rotations and is disjoint. The subgroup with $\det R = 1$ is the group $SO(3)$. An $SO(3)$ element near the identity $R = I + \omega$ must satisfy

$$(I+\omega)^{\mathsf{T}}(I+\omega) = I. \tag{11.91}$$

Neglecting the tiny quadratic term, we find that the infinitesimal matrix ω is antisymmetric

$$\omega^{\mathsf{T}} = -\,\omega. \tag{11.92}$$

One complete set of real 3×3 antisymmetric matrices is

$$\omega_1 = \begin{pmatrix} 0 & 0 & 0 \\ 0 & 0 & -1 \\ 0 & 1 & 0 \end{pmatrix}, \quad \omega_2 = \begin{pmatrix} 0 & 0 & 1 \\ 0 & 0 & 0 \\ -1 & 0 & 0 \end{pmatrix}, \quad \omega_3 = \begin{pmatrix} 0 & -1 & 0 \\ 1 & 0 & 0 \\ 0 & 0 & 0 \end{pmatrix} \tag{11.93}$$

which we may write as

$$[\omega_b]_{ac} = \epsilon_{abc} \tag{11.94}$$

in which ϵ_{abc} is the **Levi-Civita symbol** which is totally antisymmetric with $\epsilon_{123} = 1$ (Tullio Levi-Civita 1873–1941). The ω_b are antihermitian, but we make them hermitian by multiplying by i

$$t_b = i\,\omega_b \quad \text{so that} \quad [t_b]_{ac} = i\epsilon_{abc} \tag{11.95}$$

and $R = I - i\theta_b\, t_b$.

The three hermitian generators t_a satisfy (Exercise 11.15) the commutation relations

$$[t_a, t_b] = i\, f_{abc}\, t_c \tag{11.96}$$

in which the structure constants are given by the Levi-Civita symbol ϵ_{abc}

$$f_{abc} = \epsilon_{abc} \tag{11.97}$$

so that

$$[t_a, t_b] = i\, \epsilon_{abc}\, t_c. \tag{11.98}$$

They are the generators of the **defining representation** of $SO(3)$ (and also of the **adjoint representation** of $SU(2)$ (Section 11.25)).

Physicists usually scale the generators by $\hbar$ and define the angular-momentum generator L_a as

$$L_a = \hbar\, t_a \tag{11.99}$$

so that the eigenvalues of the angular-momentum operators are the physical values of the angular momenta. With $\hbar$, the commutation relations are

$$[L_a, L_b] = i\,\hbar\,\epsilon_{abc}\, L_c. \tag{11.100}$$

The matrix that represents a right-handed rotation (of an object) by an angle $\theta = |\boldsymbol{\theta}|$ about an axis $\boldsymbol{\theta}$ is

$$D(\boldsymbol{\theta}) = e^{-i\boldsymbol{\theta}\cdot\boldsymbol{t}} = e^{-i\boldsymbol{\theta}\cdot\boldsymbol{L}/\hbar}. \tag{11.101}$$

By using the fact (1.294) that a matrix obeys its characteristic equation, one may show (Exercise 11.17) that the 3×3 matrix $D(\boldsymbol{\theta})$ that represents a right-handed rotation of θ radians about the axis $\boldsymbol{\theta}$ is the matrix $\exp(-i\boldsymbol{\theta}\cdot\boldsymbol{t})$ (11.7) whose i, jth entry is

$$D_{ij}(\boldsymbol{\theta}) = \cos\theta\,\delta_{ij} - \sin\theta\,\epsilon_{ijk}\,\theta_k/\theta + (1-\cos\theta)\,\theta_i\theta_j/\theta^2 \tag{11.102}$$

in which a sum over $k = 1, 2, 3$ is understood.

A set of generators J_a equivalent to the antisymmetric L's and ω's (11.93) but with J_3 diagonal is

$$J_1 = \frac{1}{\sqrt{2}}\begin{pmatrix} 0 & 1 & 0 \\ 1 & 0 & 1 \\ 0 & 1 & 0 \end{pmatrix}, \quad J_2 = \frac{1}{\sqrt{2}}\begin{pmatrix} 0 & -i & 0 \\ i & 0 & -i \\ 0 & i & 0 \end{pmatrix}, \quad J_3 = \begin{pmatrix} 1 & 0 & 0 \\ 0 & 0 & 0 \\ 0 & 0 & -1 \end{pmatrix}. \tag{11.103}$$

Example 11.19 (Demonstration of commutation relations) Take a big sphere with a distinguished point and orient the sphere so that the point lies in the y-direction from the center of the sphere. Now rotate the sphere by a small angle, say 15 degrees or $\epsilon = \pi/12$, right-handedly about the x-axis, then right-handedly about the y-axis by the same angle, then left-handedly about the x-axis and then left-handedly about the y-axis. Using the approximations (11.65 and 11.67) for the product of these four rotation matrices and the definitions (11.99) of the generators and of their structure constants (11.100), we have $\hbar t_a = L_1 = L_x$, $\hbar t_b = L_2 = L_y$, $\hbar f_{abc} t_c = \epsilon_{12c} L_c = L_3 = L_z$, and

$$e^{i\epsilon L_y/\hbar}\, e^{i\epsilon L_x/\hbar}\, e^{-i\epsilon L_y/\hbar}\, e^{-i\epsilon L_x/\hbar} \approx 1 + \frac{\epsilon^2}{\hbar^2}[L_x, L_y] = 1 + i\frac{\epsilon^2}{\hbar}L_z \approx e^{i\epsilon^2 L_z/\hbar} \tag{11.104}$$

which is a left-handed rotation about the (vertical) z-axis. The magnitude of that rotation should be about $\epsilon^2 = (\pi/12)^2 \approx 0.069$ or about 3.9 degrees. Photographs of an actual demonstration are displayed in Fig. 11.1.

The demonstrated equation (11.104) shows (Exercise 11.16) that the generators L_x and L_y satisfy the commutation relation

$$[L_x, L_y] = i\hbar L_z \tag{11.105}$$

of the rotation group.

Physical demonstration of commutation relations

Figure 11.1 Demonstration of Equation (11.104) and the commutation relation (11.105). Upper left: black ball with a white stick pointing in the y-direction; the x-axis is to the reader's left, the z-axis is vertical. Upper right: ball after a small right-handed rotation about the x-axis. Center left: ball after that rotation is followed by a small right-handed rotation about the y-axis. Center right: ball after these rotations are followed by a small left-handed rotation about the x-axis. Bottom: ball after these rotations are followed by a small left-handed rotation about the y-axis. The net effect is approximately a small left-handed rotation about the z-axis.

11.19 Rotations and Reflections in 2n Dimensions

The orthogonal group $O(2n)$ of rotations and reflections in $2n$ dimensions is the group of all real $2n \times 2n$ matrices O whose transposes O^{T} are their inverses

$$O^{\mathsf{T}} O = O\, O^{\mathsf{T}} = I \tag{11.106}$$

in which I is the $2n \times 2n$ identity matrix. These orthogonal matrices leave unchanged the distances from the origin of points in $2n$ dimensions. Those with unit determinant, $\det O = 1$, constitute the subgroup $SO(2n)$ of rotations in $2n$ dimensions.

A symmetric sum $\{A, B\} = AB + BA$ is called an **anticommutator**. Complex fermionic variables ψ_i obey the anticommutation relations

$$\{\psi_i, \psi_k^\dagger\} = \hbar\, \delta_{ik}, \quad \{\psi_i, \psi_k\} = 0, \quad \text{and} \quad \{\psi_i^\dagger, \psi_k^\dagger\} = 0. \tag{11.107}$$

Their real x_i and imaginary y_i parts

$$x_i = \frac{1}{\sqrt{2}}(\psi_i + \psi_i^\dagger) \quad \text{and} \quad y_i = \frac{1}{i\sqrt{2}}(\psi_i - \psi_i^\dagger) \tag{11.108}$$

obey the anticommutation relations

$$\{x_i, x_k\} = \hbar\, \delta_{ik}, \quad \{y_i, y_k\} = \hbar\, \delta_{ik}, \quad \text{and} \quad \{x_i, y_k\} = 0. \tag{11.109}$$

More simply, the anticommutation relations of these $2n$ hermitian variables $v = (x_1, \ldots, x_n, y_1, \ldots, y_n)$ are

$$\{v_i, v_k\} = \hbar\, \delta_{ik}. \tag{11.110}$$

If the real linear transformation $v_i' = L_{i1}\, v_1 + L_{i2}\, v_2 + \cdots + L_{i2n}\, v_{2n}$ preserves these anticommutation relations, then the matrix L must satisfy

$$\hbar\, \delta_{ik} = \{v_i', v_k'\} = L_{ij}\, L_{k\ell} \{v_j, v_\ell\} = L_{ij}\, L_{k\ell}\, \hbar\, \delta_{j\ell} = \hbar L_{ij}\, L_{kj} \tag{11.111}$$

which is the statement that it is orthogonal, $LL^{\mathsf{T}} = I$. Thus the group $O(2n)$ is the largest group of linear transformations that preserve the anticommutation relations of the $2n$ hermitian real and imaginary parts v_i of n complex fermionic variables ψ_i.

11.20 Defining Representation of SU(2)

The smallest positive value of angular momentum is $\hbar/2$. The spin-one-half angular-momentum operators are represented by three 2×2 matrices

$$S_a = \frac{\hbar}{2}\, \sigma_a \tag{11.112}$$

in which the σ_a are the **Pauli matrices**

$$\sigma_1 = \begin{pmatrix} 0 & 1 \\ 1 & 0 \end{pmatrix}, \quad \sigma_2 = \begin{pmatrix} 0 & -i \\ i & 0 \end{pmatrix}, \quad \text{and} \quad \sigma_3 = \begin{pmatrix} 1 & 0 \\ 0 & -1 \end{pmatrix} \tag{11.113}$$

which obey the multiplication law

$$\sigma_i \, \sigma_j = \delta_{ij} + i \epsilon_{ijk} \, \sigma_k \tag{11.114}$$

summed over k from 1 to 3. Since the symbol ϵ_{ijk} is totally antisymmetric in i, j, and k, the Pauli matrices obey the commutation and anticommutation relations

$$\begin{aligned} [\sigma_i, \sigma_j] &\equiv \sigma_i \sigma_j - \sigma_j \sigma_i = 2i \epsilon_{ijk} \, \sigma_k \\ \{\sigma_i, \sigma_j\} &\equiv \sigma_i \, \sigma_j + \sigma_j \, \sigma_i = 2\delta_{ij}. \end{aligned} \tag{11.115}$$

The Pauli matrices divided by 2 satisfy the commutation relations (11.98) of the rotation group

$$\left[\frac{1}{2}\sigma_a, \frac{1}{2}\sigma_b \right] = i \, \epsilon_{abc} \, \frac{1}{2}\sigma_c \tag{11.116}$$

and generate the elements of the group $SU(2)$

$$\exp\left(i \, \boldsymbol{\theta} \cdot \frac{\boldsymbol{\sigma}}{2} \right) = I \, \cos\frac{\theta}{2} + i \, \hat{\boldsymbol{\theta}} \cdot \boldsymbol{\sigma} \, \sin\frac{\theta}{2} \tag{11.117}$$

in which I is the 2×2 identity matrix, $\theta = \sqrt{\boldsymbol{\theta}^2}$ and $\hat{\boldsymbol{\theta}} = \boldsymbol{\theta}/\theta$.

It follows from (11.116) that the spin operators (11.112) satisfy

$$[S_a, S_b] = i \, \hbar \, \epsilon_{abc} \, S_c. \tag{11.118}$$

11.21 The Lie Algebra and Representations of *SU(2)*

The three generators of $SU(2)$ in its 2×2 defining representation are the Pauli matrices divided by 2, $t_a = \sigma_a/2$. The structure constants of $SU(2)$ are $f_{abc} = \epsilon_{abc}$ which is totally antisymmetric with $\epsilon_{123} = 1$

$$[t_a, t_b] = i f_{abc} t_c = \left[\frac{1}{2}\sigma_a, \frac{1}{2}\sigma_b \right] = i\epsilon_{abc} \, \frac{1}{2}\sigma_c. \tag{11.119}$$

For every half-integer

$$j = \frac{n}{2} \quad \text{for} \quad n = 0, 1, 2, 3, \ldots \tag{11.120}$$

there is an irreducible representation of $SU(2)$

$$D^{(j)}(\boldsymbol{\theta}) = e^{-i\boldsymbol{\theta}\cdot \boldsymbol{J}^{(j)}} \tag{11.121}$$

in which the three generators $t_a^{(j)} \equiv J_a^{(j)}$ are $(2j+1)\times(2j+1)$ square hermitian matrices. In a basis in which $J_3^{(j)}$ is diagonal, the matrix elements of the complex linear combinations $J_\pm^{(j)} \equiv J_1^{(j)} \pm i J_2^{(j)}$ are

$$\left[J_1^{(j)} \pm i J_2^{(j)}\right]_{s',s} = \delta_{s',s\pm1}\sqrt{(j \mp s)(j \pm s + 1)} \tag{11.122}$$

where s and s' run from $-j$ to j in integer steps and those of $J_3^{(j)}$ are

$$\left[J_3^{(j)}\right]_{s',s} = s\,\delta_{s',s}. \tag{11.123}$$

Borrowing a trick from Section 11.26, one may show that the commutator of the square $\boldsymbol{J}^{(j)} \cdot \boldsymbol{J}^{(j)}$ of the angular momentum matrix commutes with every generator $J_a^{(j)}$. Thus $\boldsymbol{J}^{(j)2}$ commutes with $D^{(j)}(\boldsymbol{\theta})$ for every element of the group. Part 2 of Schur's lemma (Section 11.7) then implies that $\boldsymbol{J}^{(j)2}$ must be a multiple of the $(2j+1)\times(2j+1)$ identity matrix. The coefficient turns out to be $j(j+1)$

$$\boldsymbol{J}^{(j)} \cdot \boldsymbol{J}^{(j)} = j(j+1)\,I. \tag{11.124}$$

Combinations of generators that are a multiple of the identity are called **Casimir operators**.

Example 11.20 (Spin 2) For $j = 2$, the spin-two matrices $J_+^{(2)}$ and $J_3^{(2)}$ are

$$J_+^{(2)} = \begin{pmatrix} 0 & 2 & 0 & 0 & 0 \\ 0 & 0 & \sqrt{6} & 0 & 0 \\ 0 & 0 & 0 & \sqrt{6} & 0 \\ 0 & 0 & 0 & 0 & 2 \\ 0 & 0 & 0 & 0 & 0 \end{pmatrix} \quad \text{and} \quad J_3^{(2)} = \begin{pmatrix} 2 & 0 & 0 & 0 & 0 \\ 0 & 1 & 0 & 0 & 0 \\ 0 & 0 & 0 & 0 & 0 \\ 0 & 0 & 0 & -1 & 0 \\ 0 & 0 & 0 & 0 & -2 \end{pmatrix} \tag{11.125}$$

and $J_- = \left(J_+^{(2)}\right)^\dagger$.

The tensor product of any two irreducible representations $D^{(j)}$ and $D^{(k)}$ of $SU(2)$ is equivalent to the direct sum of all the irreducible representations $D^{(\ell)}$ for $|j-k| \le \ell \le j+k$

$$D^{(j)} \otimes D^{(k)} = \bigoplus_{\ell=|j-k|}^{j+k} D^{(\ell)} \tag{11.126}$$

each $D^{(\ell)}$ occurring once.

Example 11.21 (Addition theorem) The spherical harmonics $Y_{\ell m}(\theta,\phi) = \langle\theta,\phi|\ell,m\rangle$ of Section 9.13 transform under the $(2\ell+1)$-dimensional representation D^ℓ of the rotation group. If a rotation R takes θ,ϕ into the vector θ',ϕ', so that $|\theta',\phi'\rangle = U(R)|\theta,\phi\rangle$, then summing over m' from $-\ell$ to ℓ, we get

$$Y^*_{\ell,m}(\theta',\phi') = \langle\ell,m|\theta',\phi'\rangle = \langle\ell,m|U(R)|\theta,\phi\rangle$$

$$= \langle\ell,m|U(R)|\ell,m'\rangle\langle\ell,m'|\theta,\phi\rangle = D^\ell(R)_{m,m'}Y^*_{\ell,m'}(\theta,\phi).$$

Suppose now that a rotation R maps $|\theta_1,\phi_1\rangle$ and $|\theta_2,\phi_2\rangle$ into $|\theta_1',\phi_1'\rangle = U(R)\,|\theta_1,\phi_1\rangle$ and $|\theta_2',\phi_2'\rangle = U(R)\,|\theta_2,\phi_2\rangle$. Then summing over the repeated indices m, m', and m'' from $-\ell$ to ℓ, we find

$$Y_{\ell,m}(\theta_1',\phi_1')Y^*_{\ell,m}(\theta_2',\phi_2') = D^\ell(R)^*_{m,m'}Y_{\ell,m'}(\theta_1,\phi_1)D^\ell(R)_{m,m''}Y^*_{\ell,m''}(\theta_2,\phi_2).$$

In this equation, the matrix element $D^\ell(R)^*_{m,m'}$ is

$$D^\ell(R)^*_{m,m'} = \langle\ell,m|U(R)|\ell,m'\rangle^* = \langle\ell,m'|U^\dagger(R)|\ell,m\rangle = D^\ell(R^{-1})_{m',m}.$$

Thus since D^ℓ is a representation of the rotation group, the product of the two D^ℓ's in example (11.21) is

$$D^\ell(R)^*_{m,m'}D^\ell(R)_{m,m''} = D^\ell(R^{-1})_{m',m}D^\ell(R)_{m,m''}$$

$$= D^\ell(R^{-1}R)_{m',m''} = D^\ell(I)_{m',m''} = \delta_{m',m''}.$$

So as long as the same rotation R maps θ_1,ϕ_1 into θ_1',ϕ_1' and θ_2,ϕ_2 into θ_2',ϕ_2', then we have

$$\sum_{m=-\ell}^{\ell} Y_{\ell,m}(\theta_1',\phi_1')Y^*_{\ell,m}(\theta_2',\phi_2') = \sum_{m=-\ell}^{\ell} Y_{\ell,m}(\theta_1,\phi_1)Y^*_{\ell,m}(\theta_2,\phi_2).$$

We choose the rotation R as the product of a rotation that maps the unit vector $\hat{n}(\theta_2,\phi_2)$ into $\hat{n}(\theta_2',\phi_2') = \hat{z} = (0,0,1)$ and a rotation about the z axis that maps $\hat{n}(\theta_1,\phi_1)$ into $\hat{n}(\theta_1',\phi_1') = (\sin\theta,0,\cos\theta)$ in the x-z plane where it makes an angle θ with $\hat{n}(\theta_2',\phi_2') = \hat{z}$. We then have $Y^*_{\ell,m}(\theta_2',\phi_2') = Y^*_{\ell,m}(0,0)$ and $Y_{\ell,m}(\theta_1',\phi_1') = Y_{\ell,m}(\theta,0)$ in which θ is the angle between the unit vectors $\hat{n}(\theta_1',\phi_1')$ and $\hat{n}(\theta_2',\phi_2')$, which is the same as the angle between the unit vectors $\hat{n}(\theta_1,\phi_1)$ and $\hat{n}(\theta_2,\phi_2)$. The vanishing (9.108) at $\theta = 0$ of the associated Legendre functions $P_{\ell,m}$ for $m \neq 0$ and the definitions (9.4, 9.101, and 9.112–9.114) say that $Y^*_{\ell,m}(0,0) = \sqrt{(2\ell+1)/4\pi}\,\delta_{m,0}$, and that $Y_{\ell,0}(\theta,0) = \sqrt{(2\ell+1)/4\pi}\,P_\ell(\cos\theta)$. Thus an identity of example (11.21) gives us the spherical harmonics **addition theorem** (9.123)

$$P_\ell(\cos\theta) = \frac{2\ell+1}{4\pi}\sum_{m=-\ell}^{\ell} Y_{\ell,m}(\theta_1,\phi_1)Y^*_{\ell,m}(\theta_2,\phi_2).$$

11.22 How a Field Transforms Under a Rotation

Under a rotation R, a field $\psi_s(x)$ that transforms under the $D^{(j)}$ representation of $SU(2)$ responds as

$$U(R)\,\psi_s(x)\,U^{-1}(R) = D^{(j)}_{s,s'}(R^{-1})\,\psi_{s'}(Rx). \tag{11.127}$$

Example 11.22 (Spin and statistics) Suppose $|a,m\rangle$ and $|b,m\rangle$ are any eigenstates of the rotation operator J_3 with eigenvalue m (in units with $\hbar = c = 1$). If u and v are any two space-like points, then in some Lorentz frame they have spacetime coordinates $u = (t,x,0,0)$ and $v = (t,-x,0,0)$. Let U be the unitary operator that represents a right-handed rotation by π about the 3-axis or z-axis of this Lorentz frame. Then

$$U|a,m\rangle = e^{-im\pi}|a,m\rangle \quad \text{and} \quad \langle b,m|U^{-1} = \langle b,m|e^{im\pi}. \tag{11.128}$$

And by (11.127), U transforms a field ψ of spin j with $\boldsymbol{x} \equiv (x,0,0)$ to

$$U(R)\,\psi_s(t,\boldsymbol{x})\,U^{-1}(R) = D^{(j)}_{ss'}(R^{-1})\,\psi_{s'}(t,-\boldsymbol{x}) = e^{i\pi s}\psi_s(t,-\boldsymbol{x}). \tag{11.129}$$

Thus by inserting the identity operator in the form $I = U^{-1}U$ and using both (11.128) and (11.129), we find, since the phase factors $\exp(-im\pi)$ and $\exp(im\pi)$ cancel,

$$\begin{aligned}\langle b,m|\psi_s(t,\boldsymbol{x})\,\psi_s(t,-\boldsymbol{x})|a,m\rangle &= \langle b,m|U\psi_s(t,\boldsymbol{x})U^{-1}U\psi_s(t,-\boldsymbol{x})U^{-1}|a,m\rangle \\ &= e^{2i\pi s}\,\langle b,m|\psi_s(t,-\boldsymbol{x})\psi_s(t,\boldsymbol{x})|a,m\rangle.\end{aligned} \tag{11.130}$$

Now if j is an integer, then so is s, and the phase factor $\exp(2i\pi s) = 1$ is unity. In this case, we find that the mean value of the equal-time commutator vanishes

$$\langle b,m|[\psi_s(t,\boldsymbol{x}),\psi_s(t,-\boldsymbol{x})]|a,m\rangle = 0 \tag{11.131}$$

which suggests that fields of integral spin commute at space-like separations. They represent bosons. On the other hand, if j is half an odd integer, that is, $j = (2n+1)/2$, where n is an integer, then the phase factor $\exp(2i\pi s) = -1$ is minus one. In this case, the mean value of the equal-time anticommutator vanishes

$$\langle b,m|\{\psi_s(t,\boldsymbol{x}),\psi_s(t,-\boldsymbol{x})\}|a,m\rangle = 0 \tag{11.132}$$

which suggests that fields of half-odd-integral spin anticommute at space-like separations. They represent fermions. This argument shows that the behavior of fields under rotations is related to their equal-time commutation or anticommutation relations

$$\psi_s(t,\boldsymbol{x})\psi_{s'}(t,\boldsymbol{x}') + (-1)^{2j}\psi_{s'}(t,\boldsymbol{x}')\psi_s(t,\boldsymbol{x}) = 0 \tag{11.133}$$

and their statistics.

11.23 Addition of Two Spin-One-Half Systems

The spin operators (11.112)

$$S_a = \frac{\hbar}{2}\sigma_a \tag{11.134}$$

obey the commutation relation (11.118)

$$[S_a, S_b] = i\,\hbar\,\epsilon_{abc}\,S_c. \tag{11.135}$$

The raising and lowering operators

$$S_\pm = S_1 \pm i S_2 \tag{11.136}$$

have simple commutators with S_3

$$[S_3, S_\pm] = \pm\hbar\,S_\pm. \tag{11.137}$$

This relation implies that if the state $|\frac{1}{2}, m\rangle$ is an eigenstate of S_3 with eigenvalue $\hbar m$, then the states $S_\pm|\frac{1}{2}, m\rangle$ either vanish or are eigenstates of S_3 with eigenvalues $\hbar(m \pm 1)$

$$S_3 S_\pm|\frac{1}{2}, m\rangle = S_\pm S_3|\frac{1}{2}, m\rangle \pm \hbar S_\pm|\frac{1}{2}, m\rangle = \hbar(m \pm 1) S_\pm|\frac{1}{2}, m\rangle. \tag{11.138}$$

Thus the raising and lowering operators raise and lower the eigenvalues of S_3. The eigenvalues of $S_3 = \hbar\sigma_3/2$ are $\pm\hbar/2$. So with the usual sign and normalization conventions

$$S_+|-\rangle = \hbar|+\rangle \quad \text{and} \quad S_-|+\rangle = \hbar|-\rangle \tag{11.139}$$

while

$$S_+|+\rangle = 0 \quad \text{and} \quad S_-|-\rangle = 0. \tag{11.140}$$

The square of the total spin operator is simply related to the raising and lowering operators and to S_3

$$\boldsymbol{S}^2 = S_1^2 + S_2^2 + S_3^2 = \frac{1}{2}S_+S_- + \frac{1}{2}S_-S_+ + S_3^2. \tag{11.141}$$

But the squares of the Pauli matrices are unity, and so $S_a^2 = (\hbar/2)^2$ for all three values of a. Thus

$$\boldsymbol{S}^2 = \frac{3}{4}\,\hbar^2 \tag{11.142}$$

is a Casimir operator (11.124) for a spin-one-half system.

Consider two spin operators $\boldsymbol{S}^{(1)}$ and $\boldsymbol{S}^{(2)}$ as defined by (11.112) acting on two spin-one-half systems. Let the tensor-product states

$$|\pm, \pm\rangle = |\pm\rangle_1 |\pm\rangle_2 = |\pm\rangle_1 \otimes |\pm\rangle_2 \tag{11.143}$$

be eigenstates of $S_3^{(1)}$ and $S_3^{(2)}$ so that

$$\begin{aligned} S_3^{(1)}|+,\pm\rangle &= \frac{\hbar}{2}\,|+,\pm\rangle \quad \text{and} \quad S_3^{(2)}|\pm,+\rangle = \frac{\hbar}{2}\,|\pm,+\rangle \\ S_3^{(1)}|-,\pm\rangle &= -\,\frac{\hbar}{2}\,|-,\pm\rangle \quad \text{and} \quad S_3^{(2)}|\pm,-\rangle = -\,\frac{\hbar}{2}\,|\pm,-\rangle. \end{aligned} \tag{11.144}$$

The total spin of the system is the sum of the two spins $\boldsymbol{S} = \boldsymbol{S}^{(1)} + \boldsymbol{S}^{(2)}$, so

$$\boldsymbol{S}^2 = \left(\boldsymbol{S}^{(1)} + \boldsymbol{S}^{(2)}\right)^2 \quad \text{and} \quad S_3 = S_3^{(1)} + S_3^{(2)}. \tag{11.145}$$

The state $|+,+\rangle$ is an eigenstate of S_3 with eigenvalue $\hbar$

$$S_3|+,+\rangle = S_3^{(1)}|+,+\rangle + S_3^{(2)}|+,+\rangle = \frac{\hbar}{2}|+,+\rangle + \frac{\hbar}{2}|+,+\rangle = \hbar|+,+\rangle. \tag{11.146}$$

So the state of angular momentum $\hbar$ in the 3-direction is $|1,1\rangle = |+,+\rangle$. Similarly, the state $|-,-\rangle$ is an eigenstate of S_3 with eigenvalue $-\hbar$

$$S_3|-,-\rangle = S_3^{(1)}|-,-\rangle + S_3^{(2)}|-,-\rangle = -\,\frac{\hbar}{2}|-,-\rangle - \frac{\hbar}{2}|-,-\rangle = -\,\hbar|-,-\rangle \tag{11.147}$$

and so the state of angular momentum $\hbar$ in the negative 3-direction is $|1,-1\rangle = |-,-\rangle$. The states $|+,-\rangle$ and $|-,+\rangle$ are eigenstates of S_3 with eigenvalue 0

$$\begin{aligned} S_3|+,-\rangle &= S_3^{(1)}|+,-\rangle + S_3^{(2)}|+,-\rangle = \frac{\hbar}{2}|+,-\rangle - \frac{\hbar}{2}|+,-\rangle = 0 \\ S_3|-,+\rangle &= S_3^{(1)}|-,+\rangle + S_3^{(2)}|-,+\rangle = -\,\frac{\hbar}{2}|-,+\rangle + \frac{\hbar}{2}|-,+\rangle = 0. \end{aligned} \tag{11.148}$$

To see which states are eigenstates of $\boldsymbol{S}^2$, we use the lowering operator for the combined system $S_- = S_-^{(1)} + S_-^{(2)}$ and the rules (11.122, 11.139, and 11.140) to lower the state $|1,1\rangle$

$$S_-|+,+\rangle = \left(S_-^{(1)} + S_-^{(2)}\right)|+,+\rangle = \hbar\left(|-,+\rangle + |+,-\rangle\right) = \hbar\sqrt{2}\,|1,0\rangle.$$

Thus the state $|1,0\rangle$ is

$$|1,0\rangle = \frac{1}{\sqrt{2}}\left(|+,-\rangle + |-,+\rangle\right). \tag{11.149}$$

The orthogonal and normalized combination of $|+,-\rangle$ and $|-,+\rangle$ must be the state of spin zero

$$|0,0\rangle = \frac{1}{\sqrt{2}}\left(|+,-\rangle - |-,+\rangle\right) \tag{11.150}$$

with the usual sign convention.

To check that the states $|1,0\rangle$ and $|0,0\rangle$ really are eigenstates of $\boldsymbol{S}^2$, we use (11.141 and 11.142) to write $\boldsymbol{S}^2$ as

$$\begin{aligned}\boldsymbol{S}^2 = \left(\boldsymbol{S}^{(1)} + \boldsymbol{S}^{(2)}\right)^2 &= \frac{3}{2}\,\hbar^2 + 2\boldsymbol{S}^{(1)} \cdot \boldsymbol{S}^{(2)} \\ &= \frac{3}{2}\,\hbar^2 + S_+^{(1)} S_-^{(2)} + S_-^{(1)} S_+^{(2)} + 2S_3^{(1)} S_3^{(2)}. \end{aligned} \tag{11.151}$$

Now the sum $S_+^{(1)} S_-^{(2)} + S_-^{(1)} S_+^{(2)}$ merely interchanges the states $|+,-\rangle$ and $|-,+\rangle$ and multiplies them by $\hbar^2$, so

$$\begin{aligned}\boldsymbol{S}^2|1,0\rangle &= \frac{3}{2}\,\hbar^2|1,0\rangle + \hbar^2|1,0\rangle - \frac{2}{4}\hbar^2|1,0\rangle \\ &= 2\hbar^2|1,0\rangle = s(s+1)\hbar^2|1,0\rangle \end{aligned} \tag{11.152}$$

which confirms that $s = 1$. Because of the relative minus sign in formula (11.150) for the state $|0,0\rangle$, we have

$$\begin{aligned}\boldsymbol{S}^2|0,0\rangle &= \frac{3}{2}\,\hbar^2|0,0\rangle - \hbar^2|1,0\rangle - \frac{1}{2}\hbar^2|1,0\rangle \\ &= 0\hbar^2|1,0\rangle = s(s+1)\hbar^2|1,0\rangle \end{aligned} \tag{11.153}$$

which confirms that $s = 0$.

Example 11.23 (Two equivalent representations of $SU(2)$) The identity

$$\left[\exp\left(i\boldsymbol{\theta}\cdot\frac{\boldsymbol{\sigma}}{2}\right)\right]^* = \sigma_2 \exp\left(i\boldsymbol{\theta}\cdot\frac{\boldsymbol{\sigma}}{2}\right)\sigma_2 \tag{11.154}$$

shows that the defining representation of $SU(2)$ (Section 11.20) and its complex conjugate are equivalent (11.8) representations. To prove this identity, we expand the exponential on the right-hand side in powers of its argument

$$\sigma_2 \exp\left(i\boldsymbol{\theta}\cdot\frac{\boldsymbol{\sigma}}{2}\right)\sigma_2 = \sigma_2\left[\sum_{n=0}^{\infty}\frac{1}{n!}\left(i\boldsymbol{\theta}\cdot\frac{\boldsymbol{\sigma}}{2}\right)^n\right]\sigma_2 \tag{11.155}$$

and use the fact that σ_2 is its own inverse to get

$$\sigma_2 \exp\left(i\boldsymbol{\theta}\cdot\frac{\boldsymbol{\sigma}}{2}\right)\sigma_2 = \sum_{n=0}^{\infty}\frac{1}{n!}\left[\sigma_2\left(i\boldsymbol{\theta}\cdot\frac{\boldsymbol{\sigma}}{2}\right)\sigma_2\right]^n. \tag{11.156}$$

Since the Pauli matrices obey the anticommutation relation (11.115), and since both σ_1 and σ_3 are real, while σ_2 is imaginary, we can write the 2×2 matrix within the square brackets as

$$\sigma_2\left(i\boldsymbol{\theta}\cdot\frac{\boldsymbol{\sigma}}{2}\right)\sigma_2 = -i\theta_1\frac{\sigma_1}{2} + i\theta_2\frac{\sigma_2}{2} - i\theta_3\frac{\sigma_3}{2} = \left(i\boldsymbol{\theta}\cdot\frac{\boldsymbol{\sigma}}{2}\right)^* \tag{11.157}$$

which implies the identity (11.154)

$$\sigma_2 \exp\left(i\boldsymbol{\theta}\cdot\frac{\boldsymbol{\sigma}}{2}\right)\sigma_2 = \sum_{n=0}^{\infty}\frac{1}{n!}\left[\left(i\boldsymbol{\theta}\cdot\frac{\boldsymbol{\sigma}}{2}\right)^*\right]^n = \left[\exp\left(i\boldsymbol{\theta}\cdot\frac{\boldsymbol{\sigma}}{2}\right)\right]^*. \tag{11.158}$$

11.24 Jacobi Identity

Any three square matrices A, B, and C satisfy the commutator-product rule

$$\begin{aligned}[A, BC] &= ABC - BCA = ABC - BAC + BAC - BCA \\ &= [A, B]C + B[A, C].\end{aligned} \tag{11.159}$$

Interchanging B and C gives

$$[A, CB] = [A, C]B + C[A, B]. \tag{11.160}$$

Subtracting the second equation from the first, we get the Jacobi identity

$$[A, [B, C]] = [[A, B], C] + [B, [A, C]] \tag{11.161}$$

and its equivalent cyclic form

$$[A, [B, C]] + [B, [C, A]] + [C, [A, B]] = 0. \tag{11.162}$$

Another Jacobi identity uses the anticommutator $\{A, B\} \equiv AB + BA$

$$\{[A, B], C\} + \{[A, C], B\} + [\{B, C\}, A] = 0. \tag{11.163}$$

11.25 Adjoint Representations

Any three generators t_a, t_b, and t_c satisfy the Jacobi identity (11.162)

$$[t_a, [t_b, t_c]] + [t_b, [t_c, t_a]] + [t_c, [t_a, t_b]] = 0. \tag{11.164}$$

By using the structure-constant formula (11.84), we may express each of these double commutators as a linear combination of the generators

$$\begin{aligned}[t_a, [t_b, t_c]] &= [t_a, i f_{bc}^d t_d] = - f_{bc}^d f_{ad}^e t_e \\ [t_b, [t_c, t_a]]] &= [t_b, i f_{ca}^d t_d] = - f_{ca}^d f_{bd}^e t_e \\ [t_c, [t_a, t_b]] &= [t_c, i f_{ab}^d t_d] = - f_{ab}^d f_{cd}^e t_e.\end{aligned} \tag{11.165}$$

So the Jacobi identity (11.164) implies that

$$\left(f_{bc}^d f_{ad}^e + f_{ca}^d f_{bd}^e + f_{ab}^d f_{cd}^e \right) t_e = 0 \tag{11.166}$$

or since the generators are linearly independent

$$f_{bc}^d f_{ad}^e + f_{ca}^d f_{bd}^e + f_{ab}^d f_{cd}^e = 0. \tag{11.167}$$

If we define a set of matrices T_a by

$$(T_b)_{ac} = i f_{ab}^c \tag{11.168}$$

then, since the structure constants are antisymmetric in their lower indices, we may write the three terms in the preceding equation (11.167) as

$$f^d_{bc} f^e_{ad} = f^d_{cb} f^e_{da} = (-T_b T_a)_{ce} \tag{11.169}$$

$$f^d_{ca} f^e_{bd} = -f^d_{ca} f^e_{db} = (T_a T_b)_{ce} \tag{11.170}$$

and

$$f^d_{ab} f^e_{cd} = -i f^d_{ab} (T_d)_{ce} \tag{11.171}$$

or in matrix notation

$$[T_a, T_b] = i f^c_{ab} T_c. \tag{11.172}$$

So the matrices T_a, which we made out of the structure constants by the rule $(T_b)_{ac} = i f^c_{ab}$ (11.168), obey the same algebra (11.68) as do the generators t_a. They are the **generators in the adjoint representation** of the Lie algebra. If the Lie algebra has N generators t_a, then the N generators T_a in the adjoint representation are $N \times N$ matrices.

11.26 Casimir Operators

For any compact Lie algebra, the sum of the squares of all the generators

$$C = \sum_{a=1}^{N} t_a t_a \equiv t_a t_a \tag{11.173}$$

commutes with every generator t_b

$$\begin{aligned}[C, t_b] &= [t_a t_a, t_b] = [t_a, t_b] t_a + t_a [t_a, t_b] \\ &= i f_{abc} t_c t_a + t_a i f_{abc} t_c = i \left(f_{abc} + f_{cba} \right) t_c t_a = 0\end{aligned} \tag{11.174}$$

because of the total antisymmetry (11.82) of the structure constants. This sum, called a **Casimir operator**, commutes with every matrix

$$[C, D(\alpha)] = [C, \exp(i\alpha_a t_a)] = 0 \tag{11.175}$$

of the representation generated by the t_a's. Thus by part 2 of Schur's lemma (Section 11.7), it must be a multiple of the identity matrix

$$C = t_a t_a = cI. \tag{11.176}$$

The constant c depends upon the representation $D(\alpha)$ and is called the **quadratic Casimir**

$$C_2(D) = \text{Tr}\left(t_a^2\right) / \text{Tr} I \ . \tag{11.177}$$

Example 11.24 (Quadratic Casimirs of $SU(2)$) The quadratic Casimir $C(\mathbf{2})$ of the defining representation of $SU(2)$ with generators $t_a = \sigma_a/2$ (11.113) is

$$C_2(\mathbf{2}) = \mathrm{Tr}\left(\sum_{a=1}^{3}\left(\frac{1}{2}\sigma_a\right)^2\right)/\mathrm{Tr}\,(I) = \frac{3\cdot 2\cdot\left(\frac{1}{2}\right)^2}{2} = \frac{3}{4}. \tag{11.178}$$

That of the adjoint representation (11.95) is

$$C_2(\mathbf{3}) = \mathrm{Tr}\left(\sum_{b=1}^{3} t_b^2\right)/\mathrm{Tr}\,(I) = \sum_{a,b,c=1}^{3}\frac{i\epsilon_{abc}i\epsilon_{cba}}{3} = 2. \tag{11.179}$$

The generators of some noncompact groups come in pairs t_a and it_a, and so the sum of the squares of these generators vanishes, $C = t_a t_a - t_a t_a = 0$.

11.27 Tensor Operators for the Rotation Group

Suppose $A_m^{(j)}$ is a set of $2j+1$ operators whose commutation relations with the generators J_i of rotations are

$$[J_i, A_m^{(j)}] = A_s^{(j)}(J_i^{(j)})_{sm} \tag{11.180}$$

in which the sum over s runs from $-j$ to j. Then $A^{(j)}$ is said to be a **spin-j tensor operator** for the group $SU(2)$.

Example 11.25 (A spin-one tensor operator) For instance, if $j=1$, then $(J_i^{(1)})_{sm} = i\hbar\epsilon_{sim}$, and so a spin-1 tensor operator of $SU(2)$ is a vector $A_m^{(1)}$ that transforms as

$$[J_i, A_m^{(1)}] = A_s^{(1)}\, i\,\hbar\,\epsilon_{sim} = i\,\hbar\,\epsilon_{ims}\, A_s^{(1)} \tag{11.181}$$

under rotations.

Let's rewrite the definition (11.180) as

$$J_i A_m^{(j)} = A_s^{(j)}(J_i^{(j)})_{sm} + A_m^{(j)} J_i \tag{11.182}$$

and specialize to the case $i=3$ so that $(J_3^{(j)})_{sm}$ is diagonal, $(J_3^{(j)})_{sm} = \hbar m\delta_{sm}$

$$J_3 A_m^{(j)} = A_s^{(j)}(J_3^{(j)})_{sm} + A_m^{(j)} J_3 = A_s^{(j)}\hbar m\delta_{sm} + A_m^{(j)} J_3 = A_m^{(j)}\left(\hbar m + J_3\right). \tag{11.183}$$

Thus if the state $|j, s, E\rangle$ is an eigenstate of J_3 with eigenvalue $\hbar s$, then the state $A_m^{(j)}|j, s, E\rangle$ is an eigenstate of J_3 with eigenvalue $\hbar(m+s)$

$$J_3 A_m^{(j)}|j, s, E\rangle = A_m^{(j)} (\hbar m + J_3)\,|j, s, E\rangle = \hbar\,(m+s)\,A_m^{(j)}|j, s, E\rangle. \tag{11.184}$$

The J_3 eigenvalues of the tensor operator $A_m^{(j)}$ and the state $|j, s, E\rangle$ add.

11.28 Simple and Semisimple Lie Algebras

An **invariant subalgebra** is a set of generators $t_a^{(i)}$ whose commutator with every generator t_b of the group is a linear combination of the generators $t_c^{(i)}$ of the invariant subalgebra

$$[t_a^{(i)}, t_b] = i f_{abc} t_c^{(i)}. \tag{11.185}$$

The whole algebra and the null algebra are trivial invariant subalgebras.

An algebra with no nontrivial invariant subalgebras is a **simple** algebra. A simple algebra generates a **simple group**. An algebra that has no nontrivial abelian invariant subalgebras is a **semisimple** algebra. A semisimple algebra generates a **semisimple group**.

Example 11.26 (Some simple Lie groups) The groups of unitary matrices of unit determinant $SU(2)$, $SU(3)$, ... are simple. So are the groups of orthogonal matrices of unit determinant $SO(n)$ (except $SO(4)$, which is semisimple) and the groups of symplectic matrices $Sp(2n)$ (Section 11.33).

Example 11.27 (Unification and grand unification) The symmetry group of the **standard model of particle physics** is a **direct product** of an $SU(3)$ group that acts on colored fields, an $SU(2)$ group that acts on **left-handed** quark and lepton fields, and a $U(1)$ group that acts on fields that carry hypercharge. Each of these three groups, is an invariant subgroup of the full symmetry group $SU(3)_c \otimes SU(2)_\ell \otimes U(1)_Y$, and the last one is abelian. Thus the symmetry group of the standard model is neither simple nor semisimple. In theories of **grand unification**, the strong and electroweak interactions unify at very high energies and are described by a simple group which makes all its charges simple multiples of each other. Georgi and Glashow suggested the group $SU(5)$ in 1976 (Howard Georgi, 1947–; Sheldon Glashow, 1932–). Others have proposed $SO(10)$ and even bigger groups.

11.29 SU(3)

The Gell-Mann matrices are

$$\lambda_1 = \begin{pmatrix} 0 & 1 & 0 \\ 1 & 0 & 0 \\ 0 & 0 & 0 \end{pmatrix}, \quad \lambda_2 = \begin{pmatrix} 0 & -i & 0 \\ i & 0 & 0 \\ 0 & 0 & 0 \end{pmatrix}, \quad \lambda_3 = \begin{pmatrix} 1 & 0 & 0 \\ 0 & -1 & 0 \\ 0 & 0 & 0 \end{pmatrix},$$

$$\lambda_4 = \begin{pmatrix} 0 & 0 & 1 \\ 0 & 0 & 0 \\ 1 & 0 & 0 \end{pmatrix}, \quad \lambda_5 = \begin{pmatrix} 0 & 0 & -i \\ 0 & 0 & 0 \\ i & 0 & 0 \end{pmatrix}, \quad \lambda_6 = \begin{pmatrix} 0 & 0 & 0 \\ 0 & 0 & 1 \\ 0 & 1 & 0 \end{pmatrix},$$

$$\lambda_7 = \begin{pmatrix} 0 & 0 & 0 \\ 0 & 0 & -i \\ 0 & i & 0 \end{pmatrix}, \quad \text{and} \quad \lambda_8 = \frac{1}{\sqrt{3}} \begin{pmatrix} 1 & 0 & 0 \\ 0 & 1 & 0 \\ 0 & 0 & -2 \end{pmatrix}. \tag{11.186}$$

The generators t_a of the 3×3 defining representation of $SU(3)$ are these Gell-Mann matrices divided by 2

$$t_a = \lambda_a / 2 \tag{11.187}$$

(Murray Gell-Mann, 1929–).

The eight generators t_a are orthogonal with $k = 1/2$

$$\text{Tr}\,(t_a t_b) = \frac{1}{2} \delta_{ab} \tag{11.188}$$

and satisfy the commutation relation

$$[t_a, t_b] = i f_{abc}\, t_c. \tag{11.189}$$

The trace formula (11.72) gives us the ***SU*(3) structure constants** as

$$f_{abc} = -2i\, \text{Tr}\,([t_a, t_b] t_c)\,. \tag{11.190}$$

They are real and totally antisymmetric with $f_{123} = 1$, $f_{458} = f_{678} = \sqrt{3}/2$, and $f_{147} = -f_{156} = f_{246} = f_{257} = f_{345} = -f_{367} = 1/2$.

While no two generators of $SU(2)$ commute, two generators of $SU(3)$ do. In the representation (11.186,11.187), t_3 and t_8 are diagonal and so commute

$$[t_3, t_8] = 0. \tag{11.191}$$

They generate the **Cartan subalgebra** (11.199) of $SU(3)$.

11.30 SU(3) and Quarks

The generators (11.186 and 11.187) give us the 3×3 representation

$$D(\alpha) = \exp\,(i \alpha_a t_a) \tag{11.192}$$

in which the sum $a = 1, 2, \ldots, 8$ is over the eight generators t_a. This representation acts on complex 3-vectors and is called the **3**.

Note that if

$$D(\alpha_1)D(\alpha_2) = D(\alpha_3) \tag{11.193}$$

then the complex conjugates of these matrices obey the same multiplication rule

$$D^*(\alpha_1)D^*(\alpha_2) = D^*(\alpha_3) \tag{11.194}$$

and so form another representation of $SU(3)$. It turns out that (unlike in $SU(2)$) this representation is inequivalent to the **3**; it is the $\overline{\mathbf{3}}$.

There are three quarks with masses less than about 100 MeV/c^2 – the u, d, and s quarks. The other three quarks c, b, and t are more massive; $m_c = 1.28$ GeV, $m_b = 4.18$ GeV, and $m_t = 173.1$ GeV. Nobody knows why. Gell-Mann and Zweig suggested that the low-energy strong interactions were approximately invariant under unitary transformations of the three light quarks, which they represented by a **3**, and of the three light antiquarks, which they represented by a $\overline{\mathbf{3}}$. They imagined that the eight light pseudoscalar mesons, that is, the three pions π^-, π^0, π^+, the neutral η, and the four kaons K^0, K^+, K^- $\overline{K}^0$, were composed of a quark and an antiquark. So they should transform as the tensor product

$$\mathbf{3} \otimes \overline{\mathbf{3}} = \mathbf{8} \oplus \mathbf{1}. \tag{11.195}$$

They put the eight pseudoscalar mesons into an **8**.

They imagined that the eight light baryons – the two nucleons N and P, the three sigmas Σ^-, Σ^0, Σ^+, the neutral lambda Λ, and the two cascades Ξ^- and Ξ^0 were each made of three quarks. They should transform as the tensor product

$$\mathbf{3} \otimes \mathbf{3} \otimes \mathbf{3} = \mathbf{10} \oplus \mathbf{8} \oplus \mathbf{8} \oplus \mathbf{1}. \tag{11.196}$$

They put the eight light baryons into one of these **8**'s. When they were writing these papers, there were nine spin-3/2 resonances with masses somewhat heavier than 1200 MeV/c^2 – four Δ's, three Σ^*'s, and two Ξ^*'s. They put these into the **10** and predicted the tenth and its mass. In 1964, a tenth spin-3/2 resonance, the Ω^-, was found with a mass close to their prediction of 1680 MeV/c^2, and by 1973 an MIT-SLAC team had discovered quarks inside protons and neutrons. (George Zweig, 1937–)

11.31 Fierz Identity for $SU(n)$

In terms of the $n \times n$ matrices t^a that are the $n^2 - 1$ hermitian generators of the fundamental (defining) representation of $SU(n)$ with

$$\mathrm{Tr}\left(t^a\, t^b\right) = k\, \delta_{ab}, \tag{11.197}$$

Fierz's identity for $SU(n)$ (Nishi, 2005)

$$\frac{1}{k}\sum_{a=1}^{n^2-1} t^a_{ij}\, t^a_{k\ell} + \frac{1}{n}\,\delta_{ij}\,\delta_{k\ell} = \delta_{i\ell}\,\delta_{kj} \tag{11.198}$$

follows from his identity (1.171) for the generators of $U(n)$.

11.32 Cartan Subalgebra

In any Lie group, the maximum set of mutually commuting generators H_a generate the **Cartan subalgebra**

$$[H_a, H_b] = 0 \tag{11.199}$$

which is an abelian subalgebra. The number of generators in the Cartan subalgebra is the **rank** of the Lie algebra. The Cartan generators H_a can be simultaneously diagonalized, and their eigenvalues or diagonal elements are the **weights**

$$H_a|\mu, x, D\rangle = \mu_a|\mu, x, D\rangle \tag{11.200}$$

in which D labels the representation and x whatever other variables are needed to specify the state. The vector μ is the **weight vector**. The **roots** are the weights of the adjoint representation.

11.33 Symplectic Group *Sp(2n)*

The real symplectic group $Sp(2n, R)$ is the group of real linear transformations $R_{i\ell}$ that preserve the canonical commutation relations of quantum mechanics

$$[q_i, p_k] = i\hbar\delta_{ik}, \quad [q_i, q_k] = 0, \quad \text{and} \quad [p_i, p_k] = 0 \tag{11.201}$$

for $i, k = 1, \ldots, n$. In terms of the $2n$ vector $v = (q_1, \ldots, q_n, p_1, \ldots, p_n)$ of quantum variables, these commutation relations are $[v_i, v_k] = i\hbar J_{ik}$ where J is the $2n \times 2n$ real matrix

$$J = \begin{pmatrix} 0 & I \\ -I & 0 \end{pmatrix} \tag{11.202}$$

in which I is the $n \times n$ identity matrix. The real linear transformation

$$v'_i = \sum_{\ell=1}^{2n} R_{i\ell}\, v_\ell \tag{11.203}$$

will preserve the quantum-mechanical commutation relations (11.201) if

$$[v'_i, v'_k] = \left[\sum_{\ell=1}^{2n} R_{i\ell}\, v_\ell, \sum_{m=1}^{2n} R_{km}\, v_m\right] = i\hbar \sum_{\ell,k=1}^{2n} R_{i\ell}\, J_{\ell m}\, R_{km} = i\hbar J_{ik} \tag{11.204}$$

which in matrix notation is just the condition

$$R\,J\,R^{\mathsf{T}} = J \tag{11.205}$$

that the matrix R be in the real symplectic group $Sp(2n,\mathbb{R})$. The transpose and product rules (1.205 and 1.222) for determinants imply that $\det(R) = \pm 1$, but the condition (11.205) itself implies that $\det(R) = 1$ (Zee, 2016, p. 281).

In terms of the matrix J and the hamiltonian $H(v) = H(q,p)$, Hamilton's equations have the symplectic form

$$\dot{q}_i = \frac{\partial H(q,p)}{\partial p_i} \quad \text{and} \quad \dot{p}_i = -\frac{\partial H(q,p)}{\partial q_i} \quad \text{or} \quad \frac{dv_i}{dt} = \sum_{\ell=1}^{2n} J_{i\ell}\,\frac{\partial H(v)}{\partial v_\ell}. \tag{11.206}$$

A matrix $R = e^t$ obeys the defining condition (11.205) if $t\,J = -J\,t^{\mathsf{T}}$ (Exercise 11.22) or equivalently if $JtJ = t^{\mathsf{T}}$. It follows (Exercise 11.23) that the generator t must be

$$t = \begin{pmatrix} b & s_1 \\ s_2 & -b^{\mathsf{T}} \end{pmatrix} \tag{11.207}$$

in which the matrices b, s_1, s_2 are real, and both s_1 and s_2 are symmetric.

The group $Sp(2n, R)$ is noncompact.

Example 11.28 (Squeezed states) A coherent state $|\alpha\rangle$ is an eigenstate of the annihilation operator $a = (\lambda q + ip/\lambda)/\sqrt{2\hbar}$ with a complex eigenvalue α (3.146), $a|\alpha\rangle = \alpha|\alpha\rangle$. In a coherent state with $\lambda = \sqrt{m\omega}$, the variances are $(\Delta q)^2 = \langle\alpha|(q-\bar{q})^2|\alpha\rangle = \hbar/(2m\omega)$ and $(\Delta p)^2 = \langle\alpha|(p-\bar{p})^2|\alpha\rangle = \hbar m\omega/2$. Thus coherent states have minimum uncertainty, $\Delta q\,\Delta p = \hbar/2$.

A squeezed state $|\alpha\rangle'$ is an eigenstate of $a' = (\lambda q' + ip'/\lambda)/\sqrt{2\hbar}$ in which q' and p' are related to q and p by an $Sp(2)$ transformation

$$\begin{pmatrix} q' \\ p' \end{pmatrix} = \begin{pmatrix} a & b \\ c & d \end{pmatrix}\begin{pmatrix} q \\ p \end{pmatrix} \quad \text{with inverse} \quad \begin{pmatrix} q \\ p \end{pmatrix} = \begin{pmatrix} d & -b \\ -c & a \end{pmatrix}\begin{pmatrix} q' \\ p' \end{pmatrix} \tag{11.208}$$

in which $ad - bc = 1$. The standard deviations of the variables q and p in the squeezed state $|\alpha'\rangle$ are

$$\Delta q = \sqrt{\frac{\hbar}{2}}\sqrt{\frac{d^2}{m\omega} + m\omega\,b^2} \quad \text{and} \quad \Delta p = \sqrt{\frac{\hbar}{2}}\sqrt{\frac{c^2}{m\omega} + m\omega\,a^2}. \tag{11.209}$$

Thus by making b and d tiny, one can reduce the uncertainty Δq by any factor, but then Δp will increase by the same factor since the determinant of the $Sp(2)$ transformation must remain equal to unity, $ad - bc = 1$.

Example 11.29 ($Sp(2, R)$) The matrices (Exercise 11.27)

$$T = \pm \begin{pmatrix} \cosh\theta & \sinh\theta \\ \sinh\theta & \cosh\theta \end{pmatrix} \tag{11.210}$$

are elements of the noncompact symplectic group $Sp(2, R)$.

A dynamical map $\mathcal{M}$ that takes a $2n$ vector $v = (q_1, \dots, q_n, p_1, \dots, p_n)$ from $v(t_1)$ to $v(t_2)$ has a jacobian (section 1.21)

$$M_{ab} = \frac{\partial z_a(t_2)}{\partial z_b(t_1)} \tag{11.211}$$

in $Sp(2n, R)$ if and only if its dynamics are hamiltonian (11.206, Section 18.1) (Carl Jacobi 1804–1851, William Hamilton 1805–1865).

The complex symplectic group $Sp(2n, C)$ consists of all $2n \times 2n$ complex matrices C that satisfy the condition

$$C\, J\, C^{\mathsf{T}} = J. \tag{11.212}$$

The group $Sp(2n, C)$ also is noncompact.

The unitary symplectic group $USp(2n)$ consists of all $2n \times 2n$ unitary matrices U that satisfy the condition

$$U\, J\, U^{\mathsf{T}} = J. \tag{11.213}$$

It is compact.

11.34 Quaternions

If z and w are any two complex numbers, then the 2×2 matrix

$$q = \begin{pmatrix} z & w \\ -w^* & z^* \end{pmatrix} \tag{11.214}$$

is a quaternion. The quaternions are closed under addition and multiplication and under multiplication by a real number (Exercise 11.21), but not under multiplication by an arbitrary complex number. The squared norm of q is its determinant

$$\|q\|^2 = |z|^2 + |w|^2 = \det q. \tag{11.215}$$

The matrix products $q^\dagger q$ and $q\, q^\dagger$ are the squared norm $\|q\|^2$ multiplied by the 2×2 identity matrix

$$q^\dagger q = q\, q^\dagger = \|q\|^2\, I. \tag{11.216}$$

The 2×2 matrix

$$i\sigma_2 = \begin{pmatrix} 0 & 1 \\ -1 & 0 \end{pmatrix} \tag{11.217}$$

provides another expression for $\|q\|^2$ in terms of q and its transpose q^{T}

$$q^{\mathsf{T}} i\sigma_2 \, q = \|q\|^2 \, i\sigma_2. \tag{11.218}$$

Clearly $\|q\| = 0$ implies $q = 0$. The norm of a product of quaternions is the product of their norms

$$\|q_1 q_2\| = \sqrt{\det(q_1 q_2)} = \sqrt{\det q_1 \, \det q_2} = \|q_1\| \, \|q_2\|. \tag{11.219}$$

The quaternions therefore form an **associative division algebra** (over the real numbers); the only others are the real numbers and the complex numbers; the **octonions** are a nonassociative division algebra.

One may use the Pauli matrices to define for any real 4-vector x a quaternion $q(x)$ as

$$\begin{aligned} q(x) &= x_0 + i\,\sigma_k\, x_k = x_0 + i\,\boldsymbol{\sigma} \cdot \boldsymbol{x} \\ &= \begin{pmatrix} x_0 + i\,x_3 & x_2 + i\,x_1 \\ -x_2 + i\,x_1 & x_0 - i\,x_3 \end{pmatrix} \end{aligned} \tag{11.220}$$

with squared norm

$$\|q(x)\|^2 = x_0^2 + x_1^2 + x_2^2 + x_3^2. \tag{11.221}$$

The product rule (11.114) for the Pauli matrices tells us that the product of two quaternions is

$$\begin{aligned} q(x)\,q(y) &= (x_0 + i\,\boldsymbol{\sigma} \cdot \boldsymbol{x})(y_0 + i\,\boldsymbol{\sigma} \cdot \boldsymbol{y}) \\ &= x_0 y_0 + i\,\boldsymbol{\sigma} \cdot (y_0\,\boldsymbol{x} + x_0\,\boldsymbol{y}) - i\,(\boldsymbol{x} \times \boldsymbol{y}) \cdot \boldsymbol{\sigma} - \boldsymbol{x} \cdot \boldsymbol{y} \end{aligned} \tag{11.222}$$

so their commutator is

$$[q(x), q(y)] = -2i\,(\boldsymbol{x} \times \boldsymbol{y}) \cdot \boldsymbol{\sigma}. \tag{11.223}$$

Example 11.30 (Lack of analyticity) One may define a function $f(q)$ of a quaternionic variable and then ask what functions are analytic in the sense that the (one-sided) derivative

$$f'(q) = \lim_{q' \to 0} \left[f(q + q') - f(q) \right] q'^{-1} \tag{11.224}$$

exists and is independent of the direction through which $q' \to 0$. This space of functions is extremely limited and does not even include the function $f(q) = q^2$ (exercise 11.24).

11.35 Quaternions and Symplectic Groups

This section is optional on a first reading.

One may regard the unitary symplectic group $USp(2n)$ as made of $2n \times 2n$ unitary matrices W that map n-tuples q of quaternions into n-tuples $q' = Wq$ of quaternions with the same value of the quadratic quaternionic form

$$\|q'\|^2 = \|q'_1\|^2 + \|q'_2\|^2 + \cdots + \|q'_n\|^2 = \|q_1\|^2 + \|q_2\|^2 + \cdots + \|q_n\|^2 = \|q\|^2. \tag{11.225}$$

By (11.216), the quadratic form $\|q'\|^2$ times the 2×2 identity matrix I is equal to the hermitian form $q'^\dagger q'$

$$\|q'\|^2 I = q'^\dagger q' = q_1'^\dagger q_1' + \cdots + q_n'^\dagger q_n' = q^\dagger W^\dagger W q \tag{11.226}$$

and so any matrix W that is both a $2n \times 2n$ unitary matrix and an $n \times n$ matrix of quaternions keeps $\|q'\|^2 = \|q\|^2$

$$\|q'\|^2 I = q^\dagger W^\dagger W q = q^\dagger q = \|q\|^2 I. \tag{11.227}$$

The group $USp(2n)$ thus consists of all $2n \times 2n$ unitary matrices that also are $n \times n$ matrices of quaternions. (This last requirement is needed so that $q' = Wq$ is an n-tuple of quaternions.)

The generators t_a of the symplectic group $USp(2n)$ are $2n \times 2n$ direct-product matrices of the form

$$I \otimes A, \quad \sigma_1 \otimes S_1, \quad \sigma_2 \otimes S_2, \quad \text{and} \quad \sigma_3 \otimes S_3 \tag{11.228}$$

in which I is the 2×2 identity matrix, the three σ_i's are the Pauli matrices, A is an imaginary $n \times n$ antisymmetric matrix, and the S_i are $n \times n$ real symmetric matrices. These generators t_a close under commutation

$$[t_a, t_b] = i f_{abc} t_c. \tag{11.229}$$

Any imaginary linear combination $i\alpha_a t_a$ of these generators is not only a $2n \times 2n$ antihermitian matrix but also an $n \times n$ matrix of quaternions. Thus the matrices

$$D(\alpha) = e^{i\alpha_a t_a} \tag{11.230}$$

are both unitary $2n \times 2n$ matrices and $n \times n$ quaternionic matrices and so are elements of the group $Sp(2n)$.

Example 11.31 ($USp(2) \cong SU(2)$) There is no 1×1 antisymmetric matrix, and there is only one 1×1 symmetric matrix. So the generators t_a of the group $Sp(2)$ are the Pauli matrices $t_a = \sigma_a$, and $Sp(2) = SU(2)$. The elements $g(\alpha)$ of

$SU(2)$ are quaternions of unit norm (Exercise 11.20), and so the product $g(\alpha)q$ is a quaternion

$$\|g(\alpha)q\|^2 = \det(g(\alpha)q) = \det(g(\alpha)) \det q = \det q = \|q\|^2 \tag{11.231}$$

with the same squared norm.

Example 11.32 ($SO(4) \cong SU(2) \otimes SU(2)$) If g and h are any two elements of the group $SU(2)$, then the squared norm (11.221) of the quaternion $q(x) = x_0 + i\boldsymbol{\sigma} \cdot \boldsymbol{x}$ is invariant under the transformation $q(x') = g\, q(x)\, h^{-1}$, that is, $x_0'^2 + x_1'^2 + x_2'^2 + x_3'^2 = x_0^2 + x_1^2 + x_2^2 + x_3^2$. So $x \to x'$ is an $SO(4)$ rotation of the 4-vector x. The Lie algebra of $SO(4)$ thus contains two commuting invariant $SU(2)$ subalgebras and so is semisimple.

Example 11.33 ($USp(4) \cong SO(5)$) Apart from scale factors, there are three real symmetric 2×2 matrices $S_1 = \sigma_1$, $S_2 = I$, and $S_3 = \sigma_3$ and one imaginary antisymmetric 2×2 matrix $A = \sigma_2$. So there are 10 generators of $USp(4) = SO(5)$

$$\begin{aligned} t_1 = I \otimes \sigma_2 &= \begin{pmatrix} 0 & -iI \\ iI & 0 \end{pmatrix}, & t_{k1} = \sigma_k \otimes \sigma_1 &= \begin{pmatrix} 0 & \sigma_k \\ \sigma_k & 0 \end{pmatrix} \\ t_{k2} = \sigma_k \otimes I &= \begin{pmatrix} \sigma_k & 0 \\ 0 & \sigma_k \end{pmatrix}, & t_{k3} = \sigma_k \otimes \sigma_3 &= \begin{pmatrix} \sigma_k & 0 \\ 0 & -\sigma_k \end{pmatrix} \end{aligned} \tag{11.232}$$

where k runs from 1 to 3.

Another way of looking at $USp(2n)$ is to use (11.218) to write the quadratic form $\|q\|^2$ as

$$\|q\|^2 \mathcal{J} = q^{\mathsf{T}} \mathcal{J} q \tag{11.233}$$

in which the $2n \times 2n$ matrix $\mathcal{J}$ has n copies of $i\sigma_2$ on its 2×2 diagonal

$$\mathcal{J} = \begin{pmatrix} i\sigma_2 & 0 & 0 & 0 & \dots & 0 \\ 0 & i\sigma_2 & 0 & 0 & \dots & 0 \\ 0 & 0 & i\sigma_2 & 0 & \dots & 0 \\ 0 & 0 & 0 & i\sigma_2 & \dots & 0 \\ \vdots & \vdots & \vdots & \vdots & \ddots & 0 \\ 0 & 0 & 0 & 0 & 0 & i\sigma_2 \end{pmatrix} \tag{11.234}$$

and is the matrix J (11.202) in a different basis. Thus any $n \times n$ matrix of quaternions W that satisfies

$$W^{\mathsf{T}} \mathcal{J} W = \mathcal{J} \tag{11.235}$$

also satisfies

$$\|Wq\|^2 \mathcal{J} = q^{\mathsf{T}} W^{\mathsf{T}} \mathcal{J} W q = q^{\mathsf{T}} \mathcal{J} q = \|q\|^2 \mathcal{J} \tag{11.236}$$

and so leaves invariant the quadratic form (11.225). The group $USp(2n)$ therefore consists of all $2n \times 2n$ unitary matrices W that satisfy (11.235) and that also are $n \times n$ matrices of quaternions.

11.36 Compact Simple Lie Groups

Élie Cartan (1869–1951) showed that all compact, simple Lie groups fall into four infinite classes and five discrete cases. For $n = 1, 2, \ldots$, his four classes are:

- $A_n = SU(n+1)$ which are $(n+1) \times (n+1)$ unitary matrices with unit determinant,
- $B_n = SO(2n+1)$ which are $(2n+1) \times (2n+1)$ orthogonal matrices with unit determinant,
- $C_n = USp(2n)$ which are the unitary $2n \times 2n$ symplectic matrices, and
- $D_n = SO(2n)$ which are $2n \times 2n$ orthogonal matrices with unit determinant.

The five discrete cases are the **exceptional groups** G_2, F_4, E_6, E_7, and E_8.

The exceptional groups are associated with the **octonians**

$$a + b_\alpha i_\alpha \tag{11.237}$$

where the α-sum runs from 1 to 7; the eight numbers a and b_α are real; and the seven i_α's obey the multiplication law

$$i_\alpha \, i_\beta = -\delta_{\alpha\beta} + g_{\alpha\beta\gamma} \, i_\gamma \tag{11.238}$$

in which $g_{\alpha\beta\gamma}$ is totally antisymmetric with

$$g_{123} = g_{247} = g_{451} = g_{562} = g_{634} = g_{375} = g_{716} = 1. \tag{11.239}$$

Like the quaternions and the complex numbers, the octonians form a **division algebra** with an absolute value

$$|a + b_\alpha i_\alpha| = \left(a^2 + b_\alpha^2\right)^{1/2} \tag{11.240}$$

that satisfies

$$|AB| = |A||B| \tag{11.241}$$

but they lack associativity.

The group G_2 is the subgroup of $SO(7)$ that leaves the $g_{\alpha\beta\gamma}$'s of (11.238) invariant.

11.37 Group Integration

Suppose we need to integrate some function $f(g)$ over a group. Naturally, we want to do so in a way that gives equal weight to every element of the group. In particular, if g' is any group element, we want the integral of the shifted function $f(g'g)$ to be the same as the integral of $f(g)$

$$\int f(g)\, dg = \int f(g'g)\, dg. \tag{11.242}$$

Such a measure dg is said to be **left invariant** (Creutz, 1983, chap. 8).

Let's use the letters $a = a_1, \ldots, a_n$, $b = b_1, \ldots, b_n$, and so forth to label the elements $g(a)$, $g(b)$, so that an integral over the group is

$$\int f(g)\, dg = \int f(g(a))\, m(a)\, d^n a \tag{11.243}$$

in which $m(a)$ is the left-invariant measure and the integration is over the n-space of a's that label all the elements of the group.

To find the left-invariant measure $m(a)$, we use the multiplication law of the group

$$g(a(c,b)) \equiv g(c)\, g(b) \tag{11.244}$$

and impose the requirement (11.242) of left invariance with $g' \equiv g(c)$

$$\int f(g(b))\, m(b)\, d^n b = \int f(g(c)g(b))\, m(b)\, d^n b = \int f(g(a(c,b)))\, m(b)\, d^n b. \tag{11.245}$$

We change variables from b to $a = a(c,b)$ by using the jacobian $\det(\partial b/\partial a)$ which gives us $d^n b = \det(\partial b/\partial a)\, d^n a$

$$\int f(g(b))\, m(b)\, d^n b = \int f(g(a))\, \det(\partial b/\partial a)\, m(b)\, d^n a. \tag{11.246}$$

Replacing b by $a = a(c,b)$ on the left-hand side of this equation, we find

$$m(a) = \det(\partial b/\partial a)\, m(b) \tag{11.247}$$

or since $\det(\partial b/\partial a) = 1/\det(\partial a(c,b)/\partial b)$

$$m(a(c,b)) = m(b)/\det(\partial a(c,b)/\partial b). \tag{11.248}$$

So if we let $g(b) \to g(0) = e$, the identity element of the group, and set $m(e) = 1$, then we find for the measure

$$m(a) = m(c) = m(a(c,b))|_{b=0} = 1/\det(\partial a(c,b)/\partial b)|_{b=0}\,. \tag{11.249}$$

Example 11.34 (The invariant measure for SU(2)) A general element of the group $SU(2)$ is given by (11.117) as

$$\exp\left(i\, \boldsymbol{\theta} \cdot \frac{\boldsymbol{\sigma}}{2}\right) = I \cos\frac{\theta}{2} + i\, \hat{\boldsymbol{\theta}} \cdot \boldsymbol{\sigma} \sin\frac{\theta}{2}. \tag{11.250}$$

Setting $a_0 = \cos(\theta/2)$ and $\boldsymbol{a} = \hat{\boldsymbol{\theta}} \sin(\theta/2)$, we have

$$g(a) = a_0 + i\, \boldsymbol{a} \cdot \boldsymbol{\sigma} \tag{11.251}$$

in which $a^2 \equiv a_0^2 + \boldsymbol{a} \cdot \boldsymbol{a} = 1$. Thus, the parameter space for $SU(2)$ is the unit sphere S_3 in 4 dimensions. Its invariant measure is

$$\int \delta(1 - a^2)\, d^4a = \int \delta(1 - a_0^2 - \boldsymbol{a}^2)\, d^4a = \int (1 - \boldsymbol{a}^2)^{-1/2}\, d^3a \tag{11.252}$$

or

$$m(\boldsymbol{a}) = (1 - \boldsymbol{a}^2)^{-1/2} = \frac{1}{|\cos(\theta/2)|}. \tag{11.253}$$

We also can write the arbitrary element (11.251) of $SU(2)$ as

$$g(\boldsymbol{a}) = \pm\sqrt{1 - \boldsymbol{a}^2} + i\, \boldsymbol{a} \cdot \boldsymbol{\sigma} \tag{11.254}$$

and the group-multiplication law (11.244) as

$$\sqrt{1 - \boldsymbol{a}^2} + i\, \boldsymbol{a} \cdot \boldsymbol{\sigma} = \left(\sqrt{1 - \boldsymbol{c}^2} + i\, \boldsymbol{c} \cdot \boldsymbol{\sigma}\right)\left(\sqrt{1 - \boldsymbol{b}^2} + i\, \boldsymbol{b} \cdot \boldsymbol{\sigma}\right). \tag{11.255}$$

Thus, by multiplying both sides of this equation by σ_i and taking the trace, we find (Exercise 11.28) that the parameters $\boldsymbol{a}(\boldsymbol{c}, \boldsymbol{b})$ that describe the product $g(\boldsymbol{c})\, g(\boldsymbol{b})$ are

$$\boldsymbol{a}(\boldsymbol{c}, \boldsymbol{b}) = \sqrt{1 - \boldsymbol{c}^2}\, \boldsymbol{b} + \sqrt{1 - \boldsymbol{b}^2}\, \boldsymbol{c} - \boldsymbol{c} \times \boldsymbol{b}. \tag{11.256}$$

To compute the jacobian of our formula (11.249) for the invariant measure, we differentiate this expression (11.256) at $\boldsymbol{b} = \boldsymbol{0}$ and so find (Exercise 11.29)

$$m(\boldsymbol{a}) = 1/\det(\partial a(c, b)/\partial b)|_{b=0} = (1 - \boldsymbol{a}^2)^{-1/2} \tag{11.257}$$

as the left-invariant measure in agreement with (11.253).

11.38 Lorentz Group

The Lorentz group $O(3,1)$ is the set of all linear transformations L that leave invariant the Minkowski inner product

$$xy \equiv \boldsymbol{x} \cdot \boldsymbol{y} - x^0 y^0 = x^{\mathsf{T}} \eta y \tag{11.258}$$

in which η is the diagonal matrix

$$\eta = \begin{pmatrix} -1 & 0 & 0 & 0 \\ 0 & 1 & 0 & 0 \\ 0 & 0 & 1 & 0 \\ 0 & 0 & 0 & 1 \end{pmatrix}. \tag{11.259}$$

So L is in $O(3,1)$ if for all 4-vectors x and y

$$(Lx)^{\mathsf{T}} \eta L\, y = x^{\mathsf{T}} L^{\mathsf{T}} \eta\, Ly = x^{\mathsf{T}} \eta\, y. \tag{11.260}$$

Since x and y are arbitrary, this condition amounts to

$$L^{\mathsf{T}} \eta \, L = \eta \quad \text{or} \quad L^a{}_b \, \eta_{ac} \, L^c{}_d = \eta_{bd}. \tag{11.261}$$

Taking the determinant of both sides and using the transpose (1.205) and product (1.225) rules, we have

$$(\det L)^2 = 1. \tag{11.262}$$

So $\det L = \pm 1$, and every Lorentz transformation L has an inverse. Multiplying (11.261) by η, we get

$$\eta L^{\mathsf{T}} \eta L = \eta^2 = I \quad \text{or} \quad \eta^{eb} \, L^a{}_b \, \eta_{ac} \, L^c{}_d = \eta^{eb} \, \eta_{bd} = \delta^e_d \tag{11.263}$$

which identifies L^{-1} as

$$L^{-1} = \eta L^{\mathsf{T}} \eta \quad \text{or} \quad L^{-1e}{}_c = \eta^{eb} \, L^a{}_b \, \eta_{ac} = L_c{}^e. \tag{11.264}$$

The subgroup of $O(3,1)$ with $\det L = 1$ is the proper Lorentz group $SO(3,1)$. The subgroup of $SO(3,1)$ that leaves invariant the sign of the time component of timelike vectors is the **proper orthochronous Lorentz group** $SO^+(3,1)$.

To find the Lie algebra of $SO^+(3,1)$, we take a Lorentz matrix $L = I + \omega$ that differs from the identity matrix I by a tiny matrix ω and require L to obey the condition (11.261) for membership in the Lorentz group

$$\left(I + \omega^{\mathsf{T}}\right) \eta \, (I + \omega) = \eta + \omega^{\mathsf{T}} \eta + \eta \, \omega + \omega^{\mathsf{T}} \omega = \eta. \tag{11.265}$$

Neglecting $\omega^{\mathsf{T}} \omega$, we have $\omega^{\mathsf{T}} \eta = -\eta \, \omega$ or since $\eta^2 = I$

$$\omega^{\mathsf{T}} = - \, \eta \, \omega \, \eta. \tag{11.266}$$

This equation implies that the matrix ω_{ab} is antisymmetric when both indexes are down

$$\omega_{ab} = -\omega_{ba}. \tag{11.267}$$

To see why, we write it (11.266) as $\omega^e{}_a = - \, \eta_{ab} \, \omega^b{}_c \, \eta^{ce}$ and then multiply both sides by η_{de} so as to get $\omega_{da} = \eta_{de} \, \omega^e{}_a = -\eta_{ab} \, \omega^b{}_c \, \eta^{ce} \, \eta_{de} = -\omega_{ac} \, \delta^c_d = -\omega_{ad}$.

The key equation (11.266) also tells us (Exercise 11.31) that under transposition the time–time and space–space elements of ω change sign, while the time–space and spacetime elements do not. That is, the tiny matrix ω is for infinitesimal $\boldsymbol{\theta}$ and $\boldsymbol{\lambda}$ a linear combination

$$\omega = \boldsymbol{\theta} \cdot \boldsymbol{R} + \boldsymbol{\lambda} \cdot \boldsymbol{B} \tag{11.268}$$

of three antisymmetric space–space matrices

$$R_1 = \begin{pmatrix} 0 & 0 & 0 & 0 \\ 0 & 0 & 0 & 0 \\ 0 & 0 & 0 & -1 \\ 0 & 0 & 1 & 0 \end{pmatrix} \quad R_2 = \begin{pmatrix} 0 & 0 & 0 & 0 \\ 0 & 0 & 0 & 1 \\ 0 & 0 & 0 & 0 \\ 0 & -1 & 0 & 0 \end{pmatrix} \quad R_3 = \begin{pmatrix} 0 & 0 & 0 & 0 \\ 0 & 0 & -1 & 0 \\ 0 & 1 & 0 & 0 \\ 0 & 0 & 0 & 0 \end{pmatrix} \tag{11.269}$$

and of three symmetric time–space matrices

$$B_1 = \begin{pmatrix} 0 & 1 & 0 & 0 \\ 1 & 0 & 0 & 0 \\ 0 & 0 & 0 & 0 \\ 0 & 0 & 0 & 0 \end{pmatrix} \quad B_2 = \begin{pmatrix} 0 & 0 & 1 & 0 \\ 0 & 0 & 0 & 0 \\ 1 & 0 & 0 & 0 \\ 0 & 0 & 0 & 0 \end{pmatrix} \quad B_3 = \begin{pmatrix} 0 & 0 & 0 & 1 \\ 0 & 0 & 0 & 0 \\ 0 & 0 & 0 & 0 \\ 1 & 0 & 0 & 0 \end{pmatrix} \tag{11.270}$$

all of which satisfy condition (11.266). The three R_ℓ are 4×4 versions of the rotation generators (11.93); the three B_ℓ generate Lorentz boosts.

If we write $L = I + \omega$ as

$$L = I - i\theta_\ell \, i R_\ell - i\lambda_\ell \, i B_\ell \equiv I - i\theta_\ell J_\ell - i\lambda_\ell K_\ell \tag{11.271}$$

then the three matrices $J_\ell = i R_\ell$ are imaginary and antisymmetric, and therefore hermitian. But the three matrices $K_\ell = i B_\ell$ are imaginary and symmetric, and so are antihermitian. The 4×4 matrix $L = \exp(i\theta_\ell J_\ell - i\lambda_\ell K_\ell)$ is **not unitary** because the Lorentz group is **not compact**.

One may verify (Exercise 11.32) that the six generators J_ℓ and K_ℓ satisfy three sets of commutation relations:

$$[J_i, J_j] = i\epsilon_{ijk} J_k \tag{11.272}$$

$$[J_i, K_j] = i\epsilon_{ijk} K_k \tag{11.273}$$

$$[K_i, K_j] = -i\epsilon_{ijk} J_k. \tag{11.274}$$

The first (11.272) says that the three J_ℓ generate the rotation group $SO(3)$; the second (11.273) says that the three boost generators transform as a 3-vector under $SO(3)$; and the third (11.274) implies that four cancelling infinitesimal boosts can amount to a rotation. These three sets of commutation relations form the Lie algebra of the Lorentz group $SO(3,1)$. Incidentally, one may show (Exercise 11.33) that if $\boldsymbol{J}$ and $\boldsymbol{K}$ satisfy these commutation relations (11.272–11.274), then so do

$$\boldsymbol{J} \quad \text{and} \quad -\boldsymbol{K}. \tag{11.275}$$

The infinitesimal Lorentz transformation (11.271) is the 4×4 matrix

$$L = I + \omega = I + \theta_\ell R_\ell + \lambda_\ell B_\ell = \begin{pmatrix} 1 & \lambda_1 & \lambda_2 & \lambda_3 \\ \lambda_1 & 1 & -\theta_3 & \theta_2 \\ \lambda_2 & \theta_3 & 1 & -\theta_1 \\ \lambda_3 & -\theta_2 & \theta_1 & 1 \end{pmatrix}. \tag{11.276}$$

It moves any 4-vector x to $x' = L\,x$ or in components $x'^a = L^a{}_b\, x^b$

$$\begin{aligned}
x'^0 &= x^0 + \lambda_1 x^1 + \lambda_2 x^2 + \lambda_3 x^3 \\
x'^1 &= \lambda_1 x^0 + x^1 - \theta_3 x^2 + \theta_2 x^3 \\
x'^2 &= \lambda_2 x^0 + \theta_3 x^1 + x^2 - \theta_1 x^3 \\
x'^3 &= \lambda_3 x^0 - \theta_2 x^1 + \theta_1 x^2 + x^3.
\end{aligned} \tag{11.277}$$

More succinctly with $t = x^0$, this is

$$\begin{aligned}
t' &= t + \boldsymbol{\lambda} \cdot \boldsymbol{x} \\
\boldsymbol{x}' &= \boldsymbol{x} + t\boldsymbol{\lambda} + \boldsymbol{\theta} \wedge \boldsymbol{x}
\end{aligned} \tag{11.278}$$

in which $\wedge \equiv \times$ means cross-product.

For arbitrary real $\boldsymbol{\theta}$ and $\boldsymbol{\lambda}$, the matrices

$$L = e^{-i\boldsymbol{\theta}\cdot\boldsymbol{J} - i\boldsymbol{\lambda}\cdot\boldsymbol{K}} \tag{11.279}$$

form the subgroup of $O(3,1)$ that is connected to the identity matrix I. The matrices of this subgroup have unit determinant and preserve the sign of the time of time-like vectors, that is, if $x^2 < 0$, and $y = Lx$, then $y^0 x^0 > 0$. This is the proper orthochronous Lorentz group $SO^+(3,1)$. The rest of the (homogeneous) Lorentz group can be obtained from it by space $\mathcal{P}$, time $\mathcal{T}$, and spacetime $\mathcal{PT}$ reflections.

The task of finding all the finite-dimensional irreducible representations of the proper orthochronous homogeneous Lorentz group becomes vastly simpler when we write the commutation relations (11.272–11.274) in terms of the hermitian matrices

$$J^{\pm}_{\ell} = \frac{1}{2}\left(J_\ell \pm i K_\ell\right) \tag{11.280}$$

which generate two independent rotation groups

$$\begin{aligned}
[J_i^+, J_j^+] &= i\epsilon_{ijk} J_k^+ \\
[J_i^-, J_j^-] &= i\epsilon_{ijk} J_k^- \\
[J_i^+, J_j^-] &= 0.
\end{aligned} \tag{11.281}$$

Thus the Lie algebra of the Lorentz group is equivalent to two copies of the Lie algebra (11.119) of $SU(2)$.

The hermitian generators of the rotation subgroup $SU(2)$ are by (11.280)

$$\boldsymbol{J} = \boldsymbol{J}^+ + \boldsymbol{J}^-. \tag{11.282}$$

The antihermitian generators of the boosts are (also by 11.280)

$$\boldsymbol{K} = -i\left(\boldsymbol{J}^+ - \boldsymbol{J}^-\right). \tag{11.283}$$

Since $\boldsymbol{J}^+$ and $\boldsymbol{J}^-$ commute, the finite-dimensional irreducible representations of the Lorentz group are the direct products

$$
\begin{aligned}
D(j,j')(\boldsymbol{\theta},\boldsymbol{\lambda}) &= e^{-i\boldsymbol{\theta}\cdot\boldsymbol{J}-i\boldsymbol{\lambda}\cdot\boldsymbol{K}} = e^{(-i\boldsymbol{\theta}-\boldsymbol{\lambda})\cdot\boldsymbol{J}^+ + (-i\boldsymbol{\theta}+\boldsymbol{\lambda})\cdot\boldsymbol{J}^-} \\
&= e^{(-i\boldsymbol{\theta}-\boldsymbol{\lambda})\cdot\boldsymbol{J}^+}\, e^{(-i\boldsymbol{\theta}+\boldsymbol{\lambda})\cdot\boldsymbol{J}^-}
\end{aligned}
\tag{11.284}
$$

of the nonunitary representations

$$
D^{(j,0)}(\boldsymbol{\theta},\boldsymbol{\lambda}) = e^{(-i\boldsymbol{\theta}-\boldsymbol{\lambda})\cdot\boldsymbol{J}^+} \quad \text{and} \quad D^{(0,j')}(\boldsymbol{\theta},\boldsymbol{\lambda}) = e^{(-i\boldsymbol{\theta}+\boldsymbol{\lambda})\cdot\boldsymbol{J}^-}
\tag{11.285}
$$

generated by the three $(2j+1)\times(2j+1)$ matrices J_ℓ^+ and by the three $(2j'+1)\times(2j'+1)$ matrices J_ℓ^-.

Under a Lorentz transformation L, a field $\psi_{m,m'}^{(j,j')}(x)$ that transforms under the $D^{(j,j')}$ representation of the Lorentz group responds as

$$
U(L)\,\psi_{m,m'}^{(j,j')}(x)\,U^{-1}(L) = D_{mm''}^{(j,0)}(L^{-1})\,D_{m'm'''}^{(0,j')}(L^{-1})\,\psi_{m'',m'''}^{(j,j')}(Lx).
\tag{11.286}
$$

The representation $D^{(j,j')}$ describes objects of the spins s that can arise from the direct product of spin-j with spin-j' (Weinberg, 1995, p. 231)

$$
s = j+j',\ j+j'-1,\ldots,\ |j-j'|.
\tag{11.287}
$$

For instance, $D^{(0,0)}$ describes a spinless field or particle, while $D^{(1/2,0)}$ and $D^{(0,1/2)}$ respectively describe left-handed and right-handed spin-1/2 fields or particles. The representation $D^{(1/2,1/2)}$ describes objects of spin 1 and spin 0 – the spatial and time components of a 4-vector. The interchange of $\boldsymbol{J}^+$ and $\boldsymbol{J}^-$ replaces the generators $\boldsymbol{J}$ and $\boldsymbol{K}$ with $\boldsymbol{J}$ and $-\boldsymbol{K}$, a substitution that we know (11.275) is legitimate.

11.39 Left-Handed Representation of the Lorentz Group

The generators of the 2-dimensional representation $D^{(1/2,0)}$ with $j = 1/2$ and $j' = 0$ are given by (11.282 and 11.283) with $\boldsymbol{J}^+ = \boldsymbol{\sigma}/2$ and $\boldsymbol{J}^- = 0$. They are

$$
\boldsymbol{J} = \frac{1}{2}\boldsymbol{\sigma} \quad \text{and} \quad \boldsymbol{K} = -i\frac{1}{2}\boldsymbol{\sigma}.
\tag{11.288}
$$

The 2×2 matrix $D^{(1/2,0)}$ that represents the Lorentz transformation (11.279)

$$
L = e^{-i\boldsymbol{\theta}\cdot\boldsymbol{J}-i\boldsymbol{\lambda}\cdot\boldsymbol{K}}
\tag{11.289}
$$

is

$$
D^{(1/2,0)}(\boldsymbol{\theta},\boldsymbol{\lambda}) = \exp\left(-i\boldsymbol{\theta}\cdot\boldsymbol{\sigma}/2 - \boldsymbol{\lambda}\cdot\boldsymbol{\sigma}/2\right).
\tag{11.290}
$$

And so the generic $D^{(1/2,0)}$ matrix is

$$
D^{(1/2,0)}(\boldsymbol{\theta},\boldsymbol{\lambda}) = e^{-\boldsymbol{z}\cdot\boldsymbol{\sigma}/2}
\tag{11.291}
$$

with $\boldsymbol{\lambda} = \mathrm{Re}\,\boldsymbol{z}$ and $\boldsymbol{\theta} = \mathrm{Im}\,\boldsymbol{z}$. It is nonunitary and of unit determinant; it is a member of the group $SL(2, C)$ of complex unimodular 2×2 matrices. The (covering) group $SL(2, C)$ relates to the Lorentz group $SO(3, 1)$ as $SU(2)$ relates to the rotation group $SO(3)$.

Example 11.35 (The standard left-handed boost) For a particle of mass $m > 0$, the standard boost that takes the 4-vector $k = (m, \mathbf{0})$ to $p = (p^0, \boldsymbol{p})$, where $p^0 = \sqrt{m^2 + \boldsymbol{p}^2}$ is a boost in the $\hat{\boldsymbol{p}}$ direction. It is the 4×4 matrix

$$B(p) = R(\hat{\boldsymbol{p}})\, B_3(p^0)\, R^{-1}(\hat{\boldsymbol{p}}) = \exp\left(\alpha\, \hat{\boldsymbol{p}} \cdot \boldsymbol{B}\right) \tag{11.292}$$

in which $\cosh\alpha = p^0/m$ and $\sinh\alpha = |\boldsymbol{p}|/m$, as one may show by expanding the exponential (Exercise 11.35). This standard boost is represented by $D^{(1/2,0)}(\mathbf{0}, \boldsymbol{\lambda})$, the 2×2 matrix (11.289), with $\boldsymbol{\lambda} = \alpha\, \hat{\boldsymbol{p}}$. The power-series expansion of this matrix is (Exercise 11.36)

$$\begin{aligned} D^{(1/2,0)}(\mathbf{0}, \alpha\, \hat{\boldsymbol{p}}) &= e^{-\alpha \hat{\boldsymbol{p}}\cdot\boldsymbol{\sigma}/2} = I\, \cosh(\alpha/2) - \hat{\boldsymbol{p}} \cdot \boldsymbol{\sigma}\, \sinh(\alpha/2) \\ &= I\, \sqrt{(p^0 + m)/(2m)} - \hat{\boldsymbol{p}} \cdot \boldsymbol{\sigma}\, \sqrt{(p^0 - m)/(2m)} \\ &= \frac{(p^0 + m)I - \boldsymbol{p} \cdot \boldsymbol{\sigma}}{\sqrt{2m(p^0 + m)}} \end{aligned} \tag{11.293}$$

in which I is the 2×2 identity matrix.

Under $D^{(1/2,0)}$, the vector $(-I, \boldsymbol{\sigma})$ transforms like a 4-vector. For tiny $\boldsymbol{\theta}$ and $\boldsymbol{\lambda}$, one may show (Exercise 11.38) that the vector $(-I, \boldsymbol{\sigma})$ transforms as

$$\begin{aligned} D^{\dagger(1/2,0)}(\boldsymbol{\theta}, \boldsymbol{\lambda})(-I)D^{(1/2,0)}(\boldsymbol{\theta}, \boldsymbol{\lambda}) &= -I + \boldsymbol{\lambda} \cdot \boldsymbol{\sigma} \\ D^{\dagger(1/2,0)}(\boldsymbol{\theta}, \boldsymbol{\lambda})\, \boldsymbol{\sigma}\, D^{(1/2,0)}(\boldsymbol{\theta}, \boldsymbol{\lambda}) &= \boldsymbol{\sigma} + (-I)\boldsymbol{\lambda} + \boldsymbol{\theta} \wedge \boldsymbol{\sigma} \end{aligned} \tag{11.294}$$

which is how the 4-vector $(t, \boldsymbol{x})$ transforms (11.278). Under a finite Lorentz transformation L, the 4-vector $S^a \equiv (-I, \boldsymbol{\sigma})$ goes to

$$D^{\dagger(1/2,0)}(L)\, S^a\, D^{(1/2,0)}(L) = L^a{}_b S^b. \tag{11.295}$$

A massless field $u(x)$ that responds to a unitary Lorentz transformation $U(L)$ like

$$U(L)\, u(x)\, U^{-1}(L) = D^{(1/2,0)}(L^{-1})\, u(Lx) \tag{11.296}$$

is called a **left-handed Weyl spinor**. The action density

$$\mathcal{L}_\ell(x) = i\, u^\dagger(x)\, (\partial_0 I - \nabla \cdot \boldsymbol{\sigma})\, u(x) \tag{11.297}$$

is Lorentz covariant, that is

$$U(L)\, \mathcal{L}_\ell(x)\, U^{-1}(L) = \mathcal{L}_\ell(Lx). \tag{11.298}$$

Example 11.36 (Why $\mathcal{L}_\ell$ is Lorentz covariant) We first note that the derivatives ∂'_b in $\mathcal{L}_\ell(Lx)$ are with respect to $x' = Lx$. Since the inverse matrix L^{-1} takes x' back to $x = L^{-1}x'$ or in tensor notation $x^a = L^{-1a}{}_b\, x'^b$, the derivative ∂'_b is

$$\partial'_b = \frac{\partial}{\partial x'^b} = \frac{\partial x^a}{\partial x'^b}\frac{\partial}{\partial x^a} = L^{-1a}{}_b \frac{\partial}{\partial x^a} = \partial_a\, L^{-1a}{}_b. \tag{11.299}$$

Now using the abbreviation $\partial_0 I - \nabla \cdot \boldsymbol{\sigma} \equiv -\partial_a S^a$ and the transformation laws (11.295 and 11.296), we have

$$\begin{aligned} U(L)\,\mathcal{L}_\ell(x)\,U^{-1}(L) &= i\,u^\dagger(Lx)D^{(1/2,0)\dagger}(L^{-1})(-\,\partial_a S^a)D^{(1/2,0)}(L^{-1})\,u(Lx) \\ &= i\,u^\dagger(Lx)(-\,\partial_a L^{-1a}{}_b S^b)\,u(Lx) \\ &= i\,u^\dagger(Lx)(-\,\partial'_b S^b)\,u(Lx) = \mathcal{L}_\ell(Lx) \end{aligned} \tag{11.300}$$

which shows that $\mathcal{L}_\ell$ is Lorentz covariant.

Incidentally, the rule (11.299) ensures, among other things, that the divergence $\partial_a V^a$ is invariant

$$\left(\partial_a V^a\right)' = \partial'_a V'^a = \partial_b\, L^{-1b}{}_a\, L^a{}_c V^c = \partial_b\, \delta^b{}_c\, V^c = \partial_b\, V^b. \tag{11.301}$$

Example 11.37 (Why u is left handed) The spacetime integral S of the action density $\mathcal{L}_\ell$ is stationary when $u(x)$ satisfies the wave equation

$$(\partial_0 I - \nabla \cdot \boldsymbol{\sigma})\,u(x) = 0 \tag{11.302}$$

or in momentum space

$$(E + \boldsymbol{p} \cdot \boldsymbol{\sigma})\,u(p) = 0. \tag{11.303}$$

Multiplying from the left by $(E - \boldsymbol{p} \cdot \boldsymbol{\sigma})$, we see that the energy of a particle created or annihilated by the field u is the same as its momentum, $E = |\boldsymbol{p}|$. The particles of the field u are massless because the action density $\mathcal{L}_\ell$ has no mass term. The spin of the particle is represented by the matrix $\boldsymbol{J} = \boldsymbol{\sigma}/2$, so the momentum-space relation (11.303) says that $u(p)$ is an eigenvector of $\hat{\boldsymbol{p}} \cdot \boldsymbol{J}$ with eigenvalue $-1/2$

$$\hat{\boldsymbol{p}} \cdot \boldsymbol{J}\,u(p) = -\frac{1}{2}\,u(p). \tag{11.304}$$

A particle whose spin is opposite to its momentum is said to have **negative helicity** or to be **left handed**. Nearly massless neutrinos are nearly left handed.

One may add to this action density the **Majorana mass term**

$$\mathcal{L}_M(x) = -\frac{1}{2} m\, u^\dagger(x)\,\sigma_2\, u^*(x) - \frac{1}{2} m^*\, u^{\mathsf{T}}(x)\,\sigma_2\, u(x) \tag{11.305}$$

which is Lorentz covariant because the matrices σ_1 and σ_3 anticommute with σ_2 which is antisymmetric (Exercise 11.41). This term would vanish if $u_1 u_2$ were equal to $u_2 u_1$. Since charge is conserved, only neutral fields like neutrinos can have Majorana mass terms. The action density of a left-handed field of mass m is the sum $\mathcal{L} = \mathcal{L}_\ell + \mathcal{L}_M$ of the kinetic one (11.297) and the Majorana mass term (11.305). The resulting equations of motion

$$\begin{aligned} 0 &= i\,(\partial_0 - \nabla\cdot\boldsymbol{\sigma})\,u - m\sigma_2 u^* \\ 0 &= \left(\partial_0^2 - \nabla^2 + |m|^2\right) u \end{aligned} \tag{11.306}$$

show that the field u represents particles of mass $|m|$.

11.40 Right-Handed Representation of the Lorentz Group

The generators of the 2-dimensional representation $D^{(0,1/2)}$ with $j = 0$ and $j' = 1/2$ are given by (11.282 and 11.283) with $\boldsymbol{J}^+ = 0$ and $\boldsymbol{J}^- = \boldsymbol{\sigma}/2$; they are

$$\boldsymbol{J} = \frac{1}{2}\boldsymbol{\sigma} \quad \text{and} \quad \boldsymbol{K} = i\frac{1}{2}\boldsymbol{\sigma}. \tag{11.307}$$

Thus 2×2 matrix $D^{(0,1/2)}(\boldsymbol{\theta},\boldsymbol{\lambda})$ that represents the Lorentz transformation (11.279)

$$L = e^{-i\boldsymbol{\theta}\cdot\boldsymbol{J} - i\boldsymbol{\lambda}\cdot\boldsymbol{K}} \tag{11.308}$$

is

$$D^{(0,1/2)}(\boldsymbol{\theta},\boldsymbol{\lambda}) = \exp\left(-i\boldsymbol{\theta}\cdot\boldsymbol{\sigma}/2 + \boldsymbol{\lambda}\cdot\boldsymbol{\sigma}/2\right) = D^{(1/2,0)}(\boldsymbol{\theta}, -\boldsymbol{\lambda}) \tag{11.309}$$

which differs from $D^{(1/2,0)}(\boldsymbol{\theta},\boldsymbol{\lambda})$ only by the sign of $\boldsymbol{\lambda}$. The generic $D^{(0,1/2)}$ matrix is the complex unimodular 2×2 matrix

$$D^{(0,1/2)}(\boldsymbol{\theta},\boldsymbol{\lambda}) = e^{\boldsymbol{z}^*\cdot\boldsymbol{\sigma}/2} \tag{11.310}$$

with $\boldsymbol{\lambda} = \mathrm{Re}\boldsymbol{z}$ and $\boldsymbol{\theta} = \mathrm{Im}\boldsymbol{z}$.

Example 11.38 (The standard right-handed boost) For a particle of mass $m > 0$, the "standard" boost (11.292) that transforms $k = (m, \boldsymbol{0})$ to $p = (p^0, \boldsymbol{p})$ is the 4×4 matrix $B(p) = \exp\left(\alpha\,\hat{\boldsymbol{p}}\cdot\boldsymbol{B}\right)$ in which $\cosh\alpha = p^0/m$ and $\sinh\alpha = |\boldsymbol{p}|/m$. This Lorentz transformation with $\boldsymbol{\theta} = \boldsymbol{0}$ and $\boldsymbol{\lambda} = \alpha\,\hat{\boldsymbol{p}}$ is represented by the matrix (Exercise 11.37)

$$\begin{aligned} D^{(0,1/2)}(\boldsymbol{0},\alpha\,\hat{\boldsymbol{p}}) &= e^{\alpha\hat{\boldsymbol{p}}\cdot\boldsymbol{\sigma}/2} = I\,\cosh(\alpha/2) + \hat{\boldsymbol{p}}\cdot\boldsymbol{\sigma}\,\sinh(\alpha/2) \\ &= I\sqrt{(p^0+m)/(2m)} + \hat{\boldsymbol{p}}\cdot\boldsymbol{\sigma}\sqrt{(p^0-m)/(2m)} \\ &= \frac{p^0 + m + \boldsymbol{p}\cdot\boldsymbol{\sigma}}{\sqrt{2m(p^0+m)}} \end{aligned} \tag{11.311}$$

in the third line of which the 2×2 identity matrix I is suppressed.

Under $D^{(0,1/2)}$, the vector $(I, \boldsymbol{\sigma})$ transforms as a 4-vector; for tiny z

$$\begin{aligned} D^{\dagger(0,1/2)}(\boldsymbol{\theta},\boldsymbol{\lambda})\, I\, D^{(0,1/2)}(\boldsymbol{\theta},\boldsymbol{\lambda}) &= I + \boldsymbol{\lambda}\cdot\boldsymbol{\sigma} \\ D^{\dagger(0,1/2)}(\boldsymbol{\theta},\boldsymbol{\lambda})\, \boldsymbol{\sigma}\, D^{(0,1/2)}(\boldsymbol{\theta},\boldsymbol{\lambda}) &= \boldsymbol{\sigma} + I\boldsymbol{\lambda} + \boldsymbol{\theta}\wedge\boldsymbol{\sigma} \end{aligned} \tag{11.312}$$

as in (11.278).

A massless field $v(x)$ that responds to a unitary Lorentz transformation $U(L)$ as

$$U(L)\, v(x)\, U^{-1}(L) = D^{(0,1/2)}(L^{-1})\, v(Lx) \tag{11.313}$$

is called a **right-handed Weyl spinor**. One may show (Exercise 11.40) that the action density

$$\mathcal{L}_r(x) = i\, v^\dagger(x)\, (\partial_0 I + \nabla\cdot\boldsymbol{\sigma})\, v(x) \tag{11.314}$$

is Lorentz covariant

$$U(L)\, \mathcal{L}_r(x)\, U^{-1}(L) = \mathcal{L}_r(Lx). \tag{11.315}$$

Example 11.39 (Why v is right handed) An argument like that of Example 11.37 shows that the field $v(x)$ satisfies the wave equation

$$(\partial_0 I + \nabla\cdot\boldsymbol{\sigma})\, v(x) = 0 \tag{11.316}$$

or in momentum space

$$(E - \boldsymbol{p}\cdot\boldsymbol{\sigma})\, v(p) = 0. \tag{11.317}$$

Thus, $E = |\boldsymbol{p}|$, and $v(p)$ is an eigenvector of $\hat{\boldsymbol{p}}\cdot\boldsymbol{J}$

$$\hat{\boldsymbol{p}}\cdot\boldsymbol{J}\, v(p) = \frac{1}{2}\, v(p) \tag{11.318}$$

with eigenvalue $1/2$. A particle whose spin is parallel to its momentum is said to have **positive helicity** or to be **right handed**. Nearly massless antineutrinos are nearly right handed.

The Majorana mass term

$$\mathcal{L}_M(x) = -\frac{1}{2} m\, v^\dagger(x)\, \sigma_2\, v^*(x) - \frac{1}{2} m^*\, v^{\mathsf{T}}(x)\, \sigma_2\, v(x) \tag{11.319}$$

like (11.305) is Lorentz covariant. The action density of a right-handed field of mass m is the sum $\mathcal{L} = \mathcal{L}_r + \mathcal{L}_M$ of the kinetic one (11.314) and this Majorana mass term (11.319). The resulting equations of motion

$$\begin{aligned} 0 &= i\, (\partial_0 + \nabla\cdot\boldsymbol{\sigma})\, v - m\sigma_2 v^* \\ 0 &= \left(\partial_0^2 - \nabla^2 + |m|^2\right) v \end{aligned} \tag{11.320}$$

show that the field v represents particles of mass $|m|$.

11.41 Dirac's Representation of the Lorentz Group

Dirac's representation of $SO(3,1)$ is the direct sum $D^{(1/2,0)} \oplus D^{(0,1/2)}$ of $D^{(1/2,0)}$ and $D^{(0,1/2)}$. Its generators are the 4×4 matrices

$$\boldsymbol{J} = \frac{1}{2}\begin{pmatrix} \boldsymbol{\sigma} & 0 \\ 0 & \boldsymbol{\sigma} \end{pmatrix} \quad \text{and} \quad \boldsymbol{K} = \frac{i}{2}\begin{pmatrix} -\boldsymbol{\sigma} & 0 \\ 0 & \boldsymbol{\sigma} \end{pmatrix}. \tag{11.321}$$

Dirac's representation uses the **Clifford algebra** of the gamma matrices γ^a which satisfy the anticommutation relation

$$\{\gamma^a, \gamma^b\} \equiv \gamma^a \gamma^b + \gamma^b \gamma^a = 2\eta^{ab} I \tag{11.322}$$

in which η is the 4×4 diagonal matrix (11.259) with $\eta^{00} = -1$ and $\eta^{jj} = 1$ for $j = 1, 2$, and 3, and I is the 4×4 identity matrix.

Remarkably, the generators of the Lorentz group

$$J^{ij} = \epsilon_{ijk} J_k \quad \text{and} \quad J^{0j} = K_j \tag{11.323}$$

may be represented as commutators of gamma matrices

$$J^{ab} = -\frac{i}{4}[\gamma^a, \gamma^b]. \tag{11.324}$$

They transform the gamma matrices as a 4-vector

$$[J^{ab}, \gamma^c] = -i\gamma^a \eta^{bc} + i\gamma^b \eta^{ac} \tag{11.325}$$

(Exercise 11.42) and satisfy the commutation relations

$$i[J^{ab}, J^{cd}] = \eta^{bc} J^{ad} - \eta^{ac} J^{bd} - \eta^{da} J^{cb} + \eta^{db} J^{ca} \tag{11.326}$$

of the Lorentz group (Weinberg, 1995, pp. 213–217) (Exercise 11.43).

The gamma matrices γ^a are not unique; if S is any 4×4 matrix with an inverse, then the matrices $\gamma'^a \equiv S\gamma^a S^{-1}$ also satisfy the definition (11.322). The choice

$$\gamma^0 = -i\begin{pmatrix} 0 & 1 \\ 1 & 0 \end{pmatrix} \quad \text{and} \quad \boldsymbol{\gamma} = -i\begin{pmatrix} 0 & \boldsymbol{\sigma} \\ -\boldsymbol{\sigma} & 0 \end{pmatrix} \tag{11.327}$$

makes $\boldsymbol{J}$ and $\boldsymbol{K}$ block diagonal (11.321) and lets us assemble a left-handed spinor u and a right-handed spinor v neatly into a 4-component spinor

$$\psi = \begin{pmatrix} u \\ v \end{pmatrix}. \tag{11.328}$$

Dirac's action density for a 4-spinor is

$$\mathcal{L} = -\overline{\psi}\left(\gamma^a \partial_a + m\right)\psi \equiv -\overline{\psi}\left(\not{\partial} + m\right)\psi \tag{11.329}$$

in which

$$\overline{\psi} \equiv i\psi^\dagger \gamma^0 = \psi^\dagger \begin{pmatrix} 0 & 1 \\ 1 & 0 \end{pmatrix} = \begin{pmatrix} v^\dagger & u^\dagger \end{pmatrix}. \tag{11.330}$$

The kinetic part is the sum of the left-handed $\mathcal{L}_\ell$ and right-handed $\mathcal{L}_r$ action densities (11.297 and 11.314)

$$- \overline{\psi}\, \gamma^a \partial_a \psi = i u^\dagger \left(\partial_0 I - \nabla \cdot \boldsymbol{\sigma}\right) u + i\, v^\dagger \left(\partial_0 I + \nabla \cdot \boldsymbol{\sigma}\right) v. \tag{11.331}$$

If u is a left-handed spinor transforming as (11.296), then the spinor

$$v = \sigma_2\, u^* \equiv \begin{pmatrix} v_1 \\ v_2 \end{pmatrix} = \begin{pmatrix} 0 & -i \\ i & 0 \end{pmatrix} \begin{pmatrix} u_1^\dagger \\ u_2^\dagger \end{pmatrix} \tag{11.332}$$

transforms as a right-handed spinor (11.313), that is (Exercise 11.44)

$$e^{\boldsymbol{z}^* \cdot \boldsymbol{\sigma}/2} \sigma_2\, u^* = \sigma_2 \left(e^{-\boldsymbol{z} \cdot \boldsymbol{\sigma}/2} u\right)^*. \tag{11.333}$$

Similarly, if v is right handed, then $u = -\,\sigma_2\, v^*$ is left handed.

A Majorana 4-spinor obeys the Majorana condition

$$\psi_M = \begin{pmatrix} u \\ \sigma_2 u^* \end{pmatrix} = \begin{pmatrix} -\,\sigma_2 v^* \\ v \end{pmatrix} = -i\gamma^2 \psi_M^*. \tag{11.334}$$

Its particles are the same as its antiparticles.

If two Majorana spinors $\psi_M^{(1)}$ and $\psi_M^{(2)}$ have the same mass, then one may combine them into a Dirac spinor

$$\psi_D = \frac{1}{\sqrt{2}} \left(\psi_M^{(1)} + i\psi_M^{(2)}\right) = \frac{1}{\sqrt{2}} \begin{pmatrix} u^{(1)} + i u^{(2)} \\ v^{(1)} + i v^{(2)} \end{pmatrix} = \begin{pmatrix} u_D \\ v_D \end{pmatrix}. \tag{11.335}$$

The **Dirac mass term**

$$-\, m\, \overline{\psi}_D \psi_D = -\, m \left(v_D^\dagger u_D + u_D^\dagger v_D\right) \tag{11.336}$$

conserves charge, and since $\exp(\boldsymbol{z}^* \cdot \boldsymbol{\sigma}/2)^\dagger \exp(-\boldsymbol{z} \cdot \boldsymbol{\sigma}/2) = I$ it also is Lorentz invariant. For a Majorana field, it reduces to

$$\begin{aligned} -\frac{1}{2} m \overline{\psi}_M \psi_M &= -\,\frac{1}{2} m \left(v^\dagger u + u^\dagger v\right) = -\frac{1}{2} m \left(u^\dagger \sigma_2 u^* + u^\mathsf{T} \sigma_2 u\right) \\ &= -\,\frac{1}{2} m \left(v^\dagger \sigma_2 v^* + v^\mathsf{T} \sigma_2 v\right) \end{aligned} \tag{11.337}$$

a Majorana mass term (11.305 or 11.319).

11.42 Poincaré Group

The elements of the Poincaré group are products of Lorentz transformations and translations in space and time. The Lie algebra of the Poincaré group therefore includes the generators $\boldsymbol{J}$ and $\boldsymbol{K}$ of the Lorentz group as well as the hamiltonian H and the momentum operator $\boldsymbol{P}$ which respectively generate translations in time and space.

Suppose $T(y)$ is a translation that takes a 4-vector x to $x + y$ and $T(z)$ is a translation that takes a 4-vector x to $x + z$. Then $T(z)T(y)$ and $T(y)T(z)$ both take x to $x + y + z$. So if a translation $T(y) = T(t, \boldsymbol{y})$ is represented by a unitary operator $U(t, \boldsymbol{y}) = \exp(iHt - i\boldsymbol{P} \cdot \boldsymbol{y})$, then the hamiltonian H and the momentum operator $\boldsymbol{P}$ commute with each other

$$[H, P^j] = 0 \quad \text{and} \quad [P^i, P^j] = 0. \tag{11.338}$$

We can figure out the commutation relations of H and $\boldsymbol{P}$ with the angular-momentum $\boldsymbol{J}$ and boost $\boldsymbol{K}$ operators by realizing that $P^a = (H, \boldsymbol{P})$ is a 4-vector. Let

$$U(\boldsymbol{\theta}, \boldsymbol{\lambda}) = e^{-i\boldsymbol{\theta}\cdot\boldsymbol{J} - i\boldsymbol{\lambda}\cdot\boldsymbol{K}} \tag{11.339}$$

be the (infinite-dimensional) unitary operator that represents (in Hilbert space) the infinitesimal Lorentz transformation

$$L = I + \boldsymbol{\theta} \cdot \boldsymbol{R} + \boldsymbol{\lambda} \cdot \boldsymbol{B} \tag{11.340}$$

where $\boldsymbol{R}$ and $\boldsymbol{B}$ are the six 4×4 matrices (11.269 and 11.270). Then because P is a 4-vector under Lorentz transformations, we have

$$U^{-1}(\boldsymbol{\theta}, \boldsymbol{\lambda}) P U(\boldsymbol{\theta}, \boldsymbol{\lambda}) = e^{+i\boldsymbol{\theta}\cdot\boldsymbol{J} + i\boldsymbol{\lambda}\cdot\boldsymbol{K}} P e^{-i\boldsymbol{\theta}\cdot\boldsymbol{J} - i\boldsymbol{\lambda}\cdot\boldsymbol{K}} = (I + \boldsymbol{\theta} \cdot \boldsymbol{R} + \boldsymbol{\lambda} \cdot \boldsymbol{B})\, P \tag{11.341}$$

or using (11.312)

$$(I + i\boldsymbol{\theta} \cdot \boldsymbol{J} + i\boldsymbol{\lambda} \cdot \boldsymbol{K})\, H\, (I - i\boldsymbol{\theta} \cdot \boldsymbol{J} - i\boldsymbol{\lambda} \cdot \boldsymbol{K}) = H + \boldsymbol{\lambda} \cdot \boldsymbol{P} \tag{11.342}$$
$$(I + i\boldsymbol{\theta} \cdot \boldsymbol{J} + i\boldsymbol{\lambda} \cdot \boldsymbol{K})\, \boldsymbol{P}\, (I - i\boldsymbol{\theta} \cdot \boldsymbol{J} - i\boldsymbol{\lambda} \cdot \boldsymbol{K}) = \boldsymbol{P} + H\boldsymbol{\lambda} + \boldsymbol{\theta} \wedge \boldsymbol{P}.$$

Thus one finds (Exercise 11.44) that H is invariant under rotations, while $\boldsymbol{P}$ transforms as a 3-vector

$$[J_i, H] = 0 \quad \text{and} \quad [J_i, P_j] = i\epsilon_{ijk} P_k \tag{11.343}$$

and that

$$[K_i, H] = -iP_i \quad \text{and} \quad [K_i, P_j] = -i\delta_{ij} H. \tag{11.344}$$

By combining these equations with (11.326), one may write (Exercise 11.46) the Lie algebra of the Poincaré group as

$$
\begin{aligned}
i[J^{ab}, J^{cd}] &= \eta^{bc} J^{ad} - \eta^{ac} J^{bd} - \eta^{da} J^{cb} + \eta^{db} J^{ca} \\
i[P^a, J^{bc}] &= \eta^{ab} P^c - \eta^{ac} P^b \\
[P^a, P^b] &= 0.
\end{aligned} \tag{11.345}
$$

11.43 Homotopy Groups

Two paths $f(s)$ and $g(s)$ that map the unit interval $I = [0, 1]$ into a space X are **homotopic** if there is a continuous function $F(s, t)$ into X defined for $0 \le s, t \le 1$ such that $F(s, 0) = f(s)$ and $F(s, 1) = g(s)$ as well as $F(0, t) = x_0$ and $F(1, t) = x_1$. All paths homotopic to f form an equivalence class called the **homotopy class** $[f]$ of f. The **product** $f \cdot g$ of two paths with $f(1) = g(0)$ is $f \cdot g(s) = f(2s)$ for $0 \le s \le 1/2$ and $f \cdot g(s) = g(2s - 1)$ for $1/2 \le s \le 1$. With this kind of multiplication, the set of all homotopy classes $[f]$ of paths f that are loops, $f(0) = f(1) = x_0$, is the **fundamental group** $\pi_1(X, x_0)$ of the space X with basepoint x_0. This construction extended to paths that map the n-cube I^n into a space X and map the boundary of I^n to the same point x_0 defines the nth homotopy group $\pi_n(X, x_0)$ of X.

Example 11.40 (Some homotopy groups) The fundamental groups of the circle S^1 and the torus T^2 are $\pi_1(S^1) = \mathbb{Z}$ and $\pi_1(T^2) = \mathbb{Z}^2$. Some higher homotopy groups are $\pi_n(S^n) = \mathbb{Z}$ and $\pi_i(S^n) = 0$ for $i < n$.

Further Reading

Group theory in a Nutshell for Physicists (Zee, 2016), *Lie Algebras in Particle Physics* (Georgi, 1999), *Group theory and Its Application to the Quantum Mechanics of Atomic Spectra* (Wigner, 1964), *Unitary Symmetry and Elementary Particles* (Lichtenberg, 1978), *Chemical Applications of Group theory* (Cotton, 1990), *Group theory and Quantum Mechanics* (Tinkham, 2003), and `pi.math.cornell.edu/~hatcher/AT/ATch1.pdf`.

Exercises

11.1 Show that all $n \times n$ (real) orthogonal matrices O leave invariant the quadratic form $x_1^2 + x_2^2 + \cdots + x_n^2$, that is, that if $x' = Ox$, then $x'^2 = x^2$.

11.2 Show that the set of all $n \times n$ orthogonal matrices forms a group.

11.3 Show that all $n \times n$ unitary matrices U leave invariant the quadratic form $|x_1|^2 + |x_2|^2 + \cdots + |x_n|^2$, that is, that if $x' = Ux$, then $|x'|^2 = |x|^2$.

11.4 Show that the set of all $n \times n$ unitary matrices forms a group.

11.5 Show that the set of all $n \times n$ unitary matrices with unit determinant forms a group.

11.6 Show that the matrix $D^{(j)}_{m'm}(g) = \langle j, m'|U(g)|j, m\rangle$ is unitary because the rotation operator $U(g)$ is unitary $\langle j, m'|U^\dagger(g)U(g)|j, m\rangle = \delta_{m'm}$.

11.7 Invent a group of order 3 and compute its multiplication table. For extra credit, prove that the group is unique.

11.8 Show that the relation (11.23) between two equivalent representations is an isomorphism.

11.9 Suppose that D_1 and D_2 are equivalent, finite-dimensional, irreducible representations of a group G so that $D_2(g) = SD_1(g)S^{-1}$ for all $g \in G$. What can you say about a matrix A that satisfies $D_2(g)\, A = A\, D_1(g)$ for all $g \in G$?

11.10 Find all components of the matrix $\exp(i\alpha A)$ in which

$$A = \begin{pmatrix} 0 & 0 & -i \\ 0 & 0 & 0 \\ i & 0 & 0 \end{pmatrix}. \tag{11.346}$$

11.11 If $[A, B] = B$, find $e^{i\alpha A} B e^{-i\alpha A}$. Hint: what are the α-derivatives of this expression?

11.12 Show that the direct-product matrix (11.34) of two representations D and D' is a representation.

11.13 Find a 4×4 matrix S that relates the direct-product representation $D^{(1/2,1/2)}$ to the direct sum $D^{(1)} \oplus D^{(0)}$.

11.14 Find the generators in the adjoint representation of the group with structure constants $f_{abc} = \epsilon_{abc}$ where a, b, c run from 1 to 3. *Hint:* The answer is three 3×3 matrices t_a, often written as L_a.

11.15 Show that the generators (11.95) satisfy the commutation relations (11.98).

11.16 Show that the demonstrated equation (11.104) implies the commutation relation (11.105).

11.17 Use the Cayley–Hamilton theorem (1.294) to show that the 3×3 matrix (11.101) that represents a right-handed rotation of θ radians about the axis $\boldsymbol{\theta}$ is given by (11.102).

11.18 Verify the mixed Jacobi identity (11.163).

11.19 For the group $SU(3)$, find the structure constants f_{123} and f_{231}.

11.20 Show that every 2×2 unitary matrix of unit determinant is a quaternion of unit norm.

11.21 Show that the quaternions as defined by (11.214) are closed under addition and multiplication and that the product xq is a quaternion if x is real and q is a quaternion.

11.22 Show that the square of the matrix (11.202) is $J^2 = -I$, where I is the $2n \times 2n$ identity matrix. Then by setting $R = \exp(\epsilon t)$ with $0 < \epsilon \ll 1$, show that $RJR^{\mathsf{T}} = J$ if and only if $JtJ = t^{\mathsf{T}}$.

11.23 Show that $JtJ = t^{\mathsf{T}}$ implies that t is given by (11.207).

11.24 Show that the one-sided derivative $f'(q)$ (11.224) of the quaternionic function $f(q) = q^2$ depends upon the direction along which $q' \to 0$.

11.25 Show that the generators (11.228) of $Sp(2n)$ obey commutation relations of the form (11.229) for some real structure constants f_{abc} and a suitably extended set of matrices $A, A', \ldots$ and $S_k, S'_k, \ldots$.

11.26 Show that for $0 < \epsilon \ll 1$, the real $2n \times 2n$ matrix $T = \exp(\epsilon JS)$ in which S is symmetric satisfies $T^{\mathsf{T}} JT = J$ (at least up to terms of order ϵ^2) and so is in $Sp(2n, R)$.

11.27 Show that the matrix T of (11.210) is in $Sp(2, R)$.

11.28 Use the parametrization (11.254) of the group $SU(2)$ to show that the parameters $\boldsymbol{a}(\boldsymbol{c}, \boldsymbol{b})$ that describe the product $g(\boldsymbol{a}(\boldsymbol{c}, \boldsymbol{b})) = g(\boldsymbol{c})\, g(\boldsymbol{b})$ are those of (11.256).

11.29 Use formulas (11.256) and (11.249) to show that the left-invariant measure for $SU(2)$ is given by (11.257).

11.30 In tensor notation, which is explained in Chapter 13, the condition (11.266) that $I + \omega$ be an infinitesimal Lorentz transformation reads $\left(\omega^{\mathsf{T}}\right)_b{}^a = \omega^a{}_b = -\,\eta_{bc}\,\omega^c{}_d\,\eta^{da}$ in which sums over c and d from 0 to 3 are understood. In this notation, the matrix η_{ef} lowers indices and η^{gh} raises them, so that $\omega_b{}^a = -\,\omega_{bd}\,\eta^{da}$. (Both η_{ef} and η^{gh} are numerically equal to the matrix η displayed in equation (11.259).) Multiply both sides of the condition (11.266) by $\eta_{ae} = \eta_{ea}$ and use the relation $\eta^{da}\,\eta_{ae} = \eta^d{}_e \equiv \delta^d{}_e$ to show that the matrix ω_{ab} with both indices lowered (or raised) is antisymmetric, that is,

$$\omega_{ba} = -\,\omega_{ab} \quad \text{and} \quad \omega^{ba} = -\,\omega^{ab}. \tag{11.347}$$

11.31 Show that the six matrices (11.269) and (11.270) satisfy the $SO(3, 1)$ condition (11.266).

11.32 Show that the six generators $\boldsymbol{J}$ and $\boldsymbol{K}$ obey the commutation relations (11.272–11.274).

11.33 Show that if $\boldsymbol{J}$ and $\boldsymbol{K}$ satisfy the commutation relations (11.272–11.274) of the Lie algebra of the Lorentz group, then so do $\boldsymbol{J}$ and $-\boldsymbol{K}$.

11.34 Show that if the six generators $\boldsymbol{J}$ and $\boldsymbol{K}$ obey the commutation relations (11.272–11.274), then the six generators $\boldsymbol{J}^+$ and $\boldsymbol{J}^-$ obey the commutation relations (11.281).

11.35 Relate the parameter α in the definition (11.292) of the standard boost $B(p)$ to the 4-vector p and the mass m.

11.36 Derive the formulas for $D^{(1/2,0)}(\boldsymbol{0}, \alpha\,\hat{\boldsymbol{p}})$ given in Equation (11.293).

11.37 Derive the formulas for $D^{(0,1/2)}(\boldsymbol{0}, \alpha\,\hat{\boldsymbol{p}})$ given in Equation (11.311).

11.38 For infinitesimal complex z, derive the 4-vector properties (11.294 and 11.312) of $(-I, \boldsymbol{\sigma})$ under $D^{(1/2,0)}$ and of $(I, \boldsymbol{\sigma})$ under $D^{(0,1/2)}$.

11.39 Show that under the unitary Lorentz transformation (11.296), the action density (11.297) is Lorentz covariant (11.298).

11.40 Show that under the unitary Lorentz transformation (11.313), the action density (11.314) is Lorentz covariant (11.315).

11.41 Show that under the unitary Lorentz transformations (11.296 and 11.313), the Majorana mass terms (11.305 and 11.319) are Lorentz covariant.

11.42 Show that the definitions of the gamma matrices (11.322) and of the generators (11.324) imply that the gamma matrices transform as a 4-vector under Lorentz transformations (11.325).

11.43 Show that (11.324) and (11.325) imply that the generators J^{ab} satisfy the commutation relations (11.326) of the Lorentz group.

11.44 Show that the spinor $v = \sigma_2 u^*$ defined by (11.332) is right handed (11.313) if u is left handed (11.296).

11.45 Use (11.342) to get (11.343 and 11.344).

11.46 Derive (11.345) from (11.326, 11.338, 11.343, and 11.344).

12

Special Relativity

12.1 Inertial Frames and Lorentz Transformations

An **inertial reference frame** is a system of coordinates in which free particles move in straight lines at constant speeds. Our spacetime has one time dimension $x^0 = ct$ and three space dimensions $\boldsymbol{x}$. Its physical points are labeled by four coordinates, $p = (x^0, x^1, x^2, x^3)$. The quadratic difference between two infinitesimally separated points p and $p + dp$ whose coordinates differ by dx^0, dx^1, dx^2, dx^3 is

$$ds^2 = -c^2 dt^2 + (dx^1)^2 + (dx^2)^2 + (dx^3)^2. \tag{12.1}$$

In the absence of gravity, ds^2 is the **physical** quadratic distance between the points p and $p + dp$. Since it is physical, it must not change when we change coordinates. Changes of coordinates $x \to x'$ that leave ds^2 invariant are called **Lorentz transformations**.

If we adopt the **summation convention** in which an index is summed from 0 to 3 if it occurs both raised and lowered in the same monomial, then we can write a Lorentz transformation as

$$x'^i = \sum_{k=0}^{3} L^i{}_k \, x^k = L^i{}_k \, x^k. \tag{12.2}$$

Lorentz transformations change coordinate differences dx^k to $dx'^i = L^i{}_k \, dx^k$.

The Minkowski-space metric $\eta_{ik} = \eta^{ik}$ is the 4×4 matrix

$$(\eta_{ik}) = \begin{pmatrix} -1 & 0 & 0 & 0 \\ 0 & 1 & 0 & 0 \\ 0 & 0 & 1 & 0 \\ 0 & 0 & 0 & 1 \end{pmatrix} = (\eta^{ik}). \tag{12.3}$$

It is its own inverse: $\eta^2 = I$ or $\eta_{ik} \, \eta^{k\ell} = \delta_i^\ell$.

In terms of the metric η, the formula (12.1) for the quadratic distance is $ds^2 = dx^i\, \eta_{ik}\, dx^k$. A Lorentz transformation (12.2) must preserve this quadratic distance, so

$$ds^2 = dx'^i\, \eta_{ik}\, dx'^k = L^i{}_\ell\, dx^\ell\, \eta_{ik}\, L^k{}_j\, dx^j = dx^\ell\, \eta_{\ell j}\, dx^j. \tag{12.4}$$

By differentiating both sides with respect to dx^ℓ and dx^j, one may show (Exercise 12.1) that the matrix L must obey the equation

$$L^i{}_\ell\, \eta_{ik}\, L^k{}_j = \eta_{\ell j} \tag{12.5}$$

in order to preserve quadratic distances.

In matrix notation, left indexes label rows; right indexes label columns; and transposition interchanges rows and columns. Thus in matrix notation the condition (12.5) that defines a Lorentz transformation is $L^{\mathsf{T}}\,\eta\, L = \eta$ which since $\eta^2 = I$, implies that $\eta\, L^{\mathsf{T}}\,\eta\, L = \eta^2 = I$. Thus the inverse (11.264) of a Lorentz transformation is $L^{-1} = \eta L^{\mathsf{T}} \eta$ or $L^{-1\mathsf{T}} = \eta\, L\, \eta$ or more explcitly

$$L^{-1\ell}{}_j = \eta^{\ell k}\, L^i{}_k\, \eta_{ij} \quad \text{and} \quad L^{-1\ell}{}_j = \eta^{\ell k}\, L^{\mathsf{T}}{}_k{}^i\, \eta_{ij}. \tag{12.6}$$

Example 12.1 (Lorentz transformations) If we change coordinates to

$$\begin{aligned} x'^0 &= \cosh\theta\, x^0 + \sinh\theta\, x^1 \\ x'^1 &= \sinh\theta\, x^0 + \cosh\theta\, x^1 \\ x'^2 &= x^2 \quad \text{and} \quad x'^3 = x^3, \end{aligned} \tag{12.7}$$

then the matrix L of the Lorentz transformation is (Exercise 12.2)

$$L = \begin{pmatrix} \cosh\theta & \sinh\theta & 0 & 0 \\ \sinh\theta & \cosh\theta & 0 & 0 \\ 0 & 0 & 1 & 0 \\ 0 & 0 & 0 & 1 \end{pmatrix} \tag{12.8}$$

which is a boost in the x-direction to speed $v/c = \tanh\theta$ and **rapidity** θ. One may check (Exercise 12.3) that $L^{\mathsf{T}}\,\eta\, L = \eta$. In the new coordinates, the point $p = (x^0, x^1, x^2, x^3)$ is

$$p = (\cosh\theta\, x'^0 - \sinh\theta\, x'^1, \cosh\theta\, x'^1 - \sinh\theta\, x'^0, x'^2, x'^3). \tag{12.9}$$

Example 12.2 (Spacelike points) Points p and q with $(p-q)\cdot(p-q) = (\boldsymbol{p}-\boldsymbol{q})^2 - (p^0-q^0)^2 > 0$ are **spacelike**. Spacelike events occur at the same time in some Lorentz frames. Let the coordinates of p and q be $(0, \mathbf{0})$ and $(ct, L, 0, 0)$ with $|ct/L| < 1$ so that $(p-q)^2 > 0$. The Lorentz transformation (12.8) leaves the coordinates of p unchanged but takes those of q to $(ct \cosh\theta + L\sinh\theta, ct\sinh\theta + L\cosh\theta, 0, 0)$. So if $v/c = \tanh\theta = -ct/L$, then p and q occur at time 0 frame with $|v/c| = |\tanh\theta| < 1$.

Example 12.3 (Timelike points) Points p and q with $(p - q) \cdot (p - q) < 0$ are **timelike**. Timelike events occur at the same place in some Lorentz frames. We can use the same coordinates as in the previous example (12.2) but with $|ct/L| > 1$ so that $(p - q)^2 < 0$. The Lorentz transformation (12.8) leaves the coordinates of p unchanged but takes those of q to $(ct \cosh\theta + L \sinh\theta, ct \sinh\theta + L \cosh\theta, 0, 0)$. So if $v/c = \tanh\theta = -L/(ct)$, p and q occur at the same place $\mathbf{0}$ in the e' frame with $|v/c| = |\tanh\theta| < 1$.

In special relativity, the spacetime coordinates $x^0 = ct$ and $(x^1, x^2, x^3) = \boldsymbol{x}$ of a point have upper indexes and transform contravariantly $x'^k = L^k{}_\ell x^\ell$ under Lorentz transformations. Spacetime derivatives $\partial_0 = \partial/\partial x^0$ and $\nabla = (\partial_1, \partial_2, \partial_3)$ have lower indexes and transform covariantly under Lorentz transformations. That is, since $x^\ell = L^{-1\ell}{}_j x'^j$, derivatives transform as

$$\frac{\partial}{\partial x'^j} = \frac{\partial x^\ell}{\partial x'^j}\frac{\partial}{\partial x^\ell} = L^{-1\ell}{}_j \frac{\partial}{\partial x^\ell} = \eta^{\ell k} L^i{}_k \eta_{ij} \frac{\partial}{\partial x^\ell} \tag{12.10}$$

or equivalently because $\eta^{mj}\eta_{ij} = \delta^m_i$ as

$$\eta^{mj}\frac{\partial}{\partial x'^j} = L^m{}_k \eta^{\ell k} \frac{\partial}{\partial x^\ell} \quad \text{or} \quad \partial'^m = L^m{}_k \partial^\ell . \tag{12.11}$$

This last equation illustrates a general rule: the metric η^{ik} raises indexes turning covariant vectors into contravariant vectors $\eta^{ik} A_k = A^i$, and the metric η_{ik} lowers indexes turning contravariant vectors into covariant vectors $\eta_{ik} A^k = A_i$. Thus another way to write the inverse (12.6) of a Lorentz transformation is $L^{-1\ell}{}_j = \eta^{\ell k} L^i{}_k \eta_{ij} = L_j{}^\ell$.

12.2 Special Relativity

The spacetime of special relativity is flat, 4-dimensional Minkowski space. In the absence of gravity, the inner product $(p - q) \cdot (p - q)$

$$(p - q) \cdot (p - q) = (\boldsymbol{p} - \boldsymbol{q})^2 - (p^0 - q^0)^2 = (p - q)^i \, \eta_{ik} \, (p - q)^k \tag{12.12}$$

is physical and the same in all Lorentz frames. If the points p and q are close neighbors with coordinates $x^i + dx^i$ for p and x^i for q, then that invariant inner product is $ds^2 = dx^i \, \eta_{ij} \, dx^j = \boldsymbol{dx}^2 - (dx^0)^2$.

If the points p and q are on the trajectory of a massive particle moving at velocity $\boldsymbol{v}$, then this invariant quantity is the square of the **invariant distance**

$$ds^2 = \boldsymbol{dx}^2 - c^2 dt^2 = \left(\boldsymbol{v}^2 - c^2\right) dt^2 \tag{12.13}$$

which is negative since $v < c$. The time in the rest frame of the particle is the **proper time** τ, and

$$d\tau^2 = -ds^2/c^2 = \left(1 - \boldsymbol{v}^2/c^2\right) dt^2. \tag{12.14}$$

A particle of mass zero moves at the speed of light, so its element $d\tau$ of proper time is zero. But for a particle of mass $m > 0$ moving at speed v, the element of proper time $d\tau$ is smaller than the corresponding element of laboratory time dt by the factor $\sqrt{1 - \boldsymbol{v}^2/c^2}$. The proper time is the time in the rest frame of the particle where $\boldsymbol{v} = 0$. So if $T(0)$ is the lifetime of a particle at rest, then the apparent lifetime $T(v)$ when the particle is moving at speed v is

$$T(v) = dt = \frac{d\tau}{\sqrt{1 - v^2/c^2}} = \frac{T(0)}{\sqrt{1 - v^2/c^2}} \tag{12.15}$$

which is longer than $T(0)$ since $1 - v^2/c^2 \le 1$, an effect known as **time dilation**.

Example 12.4 (Time dilation in muon decay) A muon at rest has a mean life of $T(0) = 2.2 \times 10^{-6}$ seconds. Cosmic rays hitting nitrogen and oxygen nuclei make pions high in the Earth's atmosphere. The pions rapidly decay into muons in 2.6×10^{-8} s. A muon moving at the speed of light from 10 km takes at least $t = 10\,\text{km}/300,000\,(\text{km/sec}) = 3.3 \times 10^{-5}$ s to hit the ground. Were it not for time dilation, the probability P of such a muon reaching the ground as a muon would be

$$P = e^{-t/T(0)} = \exp(-33/2.2) = e^{-15} = 2.6 \times 10^{-7}. \tag{12.16}$$

The mass of a muon is 105.66 MeV. So a muon of energy $E = 749$ MeV has by (12.23) a time-dilation factor of

$$\frac{1}{\sqrt{1 - v^2/c^2}} = \frac{E}{mc^2} = \frac{749}{105.7} = 7.089 = \frac{1}{\sqrt{1 - (0.99)^2}}. \tag{12.17}$$

So a muon moving at a speed of $v = 0.99\,c$ has an apparent mean life $T(v)$ given by equation (12.15) as

$$T(v) = \frac{E}{mc^2}\,T(0) = \frac{T(0)}{\sqrt{1 - v^2/c^2}} = \frac{2.2 \times 10^{-6}\,\text{s}}{\sqrt{1 - (0.99)^2}} = 1.6 \times 10^{-5}\,\text{s}. \tag{12.18}$$

The probability of survival with time dilation is

$$P = e^{-t/T(v)} = \exp(-33/16) = 0.12 \tag{12.19}$$

so that 12% survive. Time dilation increases the chance of survival by a factor of 460,000 – no small effect.

12.3 Kinematics

From the scalar $d\tau$, and the contravariant vector dx^i, we can make the 4-vector

$$u^i = \frac{dx^i}{d\tau} = \frac{dt}{d\tau}\left(\frac{dx^0}{dt}, \frac{d\boldsymbol{x}}{dt}\right) = \frac{1}{\sqrt{1 - v^2/c^2}}\,(c, \boldsymbol{v}) \tag{12.20}$$

in which $u^0 = c\,dt/d\tau = c/\sqrt{1 - v^2/c^2}$ and $\boldsymbol{u} = u^0\,\boldsymbol{v}/c$. The product mu^i is the **energy–momentum 4-vector** p^i

$$\begin{aligned} p^i &= m\,u^i = m\,\frac{dx^i}{d\tau} = m\,\frac{dt}{d\tau}\frac{dx^i}{dt} = \frac{m}{\sqrt{1 - v^2/c^2}}\,\frac{dx^i}{dt} \\ &= \frac{m}{\sqrt{1 - v^2/c^2}}\,(c, \boldsymbol{v}) = \left(\frac{E}{c}, \boldsymbol{p}\right). \end{aligned} \tag{12.21}$$

Its invariant inner product is a constant characteristic of the particle and proportional to the square of its mass

$$c^2\,p^i\,p_i = mc\,u^i\,mc\,u_i = -E^2 + c^2\,\boldsymbol{p}^{\,2} = -m^2\,c^4. \tag{12.22}$$

Note that the time-dilation factor is the ratio of the energy of a particle to its rest energy

$$\frac{1}{\sqrt{1 - v^2/c^2}} = \frac{E}{mc^2} \tag{12.23}$$

and the velocity of the particle is its momentum divided by its equivalent mass E/c^2

$$\boldsymbol{v} = \frac{\boldsymbol{p}}{E/c^2}. \tag{12.24}$$

The analog of $\boldsymbol{F} = m\,\boldsymbol{a}$ is

$$m\,\frac{d^2x^i}{d\tau^2} = m\,\frac{du^i}{d\tau} = \frac{dp^i}{d\tau} = f^i \tag{12.25}$$

in which $p^0 = E/c$, and f^i is a 4-vector force.

Example 12.5 (Time dilation and proper time) In the frame of a laboratory, a particle of mass m with 4-momentum $p^i_{lab} = (E/c, p, 0, 0)$ travels a distance L in a time t for a 4-vector displacement of $x^i_{lab} = (ct, L, 0, 0)$. In its own rest frame, the particle's 4-momentum and 4-displacement are $p^i_{rest} = (mc, 0, 0, 0)$ and $x^i_{rest} = (c\tau, 0, 0, 0)$. Since the Minkowski inner product of two 4-vectors is Lorentz invariant, we have

$$\left(p^i x_i\right)_{rest} = \left(p^i x_i\right)_{lab} \quad \text{or} \quad pL - Et = -mc^2\tau = -mc^2 t\sqrt{1 - v^2/c^2}. \tag{12.26}$$

So a massive particle's phase $\exp(ip^i x_i/\hbar)$ is $\exp(-imc^2\tau/\hbar)$.

Example 12.6 ($p + p \to 3p + \bar{p}$) Conservation of the energy–momentum 4-vector gives $p + p_0 = 3p' + \bar{p}'$. We set $c = 1$ and use this equality in the invariant form $(p+p_0)^2 = (3p'+\bar{p}')^2$. We compute $(p+p_0)^2 = p^2+p_0^2+2p\cdot p_0 = -2m_p^2+2p\cdot p_0$ in the laboratory frame in which $p_0 = (m, \mathbf{0})$. Thus $(p+p_0)^2 = -2m_p^2 - 2E_p m_p$. We compute $(3p' + \bar{p}')^2$ in the frame in which each of the three protons and the antiproton has zero spatial momentum. There $(3p' + \bar{p}')^2 = (4m, \mathbf{0})^2 = -16m_p^2$. We get $E_p = 7m_p$ of which $6m_p = 5.63$ GeV is the threshold kinetic energy of the proton. In 1955, when the group led by Owen Chamberlain and Emilio Segrè discovered the antiproton, the nominal maximum energy of the protons in the Bevatron was 6.2 GeV.

12.4 Electrodynamics

In electrodynamics and in MKSA (SI) units, the 3-dimensional vector potential $\boldsymbol{A}$ and the scalar potential ϕ form a covariant 4-vector potential

$$A_i = \left(\frac{-\phi}{c}, \boldsymbol{A}\right). \tag{12.27}$$

The contravariant 4-vector potential is $A^i = (\phi/c, \boldsymbol{A})$. The **magnetic induction** is

$$\boldsymbol{B} = \nabla \times \boldsymbol{A} \quad \text{or} \quad B_i = \epsilon_{ijk}\partial_j A_k \tag{12.28}$$

in which $\partial_j = \partial/\partial x^j$, the sum over the repeated indices j and k runs from 1 to 3, and ϵ_{ijk} is totally antisymmetric with $\epsilon_{123} = 1$. The electric field is for $i = 1, 2, 3$

$$E_i = c\left(\frac{\partial A_0}{\partial x^i} - \frac{\partial A_i}{\partial x^0}\right) = -\frac{\partial \phi}{\partial x^i} - \frac{\partial A_i}{\partial t} \tag{12.29}$$

where $x^0 = ct$. In 3-vector notation, $\boldsymbol{E}$ is given by the gradient of ϕ and the time-derivative of $\boldsymbol{A}$

$$\boldsymbol{E} = -\nabla\phi - \dot{\boldsymbol{A}}. \tag{12.30}$$

The second-rank, antisymmetric Faraday field-strength tensor is for $i, j = 0, 1, 2, 3$

$$F_{ij} = \frac{\partial A_j}{\partial x^i} - \frac{\partial A_i}{\partial x^j} = -F_{ji}. \tag{12.31}$$

In terms of it, the electric field is $E_i = c\, F_{i0}$, and the magnetic field B_i is

$$B_i = \frac{1}{2}\epsilon_{ijk}\, F_{jk} = \frac{1}{2}\epsilon_{ijk}\left(\frac{\partial A_k}{\partial x^j} - \frac{\partial A_j}{\partial x^k}\right) = (\nabla \times \boldsymbol{A})_i \tag{12.32}$$

where the sum over repeated indexes runs from 1 to 3. The inverse equation $F_{jk} = \epsilon_{jki} B_i$ for spatial j and k follows from the Levi-Civita identity (1.497)

$$\begin{aligned}\epsilon_{jki} B_i &= \frac{1}{2}\epsilon_{jki}\epsilon_{inm} F_{nm} = \frac{1}{2}\epsilon_{ijk}\epsilon_{inm} F_{nm} \\ &= \frac{1}{2}\left(\delta_{jn}\,\delta_{km} - \delta_{jm}\,\delta_{kn}\right) F_{nm} = \frac{1}{2}\left(F_{jk} - F_{kj}\right) = F_{jk}. \end{aligned} \tag{12.33}$$

In 3-vector notation and MKSA (SI) units, Maxwell's equations are a ban on magnetic monopoles and **Faraday's law**, both homogeneous,

$$\nabla \cdot \boldsymbol{B} = 0 \quad \text{and} \quad \nabla \times \boldsymbol{E} + \dot{\boldsymbol{B}} = 0 \tag{12.34}$$

and **Gauss's law** and the **Maxwell–Ampère law**, both inhomogeneous,

$$\nabla \cdot \boldsymbol{D} = \rho_f \quad \text{and} \quad \nabla \times \boldsymbol{H} = \boldsymbol{j}_f + \dot{\boldsymbol{D}}. \tag{12.35}$$

Here ρ_f is the density of **free** charge and $\boldsymbol{j}_f$ is the **free current density**. By *free*, we understand charges and currents that do not arise from polarization and are not restrained by chemical bonds. The divergence of $\nabla \times \boldsymbol{H}$ vanishes (like that of any curl), and so the Maxwell–Ampère law and Gauss's law imply that free charge is conserved

$$0 = \nabla \cdot (\nabla \times \boldsymbol{H}) = \nabla \cdot \boldsymbol{j}_f + \nabla \cdot \dot{\boldsymbol{D}} = \nabla \cdot \boldsymbol{j}_f + \dot{\rho}_f. \tag{12.36}$$

If we use this **continuity equation** to replace $\nabla \cdot \boldsymbol{j}_f$ with $-\,\dot{\rho}_f$ in its middle form $0 = \nabla \cdot \boldsymbol{j}_f + \nabla \cdot \dot{\boldsymbol{D}}$, then we see that the Maxwell–Ampère law preserves the Gauss-law constraint in time

$$0 = \nabla \cdot \boldsymbol{j}_f + \nabla \cdot \dot{\boldsymbol{D}} = \frac{\partial}{\partial t}\left(-\rho_f + \nabla \cdot \boldsymbol{D}\right). \tag{12.37}$$

Similarly, Faraday's law preserves the constraint $\nabla \cdot \boldsymbol{B} = 0$

$$0 = -\,\nabla \cdot (\nabla \times \boldsymbol{E}) = \frac{\partial}{\partial t}\nabla \cdot \boldsymbol{B} = 0. \tag{12.38}$$

In a **linear, isotropic** medium, the **electric displacement** $\boldsymbol{D}$ is related to the electric field $\boldsymbol{E}$ by the **permittivity** ϵ, $\boldsymbol{D} = \epsilon \boldsymbol{E}$, and the **magnetic** or **magnetizing** field $\boldsymbol{H}$ differs from the magnetic induction $\boldsymbol{B}$ by the **permeability** μ, $\boldsymbol{H} = \boldsymbol{B}/\mu$.

On a sub-nanometer scale, the microscopic form of Maxwell's equations applies. On this scale, the homogeneous equations (12.34) are unchanged, but the inhomogeneous ones are

$$\nabla \cdot \boldsymbol{E} = \frac{\rho}{\epsilon_0} \quad \text{and} \quad \nabla \times \boldsymbol{B} = \mu_0\, \boldsymbol{j} + \epsilon_0\, \mu_0\, \dot{\boldsymbol{E}} = \mu_0\, \boldsymbol{j} + \frac{\dot{\boldsymbol{E}}}{c^2} \tag{12.39}$$

in which ρ and $\boldsymbol{j}$ are the total charge and current densities, and $\epsilon_0 = 8.854 \times 10^{-12}$ F/m and $\mu_0 = 4\pi \times 10^{-7}$ N/A^2 are the **electric** and **magnetic constants,** whose product is the inverse of the square of the speed of light, $\epsilon_0\mu_0 = 1/c^2$. Gauss's law and the Maxwell–Ampère law (12.39) imply (Exercise 12.10) that the microscopic (total) current-density 4-vector $j = (c\rho, \boldsymbol{j})$ obeys the continuity equation $\dot{\rho} + \nabla \cdot \boldsymbol{j} = 0$. Electric charge is conserved.

In vacuum, $\rho = \boldsymbol{j} = 0$, $\boldsymbol{D} = \epsilon_0 \boldsymbol{E}$, and $\boldsymbol{H} = \boldsymbol{B}/\mu_0$, and Maxwell's equations become

$$\begin{aligned} \nabla \cdot \boldsymbol{B} = 0 \quad &\text{and} \quad \nabla \times \boldsymbol{E} + \dot{\boldsymbol{B}} = 0 \\ \nabla \cdot \boldsymbol{E} = 0 \quad &\text{and} \quad \nabla \times \boldsymbol{B} = \frac{1}{c^2}\dot{\boldsymbol{E}}. \end{aligned} \tag{12.40}$$

Two of these equations $\nabla \cdot \boldsymbol{B} = 0$ and $\nabla \cdot \boldsymbol{E} = 0$ are constraints. Taking the curl of the other two equations, we find

$$\nabla \times (\nabla \times \boldsymbol{E}) = -\frac{1}{c^2}\ddot{\boldsymbol{E}} \quad \text{and} \quad \nabla \times (\nabla \times \boldsymbol{B}) = -\frac{1}{c^2}\ddot{\boldsymbol{B}}. \tag{12.41}$$

One may use the Levi-Civita identity (1.497) to show (Exercise 12.13) that

$$\nabla \times (\nabla \times \boldsymbol{E}) = \nabla(\nabla \cdot \boldsymbol{E}) - \Delta \boldsymbol{E} \text{ and } \nabla \times (\nabla \times \boldsymbol{B}) = \nabla(\nabla \cdot \boldsymbol{B}) - \Delta \boldsymbol{B} \tag{12.42}$$

in which $\Delta \equiv \nabla^2$. Since in vacuum the divergence of $\boldsymbol{E}$ vanishes, and since that of $\boldsymbol{B}$ always vanishes, these identities and the curl–curl equations (12.41) tell us that waves of $\boldsymbol{E}$ and $\boldsymbol{B}$ move at the speed of light

$$\frac{1}{c^2}\ddot{\boldsymbol{E}} - \Delta \boldsymbol{E} = 0 \quad \text{and} \quad \frac{1}{c^2}\ddot{\boldsymbol{B}} - \Delta \boldsymbol{B} = 0. \tag{12.43}$$

We may write the two homogeneous Maxwell equations (12.34) as

$$\begin{aligned} \partial_i F_{jk} + \partial_k F_{ij} + \partial_j F_{ki} &= \partial_i \left(\partial_j A_k - \partial_k A_j\right) + \partial_k \left(\partial_i A_j - \partial_j A_i\right) \\ &\quad + \partial_j \left(\partial_k A_i - \partial_i A_k\right) = 0 \end{aligned} \tag{12.44}$$

(exercise 12.11). This relation, known as the **Bianchi identity**, actually is a generally covariant tensor equation

$$\epsilon^{\ell ijk} \partial_i F_{jk} = 0 \tag{12.45}$$

in which $\epsilon^{\ell ijk}$ is totally antisymmetric, as explained in Section 13.20. There are four versions of this identity (corresponding to the four ways of choosing three different indices i, j, k from among four and leaving out one, ℓ). The $\ell = 0$ case gives the scalar equation $\nabla \cdot \boldsymbol{B} = 0$, and the three that have $\ell \neq 0$ give the vector equation $\nabla \times \boldsymbol{E} + \dot{\boldsymbol{B}} = 0$.

In tensor notation, the microscopic form of the two inhomogeneous equations (12.39) – the laws of Gauss and Ampère – are a single equation

$$\partial_i F^{ki} = \mu_0 \, j^k \quad \text{in which} \quad j^k = (c\rho, \boldsymbol{j}) \tag{12.46}$$

is the current 4-vector.

The **Lorentz force law** for a particle of charge q is

$$m \frac{d^2x^i}{d\tau^2} = m \frac{du^i}{d\tau} = \frac{dp^i}{d\tau} = f^i = q\, F^{ij} \frac{dx_j}{d\tau} = q\, F^{ij}\, u_j. \tag{12.47}$$

We may cancel a factor of $dt/d\tau$ from both sides and find for $i = 1, 2, 3$

$$\frac{dp^i}{dt} = q\left(-F^{i0} + \epsilon_{ijk} B_k v_j\right) \quad \text{or} \quad \frac{d\boldsymbol{p}}{dt} = q\,(\boldsymbol{E} + \boldsymbol{v} \times \boldsymbol{B}) \tag{12.48}$$

and for $i = 0$

$$\frac{dE}{dt} = q\, \boldsymbol{E} \cdot \boldsymbol{v} \tag{12.49}$$

which shows that only the electric field does work. The only special-relativistic correction needed in Maxwell's electrodynamics is a factor of $1/\sqrt{1 - v^2/c^2}$ in these equations. That is, we use $\boldsymbol{p} = m\boldsymbol{u} = m\boldsymbol{v}/\sqrt{1 - v^2/c^2}$ not $\boldsymbol{p} = m\boldsymbol{v}$ in (12.48), and we use the total energy E not the kinetic energy in (12.49). The reason why so little of classical electrodynamics was changed by special relativity is that electric and magnetic effects were accessible to measurement during the 1800s. Classical electrodynamics was almost perfect.

Keeping track of factors of the speed of light is a lot of trouble and a distraction; in what follows, we'll often use units with $c = 1$.

12.5 Principle of Stationary Action in Special Relativity

The action for a free particle of mass m in special relativity is

$$S = -m \int_{\tau_1}^{\tau_2} d\tau = -\int_{t_1}^{t_2} m\sqrt{1 - \dot{\boldsymbol{x}}^2}\, dt \tag{12.50}$$

where $c = 1$ and $\dot{\boldsymbol{x}} = d\boldsymbol{x}/dt$. The requirement of stationary action is

$$0 = \delta S = -\delta \int_{t_1}^{t_2} m\sqrt{1 - \dot{\boldsymbol{x}}^2}\, dt = m \int_{t_1}^{t_2} \frac{\dot{\boldsymbol{x}} \cdot \delta\dot{\boldsymbol{x}}}{\sqrt{1 - \dot{\boldsymbol{x}}^2}}\, dt. \tag{12.51}$$

But $1/\sqrt{1 - \dot{\boldsymbol{x}}^2} = dt/d\tau$ and so

$$\begin{aligned} 0 = \delta S &= m \int_{t_1}^{t_2} \frac{d\boldsymbol{x}}{dt} \cdot \frac{d\delta\boldsymbol{x}}{dt} \frac{dt}{d\tau}\, dt = m \int_{\tau_1}^{\tau_2} \frac{d\boldsymbol{x}}{dt} \cdot \frac{d\delta\boldsymbol{x}}{dt} \frac{dt}{d\tau} \frac{dt}{d\tau}\, d\tau \\ &= m \int_{\tau_1}^{\tau_2} \frac{d\boldsymbol{x}}{d\tau} \cdot \frac{d\delta\boldsymbol{x}}{d\tau}\, d\tau. \end{aligned} \tag{12.52}$$

So integrating by parts, keeping in mind that $\delta\boldsymbol{x}(\tau_2) = \delta\boldsymbol{x}(\tau_1) = \boldsymbol{0}$, we have

$$0 = \delta S = m \int_{\tau_1}^{\tau_2} \left[\frac{d}{d\tau}\left(\frac{d\boldsymbol{x}}{d\tau} \cdot \delta\boldsymbol{x} \right) - \frac{d^2\boldsymbol{x}}{d\tau^2} \cdot \delta\boldsymbol{x} \right] d\tau = -m \int_{\tau_1}^{\tau_2} \frac{d^2\boldsymbol{x}}{d\tau^2} \cdot \delta\boldsymbol{x}\, d\tau. \tag{12.53}$$

To have this hold for arbitrary $\delta \boldsymbol{x}$, we need

$$\frac{d^2 \boldsymbol{x}}{d\tau^2} = \boldsymbol{0} \tag{12.54}$$

which is the equation of motion for a free particle in special relativity.

What about a charged particle in an electromagnetic field A_i? Its action is

$$S = -m \int_{\tau_1}^{\tau_2} d\tau + q \int_{x_1}^{x_2} A_i(x)\, dx^i = \int_{\tau_1}^{\tau_2} \left(-m + q A_i(x) \frac{dx^i}{d\tau} \right) d\tau. \tag{12.55}$$

We now treat the first term in a 4-dimensional manner

$$\delta d\tau = \delta \sqrt{-\eta_{ik} dx^i dx^k} = \frac{-\eta_{ik} dx^i \delta dx^k}{\sqrt{-\eta_{ik} dx^i dx^k}} = -u_k \delta dx^k = -u_k d\delta x^k \tag{12.56}$$

in which $u_k = dx_k/d\tau$ is the 4-velocity (12.20) and η is the Minkowski metric (12.3) of flat spacetime. The variation of the other term is

$$\delta \left(A_i\, dx^i \right) = (\delta A_i)\, dx^i + A_i\, \delta dx^i = A_{i,k} \delta x^k\, dx^i + A_i\, d\delta x^i. \tag{12.57}$$

Putting them together, we get for δS

$$\delta S = \int_{\tau_1}^{\tau_2} \left(m u_k \frac{d\delta x^k}{d\tau} + q A_{i,k} \delta x^k \frac{dx^i}{d\tau} + q A_i \frac{d\delta x^i}{d\tau} \right) d\tau. \tag{12.58}$$

After integrating by parts the last term, dropping the boundary terms, and changing a dummy index, we get

$$\begin{aligned} \delta S &= \int_{\tau_1}^{\tau_2} \left(-m \frac{du_k}{d\tau} \delta x^k + q A_{i,k} \delta x^k \frac{dx^i}{d\tau} - q \frac{dA_k}{d\tau} \delta x^k \right) d\tau \\ &= \int_{\tau_1}^{\tau_2} \left[-m \frac{du_k}{d\tau} + q \left(A_{i,k} - A_{k,i} \right) \frac{dx^i}{d\tau} \right] \delta x^k\, d\tau. \end{aligned} \tag{12.59}$$

If this first-order variation of the action is to vanish for arbitrary δx^k, then the particle must follow the path

$$0 = -m \frac{du_k}{d\tau} + q \left(A_{i,k} - A_{k,i} \right) \frac{dx^i}{d\tau} \quad \text{or} \quad \frac{dp^k}{d\tau} = q F^{ki} u_i \tag{12.60}$$

which is the Lorentz force law (12.47).

12.6 Differential Forms

By (13.9 and 13.4), a covariant vector field contracted with contravariant coordinate differentials is invariant under arbitrary coordinate transformations

$$A' = A'_i\, dx'^i = \frac{\partial x^j}{\partial x'^i} A_j \frac{\partial x'^i}{\partial x^k} dx^k = \delta^j_k\, A_j\, dx^k = A_k\, dx^k = A. \tag{12.61}$$

This invariant quantity $A = A_k\, dx^k$ is a called a **1-form** in the language of **differential forms** introduced about a century ago by Élie Cartan, son of a blacksmith (1869–1951).

The **wedge product** $dx \wedge dy$ of two coordinate differentials is the directed area spanned by the two differentials and is defined to be antisymmetric

$$dx \wedge dy = -\,dy \wedge dx \quad \text{and} \quad dx \wedge dx = dy \wedge dy = 0 \tag{12.62}$$

so as to transform correctly under a change of coordinates. In terms of the coordinates $u = u(x, y)$ and $v = v(x, y)$, the new element of area is

$$du \wedge dv = \left(\frac{\partial u}{\partial x}dx + \frac{\partial u}{\partial y}dy\right) \wedge \left(\frac{\partial v}{\partial x}dx + \frac{\partial v}{\partial y}dy\right). \tag{12.63}$$

Labeling partial derivatives by subscripts (2.7) and using the antisymmetry (12.62) of the wedge product, we see that $du \wedge dv$ is the old area $dx \wedge dy$ multiplied by the Jacobian (Section 1.21) of the transformation $x, y \to u, v$

$$\begin{aligned} du \wedge dv &= \left(u_x dx + u_y dy\right) \wedge \left(v_x dx + v_y dy\right) \\ &= u_x\, v_x\, dx \wedge dx + u_x\, v_y\, dx \wedge dy + u_y\, v_x\, dy \wedge dx + u_y\, v_y\, dy \wedge dy \\ &= \left(u_x\, v_y - u_y v_x\right)\, dx \wedge dy \\ &= \begin{vmatrix} u_x & u_y \\ v_x & v_y \end{vmatrix}\, dx \wedge dy = J(u, v; x, y)\, dx \wedge dy. \end{aligned} \tag{12.64}$$

A contraction $H = \frac{1}{2} H_{ik}\, dx^i \wedge dx^k$ of a second-rank covariant tensor with a wedge product of two differentials is a 2-form. A ***p*-form** is a rank-p covariant tensor contracted with a wedge product of p differentials

$$K = \frac{1}{p!}\, K_{i_1 \dots i_p}\, dx^{i_1} \wedge \dots dx^{i_p}. \tag{12.65}$$

The **exterior derivative** d differentiates and adds a differential. It turns a p-form into a $(p+1)$-form. It turns a function f, which is a **0-form**, into a 1-form

$$df = \frac{\partial f}{\partial x^i}\, dx^i \tag{12.66}$$

and a 1-form $A = A_j\, dx^j$ into a 2-form $dA = d(A_j\, dx^j) = (\partial_i A_j)\, dx^i \wedge dx^j$.

Example 12.7 (The curl) The exterior derivative of the 1-form

$$A = A_x\, dx + A_y\, dy + A_z\, dz \tag{12.67}$$

is a 2-form that contains the curl (2.42) of $\boldsymbol{A}$

$$\begin{aligned} dA &= \partial_y A_x\, dy \wedge dx + \partial_z A_x\, dz \wedge dx \\ &\quad + \partial_x A_y\, dx \wedge dy + \partial_z A_y\, dz \wedge dy \\ &\quad + \partial_x A_z\, dx \wedge dz + \partial_y A_z\, dy \wedge dz \end{aligned}$$

$$
\begin{aligned}
&= \left(\partial_y A_z - \partial_z A_y\right) dy \wedge dz \\
&\quad + \left(\partial_z A_x - \partial_x A_z\right) dz \wedge dx \\
&\quad + \left(\partial_x A_y - \partial_y A_x\right) dx \wedge dy \\
&= (\nabla \times A)_x \, dy \wedge dz + (\nabla \times A)_y \, dz \wedge dx + (\nabla \times A)_z \, dx \wedge dy.
\end{aligned} \tag{12.68}
$$

The exterior derivative of the electromagnetic 1-form $A = A_j \, dx^j$ made from the 4-vector potential or gauge field A_j is the Faraday 2-form (12.31), the tensor F_{ij}

$$
dA = d\left(A_j \, dx^j\right) = \partial_i A_j \, dx^i \wedge dx^j = \frac{1}{2} F_{ij} \, dx^i \wedge dx^j = F \tag{12.69}
$$

in which $\partial_i = \partial/\partial x^i$.

The square dd of the exterior derivative vanishes in the sense that dd applied to any p-form Q is zero

$$
\begin{aligned}
d\left[d\left(Q_{i\ldots} dx^i \wedge \ldots\right)\right] &= d\left[\left(\partial_r Q_{i\ldots}\right) dx^r \wedge dx^i \wedge \ldots\right] \\
&= \left(\partial_s \partial_r Q_{i\ldots}\right) dx^s \wedge dx^r \wedge dx^i \wedge \ldots = 0
\end{aligned} \tag{12.70}
$$

because $\partial_s \partial_r Q$ is symmetric in r and s while $dx^s \wedge dx^r$ is antisymmetric.

If M_{ik} is a covariant second-rank tensor with no particular symmetry, then (Exercise 12.12) only its antisymmetric part contributes to the 2-form $M_{ik} \, dx^i \wedge dx^k$ and only its symmetric part contributes to $M_{ik} \, dx^i dx^k$.

Example 12.8 (The homogeneous Maxwell equations) The exterior derivative d applied to the Faraday 2-form $F = dA$ gives the homogeneous Maxwell equations

$$
0 = ddA = dF = dF_{ik} \, dx^i \wedge dx^k = \partial_\ell F_{ik} \, dx^\ell \wedge dx^i \wedge dx^k \tag{12.71}
$$

an equation known as the Bianchi identity (12.45).

A p-form H is **closed** if $dH = 0$. By (12.71), the Faraday 2-form is closed, $dF = 0$. A p-form H is **exact** if it is the differential $H = dK$ of a $(p-1)$-form K. The identity (12.70) or $dd = 0$ implies that **every exact form is closed.** A lemma (Section 14.5) due to Poincaré shows that **every closed form is locally exact**.

If the A_i in the 1-form $A = A_i dx^i$ commute with each other, then the 2-form $A \wedge A$ is identically zero. But if the A_i don't commute because they are matrices, operators, or Grassmann variables, then $A \wedge A = \frac{1}{2}[A_i, A_j] \, dx^i \wedge dx^j$ need not vanish.

Example 12.9 (If $\dot{B} = 0$, the electric field is closed and exact) If $\dot{\boldsymbol{B}} = 0$, then by Faraday's law (12.34) the curl of the electric field vanishes, $\nabla \times \boldsymbol{E} = 0$. In terms of the 1-form $E = E_i\, dx^i$ for $i = 1, 2, 3$, the vanishing of its curl $\nabla \times \boldsymbol{E}$ is

$$dE = \partial_j E_i\, dx^j \wedge dx^i = \frac{1}{2}\left(\partial_j E_i - \partial_i E_j\right) dx^j \wedge dx^i = 0. \tag{12.72}$$

So E is closed. It also is exact because we can define a quantity $V(\boldsymbol{x})$ whose gradient is $\boldsymbol{E} = -\nabla V$. We first define $V_P(\boldsymbol{x})$ as a line integral of the 1-form E along an arbitrary path P from some starting point $\boldsymbol{x}_0$ to $\boldsymbol{x}$

$$V_P(\boldsymbol{x}) = -\int_{P,\,\boldsymbol{x}_0}^{\boldsymbol{x}} E_i\, dx^i = -\int_P E. \tag{12.73}$$

The potential $V_P(\boldsymbol{x})$ might seem to depend on the path P. But the difference $V_{P'}(\boldsymbol{x}) - V_P(\boldsymbol{x})$ is a line integral of E from $\boldsymbol{x}_0$ to $\boldsymbol{x}$ along the path P' and then back to $\boldsymbol{x}_0$ along the path P. And by Stokes's theorem (2.50), the integral of $\boldsymbol{E}$ around such a closed loop is an integral of the curl $\nabla \times \boldsymbol{E}$ of $\boldsymbol{E}$ over any surface S whose boundary is that closed loop.

$$V_{P'}(\boldsymbol{x}) - V_P(\boldsymbol{x}) = \oint_{P-P'} E_i\, dx^i = \int_S (\nabla \times \boldsymbol{E}) \cdot da = 0. \tag{12.74}$$

In the notation of forms, this is

$$V_{P'}(\boldsymbol{x}) - V_P(\boldsymbol{x}) = \int_{\partial S} E = \int_S dE = 0. \tag{12.75}$$

Thus the potential $V_P(\boldsymbol{x}) = V(\boldsymbol{x})$ is independent of the path, and $\boldsymbol{E} = -\nabla V(\boldsymbol{x})$, and so the 1-form $E = E_i\, dx^i = -\partial_i V\, dx^i = -dV$ is exact.

The general form of Stokes's theorem is that the integral of any p-form H over the boundary ∂R of any $(p+1)$-dimensional, simply connected, orientable region R is equal to the integral of the $(p+1)$-form dH over R

$$\int_{\partial R} H = \int_R dH. \tag{12.76}$$

Equation (12.75) is the $p = 1$ case (George Stokes, 1819–1903).

Example 12.10 (Stokes's theorem for 0-forms) When $p = 0$, the region $R = [a, b]$ is 1-dimensional, H is a 0-form, and Stokes's theorem is the formula of elementary calculus

$$H(b) - H(a) = \int_{\partial R} H = \int_R dH = \int_a^b dH(x) = \int_a^b H'(x)\, dx. \tag{12.77}$$

Example 12.11 (Exterior derivatives anticommute with differentials) The exterior derivative acting on the wedge product of two 1-forms $A = A_i dx^i$ and $B = B_\ell dx^\ell$ is

$$\begin{aligned} d(A \wedge B) &= d(A_i dx^i \wedge B_\ell dx^\ell) = \partial_k (A_i B_\ell)\, dx^k \wedge dx^i \wedge dx^\ell \\ &= (\partial_k A_i)\, B_\ell\, dx^k \wedge dx^i \wedge dx^\ell + A_i\, (\partial_k B_\ell)\, dx^k \wedge dx^i \wedge dx^\ell \\ &= (\partial_k A_i)\, B_\ell\, dx^k \wedge dx^i \wedge dx^\ell - A_i\, (\partial_k B_\ell)\, dx^i \wedge dx^k \wedge dx^\ell \\ &= (\partial_k A_i)\, dx^k \wedge dx^i \wedge B_\ell dx^\ell - A_i dx^i \wedge (\partial_k B_\ell)\, dx^k \wedge dx^\ell \\ &= dA \wedge B - A \wedge dB. \end{aligned} \tag{12.78}$$

If A is a p-form, then $d(A \wedge B) = dA \wedge B + (-1)^p A \wedge dB$ (Exercise 12.14).

Although I have avoided gravity in this chapter, special relativity is not in conflict with general relativity. In fact, Einstein's **equivalence principle** says that special relativity applies in a suitably small neighborhood of any point in any inertial reference frame that has free-fall coordinates.

Exercises

12.1 Show that (12.4) implies (12.5).

12.2 Show that the matrix form of the Lorentz transformation (12.7) is the x boost (12.8).

12.3 Show that the Lorentz matrix (12.8) satisfies $L^{\mathsf{T}}\, \eta\, L = \eta$.

12.4 The basis vectors at a point p are the derivatives of the point with respect to the coordinates. Find the basis vectors $e_i = \partial p / \partial x^i$ at the point $p = (x^0, x^1, x^2, x^3)$. What are the basis vectors e'_i in the coordinates x' (12.9)?

12.5 The basis vectors e^i that are dual to the basis vectors e_k are defined by $e^i = \eta^{ik}\, e_k$. (a) Show that they obey $e^i \cdot e_k = \delta^i_k$. (b) In the two coordinate systems described in Example 12.1, the vectors $e_k\, x^k$ and $e'_i\, x'^i$ represent the same point, so $e_k\, x^k = e'_i\, x'^i$. Find a formula for $e'^i \cdot e_k$. (c) Relate $e'^i \cdot e_k$ to the Lorentz matrix (12.8).

12.6 Show that the equality of the inner products $x^i \eta_{ik} x^k = x'^j \eta_{j\ell} x'^\ell$ means that the matrix $L^i{}_k = e'^i \cdot e_k$ that relates the coordinates $x'^i = L^i{}_k\, x^k$ to the coordinates x^k must obey the relation $\eta_{ik} = L^i{}_k\, \eta_{i\ell}\, L^\ell{}_k$ which is $\eta = L^{\mathsf{T}} \eta L$ in matrix notation. Hint: First doubly differentiate the equality with respect to x^k and to x^ℓ for $k \neq \ell$. Then differentiate it twice with respect to x^k.

12.7 The relations $x'^i = e'^i \cdot e_j\, x^j$ and $x^\ell = e^\ell \cdot e'_k\, x^k$ imply (for fixed basis vectors e and e') that

$$\frac{\partial x'^i}{\partial x^j} = e'^i \cdot e_j = e_j \cdot e'^i = \eta_{j\ell}\, \eta^{ik} e^\ell \cdot e'_k = \eta_{j\ell}\, \eta^{ik} \frac{\partial x^\ell}{\partial x'^k}.$$

Use this equation to show that if A^i transforms (13.6) as a contravariant vector

$$A'^i = \frac{\partial x'^i}{\partial x^j} A^j, \tag{12.79}$$

then $A_\ell = \eta_{\ell j} A^j$ transforms covariantly (13.9)

$$A'_s = \frac{\partial x^\ell}{\partial x'^s} A_\ell.$$

The metric η also turns a covariant vector A_ℓ into its contravariant form $\eta^{k\ell} A_\ell = \eta^{k\ell} \eta_{\ell j} A^j = \delta^k_j A^j = A^k$.

12.8 The LHC is designed to collide 7 TeV protons against 7 TeV protons for a total collision energy of 14 TeV. Suppose one used a linear accelerator to fire a beam of protons at a target of protons at rest at one end of the accelerator. What energy would you need to see the same physics as at the LHC?

12.9 What is the minimum energy that a beam of pions must have to produce a sigma hyperon and a kaon by striking a proton at rest? The relevant masses (in MeV) are $m_{\Sigma^+} = 1189.4$, $m_{K^+} = 493.7$, $m_p = 938.3$, and $m_{\pi^+} = 139.6$.

12.10 Use Gauss's law and the Maxwell–Ampère law (12.39) to show that the microscopic (total) current-density 4-vector $j = (c\rho, \boldsymbol{j})$ obeys the continuity equation $\dot{\rho} + \nabla \cdot \boldsymbol{j} = 0$.

12.11 Derive the Bianchi identity (12.44) from the definition (12.31) of the Faraday field-strength tensor, and show that it implies the two homogeneous Maxwell equations (12.34).

12.12 Show that if M_{ik} is a covariant second-rank tensor with no particular symmetry, then only its antisymmetric part contributes to the 2-form $M_{ik}\, dx^i \wedge dx^k$ and only its symmetric part contributes to the quantity $M_{ik}\, dx^i dx^k$.

12.13 In rectangular coordinates, use the Levi-Civita identity (1.497) to derive the curl–curl equations (12.42).

12.14 Show that if A is a p-form, then $d(AB) = dA \wedge B + (-1)^p A \wedge dB$.

12.15 Show that if $\omega = a_{ij} dx^i \wedge dx^j/2$ with $a_{ij} = -a_{ji}$, then

$$d\omega = \frac{1}{3!} \left(\partial_k a_{ij} + \partial_i a_{jk} + \partial_j a_{ki} \right) dx^i \wedge dx^j \wedge dx^k. \tag{12.80}$$

13

General Relativity

13.1 Points and Their Coordinates

We use coordinates to label the physical points of a spacetime and the mathematical points of an abstract object. For example, we may label a point on a sphere by its latitude and longitude with respect to a polar axis and meridian. If we use a different axis and meridian, our coordinates for the point will change, but the point remains as it was. **Physical and mathematical points exist independently of the coordinates we use to talk about them. When we change our system of coordinates, we change our labels for the points, but the points remain as they were.**

At each point p, we can set up various coordinate systems that assign unique coordinates $x^i(p)$ and $x'^i(p)$ to p and to points near it. For instance, polar coordinates (θ, ϕ) are unique for all points on a sphere – except the north and south poles which are labeled by $\theta = 0$ and $\theta = \pi$ and all $0 \le \phi < 2\pi$. By using a second coordinate system with $\theta' = 0$ and $\theta' = \pi$ on the equator in the (θ, ϕ) system, we can assign unique coordinates to the north and south poles in that system. Embedding simplifies labeling. In a 3-dimensional euclidian space and in the 4-dimensional Minkowski spacetime in which the sphere is a surface, each point of the sphere has unique coordinates, (x, y, z) and (t, x, y, z).

We will use coordinate systems that represent the points of a space or spacetime uniquely and smoothly at least in local patches, so that the maps

$$\begin{aligned} x^i &= x^i(p) = x^i(p(x')) = x^i(x') \\ x'^i &= x'^i(p) = x'^i(p(x)) = x'^i(x) \end{aligned} \tag{13.1}$$

are well defined, differentiable, and one to one in the patches. We'll often group the n coordinates x^i together and write them collectively as x without superscripts. Since the coordinates $x(p)$ label the point p, we sometimes will call them "the point x." But p and x are different. The point p is unique with infinitely many coordinates x, x', x'',... in infinitely many coordinate systems.

We begin this chapter by noticing carefully how things change as we change our coordinates. Our goal is to write physical theories so their equations look the same in all systems of coordinates as Einstein taught us.

13.2 Scalars

A **scalar** is a quantity B that is the same in all coordinate systems

$$B' = B. \tag{13.2}$$

If it also depends upon the coordinates of the spacetime point $p(x) = p(x')$, then it is a **scalar field**, and

$$B'(x') = B(x). \tag{13.3}$$

13.3 Contravariant Vectors

By the chain rule, the change in dx'^i due to changes in the unprimed coordinates is

$$dx'^i = \sum_k \frac{\partial x'^i}{\partial x^k} dx^k. \tag{13.4}$$

This transformation defines **contravariant vectors**: a quantity A^i is a component of a contravariant vector if it transforms like dx^i

$$A'^i = \sum_k \frac{\partial x'^i}{\partial x^k} A^k. \tag{13.5}$$

The coordinate differentials dx^i form a contravariant vector. A contravariant vector $A^i(x)$ that depends on the coordinates is a **contravariant vector field** and transforms as

$$A'^i(x') = \sum_k \frac{\partial x'^i}{\partial x^k} A^k(x). \tag{13.6}$$

13.4 Covariant Vectors

The chain rule for partial derivatives

$$\frac{\partial}{\partial x'^i} = \sum_k \frac{\partial x^k}{\partial x'^i} \frac{\partial}{\partial x^k} \tag{13.7}$$

defines **covariant vectors**: a quantity C_i that transforms like a partial derivative

$$C'_i = \sum_k \frac{\partial x^k}{\partial x'^i} C_k \tag{13.8}$$

is a **covariant vector**. A covariant vector $C_i(x)$ that depends on the coordinates and transforms as

$$C'_i(x') = \sum_k \frac{\partial x^k}{\partial x'^i} C_k(x) \tag{13.9}$$

is a **covariant vector field**.

Example 13.1 (Gradient of a scalar) The derivatives of a scalar field $B'(x') = B(x)$ form a covariant vector field because

$$\frac{\partial B'(x')}{\partial x'^i} = \frac{\partial B(x)}{\partial x'^i} = \sum_k \frac{\partial x^k}{\partial x'^i} \frac{\partial B(x)}{\partial x^k}, \tag{13.10}$$

which shows that the gradient $\partial B(x)/\partial x^k$ fits the definition (13.9) of a covariant vector field.

13.5 Tensors

Tensors are structures that transform like products of vectors. A rank-zero tensor is a scalar. A rank-one tensor is a covariant or contravariant vector. Second-rank tensors are distinguished by how they transform under changes of coordinates:

$$\begin{aligned}
\text{covariant} \quad F'_{ij} &= \frac{\partial x^k}{\partial x'^i} \frac{\partial x^l}{\partial x'^j} F_{kl} \\
\text{contravariant} \quad M'^{ij} &= \frac{\partial x'^i}{\partial x^k} \frac{\partial x'^j}{\partial x^l} M^{kl} \\
\text{mixed} \quad N'^i_j &= \frac{\partial x'^i}{\partial x^k} \frac{\partial x^l}{\partial x'^j} N^k_l .
\end{aligned} \tag{13.11}$$

We can define tensors of higher rank by extending these definitions to quantities with more indices. The rank of a tensor also is called its order and its degree.

If $S(x)$ is a scalar field, then its derivatives with respect to the coordinates are covariant vectors (13.10) and tensors

$$V_i = \frac{\partial S}{\partial x^i}, \quad T_{ik} = \frac{\partial^2 S}{\partial x^i \partial x^k}, \quad \text{and} \quad U_{ik\ell} = \frac{\partial^3 S}{\partial x^i \partial x^k \partial x^\ell}. \tag{13.12}$$

Example 13.2 (Rank-2 tensors) If A_k and B_ℓ are covariant vectors, and C^m and D^n are contravariant vectors, then the product $A_k\, B_\ell$ is a second-rank covariant tensor; $C^m\, D^n$ is a second-rank contravariant tensor; and $A_k\, C^m$, $A_k\, D^n$, $B_k\, C^m$, and $B_k\, D^n$ are second-rank mixed tensors.

Since the transformation laws that define tensors are linear, any linear combination (with constant coefficients) of tensors of a given rank and kind is a tensor of that rank and kind. Thus if F_{ij} and G_{ij} are both second-rank covariant tensors, so is their sum $H_{ij} = F_{ij} + G_{ij}$.

13.6 Summation Convention and Contractions

An index that appears in the same monomial once as a covariant subscript and once as a contravariant superscript, is a dummy index that is summed over

$$A_i\, B^i \equiv \sum_i A_i\, B^i \tag{13.13}$$

usually from 0 to 3. Such a sum in which an index is repeated once covariantly and once contravariantly is a **contraction**. The **rank** of a tensor is the number of its uncontracted indices.

Although the product $A_k\, C^\ell$ is a mixed second-rank tensor, the contraction $A_k\, C^k$ is a scalar because

$$A'_k\, C'^k = \frac{\partial x^\ell}{\partial x'^k}\frac{\partial x'^k}{\partial x^m} A_\ell\, C^m = \frac{\partial x^\ell}{\partial x^m} A_\ell\, C^m = \delta^\ell_m\, A_\ell\, C^m = A_\ell\, C^\ell. \tag{13.14}$$

Similarly, the doubly contracted product $F^{ik}\, F_{ik}$ is a scalar.

Example 13.3 (Kronecker delta) The summation convention and the chain rule imply that

$$\frac{\partial x'^i}{\partial x^k}\frac{\partial x^k}{\partial x'^\ell} = \frac{\partial x'^i}{\partial x'^\ell} = \delta^i_\ell = \begin{cases} 1 & \text{if } i = \ell \\ 0 & \text{if } i \neq \ell. \end{cases} \tag{13.15}$$

The repeated index k has disappeared in this contraction. The **Kronecker delta** δ^i_ℓ is a mixed second-rank tensor; it transforms as

$$\delta'^i_\ell = \frac{\partial x'^i}{\partial x^k}\frac{\partial x^j}{\partial x'^\ell}\,\delta^k_j = \frac{\partial x'^i}{\partial x^k}\frac{\partial x^k}{\partial x'^\ell} = \frac{\partial x'^i}{\partial x'^\ell} = \delta^i_\ell \tag{13.16}$$

and is **invariant** under changes of coordinates.

13.7 Symmetric and Antisymmetric Tensors

A covariant tensor is **symmetric** if it is independent of the order of its indices. That is, if $S_{ik} = S_{ki}$, then S is symmetric. Similarly a contravariant tensor $S^{k\ell m}$ is symmetric if permutations of its indices k, ℓ, m leave it unchanged. The metric of spacetime $g_{ik}(x) = g_{ki}(x)$ is symmetric because its whole role is to express infinitesimal distances as $ds^2 = g_{ik}(x)dx^i dx^k$ which is symmetric in i and k.

A covariant or contravariant tensor is **antisymmetric** if it changes sign when any two of its indices are interchanged. The Maxwell field strength $F_{k\ell}(x) = -F_{\ell k}(x)$ is an antisymmetric rank-2 covariant tensor.

If $T^{ik}\,\epsilon_{ik} = 0$ where $\epsilon_{12} = -\epsilon_{21} = 1$ is antisymmetric, then $T^{12} - T^{21} = 0$. Thus $T^{ik}\,\epsilon_{ik} = 0$ means that the tensor T^{ik} is symmetric.

13.8 Quotient Theorem

Suppose that B has unknown transformation properties, but that its product BA with all tensors A a given rank and kind is a tensor. Then B must be a tensor.

The simplest example is when $B_i A^i$ is a scalar for all contravariant vectors A^i

$$B'_i A'^i = B_j A^j. \tag{13.17}$$

Then since A^i is a contravariant vector

$$B'_i A'^i = B'_i \frac{\partial x'^i}{\partial x^j} A^j = B_j A^j \tag{13.18}$$

or

$$\left(B'_i \frac{\partial x'^i}{\partial x^j} - B_j \right) A^j = 0. \tag{13.19}$$

Since this equation holds for all vectors A, we may promote it to the level of a vector equation

$$B'_i \frac{\partial x'^i}{\partial x^j} - B_j = 0. \tag{13.20}$$

Multiplying both sides by $\partial x^j/\partial x'^k$ and summing over j, we get

$$B'_i \frac{\partial x'^i}{\partial x^j} \frac{\partial x^j}{\partial x'^k} = B_j \frac{\partial x^j}{\partial x'^k} \tag{13.21}$$

which shows that the unknown quantity B_i transforms as a covariant vector

$$B'_k = \frac{\partial x^j}{\partial x'^k} B_j. \tag{13.22}$$

The quotient rule works for tensors A and B of arbitrary rank and kind. The proof in each case is similar to the one given here.

13.9 Tensor Equations

Maxwell's homogeneous equations (12.45) relate the derivatives of the field-strength tensor to each other as

$$0 = \partial_i F_{jk} + \partial_k F_{ij} + \partial_j F_{ki}. \tag{13.23}$$

They are generally covariant **tensor equations** (Sections 13.19 and 13.20). They follow from the Bianchi identity (12.71)

$$dF = ddA = 0. \tag{13.24}$$

Maxwell's inhomegneous equations (12.46) relate the derivatives of the field-strength tensor to the current density j^i and to the square root of the modulus g of the determinant of the metric tensor g_{ij} (Section 13.12)

$$\frac{\partial(\sqrt{g}\, F^{ik})}{\partial x^k} = \mu_0 \sqrt{g}\, j^i. \tag{13.25}$$

They are generally covariant tensor equations. We'll write them as the divergence of a contravariant vector in section 13.29, derive them from an action principle in Section 13.31, and write them as invariant forms in Section 14.7.

If we can write a physical law in one coordinate system as a tensor equation $G^{j\ell}(x) = 0$, then in any other coordinate system the corresponding tensor equation $G'^{ik}(x') = 0$ is valid because

$$G'^{ik}(x') = \frac{\partial x'^i}{\partial x^j} \frac{\partial x'^k}{\partial x^\ell} G^{j\ell}(x) = 0. \tag{13.26}$$

Physical laws also remain the same if expressed in terms of invariant forms. **A theory written in terms of tensors or forms has equations that are true in all coordinate systems if they are true in any coordinate system**. Only such generally covariant theories have a chance at being right because we can't be sure that our particular coordinate system is the correct one. One can make a theory the same in all coordinate systems by applying the principle of stationary action (Section 13.31) to an action that is invariant under all coordinate transformations.

13.10 Comma Notation for Derivatives

Commas are used to denote derivatives. If $f(\theta, \phi)$ is a function of θ and ϕ, we can write its derivatives with respect to these coordinates as

$$f_{,\theta} = \partial_\theta f = \frac{\partial f}{\partial \theta} \quad \text{and} \quad f_{,\phi} = \partial_\phi f = \frac{\partial f}{\partial \phi}. \tag{13.27}$$

And we can write its double derivatives as

$$f_{,\theta\theta} = \frac{\partial^2 f}{\partial \theta^2}, \quad f_{,\theta\phi} = \frac{\partial^2 f}{\partial \theta \partial \phi}, \quad \text{and} \quad f_{,\phi\phi} = \frac{\partial^2 f}{\partial \phi^2}. \tag{13.28}$$

If we use indices $i, k, \ldots$ to label the coordinates x^i, x^k, then we can write the derivatives of a scalar f as

$$f_{,i} = \partial_i f = \frac{\partial f}{\partial x^i} \quad \text{and} \quad f_{,ik} = \partial_k \partial_i f = \frac{\partial^2 f}{\partial x^k \partial x^i} \tag{13.29}$$

and those of tensors T^{ik} and F_{ik} as

$$T^{ik}_{\ \ ,j\ell} = \frac{\partial^2 T^{ik}}{\partial x^j \partial x^\ell} \quad \text{and} \quad F_{ik,j\ell} = \frac{\partial^2 F_{ik}}{\partial x^j \partial x^\ell} \tag{13.30}$$

and so forth.

Semicolons are used to denote covariant derivatives (Section 13.15).

13.11 Basis Vectors and Tangent Vectors

A point $p(x)$ in a space or spacetime with coordinates x is a scalar (13.3) because it is the same point $p'(x') = p(x') = p(x)$ in any other system of coordinates x'. Thus its derivatives with respect to the coordinates

$$\frac{\partial p(x)}{\partial x^i} = e_i(x) \tag{13.31}$$

form a **covariant vector** $e_i(x)$

$$e'_i(x') = \frac{\partial p'(x')}{\partial x'^i} = \frac{\partial p(x)}{\partial x'^i} = \frac{\partial x^k}{\partial x'^i}\frac{\partial p(x)}{\partial x^k} = \frac{\partial x^k}{\partial x'^i} e_k(x). \tag{13.32}$$

Small changes dx^i in the coordinates (in any fixed system of coordinates) lead to small changes in the point $p(x)$

$$dp(x) = e_i(x)\, dx^i. \tag{13.33}$$

The covariant vectors $e_i(x)$ therefore form a basis (1.49) for the space or spacetime at the point $p(x)$. These **basis vectors** $e_i(x)$ are tangent to the curved space or spacetime at the point x and so are called **tangent vectors**. Although complex and fermionic manifolds may be of interest, the manifolds, points, and vectors of this chapter are assumed to be real.

13.12 Metric Tensor

A **Riemann manifold** of dimension d is a space that locally looks like d-dimensional euclidian space $\mathbb{E}^d$ and that is smooth enough for the derivatives

(13.31) that define tangent vectors to exist. The surface of the Earth, for example, looks flat at horizontal distances of less than a kilometer.

Just as the surface of a sphere can be embedded in flat 3-dimensional space, so too every Riemann manifold can be embedded without change of shape (isometrically) in a euclidian space $\mathbb{E}^n$ of suitably high dimension (Nash, 1956). In particular, every Riemann manifold of dimension $d = 3$ (or 4) can be isometrically embedded in a euclidian space of at most $n = 14$ (or 19) dimensions, $\mathbb{E}^{14}$ or $\mathbb{E}^{19}$ (Günther, 1989).

The euclidian dot products (Example 1.15) of the tangent vectors (13.31) define the metric of the manifold

$$g_{ik}(x) = e_i(x) \cdot e_k(x) = \sum_{\alpha=1}^{n} e_i^\alpha(x)\, e_k^\alpha(x) = e_k(x) \cdot e_i(x) = g_{ki}(x) \tag{13.34}$$

which is symmetric, $g_{ik}(x) = g_{ki}(x)$. Here $1 \le i, k \le d$ and $1 \le \alpha \le n$. The dot product of this equation is the dot product of the n-dimensional euclidian embedding space $\mathbb{E}^n$.

Because the tangent vectors $e_i(x)$ are covariant vectors, the metric tensor transforms as a covariant tensor if we change coordinates from x to x'

$$g'_{ik}(x') = \frac{\partial x^j}{\partial x'^i} \frac{\partial x^\ell}{\partial x'^k}\, g_{j\ell}(x). \tag{13.35}$$

The squared distance ds^2 between two nearby points is the dot product of the small change $dp(x)$ (13.33) with itself

$$\begin{aligned} ds^2 &= dp(x) \cdot dp(x) = (e_i(x)\, dx^i) \cdot (e_i(x)\, dx^i) \\ &= e_i(x) \cdot e_i(x)\, dx^i dx^k = g_{ik}(x)\, dx^i dx^k. \end{aligned} \tag{13.36}$$

So by measuring the distances ds between nearby points, one can determine the metric $g_{ik}(x)$ of a Riemann space.

Example 13.4 (The sphere S^2 in $\mathbb{E}^3$) In polar coordinates, a point $\boldsymbol{p}$ on the 2-dimensional surface of a sphere of radius R has coordinates $\boldsymbol{p} = R(\sin\theta\cos\phi, \sin\theta\sin\phi, \cos\theta)$ in an embedding space $\mathbb{E}^3$. The tangent space $\mathbb{E}^2$ at $\boldsymbol{p}$ is spanned by the tangent vectors

$$\begin{aligned} \boldsymbol{e}_\theta &= \boldsymbol{p}_{,\theta} = \frac{\partial \boldsymbol{p}}{\partial\theta} = R\,(\cos\theta\cos\phi, \cos\theta\sin\phi, -\sin\theta) \\ \boldsymbol{e}_\phi &= \boldsymbol{p}_{,\phi} = \frac{\partial \boldsymbol{p}}{\partial\phi} = R\,(-\sin\theta\sin\phi, \sin\theta\cos\phi, 0). \end{aligned} \tag{13.37}$$

The dot products of these tangent vectors are easy to compute in the embedding space $\mathbb{E}^3$. They form the metric tensor of the sphere

$$g_{ik} = \begin{pmatrix} g_{\theta\theta} & g_{\theta\phi} \\ g_{\phi\theta} & g_{\phi\phi} \end{pmatrix} = \begin{pmatrix} \boldsymbol{e}_\theta \cdot \boldsymbol{e}_\theta & \boldsymbol{e}_\theta \cdot \boldsymbol{e}_\phi \\ \boldsymbol{e}_\phi \cdot \boldsymbol{e}_\theta & \boldsymbol{e}_\phi \cdot \boldsymbol{e}_\phi \end{pmatrix} = \begin{pmatrix} R^2 & 0 \\ 0 & R^2 \sin^2\theta \end{pmatrix}. \tag{13.38}$$

Its determinant is $\det(g_{ik}) = R^4 \sin^2\theta$. Since $\boldsymbol{e}_\theta \cdot \boldsymbol{e}_\phi = 0$, the squared infinitesimal distance (13.36) is

$$ds^2 = \boldsymbol{e}_\theta \cdot \boldsymbol{e}_\theta \, d\theta^2 + \boldsymbol{e}_\phi \cdot \boldsymbol{e}_\phi \, d\phi^2 = R^2 d\theta^2 + R^2 \sin^2\theta \, d\phi^2. \tag{13.39}$$

We change coordinates from the angle θ to a radius $r = R \sin\theta / a$ in which a is a dimensionless scale factor. Then $R^2 d\theta^2 = a^2 dr^2 / \cos^2\theta$, and $\cos^2\theta = 1 - \sin^2\theta = 1 - a^2 r^2 / R^2 = 1 - kr^2$ where $k = (a/R)^2$. In these coordinates, the squared distance (13.39) is

$$ds^2 = \frac{a^2}{1 - kr^2} dr^2 + a^2 r^2 \, d\phi^2 \tag{13.40}$$

and the r, ϕ metric of the sphere and its inverse are

$$g_{ik} = a^2 \begin{pmatrix} (1 - kr^2)^{-1} & 0 \\ 0 & r^2 \end{pmatrix} \quad \text{and} \quad g^{ik} = a^{-2} \begin{pmatrix} 1 - kr^2 & 0 \\ 0 & r^{-2} \end{pmatrix}. \tag{13.41}$$

The sphere is a **maximally symmetric space** (Section 13.24).

Example 13.5 (Graph paper) Imagine a piece of slightly crumpled graph paper with horizontal and vertical lines. The lines give us a 2-dimensional coordinate system (x^1, x^2) that labels each point $p(x)$ on the paper. The vectors $e_1(x) = \partial_1 p(x)$ and $e_2(x) = \partial_2 p(x)$ define how a point moves $dp(x) = e_i(x) \, dx^i$ when we change its coordinates by dx^1 and dx^2. The vectors $e_1(x)$ and $e_2(x)$ span a different tangent space at the intersection of every horizontal line with every vertical line. Each tangent space is like the tiny square of the graph paper at that intersection. We can think of the two vectors $e_i(x)$ as three-component vectors in the 3-dimensional embedding space we live in. The squared distance between any two nearby points separated by $dp(x)$ is $ds^2 \equiv dp^2(x) = e_1^2(x)(dx^1)^2 + 2e_1(x) \cdot e_2(x) \, dx^1 dx^2 + e_2^2(x)(dx^2)^2$ in which the inner products $g_{ij} = e_i(x) \cdot e_j(x)$ are defined by the euclidian metric of the embedding euclidian space $\mathbb{R}^3$.

But our universe has time. A **semi-euclidian** spacetime $\mathbb{E}^{(p,d-p)}$ of dimension d is a flat spacetime with a dot product that has p minus signs and $q = d - p$ plus signs. A **semi-riemannian** manifold of dimension d is a spacetime that locally looks like a **semi-euclidian** spacetime $\mathbb{E}^{(p,d-p)}$ and that is smooth enough for the derivatives (13.31) that define its tangent vectors to exist.

Every semi-riemannian manifold can be embedded without change of shape (isometrically) in a semi-euclidian spacetime $\mathbb{E}^{(u,n-u)}$ for sufficiently large u and n (Greene, 1970; Clarke, 1970). Every physically reasonable (globally hyperbolic) semi-riemannian manifold with 1 dimension of time and 3 dimensions of space can be embedded without change of shape (isometrically) in a flat semi-euclidian spacetime of 1 temporal and at most 19 spatial dimensions $\mathbb{E}^{(1,19)}$ (Müller and Sánchez, 2011; Aké et al., 2018).

The semi-euclidian dot products of the tangent vectors of a semi-riemannian manifold of d dimensions define its metric as

$$g_{ik}(x) = e_i(x) \cdot e_k(x) = -\sum_{\alpha=1}^{u} e_i^\alpha(x)\, e_k^\alpha(x) + \sum_{\alpha=u+1}^{n} e_i^\alpha(x)\, e_k^\alpha(x) \tag{13.42}$$

for $0 \le i, k \le d-1$. The metric (13.42) is symmetric $g_{ik}(x) = g_{ki}$. In an extended summation convention, the dot product (13.42) is $g_{ik}(x) = e_{i\alpha}(x)\, e_k^\alpha(x)$.

The squared pseudo-distance or **line element** ds^2 between two nearby points is the inner product of the small change $dp(x)$ (13.33) with itself

$$\begin{aligned} ds^2 &= dp(x) \cdot dp(x) = (e_i(x)\, dx^i) \cdot (e_i(x)\, dx^i) \\ &= e_i(x) \cdot e_i(x)\, dx^i dx^k = g_{ik}(x)\, dx^i dx^k. \end{aligned} \tag{13.43}$$

Thus measurements of line elements ds^2 determine the metric $g_{ik}(x)$ of the spacetime.

Some Riemann spaces have natural embeddings in semi-euclidian spaces. One example is the hyperboloid H^2.

Example 13.6 (The hyperboloid H^2) If we embed a hyperboloid H^2 of radius R in a semi-euclidian spacetime $\mathbb{E}^{(1,2)}$, then a point $\boldsymbol{p} = (x, y, z)$ on the 2-dimensional surface of H^2 obeys the equation $R^2 = x^2 - y^2 - z^2$ and has polar coordinates $\boldsymbol{p} = R(\cosh\theta, \sinh\theta\cos\phi, \sinh\theta\sin\phi)$. The tangent vectors are

$$\begin{aligned} \boldsymbol{e}_\theta &= \boldsymbol{p}_{,\theta} = \frac{\partial \boldsymbol{p}}{\partial \theta} = R\,(\sinh\theta, \cosh\theta\cos\phi, \cosh\theta\sin\phi) \\ \boldsymbol{e}_\phi &= \boldsymbol{p}_{,\phi} = \frac{\partial \boldsymbol{p}}{\partial \phi} = R\,(0, -\sinh\theta\sin\phi, \sinh\theta\cos\phi). \end{aligned} \tag{13.44}$$

The line element $d\boldsymbol{p}^2 = ds^2$ between nearby points is

$$ds^2 = \boldsymbol{e}_\theta \cdot \boldsymbol{e}_\theta\, d\theta^2 + \boldsymbol{e}_\phi \cdot \boldsymbol{e}_\phi\, d\phi^2. \tag{13.45}$$

The metric of $\mathbb{E}^{(1,2)}$ is $(-1, 1, 1)$, so the metric and line element (13.45) of H^2 are

$$R^2 \begin{pmatrix} 1 & 0 \\ 0 & \sinh^2\theta \end{pmatrix} \quad \text{and} \quad ds^2 = R^2\, d\theta^2 + R^2 \sinh^2\theta\, d\phi^2. \tag{13.46}$$

We change coordinates from the angle θ to a radius $r = R\sinh\theta/a$ in which a is a dimensionless scale factor. Then in terms of the parameter $k = (a/R)^2$, the metric and line element (13.46) are (Exercise 13.7)

$$a^2 \begin{pmatrix} (1+r^2)^{-1} & 0 \\ 0 & r^2 \end{pmatrix} \quad \text{and} \quad ds^2 = a^2 \left(\frac{dr^2}{1+kr^2} + r^2\, d\phi^2 \right) \tag{13.47}$$

which describe one of only three maximally symmetric (Section 13.24) two-dimensional spaces. The other two are the sphere S^2 (13.40) and the plane.

13.13 Inverse of Metric Tensor

The metric g_{ik} is a nonsingular matrix (exercise 13.4), and so it has an inverse g^{ik} that satisfies

$$g^{ik} g_{k\ell} = \delta^i_\ell = g'^{ik} g'_{k\ell} \tag{13.48}$$

in all coordinate systems. The inverse metric g^{ik} is a rank-2 contravariant tensor (13.11) because the metric $g_{k\ell}$ is a rank-2 covariant tensor (13.35). To show this, we combine the transformation law (13.35) with the definition (13.48) of the inverse of the metric tensor

$$\delta^i_\ell = g'^{ik} g'_{k\ell} = g'^{ik} \frac{\partial x^r}{\partial x'^k} \frac{\partial x^s}{\partial x'^\ell} g_{rs} \tag{13.49}$$

and multiply both sides by

$$g^{tu} \frac{\partial x'^\ell}{\partial x^t} \frac{\partial x'^v}{\partial x^u}. \tag{13.50}$$

Use of the Kronecker-delta chain rule (13.15) now leads (Exercise 13.5) to

$$g'^{iv}(x') = \frac{\partial x'^i}{\partial x^t} \frac{\partial x'^v}{\partial x^u} g^{tu}(x) \tag{13.51}$$

which shows that the inverse metric g^{ik} transforms as a rank-2 contravariant tensor.

The contravariant vector A^i associated with any covariant vector A_k **is defined as** $A^i = g^{ik} A_k$ which ensures that A^i transforms contravariantly (Exercise 13.6). This is called **raising an index**. It follows that the covariant vector corresponding to the contravariant vector A^i is $A_k = g_{ki} A^i = g_{ki} g^{i\ell} A_\ell = \delta^\ell_k A_\ell = A_k$ which is called **lowering an index**. These definitions apply to all tensors, so $T^{ik\ell} = g^{ij} g^{km} g^{\ell n} T_{jmn}$, and so forth.

Example 13.7 (Making scalars) Fully contracted products of vectors and tensors are scalars. Two contravariant vectors A^i and B^k contracted with the metric tensor form the scalar $g_{ik} A^i B^k = A_k B^k$. Similarly, $g^{ik} A_i B_k = A^k B_k$. Derivatives of scalar fields with respect to the coordinates are covariant vectors $S_{,i}$ (Example 13.1) and covariant tensors $S_{,ik}$ (Section 13.5). If S is a scalar, then $S_{,i}$ is a covariant vector, $g^{ik} S_{,k}$ is a contravariant vector, and the contraction $g^{ik} S_{,i}\, S_{,k}$ is a scalar.

In what follows, I will often use *space* to mean either *space* or *spacetime*.

13.14 Dual Vectors, Cotangent Vectors

Since the inverse metric g^{ik} is a rank-2 contravariant tensor, **dual vectors**

$$e^i = g^{ik} e_k \tag{13.52}$$

are contravariant vectors. They are orthonormal to the tangent vectors e_ℓ because

$$e^i \cdot e_\ell = g^{ik} e_k \cdot e_\ell = g^{ik}\, g_{k\ell} = \delta^i_\ell. \tag{13.53}$$

Here the dot product is that (13.34) of the euclidian space or embedding space or that (13.42) of the semi-euclidian space or embedding space. The dual vectors e^i are called **cotangent vectors** or **tangent covectors**. The tangent vector e_k is the sum $e_k = g_{ki}\, e^i$ because

$$e_k = g_{ki}\, e^i = g_{ki}\, g^{i\ell}\, e_\ell = \delta^\ell_k\, e_\ell = e_k. \tag{13.54}$$

The definition (13.52) of the dual vectors and their orthonormality (13.53) to the tangent vectors imply that their inner products are the matrix elements of the inverse of the metric tensor

$$e^i \cdot e^\ell = g^{ik}\, e_k \cdot e^\ell = g^{ik}\, \delta^\ell_k = g^{i\ell}. \tag{13.55}$$

The outer product of a tangent vector with its cotangent vector $P = e_k e^k$ (summed over the dimensions of the space) is both a projection matrix P from the embedding space onto the tangent space and an identity matrix for the tangent space because $P e_i = e_i$. Its transpose $P^\mathsf{T} = e^k e_k$ is both a projection matrix P from the embedding space onto the cotangent space and an identity matrix for the cotangent space because $P^\mathsf{T} e^i = e^i$. So

$$P = e_k\, e^k = I_t \quad \text{and} \quad P^\mathsf{T} = e^k e_k = I_{ct}. \tag{13.56}$$

Details and examples are in the file tensors.pdf in Tensors_and_general_relativity at `github.com/kevinecahill`.

13.15 Covariant Derivatives of Contravariant Vectors

The **covariant derivative** $D_\ell V^k$ of a contravariant vector V^k is a derivative of V^k that transforms like a mixed rank-2 tensor. An easy way to make such a derivative is to note that the invariant description $V(x) = V^i(x)\, e_i(x)$ of a contravariant vector field $V^i(x)$ in terms of tangent vectors $e_i(x)$ is a scalar. Its derivative

$$\frac{\partial V}{\partial x^\ell} = \frac{\partial V^i}{\partial x^\ell}\, e_i + V^i\, \frac{\partial e_i}{\partial x^\ell} \tag{13.57}$$

is therefore a covariant vector. And the inner product of that covariant vector $V_{,\ell}$ with a contravariant tangent vector e^k is a mixed rank-2 tensor

$$\begin{aligned} D_\ell V^k &= e^k \cdot V_{,\ell} = e^k \cdot \left(V^i_{,\ell} e_i + e_{i,\ell} V^i\right) = \delta^k_i V^i_{,\ell} + e^k \cdot e_{i,\ell} V^i \\ &= V^k_{,\ell} + e^k \cdot e_{i,\ell}\, V^i . \end{aligned} \tag{13.58}$$

The inner product $e^k \cdot e_{i,\ell}$ is usually written as

$$e^k \cdot e_{i,\ell} = e^k \cdot \frac{\partial e_i}{\partial x^\ell} \equiv \Gamma^k{}_{i\ell} \tag{13.59}$$

and is variously called an **affine connection** (it relates tangent spaces lacking a common origin), a **Christoffel connection**, and a **Christoffel symbol** of the second kind. The covariant derivative itself often is written with a semicolon, thus

$$D_\ell V^k = V^k_{;\ell} = V^k_{,\ell} + e^k \cdot e_{i,\ell}\, V^i = V^k_{,\ell} + \Gamma^k{}_{i\ell}\, V^i . \tag{13.60}$$

Example 13.8 (Covariant derivatives of cotangent vectors) Using the identity

$$0 = \delta^k_{i,\ell} = (e^k \cdot e_i)_{,\ell} = e^k_{,\ell} \cdot e_i + e^k \cdot e_{i,\ell} \tag{13.61}$$

and the projection matrix (13.56), we find that

$$D_\ell e^k = e^k_{,\ell} + e^k \cdot e_{i,\ell}\, e^i = e^k_{,\ell} - e^k_{,\ell} \cdot e_i\, e^i = e^k_{,\ell} - e^k_{,\ell} = 0 \tag{13.62}$$

the covariant derivatives of cotangent vectors vanish.

Under general coordinate transformations, $D_\ell V^k$ transforms as a rank-2 mixed tensor

$$\left(D_\ell V^k\right)'(x') = \left(V^k_{;\ell}\right)'(x') = \frac{\partial x'^k}{\partial x^p} \frac{\partial x^m}{\partial x'^\ell}\, V^p_{;m}(x) = x'^k_{,p} x^m_{,\ell'}\, V^p_{;m}(x). \tag{13.63}$$

Tangent basis vectors e_i are derivatives (13.31) of the spacetime point p with respect to the coordinates x^i, and so $e_{i,\ell} = e_{\ell,i}$ because partial derivatives commute

$$e_{i,\ell} = \frac{\partial e_i}{\partial x^\ell} = \frac{\partial^2 p}{\partial x^\ell \partial x^i} = \frac{\partial^2 p}{\partial x^i \partial x^\ell} = e_{\ell,i}. \tag{13.64}$$

Thus the affine connection (13.59) is symmetric in its lower indices

$$\Gamma^k{}_{i\ell} = e^k \cdot e_{i,\ell} = e^k \cdot e_{\ell,i} = \Gamma^k{}_{\ell i}. \tag{13.65}$$

Although the covariant derivative $V^i_{;\ell}$ (13.60) is a rank-2 mixed tensor, the affine connection $\Gamma^k{}_{i\ell}$ transforms inhomogeneously (Exercise 13.8)

$$\Gamma'^k{}_{i\ell} = e'^k \cdot \frac{\partial e'_i}{\partial x'^\ell} = \frac{\partial x'^k}{\partial x^p} \frac{\partial x^m}{\partial x'^\ell} \frac{\partial x^n}{\partial x'^i} \Gamma^p{}_{nm} + \frac{\partial x'^k}{\partial x^p} \frac{\partial^2 x^p}{\partial x'^\ell \partial x'^i}$$
$$= x'^k_{,p}\, x^m_{,\ell'}\, x^n_{,i'}\, \Gamma^p{}_{nm} + x'^k_{,p}\, x^p_{,\ell' i'} \tag{13.66}$$

and so is not a tensor. Its variation $\delta\Gamma^k{}_{i\ell} = \Gamma'^k{}_{i\ell} - \Gamma^k{}_{i\ell}$ is a tensor, however, because the inhomogeneous terms in the difference cancel.

Since the affine connection $\Gamma^k{}_{i\ell}$ is symmetric in i and ℓ, in four-dimensional spacetime, there are 10 Γ's for each k, or 40 in all. The 10 correspond to 3 rotations, 3 boosts, and 4 translations.

13.16 Covariant Derivatives of Covariant Vectors

The derivative of the scalar $V = V_k\, e^k$ is the covariant vector

$$V_{,\ell} = (V_k\, e^k)_{,\ell} = V_{k,\ell}\, e^k + V_k\, e^k_{,\ell}. \tag{13.67}$$

Its inner product with the covariant vector e_i transforms as a rank-2 covariant tensor. Thus using again the identity (13.61), we see that the covariant derivative of a covariant vector is

$$D_\ell V_i = V_{i;\ell} = e_i \cdot V_{,\ell} = e_i \cdot \left(V_{k,\ell}\, e^k + V_k\, e^k_{,\ell}\right) = \delta^k_i\, V_{k,\ell} + e_i \cdot e^k_{,\ell}\, V_k$$
$$= V_{i,\ell} - e_{i,\ell} \cdot e^k\, V_k = V_{i,\ell} - \Gamma^k{}_{i\ell}\, V_k. \tag{13.68}$$

$D_\ell V_i$ transforms as a rank-2 covariant tensor because it is the inner product of a covariant tangent vector e_i with the derivative $V_{,\ell}$ of a scalar. Note that $\Gamma^k_{i\ell}$ appears with a minus sign in $V_{i;\ell}$ and a plus sign in $V^k_{;\ell}$.

Example 13.9 (Covariant derivatives of tangent vectors) Using again the projection matrix (13.56), we find that

$$D_\ell e_i = e_{i\ell} = e_{i,\ell} - e_{i,\ell} \cdot e^k e_k = e_{i,\ell} - e_{i,\ell} = 0 \tag{13.69}$$

covariant derivatives of tangent vectors vanish.

13.17 Covariant Derivatives of Tensors

Tensors transform like products of vectors. So we can make the derivative of a tensor transform covariantly by using Leibniz's rule (5.49) to differentiate products of vectors and by turning the derivatives of the vectors into their covariant derivatives (13.60) and (13.68).

Example 13.10 (Covariant derivative of a rank-2 contravariant tensor) An arbitrary rank-2 contravariant tensor T^{ik} transforms like the product of two contravariant vectors $A^i\, B^k$. So its derivative $\partial_\ell T^{ik}$ transforms like the derivative of the product of the vectors $A^i\, B^k$

$$\partial_\ell (A^i\, B^k) = (\partial_\ell A^i)\, B^k + A^i\, \partial_\ell B^k. \tag{13.70}$$

By using twice the formula (13.60) for the covariant derivative of a contravariant vector, we can convert these two ordinary derivatives $\partial_\ell A^i$ and $\partial_\ell B^k$ into tensors

$$\begin{aligned} D_\ell (A^i\, B^k) = (A^i\, B^k)_{;\ell} &= (A^i_{,\ell} + \Gamma^i{}_{j\ell}\, A^j) B^k + A^i (B^k_{,\ell} + \Gamma^k{}_{j\ell}\, B^j) \\ &= (A^i\, B^k)_{,\ell} + \Gamma^i{}_{j\ell}\, A^j\, B^k + \Gamma^k{}_{j\ell}\, A^i\, B^j. \end{aligned} \tag{13.71}$$

Thus the covariant derivative of a rank-2 contravariant tensor is

$$D_\ell T^{ik} = T^{ik}{}_{;\ell} = T^{ik}{}_{,\ell} + \Gamma^i{}_{j\ell}\, T^{jk} + \Gamma^k{}_{j\ell}\, T^{ij}. \tag{13.72}$$

It transforms as a rank-3 tensor with one covariant index.

Example 13.11 (Covariant derivative of a rank-2 mixed tensor) A rank-2 mixed tensor $T^i{}_k$ transforms like the product $A^i\, B_k$ of a contravariant vector A^i and a covariant vector B_k. Its derivative $\partial_\ell T^i{}_k$ transforms like the derivative of the product of the vectors $A^i\, B_k$

$$\partial_\ell (A^i\, B_k) = (\partial_\ell A^i)\, B_k + A^i\, \partial_\ell B_k. \tag{13.73}$$

We can make these derivatives transform like tensors by using the formulas (13.60) and (13.68)

$$\begin{aligned} D_\ell (A^i\, B_k) = (A^i\, B_k)_{;\ell} &= (A^i_{,\ell} + \Gamma^i{}_{j\ell}\, A^j) B_k + A^i (B_{k,\ell} - \Gamma^j{}_{k\ell}\, B_j) \\ &= (A^i\, B_k)_{,\ell} + \Gamma^i{}_{j\ell}\, A^j\, B_k - \Gamma^j{}_{k\ell}\, A^i\, B_j. \end{aligned} \tag{13.74}$$

Thus the covariant derivative of a mixed rank-2 tensor is

$$D_\ell\, T^i{}_k = T^i{}_{k;\ell} = T^i{}_{k,\ell} + \Gamma^i{}_{j\ell}\, T^j{}_k - \Gamma^j{}_{k\ell}\, T^i{}_j. \tag{13.75}$$

It transforms as a rank-3 tensor with two covariant indices.

Example 13.12 (Covariant derivative of a rank-2 covariant tensor) A rank-2 covariant tensor T_{ik} transforms like the product $A_i\, B_k$ of two covariant vectors A_i and B_k. Its derivative $\partial_\ell T_{ik}$ transforms like the derivative of the product of the vectors $A_i\, B_k$

$$\partial_\ell (A_i\, B_k) = (\partial_\ell A_i)\, B_k + A_i\, \partial_\ell B_k. \tag{13.76}$$

We can make these derivatives transform like tensors by twice using the formula (13.68)

$$\begin{aligned} D_\ell (A_i\, B_k) = (A_i\, B_k)_{;\ell} &= A_{i;\ell}\, B_k + A_i\, B_{k\ell} \\ &= (A_{i,\ell} - \Gamma^j{}_{i\ell}\, A_j) B_k + A_i (B_{k,\ell} - \Gamma^j{}_{k\ell}\, B_j) \\ &= (A_i\, B_k)_{,\ell} - \Gamma^j{}_{i\ell}\, A_j\, B_k - \Gamma^j{}_{k\ell}\, A_i\, B_j. \end{aligned} \tag{13.77}$$

Thus the covariant derivative of a rank-2 covariant tensor T_{ik} is

$$D_\ell \, T_{ik} = T_{ik;\ell} = T_{ik,\ell} - \Gamma^j{}_{i\ell} \, T_{jk} - \Gamma^j{}_{k\ell} \, T_{ij}. \tag{13.78}$$

It transforms as a rank-3 covariant tensor.

Another way to derive the same result is to note that the scalar form of a rank-2 covariant tensor T_{ik} is $T = e^i \otimes e^k \, T_{ik}$. So its derivative is a covariant vector

$$T_{,\ell} = e^i \otimes e^k \, T_{ik,\ell} + e^i_{,\ell} \otimes e^k \, T_{ik} + e^i \otimes e^k_{,\ell} \, T_{ik}. \tag{13.79}$$

Using the projector $P_t = e^j e_j$ (13.56), the duality $e^i \cdot e_n = \delta^i_n$ of tangent and cotangent vectors (13.53), and the relation $e_j \cdot e^k_{,\ell} = - e^k \cdot e_{j,\ell} = -\Gamma^k_{j\ell}$ (13.59 and 13.61), we can project this derivative onto the tangent space and find after shuffling some indices

$$\begin{aligned}
(e^n e_n \otimes e^j e_j) \, T_{,\ell} &= e^i \otimes e^k \, T_{ik,\ell} + e^n \otimes e^k \, (e_n \cdot e^i_{,\ell}) T_{ik} + e^i \otimes e^j \, (e_j \cdot e^k_{,\ell}) \, T_{ik} \\
&= e^i \otimes e^k \, T_{ik,\ell} - e^n \otimes e^k \, \Gamma^i{}_{n\ell} \, T_{ik} - e^i \otimes e^j \, \Gamma^k{}_{j\ell} \, T_{ik} \\
&= (e^i \otimes e^k) \left(T_{ik,\ell} - \Gamma^j{}_{i\ell} \, T_{jk} - \Gamma^j{}_{k\ell} \, T_{ij} \right)
\end{aligned}$$

which again gives us the formula (13.78).

As in these examples, covariant derivatives are **derivations**;

$$D_k(AB) = (AB)_{;k} = A_{;k} \, B + A \, B_{;k} = (D_k A) \, B + A \, D_k B. \tag{13.80}$$

The rule for a general tensor is to treat every contravariant index as in (13.60) and every covariant index as in (13.68). The covariant derivative of a mixed rank-4 tensor, for instance, is

$$T^{ab}_{xy;k} = T^{ab}_{xy,k} + T^{jb}_{xy} \Gamma^a{}_{jk} + T^{am}_{xy} \Gamma^b{}_{mk} - T^{ab}_{jy} \Gamma^j{}_{xk} - T^{ab}_{xm} \Gamma^m{}_{yk}. \tag{13.81}$$

13.18 The Covariant Derivative of the Metric Tensor Vanishes

The metric tensor is the inner product (13.42) of tangent basis vectors

$$g_{ik} = e^\alpha_i \, \eta_{\alpha\beta} \, e^\beta_k \tag{13.82}$$

in which α and β are summed over the dimensions of the embedding space. Thus by the product rule (13.77), the covariant derivative of the metric

$$D_\ell \, g_{ik} = g_{ik;\ell} = D_\ell \, (e^\alpha_i \, \eta_{\alpha\beta} \, e^\beta_k) = (D_\ell \, e^\alpha_i) \, \eta_{\alpha\beta} \, e^\beta_k + e^\alpha_i \, \eta_{\alpha\beta} \, D_\ell \, e^\beta_k = 0 \tag{13.83}$$

vanishes because the covariant derivatives of tangent vectors vanish (13.69), $D_\ell \, e^\alpha_i = e^\alpha_{i;\ell} = 0$ and $D_\ell \, e^\beta_k = e^\beta_{k;\ell} = 0$.

13.19 Covariant Curls

Because the connection $\Gamma^k{}_{i\ell}$ is symmetric (13.65) in its lower indices, the covariant curl of a covariant vector V_i is simply its ordinary curl

$$V_{\ell;i} - V_{i;\ell} = V_{\ell,i} - V_k\,\Gamma^k{}_{\ell i} - V_{i,\ell} + V_k\,\Gamma^k{}_{i\ell} = V_{\ell,i} - V_{i,\ell}. \tag{13.84}$$

Thus the Faraday field-strength tensor $F_{i\ell} = A_{\ell,i} - A_{i,\ell}$ being the curl of the covariant vector field A_i is a generally covariant second-rank tensor.

13.20 Covariant Derivatives and Antisymmetry

The covariant derivative (13.78) $A_{i\ell;k}$ is $A_{i\ell;k} = A_{i\ell,k} - A_{m\ell}\,\Gamma^m{}_{ik} - A_{im}\,\Gamma^m{}_{\ell k}$. If the tensor A is antisymmetric $A_{i\ell} = -A_{\ell i}$, then by adding together the three cyclic permutations of the indices $i\ell k$, we find that the antisymmetry of the tensor and the symmetry (13.65) of the affine connection $\Gamma^m{}_{ik} = \Gamma^m{}_{ki}$ conspire to cancel the terms with Γs

$$\begin{aligned} A_{i\ell;k} + A_{ki;\ell} + A_{\ell k;i} = {} & A_{i\ell,k} - A_{m\ell}\,\Gamma^m{}_{ik} - A_{im}\,\Gamma^m{}_{\ell k} \\ & + A_{ki,\ell} - A_{mi}\,\Gamma^m{}_{k\ell} - A_{km}\,\Gamma^m{}_{i\ell} \\ & + A_{\ell k,i} - A_{mk}\,\Gamma^m{}_{\ell i} - A_{\ell m}\,\Gamma^m{}_{ki} \\ = {} & A_{i\ell,k} + A_{ki,\ell} + A_{\ell k,i} \end{aligned} \tag{13.85}$$

an identity named after Luigi Bianchi (1856–1928).

The Maxwell field-strength tensor $F_{i\ell}$ is antisymmetric by construction ($F_{i\ell} = A_{\ell,i} - A_{i,\ell}$), and so Maxwell's homogeneous equations

$$\begin{aligned} \frac{1}{2}\epsilon^{ijk\ell}\,F_{jk,\ell} &= F_{jk,\ell} + F_{k\ell,j} + F_{\ell j,k} \\ &= A_{k,j\ell} - A_{j,k\ell} + A_{\ell,kj} - A_{k,\ell j} + A_{j,\ell k} - A_{\ell,jk} = 0 \end{aligned} \tag{13.86}$$

are tensor equations valid in all coordinate systems.

13.21 What is the Affine Connection?

We insert the identity matrix (13.56) of the tangent space in the form $e^j\,e_j$ into the formula (13.59) for the affine connection $\Gamma^k{}_{i\ell} = e^k \cdot e_{i,\ell}$. In the resulting combination $\Gamma^k{}_{i\ell} = e^k \cdot e^j\;e_j \cdot e_{\ell,i}$ we recognize $e^k \cdot e^j$ as the inverse (13.55) of the metric tensor $e^k \cdot e^j = g^{kj}$. Repeated use of the relation $e_{i,k} = e_{k,i}$ (13.64) then leads to a formula for the affine connection

$$\begin{aligned} \Gamma^k{}_{i\ell} &= e^k \cdot e_{i,\ell} = e^k \cdot e^j\;e_j \cdot e_{i,\ell} = e^k \cdot e^j\;e_j \cdot e_{\ell,i} = \frac{1}{2}\,g^{kj}\left(e_j \cdot e_{i,\ell} + e_j \cdot e_{\ell,i}\right) \\ &= \frac{1}{2}\,g^{kj}\left((e_j \cdot e_i)_{,\ell} - e_{j,\ell} \cdot e_i + (e_j \cdot e_\ell)_{,i} - e_{j,i} \cdot e_\ell\right) \end{aligned}$$

$$= \frac{1}{2} g^{kj} \left(g_{ji,\ell} + g_{j\ell,i} - e_{j,\ell} \cdot e_i - e_{j,i} \cdot e_\ell \right) \tag{13.87}$$
$$= \frac{1}{2} g^{kj} \left(g_{ji,\ell} + g_{j\ell,i} - e_{\ell,j} \cdot e_i - e_{i,j} \cdot e_\ell \right)$$
$$= \frac{1}{2} g^{kj} \left(g_{ji,\ell} + g_{j\ell,i} - (e_i \cdot e_\ell)_{,j} \right) = \frac{1}{2} g^{kj} \left(g_{ji,\ell} + g_{j\ell,i} - g_{i\ell,j} \right)$$

in terms of the inverse of the metric tensor and a combination of its derivatives. The metric g_{ik} determines the affine connection ${\Gamma^k}_{i\ell}$.

The affine connection with all lowerindices is

$$\Gamma_{ni\ell} = g_{nk} {\Gamma^k}_{i\ell} = \frac{1}{2} \left(g_{ni,\ell} + g_{n\ell,i} - g_{i\ell,n} \right). \tag{13.88}$$

13.22 Parallel Transport

The movement of a vector along a curve on a manifold so that its length and direction in successive tangent spaces do not change is called **parallel transport**. In parallel transport, a vector $V = V^k e_k = V_k e^k$ may change $dV = V_{,\ell} dx^\ell$, but the projection of the change $P dV = e^i e_i dV = e_i e^i dV$ into the tangent space must vanish, $P dV = 0$. In terms of its contravariant components $V = V^k e_k$, this condition for parallel transport is just the vanishing of its covariant derivative (13.60)

$$\begin{aligned} 0 = e^i dV &= e^i V_{,\ell} dx^\ell = e^i (V^k e_k)_{,\ell} dx^\ell = e^i \left(V^k_{,\ell} e_k + V^k e_{k,\ell} \right) dx^\ell \\ &= \left(\delta^i_k V^k_{,\ell} + e^i \cdot e_{k,\ell} V^k \right) dx^\ell = \left(V^i_{,\ell} + {\Gamma^i}_{k\ell} V^k \right) dx^\ell. \end{aligned} \tag{13.89}$$

In terms of its covariant components $V = V_k e^k$, the condition of parallel transport is also the vanishing of its covariant derivative (13.68)

$$\begin{aligned} 0 = e_i dV &= e_i V_{,\ell} dx^\ell = e_i (V_k e^k)_{,\ell} dx^\ell = e_i \left(V_{k,\ell} e^k + V_k e^k_{,\ell} \right) dx^\ell \\ &= \left(\delta^k_i V_{k,\ell} + e_i \cdot e^k_{,\ell} V^k \right) dx^\ell = \left(V_{i,\ell} - {\Gamma^k}_{i\ell} V_k \right) dx^\ell. \end{aligned} \tag{13.90}$$

If the curve is $x^\ell(u)$, then these conditions (13.89 and 13.90) for parallel transport are

$$\frac{dV^i}{du} = V^i_{,\ell} \frac{dx^\ell}{du} = - {\Gamma^i}_{k\ell} V^k \frac{dx^\ell}{du} \quad \text{and} \quad \frac{dV_i}{du} = V_{i,\ell} \frac{dx^\ell}{du} = {\Gamma^k}_{i\ell} V_k \frac{dx^\ell}{du}. \tag{13.91}$$

Example 13.13 (Parallel transport on a sphere) We parallel-transport the vector $\boldsymbol{v} = \boldsymbol{e}_\phi = (0, 1, 0)$ up from the equator along the line of longitude $\phi = 0$. Along this path, the vector $\boldsymbol{v} = (0, 1, 0) = \boldsymbol{e}_\phi$ is constant, so $\partial_\theta \boldsymbol{v} = 0$ and so both $e^\theta \cdot \boldsymbol{e}_{\phi,\theta} = 0$ and $e^\phi \cdot \boldsymbol{e}_{\phi,\theta} = 0$. Thus $D_\theta v^k = v^k_{;\theta} = 0$ between the equator and the north pole. As $\theta \to 0$ along the meridian $\phi = 0$, the vector $\boldsymbol{v} = (0, 1, 0)$ approaches the vector $\boldsymbol{e}_\theta$ of the $\phi = \pi/2$ meridian. We then parallel-transport $\boldsymbol{v} = \boldsymbol{e}_\theta$ down from the north pole along that meridian to the equator. Along this path, the vector $\boldsymbol{v} = \boldsymbol{e}_\theta/r = (0, \cos\theta, -\sin\theta)$ obeys the parallel-transport condition (13.90) because its θ-derivative is

$$\boldsymbol{v}_{,\theta} = r^{-1}\boldsymbol{e}_{,\theta} = (0, \cos\theta, -\sin\theta)_{,\theta} = -(0, \sin\theta, \cos\theta) = -\hat{\boldsymbol{r}}\big|_{\phi=\pi/2}. \quad (13.92)$$

So $\boldsymbol{v}_{,\theta}$ is perpendicular to the tangent vectors $\boldsymbol{e}_\theta$ and $\boldsymbol{e}_\phi$ along the curve $\phi = \pi/2$. Thus $e^k \cdot \boldsymbol{v}_{,\theta} = 0$ for $k = \theta$ and $k = \phi$ and so $\boldsymbol{v}_{;\theta} = 0$, along the meridian $\phi = \pi/2$. When $\boldsymbol{e}_\theta$ reaches the equator, it is $\boldsymbol{e}_\theta = (0, 0, -1)$. Finally, we parallel-transport $\boldsymbol{v}$ along the equator back to the starting point $\phi = 0$. Along this path, the vector $\boldsymbol{v} = (0, 0, -1) = e_\theta$ is constant, so $\boldsymbol{v}_{,\phi} = 0$ and $\boldsymbol{v}_{;\phi} = 0$. The change from $\boldsymbol{v} = (0, 1, 0)$ to $\boldsymbol{v} = (0, 0, -1)$ is due to the curvature of the sphere.

13.23 Curvature

To find the curvature at a point $p(x_0)$, we parallel-transport a vector V_i along a curve x^ℓ that runs around a tiny square about the point $p(x_0)$. We then measure the change in the vector

$$\Delta V_i = \oint \Gamma^k{}_{i\ell}\, V_k\, dx^\ell. \quad (13.93)$$

On the curve x^ℓ, we approximate $\Gamma^k{}_{i\ell}(x)$ and $V_k(x)$ as

$$\begin{aligned} \Gamma^k{}_{i\ell}(x) &= \Gamma^k{}_{i\ell}(x_0) + \Gamma^k{}_{i\ell,n}(x_0)\,(x - x_0)^n \\ V_k(x) &= V_k(x_0) + \Gamma^m{}_{kn}(x_0)\, V_m(x_0)\,(x - x_0)^n. \end{aligned} \quad (13.94)$$

So keeping only terms linear in $(x - x_0)^n$, we have

$$\begin{aligned} \Delta V_i &= \oint \Gamma^k{}_{i\ell}\, V_k\, dx^\ell \qquad (13.95) \\ &= \left[\Gamma^k{}_{i\ell,n}(x_0)\, V_k(x_0) + \Gamma^k{}_{i\ell}(x_0)\, \Gamma^m{}_{kn}(x_0)\, V_m(x_0)\right] \oint (x - x_0)^n\, dx^\ell \\ &= \left[\Gamma^k{}_{i\ell,n}(x_0)\, V_k(x_0) + \Gamma^m{}_{i\ell}(x_0)\, \Gamma^k{}_{mn}(x_0)\, V_k(x_0)\right] \oint (x - x_0)^n\, dx^\ell \end{aligned}$$

after interchanging the dummy indices k and m in the second term within the square brackets. The integral around the square is antisymmetric in n and ℓ and equal in absolute value to the area a^2 of the tiny square

$$\oint (x - x_0)^n\, dx^\ell = \pm a^2\, \epsilon_{n\ell}. \quad (13.96)$$

The overall sign depends upon whether the integral is clockwise or counterclockwise, what n and ℓ are, and what we mean by positive area. The integral picks out the part of the term between the brackets in the formula (13.95) that is antisymmetric in n and ℓ. We choose minus signs in (13.96) so that the change in the vector is

$$\Delta V_i = a^2 \left[\Gamma^k{}_{in,\ell} - \Gamma^k{}_{i\ell,n} + \Gamma^k{}_{\ell m}\,\Gamma^m{}_{in} - \Gamma^k{}_{nm}\,\Gamma^m{}_{i\ell}\right] V_k. \tag{13.97}$$

The quantity between the brackets is **Riemann's curvature tensor**

$$R^k{}_{i\ell n} = \Gamma^k{}_{ni,\ell} - \Gamma^k{}_{\ell i,n} + \Gamma^k{}_{\ell m}\,\Gamma^m{}_{ni} - \Gamma^k{}_{nm}\,\Gamma^m{}_{\ell i}. \tag{13.98}$$

The sign convention is that of (Zee, 2013; Misner et al., 1973; Carroll, 2003; Schutz, 2009; Hartle, 2003; Cheng, 2010; Padmanabhan, 2010). Weinberg (Weinberg, 1972) uses the opposite sign. The covariant form $R_{ijk\ell}$ of Riemann's tensor is related to $R^k{}_{i\ell n}$ by

$$R_{ijk\ell} = g_{in} R^n{}_{jk\ell} \quad \text{and} \quad R^i{}_{jk\ell} = g^{in} R_{njk\ell}. \tag{13.99}$$

The Riemann curvature tensor is the commutator of two covariant derivatives. To see why, we first use the formula (13.78) for the covariant derivative $D_n D_\ell V_i$ of the second-rank covariant tensor $D_\ell V_i$

$$\begin{aligned}
D_n D_\ell V_i &= D_n \left(V_{i,\ell} - \Gamma^k{}_{\ell i}\, V_k\right) \\
&= V_{i,\ell n} - \Gamma^k{}_{\ell i,n}\, V_k - \Gamma^k{}_{\ell i}\, V_{k,n} \\
&\quad - \Gamma^j{}_{ni}\left(V_{j,\ell} - \Gamma^m{}_{\ell j}\, V_m\right) - \Gamma^m{}_{\ell n}\left(V_{i,m} - \Gamma^q{}_{im}\, V_q\right).
\end{aligned} \tag{13.100}$$

Subtracting $D_\ell D_n V_i$, we find the commutator $[D_n, D_\ell]V_i$ to be the contraction of the curvature tensor $R^k{}_{i\ell n}$ (13.98) with the covariant vector V_k

$$[D_n, D_\ell]V_i = \left(\Gamma^k{}_{ni,\ell} - \Gamma^k{}_{\ell i,n} + \Gamma^k{}_{\ell j}\,\Gamma^j{}_{ni} - \Gamma^k{}_{nj}\,\Gamma^j{}_{\ell i}\right) V_k = R^k{}_{i\ell n}\, V_k. \tag{13.101}$$

Since $[D_n, D_\ell]V_i$ is a rank-3 covariant tensor and V_k is an arbitrary covariant vector, the quotient theorem (Section 13.8) implies that the curvature tensor is a rank-4 tensor with one contravariant index.

If we define the matrix Γ_ℓ with row index k and column index i as $\Gamma^k{}_{i\ell}$

$$\Gamma_\ell = \begin{pmatrix} \Gamma^0{}_{\ell 0} & \Gamma^0{}_{\ell 1} & \Gamma^0{}_{\ell 2} & \Gamma^0{}_{\ell 3} \\ \Gamma^1{}_{\ell 0} & \Gamma^1{}_{\ell 1} & \Gamma^1{}_{\ell 2} & \Gamma^1{}_{\ell 3} \\ \Gamma^2{}_{\ell 0} & \Gamma^2{}_{\ell 1} & \Gamma^2{}_{\ell 2} & \Gamma^2{}_{\ell 3} \\ \Gamma^3{}_{\ell 0} & \Gamma^3{}_{\ell 1} & \Gamma^3{}_{\ell 2} & \Gamma^3{}_{\ell 3} \end{pmatrix}, \tag{13.102}$$

then we may write the covariant derivatives appearing in the curvature tensor $R^k{}_{i\ell n}$ as $D_\ell = \partial_\ell + \Gamma_\ell$ and $D_n = \partial_n + \Gamma_n$. In these terms, the curvature tensor is the i, k matrix element of their commutator

$$R^k{}_{i\ell n} = [\partial_\ell + \Gamma_\ell, \partial_n + \Gamma_n]^k{}_i = [D_\ell, D_n]^k{}_i\,. \tag{13.103}$$

The curvature tensor is therefore antisymmetric in its last two indexes

$$R^k{}_{i\ell n} = -\, R^k{}_{in\ell}. \tag{13.104}$$

The curvature tensor with all lower indices shares this symmetry

$$R_{ji\ell n} = g_{jk} R^k{}_{i\ell n} = -\, g_{jk} R^k{}_{in\ell} = -\, R_{jin\ell} \tag{13.105}$$

and has three others. In Riemann normal coordinates the derivatives of the metric vanish at any particular point x_*. In these coordinates, the Γ's all vanish, and the curvature tensor in terms of the Γ's with all lower indices (13.88) is after a cancellation

$$R_{ki\ell n} = \Gamma_{kni,\ell} - \Gamma_{k\ell i,n} = \frac{1}{2}\left(g_{kn,i\ell} - g_{ni,k\ell} - g_{k\ell,in} + g_{\ell i,kn}\right). \tag{13.106}$$

In these coordinates and therefore in all coordinates, $R_{ki\ell n}$ is antisymmetric in its first two indexes and symmetric under the interchange of its first and second pairs of indexes

$$R_{ijk\ell} = -\, R_{jik\ell} \quad \text{and} \quad R_{ijk\ell} = R_{k\ell ij}. \tag{13.107}$$

Cartan's equations of structure (13.328 and 13.330) imply (13.342) that the curvature tensor is antisymmetric in its last three indexes

$$0 = R^j{}_{[ik\ell]} = \frac{1}{3!}\left(R^j{}_{ik\ell} + R^j{}_{\ell ik} + R^j{}_{k\ell i} - R^j{}_{ki\ell} - R^j{}_{i\ell k} - R^j{}_{\ell ki}\right) \tag{13.108}$$

and obeys the cyclic identity

$$0 = R^j{}_{ik\ell} + R^j{}_{\ell ik} + R^j{}_{k\ell i}. \tag{13.109}$$

The vanishing (13.108) of $R_{i[jk\ell]}$ implies that the completely antisymmetric part of the Riemann tensor also vanishes

$$0 = R_{[ijk\ell]} = \frac{1}{4!}\left(R_{ijk\ell} - R_{jik\ell} - R_{ikj\ell} - R_{ij\ell k} + R_{jki\ell}\cdots\right). \tag{13.110}$$

The Riemann tensor also satisfies a Bianchi identity

$$0 = R^i{}_{j[k\ell;m]}. \tag{13.111}$$

These symmetries reduce 256 different functions $R_{ijk\ell}(x)$ to 20.

The **Ricci tensor** is the contraction

$$R_{in} = R^k{}_{ikn}. \tag{13.112}$$

The **curvature scalar** is the further contraction

$$R = g^{ni}\, R_{in}. \tag{13.113}$$

Example 13.14 (Curvature of the sphere S^2) While in 4-dimensional spacetime indices run from 0 to 3, on the everyday sphere S^2 (Example 13.4) they are just θ and ϕ. There are only eight possible affine connections, and because of the symmetry (13.65) in their lower indices $\Gamma^i{}_{\theta\phi} = \Gamma^i{}_{\phi\theta}$, only six are independent.

In the euclidian embedding space $\mathbb{E}^3$, the point $\boldsymbol{p}$ on a sphere of radius L has cartesian coordinates $\boldsymbol{p} = L\,(\sin\theta\cos\phi,\ \sin\theta\sin\phi,\ \cos\theta)$, so the two tangent 3-vectors are (13.37)

$$\begin{aligned} \boldsymbol{e}_\theta &= \boldsymbol{p}_{,\theta} = L\,(\cos\theta\,\cos\phi,\ \cos\theta\,\sin\phi,\ -\sin\theta) = L\,\hat{\boldsymbol{\theta}} \\ \boldsymbol{e}_\phi &= \boldsymbol{p}_{,\phi} = L\sin\theta\,(-\sin\phi,\ \cos\phi,\ 0) = L\sin\theta\,\hat{\boldsymbol{\phi}}. \end{aligned} \tag{13.114}$$

Their dot products form the metric (13.38)

$$g_{ik} = \begin{pmatrix} g_{\theta\theta} & g_{\theta\phi} \\ g_{\phi\theta} & g_{\phi\phi} \end{pmatrix} = \begin{pmatrix} \boldsymbol{e}_\theta\cdot\boldsymbol{e}_\theta & \boldsymbol{e}_\theta\cdot\boldsymbol{e}_\phi \\ \boldsymbol{e}_\phi\cdot\boldsymbol{e}_\theta & \boldsymbol{e}_\phi\cdot\boldsymbol{e}_\phi \end{pmatrix} = \begin{pmatrix} L^2 & 0 \\ 0 & L^2\sin^2\theta \end{pmatrix} \tag{13.115}$$

which is diagonal with $g_{\theta\theta} = L^2$ and $g_{\phi\phi} = L^2\sin^2\theta$. Differentiating the vectors $\boldsymbol{e}_\theta$ and $\boldsymbol{e}_\phi$, we find

$$\begin{aligned} e_{\theta,\theta} &= -\,L\,(\sin\theta\cos\phi,\ \sin\theta\sin\phi,\ \cos\theta) = -L\,\hat{\boldsymbol{r}} \\ e_{\theta,\phi} &= L\cos\theta\,(-\sin\phi,\ \cos\phi,\ 0) = L\cos\theta\,\hat{\boldsymbol{\phi}} \\ e_{\phi,\theta} &= e_{\theta,\phi} \\ e_{\phi,\phi} &= -\,L\sin\theta\,(\cos\phi,\ \sin\phi,\ 0)\,. \end{aligned} \tag{13.116}$$

The metric with upper indices g^{ij} is the inverse of the metric g_{ij}

$$(g^{ij}) = \begin{pmatrix} L^{-2} & 0 \\ 0 & L^{-2}\sin^{-2}\theta \end{pmatrix}, \tag{13.117}$$

so the dual vectors $\boldsymbol{e}^i = g^{ik} e_k$ are

$$\begin{aligned} \boldsymbol{e}^\theta &= L^{-1}\,(\cos\theta\,\cos\phi,\ \cos\theta\,\sin\phi,\ -\sin\theta) = L^{-1}\hat{\boldsymbol{\theta}} \\ \boldsymbol{e}^\phi &= \frac{1}{L\sin\theta}\,(-\sin\phi,\ \cos\phi,\ 0) = \frac{1}{L\sin\theta}\,\hat{\boldsymbol{\phi}}. \end{aligned} \tag{13.118}$$

The affine connections are given by (13.59) as

$$\Gamma^i{}_{jk} = \Gamma^i{}_{kj} = \boldsymbol{e}^i\cdot\boldsymbol{e}_{j,k}. \tag{13.119}$$

Since both $\boldsymbol{e}^\theta$ and $\boldsymbol{e}^\phi$ are perpendicular to $\hat{r}$, the affine connections $\Gamma^\theta{}_{\theta\theta}$ and $\Gamma^\phi{}_{\theta\theta}$ both vanish. Also, $e_{\phi,\phi}$ is orthogonal to $\hat{\phi}$, so $\Gamma^\phi{}_{\phi\phi} = 0$ as well. Similarly, $e_{\theta,\phi}$ is perpendicular to $\hat{\theta}$, so $\Gamma^\theta{}_{\theta\phi} = \Gamma^\theta{}_{\phi\theta}$ also vanishes.

The two nonzero affine connections are

$$\Gamma^\phi{}_{\theta\phi} = \boldsymbol{e}^\phi\cdot\boldsymbol{e}_{\theta,\phi} = L^{-1}\sin^{-1}\theta\,\hat{\boldsymbol{\phi}}\cdot L\cos\theta\,\hat{\boldsymbol{\phi}} = \cot\theta \tag{13.120}$$

and

$$\Gamma^{\theta}{}_{\phi\phi} = \boldsymbol{e}^{\theta} \cdot \boldsymbol{e}_{\phi,\phi} = -\sin\theta\,(\cos\theta\,\cos\phi,\,\cos\theta\,\sin\phi,\,-\sin\theta)\cdot(\cos\phi,\,\sin\phi,\,0)$$
$$= -\sin\theta\cos\theta. \tag{13.121}$$

The nonzero connections are $\Gamma^{\phi}_{\theta\phi} = \cot\theta$ and $\Gamma^{\theta}_{\phi\phi} = -\sin\theta\cos\theta$. So the matrices Γ_θ and Γ_ϕ, the derivative $\Gamma_{\phi,\theta}$, and the commutator $[\Gamma_\theta, \Gamma_\phi]$ are

$$\Gamma_\theta = \begin{pmatrix} 0 & 0 \\ 0 & \cot\theta \end{pmatrix} \quad \text{and} \quad \Gamma_\phi = \begin{pmatrix} 0 & -\sin\theta\cos\theta \\ \cot\theta & 0 \end{pmatrix} \tag{13.122}$$
$$\Gamma_{\phi,\theta} = \begin{pmatrix} 0 & \sin^2\theta - \cos^2\theta \\ -\csc^2\theta & 0 \end{pmatrix} \quad \text{and} \quad [\Gamma_\theta, \Gamma_\phi] = \begin{pmatrix} 0 & \cos^2\theta \\ \cot^2\theta & 0 \end{pmatrix}.$$

Both $[\Gamma_\theta, \Gamma_\theta]$ and $[\Gamma_\phi, \Gamma_\phi]$ vanish. So the commutator formula (13.103) gives for Riemann's curvature tensor

$$\begin{aligned}
R^{\theta}{}_{\theta\theta\theta} &= [\partial_\theta + \Gamma_\theta, \partial_\theta + \Gamma_\theta]^{\theta}{}_{\theta} = 0 \\
R^{\phi}{}_{\theta\phi\theta} &= [\partial_\phi + \Gamma_\phi, \partial_\theta + \Gamma_\theta]^{\phi}{}_{\theta} = \left(\Gamma_{\theta,\phi}\right)^{\phi}{}_{\theta} + [\Gamma_\phi, \Gamma_\theta]^{\phi}{}_{\theta} = 1 \\
R^{\theta}{}_{\phi\theta\phi} &= [\partial_\theta + \Gamma_\theta, \partial_\phi + \Gamma_\phi]^{\theta}{}_{\phi} = -\left(\Gamma_{\theta,\phi}\right)^{\theta}{}_{\phi} + [\Gamma_\theta, \Gamma_\phi]^{\theta}{}_{\phi} = \sin^2\theta \\
R^{\phi}{}_{\phi\phi\phi} &= [\partial_\phi + \Gamma_\phi, \partial_\phi + \Gamma_\phi]^{\phi}{}_{\phi} = 0.
\end{aligned} \tag{13.123}$$

The Ricci tensor (13.112) is the contraction $R_{mk} = R^{n}_{mnk}$, and so

$$\begin{aligned}
R_{\theta\theta} &= R^{\theta}{}_{\theta\theta\theta} + R^{\phi}{}_{\theta\phi\theta} = 1 \\
R_{\phi\phi} &= R^{\theta}{}_{\phi\theta\phi} + R^{\phi}{}_{\phi\phi\phi} = \sin^2\theta.
\end{aligned} \tag{13.124}$$

The curvature scalar (13.113) is the contraction $R = g^{km} R_{mk}$, and so since $g^{\theta\theta} = L^{-2}$ and $g^{\phi\phi} = L^{-2}\sin^{-2}\theta$, it is

$$R = g^{\theta\theta} R_{\theta\theta} + g^{\phi\phi} R_{\phi\phi} = L^{-2} + L^{-2} = \frac{2}{L^2} \tag{13.125}$$

for a 2-sphere of radius L. The scalar curvature is a constant because the sphere is a maximally symmetric space (Section 13.24).

Gauss invented a formula for the curvature K of a surface; for all two-dimensional surfaces, his $K = R/2$.

Example 13.15 (Curvature of a cylindrical hyperboloid) The points of a cylindrical hyperboloid in 3-space satisfy $L^2 = -x^2 - y^2 + z^2$ and may be parametrized as $p = L(\sinh\theta\cos\phi, \sinh\theta\sin\phi, \cosh\theta)$. The (orthogonal) coordinate basis vectors are

$$\begin{aligned}
\boldsymbol{e}_\theta &= p_{,\theta} = L(\cosh\theta\cos\phi, \cosh\theta\sin\phi, \sinh\theta) \\
\boldsymbol{e}_\phi &= p_{,\phi} = L(-\sinh\theta\sin\phi, \sinh\theta\cos\phi, 0).
\end{aligned} \tag{13.126}$$

The squared distance ds^2 between nearby points is

$$ds^2 = \boldsymbol{e}_\theta \cdot \boldsymbol{e}_\theta\, d\theta^2 + \boldsymbol{e}_\phi \cdot \boldsymbol{e}_\phi\, d\phi^2. \tag{13.127}$$

If the embedding metric is $m = \text{diag}(1, 1, -1)$, then ds^2 is

$$ds^2 = L^2\, d\theta^2 + L^2 \sinh^2\theta\, d\phi^2 \tag{13.128}$$

and

$$(g_{ij}) = L^2 \begin{pmatrix} 1 & 0 \\ 0 & \sinh^2\theta \end{pmatrix}. \tag{13.129}$$

The Mathematica scripts GREAT.m and cylindrical_hyperboloid.nb compute the scalar curvature as $R = -2/L^2$. The surface is maximally symmetric with constant negative curvature. This chapter's programs and scripts are in Tensors_and_general_relativity at `github.com/kevinecahill`.

Example 13.16 (Curvature of the sphere S^3) The 3-dimensional sphere S^3 may be embedded isometrically in 4-dimensional flat euclidian space $\mathbb{E}^4$ as the set of points $p = (x, y, z, w)$ that satisfy $L^2 = x^2 + y^2 + z^2 + w^2$. If we label its points as

$$p(\chi, \theta, \phi) = L(\sin\chi \sin\theta \cos\phi, \sin\chi \sin\theta \sin\phi, \sin\chi \cos\theta, \cos\chi), \tag{13.130}$$

then its coordinate basis vectors are

$$\begin{aligned} e_\chi &= p_{,\chi} = L(\cos\chi \sin\theta \cos\phi, \cos\chi \sin\theta \sin\phi, \cos\chi \cos\theta, -\sin\chi) \\ e_\theta &= p_{,\theta} = L(\sin\chi \cos\theta \cos\phi, \sin\chi \cos\theta \sin\phi, -\sin\chi \sin\theta, 0) \\ e_\phi &= p_{,\phi} = L(-\sin\chi \sin\theta \sin\phi, \sin\chi \sin\theta \cos\phi, 0, 0). \end{aligned} \tag{13.131}$$

The inner product of $\mathbb{E}^4$ is the 4-dimensional dot-product. The basis vectors are orthogonal. In terms of the radial variable $r = L\sin\chi$, the squared distance ds^2 between two nearby points is

$$\begin{aligned} ds^2 &= e_\chi \cdot e_\chi d\chi^2 + e_\theta \cdot e_\theta d\theta^2 + e_\phi \cdot e_\phi d\phi^2 \\ &= L^2 \left(d\chi^2 + \sin^2\chi\, d\theta^2 + \sin^2\chi \sin^2\theta\, d\phi^2\right) \\ &= \frac{dr^2}{1 - \sin^2\chi} + r^2 d\theta^2 + r^2 \sin^2\theta d\phi^2 = \frac{dr^2}{1 - (r/L)^2} + r^2 d\Omega^2 \end{aligned} \tag{13.132}$$

where $d\Omega^2 = d\theta^2 + \sin^2\theta\, d\phi^2$. In these coordinates, r, θ, ϕ, the metric is

$$g_{ik} = \begin{pmatrix} 1/(1 - (r/L)^2) & 0 & 0 \\ 0 & r^2 & 0 \\ 0 & 0 & r^2 \sin^2\theta \end{pmatrix}. \tag{13.133}$$

The Mathematica scripts GREAT.m and sphere_S3.nb compute the scalar curvature as

$$R = \frac{6}{L^2} \tag{13.134}$$

which is a constant because S^3 is maximally symmetric (Section 13.24).

Example 13.17 (Curvature of the hyperboloid H^3) The hyperboloid H^3 is a 3-dimensional surface that can be isometrically embedded in the semi-euclidian space-time $\mathbb{E}^{(1,3)}$ in which distances are $ds^2 = dx^2 + dy^2 + dz^2 - dw^2$, and w is a time coordinate. The points of H^3 satisfy $L^2 = -x^2 - y^2 - z^2 + w^2$. If we label them as

$$p(\chi, \theta, \phi) = L\,(\sinh\chi\,\sin\theta\cos\phi, \sinh\chi\,\sin\theta\sin\phi, \sinh\chi\,\cos\theta, \cosh\chi) \quad (13.135)$$

then the coordinate basis vectors or tangent vectors of H^3 are

$$\begin{aligned} e_\chi &= p_{,\chi} = L(\cosh\chi\,\sin\theta\cos\phi, \cosh\chi\,\sin\theta\sin\phi, \cosh\chi\,\cos\theta, \sinh\chi) \\ e_\theta &= p_{,\theta} = L(\sinh\chi\,\cos\theta\cos\phi, \sinh\chi\,\cos\theta\sin\phi, -\sinh\chi\,\sin\theta, 0) \\ e_\phi &= p_{,\phi} = L(-\sinh\chi\,\sin\theta\sin\phi, \sinh\chi\,\sin\theta\cos\phi, 0, 0). \end{aligned} \quad (13.136)$$

The basis vectors are orthogonal. In terms of the radial variable $r = L\sinh\chi/a$, the squared distance ds^2 between two nearby points is

$$\begin{aligned} ds^2 &= e_\chi \cdot e_\chi d\chi^2 + e_\theta \cdot e_\theta d\theta^2 + e_\phi \cdot e_\phi d\phi^2 \\ &= L^2\left(d\chi^2 + \sinh^2\chi\,d\theta^2 + \sinh^2\chi\,\sin^2\theta\,d\phi^2\right) \\ &= \frac{dr^2}{1+\sinh^2\chi} + r^2 d\theta^2 + r^2\sin^2\theta\,d\phi^2 = \frac{dr^2}{1+(r/L)^2} + r^2 d\Omega^2. \end{aligned} \quad (13.137)$$

The Mathematica scripts GREAT.m and hyperboloid_H3.nb compute the scalar curvature of H^3 as

$$R = -\frac{6}{L^2}. \quad (13.138)$$

Its curvature is a constant because H^3 is maximally symmetric (Section 13.24).

The only maximally symmetric 3-dimensional manifolds are S^3, H^3, and euclidian space $\mathbb{E}^3$ whose line element is $ds^2 = dr^2 + r^2 d\Omega^2$. They are the spatial parts of Friedmann–Lemaître–Robinson–Walker cosmologies (Section 13.42).

13.24 Maximally Symmetric Spaces

The spheres S^2 and S^3 (Examples 13.4 and 13.16) and the hyperboloids H^2 and H^3 (Examples 13.6 and 13.17) are maximally symmetric spaces. A space described by a metric $g_{ik}(x)$ is symmetric under a transformation $x \to x'$ if the distances $g_{ik}(x')dx'^i dx'^k$ and $g_{ik}(x)dx^i dx^k$ are the same. To see what this symmetry condition means, we consider the infinitesimal transformation $x'^\ell = x^\ell + \epsilon y^\ell(x)$ under which to lowest order $g_{ik}(x') = g_{ik}(x) + g_{ik,\ell}\epsilon y^\ell$ and $dx'^i = dx^i + \epsilon y^i_{,j} dx^j$. The symmetry condition requires

$$g_{ik}(x)dx^i dx^k = (g_{ik}(x) + g_{ik,\ell}\epsilon y^\ell)(dx^i + \epsilon y^i_{,j}dx^j)(dx^k + \epsilon y^k_{,m}dx^m) \quad (13.139)$$

or

$$0 = g_{ik,\ell}\, y^\ell + g_{im}\, y^m_{,k} + g_{jk}\, y^j_{,i}. \tag{13.140}$$

The vector field $y^i(x)$ must satisfy this condition if $x'^i = x^i + \epsilon y^i(x)$ is to be a symmetry of the metric $g_{ik}(x)$. By using the vanishing (13.83) of the covariant derivative of the metric tensor, we may write the condition on the symmetry vector $y^\ell(x)$ as (Exercise 13.9)

$$0 = y_{i;k} + y_{k;i}. \tag{13.141}$$

The symmetry vector y^ℓ is a **Killing** vector (Wilhelm Killing, 1847–1923). We may use symmetry conditions (13.140) and (13.141) either to find the symmetries of a space with a known metric or to find metrics with a particular symmetry.

Example 13.18 (Killing vectors of the sphere S^2) The first Killing vector is $(y_1^\theta, y_1^\phi) = (0, 1)$. Since the components of y_1 are constants, the symmetry condition (13.140) says $g_{ik,\phi} = 0$ which tells us that the metric is independent of ϕ. The other two Killing vectors are $(y_2^\theta, y_2^\phi) = (\sin\phi, \cot\theta\cos\phi)$ and $(y_3^\theta, y_3^\phi) = (\cos\phi, -\cot\theta\sin\phi)$. The symmetry condition (13.140) for $i = k = \theta$ and Killing vectors y_2 and y_3 tell us that $g_{\theta\phi} = 0$ and that $g_{\theta\theta,\theta} = 0$. So $g_{\theta\theta}$ is a constant, which we set equal to unity. Finally, the symmetry condition (13.140) for $i = k = \phi$ and the Killing vectors y_2 and y_3 tell us that $g_{\phi\phi,\theta} = 2\cot\theta\, g_{\phi\phi}$ which we integrate to $g_{\phi\phi} = \sin^2\theta$. The 2-dimensional space with Killing vectors y_1, y_2, y_3 therefore has the metric (13.115) of the sphere S^2.

Example 13.19 (Killing vectors of the hyperboloid H^2) The metric (13.46) of the hyperboloid H^2 is diagonal with $g_{\theta\theta} = R^2$ and $g_{\phi\phi} = R^2\sinh^2\theta$. The Killing vector $(y_1^\theta, y_1^\phi) = (0, 1)$ satisfies the symmetry condition (13.140). Since $g_{\theta\theta}$ is independent of θ and ϕ, the $\theta\theta$ component of (13.140) implies that $y^\theta_{,\theta} = 0$. Since $g_{\phi\phi} = R^2\sinh^2\theta$, the $\phi\phi$ component of (13.140) says that $y^\phi_{,\phi} = -\coth\theta\, y^\theta$. The $\theta\phi$ and $\phi\theta$ components of (13.140) give $y^\theta_{,\phi} = -\sinh^2\theta\, y^\phi_{,\theta}$. The vectors $y_2 = (y_2^\theta, y_2^\phi) = (\sin\phi,\ \coth\theta\ \sin\phi)$ and $y_3 = (y_3^\theta, y_3^\phi) = (\cos\phi,\ -\coth\theta\ \sin\phi)$ satisfy both of these equations.

The **Lie derivative** $\mathcal{L}_y$ of a scalar field A is defined in terms of a vector field $y^\ell(x)$ as $\mathcal{L}_y A = y^\ell A_{,\ell}$. The Lie derivative $\mathcal{L}_y$ of a contravariant vector F^i is

$$\mathcal{L}_y F^i = y^\ell F^i_{,\ell} - F^\ell y^i_{,\ell} = y^\ell F^i_{;\ell} - F^\ell y^i_{;\ell} \tag{13.142}$$

in which the second equality follows from $y^\ell \Gamma^i_{\ell k} F^k = F^\ell \Gamma^i_{\ell k} y^k$. The Lie derivative $\mathcal{L}_y$ of a covariant vector V_i is

$$\mathcal{L}_y V_i = y^\ell V_{i,\ell} + V_\ell y^\ell_{,i} = y^\ell V_{i;\ell} + V_\ell y^\ell_{;i}. \tag{13.143}$$

Similarly, the Lie derivative $\mathcal{L}_y$ of a rank-2 covariant tensor T_{ik} is

$$\mathcal{L}_y T_{ik} = y^\ell T_{ik,\ell} + T_{\ell k} y^\ell_{,i} + T_{i\ell} y^\ell_{,k}. \tag{13.144}$$

We see now that the condition (13.140) that a vector field y^ℓ be a symmetry of a metric g_{jm} is that its Lie derivative

$$\mathcal{L}_y g_{ik} = g_{ik,\ell}\, y^\ell + g_{im}\, y^m_{,k} + g_{jk}\, y^j_{,i} = 0 \tag{13.145}$$

must vanish.

A maximally symmetric space (or spacetime) in d dimensions has d translation symmetries and $d(d-1)/2$ rotational symmetries which gives a total of $d(d+1)/2$ symmetries associated with $d(d+1)/2$ Killing vectors. Thus for $d = 2$, there is one rotation and two translations. For $d = 3$, there are three rotations and three translations. For $d = 4$, there are six rotations and four translations.

A maximally symmetric space has a curvature tensor (13.99) that is simply related to its metric tensor

$$R_{ijk\ell} = c\,(g_{ik}g_{j\ell} - g_{i\ell}g_{jk}) \tag{13.146}$$

where c is a constant (Zee, 2013, IX.6). Since $g^{ki} g_{ik} = g^k_k = d$ is the number of dimensions of the space(time), the Ricci tensor (13.112) and the curvature scalar (13.113) of a maximally symmetric space are

$$R_{j\ell} = g^{ki} R_{ijk\ell} = c\,(d-1)\, g_{j\ell} \quad \text{and} \quad R = g^{\ell j} R_{j\ell} = c\, d(d-1). \tag{13.147}$$

13.25 Principle of Equivalence

Since the metric tensor $g_{ij}(x)$ is real and symmetric, it can be diagonalized at any point $p(x)$ by a 4×4 orthogonal matrix $O(x)$

$$O^{\mathsf{T}}{}_i{}^k\, g_{k\ell}\, O^\ell{}_j = \begin{pmatrix} e_0 & 0 & 0 & 0 \\ 0 & e_1 & 0 & 0 \\ 0 & 0 & e_2 & 0 \\ 0 & 0 & 0 & e_3 \end{pmatrix} \tag{13.148}$$

which arranges the four real eigenvalues e_i of the matrix $g_{ij}(x)$ in the order $e_0 \le e_1 \le e_2 \le e_3$. Thus the coordinate transformation

$$\frac{\partial x^k}{\partial x'^i} = \frac{O^{\mathsf{T}}{}_i{}^k}{\sqrt{|e_i|}} \tag{13.149}$$

takes any spacetime metric $g_{k\ell}(x)$ with one negative and three positive eigenvalues into the Minkowski metric η_{ij} of flat spacetime

$$g_{k\ell}(x)\,\frac{\partial x^k}{\partial x'^i}\,\frac{\partial x^\ell}{\partial x'^j} = g'_{ij}(x') = \eta_{ij} = \begin{pmatrix} -1 & 0 & 0 & 0 \\ 0 & 1 & 0 & 0 \\ 0 & 0 & 1 & 0 \\ 0 & 0 & 0 & 1 \end{pmatrix} \tag{13.150}$$

at the point $p(x) = p(x')$.

The **principle of equivalence** says that in these free-fall coordinates x', the physical laws of gravity-free special relativity apply in a suitably small region about the point $p(x) = p(x')$. It follows from this principle that the metric g_{ij} of spacetime accounts for all the effects of gravity.

In the x' coordinates, the invariant squared separation dp^2 is

$$\begin{aligned} dp^2 = g'_{ij}\,dx'^i dx'^j &= e'_i(x') \cdot e'_j(x')\,dx'^i dx'^j \\ &= e'^a_i(x')\eta_{ab}e'^b_j(x')\,dx'^i dx'^j = \delta^a_i\,\eta_{ab}\delta^b_j\,dx'^i dx'^j \\ &= \eta_{ij}\,dx'^i dx'^j = (\mathbf{dx}')^2 - (dx'^0)^2 = ds^2. \end{aligned} \tag{13.151}$$

If $\mathbf{dx}' = 0$, then $dt' = \sqrt{-ds^2}/c$ is the **proper time** elapsed between events p and $p + dp$. If $dt' = 0$, then ds is the **proper distance** between the events.

The x' coordinates are not unique because every Lorentz transformation (Section 12.1) leaves the metric η invariant. Coordinate systems in which $g_{ij}(x') = \eta_{ij}$ are called **Lorentz**, **inertial**, or **free-fall** coordinate systems.

The congruency transformation (1.351 and 13.148–13.150) preserves the signs of the eigenvalues e_i which make up the **signature** $(-1, 1, 1, 1)$ of the metric tensor.

13.26 Tetrads

We defined the metric tensor as the dot product (13.34) or (13.42) of tangent vectors, $g_{k\ell}(x) = e_k(x) \cdot e_\ell(x)$. If instead we invert the equation (13.150) that relates the metric tensor to the flat metric

$$g_{k\ell}(x) = \frac{\partial x'^a}{\partial x^k}\,\eta_{ab}\,\frac{\partial x'^b}{\partial x^\ell} \tag{13.152}$$

then we can express the metric in terms of four 4-vectors

$$c^a_k(x) = \frac{\partial x'^a}{\partial x^k} \quad \text{as} \quad g_{k\ell}(x) = c^a_k(x)\,\eta_{ab}\,c^b_\ell(x) \tag{13.153}$$

in which η_{ij} is the 4×4 metric (13.150) of flat Minkowski space. Cartan's four 4-vectors $c^a_i(x)$ are called a **moving frame**, a **tetrad**, and a **vierbein**.

Because $L^a{}_c(x)\,\eta_{ab}\,L^b{}_d(x) = \eta_{cd}$, every spacetime-dependent Lorentz transformation $L(x)$ maps one set of tetrads $c^c_k(x)$ to another set of tetrads $c'^a_k(x) = L^a{}_c(x)\,c^c_k(x)$ that represent the same metric

$$\begin{aligned} c'^a_k(x)\,\eta_{ab}\,c'^b_\ell(x) &= L^a{}_c(x)\,c^c_k(x)\,\eta_{ab}\,L^b{}_d(x)\,c^d_\ell(x) \\ &= c^c_k(x)\,\eta_{cd}\,c^d_\ell(x) = g_{k\ell}(x). \end{aligned} \tag{13.154}$$

Cartan's tetrad is four 4-vectors c_i that give the metric tensor as $g_{ik} = c_i \cdot c_k = \vec{c}_i \cdot \vec{c}_k - c^0_i c^0_k$. The dual tetrads $c^i_a = g^{ik}\eta_{ab}c^b_k$ satisfy

$$c^i_a c^a_k = \delta^i_k \quad \text{and} \quad c^i_a c^b_i = \delta^b_a. \tag{13.155}$$

The metric $g_{k\ell}(x)$ is symmetric, $g_{k\ell}(x) = g_{\ell k}(x)$, so it has 10 independent components at each spacetime point x. The four 4-vectors c^a_k have 16 components, but a Lorentz transformation $L(x)$ has 6 components. So the tetrads have $16 - 6 = 10$ independent components at each spacetime point.

The distinction between tetrads and tangent basis vectors is that each tetrad $c^a{}_k$ has 4 components, $a = 0, 1, 2, 3$, while each basis vector $e^\alpha{}_k(x)$ has as many components $\alpha = 0, 1, 2, 3, \ldots$ as there are dimensions in the minimal semi-euclidian embedding space $\mathbb{E}^{1,n}$ where $n \le 19$ (Aké et al., 2018). Both represent the metric

$$g_{k\ell}(x) = \sum_{a,b=0}^{3} c^a{}_k(x)\,\eta_{ab}\,c^b{}_\ell(x) = \sum_{\alpha,\beta=0}^{n} e^\alpha{}_k(x)\,\eta'_{\alpha\beta}\,e^\beta{}_\ell(x) \tag{13.156}$$

in which η' is like η but with n diagonal elements that are unity (Élie Cartan, 1869–1951).

13.27 Scalar Densities and $g = |\det(g_{ik})|$

Let g be the absolute value of the determinant of the metric tensor g_{ik}

$$g = g(x) = |\det(g_{ik}(x))|. \tag{13.157}$$

Under a coordinate transformation, $\sqrt{g}$ becomes

$$\sqrt{g'} = \sqrt{g'(x')} = \sqrt{|\det(g'_{ik}(x'))|} = \sqrt{\left|\det\left(\frac{\partial x^j}{\partial x'^i}\frac{\partial x^\ell}{\partial x'^k}\,g_{j\ell}(x)\right)\right|}. \tag{13.158}$$

The definition (1.204) of a determinant and the product rule (1.225) for determinants tell us that

$$\sqrt{g'(x')} = \sqrt{\left|\det\left(\frac{\partial x^j}{\partial x'^i}\right)\det\left(\frac{\partial x^\ell}{\partial x'^k}\right)\det(g_{j\ell})\right|} = |J(x/x')|\sqrt{g(x)} \tag{13.159}$$

where $J(x/x')$ is the jacobian (Section 1.21) of the coordinate transformation

$$J(x/x') = \det\left(\frac{\partial x^j}{\partial x'^i}\right). \tag{13.160}$$

A quantity $s(x)$ is a **scalar density** of weight w if it transforms as

$$s'(x') = [J(x'/x)]^w s(x). \tag{13.161}$$

Thus the transformation rule (13.159) says that the determinant $\det(g_{ik})$ is a scalar density of weight minus two

$$\det(g'_{ik}(x')) = = [J(x/x')]^2 g(x) = [J(x'/x)]^{-2} \det(g_{j\ell}(x)). \tag{13.162}$$

We saw in Section 1.21 that under a coordinate transformation $x \to x'$ the d-dimensional element of volume in the new coordinates $d^d x'$ is related to that in the old coordinates $d^d x$ by a jacobian

$$d^d x' = J(x'/x)\, d^d x = \det\left(\frac{\partial x'^i}{\partial x^j}\right) d^d x. \tag{13.163}$$

Thus the product $\sqrt{g}\, d^d x$ changes at most by the sign of the jacobian $J(x'/x)$ when $x \to x'$

$$\sqrt{g'}\, d^d x' = |J(x/x')|\, J(x'/x)\, \sqrt{g(x)}\, d^d x = \pm\sqrt{g(x)}\, d^d x. \tag{13.164}$$

The quantity $\sqrt{g}\, d^4 x$ is the invariant scalar $\sqrt{g}\, |d^4 x|$ so that if $L(x)$ is a scalar, then the integral over spacetime

$$\int L(x)\, \sqrt{g}\, d^4 x \tag{13.165}$$

is invariant under general coordinate transformations. The Levi-Civita tensor provides a fancier definition.

13.28 Levi-Civita's Symbol and Tensor

In 3 dimensions, Levi-Civita's **symbol** $\epsilon_{ijk} \equiv \epsilon^{ijk}$ is totally antisymmetric with $\epsilon_{123} = 1$ in all coordinate systems. In 4 space or spacetime dimensions, Levi-Civita's **symbol** $\epsilon_{ijk\ell} \equiv \epsilon^{ijk\ell}$ is totally antisymmetric with $\epsilon_{1234} = 1$ or equivalently with $\epsilon_{0123} = 1$ in all coordinate systems. In n dimensions, Levi-Civita's symbol $\epsilon_{i_1 i_2 \ldots i_n}$ is totally antisymmetric with $\epsilon_{123\ldots n} = 1$ or $\epsilon_{012\ldots n-1} = 1$.

We can turn his symbol into a pseudotensor by multiplying it by the square root of the absolute value of the determinant of a rank-2 covariant tensor. A natural choice is the metric tensor. In a right-handed coordinate system in which the tangent vector e_0 points (orthochronously) toward the future, the Levi-Civita **tensor** $\eta_{ijk\ell}$ is the totally antisymmetric rank-4 covariant tensor

$$\eta_{ijk\ell}(x) = \sqrt{g(x)}\, \epsilon_{ijk\ell} \tag{13.166}$$

in which $g(x) = |\det g_{mn}(x)|$ is (13.157) the absolute value of the determinant of the metric tensor g_{mn}. In a different system of coordinates x', the Levi-Civita tensor $\eta_{ijk\ell}(x')$ differs from (13.166) by the sign s of the jacobian $J(x'/x)$ of any coordinate transformation to x' from a right-handed, orthochronous coordinate system x

$$\eta_{ijk\ell}(x') = s(x')\sqrt{g(x')}\, \epsilon_{ijk\ell}. \tag{13.167}$$

The transformation rule (13.159) and the definition (1.204) and product rule (1.225) of determinants show that $\eta_{ijk\ell}$ transforms as a rank-4 covariant tensor

$$\begin{aligned}\eta'_{ijk\ell}(x') &= s(x')\sqrt{g'(x')}\,\epsilon_{ijk\ell} = s(x')\,|J(x/x')|\sqrt{g(x)}\,\epsilon_{ijk\ell}\\ &= J(x/x')\sqrt{g(x)}\,\epsilon_{ijk\ell} = \det\left(\frac{\partial x}{\partial x'}\right)\sqrt{g}\,\epsilon_{ijk\ell}\\ &= \frac{\partial x^t}{\partial x'^i}\frac{\partial x^u}{\partial x'^j}\frac{\partial x^v}{\partial x'^k}\frac{\partial x^w}{\partial x'^\ell}\sqrt{g}\,\epsilon_{tuvw} = \frac{\partial x^t}{\partial x'^i}\frac{\partial x^u}{\partial x'^j}\frac{\partial x^v}{\partial x'^k}\frac{\partial x^w}{\partial x'^\ell}\,\eta_{tuvw}.\end{aligned} \tag{13.168}$$

Raising the indices of η and using σ as the sign of $\det(g_{ik})$, we have

$$\begin{aligned}\eta^{ijk\ell} &= g^{it}\, g^{ju}\, g^{kv}\, g^{\ell w}\,\eta_{tuvw} = g^{it}\, g^{ju}\, g^{kv}\, g^{\ell w}\sqrt{g}\,\epsilon_{tuvw} = \sqrt{g}\,\epsilon_{ijk\ell}\det(g^{mn})\\ &= \sqrt{g}\,\epsilon_{ijk\ell}/\det(g_{mn}) = \sigma\,\epsilon_{ijk\ell}/\sqrt{g} \equiv \sigma\,\epsilon^{ijk\ell}/\sqrt{g}.\end{aligned} \tag{13.169}$$

In terms of the Hodge star (14.151), the invariant volume element is

$$\sqrt{g}\;|d^4x| = *\,1 = \frac{1}{4!}\,\eta_{ijk\ell}\, dx^i\wedge dx^j\wedge dx^k\wedge dx^\ell. \tag{13.170}$$

13.29 Divergence of a Contravariant Vector

The contracted covariant derivative of a contravariant vector is a **scalar** known as the **divergence**,

$$\nabla\cdot V = V^i_{;i} = V^i_{,i} + V^k\,\Gamma^i_{ki}. \tag{13.171}$$

Because $g_{ik} = g_{ki}$, in the sum over i of the connection (13.59)

$$\Gamma^i{}_{ki} = \frac{1}{2}g^{i\ell}\left(g_{i\ell,k} + g_{\ell k,i} - g_{ki,\ell}\right) \tag{13.172}$$

the last two terms cancel because they differ only by the interchange of the dummy indices i and ℓ

$$g^{i\ell}g_{\ell k,i} = g^{\ell i}g_{ik,\ell} = g^{i\ell}g_{ki,\ell}. \tag{13.173}$$

So the contracted connection collapses to

$$\Gamma^{i}{}_{ki} = \frac{1}{2} g^{i\ell} g_{i\ell,k}. \tag{13.174}$$

There is a nice formula for this last expression. To derive it, let $\underline{g} \equiv g_{i\ell}$ be the 4×4 matrix whose elements are those of the covariant metric tensor $g_{i\ell}$. Its determinant, like that of any matrix, is the cofactor sum (1.213) along any row or column, that is, over ℓ for fixed i or over i for fixed ℓ

$$\det(\underline{g}) = \sum_{i \text{ or } \ell} g_{i\ell}\, C_{i\ell} \tag{13.175}$$

in which the cofactor $C_{i\ell}$ is $(-1)^{i+\ell}$ times the determinant of the reduced matrix consisting of the matrix $\underline{g}$ with row i and column ℓ omitted. Thus the partial derivative of $\det \underline{g}$ with respect to the $i\ell$th element $g_{i\ell}$ is

$$\frac{\partial \det(\underline{g})}{\partial g_{i\ell}} = C_{i\ell} \tag{13.176}$$

in which we allow $g_{i\ell}$ and $g_{\ell i}$ to be independent variables for the purposes of this differentiation. The inverse $g^{i\ell}$ of the metric tensor $\underline{g}$, like the inverse (1.215) of any matrix, is the transpose of the cofactor matrix divided by its determinant $\det(\underline{g})$

$$g^{i\ell} = \frac{C_{\ell i}}{\det(\underline{g})} = \frac{1}{\det(\underline{g})} \frac{\partial \det(\underline{g})}{\partial g_{\ell i}}. \tag{13.177}$$

Using this formula and the chain rule, we may write the derivative of the determinant $\det(\underline{g})$ as

$$\det(\underline{g})_{,k} = \frac{\partial \det(\underline{g})}{\partial g_{i\ell}} g_{i\ell,k} = \det(\underline{g})\, g^{\ell i}\, g_{i\ell,k} \tag{13.178}$$

and so since $g_{i\ell} = g_{\ell i}$, the contracted connection (13.174) is

$$\Gamma^{i}{}_{ki} = \frac{1}{2} g^{i\ell} g_{i\ell,k} = \frac{\det(\underline{g})_{,k}}{2\det(\underline{g})} = \frac{|\det(\underline{g})|_{,k}}{2|\det(\underline{g})|} = \frac{g_{,k}}{2g} = \frac{(\sqrt{g}\,)_{,k}}{\sqrt{g}} \tag{13.179}$$

in which $g \equiv \left|\det(\underline{g})\right|$ is the absolute value of the determinant of the metric tensor.

Thus from (13.171 and 13.179), we arrive at our formula for the covariant divergence of a contravariant vector:

$$\nabla \cdot V = V^{i}_{;i} = V^{i}_{,i} + \Gamma^{i}_{ki} V^{k} = V^{k}_{,k} + \frac{(\sqrt{g}\,)_{,k}}{\sqrt{g}} V^{k} = \frac{(\sqrt{g}\, V^{k})_{,k}}{\sqrt{g}}. \tag{13.180}$$

Example 13.20 (Maxwell's inhomogeneous equations) An important application of this divergence formula (13.180) is the generally covariant form (14.157) of Maxwell's inhomogeneous equations

$$\frac{1}{\sqrt{g}}\left(\sqrt{g}F^{k\ell}\right)_{,\ell} = \mu_0 j^k. \tag{13.181}$$

Example 13.21 (Energy–momentum tensor) Another application is to the divergence of the symmetric energy–momentum tensor $T^{ij} = T^{ji}$

$$\begin{aligned} T^{ij}_{;i} &= T^{ij}_{,i} + \Gamma^i{}_{ki}\, T^{kj} + \Gamma^j{}_{mi}\, T^{im} \\ &= \frac{(\sqrt{g}T^{kj})_k}{\sqrt{g}} + \Gamma^j{}_{mi}\, T^{im}. \end{aligned} \tag{13.182}$$

13.30 Covariant Laplacian

In flat 3-space, we write the laplacian as $\nabla \cdot \nabla = \nabla^2$ or as $\triangle$. In euclidian coordinates, both mean $\partial_x^2 + \partial_y^2 + \partial_z^2$. In flat minkowski space, one often turns the triangle into a square and writes the 4-laplacian as $\Box = \triangle - \partial_0^2$.

The gradient $f_{,k}$ of a scalar field f is a covariant vector, and $f^{,i} = g^{ik} f_{,k}$ is its contravariant form. The **invariant laplacian** $\Box f$ of a scalar field f is the covariant divergence $f^{,i}_{;i}$. We may use our formula (13.180) for the divergence of a contravariant vector to write it in these equivalent ways

$$\Box f = f^{,i}_{;i} = (g^{ik} f_{,k})_{;i} = \frac{(\sqrt{g}\, f^{,i})_{,i}}{\sqrt{g}} = \frac{(\sqrt{g}\, g^{ik} f_{,k})_{,i}}{\sqrt{g}}. \tag{13.183}$$

13.31 Principle of Stationary Action in General Relativity

The invariant proper time for a particle to move along a path $x^i(t)$

$$T = \int_{\tau_1}^{\tau_2} d\tau = \frac{1}{c}\int \left(-\, g_{i\ell} dx^i dx^\ell\right)^{\frac{1}{2}} \tag{13.184}$$

is extremal and stationary on free-fall paths called **geodesics**. We can identify a geodesic by computing the variation $\delta d\tau$

$$\begin{aligned} c\delta d\tau &= \delta\sqrt{-g_{i\ell} dx^i dx^\ell} = \frac{-\delta(g_{i\ell}) dx^i dx^\ell - 2 g_{i\ell} dx^i \delta dx^\ell}{2\sqrt{-g_{i\ell} dx^i dx^\ell}} \\ &= -\frac{g_{i\ell,k}}{2c}\delta x^k u^i u^\ell d\tau - \frac{g_{i\ell}}{c} u^i \delta dx^\ell = -\frac{g_{i\ell,k}}{2c}\delta x^k u^i u^\ell d\tau - \frac{g_{i\ell}}{c} u^i d\delta x^\ell \end{aligned} \tag{13.185}$$

in which $u^\ell = dx^\ell/d\tau$ is the 4-velocity (12.20). The path is extremal if

$$0 = c\delta T = c\int_{\tau_1}^{\tau_2} \delta d\tau = -\frac{1}{c}\int_{\tau_1}^{\tau_2}\left(\frac{1}{2}g_{i\ell,k}\delta x^k u^i u^\ell + g_{i\ell}u^i\frac{d\delta x^\ell}{d\tau}\right)d\tau \quad (13.186)$$

which we integrate by parts keeping in mind that $\delta x^\ell(\tau_2) = \delta x^\ell(\tau_1) = 0$

$$\begin{aligned} 0 &= -\int_{\tau_1}^{\tau_2}\left(\frac{1}{2}g_{i\ell,k}\delta x^k u^i u^\ell - \frac{d(g_{i\ell}u^i)}{d\tau}\delta x^\ell\right)d\tau \\ &= -\int_{\tau_1}^{\tau_2}\left(\frac{1}{2}g_{i\ell,k}\delta x^k u^i u^\ell - g_{i\ell,k}u^i u^k\delta x^\ell - g_{i\ell}\frac{du^i}{d\tau}\delta x^\ell\right)d\tau. \quad (13.187)\end{aligned}$$

Now interchanging the dummy indices ℓ and k on the second and third terms, we have

$$0 = -\int_{\tau_1}^{\tau_2}\left(\frac{1}{2}g_{i\ell,k}u^i u^\ell - g_{ik,\ell}u^i u^\ell - g_{ik}\frac{du^i}{d\tau}\right)\delta x^k d\tau \quad (13.188)$$

or since δx^k is arbitrary

$$0 = \frac{1}{2}g_{i\ell,k}u^i u^\ell - g_{ik,\ell}u^i u^\ell - g_{ik}\frac{du^i}{d\tau}. \quad (13.189)$$

If we multiply this equation of motion by g^{rk} and note that $g_{ik,\ell}u^i u^\ell = g_{\ell k,i}u^i u^\ell$, then we find

$$0 = \frac{du^r}{d\tau} + \frac{1}{2}g^{rk}\left(g_{ik,\ell} + g_{\ell k,i} - g_{i\ell,k}\right)u^i u^\ell. \quad (13.190)$$

So using the symmetry $g_{i\ell} = g_{\ell i}$ and the formula (13.87) for $\Gamma^r{}_{i\ell}$, we get

$$0 = \frac{du^r}{d\tau} + \Gamma^r{}_{i\ell}\,u^i u^\ell \quad \text{or} \quad 0 = \frac{d^2x^r}{d\tau^2} + \Gamma^r{}_{i\ell}\frac{dx^i}{d\tau}\frac{dx^\ell}{d\tau} \quad (13.191)$$

which is the geodesic equation. In empty space, particles fall along **geodesics independently of their masses.**

One gets the same geodesic equation from the simpler action principle

$$0 = \delta\int_{\lambda_1}^{\lambda_2} g_{i\ell}(x)\frac{dx^i}{d\lambda}\frac{dx^\ell}{d\lambda}d\lambda \implies 0 = \frac{d^2x^r}{d\lambda^2} + \Gamma^r{}_{i\ell}\frac{dx^i}{d\lambda}\frac{dx^\ell}{d\lambda}. \quad (13.192)$$

The right-hand side of the geodesic equation (13.191) is a contravariant vector because (Weinberg, 1972) under general coordinate transformations, the inhomogeneous terms arising from $\ddot{x}^r$ cancel those from $\Gamma^r{}_{i\ell}\dot{x}^i\dot{x}^\ell$. Here and often in what follows we'll use dots to mean proper-time derivatives.

The action for a particle of mass m and charge q in a gravitational field $\Gamma^r{}_{i\ell}$ and an electromagnetic field A_i is

$$S = -mc\int\left(-g_{i\ell}dx^i dx^\ell\right)^{\frac{1}{2}} + \frac{q}{c}\int_{\tau_1}^{\tau_2} A_i(x)\,dx^i \quad (13.193)$$

in which the interaction $q\int A_i dx^i$ is invariant under general coordinate transformations. By (12.59 and 13.188), the first-order change in S is

$$\delta S = m \int_{\tau_1}^{\tau_2} \left[\frac{1}{2} g_{i\ell,k} u^i u^\ell - g_{ik,\ell} u^i u^\ell - g_{ik} \frac{du^i}{d\tau} + \frac{q}{mc} \left(A_{i,k} - A_{k,i} \right) u^i \right] \delta x^k d\tau \tag{13.194}$$

and so by combining the Lorentz force law (12.60) and the geodesic equation (13.191) and by writing $F^{ri} \dot{x}_i$ as $F^r{}_i \dot{x}^i$, we have

$$0 = \frac{d^2 x^r}{d\tau^2} + \Gamma^r{}_{i\ell} \frac{dx^i}{d\tau} \frac{dx^\ell}{d\tau} - \frac{q}{m} F^r{}_i \frac{dx^i}{d\tau} \tag{13.195}$$

as the equation of motion of a particle of mass m and charge q. It is striking how nearly perfect the electromagnetism of Faraday and Maxwell is.

The action of the electromagnetic field interacting with an electric current j^k in a gravitational field is

$$S = \int \left[-\tfrac{1}{4} F_{k\ell} F^{k\ell} + \mu_0 A_k j^k \right] \sqrt{g}\, d^4 x \tag{13.196}$$

in which $\sqrt{g}\, d^4 x$ is the invariant volume element. After an integration by parts, the first-order change in the action is

$$\delta S = \int \left[-\frac{\partial}{\partial x^\ell} \left(F^{k\ell} \sqrt{g} \right) + \mu_0 j^k \sqrt{g} \right] \delta A_k \, d^4 x, \tag{13.197}$$

and so the inhomogeneous Maxwell equations in a gravitational field are

$$\frac{\partial}{\partial x^\ell} \left(\sqrt{g}\, F^{k\ell} \right) = \mu_0 \sqrt{g}\, j^k. \tag{13.198}$$

The action of a scalar field ϕ of mass m in a gravitational field is

$$S = \frac{1}{2} \int \left(-\phi_{,i}\, g^{ik} \phi_{,k} - m^2 \phi^2 \right) \sqrt{g}\, d^4 x. \tag{13.199}$$

After an integration by parts, the first-order change in the action is

$$\delta S = \int \delta\phi \left[\left(\sqrt{g}\, g^{ik} \phi_{,k} \right)_{,i} - m^2 \sqrt{g}\, \phi \right] d^4 x \tag{13.200}$$

which yields the equation of motion

$$\left(\sqrt{g}\, g^{ik} \phi_{,k} \right)_{,i} - m^2 \sqrt{g}\, \phi = 0. \tag{13.201}$$

The action of the gravitational field itself is a spacetime integral of the Riemann scalar (13.113) divided by Newton's constant

$$S = \frac{c^3}{16\pi G} \int R \sqrt{g}\, d^4 x. \tag{13.202}$$

Its variation leads to Einstein'sequations (Section 13.35).

13.32 Equivalence Principle and Geodesic Equation

The **principle of equivalence** (Section 13.25) says that in any gravitational field, one may choose **free-fall coordinates** in which all physical laws take the same form as in special relativity without acceleration or gravitation – at least over a suitably small volume of spacetime. Within this volume and in these coordinates, things behave as they would at rest deep in empty space far from any matter or energy. The volume must be small enough so that the gravitational field is constant throughout it. Such free-fall coordinate systems are called **local Lorentz frames** and **local inertial frames**.

Example 13.22 (Elevators) When a modern elevator starts going down from a high floor, it accelerates downward at something less than the local acceleration of gravity. One feels less pressure on one's feet; one feels lighter. After accelerating downward for a few seconds, the elevator assumes a constant downward speed, and then one feels the normal pressure of one's weight on one's feet. The elevator seems to be slowing down for a stop, but actually it has just stopped accelerating downward.

What if the cable snapped, and a frightened passenger dropped his laptop? He could catch it very easily as it would not seem to fall because the elevator, the passenger, and the laptop would all fall at the same rate. The physics in the falling elevator would be the same as if the elevator were at rest in empty space far from any gravitational field. The laptop's clock would tick as fast as it would at rest in the absence of gravity, but to an observer on the ground it would appear slower.

What if a passenger held an electric charge? Observers in the falling elevator would see a static electric field around the charge, but observers on the ground could detect radiation from the accelerating charge.

Example 13.23 (Proper time) If the events are the ticks of a clock, then the proper time between ticks $d\tau/c$ is the time between the ticks of the clock at rest or at speed zero if the clock is accelerating. The proper lifetime $d\tau_\ell/c$ of an unstable particle is the average time it takes to decay at speed zero. In arbitrary coordinates, this proper lifetime is

$$c^2 d\tau_\ell^2 = -ds^2 = -g_{ik}(x)\, dx^i dx^k. \tag{13.203}$$

Example 13.24 (Clock hypothesis) The apparent lifetime of an unstable particle is independent of the acceleration of the particle even when the particle is subjected to centripetal accelerations of 10^{19} m/s^2(Bailey et al., 1977) and to longitudinal accelerations of 10^{16}m/s^2(Roos et al., 1980).

The transformation from arbitrary coordinates x^k to free-fall coordinates y^i changes the metric $g_{j\ell}$ to the diagonal metric η_{ik} of flat spacetime $\eta =$ diag$(-1, 1, 1, 1)$, which has two indices and is not a Levi-Civita tensor. Algebraically, this transformation is a congruence (1.353)

$$\eta_{ik} = \frac{\partial x^j}{\partial y^i} \, g_{j\ell} \, \frac{\partial x^\ell}{\partial y^k}. \tag{13.204}$$

The geodesic equation (13.191) follows from the **principle of equivalence** (Weinberg, 1972; Hobson et al., 2006). Suppose a particle is moving under the influence of gravitation alone. Then one may choose free-fall coordinates $y(x)$ so that the particle obeys the force-free equation of motion

$$\frac{d^2 y^i}{d\tau^2} = 0 \tag{13.205}$$

with $d\tau$ the proper time $d\tau^2 = -\eta_{ik}\, dy^i dy^k / c^2$. The chain rule applied to $y^i(x)$ in (13.205) gives

$$\begin{aligned} 0 &= \frac{d}{d\tau}\left(\frac{\partial y^i}{\partial x^k}\frac{dx^k}{d\tau}\right) \\ &= \frac{\partial y^i}{\partial x^k}\frac{d^2 x^k}{d\tau^2} + \frac{\partial^2 y^i}{\partial x^k \partial x^\ell}\frac{dx^k}{d\tau}\frac{dx^\ell}{d\tau}. \end{aligned} \tag{13.206}$$

We multiply by $\partial x^m/\partial y^i$ and use the identity

$$\frac{\partial x^m}{\partial y^i}\frac{\partial y^i}{\partial x^k} = \delta^m_k \tag{13.207}$$

to write the equation of motion (13.205) in the x-coordinates

$$\frac{d^2 x^m}{d\tau^2} + \Gamma^m{}_{k\ell}\,\frac{dx^k}{d\tau}\frac{dx^\ell}{d\tau} = 0. \tag{13.208}$$

This is the geodesic equation (13.191) in which the affine connection is

$$\Gamma^m{}_{k\ell} = \frac{\partial x^m}{\partial y^i}\frac{\partial^2 y^i}{\partial x^k \partial x^\ell}. \tag{13.209}$$

13.33 Weak Static Gravitational Fields

Newton's equations describe slow motion in a weak static gravitational field. Because the motion is slow, we neglect u^i compared to u^0 and simplify the geodesic equation (13.191) to

$$0 = \frac{du^r}{d\tau} + \Gamma^r{}_{00}\,(u^0)^2. \tag{13.210}$$

Because the gravitational field is static, we neglect the time derivatives $g_{k0,0}$ and $g_{0k,0}$ in the connection formula (13.87) and find for $\Gamma^r{}_{00}$

$$\Gamma^r{}_{00} = \frac{1}{2}\, g^{rk}\left(g_{0k,0} + g_{0k,0} - g_{00,k}\right) = -\frac{1}{2}\, g^{rk}\, g_{00,k} \tag{13.211}$$

with ${\Gamma^0}_{00} = 0$. Because the field is weak, the metric can differ from η_{ij} by only a tiny tensor $g_{ij} = \eta_{ij} + h_{ij}$ so that to first order in $|h_{ij}| \ll 1$ we have ${\Gamma^r}_{00} = -\frac{1}{2} h_{00,r}$ for $r = 1, 2, 3$. With these simplifications, the geodesic equation (13.191) reduces to

$$\frac{d^2x^r}{d\tau^2} = \frac{1}{2}\,(u^0)^2\, h_{00,r} \qquad \text{or} \qquad \frac{d^2x^r}{d\tau^2} = \frac{1}{2}\left(\frac{dx^0}{d\tau}\right)^2 h_{00,r}. \tag{13.212}$$

So for slow motion, the ordinary acceleration is described by Newton's law

$$\frac{d^2\boldsymbol{x}}{dt^2} = \frac{c^2}{2}\,\nabla h_{00}. \tag{13.213}$$

If ϕ is his potential, then for slow motion in weak static fields

$$g_{00} = -1 + h_{00} = -1 - 2\phi/c^2 \qquad \text{and so} \qquad h_{00} = -\,2\phi/c^2. \tag{13.214}$$

Thus, if the particle is at a distance r from a mass M, then $\phi = -\,GM/r$ and $h_{00} = -2\phi/c^2 = 2GM/rc^2$ and so

$$\frac{d^2\boldsymbol{x}}{dt^2} = -\,\nabla\phi = \nabla\,\frac{GM}{r} = -\,GM\frac{\boldsymbol{r}}{r^3}. \tag{13.215}$$

How weak are the static gravitational fields we know about? The dimensionless ratio ϕ/c^2 is 10^{-39} on the surface of a proton, 10^{-9} on the Earth, 10^{-6} on the surface of the sun, and 10^{-4} on the surface of a white dwarf.

13.34 Gravitational Time Dilation

The proper time (Example 13.23) interval $d\tau$ of a clock at rest in the weak, static gravitational field (13.210–13.215) satisfies equation (13.203)

$$c^2 d\tau^2 = -\,ds^2 = -g_{ik}(x)\,dx^i dx^k = -g_{00}\,c^2 dt^2 = (1 + 2\phi/c^2)\,c^2 dt^2. \tag{13.216}$$

So if two clocks are at rest at distances r and $r + h$ from the center of the Earth, then the times dt_r and dt_{r+h} between ticks of the clock at r and the one at $r + h$ are related by the proper time $d\tau$ of the ticks of the clock

$$(c^2 + 2\phi(r))dt_r^2 = c^2 d\tau^2 = (c^2 + 2\phi(r+h))dt_{r+h}^2. \tag{13.217}$$

Since $\phi(r) = -\,GM/r$, the potential at $r + h$ is $\phi(r + h) \approx \phi(r) + gh$, and so the ratio of the tick time dt_r of the lower clock at r to the tick time of the upper clock at $r + h$ is

$$\frac{dt_r}{dt_{r+h}} = \frac{\sqrt{c^2 + 2\phi(r+h)}}{\sqrt{c^2 + 2\phi(r)}} = \frac{\sqrt{c^2 + 2\phi(r) + 2gh}}{\sqrt{c^2 + 2\phi(r)}} \approx 1 + \frac{gh}{c^2}. \tag{13.218}$$

The lower clock is slower.

Example 13.25 (Pound and Rebka) Pound and Rebka in 1960 used the Mössbauer effect to measure the blue shift of light falling down a 22.6 m shaft. They found $(\nu_\ell - \nu_u)/\nu = gh/c^2 = 2.46 \times 10^{-15}$ (Robert Pound 1919–2010, Glen Rebka 1931–2015, `media.physics.harvard.edu/video/?id=LOEB_POUND_092591.flv`).

Example 13.26 (Redshift of the Sun) A photon emitted with frequency ν_0 at a distance r from a mass M would be observed at spatial infinity to have frequency $\nu = \nu_0\sqrt{-g_{00}} = \nu_0\sqrt{1 - 2MG/c^2r}$ for a redshift of $\Delta\nu = \nu_0 - \nu$. Since the Sun's dimensionless potential $\phi_\odot/c^2$ is $-MG/c^2r = -2.12 \times 10^{-6}$ at its surface, sunlight is shifted to the red by 2 parts per million.

13.35 Einstein's Equations

If we make an action that is a scalar, invariant under general coordinate transformations, and then apply to it the principle of stationary action, we will get tensor field equations that are invariant under general coordinate transformations. If the metric of spacetime is among the fields of the action, then the resulting theory will be a possible theory of gravity. If we make the action as simple as possible, it will be Einstein's theory.

To make the action of the gravitational field, we need a scalar. Apart from the scalar $\sqrt{g}\, d^4x = \sqrt{g}\, c\, dt\, d^3x$, where $g = |\det(g_{ik})|$, the simplest scalar we can form from the metric tensor and its first and second derivatives is the scalar curvature R which gives us the **Einstein–Hilbert action**

$$S_{EH} = \frac{c^3}{16\pi G}\int R\,\sqrt{g}\, d^4x = \frac{c^3}{16\pi G}\int g^{ik}\, R_{ik}\,\sqrt{g}\, d^4x \tag{13.219}$$

in which $G = 6.7087 \times 10^{-39}\, \hbar c\, (\text{GeV}/c^2)^{-2} = 6.6742 \times 10^{-11}\, \text{m}^3\, \text{kg}^{-1}\, \text{s}^{-2}$ is Newton's constant.

If $\delta g^{ik}(x)$ is a tiny local change in the inverse metric, then the rule $\delta \det A = \det A\, \text{Tr}(A^{-1}\delta A)$ (1.228), valid for any nonsingular, nondefective matrix A, together with the identity $0 = \delta(g^{ik} g_{k\ell}) = \delta g^{ik}\, g_{k\ell} + g^{ik}\, \delta g_{k\ell}$ and the notation $\underline{g}$ for the metric tensor $g_{j\ell}$ considered as a matrix imply that

$$\delta\sqrt{g} = \frac{\det \underline{g}}{2g\sqrt{g}}\,\delta \det \underline{g} = \frac{(\det \underline{g})^2 g^{ik}\,\delta g_{ik}}{2g\sqrt{g}} = -\frac{1}{2}\sqrt{g}\; g_{ik}\,\delta g^{ik}. \tag{13.220}$$

So the first-order change in the action density is

$$\begin{aligned}\delta\left(g^{ik}\, R_{ik}\,\sqrt{g}\right) &= R_{ik}\,\sqrt{g}\;\delta g^{ik} + g^{ik}\, R_{ik}\;\delta\sqrt{g} + g^{ik}\,\sqrt{g}\;\delta R_{ik} \\ &= \left(R_{ik} - \frac{1}{2}R\, g_{ik}\right)\sqrt{g}\;\delta g^{ik} + g^{ik}\,\sqrt{g}\;\delta R_{ik}.\end{aligned} \tag{13.221}$$

The product $g^{ik}\delta R_{ik}$ is a scalar, so we can evaluate it in any coordinate system. In a local inertial frame, where $\Gamma^a{}_{bc} = 0$ and g_{de} is constant, this invariant variation of the Ricci tensor (13.112) is

$$\begin{aligned} g^{ik}\delta R_{ik} &= g^{ik}\,\delta\left(\Gamma^n{}_{in,k} - \Gamma^n{}_{ik,n}\right) = g^{ik}\left(\partial_k\,\delta\Gamma^n{}_{in} - \partial_n\,\delta\Gamma^n{}_{ik}\right) \\ &= g^{ik}\,\partial_k\,\delta\Gamma^n{}_{in} - g^{in}\,\partial_k\,\delta\Gamma^k{}_{in} = \partial_k\left(g^{ik}\,\delta\Gamma^n{}_{in} - g^{in}\,\delta\Gamma^k{}_{in}\right). \end{aligned} \tag{13.222}$$

The transformation law (13.66) for the affine connection shows that the variations $\delta\Gamma^n{}_{in}$ and $\delta\Gamma^k{}_{in}$ are tensors although the connections themselves aren't. Thus we can evaluate this invariant variation of the Ricci tensor in any coordinate system by replacing the derivatives with covariant ones getting

$$g^{ik}\delta R_{ik} = \left(g^{ik}\,\delta\Gamma^n{}_{in} - g^{in}\,\delta\Gamma^k{}_{in}\right)_{;k} \tag{13.223}$$

which we recognize as the covariant divergence (13.180) of a contravariant vector. The last term in the first-order change (13.221) in the action density is therefore a surface term whose variation vanishes for tiny local changes δg^{ik} of the metric

$$\sqrt{g}\,g^{ik}\delta R_{ik} = \left[\sqrt{g}\left(g^{ik}\,\delta\Gamma^n{}_{in} - g^{in}\,\delta\Gamma^k{}_{in}\right)\right]_{,k}. \tag{13.224}$$

Hence the variation of S_{EH} is simply

$$\delta S_{EH} = \frac{c^3}{16\pi G}\int\left(R_{ik} - \frac{1}{2}\,g_{ik}R\right)\sqrt{g}\;\delta g^{ik}\,d^4x. \tag{13.225}$$

The principle of least action $\delta S_{EH} = 0$ now gives us **Einstein's equations for empty space**:

$$R_{ik} - \frac{1}{2}\,g_{ik}\,R = 0. \tag{13.226}$$

The tensor $G_{ik} = R_{ik} - \frac{1}{2}g_{ik}\,R$ is Einstein's tensor.

Taking the trace of Einstein's equations (13.226), we find that the scalar curvature R and the Ricci tensor R_{ik} are zero in empty space:

$$R = 0 \quad \text{and} \quad R_{ik} = 0. \tag{13.227}$$

The **energy–momentum tensor** T_{ik} is the source of the gravitational field. It is defined so that the change in the action of the matter fields due to a tiny local change $\delta g^{ik}(x)$ in the metric is

$$\delta S_m = -\frac{1}{2c}\int T_{ik}\sqrt{g}\;\delta g^{ik}\,d^4x = \frac{1}{2c}\int T^{ik}\sqrt{g}\;\delta g_{ik}\,d^4x \tag{13.228}$$

in which the identity $\delta g^{ik} = -\,g^{ij}g^{\ell k}\delta g_{j\ell}$ explains the sign change. Now the principle of least action $\delta S = \delta S_{EH} + \delta S_m = 0$ yields Einstein's equations in the presence of matter and energy

$$R_{ik} - \frac{1}{2}\,g_{ik}\,R = \frac{8\pi G}{c^4}T_{ik}. \tag{13.229}$$

Taking the trace of both sides, we get

$$R = -\frac{8\pi G}{c^4}T \quad \text{and} \quad R_{ik} = \frac{8\pi G}{c^4}\left(T_{ik} - \frac{T}{2}g_{ik}\right). \tag{13.230}$$

13.36 Energy–Momentum Tensor

The action S_m of the matter fields ϕ_i is a scalar that is invariant under general coordinate transformations. In particular, a tiny local general coordinate transformation $x'^a = x^a + \epsilon^a(x)$ leaves S_m invariant

$$0 = \delta S_m = \int \delta\left(L(\phi_i(x))\sqrt{g(x)}\right) d^4x. \tag{13.231}$$

The vanishing change $\delta S_m = \delta S_{m\phi} + \delta S_{mg}$ has a part $\delta S_{m\phi}$ due to the changes in the fields $\delta\phi_i(x)$ and a part δS_{mg} due to the change in the metric δg^{ik}. The principle of stationary action tells us that the change $\delta S_{m\phi}$ is zero as long as the fields obey the classical equations of motion. Combining the result $\delta S_{m\phi} = 0$ with the definition (13.228) of the energy–momentum tensor, we find

$$0 = \delta S_m = \delta S_{m\phi} + \delta S_{mg} = \delta S_{mg} = \frac{1}{2c}\int T^{ik}\sqrt{g}\,\delta g_{ik}\,d^4x. \tag{13.232}$$

Since $x^a = x'^a - \epsilon^a(x)$, the transformation law for rank-2 covariant tensors (13.35) gives us

$$\begin{aligned}
\delta g_{ik} &= g'_{ik}(x) - g_{ik}(x) = g'_{ik}(x') - g_{ik}(x) - (g'_{ik}(x') - g'_{ik}(x)) \\
&= (\delta_i^a - \epsilon^a_{,i})(\delta_k^b - \epsilon^b_{,k})g_{ab} - g_{ik} - \epsilon^c g_{ik,c} \\
&= -\,g_{ib}\,\epsilon^b_{,k} - g_{ak}\,\epsilon^a_{,i} - \epsilon^c\,g_{ik,c} \\
&= -\,g_{ib}(g^{bc}\,\epsilon_c)_{,k} - g_{ak}(g^{ac}\,\epsilon_c)_{,i} - \epsilon^c\,g_{ik,c} \\
&= -\,\epsilon_{i,k} - \epsilon_{k,i} - \epsilon_c\,g_{ib}\,g^{bc}_{,k} - \epsilon_c\,g_{ak}\,g^{ac}_{,i} - \epsilon^c\,g_{ik,c}.
\end{aligned} \tag{13.233}$$

Now using the identity $\partial_i(g^{ik}\,g_{k\ell}) = 0$, the definition (13.68) of the covariant derivative of a covariant vector, and the formula (13.87) for the connection in terms of the metric, we find to lowest order in the change $\epsilon^a(x)$ in x^a that the change in the metric is

$$\begin{aligned}
\delta g_{ik} &= -\,\epsilon_{i,k} - \epsilon_{k,i} + \epsilon_c\,g^{bc}\,g_{ib,k} + \epsilon_c\,g^{ac}\,g_{ak,i} - \epsilon^c\,g_{ik,c} \\
&= -\,\epsilon_{i,k} - \epsilon_{k,i} + \epsilon_c\,g^{ac}\left(g_{ia,k} + g_{ak,i} - g_{ik,a}\right) \\
&= -\,\epsilon_{i,k} - \epsilon_{k,i} + \epsilon_c\,\Gamma^c{}_{ik} + \epsilon_c\,\Gamma^c{}_{ki} = -\,\epsilon_{i;k} - \epsilon_{k;i}.
\end{aligned} \tag{13.234}$$

Combining this result (13.234) with the vanishing (13.232) of the change δS_{mg}, we have

$$0 = \int T^{ik}\sqrt{g}\,(\epsilon_{i;k} + \epsilon_{k;i})\,d^4x. \tag{13.235}$$

Since the energy–momentum tensor is symmetric, we may combine the two terms, integrate by parts, divide by $\sqrt{g}$, and so find that the covariant divergence of the energy–momentum tensor is zero

$$0 = T^{ik}_{\;;k} = T^{ik}_{\;,k} + \Gamma^k_{\;ak}\, T^{ia} + \Gamma^i_{\;ak}\, T^{ak} = \frac{1}{\sqrt{g}}(\sqrt{g}T^{ik})_{,k} + \Gamma^i_{\;ak}\, T^{ak} \tag{13.236}$$

when the fields obey their equations of motion. In a given inertial frame, only the total energy, momentum, and angular momentum of both the matter and the gravitational field are conserved.

13.37 Perfect Fluids

In many cosmological models, the energy–momentum tensor is assumed to be that of a **perfect fluid**, which is isotropic in its rest frame, does not conduct heat, and has zero viscosity. The **energy–momentum** tensor T_{ij} of a perfect fluid moving with 4-velocity u^i (12.20) is

$$T_{ij} = p\, g_{ij} + (\frac{p}{c^2} + \rho)\, u_i\, u_j \tag{13.237}$$

in which p and ρ are the pressure and mass density of the fluid in its *rest frame* and g_{ij} is the spacetime metric. Einstein's equations (13.229) then are

$$R_{ik} - \frac{1}{2}\, g_{ik}\, R = \frac{8\pi G}{c^4} T_{ik} = \frac{8\pi G}{c^4}\left[p\, g_{ij} + (\frac{p}{c^2} + \rho)\, u_i\, u_j\right]. \tag{13.238}$$

An important special case is the energy-momentum tensor due to a nonzero value of the energy density of the vacuum. In this case $p = -c^2\rho$ and the energy–momentum tensor is

$$T_{ij} = p\, g_{ij} = -c^2 \rho\, g_{ij} \tag{13.239}$$

in which $T_{00} = c^2\rho$ is the energy density of the ground state of the theory. This energy density $c^2\rho$ is a plausible candidate for the **dark-energy** density. It is equivalent to a **cosmological constant** $\Lambda = 8\pi G\rho$.

On small scales, such as that of our solar system, one may neglect matter and dark energy. So in empty space and on small scales, the energy–momentum tensor vanishes $T_{ij} = 0$ along with its trace and the scalar curvature $T = 0 = R$, and Einstein's equations (13.230) are

$$R_{ij} = 0. \tag{13.240}$$

13.38 Gravitational Waves

The nonlinear properties of Einstein's equations (13.229– 13.230) are important on large scales of distance (Sections 13.42 & 13.43) and near great masses (Sections 13.39 & 13.40). But throughout most of the mature universe, it is helpful to linearize them by writing the metric as the metric η_{ik} of empty, flat spacetime (12.3) plus a tiny deviation h_{ik}

$$g_{ik} = \eta_{ik} + h_{ik}. \tag{13.241}$$

To first order in h_{ik}, the affine connection (13.87) is

$$\Gamma^k{}_{i\ell} = \frac{1}{2} g^{kj} \left(g_{ji,\ell} + g_{j\ell,i} - g_{i\ell,j} \right) = \frac{1}{2} \eta^{kj} \left(h_{ji,\ell} + h_{j\ell,i} - h_{i\ell,j} \right) \tag{13.242}$$

and the Ricci tensor (13.112) is the contraction

$$R_{i\ell} = R^k{}_{ik\ell} = [\partial_k + \Gamma_k, \partial_\ell + \Gamma_\ell]^k{}_i = \Gamma^k{}_{\ell i,k} - \Gamma^k{}_{ki,\ell}. \tag{13.243}$$

Since $\Gamma^k{}_{i\ell} = \Gamma^k{}_{\ell i}$ and $h_{ik} = h_{ki}$, the linearized Ricci tensor is

$$\begin{aligned} R_{i\ell} &= \frac{1}{2} \eta^{kj} \left(h_{ji,\ell} + h_{j\ell,i} - h_{i\ell,j} \right)_{,k} - \frac{1}{2} \eta^{kj} \left(h_{ji,k} + h_{jk,i} - h_{ik,j} \right)_{,\ell} \\ &= \frac{1}{2} \left(h^k{}_{\ell,ik} + h_{ik,\ell}{}^{,k} - h_{i\ell,k}{}^{,k} - h^k{}_{k,i\ell} \right). \end{aligned} \tag{13.244}$$

We can simplify Einstein's equations (13.230) in empty space $R_{i\ell} = 0$ by using coordinates in which h_{ik} obeys (Exercise 13.17) de Donder's harmonic gauge condition $h^i{}_{k,i} = \frac{1}{2}(\eta^{j\ell} h_{j\ell})_{,k} \equiv \frac{1}{2} h_{,k}$. In this gauge, the linearized Einstein equations in empty space are

$$R_{i\ell} = -\frac{1}{2} h_{i\ell,k}{}^{,k} = 0 \quad \text{or} \quad \left(c^2 \nabla^2 - \partial_0^2 \right) h_{i\ell} = 0. \tag{13.245}$$

On 14 September 2015, the LIGO collaboration detected the merger of two black holes of 29 and 36 solar masses which liberated $3M_\odot c^2$ of energy. They have set an upper limit of $c^2 m_g < 2 \times 10^{-25}$ eV on the mass of the graviton, have detected 10 black-hole mergers, and are expected to detect a new black-hole merger every week in late 2019.

13.39 Schwarzschild's Solution

In 1916, Schwarzschild solved Einstein's field equations (13.240) in empty space $R_{ij} = 0$ outside a static mass M and found as the metric

$$ds^2 = -\left(1 - \frac{2MG}{c^2 r} \right) c^2 dt^2 + \left(1 - \frac{2MG}{c^2 r} \right)^{-1} dr^2 + r^2 \, d\Omega^2 \tag{13.246}$$

in which $d\Omega^2 = d\theta^2 + \sin^2\theta \, d\phi^2$.

The Mathematica scripts GREAT.m and Schwarzschild.nb give for the Ricci tensor and the scalar curvature $R_{ik} = 0$ and $R = 0$, which show that the metric obeys Einstein's equations and that the singularity in

$$g_{rr} = \left(1 - \frac{2MG}{c^2 r}\right)^{-1} \tag{13.247}$$

at the Schwarzschild radius $r_s = 2MG/c^2$ is an artifact of the coordinates. Schwarzschild.nb also computes the affine connections.

The Schwarzschild radius of the Sun $r_s = 2M_\odot G/c^2 = 2.95$ km is far less than the Sun's radius $r_\odot = 6.955 \times 10^5$ km, beyond which his metric applies (Karl Schwarzschild, 1873–1916).

13.40 Black Holes

Suppose an uncharged, spherically symmetric star of mass M has collapsed within a sphere of radius r_b less than its Schwarzschild radius or horizon $r_h = r_s = 2MG/c^2$. Then for $r > r_b$, the Schwarzschild metric (13.246) is correct. The time dt measured on a clock outside the gravitational field is related to the proper time $d\tau$ on a clock fixed at $r \geq 2MG/c^2$ by (13.216)

$$dt = d\tau/\sqrt{-g_{00}} = d\tau/\sqrt{1 - \frac{2MG}{c^2 r}}. \tag{13.248}$$

The time dt measured away from the star becomes infinite as r approaches the horizon $r_h = r_s = 2MG/c^2$. To outside observers, a clock at the horizon r_h seems frozen in time.

Due to the gravitational redshift (13.248), light of frequency ν_p emitted at $r \geq 2MG/c^2$ will have frequency ν

$$\nu = \nu_p \sqrt{-g_{00}} = \nu_p \sqrt{1 - \frac{2MG}{c^2 r}} \tag{13.249}$$

when observed at great distances. Light coming from the surface at $r_s = 2MG/c^2$ is redshifted to zero frequency $\nu = 0$. The star is black. It is a black hole with a horizon at its Schwarzschild radius $r_h = r_s = 2MG/c^2$, although there is no singularity there. If the radius of the Sun were less than its Schwarzschild radius of 2.95 km, then the Sun would be a black hole. The radius of the Sun is 6.955×10^5 km.

Black holes are not black. They often are surrounded by bright hot accretion disks, and Stephen Hawking (1942–2018) showed (Hawking, 1975) that the intense gravitational field of a black hole of mass M radiates at a temperature

$$T = \frac{\hbar c^3}{8\pi kGM} = \frac{\hbar c}{4\pi k r_h} = \frac{\hbar g}{2\pi k c} \tag{13.250}$$

in which $k = 8.617 \times 10^{-5}$ eV K^{-1} is Boltzmann's constant, $\hbar$ is Planck's constant $h = 6.626 \times 10^{-34}$ J s divided by 2π, $\hbar = h/(2\pi)$, and $g = GM/r_h^2$ is the gravitational acceleration at $r = r_h$. More generally, a detector in vacuum subject to a uniform acceleration a (in its instantaneous rest frame) sees a temperature $T = \hbar a/(2\pi kc)$ (Alsing and Milonni, 2004).

In a region of empty space where the pressure p and the chemical potentials μ_j all vanish, the change (7.111) in the internal energy $U = c^2 M$ of a black hole of mass M is $c^2 dM = T dS$ where S is its entropy. So the change dS in the entropy of a black hole of mass $M = c^2 r_h/(2G)$ and temperature $T = \hbar c/(4\pi k r_h)$ (13.250) is

$$dS = \frac{c^2}{T} dM = \frac{4\pi c^2 k r_h}{(\hbar c)} dM = \frac{4\pi c k r_h}{\hbar} \frac{c^2}{2G} dr_h = \frac{\pi c^3 k}{G\hbar} 2 r_h dr_h. \tag{13.251}$$

Integrating, we get a formula for the entropy of a black hole in terms of its area (Bekenstein, 1973; Hawking, 1975)

$$S = \frac{\pi c^3 k}{G\hbar} r_h^2 = \frac{c^3 k}{\hbar G} \frac{A}{4} \tag{13.252}$$

where $A = 4\pi r_h^2$ is the area of the horizon of the black hole.

A black hole is entirely converted into radiation after a time

$$t = \frac{5120\,\pi\, G^2}{\hbar\, c^4} M^3 \tag{13.253}$$

proportional to the cube of its mass M.

13.41 Rotating Black Holes

A half-century after Einstein invented general relativity, Roy Kerr invented the metric for a mass M rotating with angular momentum $J = GMa/c$. Two years later, Newman and others generalized the Kerr metric to one of charge Q. In Boyer–Lindquist coordinates, its line element is

$$\begin{aligned} ds^2 = & -\frac{\Delta}{\rho^2}\left(dt - a\sin^2\theta d\phi\right)^2 \\ & + \frac{\sin^2\theta}{\rho^2}\left((r^2+a^2)d\phi - a\,dt\right)^2 + \frac{\rho^2}{\Delta}dr^2 + \rho^2 d\theta^2 \end{aligned} \tag{13.254}$$

where $\rho^2 = r^2 + a^2\cos^2\theta$ and $\Delta = r^2 + a^2 - 2GMr/c^2 + Q^2$. The Mathematica script Kerr_black_hole.nb shows that the Kerr–Newman metric for the uncharged case, $Q = 0$, has $R_{ik} = 0$ and $R = 0$ and so is a solution of Einstein's equations in empty space (13.230) with zero scalar curvature.

A rotating mass drags nearby masses along with it. The daily rotation of the Earth moves satellites to the East by tens of meters per year. The **frame dragging** of extremal black holes with $J \lesssim GM^2/c$ approaches the speed of light (Ghosh et al., 2018) (Roy Kerr, 1934–).

13.42 Spatially Symmetric Spacetimes

Einstein's equations (13.230) are second-order, nonlinear partial differential equations for 10 unknown functions $g_{ik}(x)$ in terms of the energy–momentum tensor $T_{ik}(x)$ throughout the universe, which of course we don't know. The problem is not quite hopeless, however. The ability to choose arbitrary coordinates, the appeal to symmetry, and the choice of a reasonable form for T_{ik} all help.

Astrophysical observations tell us that the universe extends at least 46 billion light years in all directions; that it is **flat** or very nearly flat; and that the cosmic microwave background (CMB) radiation is isotropic to one part in 10^5 apart from a Doppler shift due the motion of the Sun at 370 km/s towards the constellation Leo. These microwave photons have been moving freely since the universe became cool enough for hydrogen atoms to be stable. Observations of clusters of galaxies reveal an expanding universe that is homogeneous on suitably large scales of distance. Thus as far as we know, the universe is **homogeneous** and **isotropic** in space, but not in time.

There are only three **maximally symmetric** 3-dimensional spaces: euclidian space $\mathbb{E}^3$, the sphere S^3 (Example 13.16), and the hyperboloid H^3 (Example 13.17). Their line elements may be written in terms of a distance L as

$$ds^2 = \frac{dr^2}{1 - k\,r^2/L^2} + r^2 d\Omega^2 \tag{13.255}$$

in which $k = 1$ for the sphere, $k = 0$ for euclidian space, and $k = -1$ for the hyperboloid. The **Friedmann–Lemaître–Robinson–Walker** (FLRW) cosmologies add to these spatially symmetric line elements a **dimensionless scale factor** $a(t)$ that describes the expansion (or contraction) of space

$$ds^2 = -\,c^2 dt^2 + a^2(t)\left(\frac{dr^2}{1 - k\,r^2/L^2} + r^2\,d\theta^2 + r^2\,\sin^2\theta\,d\phi^2\right). \tag{13.256}$$

The FLRW metric is

$$g_{ik}(t,r,\theta,\phi) = \begin{pmatrix} -\,c^2 & 0 & 0 & 0 \\ 0 & a^2/(1 - k\,r^2/L^2) & 0 & 0 \\ 0 & 0 & a^2\,r^2 & 0 \\ 0 & 0 & 0 & a^2\,r^2\,\sin^2\theta \end{pmatrix}. \tag{13.257}$$

The constant k determines whether the spatial universe is **open** $k = -1$, **flat** $k = 0$, or **closed** $k = 1$. The coordinates $x^0, x^1, x^2, x^3 \equiv t, r, \theta, \phi$ are **comoving** in that a detector at rest at r, θ, ϕ records the CMB as isotropic with no Doppler shift.

The metric (13.257) is diagonal; its inverse g^{ij} also is diagonal

$$g^{ik} = \begin{pmatrix} -c^{-2} & 0 & 0 & 0 \\ 0 & (1-k\,r^2/L^2)/a^2 & 0 & 0 \\ 0 & 0 & (a\,r)^{-2} & 0 \\ 0 & 0 & 0 & (a\,r\,\sin\theta)^{-2} \end{pmatrix}. \tag{13.258}$$

One may use the formula (13.87) to compute the affine connection in terms of the metric and its inverse as $\Gamma^k{}_{i\ell} = \frac{1}{2}g^{kj}(g_{ji,\ell} + g_{j\ell,i} - g_{\ell i,j})$. It usually is easier, however, to use the action principle (13.192) to derive the geodesic equation directly and then to read its expressions for the $\Gamma^i{}_{jk}$'s. So we require that the integral

$$0 = \delta \int \left(-c^2 t'^2 + \frac{a^2\,r'^2}{1-kr^2/L^2} + a^2\,r^2\,\theta'^2 + a^2\,r^2\,\sin^2\theta\,\phi'^2 \right) d\lambda, \tag{13.259}$$

in which a prime means derivative with respect to λ, be stationary with respect to the tiny variations $\delta t(\lambda)$, $\delta r(\lambda)$, $\delta\theta(\lambda)$, and $\delta\phi(\lambda)$. By varying $t(\lambda)$, we get the equation

$$0 = t'' + \frac{a\dot{a}}{c^2}\left(\frac{r'^2}{1-kr^2/L^2} + r^2\,\theta'^2 + r^2\,\sin^2\theta\,\phi'^2 \right) = t'' + \Gamma^t{}_{i\ell}x'^i x'^\ell \tag{13.260}$$

which tells us that the nonzero $\Gamma^t{}_{jk}$'s are

$$\Gamma^t{}_{rr} = \frac{a\dot{a}}{c^2(1-kr^2/L^2)}, \quad \Gamma^t{}_{\theta\theta} = \frac{a\dot{a}\,r^2}{c^2}, \quad \text{and} \quad \Gamma^t{}_{\phi\phi} = \frac{a\dot{a}\,r^2\sin^2\theta}{c^2}. \tag{13.261}$$

By varying $r(\lambda)$ we get (with more effort)

$$0 = r'' + \frac{rr'^2k/L^2}{(1-kr^2/L^2)} + 2\frac{\dot{a}\,t'r'}{a} - r\,(1-\frac{kr^2}{L^2})(\theta'^2 + \sin^2\theta\,\phi'^2). \tag{13.262}$$

So we find that $\Gamma^r{}_{tr} = \dot{a}/a$,

$$\Gamma^r{}_{rr} = \frac{kr}{L^2-kr^2}, \quad \Gamma^r{}_{\theta\theta} = -r + \frac{kr^3}{L^2}, \quad \text{and} \quad \Gamma^r{}_{\phi\phi} = \sin^2\theta\,\Gamma^r{}_{\theta\theta}. \tag{13.263}$$

Varying $\theta(\lambda)$ gives

$$\begin{gathered} 0 = \theta'' + 2\frac{\dot{a}}{a}t'\theta' + \frac{2}{r}\theta'r' - \sin\theta\cos\theta\phi'^2 \quad \text{and} \\ \Gamma^\theta{}_{t\theta} = \frac{\dot{a}}{a}, \quad \Gamma^\theta{}_{r\theta} = \frac{1}{r}, \quad \text{and} \quad \Gamma^\theta{}_{\phi\phi} = -\sin\theta\cos\theta. \end{gathered} \tag{13.264}$$

Finally, varying $\phi(\lambda)$ gives

$$\begin{gathered} 0 = \phi'' + 2\frac{\dot{a}}{a}t'\phi' + 2\frac{r'\phi'}{r} + 2\cot\theta\,\theta'\phi' \quad \text{and} \\ \Gamma^\phi{}_{t\phi} = \frac{\dot{a}}{a}, \quad \Gamma^\phi{}_{r\phi} = \frac{1}{r}, \quad \text{and} \quad \Gamma^\phi{}_{\theta\phi} = \cot\theta. \end{gathered} \tag{13.265}$$

Other Γ's are either zero or related by the symmetry $\Gamma^k{}_{i\ell} = \Gamma^k{}_{\ell i}$.

Our formulas for the Ricci (13.112) and curvature (13.103) tensors give

$$R_{00} = R^k{}_{0k0} = [D_k, D_0]^k{}_0 = [\partial_k + \Gamma_k, \partial_0 + \Gamma_0]^k{}_0. \tag{13.266}$$

Because $[D_0, D_0] = 0$, we need only compute $[D_1, D_0]^1{}_0$, $[D_2, D_0]^2{}_0$, and $[D_3, D_0]^3{}_0$. Using the formulas (13.261–13.265) for the Γ's and keeping in mind (13.102) that the element of row r and column c of the ℓth gamma matrix is $\Gamma^r{}_{\ell c}$, we find

$$\begin{aligned}
[D_1, D_0]^1{}_0 &= \Gamma^1{}_{00,1} - \Gamma^1{}_{10,0} + \Gamma^1{}_{1j}\Gamma^j{}_{00} - \Gamma^1{}_{0j}\Gamma^j{}_{10} = -(\dot a/a)_{,0} - (\dot a/a)^2 \\
[D_2, D_0]^2{}_0 &= \Gamma^2{}_{00,2} - \Gamma^2{}_{20,0} + \Gamma^2{}_{2j}\Gamma^j{}_{00} - \Gamma^2{}_{0j}\Gamma^j{}_{20} = -(\dot a/a)_{,0} - (\dot a/a)^2 \\
[D_3, D_0]^3{}_0 &= \Gamma^3{}_{00,3} - \Gamma^3{}_{30,0} + \Gamma^3{}_{3j}\Gamma^j{}_{00} - \Gamma^3{}_{0j}\Gamma^j{}_{30} = -(\dot a/a)_{,0} - (\dot a/a)^2 \\
R_{tt} = R_{00} &= [D_k, D_0]^k{}_0 = -3(\dot a/a)_{,0} - 3(\dot a/a)^3 = -3\ddot a/a.
\end{aligned} \tag{13.267}$$

Thus for $R_{rr} = R_{11} = R^k{}_{1k1} = [D_k, D_1]^k{}_1 = [\partial_k + \Gamma_k, \partial_1 + \Gamma_1]^k{}_1$, we get

$$R_{rr} = [D_k, D_1]^k{}_1 = \frac{a\ddot a + 2\dot a^2 + 2kc^2/L^2}{c^2(1 - kr^2/L^2)} \tag{13.268}$$

(Exercise 13.21), and for $R_{22} = R_{\theta\theta}$ and $R_{33} = R_{\phi\phi}$ we find

$$R_{\theta\theta} = [(a\ddot a + 2\dot a^2 + 2kc^2/L^2)r^2]/c^2 \quad \text{and} \quad R_{\phi\phi} = \sin^2\theta \, R_{\theta\theta} \tag{13.269}$$

(Exercises 13.22 and 13.23). And so the scalar curvature $R = g^{ab}R_{ba}$ is

$$\begin{aligned}
R = g^{ab}R_{ba} &= -\frac{R_{00}}{c^2} + \frac{(1 - kr^2/L^2)R_{11}}{a^2} + \frac{R_{22}}{a^2r^2} + \frac{R_{33}}{a^2r^2\sin^2\theta} \\
&= 6\,\frac{a\ddot a + \dot a^2 + kc^2/L^2}{c^2a^2}.
\end{aligned} \tag{13.270}$$

It is, of course, quicker to use the Mathematica script FLRW.nb.

13.43 Friedmann–Lemaître–Robinson–Walker Cosmologies

The energy–momentum tensor (13.237) of a perfect fluid moving at 4-velocity u_i is $T_{ik} = pg_{ik} + (p/c^2 + \rho)u_iu_k$ where p and ρ are the pressure and mass density of the fluid in its rest frame. In the comoving coordinates of the FLRW metric (13.257), the 4-velocity (12.20) is $u^i = (1, 0, 0, 0)$, and the energy–momentum tensor (13.237) is

$$T_{ij} = \begin{pmatrix} -c^2\rho\, g_{00} & 0 & 0 & 0 \\ 0 & p\, g_{11} & 0 & 0 \\ 0 & 0 & p\, g_{22} & 0 \\ 0 & 0 & 0 & p\, g_{33} \end{pmatrix}. \tag{13.271}$$

Its trace is

$$T = g^{ij}\, T_{ij} = -c^2 \rho + 3p. \tag{13.272}$$

Thus using our formulas (13.257) for $g_{00} = -c^2$, (13.267) for $R_{00} = -3\ddot{a}/a$, (13.271) for T_{ij}, and (13.272) for T, we can write the 00 Einstein equation (13.230) as the second-order equation

$$\frac{\ddot{a}}{a} = -\frac{4\pi G}{3}\left(\rho + \frac{3p}{c^2}\right) \tag{13.273}$$

which is nonlinear because ρ and p depend upon a. The sum $c^2\rho + 3p$ determines the acceleration $\ddot{a}$ of the scale factor $a(t)$; when it is negative, it accelerates the expansion. If we combine Einstein's formula for the scalar curvature $R = -8\pi G\, T/c^4$ (13.230) with the FLRW formulas for R (13.270) and for the trace T (13.272) of the energy–momentum tensor, we get

$$\frac{\ddot{a}}{a} + \left(\frac{\dot{a}}{a}\right)^2 + \frac{c^2 k}{L^2 a^2} = \frac{4\pi G}{3}\left(\rho - \frac{3p}{c^2}\right). \tag{13.274}$$

Using the 00-equation (13.273) to eliminate the second derivative $\ddot{a}$, we find

$$\left(\frac{\dot{a}}{a}\right)^2 = \frac{8\pi G}{3}\rho - \frac{c^2 k}{L^2 a^2} \tag{13.275}$$

which is a first-order nonlinear equation. Both this and the second-order equation (13.273) are known as the **Friedmann equations**.

The left-hand side of the first-order Friedmann equation (13.275) is the square of the **Hubble rate**

$$H = \frac{\dot{a}}{a} \tag{13.276}$$

which is an inverse time or a frequency. Its present value H_0 is the **Hubble constant**.

In terms of H, the first-order Friedmann equation (13.275) is

$$H^2 = \frac{8\pi G}{3}\rho - \frac{c^2 k}{L^2 a^2}. \tag{13.277}$$

An absolutely flat universe has $k = 0$, and therefore its density must be

$$\rho_c = \frac{3H^2}{8\pi G} \tag{13.278}$$

which is the **critical mass density**.

13.44 Density and Pressure

The 0th energy-momentum conservation law (13.236) is

$$0 = T^{0a}_{\ ;a} = \partial_a T^{0a} + \Gamma^a{}_{ca} T^{0c} + \Gamma^0{}_{ca} T^{ca}. \tag{13.279}$$

For a perfect fluid of 4-velocity u^a, the energy–momentum tensor (13.271) is $T^{ik} = (\rho + p/c^2)u^i u^k + p g^{ik}$ in which ρ and p are the mass density and pressure of the fluid in its rest frame. The comoving frame of the Friedmann–Lemaître–Robinson–Walker metric (13.257) is the rest frame of the fluid. In these coordinates, the 4-velocity u^a is $(1, 0, 0, 0)$, and the energy–momentum tensor is diagonal with $T^{00} = \rho$ and $T^{jj} = p g^{jj}$ for $j = 1, 2, 3$. Our connection formulas (13.261) tell us that $\Gamma^0{}_{00} = 0$, that $\Gamma^0{}_{jj} = \dot{a} g_{jj}/(c^2 a)$, and that $\Gamma^j{}_{0j} = 3\dot{a}/a$. Thus the conservation law (13.279) becomes for spatial j

$$\begin{aligned} 0 &= \partial_0 T^{00} + \Gamma^j{}_{0j} T^{00} + \Gamma^0{}_{jj} T^{jj} \\ &= \dot{\rho} + 3\frac{\dot{a}}{a}\rho + \sum_{j=1}^{3} \frac{\dot{a}\, g_{jj}}{c^2 a}\, p\, g^{jj} = \dot{\rho} + 3\frac{\dot{a}}{a}\left(\rho + \frac{p}{c^2}\right). \end{aligned} \tag{13.280}$$

Thus

$$\dot{\rho} = -\frac{3\dot{a}}{a}(\rho + \frac{p}{c^2}), \quad \text{and so} \quad \frac{d\rho}{da} = -\frac{3}{a}\left(\rho + \frac{p}{c^2}\right). \tag{13.281}$$

The energy density ρ is composed of fractions ρ_i each contributing its own partial pressure p_i according to its own **equation of state**

$$p_i = c^2 w_i \rho_i \tag{13.282}$$

in which w_i is a constant. The rate of change (13.282) of the density ρ_i is then

$$\frac{d\rho_i}{da} = -\frac{3}{a}\,(1 + w_i)\,\rho_i. \tag{13.283}$$

In terms of the present density ρ_{i0} and scale factor a_0, the solution is

$$\rho_i = \rho_{i0}\left(\frac{a_0}{a}\right)^{3(1+w_i)}. \tag{13.284}$$

There are three important kinds of density. The dark-energy density ρ_Λ is assumed to be like a cosmological constant Λ or like the energy density of the vacuum, so it is independent of the scale factor a and has $w_\Lambda = -1$.

A universe composed only of **dust** or **nonrelativistic collisionless matter** has no pressure. Thus $p = w\rho = 0$ with $\rho \neq 0$, and so $w = 0$. So the matter density falls inversely with the volume

$$\rho_m = \rho_{m0}\left(\frac{a_0}{a}\right)^3. \tag{13.285}$$

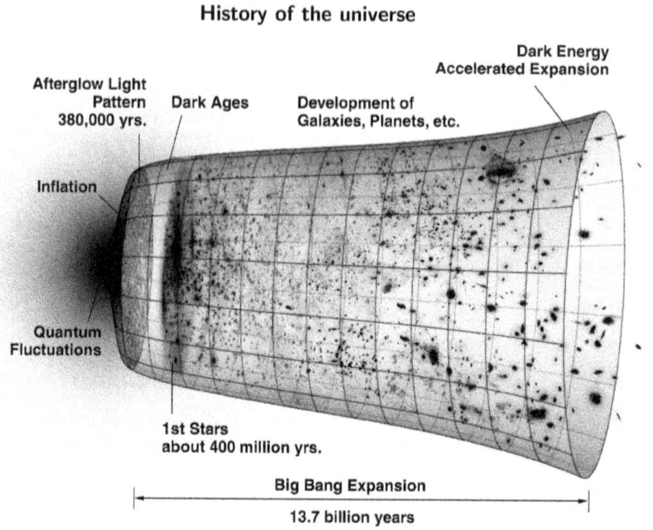

Figure 13.1 NASA/WMAP Science Team's timeline of the known universe. (Source: NASA/WMAP Science Team, https://map.gsfc.nasa.gov/media/060915/index.html)

The density of radiation ρ_r has $w_r = 1/3$ because wavelengths scale with the scale factor, and so there's an extra factor of a

$$\rho_r = \rho_{r0} \left(\frac{a_0}{a}\right)^4. \tag{13.286}$$

The total density ρ varies with a as

$$\rho = \rho_\Lambda + \rho_{m0} \left(\frac{a_0}{a}\right)^3 + \rho_{r0} \left(\frac{a_0}{a}\right)^4. \tag{13.287}$$

This mass density ρ, the Friedmann equations (13.273 and 13.275), and the physics of the standard model have caused our universe to evolve as in Fig. 13.1 over the past 14 billion years.

13.45 How the Scale Factor Evolves with Time

The first-order Friedmann equation (13.275) expresses the square of the instantaneous Hubble rate $H = \dot{a}/a$ in terms of the density ρ and the scale factor $a(t)$

$$H^2 = \left(\frac{\dot{a}}{a}\right)^2 = \frac{8\pi G}{3}\rho - \frac{c^2 k}{L^2 a^2} \tag{13.288}$$

in which $k = \pm 1$ or 0. The critical density $\rho_c = 3H^2/(8\pi G)$ (13.278) is the one that satisfies this equation for a flat ($k = 0$) universe. Its present value is

Table 13.1 *Cosmological parameters of the Planck collaboration*

H_0 (km/(s Mpc)	Ω_Λ	Ω_m	Ω_k
67.66 ± 0.42	0.6889 ± 0.0056	0.3111 ± 0.0056	0.0007 ± 0.0037

$\rho_{c0} = 3H_0^2/(8\pi G) = 8.599 \times 10^{-27}$ kg m^{-3}. Dividing Friedmann's equation by the square of the present Hubble rate H_0^2, we get

$$\frac{H^2}{H_0^2} = \frac{1}{H_0^2}\left(\frac{\dot{a}}{a}\right)^2 = \frac{1}{H_0^2}\left(\frac{8\pi G}{3}\rho - \frac{c^2 k}{a^2 L^2}\right) = \frac{\rho}{\rho_{0c}} - \frac{c^2 k}{a^2 H_0^2 L^2} \tag{13.289}$$

in which ρ is the total density (13.287)

$$\begin{aligned} \frac{H^2}{H_0^2} &= \frac{\rho_\Lambda}{\rho_{c0}} + \frac{\rho_m}{\rho_{c0}} + \frac{\rho_r}{\rho_{c0}} - \frac{c^2 k}{a^2 H_0^2 L^2} \\ &= \frac{\rho_\Lambda}{\rho_{c0}} + \frac{\rho_{m0}}{\rho_{c0}}\frac{a_0^3}{a^3} + \frac{\rho_{r0}}{\rho_{c0}}\frac{a_0^4}{a^4} - \frac{c^2 k}{a_0^2 H_0^2 L^2}\frac{a_0^2}{a^2}. \end{aligned} \tag{13.290}$$

The Planck collaboration use a model in which the energy density of the universe is due to radiation, matter, and a cosmological constant Λ. Only about 18.79% of the matter in their model is composed of baryons, $\Omega_b = 0.05845 \pm 0.0003$. Most of the matter is transparent and is called **dark matter**. They assume the dark matter is composed of particles that have masses in excess of a keV so that they are heavy enough to have been nonrelativistic or "cold" when the universe was about a year old (Peebles, 1982). The energy density of the cosmological constant Λ is known as **dark energy**. The Planck collaboration use this Λ-cold-dark-matter (ΛCDM) model and their CMB data to estimate the Hubble constant as $H_0 = 67.66$km/(s Mpc) $= 2.1927 \times 10^{-18}\,\mathrm{s}^{-1}$ and the density ratios $\Omega_\Lambda = \rho_\Lambda/\rho_{c0}$, $\Omega_m = \rho_{m0}/\rho_{c0}$, and $\Omega_k \equiv -c^2k/(a_0 H_0 L)^2$ as listed in the table (13.1) (Aghanim et al., 2018). The Riess group use the Gaia observatory to calibrate Cepheid stars and type Ia supernovas as standard candles for measuring distances to remote galaxies. The distances and redshifts of these galaxies give the Hubble constant as $H_0 = 73.48 \pm 1.66$ (Riess et al., 2018). As this book goes to press, the 9% discrepancy between the Planck and Riess H_0's is unexplained.

To estimate the ratio $\Omega_r = \rho_{r0}/\rho_{c0}$ of densities, one may use the present temperature $T_0 = 2.7255 \pm 0.0006$ K (Fixsen, 2009) of the CMB radiation and the formula (5.110) for the energy density of photons

$$\rho_\gamma = \frac{8\pi^5\,(k_B T_0)^4}{15 h^3 c^5} = 4.6451 \times 10^{-31}\ \mathrm{kg\,m^{-3}}. \tag{13.291}$$

Adding in three kinds of neutrinos and antineutrinos at $T_{0\nu} = (4/11)^{1/3}\, T_0$, we get for the present density of massless and nearly massless particles (Weinberg, 2010, section 2.1)

$$\rho_{r0} = \left[1 + 3\left(\frac{7}{8}\right)\left(\frac{4}{11}\right)^{4/3}\right] \rho_\gamma = 7.8099 \times 10^{-31}\ \text{kg m}^{-3}. \tag{13.292}$$

The fraction Ω_r of the present critical energy density that is due to radiation is then

$$\Omega_r = \frac{\rho_{r0}}{\rho_{c0}} = 9.0824 \times 10^{-5}. \tag{13.293}$$

In terms of Ω_r and of the Ω's in the table (13.1), the formula (13.290) for H^2/H_0^2 is

$$\frac{H^2}{H_0^2} = \Omega_\Lambda + \Omega_k \frac{a_0^2}{a^2} + \Omega_m \frac{a_0^3}{a^3} + \Omega_r \frac{a_0^4}{a^4}. \tag{13.294}$$

Since $H = \dot{a}/a$, one has $dt = da/(aH) = H_0^{-1}(da/a)(H_0/H)$, and so with $x = a/a_0$, the time interval dt is

$$dt = \frac{1}{H_0}\frac{dx}{x}\frac{1}{\sqrt{\Omega_\Lambda + \Omega_k\, x^{-2} + \Omega_m\, x^{-3} + \Omega_r\, x^{-4}}}. \tag{13.295}$$

Integrating and setting the origin of time $t(0) = 0$ and the scale factor at the present time equal to unity $a_0 = 1$, we find that the time $t(a)$ that $a(t)$ took to grow from 0 to $a(t)$ is

$$t(a) = \frac{1}{H_0}\int_0^a \frac{dx}{\sqrt{\Omega_\Lambda\, x^2 + \Omega_k + \Omega_m\, x^{-1} + \Omega_r\, x^{-2}}}. \tag{13.296}$$

This integral gives the age of the universe as $t(1) = 13.789$ Gyr; the Planck-collaboration value is 13.787 ± 0.020 Gyr (Aghanim et al., 2018). Figure 13.2 plots the scale factor $a(t)$ and the redshift $z(t) = 1/a - 1$ as functions of the time t (13.296) for the first 14 billion years after the time $t = 0$ of infinite redshift. A photon emitted with wavelength λ at time $t(a)$ now has wavelength $\lambda_0 = \lambda/a(t)$. The change in its wavelength is $\Delta\lambda = \lambda\, z(t) = \lambda\,(1/a - 1) = \lambda_0 - \lambda$.

13.46 The First Hundred Thousand Years

Figure 13.3 plots the scale factor $a(t)$ as given by the integral (13.296) and the densities of radiation $\rho_r(t)$ and matter $\rho_m(t)$ for the first 100,000 years after the time of infinite redshift. Because wavelengths grow with the scale factor, the radiation density (13.286) is proportional to the inverse fourth power of the scale factor $\rho_r(t) = \rho_{r0}/a^4(t)$. The density of radiation therefore was dominant at early

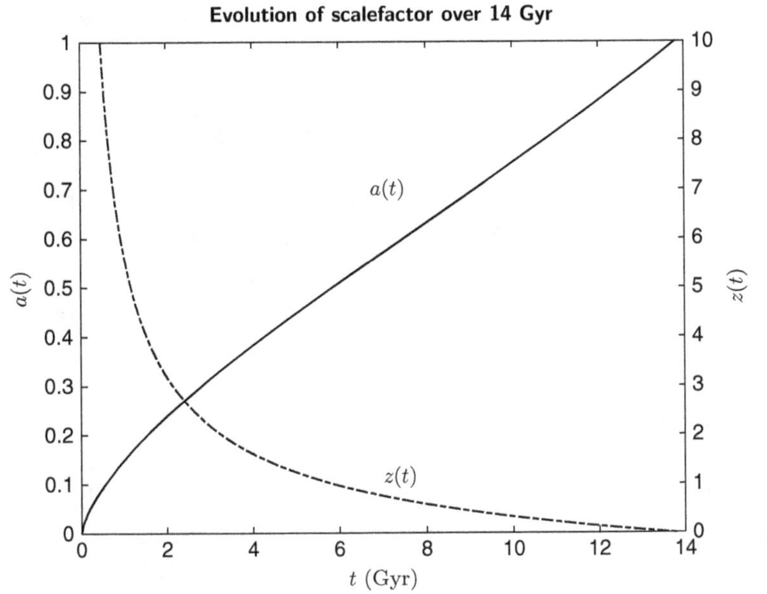

Figure 13.2 The scale factor $a(t)$ (solid, left axis) and redshift $z(t)$ (dotdash, right axis) are plotted against the time (13.296) in Gyr. This chapter's Fortran, Matlab, and Mathematica scripts are in Tensors_and_general_relativity at `github.com/kevinecahill`.

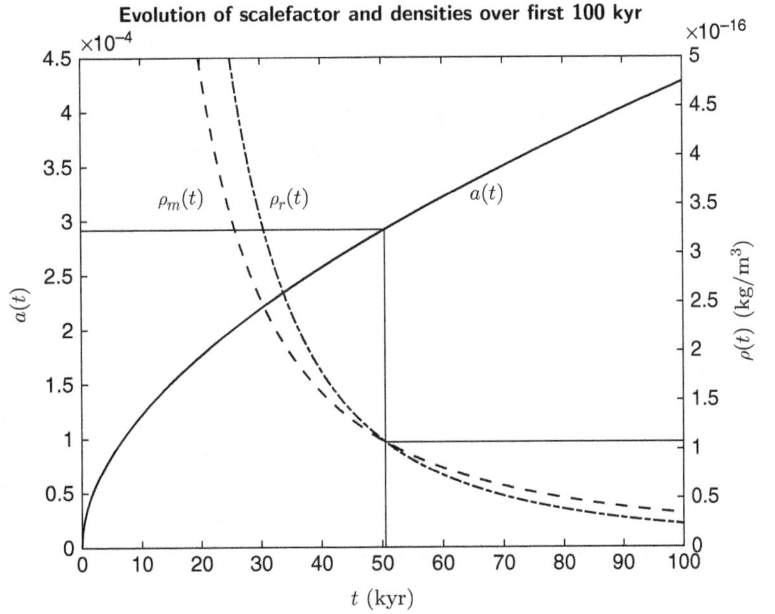

Figure 13.3 The Planck-collaboration scale factor a (solid), radiation density ρ_r (dotdash), and matter density ρ_m (dashed) are plotted as functions of the time (13.296) in kyr. The era of radiation ends at $t = 50{,}506$ years when the two densities are equal to 1.0751×10^{-16} kg/m^3, $a = 2.919 \times 10^{-4}$, and $z = 3425$.

times when the scale factor was small. Keeping only $\Omega_r = 0.6889$ in the integral (13.296), we get

$$t = \frac{a^2}{2H_0\sqrt{\Omega_r}} \quad \text{and} \quad a(t) = \Omega_r^{1/4}\sqrt{2\,H_0\,t}. \tag{13.297}$$

Since the radiation density $\rho_r(t) = \rho_{r0}/a^4(t)$ also is proportional to the fourth power of the temperature $\rho_r(t) \sim T^4$, the temperature varied as the inverse of the scale factor $T \sim 1/a(t) \sim t^{-1/2}$ during the era of radiation.

In cold-dark-matter models, when the temperature was in the range $10^{12} > T > 10^{10}$ K or $m_\mu c^2 > kT > m_e c^2$, where m_μ is the mass of the muon and m_e that of the electron, the radiation was mostly electrons, positrons, photons, and neutrinos, and the relation between the time t and the temperature T was $t \sim 0.994$ sec $\times$ $(10^{10}\,\text{K}/T)^2$ (Weinberg, 2010, ch. 3). By 10^9 K, the positrons had annihilated with electrons, and the neutrinos fallen out of equilibrium. Between 10^9 K and 10^6 K, when the energy density of nonrelativistic particles became relevant, the time–temperature relation was $t \sim 1.78$ sec $\times$ $(10^{10}\,\text{K}/T)^2$ (Weinberg, 2010, ch. 3). During the first three minutes (Weinberg, 1988) of the era of radiation, quarks and gluons formed hadrons, which decayed into protons and neutrons. As the neutrons decayed ($\tau = 877.7$ s), they and the protons formed the light elements – principally hydrogen, deuterium, and helium in a process called **big-bang nucleosynthesis**.

The density of nonrelativistic matter (13.285) falls as the third power of the scale factor $\rho_m(t) = \rho_{m0}/a^3(t)$. The more rapidly falling density of radiation $\rho_r(t)$ crosses it 50,506 years after the Big Bang as indicated by the vertical line in the figure (13.3). This time $t = 50{,}506$ yr and redshift $z = 3425$ mark the end of the **era of radiation**.

13.47 The Next Ten Billion Years

The **era of matter** began about 50,506 years after the time of infinite redshift when the matter density ρ_m first exceeded the radiation density ρ_r. Some 330,000 years later at $t \sim 380{,}000$ yr, the universe had cooled to about $T = 3000$ K or $kT = 0.26$ eV – a temperature at which less than 1% of the hydrogen was ionized. At this redshift of $z = 1090$, the plasma of ions and electrons became a **transparent** gas of neutral hydrogen and helium with trace amounts of deuterium, helium-3, and lithium-7. The photons emitted or scattered at that time as 0.26 eV or 3000 K photons have redshifted down to become the 2.7255 K photons of the cosmic microwave background (CMB) radiation. This time of last scattering and first transparency is called **recombination**, a term that makes sense if the universe is cyclic.

If we approximate time periods $t - t_m$ during the era of matter by keeping only Ω_m in the integral (13.296), then we get

$$t - t_m = \frac{2\,a^{3/2}}{3H_0\sqrt{\Omega_m}} \quad \text{and} \quad a(t) = \left(\frac{3H_0\sqrt{\Omega_m}\,(t - t_m)}{2}\right)^{2/3} \tag{13.298}$$

in which t_m is a time well inside the era of matter.

Between 10 and 17 million years after the Big Bang, the temperature of the known universe fell from 373 to 273 K. If by then the supernovas of very early, very heavy stars had produced carbon, nitrogen, and oxygen, biochemistry may have started during this period of 7 million years. Stars did form at least as early as 180 million years after the Big Bang (Bowman et al., 2018).

The era of matter lasted until the energy density of matter $\rho_m(t)$, falling as $\rho_m(t) = \rho_{m0}/a^3(t)$, had dropped to the energy density of dark energy $\rho_\Lambda = 5.9238 \times 10^{-27}\text{kg/m}^3$. This happened at $t = 10.228$ Gyr as indicated by the first vertical line in the figure (13.4).

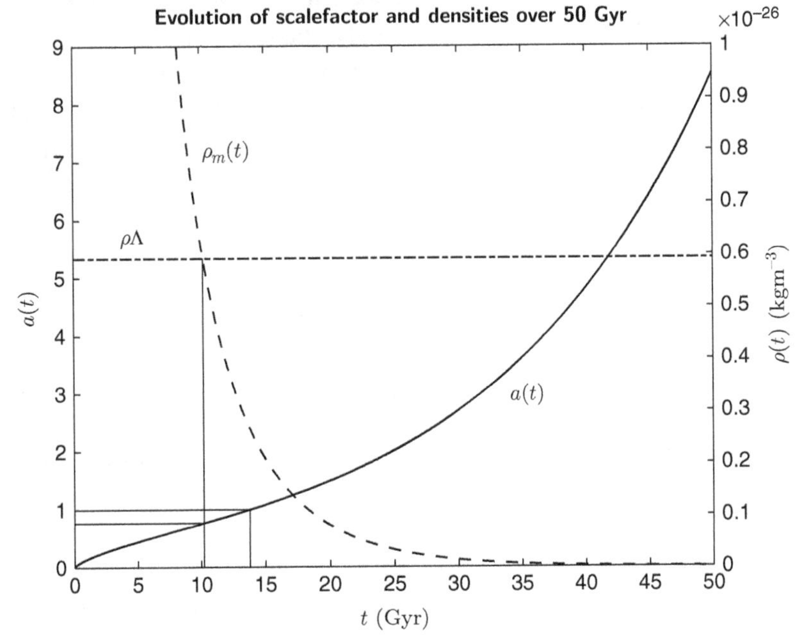

Figure 13.4 The scale factor a (solid), the vacuum density ρ_Λ (dotdash), and the matter density ρ_m (dashed) are plotted as functions of the time (13.296) in Gyr. The era of matter ends at $t = 10.228$ Gyr (first vertical line) when the two densities are equal to $5.9238 \times 10^{-27}\text{kg m}^{-3}$ and $a = 0.7672$. The present time t_0 is 13.787 Gyr (second vertical line) at which $a(t) = 1$.

13.48 Era of Dark Energy

The era of dark energy began 3.6 billion years ago at a redshift of $z = 0.3034$ when the energy density of the universe $\rho_m + \rho_\Lambda$ was twice that of empty space, $\rho = 2\rho_\Lambda = 1.185 \times 10^{-26}$ kg/m^3. The energy density of matter now is only 31.11% of the energy density of the universe, and it is falling as the cube of the scale factor $\rho_m(t) = \rho_{m0}/a^3(t)$. In another 20 billion years, the energy density of the universe will have declined almost all the way to the dark-energy density $\rho_\Lambda = 5.9238 \times 10^{-27}$ kg/m^3 or $(1.5864\ \text{meV})^4/(\hbar^3 c^5)$. At that time t_Λ and in the indefinite future, the only significant term in the integral (13.296) will be the vacuum energy. Neglecting the others and replacing $a_0 = 1$ with $a_\Lambda = a(t_\Lambda)$, we find

$$t(a/a_\Lambda) - t_\Lambda = \frac{\log(a/a_\Lambda)}{H_0\sqrt{\Omega_\Lambda}} \quad \text{or} \quad a(t) = e^{H_0\sqrt{\Omega_\Lambda}(t-t_\Lambda)}\, a(t_\Lambda) \tag{13.299}$$

in which $t_\Lambda \gtrsim 35$ Gyr.

13.49 Before the Big Bang

The ΛCDM model is remarkably successful (Aghanim et al., 2018). But it does not explain why the CMB is so isotropic, apart from a Doppler shift, and why the universe is so flat (Guth, 1981). A brief period of rapid exponential growth like that of the era of dark energy may explain the isotropy and the flatness.

Inflation occurs when the potential energy ρ dwarfs the energy of matter and radiation. The internal energy of the universe then is proportional to its volume $U = c^2\rho\, V$, and its pressure p as given by the thermodynamic relation

$$p = -\frac{\partial U}{\partial V} = -c^2\rho \tag{13.300}$$

is *negative*. The second-order Friedmann equation (13.273)

$$\frac{\ddot a}{a} = -\frac{4\pi G}{3}\left(\rho + \frac{3p}{c^2}\right) = \frac{8\pi G\rho}{3} \tag{13.301}$$

then implies exponential growth like that of the era of dark energy (13.299)

$$a(t) = e^{\sqrt{8\pi G\rho/3}\, t}\, a(0). \tag{13.302}$$

The origin of the potential-energy density ρ is unknown.

In **chaotic inflation** (Linde, 1983), a scalar field ϕ fluctuates to a mean value $\langle\phi\rangle_i$ that makes its potential-energy density ρ huge. The field remains at or close to the value $\langle\phi\rangle_i$, and the scale factor inflates rapidly and exponentially (13.302) until time t at which the potential energy of the universe is

$$E = c^2\rho\, e^{\sqrt{24\pi G\rho}\, t}\, V(0) \tag{13.303}$$

where $V(0)$ is the spatial volume in which the field ϕ held the value $\langle\phi\rangle_i$. After time t, the field returns to its mean value $\langle 0|\phi|0\rangle$ in the ground state $|0\rangle$ of the theory, and the huge energy E is released as radiation in a Big Bang. The energy E_g of the gravitational field caused by inflation is negative, $E_g = -E$, and energy is conserved. Chaotic inflation may explain why there is a universe: a quantum fluctuation made it. Regions where the field remained longer at $\langle\phi\rangle_i$ would inflate longer and make new local universes in new Big Bangs creating a **multiverse**, which also might arise from multiple quantum fluctuations.

If a quantum fluctuation gives a field ϕ a spatially constant mean value $\langle\phi\rangle_i \equiv \phi$ in an initial volume $V(0)$, then the equations for the scale factor (13.301) and for the scalar field (13.201) simplify to

$$H = \left(\frac{8\pi G\rho}{3}\right)^{1/2} \quad \text{and} \quad \ddot{\phi} = -3H\dot{\phi} - \frac{m^2c^4}{\hbar^2}\phi \tag{13.304}$$

in which ρ is the mass density of the potential energy of the field ϕ. The term $-3H\dot{\phi}$ is a kind of gravitational friction. It may explain why a field ϕ sticks at the value $\langle\phi\rangle_i$ long enough to resolve the isotropy and flatness puzzles.

Classical, de Sitter-like (Example 7.63), bouncing solutions (Steinhardt et al., 2002; Ijjas and Steinhardt, 2018) of Einstein's equations can explain the homogeneity, flatness, and isotropy of the universe as due to repeated collapses and rebirths. Experiments will tell us whether inflation or bouncing or something else actually occurred (Akrami et al., 2018).

13.50 Yang–Mills Theory

The gauge transformation of an **abelian** gauge theory like electrodynamics multiplies a single charged field by a spacetime-dependent phase factor $\phi'(x) = \exp(iq\theta(x))\,\phi(x)$. Yang and Mills generalized this gauge transformation to one that multiplies a vector ϕ of matter fields by a spacetime dependent unitary matrix $U(x)$

$$\phi'_a(x) = \sum_{b=1}^{n} U_{ab}(x)\,\phi_b(x) \quad \text{or} \quad \phi'(x) = U(x)\,\phi(x) \tag{13.305}$$

and showed how to make the action of the theory invariant under such **non-abelian** gauge transformations. (The fields ϕ are scalars for simplicity.)

Since the matrix U is unitary, inner products like $\phi^\dagger(x)\,\phi(x)$ are automatically invariant

$$\left(\phi^\dagger(x)\,\phi(x)\right)' = \phi^\dagger(x)U^\dagger(x)U(x)\phi(x) = \phi^\dagger(x)\phi(x). \tag{13.306}$$

But inner products of derivatives $\partial^i \phi^\dagger \, \partial_i \phi$ are not invariant because the derivative acts on the matrix $U(x)$ as well as on the field $\phi(x)$.

Yang and Mills made derivatives $D_i \phi$ that transform like the fields ϕ

$$(D_i \phi)' = U \, D_i \phi. \tag{13.307}$$

To do so, they introduced **gauge-field matrices** A_i that play the role of the connections Γ_i in general relativity and set

$$D_i = \partial_i + A_i \tag{13.308}$$

in which A_i like ∂_i is antihermitian. They required that under the gauge transformation (13.305), the gauge-field matrix A_i transform to A_i' in such a way as to make the derivatives transform as in (13.307)

$$(D_i \phi)' = \left(\partial_i + A_i'\right) \phi' = \left(\partial_i + A_i'\right) U \phi = U \, D_i \phi = U \left(\partial_i + A_i\right) \phi. \tag{13.309}$$

So they set

$$\left(\partial_i + A_i'\right) U \phi = U \left(\partial_i + A_i\right) \phi \quad \text{or} \quad \left(\partial_i U\right) \phi + A_i' \, U \phi = U A_i \, \phi \tag{13.310}$$

and made the gauge-field matrix A_i transform as

$$A_i' = U A_i U^{-1} - \left(\partial_i U\right) U^{-1}. \tag{13.311}$$

Thus under the gauge transformation (13.305), the derivative $D_i \phi$ transforms as in (13.307), like the vector ϕ in (13.305), and the inner product of covariant derivatives

$$\left[\left(D^i \phi\right)^\dagger D_i \phi\right]' = \left(D^i \phi\right)^\dagger U^\dagger U \, D_i \phi = \left(D^i \phi\right)^\dagger D_i \phi \tag{13.312}$$

remains invariant.

To make an invariant action density for the gauge-field matrices A_i, they used the transformation law (13.309) which implies that $D_i' \, U \phi = U D_i \, \phi$ or $D_i' = U D_i \, U^{-1}$. So they defined their generalized Faraday tensor as

$$F_{ik} = [D_i, D_k] = \partial_i A_k - \partial_k A_i + [A_i, A_k] \tag{13.313}$$

so that it transforms covariantly

$$F_{ik}' = U F_{ik} U^{-1}. \tag{13.314}$$

They then generalized the action density $F_{ik} F^{ik}$ of electrodynamics to the trace $\mathrm{Tr}\left(F_{ik} F^{ik}\right)$ of the square of the Faraday matrices which is invariant under gauge transformations since

$$\mathrm{Tr}\left(U F_{ik} U^{-1} U F^{ik} U^{-1}\right) = \mathrm{Tr}\left(U F_{ik} F^{ik} U^{-1}\right) = \mathrm{Tr}\left(F_{ik} F^{ik}\right). \tag{13.315}$$

As an action density for fermionic matter fields, they replaced the ordinary derivative in Dirac's formula $\overline{\psi}(\gamma^i \partial_i + m)\psi$ by the covariant derivative (13.308) to get $\overline{\psi}(\gamma^i D_i + m)\psi$ (Chen-Ning Yang 1922–, Robert L. Mills 1927–1999).

In an abelian gauge theory, the square of the 1-form $A = A_i\, dx^i$ vanishes $A^2 = A_i\, A_k\, dx^i \wedge dx^k = 0$, but in a nonabelian gauge theory the gauge fields are matrices, and $A^2 \neq 0$. The sum $dA + A^2$ is the Faraday 2-form

$$F = dA + A^2 = (\partial_i A_k + A_i A_k)\, dx^i \wedge dx^k \tag{13.316}$$
$$= \frac{1}{2}\,(\partial_i A_k - \partial_k A_i + [A_i,\, A_k])\, dx^i \wedge dx^k = \frac{1}{2} F_{ik}\, dx^i \wedge dx^k.$$

The scalar matter fields ϕ may have self-interactions described by a potential $V(\phi)$ such as $V(\phi) = \lambda(\phi^\dagger \phi - m^2/\lambda)^2$ which is positive unless $\phi^\dagger \phi = m^2/\lambda$. The kinetic action of these fields is $(D^i \phi)^\dagger D_i \phi$. At low temperatures, these scalar fields assume mean values $\langle 0|\phi|0\rangle = \phi_0$ in the vacuum with $\phi_0^\dagger \phi_0 = m^2/\lambda$ so as to minimize their potential energy density $V(\phi)$ and their kinetic action $(D^i \phi)^\dagger D_i \phi = (\partial^i \phi + A^i \phi)^\dagger (\partial_i \phi + A_i \phi)$ is approximately $\phi_0^\dagger A^i A_i \phi_0$. The matrix of the gauge fields $A^i_{ab} = -i t^\alpha_{ab} A^i_\alpha$ is a linear combination of the generators t^α of the gauge group. So the action of the scalar fields contains the term $\phi_0^\dagger A^i A_i \phi_0 = -M^2_{\alpha\beta} A^i_\alpha A_{i\beta}$ in which the mass-squared matrix for the gauge fields is $M^2_{\alpha\beta} = \phi_0^{*a} t^\alpha_{ab} t^\beta_{bc} \phi_0^c$. This **Higgs mechanism** gives masses to those linear combinations $b_{\beta i} A_\beta$ of the gauge fields for which $M^2_{\alpha\beta} b_{\beta i} = m_i^2 b_{\alpha i} \neq 0$.

The Higgs mechanism also gives masses to the fermions. The mass term m in the Yang–Mills–Dirac action is replaced by something like $c\phi$ in which c is a constant, different for each fermion. In the vacuum and at low temperatures, each fermion acquires as its mass $c\phi_0$. On 4 July 2012, physicists at CERN's Large Hadron Collider announced the discovery of a Higgs-like particle with a mass near 125 GeV/c^2 (Peter Higgs 1929 –).

13.51 Cartan's Spin Connection and Structure Equations

Cartan's tetrads (13.153) $c^a{}_k(x)$ are the rows and columns of the orthogonal matrix that turns the flat-space metric η_{ab} into the curved-space metric $g_{ik} = c^a_i\, \eta_{ab} c^b_k$. Early-alphabet letters $a, b, c, d, \ldots = 0, 1, 2, 3$ are Lorentz indexes, and middle-to-late letters $i, j, k, \ell, \ldots = 0, 1, 2, 3$ are spacetime indexes. Under a combined local Lorentz (13.154) and general coordinate transformation the tetrads transform as

$$c'^a{}_k(x') = L^a{}_b(x')\, \frac{\partial x^\ell}{\partial x'^k}\, c^b{}_\ell(x). \tag{13.317}$$

The covariant derivative of a tetrad $D_\ell c$ must transform as

$$(D_\ell \, c^a{}_k)'(x') = L^a{}_b(x') \, \frac{\partial x^i}{\partial x'^\ell} \frac{\partial x^j}{\partial x'^k} \, D_i \, c^b{}_j(x). \tag{13.318}$$

We can use the affine connection $\Gamma^j{}_{k\ell}$ and the formula (13.68) for the covariant derivative of a covariant vector to cope with the index j. And we can treat the Lorentz index b like an index of a nonabelian group as in Section 13.50 by introducing a gauge field $\omega^a{}_{\ell b}$

$$D_\ell \, c^a{}_k = c^a{}_{,\ell} - \Gamma^j{}_{k\ell} \, c^a{}_j + \omega^a{}_{\ell b} \, c^b{}_k. \tag{13.319}$$

The affine connection is defined so as to make the covariant derivative of the tangent basis vectors vanish

$$D_\ell \, e^\alpha_k = e^\alpha_{k,\ell} - \Gamma^j{}_{k\ell} e^\alpha_j = 0 \tag{13.320}$$

in which the Greek letter α labels the coordinates $0, 1, 2, \ldots, n$ of the embedding space. We may verify this relation by taking the inner product in the embedding space with the dual tangent vector e^i_α

$$e^i_\alpha \Gamma^j{}_{k\ell} e^\alpha_j = \delta^i_j \Gamma^j{}_{k\ell} = \Gamma^i{}_{k\ell} = e^i_\alpha e^\alpha_{k,\ell} = e^i \cdot e_{k,\ell} \tag{13.321}$$

which is the definition (13.59) of the affine connection, $\Gamma^i{}_{k\ell} = e^i \cdot e_{k,\ell}$. So too the **spin connection** $\omega^a{}_{b\ell}$ is defined so as to make the covariant derivative of the tetrad vanish

$$D_\ell \, c^a{}_k = c^a{}_{k;\ell} = c^a{}_{k,\ell} - \Gamma^j{}_{k\ell} \, c^a_j + \omega^a{}_{d\ell} \, c^d_k = 0. \tag{13.322}$$

The dual tetrads c^k_b are doubly orthonormal:

$$c^k_b \, c^b_i = \delta^k_i \quad \text{and} \quad c^k_a \, c^b_k = \delta^b_a. \tag{13.323}$$

Thus using their orthonormality, we have $\omega^a{}_{d\ell} \, c^d_k \, c^k_b = \omega^a{}_{d\ell} \, \delta^d_b = \omega^a{}_{b\ell}$, and so the spin connection is

$$\omega^a{}_{b\ell} = -\, c^k_b \big(c^a_{k,\ell} - \Gamma^j_{k\ell} \, c^a_j \big) = c^a_j \, c^k_b \, \Gamma^j{}_{k\ell} - c^a_{k,\ell} \, c^k_b. \tag{13.324}$$

In terms of the differential forms (Section 12.6)

$$c^a = c^a_k \, dx^k \quad \text{and} \quad \omega^a{}_b = \omega^a{}_{b\ell} \, dx^\ell \tag{13.325}$$

we may use the exterior derivative to express the vanishing (13.322) of the covariant derivative $c^a{}_{k;\ell}$ as

$$dc^a = c^a_{k,\ell} \, dx^\ell \wedge dx^k = \left(\Gamma^j{}_{k\ell} \, c^a_j - \omega^a{}_{b\ell} \, c^b_k \right) dx^\ell \wedge dx^k. \tag{13.326}$$

But the affine connection ${\Gamma^j}_{k\ell}$ is symmetric in k and ℓ while the wedge product $dx^\ell \wedge dx^k$ is antisymmetric in k and ℓ. Thus we have

$$dc^a = c^a_{k,\ell}\, dx^\ell \wedge dx^k = -\,{\omega^a}_{b\,\ell}\, dx^\ell \wedge c^b_k\, dx^k \tag{13.327}$$

or with $c \equiv c^a_k dx^k$ and $\omega \equiv {\omega^a}_{b\,\ell} dx^\ell$

$$dc = -\omega \wedge c \tag{13.328}$$

which is **Cartan's first equation of structure**. Cartan's curvature 2-form is

$$\begin{aligned} {R^a}_b &= \frac{1}{2}\, c^a_j\, c^i_b\, {R^j}_{ik\ell}\, dx^k \wedge dx^\ell \\ &= \frac{1}{2}\, c^a_j\, c^i_b \Big[{\Gamma^j}_{\ell i,k} - {\Gamma^j}_{ki,\ell} + {\Gamma^j}_{kn}\, {\Gamma^n}_{\ell i} - {\Gamma^j}_{\ell n}\, {\Gamma^n}_{ki} \Big] dx^k \wedge dx^\ell. \end{aligned} \tag{13.329}$$

His **second equation of structure** expresses ${R^a}_b$ as

$$ {R^a}_b = d{\omega^a}_b + {\omega^a}_c \wedge {\omega^c}_b \tag{13.330}$$

or more simply as

$$R = d\omega + \omega \wedge \omega. \tag{13.331}$$

A more compact notation, similar to that of Yang–Mills theory, uses Cartan's covariant exterior derivative

$$D \equiv d + \omega\, \wedge \tag{13.332}$$

to express his two structure equations as

$$D\,c = 0 \quad \text{and} \quad R = D\,\omega. \tag{13.333}$$

To derive Cartan's second structure equation (13.330), we let the exterior derivative act on the 1-form ${\omega^a}_b$

$$d{\omega^a}_b = d({\omega^a}_{b\,\ell}\, dx^\ell) = {\omega^a}_{b\,\ell,k}\, dx^k \wedge dx^\ell \tag{13.334}$$

and add the 2-form ${\omega^a}_c \wedge {\omega^c}_b$

$${\omega^a}_c \wedge {\omega^c}_b = {\omega^a}_{c\,k}\, {\omega^c}_{b\,\ell}\, dx^k \wedge dx^\ell \tag{13.335}$$

to get

$${S^a}_b = ({\omega^a}_{b\,\ell,k} + {\omega^a}_{c\,k}\, {\omega^c}_{b\,\ell})\, dx^k \wedge dx^\ell \tag{13.336}$$

which we want to show is Cartan's curvature 2-form ${R^a}_b$ (13.329). First we replace ${\omega^a}_{\ell b}$ with its equivalent (13.324) $c^a_j\, c^i_b\, {\Gamma^j}_{i\ell} - c^a_{i,\ell}\, c^i_b$

$$\begin{aligned} {S^a}_b = \Big[& (c^a_j\, c^i_b\, {\Gamma^j}_{i\ell} - c^a_{i,\ell}\, c^i_b)_{,k} + (c^a_j\, c^i_c\, {\Gamma^j}_{ik} - c^a_{i,k}\, c^i_c)(c^c_n\, c^p_b\, {\Gamma^n}_{p\ell} - c^c_{p,\ell}\, c^p_b) \Big] \\ & \times\, dx^k \wedge dx^\ell. \end{aligned} \tag{13.337}$$

The terms proportional to ${\Gamma^j}_{i\ell,k}$ are equal to those in the definition (13.329) of Cartan's curvature 2-form. Among the remaining terms in ${S^a}_b$, those independent of Γ are after explicit antisymmetrization

$$S_0 = c^a_{i,k}\, c^i_{b,\ell} - c^a_{i,\ell}\, c^i_{b,k} + c^a_{i,k}\, c^i_c\, c^c_{p,\ell}\, c^p_b - c^a_{i,\ell}\, c^i_c\, c^c_{p,k}\, c^p_b \tag{13.338}$$

which vanishes (Exercise 13.36) because $c^i_{b,k} = -c^i_c\, c^c_{p,k}\, c^p_b$. The terms in ${S^a}_b$ that are linear in Γ's also vanish (Exercise 13.37). Finally, the terms in ${S^a}_b$ that are quadratic in Γ's are

$$\begin{aligned} c^a_j\, c^i_c\, c^c_n\, c^p_b\, {\Gamma^j}_{ik}\, {\Gamma^n}_{p\ell}\, dx^k \wedge dx^\ell &= c^a_j\, \delta^i_n\, c^p_b\, {\Gamma^j}_{ik}\, {\Gamma^n}_{p\ell}\, dx^k \wedge dx^\ell \\ &= c^a_j\, c^p_b\, {\Gamma^j}_{nk}\, {\Gamma^n}_{p\ell}\, dx^k \wedge dx^\ell \end{aligned} \tag{13.339}$$

and these match those of Cartan's curvature 2-form ${R^a}_b$ (13.329). Since ${S^a}_b = {R^a}_b$, Cartan's second equation of structure (13.330) follows.

Example 13.27 (Cyclic identity for the curvature tensor) We can use Cartan's structure equations to derive the cyclic identity (13.109) of the curvature tensor. We apply the exterior derivative (whose square $dd = 0$) to Cartan's first equation of structure (13.328) and then use it and his second equation of structure (13.330) to write the result as

$$0 = d(dc + \omega \wedge c) = (d\omega) \wedge c - \omega \wedge dc = (d\omega + \omega \wedge \omega) \wedge c = R \wedge c. \tag{13.340}$$

The definition (13.329) of Cartan's curvature 2-form R and of his 1-form (13.325) now give

$$\begin{aligned} 0 = R \wedge c &= \frac{1}{2}\, c^a_j\, c^i_b\, {R^j}_{ik\ell}\, dx^k \wedge dx^\ell \wedge c^b_m dx^m \\ &= \frac{1}{2}\, c^a_j\, {R^j}_{ik\ell}\, dx^k \wedge dx^\ell \wedge dx^i \end{aligned} \tag{13.341}$$

which implies that

$$0 = {R^j}_{[ik\ell]} = \frac{1}{3!}\left({R^j}_{ik\ell} + {R^j}_{\ell ik} + {R^j}_{k\ell i} - {R^j}_{ki\ell} - {R^j}_{i\ell k} - {R^j}_{\ell ki}\right). \tag{13.342}$$

But since Riemann's tensor is antisymmetric in its last two indices (13.104), we can write this result more simply as the cyclic identity (13.109) for the curvature tensor

$$0 = {R^j}_{ik\ell} + {R^j}_{\ell ik} + {R^j}_{k\ell i}. \tag{13.343}$$

The vanishing of the covariant derivative of the flat-space metric

$$0 = \eta_{ab;k} = \eta_{ab,k} - {\omega^c}_{ak}\eta_{cb} - {\omega^c}_{bk}\eta_{ac} = -\omega_{bak} - \omega_{abk} \tag{13.344}$$

shows that the spin connection is antisymmetric in its Lorentz indexes

$$\omega_{abk} = -\,\omega_{bak} \quad \text{and} \quad \omega^{ab}{}_{k} = -\,\omega^{ba}{}_{k}. \tag{13.345}$$

Under a general coordinate transformation and a local Lorentz transformation, the spin connection (13.324) transforms as

$$\omega'^{a}{}_{b\ell} = \frac{\partial x^i}{\partial x'^\ell}\left[L^a{}_d\,\omega^d{}_{ei} - (\partial_i L^a{}_e)\right] L^{-1e}{}_b. \tag{13.346}$$

13.52 Spin-One-Half Fields in General Relativity

The action density (11.329) of a free Dirac field is $L = -\bar\psi\,(\gamma^a\partial_a + m)\,\psi$ in which $a = 0, 1, 2, 3$; ψ is a 4-component Dirac field; $\bar\psi = \psi^\dagger\beta = i\psi^\dagger\gamma^0$; and m is a mass. Tetrads $c^a_k(x)$ turn the flat-space indices a into curved-space indices i, so one first replaces $\gamma^a\partial_a$ by $\gamma^a\, c^\ell_a\partial_\ell$. The next step is to use the spin connection (13.324) to correct for the effect of the derivative ∂_ℓ on the field ψ. The fully covariant derivative is $D_\ell = \partial_\ell - \frac{1}{8}\,\omega^{ab}{}_\ell\,[\gamma_a, \gamma_b]$ where $\omega^{ab}{}_\ell = \omega^a{}_{c\ell}\,\eta^{bc}$, and the action density is $L = -\,\bar\psi(\gamma^a c^\ell_a\, D_\ell + m)\psi$.

13.53 Gauge Theory and Vectors

This section is optional on a first reading.

We can formulate Yang–Mills theory in terms of vectors as we did relativity. To accomodate noncompact groups, we generalize the unitary matrices $U(x)$ of the Yang–Mills gauge group to nonsingular matrices $V(x)$ that act on n matter fields $\psi^a(x)$ as $\psi'^a(x) = V^a_b(x)\,\psi^b(x)$. The field $\Psi(x) = e_a(x)\,\psi^a(x)$ will be gauge invariant $\Psi'(x) = \Psi(x)$ if the vectors $e_a(x)$ transform as $e'_a(x) = e_b(x)\,V^{-1b}{}_a(x)$. We are summing over repeated indices from 1 to n and often will suppress explicit mention of the spacetime coordinates. In this compressed notation, the field Ψ is gauge invariant because

$$\Psi' = e'_a\,\psi'^a = e_b\,V^{-1b}{}_a\,V^a{}_c\,\psi^c = e_b\,\delta^b{}_c\,\psi^c = e_b\,\psi^b = \Psi \tag{13.347}$$

which is $e'^{\mathsf T}\psi' = e^{\mathsf T}V^{-1}V\psi = e^{\mathsf T}\psi$ in matrix notation.

The inner product of two basis vectors is an internal "metric tensor"

$$e^*_a \cdot e_b = \sum_{\alpha=1}^{N} e^{\alpha *}_a\, e^\alpha_b = g_{ab} \tag{13.348}$$

in which for simplicity I used the the N-dimensional identity matrix as the metric of the embedding space. As in relativity, we'll assume the matrix g_{ab} to be nonsingular. We then can use its inverse to construct dual vectors $e^a = g^{ab}e_b$ that satisfy $e^{a\dagger}\cdot e_b = \delta^a_b$.

The free Dirac action density of the invariant field Ψ

$$\overline{\Psi}(\gamma^i \partial_i + m)\Psi = \overline{\psi}_a e^{a\dagger}(\gamma^i \partial_i + m) e_b \psi^b = \overline{\psi}_a \left[\gamma^i (\delta^a{}_b \partial_i + e^{a\dagger} \cdot e_{b,i}) + m\delta^a{}_b\right] \psi^b \tag{13.349}$$

is the full action of the component fields ψ^b

$$\overline{\Psi}(\gamma^i \partial_i + m)\Psi = \overline{\psi}_a(\gamma^i D^a_{i\,b} + m\,\delta^a{}_b)\psi^b = \overline{\psi}_a\left[\gamma^i (\delta^a{}_b \partial_i + A^a_{i\,b}) + m\,\delta^a{}_b\right]\psi^b \tag{13.350}$$

if we identify the gauge-field matrix as $A^a_{i\,b} = e^{a\dagger} \cdot e_{b,i}$ in harmony with the definition (13.59) of the affine connection $\Gamma^k_{i\ell} = e^k \cdot e_{\ell,i}$.

Under the gauge transformation $e'_a = e_b\, V^{-1b}{}_a$, the metric matrix transforms as

$$g'_{ab} = V^{-1c*}{}_a\, g_{cd}\, V^{-1d}{}_b \qquad \text{or as} \qquad g' = V^{-1\dagger}\, g\, V^{-1} \tag{13.351}$$

in matrix notation. Its inverse goes as $g'^{-1} = V\, g^{-1}\, V^\dagger$.

The gauge-field matrix $A^a_{i\,b} = e^{a\dagger} \cdot e_{b,i} = g^{ac} e^\dagger_c \cdot e_{b,i}$ transforms as

$$A'^a_{i\,b} = g'^{ac} e'^\dagger_a \cdot e'_{b,i} = V^a{}_c A^c_{id} V^{-1d}{}_b + V^a{}_c V^{-1c}{}_{b,i} \tag{13.352}$$

or as $A'_i = V A_i V^{-1} + V \partial_i V^{-1} = V A_i V^{-1} - (\partial_i V)\, V^{-1}$.

By using the identity $e^{a\dagger} \cdot e_{c,i} = -\, e^{a\dagger}_{,i} \cdot e_c$, we may write (Exercise 13.39) the Faraday tensor as

$$F^a_{ijb} = [D_i, D_j]^a{}_b = e^{a\dagger}_{,i} \cdot e_{b,j} - e^{a\dagger}_{,i} \cdot e_c\; e^{c\dagger} \cdot e_{b,j} - e^{a\dagger}_{,j} \cdot e_{b,i} + e^{a\dagger}_{,j} \cdot e_c\; e^{c\dagger} \cdot e_{b,i}. \tag{13.353}$$

If $n = N$, then

$$\sum_{c=1}^{n} e^\alpha_c\; e^{\beta c*} = \delta^{\alpha\beta} \quad \text{and} \quad F^a_{ijb} = 0. \tag{13.354}$$

The Faraday tensor vanishes when $n = N$ because the dimension of the embedding space is too small to allow the tangent space to have different orientations at different points x of spacetime. The Faraday tensor, which represents internal curvature, therefore must vanish. One needs at least 3 dimensions in which to bend a sheet of paper. The embedding space must have $N > 2$ dimensions for $SU(2)$, $N > 3$ for $SU(3)$, and $N > 5$ for $SU(5)$.

The covariant derivative of the internal metric matrix

$$g_{;i} = g_{,i} - g A_i - A^\dagger_i g \tag{13.355}$$

does not vanish and transforms as $\left(g_{;i}\right)' = V^{-1\dagger} g_{,i} V^{-1}$. A suitable action density for it is the trace $\mathrm{Tr}(g_{;i}\, g^{-1} g^{;i}\, g^{-1})$. If the metric matrix assumes a (constant, hermitian) mean value g_0 in the vacuum at low temperatures, then its action is

$$m^2 \mathrm{Tr}\left[(g_0 A_i + A^\dagger_i g_0) g_0^{-1} (g_0 A^i + A^{i\dagger} g_0) g_0^{-1}\right] \tag{13.356}$$

which is a mass term for the matrix of gauge bosons

$$W_i = g_0^{1/2}\, A_i\, g_0^{-1/2} + g_0^{-1/2}\, A_i^\dagger\, g_0^{1/2}. \tag{13.357}$$

This mass mechanism also gives masses to the fermions. To see how, we write the Dirac action density (13.350) as

$$\overline{\psi}_a \left[\gamma^i (\delta^a{}_b \partial_i + A^a_{i\,b}) + m\, \delta^a{}_b\right] \psi^b = \overline{\psi}^a \left[\gamma^i (g_{ab}\partial_i + g_{ac} A^c_{i\,b}) + m\, g_{ab}\right] \psi^b. \tag{13.358}$$

Each fermion now gets a mass $m\, c_i$ proportional to an eigenvalue c_i of the hermitian matrix g_0.

This mass mechanism does not leave behind scalar bosons. Whether nature ever uses it is unclear.

Further Reading

Einstein Gravity in a Nutshell (Zee, 2013), *Gravitation* (Misner et al., 1973), *Gravitation and Cosmology* (Weinberg, 1972), *Cosmology* (Weinberg, 2010), *General Theory of Relativity* (Dirac, 1996), *Spacetime and Geometry* (Carroll, 2003), *Exact Space-Times in Einstein's General Relativity* (Griffiths and Podolsky, 2009), *Gravitation: Foundations and Frontiers* (Padmanabhan, 2010), *Modern Cosmology* (Dodelson, 2003), *The Primordial Density Perturbation: Cosmology, Inflation and the Origin of Structure* (Lyth and Liddle, 2009), *A First Course in General Relativity* (Schutz, 2009), *Gravity: An Introduction to Einstein's General Relativity* (Hartle, 2003), and *Relativity, Gravitation and Cosmology: A Basic Introduction* (Cheng, 2010).

Exercises

13.1 Use the flat-space formula $d\boldsymbol{p} = \hat{\boldsymbol{x}}\, dx + \hat{\boldsymbol{y}}\, dy + \hat{\boldsymbol{z}}\, dz$ to compute the change $d\boldsymbol{p}$ due to $d\rho$, $d\phi$, and dz, and so derive expressions for the orthonormal basis vectors $\hat{\boldsymbol{\rho}}$, $\hat{\boldsymbol{\phi}}$, and $\hat{\boldsymbol{z}}$ in terms of $\hat{\boldsymbol{x}}$, $\hat{\boldsymbol{y}}$, and $\hat{\boldsymbol{z}}$.

13.2 Similarly, compute the change $d\boldsymbol{p}$ due to dr, $d\theta$, and $d\phi$, and so derive expressions for the orthonormal basis vectors $\hat{\boldsymbol{r}}$, $\hat{\boldsymbol{\theta}}$, and $\hat{\boldsymbol{\phi}}$ in terms of $\hat{\boldsymbol{x}}$, $\hat{\boldsymbol{y}}$, and $\hat{\boldsymbol{z}}$.

13.3 (a) Using the formulas you found in Exercise 13.2 for the basis vectors of spherical coordinates, compute the derivatives of the unit vectors $\hat{\boldsymbol{r}}$, $\hat{\boldsymbol{\theta}}$, and $\hat{\boldsymbol{\phi}}$ with respect to the variables r, θ, and ϕ and express them in terms of the basis vectors $\hat{\boldsymbol{r}}$, $\hat{\boldsymbol{\theta}}$, and $\hat{\boldsymbol{\phi}}$. (b) Using the formulas of (a) and our expression (2.16) for the gradient in spherical coordinates, derive the formula (2.33) for the laplacian $\nabla \cdot \nabla$.

13.4 Show that for any set of basis vectors $v_1, \ldots, v_n$ and an inner product that is either positive definite (1.78–1.81) or indefinite (1.78–1.79, 1.81, and 1.84), the inner products $g_{ik} = (v_i, v_k)$ define a matrix g_{ik} that is nonsingular and that therefore has an inverse. Hint: Show that the matrix g_{ik} cannot have a zero eigenvalue without violating either the condition (1.80) that it be positive definite or the condition (1.84) that it be nondegenerate.

13.5 Show that the inverse metric (13.48) transforms as a rank-2 contravariant tensor.

13.6 Show that if A_k is a covariant vector, then $g^{ik} A_k$ is a contravariant vector.

13.7 Show that in terms of the parameter $k = (a/R)^2$, the metric and line element (13.46) are given by (13.47).

13.8 Show that the connection $\Gamma^k_{i\ell}$ transforms as (13.66) and so is not a tensor.

13.9 Use the vanishing (13.83) of the covariant derivative of the metric tensor, to write the condition (13.140) in terms of the covariant derivatives of the symmetry vector (13.141).

13.10 Embed the points $\boldsymbol{p} = R(\cosh\theta, \sinh\theta\cos\phi, \sinh\theta\sin\phi)$ with tangent vectors (13.44) and line element (13.45) in the euclidian space $\mathbb{E}^3$. Show that the line element of this embedding is

$$\begin{aligned} ds^2 &= R^2\left(\cosh^2\theta + \sinh^2\theta\right) d\theta^2 + R^2 \sinh^2\theta\, d\phi^2 \\ &= a^2\left(\frac{(1+2kr^2)dr^2}{1+kr^2} + r^2\, d\phi^2\right) \end{aligned} \tag{13.359}$$

which describes a hyperboloid that is not maximally symmetric.

13.11 If you have Mathematica, imitate Example 13.15 and find the scalar curvature R (13.113) of the line element (13.359) of the cylindrical hyperboloid embedded in euclidian 3-space $\mathbb{E}^3$.

13.12 Consider the torus with coordinates θ, ϕ labeling the arbitrary point

$$\boldsymbol{p} = (\cos\phi(R + r\sin\theta), \sin\phi(R + r\sin\theta), r\cos\theta) \tag{13.360}$$

in which $R > r$. Both θ and ϕ run from 0 to 2π. (a) Find the basis vectors e_θ and e_ϕ. (b) Find the metric tensor and its inverse.

13.13 For the same torus, (a) find the dual vectors e^θ and e^ϕ and (b) find the nonzero connections $\Gamma^i{}_{jk}$ where i, j, k take the values θ and ϕ.

13.14 For the same torus, (a) find the two Christoffel matrices Γ_θ and Γ_ϕ, (b) find their commutator $[\Gamma_\theta, \Gamma_\phi]$, and (c) find the elements $R^\theta_{\ \theta\theta\theta}$, $R^\phi_{\ \theta\phi\theta}$, $R^\theta_{\ \phi\theta\phi}$, and $R^\phi_{\ \phi\phi\phi}$ of the curvature tensor.

13.15 Find the curvature scalar R of the torus with points (13.360). Hint: In these four problems, you may imitate the corresponding calculation for the sphere in Section 13.23.

13.16 Show that $\delta g^{ik} = -g^{is} g^{kt} \delta g_{st}$ or equivalently that $dg^{ik} = -g^{is} g^{kt} dg_{st}$ by differentiating the identity $g^{ik} g_{k\ell} = \delta^i_\ell$.

13.17 Let $g_{ik} = \eta_{ik} + h_{ik}$ and $x'^n = x^n + \epsilon^n$. To lowest order in ϵ and h, (a) show that in the x' coordinates $h'_{ik} = h_{ik} - \epsilon_{i,k} - \epsilon_{k,i}$ and (b) find an equation for ϵ that puts h' in de Donder's gauge $h'^i{}_{k,i} = \frac{1}{2}(\eta^{j\ell} h'_{j\ell})_{,k}$.

13.18 Just to get an idea of the sizes involved in black holes, imagine an isolated sphere of matter of uniform density ρ that as an initial condition is all at rest within a radius r_b. Its radius will be less than its Schwarzschild radius if

$$r_b < \frac{2MG}{c^2} = 2\left(\frac{4}{3}\pi r_b^3 \rho\right)\frac{G}{c^2}. \tag{13.361}$$

If the density ρ is that of water under standard conditions (1 gram per cc), for what range of radii r_b might the sphere be or become a black hole? Same question if ρ is the density of dark energy.

13.19 Embed the points

$$p = (ct, aL\sin\chi\sin\theta\cos\phi, aL\sin\chi\sin\theta\sin\phi, aL\sin\chi\cos\theta, aL\cos\chi) \tag{13.362}$$

in the flat semi-euclidian space $\mathbb{E}^{(1,4)}$ with metric $(-1, 1, 1, 1, 1)$ and derive the metric (13.257) with $k = 1$.

13.20 For the points $p = (ct, a\sin\theta\cos\phi, a\sin\theta\sin\phi, a\cos\theta)$, derive the metric (13.257) with $k = 0$.

13.21 Show that the 11 component of Ricci's tensor R_{11} is

$$R_{11} = [D_k, D_1]^k{}_1 = \frac{a\ddot{a} + 2\dot{a}^2 + 2kc^2/L^2}{c^2(1 - kr^2/L^2)}. \tag{13.363}$$

13.22 Show that the 22 component of Ricci's tensor R_{22} is

$$R_{22} = [D_k, D_2]^k{}_2 = \frac{(a\ddot{a} + 2\dot{a}^2 + 2kc^2/L^2)r^2}{c^2}. \tag{13.364}$$

13.23 Show that the 33 component of Ricci's tensor R_{33} is

$$R_{33} = [D_k, D_3]^k{}_3 = \frac{(a\ddot{a} + 2\dot{a}^2 + 2kc^2/L^2)r^2\sin^2\theta}{c^2}. \tag{13.365}$$

13.24 Embed the points

$$p = (ct, aL\sinh\chi\sin\theta\cos\phi, aL\sinh\chi\sin\theta\sin\phi, \\ aL\sinh\chi\cos\theta, aL\cosh\chi) \tag{13.366}$$

in the flat semi-euclidian space $\mathbb{E}^{(2,3)}$ with metric $(-1, 1, 1, 1, -1)$ and derive the line element (13.256) and the metric (13.257) with $k = -1$.

13.25 Derive the second-order FLRW equation (13.273) from the formulas (13.257) for $g_{00} = -c^2$, (13.267) for $R_{00} = -3\ddot{a}/a$, (13.271) for T_{ij}, and (13.272) for T.

13.26 Derive the second-order FLRW equation (13.274) from Einstein's formula for the scalar curvature $R = -8\pi G\, T/c^4$ (13.230) and from the FLRW formulas for R (13.270) and for the trace T (13.272).

13.27 Assume there had been no inflation, no era of radiation, and no dark energy. In this case, the magnitude of the difference $|\Omega - 1|$ would have increased as $t^{2/3}$ over the past 13.8 billion years. Show explicitly how close to unity Ω would have had to have been at $t = 1$ s so as to satisfy the observational constraint $|\Omega_0 - 1| < 0.036$ on the present value of Ω.

13.28 Derive the relation (13.284) between the energy density ρ and the scale factor $a(t)$ from the conservation law (13.281) and the equation of state $p_i = w_i \rho_i$.

13.29 For constant $\rho = -p/c^2$ and $k = 1$, set $g^2 = 8\pi G\rho/3$ and use the Friedmann equations (13.273 and 13.288) and the boundary condition that the minimum of $a(t) > 0$ is at $t = 0$ to derive the formula $a(t) = c\cosh(gt)/(Lg)$.

13.30 Use the Friedmann equations (13.273 and 13.275) with $w = -1$, ρ constant, $k = -1$, and the boundary conditions $a(0) = 0$ and $\dot{a}(0) > 0$ to derive the formula $a(t) = c\sinh(gt)/(Lg)$ where again $g^2 = 8\pi G\rho/3$.

13.31 Use the Friedmann equations (13.273 and 13.275) with $w = -1$, ρ constant, and $k = 0$ to derive the formula $a(t) = a(0)\, e^{\pm gt}$.

13.32 Use the constancy of $8\pi G\rho\, a^4/3 = f^2$ for radiation ($w = 1/3$) and the Friedmann equations (13.273 and 13.275) to show that if $k = 0$, $a(0) = 0$, and $a(t) > 0$, then $a(t) = \sqrt{2ft}$ where $f > 0$.

13.33 Show that if the matrix $U(x)$ is nonsingular, then

$$(\partial_i U)\, U^{-1} = -U\, \partial_i\, U^{-1}. \tag{13.367}$$

13.34 The gauge-field matrix is a linear combination $A_k = -ig\, t^b\, A_k^b$ of the generators t^b of a representation of the gauge group. The generators obey the commutation relations

$$[t^a, t^b] = i f_{abc} t^c \tag{13.368}$$

in which the f_{abc} are the structure constants of the gauge group. Show that under a gauge transformation (13.311)

$$A_i' = U A_i U^{-1} - (\partial_i U)\, U^{-1} \tag{13.369}$$

by the unitary matrix $U = \exp(-ig\lambda^a t^a)$ in which λ^a is infinitesimal, the gauge-field matrix A_i transforms as

$$- igA_i^{\prime a} t^a = -igA_i^a t^a - ig^2 f_{abc} \lambda^a A_i^b t^c + ig \partial_i \lambda^a t^a. \tag{13.370}$$

Show further that the gauge field transforms as

$$A_i^{\prime a} = A_i^a - \partial_i \lambda^a - g f_{abc} A_i^b \lambda^c. \tag{13.371}$$

13.35 Show that if the vectors $e_a(x)$ are orthonormal, then $e^{a\dagger} \cdot e_{c,i} = -e^{a\dagger}_{,i} \cdot e_c$.

13.36 Use the equation $0 = \delta^a_{b,k} = (c^a_i \, c^i_b)_{,k}$ to show that $c^i_{b,k} = -c^i_c \, c^c_{p,k} \, c^p_b$. Then use this result to show that the Γ-free terms S_0 (13.338) vanish.

13.37 Show that terms in $S^a{}_b$ (13.337) linear in the Γ's vanish.

13.38 Derive the formula (13.346) for how the spin connection (13.324) changes under a Lorentz transformation and a general change of coordinates.

13.39 Use the identity of Exercise 13.35 to derive the formula (13.353) for the nonabelian Faraday tensor.

13.40 Show that the dual tetrads $c^i_a = g^{ik} \eta_{ab} c^b_k$ are dual (13.155).

13.41 Write Dirac's action density in the explicitly hermitian form $L_D = -\frac{1}{2}\overline{\psi}\gamma^i \partial_i \psi - \frac{1}{2}\left[\overline{\psi}\gamma^i \partial_i \psi\right]^\dagger$ in which the field ψ has the invariant form $\psi = e_a \psi_a$ and $\overline{\psi} = i\psi^\dagger \gamma^0$. Use the identity $\left[\overline{\psi}_a \gamma^i \psi_b\right]^\dagger = -\overline{\psi}_b \gamma^i \psi_a$ to show that the gauge-field matrix A_i defined as the coefficient of $\overline{\psi}_a \gamma^i \psi_b$ as in $\overline{\psi}_a \gamma^i (\partial_i + i A_{iab}) \psi_b$ is hermitian $A^*_{iab} = A_{iba}$.

14

Forms

14.1 Exterior Forms

1-Forms: A **1-form** is a linear function ω that maps vectors into numbers. Thus, if A and B are vectors in $\mathbb{R}^n$ and z and w are numbers, then

$$\omega(zA + wB) = z\,\omega(A) + w\,\omega(B). \tag{14.1}$$

The n coordinates $x_1, \ldots, x_n$ are 1-forms; they map a vector A into its coordinates: $x_1(A) = A_1, \ldots, x_n(A) = A_n$. Every 1-form may be expanded in terms of these **basic** 1-forms as

$$\omega = B_1 x_1 + \cdots + B_n x_n \tag{14.2}$$

so that

$$\begin{aligned} \omega(A) &= B_1 x_1(A) + \cdots + B_n x_n(A) \\ &= B_1 A_1 + \cdots + B_n A_n \\ &= (B, A) = \boldsymbol{B} \cdot \boldsymbol{A}. \end{aligned} \tag{14.3}$$

Thus, every 1-form is associated with a (dual) vector, in this case B.

2-Forms: A **2-form** is a function that maps pairs of vectors into numbers linearly and skew-symmetrically. Thus, if A, B, and C are vectors in $\mathbb{R}^n$ and z and w are numbers, then

$$\begin{aligned} \omega^2(zA + wB, C) &= z\,\omega^2(A, C) + w\,\omega^2(B, C) \\ \omega^2(A, B) &= -\,\omega^2(B, A). \end{aligned} \tag{14.4}$$

One often drops the superscript and writes the addition of two 2-forms as

$$(\omega_1 + \omega_2)(A, B) = \omega_1(A, B) + \omega_2(A, B). \tag{14.5}$$

Example 14.1 (Parallelogram) The **oriented area** of the parallelogram defined by two 2-vectors A and B is the determinant

$$\omega(A, B) = \begin{vmatrix} A_1 & A_2 \\ B_1 & B_2 \end{vmatrix}. \tag{14.6}$$

This 2-form maps the ordered pair of vectors (A, B) into the oriented area ($\pm$ the usual area) of the parallelogram they describe. To check that this 2-form gives the area to within a sign, rotate the coordinates so that the 2-vector A runs from the origin along the x-axis. Then $A_2 = 0$, and the 2-form gives $A_1 B_2$ which is the base A_1 of the parallelogram times its height B_2.

Example 14.2 (Parallelepiped) The **triple scalar product** of three 3-vectors

$$\omega_A^2(B, C) = \boldsymbol{A} \cdot \boldsymbol{B} \times \boldsymbol{C} = \begin{vmatrix} A_1 & A_2 & A_3 \\ B_1 & B_2 & B_3 \\ C_1 & C_2 & C_3 \end{vmatrix} = \omega^3(A, B, C) \tag{14.7}$$

is both a 2-form that depends upon the vector A and also a **3-form** that maps the triplet of vectors A, B, C into the signed volume of their parallelepiped.

k-Forms: A **k-form** (or an **exterior form of degree *k***) is a linear function of k vectors that is antisymmetric. For vectors $A_1, \ldots, A_k$ and numbers z and w

$$\omega(zA_1' + wA_1'', A_2, \ldots, A_k) = z\,\omega(A_1', A_2, \ldots, A_k) + w\,\omega(A_1'', A_2, \ldots, A_k) \tag{14.8}$$

and the interchange of any two vectors makes a minus sign

$$\omega(A_2, A_1, \ldots, A_k) = -\omega(A_1, A_2, \ldots, A_k). \tag{14.9}$$

Exterior Product of Two 1-Forms: The 1-form ω_1 maps the vectors A and B into the numbers $\omega_1(A)$ and $\omega_1(B)$, and the 1-form ω_2 does the same thing with $1 \to 2$. The value of the **exterior product** $\omega_1 \wedge \omega_2$ on the two vectors A and B is the 2-form defined by the 2×2 determinant

$$\omega_1 \wedge \omega_2(A, B) = \begin{vmatrix} \omega_1(A) & \omega_2(A) \\ \omega_1(B) & \omega_2(B) \end{vmatrix} = \omega_1(A)\omega_2(B) - \omega_2(A)\omega_1(B) \tag{14.10}$$

or more formally

$$\omega_1 \wedge \omega_2 = \omega_1 \otimes \omega_2 - \omega_2 \otimes \omega_1. \tag{14.11}$$

The **most general 2-form** on $\mathbb{R}^n$ is a linear combination of the basic 2-forms $x_i \wedge x_j$

$$\omega^2 = \sum_{1 \le i < k \le n} a_{ik}\, x_i \wedge x_k. \tag{14.12}$$

If the unit vectors in the n orthogonal directions of $\mathbb{R}^n$ are $e_1, \ldots, e_n$, then $x_i(e_k) = \delta_{ik}$ and so

$$\omega^2(e_i, e_k) = a_{ik} \begin{vmatrix} x_i(e_i) & x_k(e_i) \\ x_i(e_k) & x_k(e_k) \end{vmatrix} = a_{ik} \begin{vmatrix} 1 & 0 \\ 0 & 1 \end{vmatrix} = a_{ik}. \tag{14.13}$$

Exterior Product of k 1-Forms: The **exterior product** of k 1-forms $\omega_1, \ldots, \omega_k$ maps the k n-vectors $A_1, A_2, \ldots, A_k$ to the determinant

$$\omega_1 \wedge \omega_2 \wedge \cdots \wedge \omega_k(A_1, A_2, \ldots, A_k) = \begin{vmatrix} \omega_1(A_1) & \ldots & \omega_k(A_1) \\ \vdots & \delta \ldots & \vdots \\ \omega_1(A_k) & \ldots & \omega_k(A_k) \end{vmatrix}. \tag{14.14}$$

The **most general k-form** on $\mathbb{R}^n$ is a linear combination of the various exterior products of k basic 1-forms $x_{i_1} \wedge \cdots \wedge x_{i_k}$

$$\omega^k = \sum_{1 \le i_1 < \ldots i_k \le n} a_{i_1 \ldots i_k} x_{i_1} \wedge \cdots \wedge x_{i_k}. \tag{14.15}$$

Exterior Multiplication: The **exterior multiplication** of a k-form with an ℓ-form is linear, associative, and antisymmetric

$$\omega^k \wedge \omega^\ell = (-1)^{k\ell} \omega^\ell \wedge \omega^k. \tag{14.16}$$

Restriction of Forms: A p-form ω^p is a map from the product $V \times \cdots \times V$ of p copies of some vector space V into the real numbers. The restriction $\omega^p|_U (A_1, A_2, \ldots, A_p)$ of the p-form ω^p to a subspace $U \subset V$ is the same p-form ω^p but with its domain restricted to vectors $A_i \in U$.

14.2 Differential Forms

A **manifold** is a set of points that can be labeled locally by coordinates in $\mathbb{R}^n$ in such a way that the coordinates make sense when the local regions overlap. The k-dimensional surface S^k of the unit sphere in $\mathbb{R}^{k+1}$

$$\sum_{i=1}^{k+1} y_i^2 = 1 \tag{14.17}$$

is an example of a manifold. A **smooth** function $f(x_1, \ldots, x_n)$ is one that is infinitely differentiable with respect to all combinations of its arguments $x_1, \ldots, x_n$.

There are two ways of thinking about differential forms. The Russian literature views a manifold as embedded in $\mathbb{R}^n$ and so is somewhat more straightforward. We will discuss it first.

The Russian Way: Suppose $x(t)$ is a curve with $x(0) = x$ on some **manifold** M, and $f(x(t))$ is a smooth function $f : \mathbb{R}^n \to \mathbb{R}$ that maps points $x(t)$ into numbers. Then the **differential** $df(\dot{x}(t))$ maps $\dot{x}(t)$ at x into

$$df\left(\frac{d}{dt}x(t)\right) \equiv \frac{d}{dt}f(x(t)) = \sum_{j=1}^{n} \dot{x}(t)_j \frac{\partial f(x(t))}{\partial x_j} = \dot{x}(t) \cdot \nabla f(x(t)) \quad (14.18)$$

all at $t = 0$. As physicists, we think of df as a number – the change in the function $f(x)$ when its argument x is changed by dx. Russian mathematicians think of df as a linear map of tangent vectors $\dot{x}$ at x into numbers. Since this map is linear, we may multiply the definition (14.18) by dt and arrive at the more familiar formula

$$dt\, df\left(\frac{d}{dt}x(t)\right) = df\left(dt\frac{d}{dt}x(t)\right) = df\,(dx(t)) = dx(t) \cdot \nabla f(x(t)) \quad (14.19)$$

all at $t = 0$. So

$$df(dx) = dx \cdot \nabla f \quad (14.20)$$

is the physicist's df.

Since the differential df is a linear map of vectors $\dot{x}(0)$ into numbers, it is a 1-form; since it is defined on vectors like $\dot{x}(0)$, it is a **differential 1-form**. The term *differential 1-form* underscores the fact that the actual value of the differential df depends upon the vector $\dot{x}(0)$ and the point $x = x(0)$. Mathematicians call the space of vectors $\dot{x}(0)$ at the point $x = x(0)$ the **tangent space** TM_x. They say df is a smooth map of the **tangent bundle** TM, which is the union of the tangent spaces for all points x in the manifold M, to the real line, so $df : TM \to \mathbb{R}$.

In the special case in which $f(x) = x_i(x) = x_i$, the differential $dx_i(\dot{x}(t))$ by (14.18) is

$$dx_i(\dot{x}(t)) = \sum_{j=1}^{n} \dot{x}_j(t) \frac{\partial x_i(x)}{\partial x_j} = \sum_{j=1}^{n} \dot{x}_j(t) \frac{\partial x_i}{\partial x_j} = \sum_{j=1}^{n} \dot{x}(t)_j\, \delta_{ij} = \dot{x}_i(t). \quad (14.21)$$

These dx_i's are the **basic differentials**. Using A for the vector $\dot{x}(t)$, we find from our definition (14.18) that

$$dx_i(A) = \sum_{j=1}^{n} A_j \frac{\partial x_i}{\partial x_j} = \sum_{j=1}^{n} A_j \delta_{ij} = A_i \quad (14.22)$$

as well as

$$df(A) = \sum_{j=1}^{n} A_j \frac{\partial f(x)}{\partial x_j} = \sum_{j=1}^{n} \frac{\partial f(x)}{\partial x_j} dx_j(A) \quad (14.23)$$

or

$$df = \sum_{j=1}^{n} \frac{\partial f(x)}{\partial x_j} \, dx_j. \tag{14.24}$$

Example 14.3 (dr^2) If $r^2 = x_1^2 + x_2^2$, then the differential 1-form dr^2 is

$$dr^2 = 2x_1 \, dx_1 + 2x_2 \, dx_2. \tag{14.25}$$

It takes the 2-vector A into the number

$$dr^2(A) = 2x_1 \, dx_1(A) + 2x_2 \, dx_2(A) = 2x_1 \, A_1 + 2x_2 \, A_2. \tag{14.26}$$

So if $A = (\epsilon_1, \epsilon_2)$, then $dr^2(A) = 2x_1 \, \epsilon_1 + 2x_2 \, \epsilon_2$.

The Other Way: Most American, French, and English mathematicians use a more abstract approach. They abstract from the basic definition (14.18) the rule

$$df\left(\frac{d}{dt}\right) = \frac{d}{dt} f \tag{14.27}$$

or more generally

$$df\left(\frac{\partial}{\partial x_k}\right) = \frac{\partial}{\partial x_k} f. \tag{14.28}$$

In particular, if $f(x) = x_i$, then

$$dx_i\left(\frac{\partial}{\partial x_k}\right) = \frac{\partial}{\partial x_k} x_i = \delta_{ik}. \tag{14.29}$$

In this frequently used notation, the idea is that the derivatives

$$\partial_k \equiv \frac{\partial}{\partial x_k} \tag{14.30}$$

form a set of orthonormal vectors to which the forms dx_i are dual

$$dx_i(\partial_k) = \delta_{ik}. \tag{14.31}$$

In the nonRussian literature, equations (14.27–14.31) *define* how the basic 1-forms dx_i act on the vectors ∂_k.

Change of Variables: Suppose that $x_1, \ldots, x_n$ and $y_1, \ldots, y_n$ are two systems of coordinates on $\mathbb{R}^n$, and that $dx_1, \ldots, dx_n$ and $dy_1, \ldots, dy_n$ are two sets of basic differentials. Then by applying the formula (14.24) to the function $y_k(x)$, we get

$$dy_k = \sum_{j=1}^{n} \frac{\partial y_k(x)}{\partial x_j} \, dx_j \tag{14.32}$$

which is the familiar rule for changing variables.

The **most general differential 1-form** ω on the space $\mathbb{R}^n$ with coordinates $x_1 \dots x_n$ is a linear combination of the basic differentials dx_i with coefficients $a_i(x)$ that are smooth functions of $x = (x_1, \dots, x_n)$

$$\omega = a_1(x)\,dx_1 + \cdots + a_n(x)\,dx_n. \tag{14.33}$$

The **basic differential 2-forms** are $dx_i \wedge dx_k$ defined as

$$dx_i \wedge dx_k(A, B) = \begin{vmatrix} dx_i(A) & dx_k(A) \\ dx_i(B) & dx_k(B) \end{vmatrix} = \begin{vmatrix} A_i & A_k \\ B_i & B_k \end{vmatrix} = A_i B_k - A_k B_i. \tag{14.34}$$

So in particular

$$dx_i \wedge dx_i = 0. \tag{14.35}$$

The **basic differential k-forms** $dx_1 \wedge \cdots \wedge dx_k$ are defined as

$$dx_1 \wedge \cdots dx_k(A_1, \dots, A_k) = \begin{vmatrix} dx_1(A_1) & \dots & dx_k(A_1) \\ \vdots & \ddots & \vdots \\ dx_1(A_k) & \dots & dx_k(A_k) \end{vmatrix} = \begin{vmatrix} A_{11} & \dots & A_{1k} \\ \vdots & \ddots & \vdots \\ A_{k1} & \dots & A_{kk} \end{vmatrix}. \tag{14.36}$$

Example 14.4 ($dx_3 \wedge dr^2$) If $r^2 = x_1^2 + x_2^2 + x_3^2$, then dr^2 is

$$dr^2 = 2(x_1 dx_1 + x_2\,dx_2 + x_3\,dx_3) \tag{14.37}$$

and the differential 2-form $\omega = dx_3 \wedge dr^2$ is

$$\omega = dx_3 \wedge 2(x_1 dx_1 + x_2\,dx_2 + x_3\,dx_3) = 2x_1 dx_3 \wedge dx_1 + 2x_2 dx_3 \wedge dx_2 \tag{14.38}$$

since in view of (14.35) $dx_3 \wedge dx_3 = 0$. So the value of the 2-form ω on the vectors $A = (1, 2, 3)$ and $B = (2, 1, 1)$ at the point $x = (3, 0, 3)$ is

$$\omega(A, B) = 2x_1 dx_3 \wedge dx_1(A, B) = 6 \begin{vmatrix} dx_3(A) & dx_1(A) \\ dx_3(B) & dx_1(B) \end{vmatrix} = 6 \begin{vmatrix} 3 & 1 \\ 1 & 2 \end{vmatrix} = 30. \tag{14.39}$$

On the vectors, $C = (1, 0, 0)$ and $D = (0, 0, 1)$ at $x = (2, 3, 4)$, this 2-form has the value $\omega(C, D) = -4$.

The most general **differential k-form** ω^k on the space $\mathbb{R}^n$ with coordinates $x_1, \dots, x_n$ is

$$\omega^k = \sum_{1 \le i_1 < \dots i_k \le n} a_{i_1 \dots i_k}(x)\,dx_{i_1} \wedge \cdots \wedge dx_{i_k} \tag{14.40}$$

in which the functions $a_{i_1 \dots i_k}(x)$ are smooth on $\mathbb{R}^n$.

Example 14.5 (Change of variables) If x_1, x_2, x_3 and y_1, y_2, y_3 are two coordinate systems on $\mathbb{R}^3$, then in terms of the basic 1-forms dy_k, the 2-form $\omega = X dx_2 \wedge dx_3$ is by (14.32)

$$\omega = X dx_2 \wedge dx_3 = X \left(\sum_{k=1}^{3} \frac{\partial x_2}{\partial y_j} dy_j \right) \wedge \left(\sum_{k=1}^{3} \frac{\partial x_3}{\partial y_k} dy_k \right) \tag{14.41}$$

in which jacobians appear such as

$$\frac{\partial(x_2, x_3)}{\partial(y_1, y_2)} = \frac{\partial x_2}{\partial y_1} \frac{\partial x_3}{\partial y_2} - \frac{\partial x_2}{\partial y_2} \frac{\partial x_3}{\partial y_1}. \tag{14.42}$$

In terms of these jacobians, the 2-form $\omega = X dx_2 \wedge dx_3$ is (exercise 14.2)

$$\omega = X \left(\frac{\partial(x_2, x_3)}{\partial(y_1, y_2)} dy_1 \wedge dy_2 + \frac{\partial(x_2, x_3)}{\partial(y_2, y_3)} dy_2 \wedge dy_3 + \frac{\partial(x_2, x_3)}{\partial(y_3, y_1)} dy_3 \wedge dy_1 \right). \tag{14.43}$$

On the vectors of Example 14.4, both forms of ω give $\omega(A, B) = -X$.

Example 14.6 (Euclidian three space) In orthogonal coordinates, the square ds^2 of the length of a vector $\boldsymbol{dx}$ is

$$ds^2 = h_1^2 dx_1^2 + h_2^2 dx_2^2 + h_3^2 dx_3^2 \tag{14.44}$$

and the gradient (2.15) is

$$\nabla f = \frac{1}{h_i} \frac{\partial f}{\partial x_i} \hat{e}_i. \tag{14.45}$$

Thus in cylindrical coordinates (ρ, ϕ, z), we have $h_\rho = 1$, $h_\phi = \rho$, and $h_z = 1$, while in spherical coordinates (r, θ, ϕ), we have $h_r = 1$, $h_\theta = r$, and $h_\phi = r \sin\theta$. The value of the form dx_i on the unit vector $\hat{\boldsymbol{e}}_j$ is by (14.20)

$$dx_k(\hat{\boldsymbol{e}}_j) = \hat{\boldsymbol{e}}_j \cdot \nabla x_k = \hat{\boldsymbol{e}}_j \cdot \frac{1}{h_i} \frac{\partial x_k}{\partial x_i} \hat{e}_i = \frac{1}{h_j} \frac{\partial x_k}{\partial x_j} = \frac{\delta_{kj}}{h_j}. \tag{14.46}$$

Thus $d\rho(\hat{\boldsymbol{e}}_\rho) = 1$, $d\phi(\hat{\boldsymbol{e}}_\phi) = 1/\rho$, and $dz(\hat{\boldsymbol{e}}_z) = 1$.

Example 14.7 (Three-dimensional vectors and their forms) Any 3-dimensional vector $\boldsymbol{A}$ defines a 1-form as the dot product

$$\omega_A^1(\boldsymbol{U}) = \boldsymbol{A} \cdot \boldsymbol{U} \tag{14.47}$$

and a 2-form as the triple cross-product

$$\omega_A^2(\boldsymbol{U}, \boldsymbol{V}) = \boldsymbol{A} \cdot (\boldsymbol{U} \times \boldsymbol{V}). \tag{14.48}$$

Here we assume that we have a right-handed set of basis vectors $\hat{\boldsymbol{e}}_1$, $\hat{\boldsymbol{e}}_2$, and $\hat{\boldsymbol{e}}_3$ with $\hat{\boldsymbol{e}}_1 \times \hat{\boldsymbol{e}}_2 = \hat{\boldsymbol{e}}_3$ so as to define the cross-product $\boldsymbol{U} \times \boldsymbol{V}$. Such a manifold is said to be **oriented**.

The quantity

$$A = A_1 \hat{e}_1 + A_2 \hat{e}_2 + A_3 \hat{e}_3 \tag{14.49}$$

is a vector field $A(x)$. So if we use (14.44) for the squared length ds^2, then we can write the 1-form (14.47) as

$$\omega_A^1 = A_1 \, h_1 \, dx_1 + A_2 \, h_2 \, dx_2 + A_3 \, h_3 \, dx_3 \tag{14.50}$$

because by (14.46) and summing over i and k we get

$$\omega_A^1(U) = A_i \, h_i \, dx_i \left(U_k \hat{e}_k\right) = A_i \, h_i \, U_k \, \delta_{ik}/h_k = A_i U_i = A \cdot U. \tag{14.51}$$

Similarly (Exercise 14.6), the 2-form (14.48) is

$$\omega_A^2 = A_1 \, h_2 h_3 \, dx_2 \wedge dx_3 + A_2 \, h_3 h_1 \, dx_3 \wedge dx_1 + A_3 \, h_1 h_2 \, dx_1 \wedge dx_2. \tag{14.52}$$

By analogy with the definition (14.50), the gradient 1-form $\omega_{\nabla f}^1$ is

$$\omega_{\nabla f}^1 = (\nabla f)_k \, h_k \, dx_k \tag{14.53}$$

summed over repeated indices. The relation $\omega_{\nabla f}^1 = df$ gives

$$\omega_{\nabla f}^1 = (\nabla f)_k \, h_k \, dx_k = df = \frac{\partial f}{\partial x_k} dx_k \tag{14.54}$$

according to the definition (14.24) of df. So the vector field ∇f is

$$\nabla f = \sum_{k=1}^{3} \frac{1}{h_k} \frac{\partial f}{\partial x_k} \hat{e}_k \tag{14.55}$$

which in cylindrical and spherical coordinates is

$$\nabla f = \frac{\partial f}{\partial \rho} \hat{e}_\rho + \frac{1}{\rho} \frac{\partial f}{\partial \phi} \hat{e}_\phi + \frac{\partial f}{\partial z} \hat{e}_z = \frac{\partial f}{\partial r} \hat{e}_r + \frac{1}{r} \frac{\partial f}{\partial \theta} \hat{e}_\theta + \frac{1}{r \sin\theta} \frac{\partial f}{\partial \phi} \hat{e}_\phi \tag{14.56}$$

in agreement with (2.16).

14.3 Exterior Differentiation

Exterior differentiation is nifty. The differential (14.24)

$$df = \sum_{k=1}^{n} \frac{\partial f}{\partial x_k} dx_k \tag{14.57}$$

is the **exterior derivative** of the function $f(x)$, itself a 0-form. The operator

$$d = \sum_{k=1}^{n} \frac{\partial}{\partial x_k} dx_k \tag{14.58}$$

turns the 0-form f into the differential 1-form df.

Applied to the 1-form

$$\omega^1 = \sum_{i=1}^{n} a_i(x)\, dx_i \tag{14.59}$$

the exterior derivative d generates the 2-form

$$d\omega^1 = d\left(\sum_{i=1}^{n} a_i(x)\, dx_i\right) = \sum_{i,k=1}^{n} a_{i,k}(x)\, dx_k \wedge dx_i \tag{14.60}$$

in which $a_{i,k} = \partial_k a_i$. But a second application of d gives zero:

$$\begin{aligned} dd\omega^1 &= dd\left(\sum_{i=1}^{n} a_i\, dx_i\right) = d\left(\sum_{i,k=1}^{n} a_{i,k}\, dx_k \wedge dx_i\right) \\ &= \sum_{i,k,\ell=1}^{n} a_{i,k\ell}\, dx_\ell \wedge dx_k \wedge dx_i = 0 \end{aligned} \tag{14.61}$$

because the double partial derivative $a_{i,k\ell} = \partial_\ell \partial_k a_i$ is symmetric in k and ℓ while the wedge product $dx_\ell \wedge dx_k$ is antisymmetric in these indices.

We have seen (14.40) that the most general differential k-form is

$$\omega^k = \sum_{1 \le i_1 < \ldots i_k \le n} a_{i_1 \ldots i_k}(x)\, dx_{i_1} \wedge \cdots \wedge dx_{i_k} \tag{14.62}$$

in which the functions $a_{i_1 \ldots i_k}(x)$ are smooth on $\mathbb{R}^n$. The exterior derivative operator d turns ω^k into the $(k+1)$-form

$$\begin{aligned} d\omega^k &= \sum_{1 \le i_1 < \ldots i_k \le n} d\left(a_{i_1 \ldots i_k}(x)\, dx_{i_1} \wedge \cdots \wedge dx_{i_k}\right) \\ &= \sum_{1 \le \ell, i_1 < \ldots i_k \le n} a_{i_1 \ldots i_k,\ell}(x)\, dx_\ell \wedge dx_{i_1} \wedge \cdots \wedge dx_{i_k}. \end{aligned} \tag{14.63}$$

Once again $d\, d\, \omega^k = 0$ so quite generally

$$d\, d = 0. \tag{14.64}$$

If ω is the wedge product of two 1-forms $\omega_a = a_i\, dx_i$ and $\omega_b = b_k\, dx_k$

$$\omega = \omega_a \wedge \omega_b = a_i\, dx_i \wedge b_k\, dx_k \tag{14.65}$$

then d maps it to

$$\begin{aligned} d\,\omega &= d\,(a_i\, dx_i \wedge b_k\, dx_k) = d\,(a_i\, b_k\, dx_i \wedge dx_k) \\ &= (a_i\, b_k)_{,\ell}\, dx_\ell \wedge dx_i \wedge dx_k \\ &= (a_{i,\ell}\, b_k + a_i\, b_{k,\ell})\, dx_\ell \wedge dx_i \wedge dx_k \\ &= a_{i,\ell}\, dx_\ell \wedge dx_i \wedge b_k\, dx_k + a_i\, b_{k,\ell}\, dx_\ell \wedge dx_i \wedge dx_k \end{aligned}$$

$$= (a_{i,\ell}\, dx_\ell \wedge dx_i) \wedge b_k\, dx_k - a_i\, dx_i \wedge (b_{k,\ell}\, dx_\ell \wedge dx_k)$$
$$= d\,\omega_a \wedge \omega_b - \omega_a \wedge d\,\omega_b. \tag{14.66}$$

More generally, the exterior derivative operator d maps the wedge product of a k-form ω^k and a p-form ω^p to

$$d\left(\omega^k \wedge \omega^p\right) = \left(d\omega^k\right) \wedge \omega^p + (-1)^k \omega^k \wedge \left(d\omega^p\right). \tag{14.67}$$

Example 14.8 (Phase space) If ω^1 is the 1-form

$$\omega^1 = p_1 dq_1 + \cdots + p_n dq_n = p \cdot dq \tag{14.68}$$

with coordinates $p_1, \ldots, p_n, q_1, \ldots, q_n$, then $d\,\omega^1$ is the 2-form

$$d\,\omega^1 = d\left(\sum_{i=1}^n p_i\, dq_i\right) = \sum_{i,k=1}^n (\partial_{p_k} p_i)\, dp_k \wedge dq_i = \sum_{i,k=1}^n \delta_{ik}\, dp_k \wedge dq_i$$
$$= dp_1 \wedge dq_1 + \cdots + dp_n \wedge dq_n = dp \wedge dq. \tag{14.69}$$

It follows that $d\,(dp \wedge dq) = d\,d\,\omega^1 = 0$.

Example 14.9 (A Poincaré invariant) The 1-form $\omega^1 = p \cdot dq$ maps a tiny piece $\partial_s q\, ds$ of a phase-space trajectory into a small element of action $p \cdot dq(\partial_s q\, ds) = p \cdot \partial_s q\, ds$. The sum of these pieces along a *closed* trajectory

$$A = \int_{\partial S} \omega^1 = \oint p \cdot dq = \oint \sum_{i=1}^n p_i\, dq_i = \oint p \cdot \frac{\partial q}{\partial s}\, ds \tag{14.70}$$

of a Hamilton system (Section 18.1) is a Poincaré invariant. Because the trajectory is a loop, we may integrate the second term in its time derivative by parts

$$\dot{A} = \oint \left(\dot{p} \cdot \frac{\partial q}{\partial s} + p \cdot \frac{\partial^2 q}{\partial s \partial t}\right) ds = \oint \left(\dot{p} \cdot \frac{\partial q}{\partial s} - \frac{\partial p}{\partial s} \cdot \dot{q}\right) ds \tag{14.71}$$

without acquiring an extra term. Hamilton's equations

$$\dot{p}_i = -\frac{\partial H}{\partial q_i} \quad \text{and} \quad \dot{q}_i = \frac{\partial H}{\partial p_i} \quad \text{for} \quad i = 1, \ldots, n \tag{14.72}$$

then let us write $\dot{A}$ as

$$\dot{A} = \oint \left(-\frac{\partial H}{\partial q} \cdot \frac{\partial q}{\partial s} - \frac{\partial p}{\partial s} \cdot \frac{\partial H}{\partial p}\right) ds = -\oint \frac{\partial H}{\partial s}\, ds = -\oint dH = 0 \tag{14.73}$$

because the trajectory is closed.

Example 14.10 (The Bohr Model) In 1912, Bohr considered an electron in a circular orbit around a proton, set Poincaré's invariant equal to an integral multiple of Planck's constant

$$A = \oint p\, dq = 2\pi r p = nh \tag{14.74}$$

and so quantized the orbital angular momentum as $L = rp = n\hbar$. One can derive the energy levels of the hydrogen atom from this rule (exercise 14.12). In 1924, Arnold Sommerfeld applied this trick to a more general orbit of a relativistic electron about a proton and got the energy levels that Dirac would four years later.

Example 14.11 (Constant area) Consider three nearby points in phase space (p, q), $(p + \delta p, q + \delta q)$, and $(p + \Delta p, q + \Delta q)$ that move according to Hamilton's equations (14.72). The time derivatives of the tiny displacements δp_i and δq_i are (Exercise 14.9)

$$\begin{aligned} \frac{d}{dt}\delta p_i &= \delta \dot{p}_i = \sum_{k=1}^{n} -\frac{\partial^2 H}{\partial q_i \partial q_k}\delta q_k - \frac{\partial^2 H}{\partial q_i \partial p_k}\delta p_k \\ \frac{d}{dt}\delta q_i &= \delta \dot{q}_i = \sum_{k=1}^{n} \frac{\partial^2 H}{\partial p_i \partial q_k}\delta q_k + \frac{\partial^2 H}{\partial p_i \partial p_k}\delta p_k . \end{aligned} \tag{14.75}$$

Similar equations give the derivatives of the small differences Δp_i and Δq_i.

The 2-form (14.69) maps the n pairs of 2-vectors $(\delta p, \delta q)$ and $(\Delta p, \Delta q)$ into a sum of areas of parallelograms

$$d\omega^1(\delta p, \delta q; \Delta p, \Delta q) = \begin{vmatrix} \delta p_1 & \delta q_1 \\ \Delta p_1 & \Delta q_1 \end{vmatrix} + \cdots + \begin{vmatrix} \delta p_n & \delta q_n \\ \Delta p_n & \Delta q_n \end{vmatrix} . \tag{14.76}$$

By using the time derivatives (14.75) of δp_i and δq_i and those of Δp_i and Δq_i, one may show (Exercise 14.10) that this sum of areas remains constant

$$\frac{d}{dt}\, d\omega^1(\delta p, \delta q; \Delta p, \Delta q) = 0 \tag{14.77}$$

along the trajectories in phase space (Gutzwiller, 1990, chap. 7).

Example 14.12 (Curl) We saw in Example 14.7 that the 1-form (14.50) of a vector field $\boldsymbol{A}$ is $\omega_A = A_1 h_1\, dx_1 + A_2 h_2\, dx_2 + A_3 h_3\, dx_3$ in which the h_k's are those that determine (14.44) the squared length $ds^2 = h_k^2\, dx_k^2$ of the triply orthogonal coordinate system with unit vectors $\hat{\boldsymbol{e}}_1, \hat{\boldsymbol{e}}_2, \hat{\boldsymbol{e}}_3$. So the exterior derivative of the 1-form ω_A is

$$\begin{aligned} d\omega_A &= \sum_{i,k=1}^{3} \partial_k (A_i\, h_i)\, dx_k \wedge dx_i \\ &= \left[\frac{\partial(A_3\, h_3)}{\partial x_2} - \frac{\partial(A_2\, h_2)}{\partial x_3} \right] dx_2 \wedge dx_3 \\ &\quad + \left[\frac{\partial(A_2\, h_2)}{\partial x_1} - \frac{\partial(A_1\, h_1)}{\partial x_2} \right] dx_1 \wedge dx_2 \\ &\quad + \left[\frac{\partial A_1\, h_1}{\partial x_3} - \frac{\partial(A_3\, h_3)}{\partial x_1} \right] dx_3 \wedge dx_1 \equiv \omega_{\nabla \times \boldsymbol{A}} . \end{aligned} \tag{14.78}$$

Comparison with Eq. (14.52) shows that the curl of $\boldsymbol{A}$ is

$$\nabla \times \boldsymbol{A} = \frac{1}{h_2\, h_3} \left(\frac{\partial A_3\, h_3}{\partial x_2} - \frac{\partial A_2\, h_2}{\partial_3} \right) dx_2 \wedge dx_3\, \hat{\boldsymbol{e}}_1 + \cdots$$

$$= \frac{1}{h_1\, h_2\, h_3} \begin{vmatrix} h_1 \hat{\boldsymbol{e}}_1 & h_2 \hat{\boldsymbol{e}}_2 & h_3 \hat{\boldsymbol{e}}_3 \\ \partial_1 & \partial_2 & \partial_3 \\ A_1 h_1 & A_2 h_2 & A_3 h_3 \end{vmatrix}$$

$$= \frac{1}{h_1 h_2 h_3} \sum_{i,j,k=1}^{3} \epsilon_{ijk} h_i\, \hat{\boldsymbol{e}}_i \frac{\partial (A_k h_k)}{\partial x_j} \tag{14.79}$$

as we saw in (2.44). This formula gives our earlier expressions for the curl in cylindrical and spherical coordinates (2.46 and 2.47).

Example 14.13 (Divergence) We have seen in Equations (14.48, 14.49, and 14.52) that the 2-form $\omega_A(U, V) = A \cdot (U \times V)$ of the vector field $\boldsymbol{A} = A_1 \hat{\boldsymbol{e}}_1 + A_2 \hat{\boldsymbol{e}}_2 + A_3 \hat{\boldsymbol{e}}_3$ is

$$\omega_A^2 = A_1\, h_2\, h_3\; dx_2 \wedge dx_3 + A_2\, h_3\, h_1\; dx_3 \wedge dx_1 + A_3\, h_1\, h_2\; dx_1 \wedge dx_2. \tag{14.80}$$

The exterior derivative of this 2-form is

$$d\,\omega_A = \sum_{k=1}^{3} \frac{\partial}{\partial x_k} \omega_A = \frac{\partial A_1\, h_2 h_3}{\partial x_1} dx_1 \wedge dx_2 \wedge dx_3$$

$$+ \frac{\partial A_2\, h_3\, h_1}{\partial x_2} dx_2 \wedge dx_3 \wedge dx_1 + \frac{\partial A_3\, h_1\, h_2}{\partial x_3} dx_3 \wedge dx_1 \wedge dx_2$$

$$= \left(\sum_{k=1}^{3} \frac{\partial (A_k\, h_1\, h_2\, h_3 / h_k)}{\partial x_k} \right) dx_1 \wedge dx_2 \wedge dx_3. \tag{14.81}$$

If one defines the divergence $\nabla \cdot \boldsymbol{A}$ as

$$d\omega_A = (\nabla \cdot \boldsymbol{A})\; h_1\, h_2\, h_3\, dx_1 \wedge dx_2 \wedge dx_3 \tag{14.82}$$

then $\nabla \cdot \boldsymbol{A}$ must be

$$\nabla \cdot \boldsymbol{A} = \frac{1}{h_1\, h_2\, h_3} \left(\sum_{k=1}^{3} \frac{\partial (A_k\, h_1\, h_2\, h_3 / h_k)}{\partial x_k} \right) \tag{14.83}$$

in agreement with (2.19) from which the specific formulas for cylindrical (2.21) and spherical (2.22) coordinates follow.

Example 14.14 (Divergence of a gradient) By combining our expression (14.83) for the divergence with our formula (14.55) for the gradient of a function f, we find that its laplacian $\triangle f$ in orthogonal coordinates is

$$\Delta f(\boldsymbol{x}) \equiv \nabla \cdot \nabla f(\boldsymbol{x}) = \frac{1}{h_1 h_2 h_3} \left[\sum_{k=1}^{3} \frac{\partial}{\partial x_k} \left(\frac{h_1 h_2 h_3}{h_k^2} \frac{\partial f(\boldsymbol{x})}{\partial x_k} \right) \right] \tag{14.84}$$

which agrees with (2.19) and so yields (2.21) for cylindrical coordinates and (2.22) for spherical ones.

14.4 Integration of Forms

Let's follow the Russian approach at first. Let $\gamma(t)$ be a smooth map from the unit interval $[0, 1]$ into some manifold $M \subset \mathbb{R}^n$. We divide this interval into tiny segments $[t_i, t_{i+1}]$ of length $dt = t_{i+1} - t_i$ which γ maps into vectors $d\gamma(dt_i) = \dot{\gamma}(t_i)\, dt$ that are tangent to the manifold at the point $\gamma(t_i)$. The integral of a 1-form ω along the curve γ is then the usual Riemann sum

$$\int_\gamma \omega = \lim_{dt \to 0} \sum_i \omega(\dot{\gamma}(t_i))\, dt. \tag{14.85}$$

If for example, the 1-form is $\omega_A(U) = A \cdot U$, then

$$\int_\gamma \omega = \int_\gamma \omega_A(\dot{\gamma}(t_i))\, dt = \int A \cdot \dot{\gamma}(t_i)\, dt = \int A \cdot d\gamma. \tag{14.86}$$

And if $\omega = a_k(x)\, dx_k$, then since $dt\, dx_k(\dot{\gamma}) = d\gamma_k$, the integral

$$\int_\gamma \omega = \lim_{dt \to 0} \sum_{i=1}^{n} \omega(\dot{\gamma}(t_i)\, dt) = \lim_{dt \to 0} \sum_{i=1}^{n} a_k(x)\, dx_k(\dot{\gamma}(t_i))\, dt = \int a_k(x)\, d\gamma_k \tag{14.87}$$

is a line integral on the manifold

$$\int_\gamma \omega = \int a_k\, dx_k = \int a_k\, d\gamma_k. \tag{14.88}$$

Suppose now that our 1-form ω is **exact**, that is, that $\omega = d\alpha = \alpha_{,k}\, dx_k$ where $\alpha(x)$ is a 0-form, that is, a function defined on the manifold. Then by (14.88) the integral of $d\alpha$ is

$$\int_\gamma d\alpha = \int \alpha_{,k}\, dx_k = \alpha(\gamma(1)) - \alpha(\gamma(0)). \tag{14.89}$$

The signed endpoints $\gamma(1)$ and $-\gamma(0)$ are the **boundary** of the curve $\gamma(t)$ which one writes as $\partial\gamma$. In this notation, we have

$$\int_\gamma d\alpha = \int_{\partial\gamma} \alpha. \tag{14.90}$$

Example 14.15 (Green's theorem) Let $\boldsymbol{u}$ and $\boldsymbol{\eta}$ be two infinitesimal vectors which form a parallelogram Π in the tangent space. We will compute the line integral of the 1-form $\omega = a_1(x_1, x_2)\, dx_1 + a_2(x_1, x_2)\, dx_2$ around the boundary $\partial\Pi$ this parallelogram. This boundary $\partial\Pi$ is a **chain** of four maps $t \to t\boldsymbol{u}$, $t \to \boldsymbol{u} + t\boldsymbol{\eta}$, $t \to \boldsymbol{\eta} + t\boldsymbol{u}$, and $t \to t\boldsymbol{\eta}$ of the unit interval $0 \le t \le 1$ into the plane defined by the two vectors $\boldsymbol{u}$ and $\boldsymbol{\eta}$. We assign multiplicities 1, 1, -1, and -1 to these four maps, that is, the full chain runs from a point that we'll call the origin to the point $\boldsymbol{u}$, then from $\boldsymbol{u}$ to $\boldsymbol{u} + \boldsymbol{\eta}$, and then from $\boldsymbol{u} + \boldsymbol{\eta}$ to $\boldsymbol{\eta}$, and then from $\boldsymbol{\eta}$ back to the origin. On the curve $\gamma(t) = t\boldsymbol{u}$, the differentials dx_k map $\dot{\gamma}(t)\, dt$ into $dx_k(\dot{\gamma}(t)\, dt) = dx_k(\boldsymbol{u}\, dt) = u_k\, dt$. Similarly, on the curve $\gamma(t) = t\boldsymbol{\eta}$, we have $dx_k(\dot{\gamma}(t)\, dt) = \eta_k\, dt$. So summing over $k = 1, 2$, we find

$$\int_{\partial\Pi} \omega = \int_0^1 \{[a_k(t\boldsymbol{u}) - a_k(t\boldsymbol{u} + \boldsymbol{\eta})]\, u_k - [a_k(t\boldsymbol{\eta}) - a_k(t\boldsymbol{\eta} + \boldsymbol{u})]\, \eta_k\}\, dt. \tag{14.91}$$

Since the tangent vectors $\boldsymbol{u}$ and $\boldsymbol{\eta}$ are infinitesimal, the square brackets are

$$\begin{aligned} a_k(t\boldsymbol{u}) - a_k(t\boldsymbol{u} + \boldsymbol{\eta}) &= -\eta_j \frac{\partial a_k}{\partial x_j} \\ a_k(t\boldsymbol{\eta}) - a_k(t\boldsymbol{\eta} + \boldsymbol{u}) &= -u_j \frac{\partial a_k}{\partial x_j} \end{aligned} \tag{14.92}$$

and so we have

$$\int_{\partial\Pi} \omega = \int_0^1 \left(-\eta_j \frac{\partial a_k}{\partial x_j} u_k + u_j \frac{\partial a_k}{\partial x_j} \eta_k\right) dt = \frac{\partial a_k}{\partial x_j} \left(u_j\, \eta_k - u_k\, \eta_j\right). \tag{14.93}$$

But the exterior derivative of ω is

$$d\,\omega = d\,(a_k\, dx_k) = \frac{\partial a_k}{\partial x_j} dx_j \wedge dx_k \tag{14.94}$$

so the last term in (14.93) is just the 2-form $d\,\omega$ applied to the tangent vectors $\boldsymbol{u}$ and $\boldsymbol{\eta}$

$$d\,\omega(\boldsymbol{u}, \boldsymbol{\eta}) = \frac{\partial a_k}{\partial x_j} dx_j \wedge dx_k(\boldsymbol{u}, \boldsymbol{\eta}) = \frac{\partial a_k}{\partial x_j} \left(u_j \eta_k - u_k \eta_j\right). \tag{14.95}$$

And $dx_j \wedge dx_k(\boldsymbol{u}, \boldsymbol{\eta}) = u_j \eta_k - u_k \eta_j$ is the area of the tiny parallelogram Π. So this last expression (14.95) is $d\omega$ integrated over the tiny parallelogram Π formed by the tangent vectors $\boldsymbol{u}$ and $\boldsymbol{\eta}$, and we have

$$\int_{\Pi} d\omega = \int_{\partial\Pi} \omega \tag{14.96}$$

for an infinitesimal parallelogram.

It is easy to extend this identity to an arbitrary surface S of finite extent. To do this, we tile the surface S with infinitesimal parallelograms Π_α with boundaries $\partial\Pi_\alpha$.

The integral over the finite surface S is then the sum of the integrals over the Π_α that tile S

$$\int_S d\omega = \sum_\alpha \int_{\Pi_\alpha} d\omega \tag{14.97}$$

which by (14.95) is a sum of integrals over the boundaries $\partial\Pi_\alpha$

$$\int_S d\omega = \sum_\alpha \int_{\Pi_\alpha} d\omega = \sum_\alpha \int_{\partial\Pi_\alpha} \omega. \tag{14.98}$$

In the sum of the integrals over the boundaries $\partial\Pi_\alpha$, the internal boundaries all cancel, leaving us with the integral over the boundary ∂S of the surface S. Thus we have

$$\int_S d\omega = \sum_\alpha \int_{\Pi_\alpha} d\omega = \sum_\alpha \int_{\partial\Pi_\alpha} \omega = \int_{\partial S} \omega \tag{14.99}$$

or more simply

$$\int_S d\omega = \int_{\partial S} \omega \tag{14.100}$$

which generalizes the identity (14.90) from 0-forms to 1-forms.

In the notation of ordinary vector calculus, this relation is

$$\int_S (\nabla \times \boldsymbol{A}) \cdot \boldsymbol{dS} = \oint_{\partial S} \boldsymbol{A} \cdot \boldsymbol{dx} \tag{14.101}$$

in accord (Exercise 14.13) with the curl formulas (14.78–14.79).

Example 14.16 (How electric motors and generators work) Since by (12.32) the curl of the vector potential $\boldsymbol{A}$ is the magnetic field $\boldsymbol{B}$, this last identity (14.101) implies that the magnetic flux Φ through a surface S is the line integral of the vector potential $\boldsymbol{A}$ around the edge of the surface

$$\Phi = \int_S \boldsymbol{B} \cdot \boldsymbol{dS} = \int_S (\nabla \times \boldsymbol{A}) \cdot \boldsymbol{dS} = \oint_{\partial S} \boldsymbol{A} \cdot \boldsymbol{dx}. \tag{14.102}$$

If we take the time derivative of this relation and remember (12.30) that the time derivative of the vector potential is $\dot{\boldsymbol{A}} = -\boldsymbol{E} - \nabla\phi$, then we find that the rate of change of the magnetic flux through a surface is the negative of the line integral of the electric field along the boundary of the surface

$$\dot{\Phi} = \int_S \dot{\boldsymbol{B}} \cdot \boldsymbol{dS} = \int_S (\nabla \times \dot{\boldsymbol{A}}) \cdot \boldsymbol{dS} = -\oint_{\partial S} \boldsymbol{E} \cdot \boldsymbol{dx} \tag{14.103}$$

or minus the voltage ($-\nabla\phi$ drops out because its curl vanishes).

Example 14.17 (Stokes's theorem) Suppose $\boldsymbol{u}$, $\boldsymbol{\eta}$, and $\boldsymbol{\zeta}$ form a triplet of infinitesimal vectors oriented so as to form a right-handed coordinate system, $\boldsymbol{u} \times \boldsymbol{\eta} \cdot \boldsymbol{\zeta} > 0$. These vectors form a tiny parallelepiped Π. We want to integrate the 2-form $\omega = a_{jk} dx_j \wedge dx_k$ over the surface $\partial\Pi$ of this tiny parallelepiped Π. We find

$$
\begin{aligned}
\int_{\partial\Pi} \omega = & \int_0^1 dt \int_0^1 ds \left\{ \left[a_{jk}(t\boldsymbol{u} + s\boldsymbol{\eta} + \boldsymbol{\zeta}) - a_{jk}(t\boldsymbol{u} + s\boldsymbol{\eta} \right] \right\} dx_j \wedge dx_k(\boldsymbol{u}, \boldsymbol{\eta}) \\
& + \int_0^1 dt \int_0^1 ds \left\{ \left[a_{jk}(t\boldsymbol{\eta} + s\boldsymbol{\zeta} + \boldsymbol{u}) - a_{jk}(t\boldsymbol{\eta} + s\boldsymbol{\zeta} \right] \right\} dx_j \wedge dx_k(\boldsymbol{\eta}, \boldsymbol{\zeta}) \\
& + \int_0^1 dt \int_0^1 ds \left\{ \left[a_{jk}(t\boldsymbol{\zeta} + s\boldsymbol{u} + \boldsymbol{\eta}) - a_{jk}(t\boldsymbol{\zeta} + s\boldsymbol{u} \right] \right\} dx_j \wedge dx_k(\boldsymbol{\zeta}, \boldsymbol{u}) \\
& = a_{jk,\ell}\zeta_\ell \left(u_j \eta_k - \eta_j u_k \right) + a_{jk,\ell} u_\ell \left(\eta_j \zeta_k - \zeta_j \eta_k \right) + a_{jk,\ell} \eta_\ell \left(\zeta_j u_k - u_j \zeta_k \right) \\
& = a_{jk,\ell}\, dx_\ell \wedge dx_j \wedge dx_k(\boldsymbol{u}, \boldsymbol{\eta}, \boldsymbol{\zeta}) = d\omega(\boldsymbol{u}, \boldsymbol{\eta}, \boldsymbol{\zeta}) = \int_\Pi d\omega
\end{aligned} \tag{14.104}
$$

for an infinitesimal parallelepiped Π.

It is easy to extend this identity to an arbitrary volume V of finite extent. To do this, we tile the volume V with infinitesimal parallelepipeds Π_α with boundaries $\partial\Pi_\alpha$. The integral over the finite volume V is then the sum of the integrals over the Π_α

$$
\int_V d\omega = \sum_\alpha \int_{\Pi_\alpha} d\omega \tag{14.105}
$$

which by (14.104) is a sum of the integrals over the boundaries $\partial\Pi_\alpha$

$$
\int_V d\omega = \sum_\alpha \int_{\Pi_\alpha} d\omega = \sum_\alpha \int_{\partial\Pi} \omega. \tag{14.106}
$$

In this sum over surface integrals, the internal boundaries all cancel, leaving us with the integral over the boundary ∂V of the volume V, so that

$$
\int_V d\omega = \sum_\alpha \int_{\Pi_\alpha} d\omega = \sum_\alpha \int_{\partial\Pi_\alpha} \omega = \int_{\partial V} \omega. \tag{14.107}
$$

Thus we have

$$
\int_V d\omega = \int_{\partial V} \omega \tag{14.108}
$$

which generalizes the identity (14.90) from 1-forms to 2-forms.

Before leaving this example, is may be instructive to examine the value of the integral of ω on one of the faces of the infinitesimal parallelepiped Π as well as that of the integral of $d\omega$ on the infinitesimal parallelepiped Π. The 2-form ω on the upper $\boldsymbol{u}, \boldsymbol{\eta}$ face of the surface $\partial\Pi$ of Π may be seen from Equation (14.104) to be

$$
\omega(\boldsymbol{u}, \boldsymbol{\eta}) = a_{jk}\, dx_j \wedge dx_k(\boldsymbol{u}, \boldsymbol{\eta}) = a_{jk} \left(u_j \eta_k - \eta_j u_k \right). \tag{14.109}
$$

The wedge $dx_j \wedge dx_k(\boldsymbol{u}, \boldsymbol{\eta})$ defines an area vector $\boldsymbol{S} = \boldsymbol{u} \times \boldsymbol{\eta}$ with components $S_i = \epsilon_{ijk} u_j \eta_k$. In terms of $\boldsymbol{S}$, the 2-form ω is

$$
\omega(\boldsymbol{u}, \boldsymbol{\eta}) = (a_{23} - a_{32})\, S_1 + (a_{31} - a_{13})\, S_2 + (a_{12} - a_{21})\, S_3 \tag{14.110}
$$

which suggests defining the vector field $A_k = \epsilon_{kij} a_{ij}$. In terms of $\boldsymbol{S}$ and $\boldsymbol{A}$, the 2-form ω on $\boldsymbol{u}, \boldsymbol{\eta}$ is $\omega(\boldsymbol{u}, \boldsymbol{\eta}) = \boldsymbol{A} \cdot \boldsymbol{S}$.

The integral of the 3-form $d\omega$ on the infinitesimal parallelepiped Π is

$$\int_\Pi d\omega = d\omega(\boldsymbol{u}, \boldsymbol{\eta}, \boldsymbol{\zeta}) = a_{jk,\ell}\, dx_\ell \wedge dx_j \wedge dx_k(\boldsymbol{u}, \boldsymbol{\eta}, \boldsymbol{\zeta}). \tag{14.111}$$

The wedge product is $\epsilon_{\ell jk}$ times the determinant $\det(\boldsymbol{u}, \boldsymbol{\eta}, \boldsymbol{\zeta})$ of the 3×3 matrix that has $\boldsymbol{u}$ as its first row, $\boldsymbol{\eta}$ as its second, and $\boldsymbol{\zeta}$ as its third row

$$dx_\ell \wedge dx_j \wedge dx_k(\boldsymbol{u}, \boldsymbol{\eta}, \boldsymbol{\zeta}) = \epsilon_{\ell jk} \det(\boldsymbol{u}, \boldsymbol{\eta}, \boldsymbol{\zeta}). \tag{14.112}$$

So $d\omega(\boldsymbol{u}, \boldsymbol{\eta}, \boldsymbol{\zeta}) = a_{jk,\ell}\, \epsilon_{\ell jk} \det(\boldsymbol{u}, \boldsymbol{\eta}, \boldsymbol{\zeta})$ or more explicitly

$$d\omega(\boldsymbol{u}, \boldsymbol{\eta}, \boldsymbol{\zeta}) = \left[(a_{23} - a_{32})_{,1} + (a_{31} - a_{13})_{,2} + (a_{12} - a_{21})_{,3}\right] \det(\boldsymbol{u}, \boldsymbol{\eta}, \boldsymbol{\zeta})$$

which we recognize as the divergence of the vector field $A_k = \epsilon_{ijk} a_{ij}$

$$d\omega(\boldsymbol{u}, \boldsymbol{\eta}, \boldsymbol{\zeta}) = \nabla \cdot \boldsymbol{A} \det(\boldsymbol{u}, \boldsymbol{\eta}, \boldsymbol{\zeta}). \tag{14.113}$$

Thus we have rediscovered the vector identity

$$\int_V \nabla \cdot \boldsymbol{A}\, dV = \oint_{\partial V} \boldsymbol{A} \cdot d\boldsymbol{S}. \tag{14.114}$$

Example 14.18 (Gauss's law) The divergence of the electric displacement $\boldsymbol{D}$ is the density ρ_f of free charge, $\nabla \cdot \boldsymbol{D} = \rho_f$, and so this last identity (14.114) gives

$$Q_{fV} = \int_V \rho_f\, dV = \int_V \nabla \cdot \boldsymbol{D}\, dV = \oint_{\partial V} \boldsymbol{D} \cdot d\boldsymbol{S} \tag{14.115}$$

which is the integral form of Gauss's law.

One may generalize these examples to what has been called the Newton–Leibniz–Gauss–Green–Ostrogradskii–Stokes–Poincaré theorem

$$\int_{\partial C} \omega = \int_C d\omega \tag{14.116}$$

in which ω is a k-form and C is any $(k+1)$-chain on a manifold.

Example 14.19 (Poincaré's invariant action) In Example 14.9, we saw that Poincaré's action

$$A = \int_{\partial S} \omega^1 = \oint p \cdot dq = \oint \sum_{i=1}^{n} p_i \wedge dq_i \tag{14.117}$$

does not change with time, $\dot{A} = 0$. In Example 14.11, we learned that the element of area $d\omega^1$ and therefore its surface integral

$$I = \int_S d\omega^1 \tag{14.118}$$

does not change with time. The identity (14.116) in the form

$$A = \int_{\partial S} \omega^1 = \int_S d\omega^1 = I \tag{14.119}$$

relates these two examples.

14.5 Are Closed Forms Exact?

A form ω is said to be **closed** if its exterior derivative vanishes

$$d\omega = 0. \tag{14.120}$$

A form ω is **exact** if it's the exterior derivative of another form ψ

$$\omega = d\psi. \tag{14.121}$$

We have seen in (14.64) that the exterior derivative of any exterior derivative vanishes; in effect, $dd = 0$. Thus the exterior derivative of any exact form ω must be zero

$$d\omega = dd\psi = 0. \tag{14.122}$$

So every exact form is closed.

But are closed forms exact? Poincaré's lemma provides the answer: A form that is defined and closed on a **simply connected** part of a manifold is exact there. More technically, if a form ω is defined and closed on a region U of a manifold M and if U can be mapped by a one-to-one differentiable map onto the interior of the unit ball in $\mathbb{R}^n$, then there is a form ψ such that $d\psi = \omega$ on U. The unit ball in $\mathbb{R}^n$ is the interior of the sphere S^{n-1} defined by $x_1^2 + \cdots + x_n^2 = 1$. You may find a proof of this result in section 4.19 of Schutz's book (Schutz, 1980).

Example 14.20 (Two dimensions) Suppose that the 1-form $\omega = f dx + g dy$ is closed

$$d\omega = f_{,y}\, dy \wedge dx + g_{,x}\, dx \wedge dy = \left(g_{,x} - f_{,y}\right) dx \wedge dy = 0 \tag{14.123}$$

on the real plane $\mathbb{R}^2$. Since $\mathbb{R}^2$ is simply connected, Poincaré's lemma tells us that there exists a 0-form h whose exterior derivative is ω

$$\omega = f dx + g dy = dh = h_{,x} dx + h_{,y} dy. \tag{14.124}$$

We may construct such a function $h(x, y)$ as the line integral

$$h(x, y) = \int_0^1 \left[f(u(t), v(t))\, \dot{u}(t) + g(u(t), v(t))\, \dot{v}(t)\right] dt \tag{14.125}$$

along any differentiable curve $\gamma(t) = (u(t), v(t))$ that goes from $(u(0), v(0)) = (x_0, y_0)$ to $(u(1), v(1)) = (x, y)$. Clearly the exterior derivative of this 0-form is

$$dh = h_{,x}dx + h_{,y}dy = f(x, y)\,dx + g(x, y)\,dy = \omega. \tag{14.126}$$

So the real issue here is whether the line integral (14.125)

$$h = \int_\gamma \omega \tag{14.127}$$

defines a function $h(x, y)$ that is the *same* for any two curves $\gamma_1(t)$ and $\gamma_2(t)$ that both go from (x_0, y_0) to (x, y). The difference $h_1(x, y) - h_2(x, y)$ is an integral of ω along a closed curve $\Gamma = \gamma_1 - \gamma_2$ that runs from (x_0, y_0) to (x, y) along $\gamma_1(t)$ and then from (x, y) to (x_0, y_0) backwards along $\gamma_2(t)$. The closed curve Γ is the boundary ∂S of the (plane) surface S that it encloses. Thus by Stokes's theorem (14.116)

$$h_1 - h_2 = \int_\Gamma \omega = \int_{\partial S} \omega = \int_S d\omega = 0 \tag{14.128}$$

in which we used the fact that ω is closed so that $d\omega = 0$. The two curves $\gamma_1(t)$ and $\gamma_2(t)$ define the same function $h(x, y)$, and $\omega = dh$ is exact.

What if the region in which $d\omega = 0$ is not simply connected? Consider for example the 1-form

$$\omega = -\frac{y}{x^2 + y^2}\,dx + \frac{x}{x^2 + y^2}\,dy \tag{14.129}$$

which is well defined except at the origin. The plane $\mathbb{R}^2$ minus the origin is not simply connected. One may check that ω is closed, $d\omega = 0$, except at the origin. But ω is not exact. In fact, by writing it as

$$\omega = \arctan(y/x)_{,y}\,dy \wedge dx + \arctan(y/x)_{,x}\,dx \wedge dy \tag{14.130}$$

we see that ω is almost exact. It is the exterior derivative of $\theta = \arctan(y/x)$ which would be a 0-form if it were single valued.

Example 14.21 (Some exact forms) It's easy to make lots of exact forms; one just applies the exterior derivative d to any form. For instance, taking the exterior derivative of the 0-form $\omega^0 = x\,y^2\,\exp(zw)$, we get the 1-form

$$d\omega^0 = y^2\,e^{zw}\,dx + 2x\,y\,e^{zw}\,dy + x\,y^2\,w\,e^{zw}\,dz + x\,y^2\,z\,e^{zw}\,dw \tag{14.131}$$

which is exact (and closed). Applying d to the 1-form $\omega^1 = y^2 z dx + x^3 dy$, we get the exact 2-form $d\omega^1 = (3x^2 - 2yz)\,dx \wedge dy - y^2 dx \wedge dz$. Incidentally, any n-form in n variables, such as $f(x, y, z)\,dx \wedge dy \wedge dz$, is closed.

14.6 Complex Differential Forms

Any function $f(x, y)$ of two real variables also is a function of the two complex variables $z = x + iy$ and $\bar{z} = x - iy$. For instance, $4xy = -i(z + \bar{z})(z - \bar{z})$ and $x^2 + y^2 = z\bar{z}$. We can write any 1-form $\omega = adx + bdy$ with complex coefficients a and b in terms of the complex differentials $dz = dx + idy$ and $d\bar{z} = dx - idy$ as

$$\omega = \frac{1}{2}(a - ib)\, dz + \frac{1}{2}(a + ib)\, d\bar{z}. \tag{14.132}$$

A 1-form of the variables $z_1, \ldots, z_n$ and $\bar{z}_1, \ldots, \bar{z}_n$ is a sum of their differentials $\omega = a_j dz_j + b_j d\bar{z}_j$. The expression

$$\omega^{1,1} = a\, dz_1 \wedge d\bar{z}_1 + b\, dz_1 \wedge d\bar{z}_2 + c\, dz_2 \wedge d\bar{z}_1 + d\, dz_2 \wedge d\bar{z}_2 \tag{14.133}$$

is a 1,1-form in z_1 and z_2, while 2,0- and 0,2-forms look like

$$\omega^{2,0} = e\, dz_1 \wedge dz_2 \quad \text{and} \quad \omega^{0,2} = f\, d\bar{z}_1 \wedge d\bar{z}_2. \tag{14.134}$$

The complex differentials anticommute: $d\bar{z}_j \wedge dz_k = -dz_k \wedge d\bar{z}_j$ as well as $dz_j \wedge dz_k = -dz_k \wedge dz_j$ and $d\bar{z}_j \wedge d\bar{z}_k = -d\bar{z}_k \wedge d\bar{z}_j$.

There are two exterior derivatives ∂ and $\bar{\partial}$ defined by

$$\partial = \sum_{j=1}^{n} \frac{\partial}{\partial z_j} dz_j \wedge \quad \text{and} \quad \bar{\partial} = \sum_{j=1}^{n} \frac{\partial}{\partial \bar{z}_j} d\bar{z}_j \wedge . \tag{14.135}$$

Their sum is the ordinary exterior derivative $\partial + \bar{\partial} = d$, and one has

$$\partial^2 = \bar{\partial}^2 = \partial\bar{\partial} + \bar{\partial}\partial = 0. \tag{14.136}$$

Example 14.22 ($\partial + \bar{\partial} = d$) We illustrate the rule $\partial + \bar{\partial} = d$ for the 1-form $\omega = z\bar{z}dz = (x^2 + y^2)(dx + idy)$. The sum $\partial + \bar{\partial}$ acting on ω gives

$$\begin{aligned}\left(\partial + \bar{\partial}\right) z\bar{z}\, dz &= \bar{\partial} z\bar{z}\, d\bar{z} \wedge dz = z\, d\bar{z} \wedge dz \\ &= (x + iy)(dx - idy) \wedge (dx + idy) = 2i(x + iy)dx \wedge dy\end{aligned} \tag{14.137}$$

while $d\omega$ is

$$d(x^2 + y^2)(dx + idy) = 2xidx \wedge dy + 2ydy \wedge dx = 2i(x + iy)dx \wedge dy \tag{14.138}$$

which is the same as $\left(\partial + \bar{\partial}\right) \omega$.

14.7 Hodge's Star

In 3 cartesian coordinates, the Hodge dual turns 1-forms into 2-forms

$$* \, dx = dy \wedge dz \qquad * \, dy = dz \wedge dx \qquad * \, dz = dx \wedge dy \tag{14.139}$$

and 2-forms into 1-forms

$$* \, (dx \wedge dy) = dz \qquad * \, (dy \wedge dz) = dx \qquad * \, (dz \wedge dx) = dy. \tag{14.140}$$

It also maps the 0-form 1 and the volume 3-form into each other

$$* \, 1 = dx \wedge dy \wedge dz \qquad * \, (dx \wedge dy \wedge dz) = 1 \tag{14.141}$$

(William Vallance Douglas Hodge, 1903–1975). More generally in 3-space, we define the Hodge dual, also called the Hodge star, in terms of Levi-Civita's pseudotensor (Section 13.28) as

$$\begin{aligned} *1 &= \frac{1}{3!} \eta_{\ell jk} dx^\ell \wedge dx^j \wedge dx^k \qquad * \, (dx^\ell \wedge dx^j \wedge dx^k) = g^{\ell t} g^{ju} g^{kv} \eta_{tuv} \\ * \, dx^i &= \frac{1}{2} g^{i\ell} \, \eta_{\ell jk} \, dx^j \wedge dx^k \qquad * \, (dx^i \wedge dx^j) = g^{ik} \, g^{j\ell} \, \eta_{k\ell m} \, dx^m \end{aligned} \tag{14.142}$$

and so if the sign of det g_{ij} is $s = +1$, then $**1 = 1$, $**dx^i = dx^i$, $**(dx^i \wedge dx^k) = dx^i \wedge dx^k$, and $* * \, (dx^i \wedge dx^j \wedge dx^k) = dx^i \wedge dx^j \wedge dx^k$.

Example 14.23 (Divergence and laplacian) The dual of the 1-form

$$df = \frac{\partial f}{\partial x} dx + \frac{\partial f}{\partial y} dy + \frac{\partial f}{\partial z} dz \tag{14.143}$$

is the 2-form

$$* \, df = \frac{\partial f}{\partial x} dy \wedge dz + \frac{\partial f}{\partial y} dz \wedge dx + \frac{\partial f}{\partial z} dx \wedge dy \tag{14.144}$$

and its exterior derivative is the laplacian

$$d * df = \left(\frac{\partial^2 f}{\partial x^2} + \frac{\partial^2 f}{\partial y^2} + \frac{\partial^2 f}{\partial z^2} \right) dx \wedge dy \wedge dz \tag{14.145}$$

multiplied by the volume 3-form.

Similarly, the dual of the one form $A = A_x \, dx + A_y \, dy + A_z \, dz$ is the 2-form $*A = A_x \, dy \wedge dz + A_y \, dz \wedge dx + A_z \, dx \wedge dy$, and its exterior derivative is the divergence times $dx \wedge dy \wedge dz$

$$d * A = \left(\frac{\partial A_x}{\partial x} + \frac{\partial A_y}{\partial y} + \frac{\partial A_z}{\partial z} \right) dx \wedge dy \wedge dz. \tag{14.146}$$

In flat Minkowski 4-space with $c = 1$, the Hodge dual turns 1-forms into 3-forms

$$\begin{aligned} *\,dt &= -\,dx \wedge dy \wedge dz & *\,dx &= -\,dy \wedge dz \wedge dt \\ *\,dy &= -\,dz \wedge dx \wedge dt & *\,dz &= -\,dx \wedge dy \wedge dt \end{aligned} \tag{14.147}$$

2-forms into 2-forms

$$\begin{aligned} *\,(dx \wedge dt) &= dy \wedge dz & *\,(dx \wedge dy) &= -\,dz \wedge dt \\ *(dy \wedge dt) &= dz \wedge dx & *\,(dy \wedge dz) &= -\,dx \wedge dt \\ *(dz \wedge dt) &= dx \wedge dy & *\,(dz \wedge dx) &= -\,dy \wedge dt \end{aligned} \tag{14.148}$$

3-forms into 1-forms

$$\begin{aligned} *\,(dx \wedge dy \wedge dz) &= -\,dt & *\,(dy \wedge dz \wedge dt) &= -\,dx \\ *\,(dz \wedge dx \wedge dt) &= -\,dy & *\,(dx \wedge dy \wedge dt) &= -\,dz \end{aligned} \tag{14.149}$$

and interchanges 0-forms and 4-forms

$$*\,1 = dt \wedge dx \wedge dy \wedge dz \qquad *\,(dt \wedge dx \wedge dy \wedge dz) = -\,1. \tag{14.150}$$

More generally in 4 dimensions, we define the Hodge star as

$$\begin{aligned} *\,1 &= \frac{1}{4!}\,\eta_{k\ell mn}\,dx^k \wedge dx^\ell \wedge dx^m \wedge dx^n \\ *\,dx^i &= \frac{1}{3!}\,g^{ik}\,\eta_{k\ell mn}\,dx^\ell \wedge dx^m \wedge dx^n \\ *\,(dx^i \wedge dx^j) &= \frac{1}{2}\,g^{ik}\,g^{j\ell}\,\eta_{k\ell mn}\,dx^m \wedge dx^n \\ *\,(dx^i \wedge dx^j \wedge dx^k) &= g^{it}\,g^{ju}\,g^{kv}\,\eta_{tuvw}\,dx^w \\ *\left(dx^i \wedge dx^j \wedge dx^k \wedge dx^\ell\right) &= g^{it}\,g^{ju}\,g^{kv}\,g^{\ell w}\eta_{tuvw} = \eta^{ijk\ell}. \end{aligned} \tag{14.151}$$

Thus (Exercise 14.18) if the determinant $\det g_{ij}$ of the metric is negative, then

$$\begin{aligned} &*\,*\,dx^i = dx^i \qquad *\,*\,(dx^i \wedge dx^j) = -\,dx^i \wedge dx^j \\ &*\,*\,(dx^i \wedge dx^j \wedge dx^k) = dx^i \wedge dx^j \wedge dx^k \qquad *\,*1 = -\,1. \end{aligned} \tag{14.152}$$

In n dimensions, the Hodge star turns p-forms into $n-p$-forms

$$*\left(dx^{i_1} \wedge \cdots \wedge dx^{i_p}\right) = g^{i_1 k_1} \cdots g^{i_p k_p}\,\frac{\eta_{k_1 \ldots k_p \ell_1 \ldots \ell_{n-p}}}{(n-p)!}\,dx^{\ell_1} \wedge \cdots \wedge dx^{\ell_{n-p}}. \tag{14.153}$$

Example 14.24 (The volume element $*1$) The scalar element of volume is $*1$. Even in the literature of general relativity, this scalar is written as $\sqrt{g}\,d^4x$, which is a pseudoscalar.

Example 14.25 (The inhomogeneous Maxwell equations) The homogeneous Maxwell equations (12.71) are

$$dF = ddA = 0. \tag{14.154}$$

To get the inhomogeneous Maxwell equations, we first form the dual $*F = *dA$

$$*F = \frac{1}{2}F_{ij} * \left(dx^i \wedge dx^j\right) = \tfrac{1}{4}F_{ij}g^{ik}g^{j\ell}\eta_{k\ell mn}dx^m \wedge dx^n = \tfrac{1}{4}F^{k\ell}\eta_{k\ell mn}dx^m \wedge dx^n,$$

and then apply the exterior derivative

$$d*F = \tfrac{1}{4}d\left(F^{k\ell}\eta_{k\ell mn}dx^m \wedge dx^n\right) = \tfrac{1}{4}\partial_p\left(F^{k\ell}\eta_{k\ell mn}\right)dx^p \wedge dx^m \wedge dx^n.$$

To get back to a 1-form like $j = j_k\, dx^k$, we apply a second Hodge star

$$\begin{aligned}
*d*F &= \tfrac{1}{4}\,\partial_p\left(F^{k\ell}\,\eta_{k\ell mn}\right) * \left(dx^p \wedge dx^m \wedge dx^n\right)\\
&= \tfrac{1}{4}\,\partial_p\left(F^{k\ell}\,\eta_{k\ell mn}\right) g^{ps}\,g^{mt}\,g^{nu}\,\eta_{stuv}\,dx^v\\
&= \tfrac{1}{4}\,\partial_p\left(\sqrt{g}\,F^{k\ell}\right)\epsilon_{k\ell mn}\,g^{ps}\,g^{mt}\,g^{nu}\,\sqrt{g}\,\epsilon_{stuv}\,dx^v \qquad (14.155)\\
&= \tfrac{1}{4}\,\partial_p\left(\sqrt{g}\,F^{k\ell}\right)\epsilon_{k\ell mn}\,g^{ps}\,g^{mt}\,g^{nu}\,g^{wv}\,\epsilon_{stuv}\,\sqrt{g}\,dx_w\\
&= \tfrac{1}{4}\,\partial_p\left(\sqrt{g}\,F^{k\ell}\right)\epsilon_{k\ell mn}\,\epsilon_{pmnw}\,\frac{\sqrt{g}}{\det g_{ij}}\,dx_w\\
&= \frac{s}{4\sqrt{g}}\,\partial_p\left(\sqrt{g}\,F^{k\ell}\right)\epsilon_{k\ell mn}\,\epsilon_{pmnw}\,dx_w
\end{aligned}$$

in which we used the definition (1.204) of the determinant. Levi-Civita's 4-symbol obeys the identity (Exercise 14.19)

$$\epsilon_{k\ell mn}\,\epsilon^{pwmn} = 2\left(\delta^p_k\,\delta^w_\ell - \delta^w_k\,\delta^p_\ell\right). \tag{14.156}$$

Applying it to $*d*F$, we get

$$*\,d*F = \frac{s}{2\sqrt{g}}\partial_p\left(\sqrt{g}\,F^{k\ell}\right)\left(\delta^p_k\,\delta^w_\ell - \delta^w_k\,\delta^p_\ell\right)dx_w = -\frac{s}{\sqrt{g}}\,\partial_p\left(\sqrt{g}\,F^{kp}\right)\,dx_k.$$

In our spacetime $s = -1$. Setting $*d*F$ equal to $j = j_k\,dx^k = j^k\,dx_k$ multiplied by the permeability μ_0 of the vacuum, we arrive at expressions for the microscopic inhomogeneous Maxwell equations in terms of both tensors and forms

$$\frac{1}{\sqrt{g}}\,\partial_p\left(\sqrt{g}\,F^{kp}\right) = \mu_0\,j^k \quad \text{and} \quad *\,d*F = \mu_0\,j. \tag{14.157}$$

They and the homogeneous Bianchi identity (12.45, 12.71, and 13.86)

$$\epsilon^{ijk\ell}\,\partial_\ell F_{jk} = dF = d\,dA = 0 \tag{14.158}$$

are invariant under general coordinate transformations.

Example 14.26 (Divergence of a contravariant vector) Another way to derive the formula (13.180) for the divergence of a contravariant vector is to recall that the Hodge dual (14.151) of the 1-form $V = V_i \, dx^i$ is

$$
\begin{aligned}
*V = V_i * dx^i &= V_i \, \frac{1}{3!} \, g^{ik} \, \eta_{k\ell mn} \, dx^\ell \wedge dx^m \wedge dx^n \\
&= \frac{1}{3!} \, \sqrt{g} \, V^k \, \epsilon_{k\ell mn} \, dx^\ell \wedge dx^m \wedge dx^n
\end{aligned}
\tag{14.159}
$$

in which g is the absolute value of the determinant of the metric tensor g_{ij}. The exterior derivative now gives

$$
d * V = \frac{1}{3!} \left(\sqrt{g}\, V^k\right)_{,p} \epsilon_{k\ell mn} \, dx^p \wedge dx^\ell \wedge dx^m \wedge dx^n . \tag{14.160}
$$

So using (14.151) to apply a second Hodge star, we find

$$
\begin{aligned}
* \, d * V &= \frac{1}{3!} \left(\sqrt{g}\, V^k\right)_{,p} \epsilon_{k\ell mn} * \left(dx^p \wedge dx^\ell \wedge dx^m \wedge dx^n\right) \\
&= \frac{1}{3!} \left(\sqrt{g}\, V^k\right)_{,p} \epsilon_{k\ell mn} \, g^{pt} \, g^{\ell u} \, g^{mv} \, g^{nw} \eta_{tuvw} \\
&= \frac{1}{3!} \left(\sqrt{g}\, V^k\right)_{,p} \epsilon_{k\ell mn} \, g^{pt} \, g^{\ell u} \, g^{mv} \, g^{nw} \epsilon_{tuvw} \sqrt{g} \\
&= \frac{1}{3!} \left(\sqrt{g}\, V^k\right)_{,p} \epsilon_{k\ell mn} \, \frac{\sqrt{g}}{\det g_{ij}} \, \epsilon^{p\ell mn} \\
&= \frac{s}{\sqrt{g}} \left(\sqrt{g}\, V^k\right)_{,p} \delta^p_k = \frac{s}{\sqrt{g}} \left(\sqrt{g}\, V^k\right)_{,k} .
\end{aligned}
\tag{14.161}
$$

So in our spacetime with $\det g_{ij} = -g$

$$
- * d * V = \frac{1}{\sqrt{g}} \left(\sqrt{g}\, V^k\right)_{,k} . \tag{14.162}
$$

In 3-space the Hodge star (14.142) of a 1-form $V = V_i \, dx^i$ is

$$
* \; V = V_i * dx^i = V_i \, \frac{1}{2} \, g^{i\ell} \, \eta_{\ell jk} \, dx^j \wedge dx^k = \frac{1}{2} \sqrt{g} \, V^\ell \, \epsilon_{\ell jk} \, dx^j \wedge dx^k . \tag{14.163}
$$

Applying the exterior derivative, we get the invariant form

$$
d * V = \frac{1}{2} \left(\sqrt{g}\, V^\ell\right)_{,p} \epsilon_{\ell jk} \, dx^p \wedge dx^j \wedge dx^k . \tag{14.164}
$$

We add a star by using the definition (14.142) of the Hodge dual in a 3-space with $\det g_{ij} > 0$ as well as the identity $\epsilon_{\ell jk} \, \epsilon^{pjk} = 2\, \delta^p_\ell$ (Exercise 14.20) and the definition (1.204) of the determinant

$$
\begin{aligned}
* \; d * V &= \frac{1}{2} \left(\sqrt{g}\, V^\ell\right)_{,p} \epsilon_{\ell jk} * \left(dx^p \wedge dx^j \wedge dx^k\right) \\
&= \frac{1}{2} \left(\sqrt{g}\, V^\ell\right)_{,p} \epsilon_{\ell jk} \, g^{pt} g^{ju} g^{kv} \eta_{tuv}
\end{aligned}
$$

$$= \frac{1}{2}\left(\sqrt{g}\, V^{\ell}\right)_{,p} \epsilon_{\ell jk}\, g^{pt} g^{ju} g^{kv} \epsilon_{tuv}\sqrt{g}$$

$$= \frac{1}{2}\left(\sqrt{g}\, V^{\ell}\right)_{,p} \epsilon_{\ell jk}\, \epsilon^{pjk} \frac{\sqrt{g}}{\det g_{ij}}$$

$$= \frac{1}{\sqrt{g}}\left(\sqrt{g}\, V^{\ell}\right)_{,p} \delta^{p}_{\ell} = \frac{1}{\sqrt{g}}\left(\sqrt{g}\, V^{p}\right)_{,p}. \tag{14.165}$$

Example 14.27 (Covariant laplacian) To find the laplacian $\Box f$ in terms of forms, we apply the exterior derivative to the Hodge dual (14.151) of the 1-form $df = f_{,i} dx^i$

$$\begin{aligned} d * df &= d\left(f_{,i} * dx^i\right) = d\left(\frac{1}{3!} f_{,i}\, g^{ik}\, \eta_{k\ell mn}\, dx^{\ell} \wedge dx^m \wedge dx^n\right) \\ &= \frac{1}{3!}\left(f^{,k}\sqrt{g}\right)_{,p} \epsilon_{k\ell mn}\, dx^p \wedge dx^{\ell} \wedge dx^m \wedge dx^n \end{aligned} \tag{14.166}$$

and then add a star using (14.151)

$$\begin{aligned} * d * df &= \frac{1}{3!}\left(f^{,k}\sqrt{g}\right)_{,p} \epsilon_{k\ell mn} * \left(dx^p \wedge dx^{\ell} \wedge dx^m \wedge dx^n\right) \\ &= \frac{1}{3!}\left(f^{,k}\sqrt{g}\right)_{,p} \epsilon_{k\ell mn}\, g^{pt} g^{\ell u} g^{mv} g^{nw} \sqrt{g}\, \epsilon_{tuvw}. \end{aligned} \tag{14.167}$$

The definition (1.204) of the determinant now gives (Exercise 14.20)

$$\begin{aligned} * d * df &= \frac{1}{3!}\left(f^{,k}\sqrt{g}\right)_{,p} \epsilon_{k\ell mn}\, \epsilon^{p\ell mn} \frac{\sqrt{g}}{\det g} \\ &= \left(f^{,k}\sqrt{g}\right)_{,p} \delta^p_k \frac{s}{\sqrt{g}} = \frac{s}{\sqrt{g}}\left(f^{,k}\sqrt{g}\right)_{,k}. \end{aligned} \tag{14.168}$$

In our spacetime $\det g_{ij} = sg = -g$, and so the laplacian is

$$\Box f = - * d * df = \frac{1}{\sqrt{g}}\left(f^{,k}\sqrt{g}\right)_{,k}. \tag{14.169}$$

14.8 Theorem of Frobenius

This section, optional on a first reading, begins with some definitions:

- If ω is a k-form that maps all vectors V of the tangent space T_P at the point P into numbers $\omega(V_1, \ldots, V_k)$, and $S \subset T_P$ is a subspace of the tangent space T_P, then the **restriction** ω_S of the form ω to S maps the vectors $S_i \in S$ into the numbers $\omega(S_1, \ldots, S_k)$.
- The **annihilator** (actually the annihilated) of a set of forms β_i at a point P of a manifold is the subspace of vectors X_P that every β_i maps to zero. (The vectors X_P are in the tangent space T_P.)

- The **complete ideal** of a set B of forms β_i is the set of all forms at P whose restriction to B's annihilator X_P vanishes. Note that for *any* form α and any vectors X_ℓ in the annihilator X_P, the wedge product $\alpha \wedge \beta_i(X_1, \ldots, X_n)$ vanishes, so $\alpha \wedge \beta_i$ is in the complete ideal of the set B of forms β_i.
- The complete ideal of a set B of forms β_i has a set α_i of linearly independent 1-forms that **generates** it, that is, whose complete ideal is the same as that of the set B.
- The complete ideal of a set B of fields β_i is the set of fields that map the annihilator X_P of B to zero at each point P of the manifold.
- If the exterior derivative $d\alpha_i$ is in an ideal whenever α_i is, then the ideal is a **differential ideal.**
- A set A of 1-forms α_i has a **closed ideal** if every $d\alpha_i$ is in the complete ideal generated by the α_i's. (Some authors call such a set A of 1-forms closed, but this terminology can be confusing.)

Example 14.28 (A rank-2 annihilator) Consider a manifold with coordinates $x_1, \ldots, x_n$, tangent vectors $\partial_1, \ldots, \partial_n$, and 1-forms $dx_1, \ldots, dx_n$ with $dx_i(\partial_k) = \delta_{ik}$ as in (14.31). The subspace $\{c_1\partial_1 + c_2\partial_2 \,|\, c_1, c_2 \text{ real}\}$ is the annihilator of the set $\{dx_3, \ldots, dx_n\}$ of 1-forms. Any linear combination of these 1-forms

$$\omega = \sum_{k=3}^{n} a_k(x)\, dx_k \tag{14.170}$$

is in the complete ideal of the 1-forms $dx_3, \ldots, dx_n$. So is any 2-form

$$\omega^2 = \sum_{i,k=3}^{n} a_{ik}(x)\, dx_i \wedge dx_k. \tag{14.171}$$

For any forms α_k the linear combination

$$\omega = \sum_{k=3}^{n} \alpha_k \wedge dx_k \tag{14.172}$$

is in the complete ideal of the set of 1-forms $dx_1, \ldots, dx_n$. In fact, each member of this complete ideal is such a linear combination.

Frobenius's Theorem: Let $\omega_1, \ldots, \omega_n$ be a linearly independent set of 1-form fields in an open region U of a k-dimensional manifold M. Then there exist functions $P_{\ell j}$ and Q_j for $i, j = 1, \ldots, n$ that express the n 1-forms ω_ℓ as

$$\omega_\ell = \sum_{j=1}^{n} P_{\ell j}\, dQ_j \tag{14.173}$$

if and only if every $d\omega_\ell$ is in the complete ideal generated by the ω_ℓ's (Schutz, 1980, secs. 3.8 & 4.26).

Further Reading

Three good books on differential forms are *Geometrical methods of mathematical physics* (Schutz, 1980), *Mathematical Methods of Classical Mechanics* (Arnold, 1989), and *Classical Mechanics* (Matzner and Shepley, 1991).

Exercises

14.1 Why might mathematicians prefer the definition (14.18) to (14.19)?

14.2 Show explicitly that the 2-form $\omega = X dx_2 \wedge dx_3$ is given by (14.43) in terms of the 1-forms dy_k.

14.3 Show explicitly that for the two 3-vectors of Example 14.4, the 2-form $\omega(A, B) = X dx_2 \wedge dx_3(A, B) = -X$.

14.4 In Example 14.5, let $x_1 = y_1 + y_2$, $x_2 = y_1 - y_2$, and $x_3 = y_1 - y_3$. Show explicitly that the y-version (14.43) of the 2-form $\omega(A, B) = X dx_2 \wedge dx_3$ maps the two 3-vectors of Example 14.4 into the same real number $-X$ as its x-version. Hint: In the y-version, first express $dy_k(A)$ and $dy_k(B)$ in terms of $dx_j(A)$ and $dx_j(B)$.

14.5 Compute $dr(\hat{\boldsymbol{e}}_r)$, $d\theta(\hat{\boldsymbol{e}}_\theta)$, and $d\phi(\hat{\boldsymbol{e}}_\phi)$ as well as the six off-diagonal ones $dr(\hat{\boldsymbol{e}}_\theta)$, $dr(\hat{\boldsymbol{e}}_\phi)$, and so forth.

14.6 Show that the 2-form (14.52) applied to the vectors $\boldsymbol{U}$ and $\boldsymbol{V}$ gives the triple scalar product (14.48).

14.7 Show that $d\, d\, \omega^k = 0$ for the general k-form (14.62).

14.8 Show that if α and β are both 2-forms, then $d(\alpha \wedge \beta) = (d\alpha) \wedge \beta + \alpha \wedge d\beta$.

14.9 Use Hamilton's equations (18.1) to derive the formula (14.75) for the time derivatives $\delta \dot{p}_i$ and $\delta \dot{q}_i$.

14.10 Use Hamilton's equations (18.1) to compute the time derivatives of the n pairs of tiny displacements Δp_j, Δq_j. Then use your resulting formulas and those (14.75) for the time derivatives of the n pairs of small differences δp_j, δq_j to show that the time derivative of the sum (14.76) of areas of tiny parallelograms vanishes (14.77).

14.11 In the early days of quantum mechanics, Bohr and Sommerfeld set action integrals like those of Example 14.9 equal to a multiple of Planck's constant, $A = \oint p \cdot dq = nh$. Why do you think they chose invariant quantities to quantize in this way?

14.12 Use Bohr's quantization of angular momentum $L = rp = n\hbar$ to find the energy levels of an electron in a circular orbit about a proton. Take the energy as $E = p^2/2m - Ze^2/4\pi\epsilon_0 r$ and balance radial forces $mv^2/r = Ze^2/4\pi\epsilon_0 r^2$ where $p = mv$.

14.13 Work out the details of using the curl formulas (14.78–14.79) to derive the curl formula (14.101) from the general identity (14.100). Do this in rectangular, cylindrical, and spherical coordinates.

14.14 Is $y^2 e^{zw}\, dx + 2x\, y\, e^{zw}\, dy + x\, y^2\, w\, e^{zw}\, dz + x\, y^2\, z\, e^{zw}\, dw$ closed?

14.15 Is $ze^y/xw\, dx + ze^y \ln x/w\, dy + e^y \ln x/w\, dz - ze^y \ln x/w^2\, dw$ closed?

14.16 Show that ∂ and $\bar{\partial}$ satisfy (14.136).

14.17 Use the definition (14.142) to show that in flat 3-space, the dual of the Hodge dual is the identity: $**\, dx^i = dx^i$ and $**\,(dx^i \wedge dx^k) = dx^i \wedge dx^k$.

14.18 Use the definition of the Hodge star (14.151) to derive (a) two of the four identities (14.152) and (b) the other two.

14.19 Show that Levi-Civita's 4-symbol obeys the identity (14.156).

14.20 Show that $\epsilon_{\ell mn}\, \epsilon^{pmn} = 2\, \delta^p_\ell$ and that $\epsilon_{k\ell mn}\, \epsilon^{p\ell mn} = 3!\, \delta^p_k$.

15

Probability and Statistics

15.1 Probability and Thomas Bayes

The probability $P(A)$ of an outcome in a set A is the sum of the probabilities P_j of all the different (mutually exclusive) outcomes j in A

$$P(A) = \sum_{j \in A} P_j. \tag{15.1}$$

For instance, if one throws two fair dice, then the probability that the sum is 2 is $P(1, 1) = 1/36$, while the probability that the sum is 3 is $P(1, 2) + P(2, 1) = 1/18$.

If A and B are two sets of possible outcomes, then the probability of an outcome in the **union** $A \cup B$ is the sum of the probabilities $P(A)$ and $P(B)$ minus that of their **intersection** $A \cap B$

$$P(A \cup B) = P(A) + P(B) - P(A \cap B). \tag{15.2}$$

If the outcomes are mutually exclusive, then $P(A \cap B) = 0$, and the probability of the union is the sum $P(A \cup B) = P(A) + P(B)$. The **joint probability** $P(A, B) \equiv P(A \cap B)$ is the probability of an outcome that is in both sets A and B. If the joint probability is the product $P(A \cap B) = P(A)\, P(B)$, then the outcomes in sets A and B are **statistically independent**.

The probability that a result in set B also is in set A is the **conditional probability** $P(A|B)$, the probability of A given B

$$P(A|B) = \frac{P(A \cap B)}{P(B)}. \tag{15.3}$$

Interchanging A and B, we get as the probability of B given A

$$P(B|A) = \frac{P(B \cap A)}{P(A)}. \tag{15.4}$$

Since $A \cap B = B \cap A$, the last two equations (15.3 and 15.4) tell us that

$$P(A \cap B) = P(B \cap A) = P(B|A)\, P(A) = P(A|B)\, P(B) \tag{15.5}$$

in which the last equality is **Bayes's theorem**

$$P(A|B) = \frac{P(B|A)\, P(A)}{P(B)} \quad \text{as well as} \quad P(B|A) = \frac{P(A|B)\, P(B)}{P(A)} \tag{15.6}$$

(Thomas Bayes, 1702–1761).

In a formula like $P(A \cap B) = P(B|A)\, P(A)$, the probability $P(A)$ is called an **a priori probability**, a **prior probability**, or a **prior**. If $B = C \cap D$ in (15.3), then $P(A|C \cap D) = P(A \cap C \cap D)/P(C \cap D)$.

If a set B of outcomes is contained in a union of n sets A_j that are mutually exclusive,

$$B \subset \bigcup_{j=1}^{n} A_j \quad \text{and} \quad A_i \cap A_k = \emptyset, \tag{15.7}$$

then we must sum over them

$$P(B) = \sum_{j=1}^{n} P(B|A_j)\, P(A_j). \tag{15.8}$$

If, for example, A_j were the probability of selecting an atom with Z_j protons and N_j neutrons, and if $P(B|A_j)$ were the probability that such a nucleus would decay in time t, then the probability that the nucleus of the selected atom would decay in time t would be given by the sum (15.8). In this case, if we replace A by A_k in the formula (15.5) for the probability $P(A \cap B)$, then we get $P(B \cap A_k) = P(B|A_k)\, P(A_k) = P(A_k|B)\, P(B)$. This last equality and the sum (15.8) give us these forms of Bayes's theorem

$$P(A_k|B) = \frac{P(B|A_k)\, P(A_k)}{\sum_{j=1}^{N} P(B|A_j)\, P(A_j)} \tag{15.9}$$

$$P(B|A_k) = \frac{P(A_k|B)}{P(A_k)} \sum_{j=1}^{N} P(B|A_j)\, P(A_j). \tag{15.10}$$

Example 15.1 (Was the cab blue?) A cab was involved in a hit-and-run accident at night. In the city 85% of the cabs are Green and 15% are Blue. A witness said the cab was Blue. Tests showed that the witness correctly distinguished Green and Blue cabs only 80% of the time. What is the probability that the guilty cab was Blue? (Kahneman, 2011, p.166)

The probabilities of a random cab being Blue or Green are $P(B) = 0.15$ and $P(G) = 0.85$. The probabilities that the witness would call Blue and Green cabs Blue are $P(sB|B) = 0.8$ and $P(sB|G) = 0.2$. So the probability $P(sB)$ that the witness said a random cab was Blue is

$$P(sB) = P(sB|B)P(B) + P(sB|G)P(G) = 0.29. \tag{15.11}$$

Now Bayes's theorem (15.6) gives the probability $P(B|sB)$ that a cab the witness said was Blue actually was Blue is

$$P(B|sB) = \frac{P(sB|B)P(B)}{P(sB)} = \frac{0.8(0.15)}{0.29} = 0.41 \tag{15.12}$$

which is about half the naive answer of 80%.

Example 15.2 (Low-base-rate problem) Suppose the incidence of a rare disease in a population is $P(D) = 0.001$. Suppose a test for the disease has a **sensitivity** of 99%, that is, the probability that a carrier of the disease will test positive is $P(+|D) = 0.99$. Suppose the test also is highly **selective** with a false-positive rate of only $P(+|N) = 0.005$. Then the probability that a random person in the population would test positive is by (15.8)

$$P(+) = P(+|D)\, P(D) + P(+|N)\, P(N) = 0.00599. \tag{15.13}$$

And by Bayes's theorem (15.6), the probability that a person who tests positive actually has the disease is only

$$P(D|+) = \frac{P(+|D)\, P(D)}{P(+)} = \frac{0.99 \times 0.001}{0.00599} = 0.165 \tag{15.14}$$

and the probability that a person testing positive actually is healthy is $P(N|+) = 1 - P(D|+) = 0.835$.

Even with an excellent test, screening for rare diseases is problematic. Similarly, screening for rare behaviors, such as disloyalty in the FBI, is dicey with a good test and absurd with a poor one like a polygraph.

Example 15.3 (Three-door problem) A prize lies behind one of three closed doors. A contestant gets to pick which door to open, but before the chosen door is opened, a door that does not lead to the prize and was not picked by the contestant swings open. Should the contestant switch and choose a different door?

We note that a contestant who picks the wrong door and switches always wins, so $P(W|Sw, WD) = 1$, while one who picks the right door and switches never wins $P(W|Sw, RD) = 0$. Since the probability of picking the wrong door is $P(WD) = 2/3$, the probability of winning if one switches is

$$P(W|Sw) = P(W|Sw, WD)\, P(WD) + P(W|Sw, RD)\, P(RD) = 2/3. \tag{15.15}$$

The probability of picking the right door is $P(RD) = 1/3$, and the probability of winning if one picks the right door and stays put is $P(W|Sp, RD) = 1$. So the probability of winning if one stays put is

$$P(W|Sp) = P(W|Sp, RD)\, P(RD) + P(W|Sp, WD)\, P(WD) = 1/3. \quad (15.16)$$

Thus, one should switch after the door opens.

If the set A is the interval $(x - dx/2, x + dx/2)$ of the real line, then $P(A) = P(x)\, dx$, and version (15.9) of Bayes's theorem says

$$P(x|B) = \frac{P(B|x)\, P(x)}{\int_{-\infty}^{\infty} P(B|x')\, P(x')\, dx'}. \quad (15.17)$$

Example 15.4 (A tiny poll) We ask 4 likely voters if they will vote for Nancy Pelosi, and 3 say *yes*. If the probability that a random voter will vote for her is y, then the probability that 3 in our sample of 4 will is

$$P(3|y) = 4\, y^3\, (1 - y) \quad (15.18)$$

which is the value $P_b(3, y, 4)$ of the binomial distribution (Section 15.3, 15.51) for $n = 3$, $p = y$, and $N = 4$. We don't know the **prior** probability distribution $P(y)$, so we set it equal to unity on the interval $(0, 1)$. Then the continuous form of Bayes's theorem (15.17) and our cheap poll give the probability distribution of the fraction y who will vote for her as

$$P(y|3) = \frac{P(3|y)\, P(y)}{\int_0^1 P(3|y')\, P(y')\, dy'} = \frac{P(3|y)}{\int_0^1 P(3|y')\, dy'}$$
$$= \frac{4\, y^3\, (1 - y)}{\int_0^1 4\, y'^3\, (1 - y')\, dy'} = 20\, y^3\, (1 - y). \quad (15.19)$$

Our best guess then for the probability that she will win the election is

$$\int_{1/2}^{1} P(y|3)\, dy = \int_{1/2}^{1} 20\, y^3\, (1 - y)\, dy = \frac{13}{16} \quad (15.20)$$

which is slightly higher that the naive estimate of $3/4$.

Example 15.5 (Quantum mechanics) But when are two sets A_1 and A_2 of microscopic events mutually exclusive? Suppose a photon can go from a laser through slits 1 and 2 and be detected at point B. Unless we measure which slit the photon goes through, the two passages are not mutually exclusive. So we can't compute the probability $P(B)$ that the photon is detected at point B as the sum (15.8) $P(B) = P(B|A_1)\, P(A_1) + P(B|A_2)\, P(A_2)$ in which $P(A_i)$ is the probability of its going through slit i. We must use the quantum-mechanical formula $P(B) = |\langle B|A_1\rangle + \langle B|A_2\rangle|^2$ in which $\langle B|A_i\rangle$ is the amplitude for the photon to get to B through slit i.

15.2 Mean and Variance

In roulette and many other games, N outcomes x_j can occur with probabilities P_j that sum to unity

$$\sum_{j=1}^{N} P_j = 1. \tag{15.21}$$

The **expected value** $E[x]$ of the outcome x is its **mean** μ or **average** value $\langle x \rangle = \overline{x}$

$$E[x] = \mu = \langle x \rangle = \overline{x} = \sum_{j=1}^{N} x_j \, P_j. \tag{15.22}$$

The **expected value** $E[x]$ also is called the **expectation** of x or the **expectation value** of x (and should be called the **mean value** of x).

The **ℓth moment** is

$$E[x^\ell] = \mu_\ell = \langle x^\ell \rangle = \sum_{j=1}^{N} x_j^\ell P_j \tag{15.23}$$

and the **ℓth central moment** is

$$E[(x-\mu)^\ell] = \nu_\ell = \sum_{j=1}^{N} (x_j - \mu)^\ell P_j \tag{15.24}$$

where always $\mu_0 = \nu_0 = 1$ and $\nu_1 = 0$ (Exercise 15.3).

The **variance** $V[x]$ is the second central moment ν_2

$$V[x] \equiv E[(x - \langle x \rangle)^2] = \nu_2 = \sum_{j=1}^{N} \left(x_j - \langle x \rangle\right)^2 \, P_j \tag{15.25}$$

which one may write as (Exercise 15.5)

$$V[x] = \langle x^2 \rangle - \langle x \rangle^2 \tag{15.26}$$

and the **standard deviation** σ is its square root

$$\sigma = \sqrt{V[x]}. \tag{15.27}$$

If the values of x are distributed continuously according to a **probability distribution** or **density** $P(x)$ normalized to unity

$$\int P(x) \, dx = 1 \tag{15.28}$$

then the mean value is

$$E[x] = \mu = \langle x \rangle = \int x \, P(x) \, dx \tag{15.29}$$

and the ℓth moment is

$$E[x^\ell] = \mu_\ell = \langle x^\ell \rangle = \int x^\ell \, P(x) \, dx. \tag{15.30}$$

The ℓth central moment is

$$E[(x - \mu)^\ell] = \nu_\ell = \int (x - \mu)^\ell \, P(x) \, dx. \tag{15.31}$$

The variance of the distribution is the second central moment

$$V[x] = \nu_2 = \int (x - \langle x \rangle)^2 \, P(x) \, dx = \mu_2 - \mu^2 \tag{15.32}$$

and the standard deviation σ is its square root $\sigma = \sqrt{V[x]}$.

Many authors use $f(x)$ for the probability distribution $P(x)$ and $F(x)$ for the cumulative probability $\Pr(-\infty, x)$ of an outcome in the interval $(-\infty, x)$

$$F(x) \equiv \Pr(-\infty, x) = \int_{-\infty}^{x} P(x') \, dx' = \int_{-\infty}^{x} f(x') \, dx' \tag{15.33}$$

a function that is necessarily **monotonic**

$$F'(x) = \Pr'(-\infty, x) = f(x) = P(x) \ge 0. \tag{15.34}$$

Some mathematicians reserve the term probability **distribution** for probabilities like $\Pr(-\infty, x)$ and P_j and call a continuous distribution $P(x)$ a **probability density function**. But the usage in physics of the Maxwell–Boltzmann distribution is too widespread for me to observe this distinction.

Although a probability distribution $P(x)$ is normalized (15.28), it can have **fat tails**, which are important in financial applications (Bouchaud and Potters, 2003). Fat tails can make the variance and even the **mean absolute deviation**

$$E_{\text{abs}} \equiv \int |x - \mu| \, P(x) \, dx \tag{15.35}$$

diverge.

Example 15.6 (Heisenberg's uncertainty principle) In quantum mechanics, the absolute-value squared $|\psi(x)|^2$ of a wave function $\psi(x)$ is the probability distribution $P(x) = |\psi(x)|^2$ of the position x of the particle, and $P(x)\,dx$ is the probability that the particle is found between $x - dx/2$ and $x + dx/2$. The variance $\langle (x - \langle x \rangle)^2 \rangle$ of the position operator x is written as the square $(\Delta x)^2$ of the standard deviation $\sigma = \Delta x$

which is the **uncertainty** in the position of the particle. Similarly, the square of the uncertainty in the momentum $(\Delta p)^2$ is the variance $\langle (p - \langle p \rangle)^2 \rangle$ of the momentum.

For the wave function (4.73)

$$\psi(x) = \left(\frac{2}{\pi}\right)^{1/4} \frac{1}{\sqrt{a}}\, e^{-(x/a)^2} \tag{15.36}$$

these uncertainties are $\Delta x = a/2$ and $\Delta p = \hbar/a$. They provide a saturated example $\Delta x\, \Delta p = \hbar/2$ of Heisenberg's uncertainty principle

$$\Delta x\, \Delta p \geq \frac{\hbar}{2}. \tag{15.37}$$

If x and y are two random variables that occur with a **joint distribution** $P(x, y)$, then the expected value of the linear combination $ax^n y^m + bx^p y^q$ is

$$\begin{aligned} E[ax^n y^m + bx^p y^q] &= \int (ax^n y^m + bx^p y^q)\, P(x, y)\, dxdy \\ &= a \int x^n y^m\, P(x, y)\, dxdy + b \int x^p y^q\, P(x, y)\, dxdy \\ &= a\, E[x^n y^m] + b\, E[x^p y^q]. \end{aligned} \tag{15.38}$$

This result and its analog for discrete probability distributions show that **expected values are linear**.

Example 15.7 (Jensen's inequalities) A **convex** function is one that lies above its tangents:

$$f(x) \geq f(y) + (x - y) f'(y). \tag{15.39}$$

For example, e^x lies above $1 + x$ which is its tangent at $x = 0$. Multiplying both sides of the definition (15.39) by the probability distribution $P(x)$ and integrating over x with $y = \langle x \rangle$, we find that the mean value of a convex function

$$\begin{aligned} \langle f(x) \rangle = \int f(x) P(x) dx &\geq \int \left[f(\langle x \rangle) + (x - \langle x \rangle) f'(\langle x \rangle) \right] P(x) dx \\ &= \int f(\langle x \rangle)\, P(x)\, dx = f(\langle x \rangle) \end{aligned} \tag{15.40}$$

exceeds its value at $\langle x \rangle$. Equivalently, $E[f(x)] \geq f(E[x])$.

For a **concave** function, the inequalities (15.39) and (15.40) reverse, and $f(E[x]) \geq E[f(x)]$. Thus since $\log(x)$ is concave, we have

$$\log(E[x]) = \log\left(\frac{1}{n}\sum_{i=1}^{n} x_i\right) \geq E[\log(x)] = \sum_{i=1}^{n} \frac{1}{n} \log(x_i). \tag{15.41}$$

Exponentiating both sides, we get the **inequality of arithmetic and geometric means**

$$\frac{1}{n}\sum_{i=1}^{n} x_i \geq \left(\prod_{i=1}^{n} x_i\right)^{1/n} \tag{15.42}$$

(Johan Jensen, 1859–1925).

The **correlation coefficient** or **covariance** of two variables x and y that occur with a **joint distribution** $P(x, y)$ is

$$C[x, y] \equiv \int P(x, y)(x - \overline{x})(y - \overline{y})\, dxdy = \langle (x - \overline{x})(y - \overline{y}) \rangle$$

$$= \langle x\, y \rangle - \langle x \rangle \langle y \rangle. \tag{15.43}$$

The variables x and y are said to be **independent** if

$$P(x, y) = P(x)\, P(y). \tag{15.44}$$

Independence implies that the covariance vanishes, but $C[x, y] = 0$ does not guarantee that x and y are independent (Roe, 2001, p. 9).

The variance of $x + y$

$$\langle (x + y)^2 \rangle - \langle x + y \rangle^2 = \langle x^2 \rangle - \langle x \rangle^2 + \langle y^2 \rangle - \langle y \rangle^2$$

$$+ 2\left(\langle x\, y \rangle - \langle x \rangle \langle y \rangle\right) \tag{15.45}$$

is the sum

$$V[x + y] = V[x] + V[y] + 2\, C[x, y]. \tag{15.46}$$

It follows (Exercise 15.6) that for any constants a and b the variance of $ax + by$ is

$$V[ax + by] = a^2\, V[x] + b^2\, V[y] + 2\, ab\, C[x, y]. \tag{15.47}$$

More generally (Exercise 15.7), the variance of the sum $a_1 x_1 + a_2 x_2 + \cdots + a_N x_N$ is

$$V[a_1 x_1 + \cdots + a_N x_N] = \sum_{j=1}^{N} a_j^2\, V[x_j] + \sum_{j,k=1, j<k}^{N} 2 a_j a_k\, C[x_j, x_k]. \tag{15.48}$$

If the variables x_j and x_k are independent for $j \neq k$, then their covariances vanish $C[x_j, x_k] = 0$, and the variance of the sum $a_1 x_1 + \cdots + a_N x_N$ is

$$V[a_1 x_1 + \cdots + a_N x_N] = \sum_{j=1}^{N} a_j^2\, V[x_j]. \tag{15.49}$$

15.3 Binomial Distribution

If the probability of success is p on each try, then we expect that in N tries the mean number of successes will be

$$\langle n \rangle = N\, p. \tag{15.50}$$

The probability of failure on each try is $q = 1-p$. So the probability of a particular sequence of successes and failures, such as n successes followed by $N-n$ failures is $p^n\, q^{N-n}$. There are $N!/n!\,(N-n)!$ different sequences of n successes and $N-n$ failures, all with the same probability $p^n\, q^{N-n}$. So the probability of n successes (and $N-n$ failures) in N tries is

$$P_b(n, p, N) = \frac{N!}{n!\,(N-n)!}\, p^n\, q^{N-n} = \binom{N}{n}\, p^n\,(1-p)^{N-n}. \tag{15.51}$$

This **binomial distribution** also is called **Bernoulli's distribution** (Jacob Bernoulli, 1654–1705).

The sum (5.93) of the probabilities $P_b(n, p, N)$ for $n = 0, 1, 2, \ldots, N$ is unity

$$\sum_{n=0}^{N} P_b(n, p, N) = \sum_{n=0}^{N} \binom{N}{n}\, p^n\,(1-p)^{N-n} = (p+1-p)^N = 1. \tag{15.52}$$

In Fig. 15.1, the probabilities $P_b(n, p, N)$ for $0 \le n \le 250$ and $p = 0.2$ are plotted for $N = 125$, 250, 500, and 1000 tries.

The mean number of successes

$$\mu = \langle n \rangle_B = \sum_{n=0}^{N} n\, P_b(n, p, N) = \sum_{n=0}^{N} n \binom{N}{n}\, p^n q^{N-n} \tag{15.53}$$

is a partial derivative with respect to p with q held fixed

$$\begin{aligned} \langle n \rangle_B &= p\, \frac{\partial}{\partial p} \sum_{n=0}^{N} \binom{N}{n}\, p^n q^{N-n} \\ &= p\, \frac{\partial}{\partial p}\, (p+q)^N = Np\,(p+q)^{N-1} = Np \end{aligned} \tag{15.54}$$

which verifies the estimate (15.50).

One may show (Exercise 15.9) that the variance (15.25) of the binomial distribution is

$$V_B = \langle (n - \langle n \rangle)^2 \rangle = p\,(1-p)\,N. \tag{15.55}$$

Its standard deviation (15.27) is

$$\sigma_B = \sqrt{V_B} = \sqrt{p\,(1-p)\,N}. \tag{15.56}$$

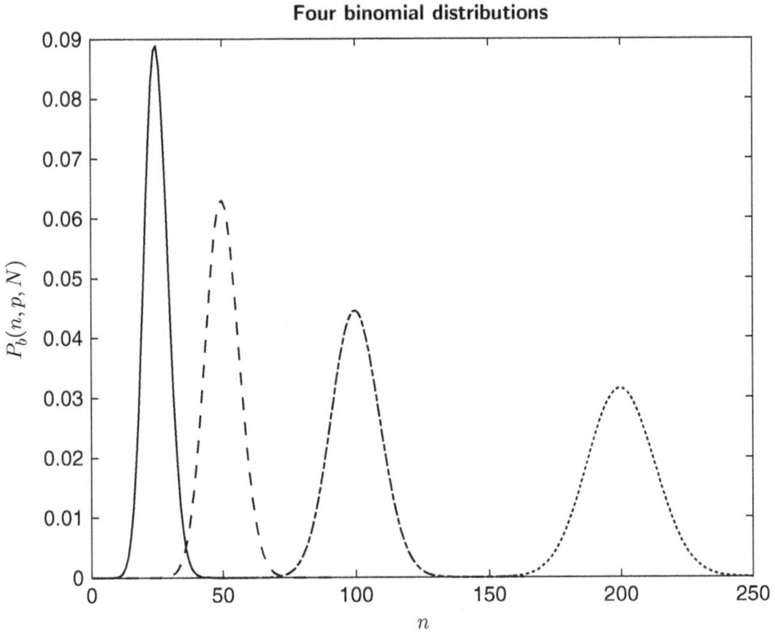

Figure 15.1 The binomial probability distribution $P_b(n, p, N)$ (15.51) is plotted here for $p = 0.2$ and $N = 125$ (solid), 250 (dashes), 500 (dot dash), and 1000 tries (dots). This chapter's codes are in Probability_and_statistics at `github.com/kevinecahill`.

The ratio of the width to the mean

$$\frac{\sigma_B}{\langle n \rangle_B} = \frac{\sqrt{p\,(1-p)\,N}}{Np} = \sqrt{\frac{1-p}{Np}} \tag{15.57}$$

decreases with N as $1/\sqrt{N}$.

Example 15.8 (Avogadro's number) A mole of gas is Avogadro's number $N_A = 6 \times 10^{23}$ of molecules. If the gas is in a cubical box, then the chance that each molecule will be in the left half of the cube is $p = 1/2$. The mean number of molecules there is $\langle n \rangle_b = pN_A = 3 \times 10^{23}$, and the uncertainty in n is $\sigma_b = \sqrt{p\,(1-p)\,N} = \sqrt{3 \times 10^{23}/4} = 3 \times 10^{11}$. So the numbers of gas molecules in the two halves of the box are equal to within $\sigma_b/\langle n \rangle_b = 10^{-12}$ or to 1 part in 10^{12}.

Example 15.9 (Counting fluorescent molecules) Molecular biologists can insert the DNA that codes for a fluorescent protein next to the DNA that codes for a specific natural protein in the genome of a bacterium. The bacterium then will make its natural protein with the fluorescent protein attached to it, and the labeled protein will produce light of a specific color when suitably illuminated by a laser. The intensity I of the light is proportional to the number n of labeled protein molecules $I = \alpha\, n$, and one can find the constant of proportionality α by measuring the light given off as the bacteria divide. When a bacterium divides, it randomly separates the total number N of fluorescent proteins inside it into its two daughter bacteria, giving one n fluorescent molecules and the other $N - n$. The variance of the difference is

$$\langle (n - (N - n))^2 \rangle = \langle (2n - N)^2 \rangle = 4\,\langle n^2 \rangle - 4\,\langle n \rangle N + N^2. \tag{15.58}$$

The mean number (15.54) is $\langle n \rangle = pN$, and our variance formula (15.55) tells us that

$$\langle n^2 \rangle = \langle n \rangle^2 + p(1-p)N = (pN)^2 + p(1-p)N. \tag{15.59}$$

Since the probability $p = 1/2$, the variance of the difference is

$$\langle (n - (N - n))^2 \rangle = (2p-1)^2 N^2 + 4p(1-p)N = N. \tag{15.60}$$

Thus the ratio of the variance of the difference of daughters' intensities to the intensity of the parent bacterium reveals the unknown constant of proportionality α (Phillips et al., 2012)

$$\frac{\langle (I_n - I_{N-n})^2 \rangle}{\langle I_N \rangle} = \frac{\alpha^2 \langle (n - (N-n))^2 \rangle}{\alpha \langle N \rangle} = \frac{\alpha^2 N}{\alpha N} = \alpha. \tag{15.61}$$

15.4 Coping with Big Factorials

Because $n!$ increases very rapidly with n, the rule

$$P_b(k+1, p, n) = \frac{p}{1-p}\,\frac{n-k}{k+1}\,P_b(k, p, n) \tag{15.62}$$

is helpful when n is big. But when n exceeds a few hundred, the formula (15.51) for $P_b(k, p, n)$ becomes unmanageable even in quadruple precision.

One solution is to work with the logarithm of the expression of interest. The Fortran function log_gamma(x), the C function lgamma(x), the Matlab function gammaln(x), and the Python function loggamma(x) all give $\log(\Gamma(x)) = \log((x-1)!)$ for real x. Using the very tame logarithm of the gamma function, one may compute $P_b(k, p, n)$ even for $n = 10^7$ as

$$\binom{n}{k} p^k q^{n-k} = \exp\big[\log(\Gamma(n+1)) - \log(\Gamma(n-k+1)) \\ - \log(\Gamma(k+1)) + k \log p + (n-k)\log q\big]. \tag{15.63}$$

Another way to cope with huge factorials is to use Stirling's formula (5.41) $n! \approx \sqrt{2\pi n}\,(n/e)^n$ or Srinivasa Ramanujan's correction (5.42) or Mermin's even more accurate approximations (5.43–5.45).

A third way to cope with the unwieldy factorials in the binomial formula $P_b(k, p, n)$ is to use its limiting forms due to Poisson and to Gauss.

15.5 Poisson's Distribution

Poisson approximated the formula (15.51) for the binomial distribution $P_b(n, p, N)$ by taking the two limits $N \to \infty$ and $p = \langle n \rangle / N \to 0$ while keeping n and the product $pN = \langle n \rangle$ constant. Using Stirling's formula $n! \approx \sqrt{2\pi n}\,(n/e)^n$ (6.331) for the two huge factorials $N!$ and $(N-n)!$, we get as $n/N \to 0$ and $\langle n \rangle / N \to 0$ with $\langle n \rangle = pN$ kept fixed

$$\begin{aligned} P_b(n, p, N) &= \binom{N}{n} p^n (1-p)^{N-n} = \frac{N!}{(N-n)!}\frac{p^n}{n!}(1-p)^{N-n} \\ &\approx \sqrt{\frac{N}{N-n}}\left(\frac{N}{e}\right)^N \left(\frac{e}{N-n}\right)^{N-n} \frac{(pN)^n}{n!}(1-p)^{N-n} \\ &\approx e^{-n}\left(1-\frac{n}{N}\right)^{-N+n} \frac{\langle n \rangle^n}{n!}\left(1-\frac{\langle n \rangle}{N}\right)^{N-n}. \end{aligned} \tag{15.64}$$

So using the definition $\exp(-x) = \lim_{N\to\infty}(1-x/N)^N$ to take the limits

$$\left(1-\frac{n}{N}\right)^{-N}\left(1-\frac{n}{N}\right)^{n} \to e^n \quad \text{and} \quad \left(1-\frac{\langle n \rangle}{N}\right)^{N}\left(1-\frac{\langle n \rangle}{N}\right)^{-n} \to e^{\langle n \rangle}, \tag{15.65}$$

we get from the binomial distribution Poisson's estimate

$$P_P(n, \langle n \rangle) = \frac{\langle n \rangle^n}{n!} e^{-\langle n \rangle} \tag{15.66}$$

of the probability of n successes in a very large number N of tries, each with a tiny chance $p = \langle n \rangle / N$ of success. (Siméon-Denis Poisson, 1781–1840. Incidentally, *poisson* means *fish* and sounds like pwahsahn.)

The Poisson distribution is normalized to unity

$$\sum_{n=0}^{\infty} P_P(n, \langle n \rangle) = \sum_{n=0}^{\infty} \frac{\langle n \rangle^n}{n!} e^{-\langle n \rangle} = e^{\langle n \rangle} e^{-\langle n \rangle} = 1. \tag{15.67}$$

Its mean μ is the parameter $\langle n \rangle = pN$ of the binomial distribution

$$\begin{aligned} \mu &= \sum_{n=0}^{\infty} n\, P_P(n, \langle n \rangle) = \sum_{n=1}^{\infty} n \frac{\langle n \rangle^n}{n!} e^{-\langle n \rangle} = \langle n \rangle \sum_{n=1}^{\infty} \frac{\langle n \rangle^{(n-1)}}{(n-1)!} e^{-\langle n \rangle} \\ &= \langle n \rangle \sum_{n=0}^{\infty} \frac{\langle n \rangle^n}{n!} e^{-\langle n \rangle} = \langle n \rangle. \end{aligned} \tag{15.68}$$

As $N \to \infty$ and $p \to 0$ with $pN = \langle n \rangle$ fixed, the variance (15.55) of the binomial distribution tends to the limit

$$V_P = \lim_{\substack{N\to\infty \\ p\to 0}} V_B = \lim_{\substack{N\to\infty \\ p\to 0}} p\,(1-p)\,N = \langle n \rangle. \tag{15.69}$$

Thus the mean and the variance of a Poisson distribution are equal

$$V_P = \langle (n - \langle n \rangle)^2 \rangle = \langle n \rangle = \mu \tag{15.70}$$

as one may show directly (Exercise 15.11).

Example 15.10 (Accuracy of Poisson's distribution) If $p = 0.0001$ and $N = 10{,}000$, then $\langle n \rangle = 1$ and Poisson's approximation to the probability that $n = 2$ is $1/2e$. The exact binomial probability (15.63) and Poisson's estimate are $P_b(2,0.01,1000)$ $= 0.18395$ and $P_P(2, 1) = 0.18394$.

Example 15.11 (Coherent states) The **coherent state** $|\alpha\rangle$ introduced in Equation (3.146)

$$|\alpha\rangle = e^{-|\alpha|^2/2} e^{\alpha a^\dagger} |0\rangle = e^{-|\alpha|^2/2} \sum_{n=0}^{\infty} \frac{\alpha^n}{\sqrt{n!}} |n\rangle \tag{15.71}$$

is an eigenstate $a|\alpha\rangle = \alpha|\alpha\rangle$ of the annihilation operator a with eigenvalue α. The probability $P(n)$ of finding n quanta in the state $|\alpha\rangle$ is the square of the absolute value of the inner product $\langle n|\alpha\rangle$

$$P(n) = |\langle n|\alpha\rangle|^2 = \frac{|\alpha|^{2n}}{n!} e^{-|\alpha|^2} \tag{15.72}$$

which is a Poisson distribution $P(n) = P_P(n, |\alpha|^2)$ with mean and variance $\mu = \langle n \rangle = V(\alpha) = |\alpha|^2$.

Example 15.12 (Radiation and cancer) If a cell becomes cancerous only after being hit N times by ionizing radiation, then the probability of cancer $P(\langle n \rangle)_N$ rises with the dose or mean number $\langle n \rangle$ of hits per cell as

$$P(\langle n \rangle)_N = \sum_{n=N}^{\infty} \frac{\langle n \rangle^n}{n!} e^{-\langle n \rangle} \tag{15.73}$$

or $P(\langle n \rangle)_N \approx \langle n \rangle^N / N!$ for $\langle n \rangle \ll 1$. As illustrated in Fig. 15.2, although the incidence of cancer $P(\langle n \rangle)_N$ rises linearly (solid) with the dose $\langle n \rangle$ of radiation if a single hit, $N = 1$, can cause a cell to become cancerous, it rises more slowly if the threshold for cancer is $N = 2$ (dot dash), 3 (dashes), or 4 (dots). Most mutations are harmless. The mean number N of harmful mutations that occur before a cell becomes cancerous is about 4, but N varies with the affected organ from 1 to 10 (Martincorena et al., 2017).

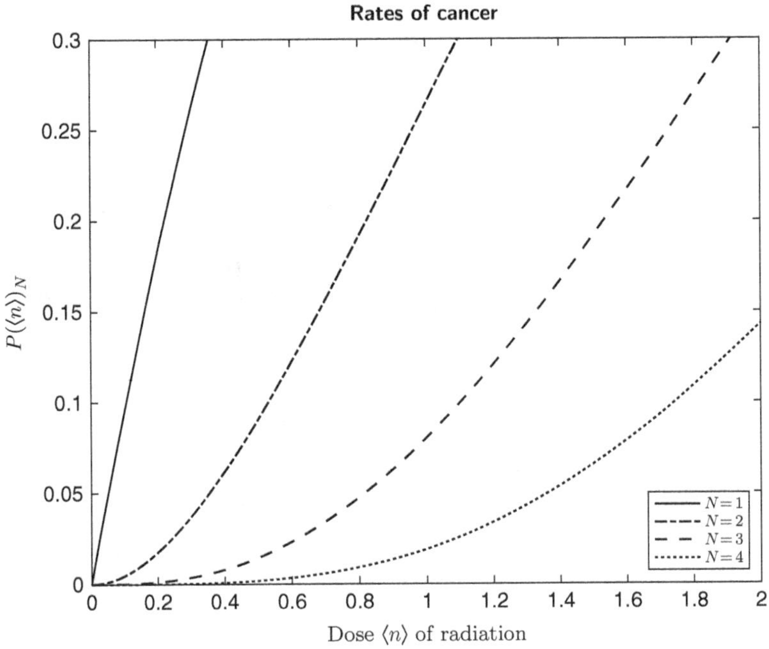

Figure 15.2 The incidence of cancer $P(\langle n\rangle)_N$ rises linearly (solid) with the dose or mean number $\langle n\rangle$ of times a cell is struck by ionizing radiation if a single hit, $N = 1$ (solid), can cause a cell to become cancerous. It rises more slowly if the threshold for cancer is $N = 2$ (dot dash), 3 (dashes), or 4 (dots) hits.

15.6 Gauss's Distribution

Gauss considered the binomial distribution in the limits $n \to \infty$ and $N \to \infty$ with the probability p fixed. In this limit, all three factorials are huge, and we may apply Stirling's formula to each of them

$$\begin{aligned} P_b(n, p, N) &= \frac{N!}{n!\,(N-n)!}\, p^n\, q^{N-n} \\ &\approx \sqrt{\frac{N}{2\pi n(N-n)}} \left(\frac{N}{e}\right)^N \left(\frac{e}{n}\right)^n \left(\frac{e}{N-n}\right)^{N-n} p^n\, q^{N-n} \\ &= \sqrt{\frac{N}{2\pi n(N-n)}} \left(\frac{pN}{n}\right)^n \left(\frac{qN}{N-n}\right)^{N-n}. \end{aligned} \tag{15.74}$$

This probability $P_b(n, p, N)$ is tiny unless n is near pN which means that $n \approx pN$ and $N - n \approx (1-p)N = qN$ are comparable. So we set $y = n - pN$ and treat y/N as small. Since $n = pN + y$ and $N - n = (1-p)N + pN - n = qN - y$, we can write the square root as

$$\sqrt{\frac{N}{2\pi\, n\,(N-n)}} = \frac{1}{\sqrt{2\pi\, N\,[(pN+y)/N]\,[(qN-y)/N]}}$$
$$= \frac{1}{\sqrt{2\pi\; pqN\,(1+y/pN)\,(1-y/qN)}}. \tag{15.75}$$

Because y remains finite as $N \to \infty$, the limit of the square root is

$$\lim_{N\to\infty} \sqrt{\frac{N}{2\pi\, n\,(N-n)}} = \frac{1}{\sqrt{2\pi\; pqN}}. \tag{15.76}$$

Substituting $pN + y$ for n and $qN - y$ for $N - n$ in (15.74), we find

$$P_b(n, p, N) \approx \frac{1}{\sqrt{2\pi\; pqN}} \left(\frac{pN}{pN+y}\right)^{pN+y} \left(\frac{qN}{qN-y}\right)^{qN-y}$$
$$= \frac{1}{\sqrt{2\pi\; pqN}} \left(1+\frac{y}{pN}\right)^{-(pN+y)} \left(1-\frac{y}{qN}\right)^{-(qN-y)} \tag{15.77}$$

which implies

$$\ln\left[P_b(n, p, N)\sqrt{2\pi\; pqN}\right] \approx -(pN+y)\ln\left[1+\frac{y}{pN}\right] - (qN-y)\ln\left[1-\frac{y}{qN}\right]. \tag{15.78}$$

The first two terms of the power series (5.101) for $\ln(1+\epsilon)$ are

$$\ln(1+\epsilon) \approx \epsilon - \frac{1}{2}\epsilon^2. \tag{15.79}$$

So applying this expansion to the two logarithms and using the relation $1/p + 1/q = (p+q)/pq = 1/pq$, we get

$$\ln\left(P_b(n, p, N)\sqrt{2\pi\; pqN}\right) \approx -(pN+y)\left[\frac{y}{pN} - \frac{1}{2}\left(\frac{y}{pN}\right)^2\right] \tag{15.80}$$
$$- (qN-y)\left[-\frac{y}{qN} - \frac{1}{2}\left(\frac{y}{qN}\right)^2\right] \approx -\frac{y^2}{2pqN}.$$

Remembering that $y = n - pN$, we get Gauss's approximation to the binomial probability distribution

$$P_{bG}(n, p, N) = \frac{1}{\sqrt{2\pi pqN}} \exp\left(-\frac{(n-pN)^2}{2pqN}\right). \tag{15.81}$$

This probability distribution is normalized

$$\sum_{n=0}^{\infty} \frac{1}{\sqrt{2\pi pqN}} \exp\left(-\frac{(n-pN)^2}{2pqN}\right) = 1 \tag{15.82}$$

almost exactly for $pN > 100$.

Extending the integer n to a continuous variable x, we have

$$P_G(x, p, N) = \frac{1}{\sqrt{2\pi pqN}} \exp\left(-\frac{(x-pN)^2}{2pqN}\right) \tag{15.83}$$

which on the real line $(-\infty, \infty)$ is (Exercise 15.12) a normalized probability distribution with mean $\langle x \rangle = \mu = pN$ and variance $\langle (x-\mu)^2 \rangle = \sigma^2 = pqN$. Replacing pN by μ and pqN by σ^2, we get the Standard form of **Gauss's distribution**

$$P_G(x, \mu, \sigma) = \frac{1}{\sigma\sqrt{2\pi}} \exp\left(-\frac{(x-\mu)^2}{2\sigma^2}\right). \tag{15.84}$$

This distribution occurs so often in mathematics and in nature that it is often called **the normal distribution**. Its odd central moments all vanish $\nu_{2n+1} = 0$, and its even ones are $\nu_{2n} = (2n-1)!!\,\sigma^{2n}$ (Exercise 15.14).

Example 15.13 (Acccuracy of Gauss's distribution) If $p = 0.1$ and $N = 10^4$, then Gauss's approximation to the probability that $n = 10^3$ is $1/(30\sqrt{2\pi})$. The exact binomial probability (15.63) and Gauss's estimate are $P_b(10^3, 0.1, 10^4) = 0.013297$ and $P_G(10^3, 0.1, 10^4) = 0.013298$.

Example 15.14 (Single-molecule super-resolution microscopy) If the wavelength of visible light were a nanometer, microscopes would yield much sharper images. Each photon from a (single-molecule) fluorophore entering the lens of a microscope would follow ray optics and be focused within a tiny circle of about a nanometer on a detector. Instead, a photon arrives not at $\boldsymbol{x} = (x_1, x_2)$ but at $\boldsymbol{y}_i = (y_{1i}, y_{2i})$ with gaussian probability

$$P(\boldsymbol{y}_i) = \frac{1}{2\pi\sigma^2} e^{-(\boldsymbol{y}_i - \boldsymbol{x})^2/2\sigma^2} \tag{15.85}$$

where $\sigma \approx 150$ nm is about a quarter of a wavelength. What to do?

In the **centroid** method, one collects $N \approx 500$ points $\boldsymbol{y}_i$ and finds the point $\boldsymbol{x}$ that maximizes the joint probability of the N image points

$$P = \prod_{i=1}^{N} P(\boldsymbol{y}_i) = d^N \prod_{i=1}^{N} e^{-(\boldsymbol{y}_i - \boldsymbol{x})^2/(2\sigma^2)} = d^N \exp\left[-\sum_{i=1}^{N} (\boldsymbol{y}_i - \boldsymbol{x})^2/(2\sigma^2)\right] \tag{15.86}$$

where $d = 1/2\pi\sigma^2$ by solving for $k = 1$ and 2 the equations

$$\frac{\partial P}{\partial x_k} = 0 = P\,\frac{\partial}{\partial x_k}\left[-\sum_{i=1}^{N} (\boldsymbol{y}_i - \boldsymbol{x})^2/(2\sigma^2)\right] = \frac{P}{\sigma^2}\sum_{i=1}^{N} (y_{ik} - x_k)\,. \tag{15.87}$$

This **maximum-likelihood** estimate of the image point $\boldsymbol{x}$ is the average of the observed points $\boldsymbol{y}_i$

$$\boldsymbol{x} = \frac{1}{N}\sum_{i=1}^{N} \boldsymbol{y}_i. \tag{15.88}$$

This method is an improvement, but it is biased by auto-fluorescence and out-of-focus fluorophores. Fang Huang and Keith Lidke use **direct stochastic optical reconstruction microscopy** (dSTORM) to locate the image point $\boldsymbol{x}$ of the fluorophore in ways that account for the finite accuracy of their pixilated detector and the randomness of photo-detection (Smith et al., 2010; Huang et al., 2011).

Actin filaments are double helices of the protein actin some 5–9 nm wide. They occur throughout a eukaryotic cell but are concentrated near its surface and determine its shape. Together with tubulin and intermediate filaments, they form a cell's cytoskeleton. The double membrane of a cell's nucleus is studded with 1000 nuclear pore complexes each of which regulates and facilitates the translocation of 1000 molecules per second. Figure 15.3 shows dSTORM images of actin filaments in a HeLa cell (left) and of the nuclear-pore-complex protein (Nup98) in the nuclear pores of a COS-7 cell (right). The finite size of the fluorophore and the motion of the molecules of living cells limit dSTORM's improvement in resolution to a factor of 10 to 20.

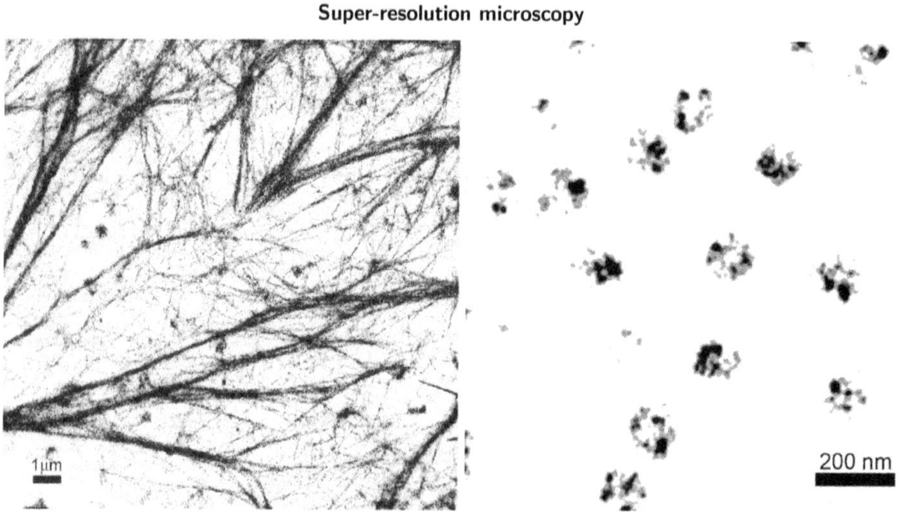

Figure 15.3 Left: dSTORM image of actin filaments in a HeLa cell, courtesy of Hanieh Mazloom Farsibaf and Keith Lidke, University of New Mexico. Right: images of the nuclear-pore-complex protein (Nup98) in the 120 nm-wide nuclear pores of a COS-7 cell, courtesy of Donghan Ma and Fang Huang, Purdue University.

15.7 The Error Function erf

The probability that a random variable x distributed according to Gauss's distribution (15.84) has a value between $\mu - \delta$ and $\mu + \delta$ is

$$P(|x-\mu| < \delta) = \int_{\mu-\delta}^{\mu+\delta} P_G(x,\mu,\sigma)\,dx = \frac{1}{\sigma\sqrt{2\pi}}\int_{\mu-\delta}^{\mu+\delta} \exp\left(-\frac{(x-\mu)^2}{2\sigma^2}\right)dx$$

$$= \frac{1}{\sigma\sqrt{2\pi}}\int_{-\delta}^{\delta} \exp\left(-\frac{x^2}{2\sigma^2}\right)dx = \frac{2}{\sqrt{\pi}}\int_0^{\delta/\sigma\sqrt{2}} e^{-t^2}\,dt. \quad (15.89)$$

The last integral is the error function

$$\operatorname{erf}(x) = \frac{2}{\sqrt{\pi}}\int_0^x e^{-t^2}dt. \quad (15.90)$$

The probability that x lies within δ of the mean μ is

$$P(|x-\mu| < \delta) = \operatorname{erf}\left(\frac{\delta}{\sigma\sqrt{2}}\right). \quad (15.91)$$

In particular, the probabilities that x falls within one, two, or three standard deviations of μ are

$$\begin{aligned} P(|x-\mu| < \sigma) &= \operatorname{erf}(1/\sqrt{2}) = 0.6827 \\ P(|x-\mu| < 2\sigma) &= \operatorname{erf}(2/\sqrt{2}) = 0.9545 \\ P(|x-\mu| < 3\sigma) &= \operatorname{erf}(3/\sqrt{2}) = 0.9973. \end{aligned} \quad (15.92)$$

The error function $\operatorname{erf}(x)$ is plotted in Fig. 15.4 in which the vertical lines are at $x = \delta/(\sigma\sqrt{2})$ for $\delta = \sigma, 2\sigma$, and 3σ.

The probability that x falls between a and b is (exercise 15.15)

$$P(a < x < b) = \frac{1}{2}\left[\operatorname{erf}\left(\frac{b-\mu}{\sigma\sqrt{2}}\right) - \operatorname{erf}\left(\frac{a-\mu}{\sigma\sqrt{2}}\right)\right]. \quad (15.93)$$

In particular, the cumulative probability $P(-\infty, x)$ that the random variable is less than x is for $\mu = 0$ and $\sigma = 1$

$$P(-\infty, x) = \frac{1}{2}\left[\operatorname{erf}\left(\frac{x}{\sqrt{2}}\right) - \operatorname{erf}\left(\frac{-\infty}{\sqrt{2}}\right)\right] = \frac{1}{2}\left[\operatorname{erf}\left(\frac{x}{\sqrt{2}}\right) + 1\right]. \quad (15.94)$$

The complement erfc of the error function is defined as

$$\operatorname{erfc}(x) = \frac{2}{\sqrt{\pi}}\int_x^{\infty} e^{-t^2}dt = 1 - \operatorname{erf}(x) \quad (15.95)$$

and is numerically useful for large x where round-off errors may occur in subtracting $\operatorname{erf}(x)$ from unity. Both erf and erfc are intrinsic functions in FORTRAN available without any effort on the part of the programmer.

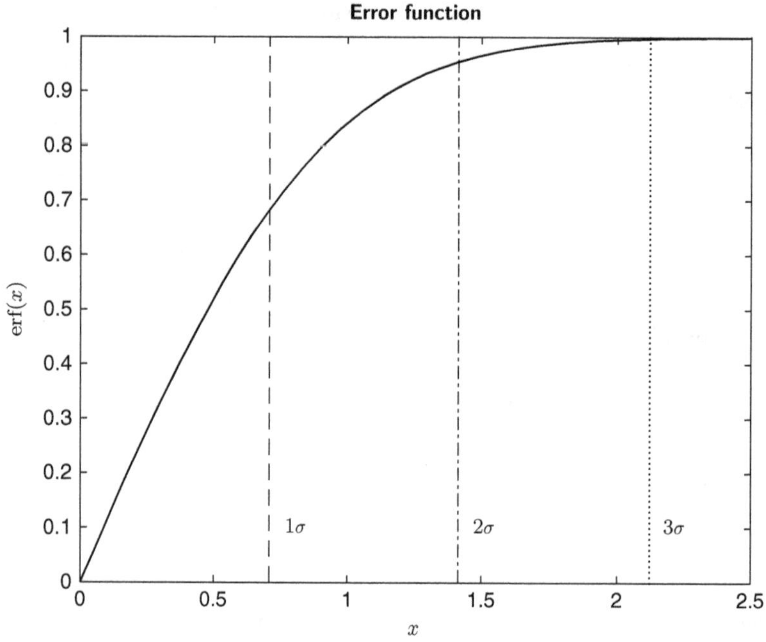

Figure 15.4 The error function erf (x) is plotted for $0 < x < 2.5$. The vertical lines are at $x = \delta/(\sigma\sqrt{2})$ for $\delta = \sigma$, 2σ, and 3σ with $\sigma = 1/\sqrt{2}$.

Example 15.15 (Summing binomial probabilities) To add up several binomial probabilities when the factorials in $P_b(n, p, N)$ are too big to handle, we first use Gauss's approximation (15.81)

$$P_b(n, p, N) = \frac{N!}{n!\,(N-n)!}\, p^n\, q^{N-n} \approx \frac{1}{\sqrt{2\pi pqN}} \exp\left(-\frac{(n-pN)^2}{2pqN}\right). \tag{15.96}$$

Then using (15.93) with $\mu = pN$, we find (Exercise 15.13)

$$P_b(n, p, N) \approx \frac{1}{2}\left[\operatorname{erf}\left(\frac{n+\frac{1}{2}-pN}{\sqrt{2pqN}}\right) - \operatorname{erf}\left(\frac{n-\frac{1}{2}-pN}{\sqrt{2pqN}}\right)\right] \tag{15.97}$$

which we can sum over the integer n to get

$$\sum_{n=n_1}^{n_2} P_b(n, p, N) \approx \frac{1}{2}\left[\operatorname{erf}\left(\frac{n_2+\frac{1}{2}-pN}{\sqrt{2pqN}}\right) - \operatorname{erf}\left(\frac{n_1-\frac{1}{2}-pN}{\sqrt{2pqN}}\right)\right] \tag{15.98}$$

which is easy to evaluate.

Example 15.16 (Polls) Suppose in a poll of 1000 likely voters, 600 have said they would vote for Nancy Pelosi. Repeating the analysis of Example 15.4, we see that if the probability that a random voter will vote for her is y, then the probability that 600 in our sample of 1000 will is by (15.96)

$$P(600|y) = P_b(600, y) = \binom{1000}{600} y^{600} (1-y)^{400}$$
$$\approx \frac{1}{10\sqrt{20\pi y(1-y)}} \exp\left(-\frac{20(3-5y)^2}{y(1-y)}\right). \qquad (15.99)$$

So if we conservatively assume that the unknown probability density $P(y)$ that a random voter will vote for her is an unknown constant which cancels, then the probability density that a random voter will vote for her, given that 600 have, is

$$P(y|600) = \frac{P(600|y)P(y)}{\int_0^1 P(600, y')\, P(y')\, dy'} = \frac{P(600|y)}{\int_0^1 P(600, y')\, dy'}$$
$$= \frac{[y(1-y)]^{-1/2} \exp\left(-\frac{20(3-5y)^2}{y(1-y)}\right)}{\int_0^1 [y'(1-y')]^{-1/2} \exp\left(-\frac{20(3-5y')^2}{y'(1-y')}\right) dy'}. \qquad (15.100)$$

So we estimate the probability that $y > 0.5$ as the ratio of the integrals

$$P(y > 0.5) \approx \frac{\int_{1/2}^1 [y(1-y)]^{-1/2} \exp\left(-\frac{20(3-5y)^2}{y(1-y)}\right) dy}{\int_0^1 [y(1-y)]^{-1/2} \exp\left(-\frac{20(3-5y)^2}{y(1-y)}\right) dy}. \qquad (15.101)$$

The Mathematica script ratio.nb gives $P(y > 1/2) \approx 0.999999999873$.

The normalized probability distribution (15.101) is negligible except for y near $3/5$ (Exercise 15.16), where it is approximately Gauss's distribution

$$P(y|600) \approx \frac{1}{\sigma\sqrt{2\pi}} \exp\left(-\frac{(y-3/5)^2}{2\sigma^2}\right) \qquad (15.102)$$

with mean $\mu = 3/5$ and variance $\sigma^2 = 3/12500 = 2.4 \times 10^{-4}$. The probability that $y > 1/2$ then is by (15.93) approximately

$$P(y > 1/2) \approx \frac{1}{2}\left[\operatorname{erf}\left(\frac{1-\mu}{\sigma\sqrt{2}}\right) - \operatorname{erf}\left(\frac{1/2-\mu}{\sigma\sqrt{2}}\right)\right]$$
$$= 0.999999999946. \qquad (15.103)$$

15.8 Error Analysis

The mean value $\bar{f} = \langle f \rangle$ of a smooth function $f(x)$ of a random variable x is

$$\bar{f} = \int f(x)\, P(x)\, dx$$
$$\approx \int \left[f(\mu) + (x-\mu)f'(\mu) + \frac{1}{2}(x-\mu)^2 f''(\mu)\right] P(x)\, dx \qquad (15.104)$$
$$= f(\mu) + \frac{1}{2}\sigma^2 f''(\mu)$$

as long as the higher central moments ν_n and the higher derivatives $f^{(n)}(\mu)$ are small. The mean value of f^2 then is

$$\begin{aligned}
\langle f^2 \rangle &= \int f^2(x)\, P(x)\, dx \\
&\approx \int \left[f(\mu) + (x-\mu) f'(\mu) + \frac{1}{2}(x-\mu)^2 f''(\mu) \right]^2 P(x)\, dx \\
&\approx \int \left[f^2(\mu) + (x-\mu)^2 f'^2(\mu) + (x-\mu)^2 f(\mu) f''(\mu) \right] P(x)\, dx \\
&= f^2(\mu) + \sigma^2 f'^2(\mu) + \sigma^2 f(\mu)\, f''(\mu).
\end{aligned} \tag{15.105}$$

Subtraction of $\bar{f}^2$ gives the variance of the variable $f(x)$

$$\sigma_f^2 = \langle (f - \bar{f})^2 \rangle = \langle f^2 \rangle - \bar{f}^2 \approx \sigma^2 f'^2(\mu). \tag{15.106}$$

A similar formula gives the variance of a smooth function $f(x_1, \ldots, x_n)$ of several independent variables $x_1, \ldots, x_n$ as

$$\sigma_f^2 = \langle (f - \bar{f})^2 \rangle = \langle f^2 \rangle - \bar{f}^2 \approx \sum_{i=1}^{n} \sigma_i^2 \left(\frac{\partial f(x)}{\partial x_i} \right)^2 \Bigg|_{x=\bar{x}} \tag{15.107}$$

in which $\bar{x}$ is the vector $(\mu_1, \ldots, \mu_n)$ of mean values, and $\sigma_i^2 = \langle (x_i - \mu_i)^2 \rangle$ is the variance of x_i.

This formula (15.107) implies that the variance of a sum $f(x, y) = c\,x + d\,y$ is

$$\sigma_{cx+dy}^2 = c^2 \sigma_x^2 + d^2 \sigma_y^2. \tag{15.108}$$

Similarly, the variance formula (15.107) gives as the variance of a product $f(x, y) = x\,y$

$$\sigma_{xy}^2 = \sigma_x^2 \mu_y^2 + \sigma_y^2 \mu_x^2 = \mu_x^2 \mu_y^2 \left(\frac{\sigma_x^2}{\mu_x^2} + \frac{\sigma_y^2}{\mu_y^2} \right) \tag{15.109}$$

and as the variance of a ratio $f(x, y) = x/y$

$$\sigma_{x/y}^2 = \frac{\sigma_x^2}{\mu_y^2} + \sigma_y^2 \frac{\mu_x^2}{\mu_y^4} = \frac{\mu_x^2}{\mu_y^2} \left(\frac{\sigma_x^2}{\mu_x^2} + \frac{\sigma_y^2}{\mu_y^2} \right). \tag{15.110}$$

The variance of a power $f(x) = x^a$ follows from the variance (15.106) of a function of a single variable

$$\sigma_{x^a}^2 = \sigma_x^2 \left(a \mu_x^{a-1} \right)^2. \tag{15.111}$$

In general, the standard deviation σ is the square root of the variance σ^2.

Example 15.17 (Photon density) The 2009 COBE/FIRAS measurement of the temperature of the cosmic microwave background (CMB) radiation is $T_0 = 2.7255 \pm 0.0006$ K. The mass density (5.110) of these photons is

$$\rho_\gamma = \sigma \frac{8\pi^5 (k_B T_0)^4}{15 h^3 c^5} = 4.6451 \times 10^{-31} \text{ kg m}^{-3}. \tag{15.112}$$

Our formula (15.111) for the variance of a power says that the standard deviation σ_ρ of the photon density is its temperature derivative times the standard deviation σ_T of the temperature

$$\sigma_\rho = \rho_\gamma \frac{4\sigma_T}{T_0} = 0.00088\, \rho_\gamma. \tag{15.113}$$

So the probability that the photon mass density lies within the range

$$\rho_\gamma = (4.6451 \pm 0.0041) \times 10^{-31} \text{ kg m}^{-3} \tag{15.114}$$

is 0.68.

15.9 Maxwell–Boltzmann Distribution

It is a small jump from Gauss's distribution (15.84) to the Maxwell–Boltzmann distribution of velocities of molecules in a gas. We start in 1 dimension and focus on a single molecule that is being knocked forward and backward with equal probabilities by other molecules. If each tiny hit increases or decreases its speed by dv, then after n hits from behind and $N - n$ hits from in front, the speed v_x of a molecule initially at rest would be

$$v_x = n dv - (N - n) dv = (2n - N) dv. \tag{15.115}$$

The probability of this speed is given by Gauss's approximation (15.81) to the binomial distribution $P_b(n, \frac{1}{2}, N)$ as

$$P_{bG}(n, \frac{1}{2}, N) = \sqrt{\frac{2}{\pi N}} \exp\left(-\frac{(2n - N)^2}{2N}\right) = \sqrt{\frac{2}{\pi N}} \exp\left(-\frac{v_x^2}{2N dv^2}\right). \tag{15.116}$$

In this formula, the product $N dv^2$ is the variance $\sigma_{v_x}^2$ which is the mean value $\langle v_x^2 \rangle$ because $\langle v_x \rangle = 0$. Kinetic theory says that this variance $\sigma_{v_x}^2 = \langle v_x^2 \rangle$ is $\langle v_x^2 \rangle = kT/m$ in which m is the mass of the molecule, k Boltzmann's constant, and T the temperature. So the probability of the molecule's having velocity v_x is the Maxwell–Boltzmann distribution

$$P_G(v_x) = \frac{1}{\sigma_v \sqrt{2\pi}} \exp\left(-\frac{v_x^2}{2\sigma_v^2}\right) = \sqrt{\frac{m}{2\pi k T}} \exp\left(-\frac{m v_x^2}{2kT}\right) \tag{15.117}$$

when normalized over the line $-\infty < v_x < \infty$.

In three space dimensions, the Maxwell–Boltzmann distribution $P_{MB}(\boldsymbol{v})$ is the product

$$P_{MB}(\boldsymbol{v})d^3v = P_G(v_x)\,P_G(v_y)\,P_G(v_z)d^3v = \left(\frac{m}{2\pi kT}\right)^{3/2} e^{-\frac{1}{2}m\boldsymbol{v}^2/(kT)}4\pi v^2 dv. \tag{15.118}$$

The mean value of the velocity of a Maxwell–Boltzmann gas vanishes

$$\langle \boldsymbol{v} \rangle = \int \boldsymbol{v}\, P_{MB}(\boldsymbol{v})d^3v = \boldsymbol{0} \tag{15.119}$$

but the mean value of the square of the velocity $\boldsymbol{v}^2 = \boldsymbol{v} \cdot \boldsymbol{v}$ is the sum of the three variances $\sigma_x^2 = \sigma_y^2 = \sigma_z^2 = kT/m$

$$\langle v^2 \rangle = V[\boldsymbol{v}^2] = \int \boldsymbol{v}^2\, P_{MB}(\boldsymbol{v})\, d^3v = 3kT/m \tag{15.120}$$

which is the familiar statement

$$\frac{1}{2}m\langle v^2 \rangle = \frac{3}{2}kT \tag{15.121}$$

that each degree of freedom gets $kT/2$ of energy.

15.10 Fermi–Dirac and Bose–Einstein Distributions

The commutation and anticommutation relations (11.133)

$$\psi_s(t,\boldsymbol{x})\psi_{s'}(t,\boldsymbol{x}') - (-1)^{2j}\psi_{s'}(t,\boldsymbol{x}')\psi_s(t,\boldsymbol{x}) = 0 \tag{15.122}$$

of Bose fields $(-1)^{2j} = 1$ and of Fermi fields $(-1)^{2j} = -1$ determine the statistics of bosons and fermions.

One can put any number N_n of noninteracting bosons, such as photons or gravitons, into any state $|n\rangle$ of energy E_n. The energy of that state is $N_n E_n$, and the states $|n, N_n\rangle$ form an independent thermodynamical system for each state $|n\rangle$. The grand canonical ensemble (1.395) gives the probability of the state $|n, N_n\rangle$ as

$$\rho_{n,N_n} = \langle n, N_n|\rho|n, N_n\rangle = \frac{e^{-\beta(E_n-\mu)N_n}}{\sum_{N_n}\langle n, N_n|\rho|n, N_n\rangle}. \tag{15.123}$$

For each state $|n\rangle$, the partition function is a geometric sum

$$Z(\beta,\mu,n) = \sum_{N_n=0}^{\infty} e^{-\beta(E_n-\mu)N_n} = \sum_{N_n=0}^{\infty}\left(e^{-\beta(E_n-\mu)}\right)^{N_n} = \frac{1}{1-e^{-\beta(E_n-\mu)}}. \tag{15.124}$$

So the probability of the state $|n, N_n\rangle$ is

$$\rho_{n,N_n} = \frac{e^{-\beta(E_n-\mu)N_n}}{1-e^{-\beta(E_n-\mu)}}, \tag{15.125}$$

and the mean number of bosons in the state $|n\rangle$ is

$$\langle N_n\rangle = \frac{1}{\beta}\frac{\partial}{\partial\mu}\ln Z(\beta,\mu,n) = \frac{1}{e^{\beta(E_n-\mu)}-1}. \tag{15.126}$$

One can put at most one fermion into a given state $|n\rangle$. If like neutrinos the fermions don't interact, then the states $|n, 0\rangle$ and $|n, 1\rangle$ form an independent thermodynamical system for each state $|n\rangle$. So for noninteracting fermions, the partition function is the sum of only two terms

$$Z(\beta,\mu,n) = \sum_{N_n=0}^{1} e^{-\beta(E_n-\mu)N_n} = 1 + e^{-\beta(E_n-\mu)}, \tag{15.127}$$

and the probability of the state $|n, N_n\rangle$ is

$$\rho_{n,N_n} = \frac{e^{-\beta(E_n-\mu)N_n}}{1+e^{-\beta(E_n-\mu)}}. \tag{15.128}$$

So the mean number of fermions in the state $|n\rangle$ is

$$\langle N_n\rangle = \;= \frac{1}{e^{\beta(E_n-\mu)}+1}. \tag{15.129}$$

15.11 Diffusion

We may apply the same reasoning as in the preceding section (15.9) to the diffusion of a gas of particles treated as a random walk with step size dx. In 1 dimension, after n steps forward and $N-n$ steps backward, a particle starting at $x=0$ is at $x=(2n-N)dx$. Thus as in (15.116), the probability of being at x is given by Gauss's approximation (15.81) to the binomial distribution $P_b(n, \frac{1}{2}, N)$ as

$$P_{bG}(n, \frac{1}{2}, N) = \sqrt{\frac{2}{\pi N}}\exp\left(-\frac{(2n-N)^2}{2N}\right) = \sqrt{\frac{2}{\pi N}}\exp\left(-\frac{x^2}{2Ndx^2}\right). \tag{15.130}$$

In terms of the diffusion constant

$$D = \frac{Ndx^2}{2t} \tag{15.131}$$

this distribution is

$$P_G(x) = \left(\frac{1}{4\pi Dt}\right)^{1/2}\exp\left(-\frac{x^2}{4Dt}\right) \tag{15.132}$$

when normalized to unity on $(-\infty, \infty)$.

In 3 dimensions, this gaussian distribution is the product

$$P(\boldsymbol{r}, t) = P_G(x)\, P_G(y)\, P_G(z) = \left(\frac{1}{4\pi D t}\right)^{3/2} \exp\left(-\frac{\boldsymbol{r}^2}{4Dt}\right). \tag{15.133}$$

The variance $\sigma^2 = 2Dt$ gives the average of the squared displacement of each of the three coordinates. Thus the mean of the squared displacement $\langle \boldsymbol{r}^2 \rangle$ rises **linearly** with the time as

$$\langle \boldsymbol{r}^2 \rangle = V[\boldsymbol{r}] = 3\sigma^2 = \int \boldsymbol{r}^2\, P(\boldsymbol{r}, t)\, d^3r = 6\, D\, t. \tag{15.134}$$

The distribution $P(\boldsymbol{r}, t)$ satisfies the **diffusion equation**

$$\dot{P}(\boldsymbol{r}, t) = D\, \nabla^2 P(\boldsymbol{r}, t) \tag{15.135}$$

in which the dot means time derivative.

15.12 Langevin's Theory of Brownian Motion

Einstein made the first theory of brownian motion in 1905, but Langevin's approach (Langevin, 1908) is simpler. A tiny particle of colloidal size and mass m in a fluid is buffeted by a force $\boldsymbol{F}(t)$ due to the 10^{21} collisions per second it suffers with the molecules of the surrounding fluid. Its equation of motion is

$$m\, \frac{d\boldsymbol{v}(t)}{dt} = \boldsymbol{F}(t). \tag{15.136}$$

Langevin suggested that the force $\boldsymbol{F}(t)$ is the sum of a viscous drag $-\boldsymbol{v}(t)/B$ and a rapidly fluctuating part $\boldsymbol{f}(t)$

$$\boldsymbol{F}(t) = -\, \boldsymbol{v}(t)/B + \boldsymbol{f}(t) \tag{15.137}$$

so that

$$m\, \frac{d\boldsymbol{v}(t)}{dt} = -\, \frac{\boldsymbol{v}(t)}{B} + \boldsymbol{f}(t). \tag{15.138}$$

The parameter B is called the **mobility**. The **ensemble average** (the average over all the particles) of the fluctuating force $\boldsymbol{f}(t)$ is zero

$$\langle \boldsymbol{f}(t) \rangle = \boldsymbol{0}. \tag{15.139}$$

Thus the ensemble average of the velocity satisfies

$$m\, \frac{d\langle \boldsymbol{v} \rangle}{dt} = -\, \frac{\langle \boldsymbol{v} \rangle}{B} \tag{15.140}$$

whose solution with $\tau = mB$ is

$$\langle \boldsymbol{v}(t) \rangle = \langle \boldsymbol{v}(\mathbf{0}) \rangle \, e^{-t/\tau}. \tag{15.141}$$

The instantaneous equation (15.138) divided by the mass m is

$$\frac{d\boldsymbol{v}(t)}{dt} = -\frac{\boldsymbol{v}(t)}{\tau} + \boldsymbol{a}(t) \tag{15.142}$$

in which $\boldsymbol{a}(t) = \boldsymbol{f}(t)/m$ is the acceleration. The ensemble average of the scalar product of the position vector $\boldsymbol{r}$ with this equation is

$$\left\langle \boldsymbol{r} \cdot \frac{d\boldsymbol{v}}{dt} \right\rangle = -\frac{\langle \boldsymbol{r} \cdot \boldsymbol{v} \rangle}{\tau} + \langle \boldsymbol{r} \cdot \boldsymbol{a} \rangle. \tag{15.143}$$

But since the ensemble average $\langle \boldsymbol{r} \cdot \boldsymbol{a} \rangle$ of the scalar product of the position vector $\boldsymbol{r}$ with the random, fluctuating part $\boldsymbol{a}$ of the acceleration vanishes, we have

$$\left\langle \boldsymbol{r} \cdot \frac{d\boldsymbol{v}}{dt} \right\rangle = -\frac{\langle \boldsymbol{r} \cdot \boldsymbol{v} \rangle}{\tau}. \tag{15.144}$$

Now

$$\frac{1}{2}\frac{d\,\boldsymbol{r}^2}{dt} = \frac{1}{2}\frac{d}{dt}\,(\boldsymbol{r} \cdot \boldsymbol{r}) = \boldsymbol{r} \cdot \boldsymbol{v} \tag{15.145}$$

and so

$$\frac{1}{2}\frac{d^2\boldsymbol{r}^2}{dt^2} = \boldsymbol{r} \cdot \frac{d\boldsymbol{v}}{dt} + \boldsymbol{v}^2. \tag{15.146}$$

The ensemble average of this equation is

$$\frac{d^2\langle \boldsymbol{r}^2 \rangle}{dt^2} = 2\left\langle \boldsymbol{r} \cdot \frac{d\boldsymbol{v}}{dt} \right\rangle + 2\langle \boldsymbol{v}^2 \rangle \tag{15.147}$$

or in view of (15.144)

$$\frac{d^2\langle \boldsymbol{r}^2 \rangle}{dt^2} = -2\frac{\langle \boldsymbol{r} \cdot \boldsymbol{v} \rangle}{\tau} + 2\langle \boldsymbol{v}^2 \rangle. \tag{15.148}$$

We now use (15.145) to replace $\langle \boldsymbol{r} \cdot \boldsymbol{v} \rangle$ with half the first time derivative of $\langle \boldsymbol{r}^2 \rangle$ so that we have

$$\frac{d^2\langle \boldsymbol{r}^2 \rangle}{dt^2} = -\frac{1}{\tau}\frac{d\langle \boldsymbol{r}^2 \rangle}{dt} + 2\langle \boldsymbol{v}^2 \rangle. \tag{15.149}$$

If the fluid is in equilibrium, then the ensemble average of $\boldsymbol{v}^2$ is given by the Maxwell–Boltzmann value (15.121)

$$\langle \boldsymbol{v}^2 \rangle = \frac{3kT}{m} \tag{15.150}$$

and so the acceleration (15.149) of $\langle \boldsymbol{r}^2 \rangle$ is

$$\frac{d^2 \langle \boldsymbol{r}^2 \rangle}{dt^2} + \frac{1}{\tau} \frac{d \langle \boldsymbol{r}^2 \rangle}{dt} = \frac{6kT}{m} \tag{15.151}$$

which we can integrate.

The general solution (7.12) to a second-order linear inhomogeneous differential equation is the sum of any particular solution to the inhomogeneous equation plus the general solution of the homogeneous equation. The function $\langle \boldsymbol{r}^2(t) \rangle_{pi} = 6kT t \tau / m$ is a particular solution of the inhomogeneous equation. The general solution to the homogeneous equation is $\langle \boldsymbol{r}^2(t) \rangle_{gh} = U + W \exp(-t/\tau)$ where U and W are constants. So $\langle \boldsymbol{r}^2(t) \rangle$ is

$$\langle \boldsymbol{r}^2(t) \rangle = U + W\, e^{-t/\tau} + 6kT \tau t / m \tag{15.152}$$

where U and W make $\langle \boldsymbol{r}^2(t) \rangle$ fit the boundary conditions. If the individual particles start out at the origin $\boldsymbol{r} = \boldsymbol{0}$, then one boundary condition is

$$\langle \boldsymbol{r}^2(0) \rangle = 0 \tag{15.153}$$

which implies that

$$U + W = 0. \tag{15.154}$$

And since the particles start out at $\boldsymbol{r} = \boldsymbol{0}$ with an isotropic distribution of initial velocities, the formula (15.145) for $\dot{r}^2$ implies that at $t = 0$

$$\left. \frac{d \, \langle \boldsymbol{r}^2 \rangle}{dt} \right|_{t=0} = 2 \langle \boldsymbol{r}(0) \cdot \boldsymbol{v}(0) \rangle = 0. \tag{15.155}$$

This boundary condition means that our solution (15.152) must satisfy

$$\left. \frac{d \, \langle \boldsymbol{r}^2(t) \rangle}{dt} \right|_{t=0} = -\frac{W}{\tau} + \frac{6kT\tau}{m} = 0. \tag{15.156}$$

Thus $W = -U = 6kT\tau^2/m$, and so our solution (15.152) is

$$\langle \boldsymbol{r}^2(t) \rangle = \frac{6kT\tau^2}{m} \left[\frac{t}{\tau} + e^{-t/\tau} - 1 \right]. \tag{15.157}$$

At times short compared to τ, the first two terms in the power series for the exponential $\exp(-t/\tau)$ cancel the terms $-1 + t/\tau$, leaving

$$\langle \boldsymbol{r}^2(t) \rangle = \frac{6kT\tau^2}{m} \left[\frac{t^2}{2\tau^2} \right] = \frac{3kT}{m} t^2 = \langle v^2 \rangle \, t^2. \tag{15.158}$$

But at times long compared to τ, the exponential vanishes, leaving

$$\langle \boldsymbol{r}^2(t) \rangle = \frac{6kT\tau}{m} t = 6\, B\, kT\, t. \tag{15.159}$$

The **diffusion constant** D is defined by

$$\langle \boldsymbol{r}^2(t) \rangle = 6\, D\, t \tag{15.160}$$

and so we arrive at **Einstein's relation**

$$D = B\, kT \tag{15.161}$$

which often is written in terms of the **viscous-friction coefficient** f_v

$$f_v \equiv \frac{1}{B} = \frac{m}{\tau} \tag{15.162}$$

as

$$f_v\, D = kT. \tag{15.163}$$

This equation expresses Boltzmann's constant k in terms of three quantities v, D, and T that were accessible to measurement in the first decade of the twentieth century. It enabled scientists to measure Boltzmann's constant k for the first time. And since Avogadro's number N_A was the known gas constant R divided by k, the number of molecules in a mole was revealed to be $N_A = 6.022 \times 10^{23}$. Chemists could then divide the mass of a mole of any pure substance by 6.022×10^{23} and find the mass of the molecules that composed it. Suddenly the masses of the molecules of chemistry became known, and molecules were recognized as real particles and not tricks for balancing chemical equations.

15.13 Einstein–Nernst Relation

If a particle of mass m carries an electric charge q and is exposed to an electric field $\boldsymbol{E}$, then in addition to viscosity $-\,\boldsymbol{v}/B$ and random buffeting $\boldsymbol{f}$, the constant force $q\boldsymbol{E}$ acts on it

$$m\,\frac{d\boldsymbol{v}}{dt} = -\frac{\boldsymbol{v}}{B} + q\boldsymbol{E} + \boldsymbol{f}. \tag{15.164}$$

The mean value of its velocity satisfies the differential equation

$$\left\langle \frac{d\boldsymbol{v}}{dt} \right\rangle = -\frac{\langle \boldsymbol{v} \rangle}{\tau} + \frac{q\boldsymbol{E}}{m} \tag{15.165}$$

where $\tau = mB$. A particular solution of this inhomogeneous equation is

$$\langle \boldsymbol{v}(t) \rangle_{pi} = \frac{q\tau\boldsymbol{E}}{m} = qB\boldsymbol{E}. \tag{15.166}$$

The general solution of its homogeneous version is $\langle \boldsymbol{v}(t) \rangle_{gh} = \boldsymbol{A} \exp(-t/\tau)$ in which the constant $\boldsymbol{A}$ is chosen to give $\langle \boldsymbol{v}(0) \rangle$ at $t = 0$. So by (7.12), the general solution $\langle \boldsymbol{v}(t)\rangle$ to Equation (15.165) is (Exercise 15.17) the sum of $\langle \boldsymbol{v}(t)\rangle_{pi}$ and $\langle \boldsymbol{v}(t)\rangle_{gh}$

$$\langle \boldsymbol{v}(t)\rangle = qB\boldsymbol{E} + \left[\langle \boldsymbol{v}(0)\rangle - qB\boldsymbol{E}\right] e^{-t/\tau}. \tag{15.167}$$

By applying the tricks of the previous section (15.12), one may show (Exercise 15.18) that the variance of the position $\boldsymbol{r}$ about its mean $\langle \boldsymbol{r}(t)\rangle$ is

$$\left\langle (\boldsymbol{r} - \langle \boldsymbol{r}(t)\rangle)^2 \right\rangle = \frac{6kT\tau^2}{m}\left(\frac{t}{\tau} - 1 + e^{-t/\tau}\right) \tag{15.168}$$

where $\langle \boldsymbol{r}(t)\rangle = (q\tau^2\boldsymbol{E}/m)\left(t/\tau - 1 + e^{-t/\tau}\right)$ if $\langle \boldsymbol{r}(0)\rangle = \langle \boldsymbol{v}(0)\rangle = 0$. So for times $t \gg \tau$, this variance is

$$\left\langle (\boldsymbol{r} - \langle \boldsymbol{r}(t)\rangle)^2 \right\rangle = \frac{6kT\tau t}{m}. \tag{15.169}$$

Since the diffusion constant D is defined by (15.160) as

$$\left\langle (\boldsymbol{r} - \langle \boldsymbol{r}(t)\rangle)^2 \right\rangle = 6\,D\,t \tag{15.170}$$

we arrive at the Einstein–Nernst relation

$$D = \frac{kT\tau}{m} = kTB = \frac{\mu}{q}kT \tag{15.171}$$

in which the electric mobility is $\mu = qB$.

15.14 Fluctuation and Dissipation

Let's look again at Langevin's equation (15.142)

$$\frac{d\boldsymbol{v}(t)}{dt} + \frac{\boldsymbol{v}(t)}{\tau} = \boldsymbol{a}(t). \tag{15.172}$$

If we multiply both sides by the exponential $\exp(t/\tau)$

$$\left(\frac{d\boldsymbol{v}}{dt} + \frac{\boldsymbol{v}}{\tau}\right) e^{t/\tau} = \frac{d}{dt}\left(\boldsymbol{v}\, e^{t/\tau}\right) = \boldsymbol{a}(t)\, e^{t/\tau} \tag{15.173}$$

and integrate from 0 to t

$$\int_0^t \frac{d}{dt'}\left(\boldsymbol{v}\, e^{t'/\tau}\right) dt' = \boldsymbol{v}(t)\, e^{t/\tau} - \boldsymbol{v}(0) = \int_0^t \boldsymbol{a}(t')\, e^{t'/\tau}\, dt' \tag{15.174}$$

then we get

$$\boldsymbol{v}(t) = e^{-t/\tau}\,\boldsymbol{v}(0) + e^{-t/\tau}\int_0^t \boldsymbol{a}(t')\, e^{t'/\tau}\, dt'. \tag{15.175}$$

Thus the ensemble average of the square of the velocity is

$$\langle \boldsymbol{v}^2(t)\rangle = e^{-2t/\tau}\,\langle \boldsymbol{v}^2(0)\rangle + 2e^{-2t/\tau}\int_0^t \langle \boldsymbol{v}(0)\cdot \boldsymbol{a}(t')\rangle\, e^{t'/\tau}\,dt' \qquad (15.176)$$
$$+\,e^{-2t/\tau}\int_0^t\int_0^t \langle \boldsymbol{a}(u_1)\cdot \boldsymbol{a}(t_2)\rangle\, e^{(u_1+t_2)/\tau}\,du_1 dt_2.$$

The second term on the RHS is zero, so we have

$$\langle \boldsymbol{v}^2(t)\rangle = e^{-2t/\tau}\,\langle \boldsymbol{v}^2(0)\rangle + e^{-2t/\tau}\int_0^t\int_0^t \langle \boldsymbol{a}(t_1)\cdot \boldsymbol{a}(t_2)\rangle\, e^{(t_1+t_2)/\tau}\,dt_1 dt_2. \qquad (15.177)$$

The ensemble average

$$C(t_1, t_2) = \langle \boldsymbol{a}(t_1)\cdot \boldsymbol{a}(t_2)\rangle \qquad (15.178)$$

is an example of an **autocorrelation function**.

All autocorrelation functions have some simple properties, which are easy to prove (Pathria, 1972, p. 458):

1. If the system is independent of time, then its autocorrelation function for any given variable $\boldsymbol{A}(t)$ depends only upon the time delay s:

$$C(t, t+s) = \langle \boldsymbol{A}(t)\cdot \boldsymbol{A}(t+s)\rangle \equiv C(s). \qquad (15.179)$$

2. The autocorrelation function for $s = 0$ is necessarily nonnegative

$$C(t, t) = \langle \boldsymbol{A}(t)\cdot \boldsymbol{A}(t)\rangle = \langle \boldsymbol{A}(t)^2\rangle \geq 0. \qquad (15.180)$$

   If the system is time independent, then $C(t, t) = C(0) \geq 0$.
3. The absolute value of $C(t_1, t_2)$ is never greater than the average of $C(t_1, t_1)$ and $C(t_2, t_2)$ because

$$\langle |\boldsymbol{A}(t_1) \pm \boldsymbol{A}(t_2)|^2\rangle = \langle \boldsymbol{A}(t_1)^2\rangle + \langle \boldsymbol{A}(t_2)^2\rangle \pm 2\langle \boldsymbol{A}(t_1)\cdot \boldsymbol{A}(t_2)\rangle \geq 0 \qquad (15.181)$$

   which implies that $-2\,C(t_1, t_2) \leq C(t_1, t_1) + C(t_2, t_2) \geq 2\,C(t_1, t_2)$ or

$$2\,|C(t_1, t_2)| \leq C(t_1, t_1) + C(t_2, t_2). \qquad (15.182)$$

   For a time-independent system, this inequality is $|C(s)| \leq C(0)$ for every time delay s.
4. If the variables $\boldsymbol{A}(t_1)$ and $\boldsymbol{A}(t_2)$ commute, then their autocorrelation function is symmetric

$$C(t_1, t_2) = \langle \boldsymbol{A}(t_1)\cdot \boldsymbol{A}(t_2)\rangle = \langle \boldsymbol{A}(t_2)\cdot \boldsymbol{A}(t_1)\rangle = C(t_2, t_1). \qquad (15.183)$$

   For a time-independent system, this symmetry is $C(s) = C(-s)$.

5. If the variable $\boldsymbol{A}(t)$ is randomly fluctuating with zero mean, then we expect both that its ensemble average vanishes

$$\langle \boldsymbol{A}(t) \rangle = \boldsymbol{0} \tag{15.184}$$

and that there is some characteristic time scale T beyond which the correlation function falls to zero:

$$\langle \boldsymbol{A}(t_1) \cdot \boldsymbol{A}(t_2) \rangle \to \langle \boldsymbol{A}(t_1) \rangle \cdot \langle \boldsymbol{A}(t_2) \rangle = 0 \tag{15.185}$$

when $|t_1 - t_2| \gg T$.

In terms of the autocorrelation function $C(t_1, t_2) = \langle \boldsymbol{a}(t_1) \cdot \boldsymbol{a}(t_2) \rangle$ of the acceleration, the variance of the velocity (15.177) is

$$\langle \boldsymbol{v}^2(t) \rangle = e^{-2t/\tau} \langle \boldsymbol{v}^2(0) \rangle + e^{-2t/\tau} \int_0^t \int_0^t C(t_1, t_2)\, e^{(t_1+t_2)/\tau}\, dt_1 dt_2. \tag{15.186}$$

Since $C(t_1, t_2)$ is big only for tiny values of $|t_2 - t_1|$, it makes sense to change variables to

$$s = t_2 - t_1 \quad \text{and} \quad w = \frac{1}{2}(t_1 + t_2). \tag{15.187}$$

The element of area then is by (14.6–14.14)

$$dt_1 \wedge dt_2 = dw \wedge ds \tag{15.188}$$

and the limits of integration are $-2w \le s \le 2w$ for $0 \le w \le t/2$ and $-2(t-w) \le s \le 2(t-w)$ for $t/2 \le w \le t$. So $\langle \boldsymbol{v}^2(t) \rangle$ is

$$\begin{aligned} \langle \boldsymbol{v}^2(t) \rangle = {} & e^{-2t/\tau} \langle \boldsymbol{v}^2(0) \rangle + e^{-2t/\tau} \int_0^{t/2} e^{2w/\tau} dw \int_{-2w}^{2w} C(s)\, ds \\ & + e^{-2t/\tau} \int_{t/2}^{t} e^{2w/\tau} dw \int_{-2(t-w)}^{2(t-w)} C(s)\, ds. \end{aligned} \tag{15.189}$$

Since by (15.185) the autocorrelation function $C(s)$ vanishes outside a narrow window of width $2T$, we may approximate each of the s-integrals by

$$C = \int_{-\infty}^{\infty} C(s)\, ds. \tag{15.190}$$

It follows then that

$$\begin{aligned} \langle \boldsymbol{v}^2(t) \rangle &= e^{-2t/\tau} \langle \boldsymbol{v}^2(0) \rangle + C\, e^{-2t/\tau} \int_0^t e^{2w/\tau} dw \\ &= e^{-2t/\tau} \langle \boldsymbol{v}^2(0) \rangle + C\, e^{-2t/\tau} \frac{\tau}{2} \left(e^{2t/\tau} - 1\right) \\ &= e^{-2t/\tau} \langle \boldsymbol{v}^2(0) \rangle + C\, \frac{\tau}{2} \left(1 - e^{-2t/\tau}\right). \end{aligned} \tag{15.191}$$

As $t \to \infty$, $\langle \boldsymbol{v}^2(t) \rangle$ must approach its equilibrium value of $3kT/m$, and so

$$\lim_{t\to\infty} \langle \boldsymbol{v}^2(t) \rangle = C\,\frac{\tau}{2} = \frac{3kT}{m} \tag{15.192}$$

which implies that

$$C = \frac{6kT}{m\tau} \quad \text{or} \quad \frac{1}{B} = \frac{m^2 C}{6kT}. \tag{15.193}$$

Our final formula for $\langle \boldsymbol{v}^2(t) \rangle$ then is

$$\langle \boldsymbol{v}^2(t) \rangle = e^{-2t/\tau}\,\langle \boldsymbol{v}^2(0) \rangle + \frac{3kT}{m}\left(1 - e^{-2t/\tau}\right). \tag{15.194}$$

Referring back to the definition (15.162) of the viscous-friction coefficient $f_v = 1/B$, we see that f_v is related to the integral

$$f_v = \frac{1}{B} = \frac{m^2}{6kT}\,C = \frac{m^2}{6kT}\int_{-\infty}^{\infty} \langle \boldsymbol{a}(0)\cdot\boldsymbol{a}(s)\rangle ds = \frac{1}{6kT}\int_{-\infty}^{\infty} \langle \boldsymbol{f}(0)\cdot\boldsymbol{f}(s)\rangle ds \tag{15.195}$$

of the autocorrelation function of the random acceleration $\boldsymbol{a}(t)$ or equivalently of the random force $\boldsymbol{f}(t)$. This equation relates the dissipation of viscous friction to the random fluctuations. It is an example of a **fluctuation–dissipation theorem**.

If we substitute our formula (15.194) for $\langle \boldsymbol{v}^2(t) \rangle$ into the expression (15.149) for the acceleration of $\langle \boldsymbol{r}^2 \rangle$, then we get

$$\frac{d^2\langle \boldsymbol{r}^2(t)\rangle}{dt^2} = -\frac{1}{\tau}\frac{d\langle \boldsymbol{r}^2(t)\rangle}{dt} + 2e^{-2t/\tau}\,\langle \boldsymbol{v}^2(0) \rangle + \frac{6kT}{m}\left(1 - e^{-2t/\tau}\right). \tag{15.196}$$

The solution with both $\langle \boldsymbol{r}^2(0)\rangle = 0$ and $d\langle \boldsymbol{r}^2(0)\rangle/dt = 0$ is (Exercise 15.19)

$$\langle \boldsymbol{r}^2(t)\rangle = \langle \boldsymbol{v}^2(0)\rangle\,\tau^2\left(1 - e^{-t/\tau}\right)^2 - \frac{3kT}{m}\,\tau^2\left(1 - e^{-t/\tau}\right)\left(3 - e^{-t/\tau}\right) + \frac{6kT\tau}{m}\,t. \tag{15.197}$$

15.15 Fokker–Planck Equation

Let $P(\boldsymbol{v}, t)$ be the probability distribution of particles in velocity space at time t, and $\psi(\boldsymbol{v}; \boldsymbol{u})$ be a normalized transition probability that the velocity changes from $\boldsymbol{v}$ to $\boldsymbol{v} + \boldsymbol{u}$ in the time interval $[t, t + \Delta t]$. We take the interval Δt to be much longer than the interval between successive particle collisions but much shorter than the time over which the velocity $\boldsymbol{v}$ changes appreciably. So $|\boldsymbol{u}| \ll |\boldsymbol{v}|$. We also assume that the successive changes in the velocities of the particles is a **Markoff stochastic process**, that is, that the changes are random and that what happens at time t depends only upon the state of the system at time t and not upon the history of the system. We then expect that the velocity distribution at time $t + \Delta t$ is related to that at time t by

$$P(\boldsymbol{v}, t + \Delta t) = \int P(\boldsymbol{v} - \boldsymbol{u}, t)\, \psi(\boldsymbol{v} - \boldsymbol{u}; \boldsymbol{u})\, d^3\boldsymbol{u}. \tag{15.198}$$

Since $|\boldsymbol{u}| \ll |\boldsymbol{v}|$, we can expand $P(\boldsymbol{v}, t + \Delta t)$, $P(\boldsymbol{v} - \boldsymbol{u}, t)$, and $\psi(\boldsymbol{v} - \boldsymbol{u}; \boldsymbol{u})$ in Taylor series in $\boldsymbol{u}$ like

$$\psi(\boldsymbol{v} - \boldsymbol{u}; \boldsymbol{u}) = \psi(\boldsymbol{v}; \boldsymbol{u}) - \boldsymbol{u} \cdot \nabla_{\boldsymbol{v}} \psi(\boldsymbol{v}; \boldsymbol{u}) + \frac{1}{2} \sum_{i,j} u_i u_j \frac{\partial^2 \psi(\boldsymbol{v}; \boldsymbol{u})}{\partial v_i \partial v_j} \tag{15.199}$$

and get

$$\begin{aligned} P(\boldsymbol{v}, t) + \Delta t\, \frac{\partial P(\boldsymbol{v}, t)}{\partial t} = \int & \left[P(\boldsymbol{v}, t) - \boldsymbol{u} \cdot \nabla_{\boldsymbol{v}} P(\boldsymbol{v}, t) + \frac{1}{2} \sum_{i,j} u_i u_j \frac{\partial^2 P(\boldsymbol{v}, t)}{\partial v_i \partial v_j} \right] \\ & \times \left[\psi(\boldsymbol{v}; \boldsymbol{u}) - \boldsymbol{u} \cdot \nabla_{\boldsymbol{v}} \psi(\boldsymbol{v}; \boldsymbol{u}) + \frac{1}{2} \sum_{i,j} u_i u_j \frac{\partial^2 \psi(\boldsymbol{v}; \boldsymbol{u})}{\partial v_i \partial v_j} \right] d^3\boldsymbol{u}. \end{aligned} \tag{15.200}$$

The normalization of the transition probability ψ and the average changes in velocity are

$$\begin{aligned} 1 &= \int \psi(\boldsymbol{v}; \boldsymbol{u})\, d^3\boldsymbol{u} \\ \langle u_i \rangle &= \int u_i\, \psi(\boldsymbol{v}; \boldsymbol{u})\, d^3\boldsymbol{u} \\ \langle u_i u_j \rangle &= \int u_i\, u_j\, \psi(\boldsymbol{v}; \boldsymbol{u})\, d^3\boldsymbol{u} \end{aligned} \tag{15.201}$$

in which the dependence of the mean values $\langle u_i \rangle$ and $\langle u_i u_j \rangle$ upon the velocity $\boldsymbol{v}$ is implicit. In these terms, the expansion (15.200) is

$$\begin{aligned} \Delta t\, \frac{\partial P(\boldsymbol{v}, t)}{\partial t} = & - \langle \boldsymbol{u} \rangle \cdot \nabla_{\boldsymbol{v}} P(\boldsymbol{v}, t) + \frac{1}{2} \sum_{i,j} \langle u_i u_j \rangle \frac{\partial^2 P(\boldsymbol{v}, t)}{\partial v_i \partial v_j} \\ & - P(\boldsymbol{v}, t)\, \nabla_{\boldsymbol{v}} \cdot \langle \boldsymbol{u} \rangle + P(\boldsymbol{v}, t)\, \frac{1}{2} \sum_{i,j} \frac{\partial^2 \langle u_i u_j \rangle}{\partial v_i \partial v_j} \\ & + \sum_{i,j} \frac{\partial P(\boldsymbol{v}, t)}{\partial v_i} \frac{\partial \langle u_i u_j \rangle}{\partial v_j}. \end{aligned} \tag{15.202}$$

Combining terms, we get the **Fokker–Planck equation** in its most general form (Chandrasekhar, 1943)

$$\Delta t \, \frac{\partial P(\boldsymbol{v},t)}{\partial t} = -\nabla_{\boldsymbol{v}} \left[P(\boldsymbol{v},t) \cdot \langle \boldsymbol{u} \rangle \right] + \frac{1}{2} \sum_{i,j} \frac{\partial^2}{\partial v_i \partial v_j} \left[P(\boldsymbol{v},t) \langle u_i u_j \rangle \right]. \quad (15.203)$$

Example 15.18 (Brownian motion) Langevin's equation (15.138) gives the change $\boldsymbol{u}$ in the velocity $\boldsymbol{v}$ as the viscous drag plus some 10^{21} random tiny accelerations per second

$$\boldsymbol{u} = -\frac{\boldsymbol{v}\,\Delta t}{mB} + \frac{\boldsymbol{f}\,\Delta t}{m} \quad (15.204)$$

in which B is the mobility of the colloidal particle. The random changes $\boldsymbol{f}\,\Delta t/m$ in velocity are gaussian, and the transition probability is

$$\psi(\boldsymbol{v};\boldsymbol{u}) = \left(\frac{\beta m^2 B}{4\pi \Delta t} \right)^{3/2} \exp\left(-\frac{\beta m^2 B}{4\Delta t} \left| \boldsymbol{u} + \frac{\boldsymbol{v}\,\Delta t}{mB} \right|^2 \right). \quad (15.205)$$

Here $\beta = 1/kT$, and Stokes's formula for the mobility B of a spherical colloidal particle of radius r in a fluid of viscosity η is $1/B = 6\pi r \eta$. The moments (15.201) of the changes $\boldsymbol{u}$ in velocity are in the limit $\Delta t \to 0$

$$\begin{aligned} \langle \boldsymbol{u} \rangle &= -\frac{\boldsymbol{v}\,\Delta t}{mB} \\ \langle u_i\, u_j \rangle &= 2\delta_{ij}\, \frac{kT}{m^2 B}\, \Delta t. \end{aligned} \quad (15.206)$$

So for Brownian motion, the Fokker–Planck equation is

$$\frac{\partial P(\boldsymbol{v},t)}{\partial t} = \frac{1}{mB} \nabla_{\boldsymbol{v}} \left[P(\boldsymbol{v},t) \cdot \boldsymbol{v} \right] + \frac{kT}{m^2 B} \nabla_{\boldsymbol{v}}^2 \, P(\boldsymbol{v},t). \quad (15.207)$$

15.16 Characteristic and Moment-Generating Functions

The Fourier transform (4.9) of a probability distribution $P(x)$ is its **characteristic function** $\tilde{P}(k)$ sometimes written as $\chi(k)$

$$\tilde{P}(k) \equiv \chi(k) \equiv E[e^{ikx}] = \int e^{ikx}\, P(x)\, dx. \quad (15.208)$$

The probability distribution $P(x)$ is the inverse Fourier transform (4.9)

$$P(x) = \int e^{-ikx}\, \tilde{P}(k)\, \frac{dk}{2\pi}. \quad (15.209)$$

Example 15.19 (Gauss) The characteristic function of the gaussian

$$P_G(x,\mu,\sigma) = \frac{1}{\sigma\sqrt{2\pi}} \exp\left(-\frac{(x-\mu)^2}{2\sigma^2}\right) \tag{15.210}$$

is by (4.19)

$$\begin{aligned} \tilde{P}_G(k,\mu,\sigma) &= \frac{1}{\sigma\sqrt{2\pi}} \int \exp\left(ikx - \frac{(x-\mu)^2}{2\sigma^2}\right) dx \\ &= \frac{e^{ik\mu}}{\sigma\sqrt{2\pi}} \int \exp\left(ikx - \frac{x^2}{2\sigma^2}\right) dx = \exp\left(i\mu k - \frac{1}{2}\sigma^2 k^2\right). \end{aligned} \tag{15.211}$$

For a discrete probability distribution P_n the characteristic function is

$$\chi(k) \equiv E[e^{ikn}] = \sum_n e^{ikn} P_n. \tag{15.212}$$

The normalization of both continuous and discrete probability distributions implies that their characteristic functions satisfy $\tilde{P}(0) = \chi(0) = 1$.

Example 15.20 (Binomial and Poisson) The characteristic function of the binomial distribution (15.51)

$$P_b(n,p,N) = \binom{N}{n} p^n (1-p)^{N-n} \tag{15.213}$$

is

$$\begin{aligned} \chi_b(k) &= \sum_{n=0}^{N} e^{ikn} \binom{N}{n} p^n (1-p)^{N-n} = \sum_{n=0}^{N} \binom{N}{n} (pe^{ik})^n (1-p)^{N-n} \\ &= \left(pe^{ik} + 1 - p\right)^N = \left[p\left(e^{ik}-1\right)+1\right]^N. \end{aligned} \tag{15.214}$$

The Poisson distribution (15.66)

$$P_P(n,\langle n\rangle) = \frac{\langle n\rangle^n}{n!} e^{-\langle n\rangle} \tag{15.215}$$

has the characteristic function

$$\chi_P(k) = \sum_{n=0}^{\infty} e^{ikn} \frac{\langle n\rangle^n}{n!} e^{-\langle n\rangle} = e^{-\langle n\rangle} \sum_{n=0}^{\infty} \frac{(\langle n\rangle e^{ik})^n}{n!} = \exp\left[\langle n\rangle\left(e^{ik}-1\right)\right]. \tag{15.216}$$

The **moment-generating function** is the characteristic function evaluated at an imaginary argument

$$M(k) \equiv E[e^{kx}] = \tilde{P}(-ik) = \chi(-ik). \tag{15.217}$$

For a continuous probability distribution $P(x)$, it is

$$M(k) = E[e^{kx}] = \int e^{kx}\, P(x)\, dx \tag{15.218}$$

and for a discrete probability distribution P_n, it is

$$M(k) = E[e^{kx}] = \sum_n e^{kx_n}\, P_n. \tag{15.219}$$

In both cases, the normalization of the probability distribution implies that $M(0) = 1$.

Derivatives of the moment-generating function and of the characteristic function give the moments μ_n

$$E[x^n] = \mu_n = \left.\frac{d^n M(k)}{dk^n}\right|_{k=0} = (-i)^n \left.\frac{d^n \tilde{P}(k)}{dk^n}\right|_{k=0}. \tag{15.220}$$

Example 15.21 (Three moment-generating functions) The characteristic functions of the binomial distribution (15.214) and those of the distributions of Poisson (15.216) and Gauss (15.210) give us the moment-generating functions

$$M_b(k, p, N) = \left[p\left(e^k - 1\right) + 1\right]^N, \quad M_P(k, \langle n\rangle) = \exp\left[\langle n\rangle\left(e^k - 1\right)\right],$$
$$\text{and } M_G(k, \mu, \sigma) = \exp\left(\mu k + \frac{1}{2}\sigma^2 k^2\right). \tag{15.221}$$

Thus by (15.220), the first three moments of these three distributions are

$$\begin{aligned} \mu_{b0} &= 1, \quad \mu_{b1} = Np, \quad \mu_{b2} = N^2 p \\ \mu_{P0} &= 1, \quad \mu_{P1} = \langle n\rangle, \quad \mu_{P2} = \langle n\rangle + \langle n\rangle^2 \\ \mu_{G0} &= 1, \quad \mu_{G1} = \mu, \quad \mu_{G2} = \mu^2 + \sigma^2 \end{aligned} \tag{15.222}$$

(Exercise 15.20).

Since the characteristic and moment-generating functions have derivatives (15.220) proportional to the moments μ_n, their Taylor series are

$$\tilde{P}(k) = E[e^{ikx}] = \sum_{n=0}^{\infty} \frac{(ik)^n}{n!} E[x^n] = \sum_{n=0}^{\infty} \frac{(ik)^n}{n!}\, \mu_n \tag{15.223}$$

and

$$M(k) = E[e^{kx}] = \sum_{n=0}^{\infty} \frac{k^n}{n!} E[x^n] = \sum_{n=0}^{\infty} \frac{k^n}{n!} \mu_n. \tag{15.224}$$

The **cumulants** c_n of a probability distribution are the derivatives of the logarithm of its moment-generating function at $k = 0$

$$c_n = \left. \frac{d^n \ln M(k)}{dk^n} \right|_{k=0} = (-i)^n \left. \frac{d^n \ln \tilde{P}(k)}{dk^n} \right|_{k=0}. \tag{15.225}$$

One may show (Exercise 15.22) that the first five cumulants of an arbitrary probability distribution are

$$c_0 = 0, \quad c_1 = \mu, \quad c_2 = \sigma^2, \quad c_3 = \nu_3, \quad \text{and} \quad c_4 = \nu_4 - 3\sigma^4 \tag{15.226}$$

where the ν's are its central moments (15.31). The third and fourth **normalized cumulants** are the **skewness** $v = c_3/\sigma^3 = \nu_3/\sigma^3$ and the **kurtosis** $\kappa = c_4/\sigma^4 = \nu_4/\sigma^4 - 3$.

Example 15.22 (Gaussian cumulants) The logarithm of the moment-generating function (15.221) of Gauss's distribution is $\mu k + \sigma^2 k^2/2$. Thus by (15.225), $P_G(x, \mu, \sigma)$ has no skewness or kurtosis, its cumulants vanish $c_{Gn} = 0$ for $n > 2$, and its fourth central moment is $\nu_4 = 3\sigma^4$.

15.17 Fat Tails

The gaussian probability distribution $P_G(x, \mu, \sigma)$ falls off for $|x - \mu| \gg \sigma$ very fast – as $\exp\left(-(x-\mu)^2/2\sigma^2\right)$. Many other probability distributions fall off more slowly; they have **fat tails**. Rare "black-swan" events – wild fluctuations, market bubbles, and crashes – lurk in their fat tails.

Gosset's distribution, which is known as **Student's t-distribution** with ν degrees of freedom

$$P_S(x, \nu, a) = \frac{1}{\sqrt{\pi}} \frac{\Gamma((1+\nu)/2)}{\Gamma(\nu/2)} \frac{a^\nu}{(a^2 + x^2)^{(1+\nu)/2}} \tag{15.227}$$

has **power-law tails**. Its even moments are

$$\mu_{2n} = (2n-1)!! \, \frac{\Gamma(\nu/2 - n)}{\Gamma(\nu/2)} \left(\frac{a^2}{2} \right)^n \tag{15.228}$$

for $2n < \nu$ and infinite otherwise. For $\nu = 1$, it coincides with the Breit–Wigner or Cauchy distribution

$$P_S(x, 1, a) = \frac{1}{\pi} \frac{a}{a^2 + x^2} \tag{15.229}$$

in which $x = E - E_0$ and $a = \Gamma/2$ is the half-width at half-maximum.

Two representative cumulative probabilities are (Bouchaud and Potters, 2003, pp. 15–16)

$$\Pr(x, \infty) = \int_x^\infty P_S(x', 3, 1)\, dx' = \frac{1}{2} - \frac{1}{\pi}\left[\arctan x + \frac{x}{1 + x^2}\right] \tag{15.230}$$

$$\Pr(x, \infty) = \int_x^\infty P_S(x', 4, \sqrt{2})\, dx' = \frac{1}{2} - \frac{3}{4}u + \frac{1}{4}u^3 \tag{15.231}$$

where $u = x/\sqrt{2 + x^2}$ and a is picked so $\sigma^2 = 1$. William Gosset (1876–1937), who worked for Guinness, wrote as Student because Guinness didn't let its employees publish.

The **log-normal** probability distribution on $(0, \infty)$

$$P_{\ln}(x) = \frac{1}{\sigma x \sqrt{2\pi}} \exp\left[-\frac{\ln^2(x/x_0)}{2\sigma^2}\right] \tag{15.232}$$

describes distributions of rates of return (Bouchaud and Potters, 2003, p. 9). Its moments are (Exercise 15.25)

$$\mu_n = x_0^n\, e^{n^2\sigma^2/2}. \tag{15.233}$$

The **exponential distribution** on $[0, \infty)$

$$P_e(x) = \alpha e^{-\alpha x} \tag{15.234}$$

has (Exercise 15.26) mean $\mu = 1/\alpha$ and variance $\sigma^2 = 1/\alpha^2$. The sum of n independent exponentially and identically distributed random variables $x = x_1 + \cdots + x_n$ is distributed on $[0, \infty)$ as (Feller, 1966, p. 10)

$$P_{n,e}(x) = \alpha \frac{(\alpha x)^{n-1}}{(n-1)!} e^{-\alpha x}. \tag{15.235}$$

The sum of the squares $x^2 = x_1^2 + \cdots + x_n^2$ of n independent normally and identically distributed random variables of zero mean and variance σ^2 gives rise to Pearson's **chi-squared distribution** on $(0, \infty)$

$$P_{n,P}(x, \sigma)dx = \frac{\sqrt{2}}{\sigma} \frac{1}{\Gamma(n/2)} \left(\frac{x}{\sigma\sqrt{2}}\right)^{n-1} e^{-x^2/(2\sigma^2)} dx \tag{15.236}$$

which for $x = v$, $n = 3$, and $\sigma^2 = kT/m$ is (Exercise 15.27) the Maxwell–Boltzmann distribution (15.118). In terms of $\chi = x/\sigma$, it is

$$P_{n,P}(\chi^2/2)\,d\chi^2 = \frac{1}{\Gamma(n/2)}\left(\frac{\chi^2}{2}\right)^{n/2-1} e^{-\chi^2/2} d\left(\chi^2/2\right). \tag{15.237}$$

It has mean and variance

$$\mu = n \quad \text{and} \quad \sigma^2 = 2n \tag{15.238}$$

and is used in the chi-squared test (Pearson, 1900). The Porter–Thomas distribution $P_{PT}(x) = e^{-x/2}/\sqrt{2\pi x}$ and the exponential distribution $P_e(x)$ (15.234) are special cases of the class (15.236) of chi-squared distributions.

Personal income, the amplitudes of catastrophes, the price changes of financial assets, and many other phenomena occur on both small and large scales. **Lévy** distributions describe such multi-scale phenomena. The characteristic function for a symmetric Lévy distribution is for $\nu \le 2$

$$\tilde{L}_\nu(k, a_\nu) = \exp\left(-a_\nu|k|^\nu\right). \tag{15.239}$$

Its inverse Fourier transform (15.209) is for $\nu = 1$ (Exercise 15.28) the **Cauchy** or **Lorentz** distribution

$$L_1(x, a_1) = \frac{a_1}{\pi(x^2 + a_1^2)} \tag{15.240}$$

and for $\nu = 2$ the gaussian

$$L_2(x, a_2) = P_G(x, 0, \sqrt{2a_2}) = \frac{1}{2\sqrt{\pi a_2}} \exp\left(-\frac{x^2}{4a_2}\right) \tag{15.241}$$

but for other values of ν no simple expression for $L_\nu(x, a_\nu)$ is available. For $0 < \nu < 2$ and as $x \to \pm\infty$, it falls off as $|x|^{-(1+\nu)}$, and for $\nu > 2$ it assumes negative values, ceasing to be a probability distribution (Bouchaud and Potters, 2003, pp. 10–13).

15.18 Central Limit Theorem and Jarl Lindeberg

We have seen in Sections 15.9 and 15.11 that unbiased fluctuations tend to distribute the position and velocity of molecules according to Gauss's distribution (15.84). Gaussian distributions occur very frequently. The **central limit theorem** suggests why they occur so often.

Let $x_1, \ldots, x_N$ be N **independent** random variables described by probability distributions $P_1(x_1), \ldots, P_N(x_N)$ with finite means μ_j and finite variances σ_j^2. The P_j's may be all different. The central limit theorem says that as $N \to \infty$ the probability distribution $P^{(N)}(y)$ for the average of the x_j's

$$y = \frac{1}{N}(x_1 + x_2 + \cdots + x_N) \tag{15.242}$$

tends to a gaussian in y quite independently of what the underlying probability distributions $P_j(x_j)$ happen to be.

Because expected values are linear (15.38), the mean value of the average y is the average of the N means

$$\begin{aligned}\mu_y = E[y] = E[(x_1 + \cdots + x_N)/N] &= \frac{1}{N}(E[x_1] + \cdots + E[x_N]) \\ &= \frac{1}{N}(\mu_1 + \cdots + \mu_N).\end{aligned} \tag{15.243}$$

The independence of the random variables $x_1, x_2, \ldots, x_N$ implies (15.44) that their joint probability distribution factorizes

$$P(x_1, \ldots, x_N) = P_1(x_1)P_2(x_2)\cdots P_N(x_N). \tag{15.244}$$

And our rule (15.49) for the variance of a linear combination of *independent* variables says that the variance of the average y is the sum of the variances

$$\sigma_y^2 = V[(x_1 + \cdots + x_N)/N] = \frac{1}{N^2}\left(\sigma_1^2 + \cdots + \sigma_N^2\right). \tag{15.245}$$

The conditional probability (15.3) $P^{(N)}(y|x_1, \ldots, x_N)$ that the average of the x's is y is the delta function (4.36)

$$P^{(N)}(y|x_1, \ldots, x_N) = \delta(y - (x_1 + x_2 + \cdots + x_N)/N). \tag{15.246}$$

Thus by (15.8) the probability distribution $P^{(N)}(y)$ for the average $y = (x_1 + x_2 + \cdots + x_N)/N$ of the x_j's is

$$\begin{aligned}P^{(N)}(y) &= \int P^{(N)}(y|x_1, \ldots, x_N)\, P(x_1, \ldots, x_N)\, d^N x \\ &= \int \delta(y - (x_1 + x_2 + \cdots + x_N)/N)\, P(x_1, \ldots, x_N)\, d^N x\end{aligned} \tag{15.247}$$

where $d^N x = dx_1 \cdots dx_N$. Its characteristic function is then

$$\begin{aligned}\tilde{P}^{(N)}(k) &= \int e^{iky}\, P^{(N)}(y)\, dy \\ &= \int e^{iky}\, \delta(y - (x_1 + x_2 + \cdots + x_N)/N)\, P(x_1, \ldots, x_N)\, d^N x\, dy \\ &= \int \exp\left[\frac{ik}{N}(x_1 + x_2 + \cdots + x_N)\right] P(x_1, \ldots, x_N)\, d^N x \\ &= \int \exp\left[\frac{ik}{N}(x_1 + x_2 + \cdots + x_N)\right] P_1(x_1)P_2(x_2)\cdots P_N(x_N)\, d^N x\end{aligned} \tag{15.248}$$

which is the product

$$\tilde{P}^{(N)}(k) = \tilde{P}_1(k/N)\, \tilde{P}_2(k/N) \cdots \tilde{P}_N(k/N) \tag{15.249}$$

of the characteristic functions

$$\tilde{P}_j(k/N) = \int e^{ikx_j/N}\, P_j(x_j)\, dx_j \tag{15.250}$$

of the probability distributions $P_1(x_1), \ldots, P_N(x_N)$.

The Taylor series (15.223) for each characteristic function is

$$\tilde{P}_j(k/N) = \sum_{n=0}^{\infty} \frac{(ik)^n}{n!\, N^n}\, \mu_{nj} \tag{15.251}$$

and so for big N we can use the approximation

$$\tilde{P}_j(k/N) \approx 1 + \frac{ik}{N}\mu_j - \frac{k^2}{2N^2}\mu_{2j} \tag{15.252}$$

in which $\mu_{2j} = \sigma_j^2 + \mu_j^2$ by the formula (15.26) for the variance. So we have

$$\tilde{P}_j(k/N) \approx 1 + \frac{ik}{N}\mu_j - \frac{k^2}{2N^2}\left(\sigma_j^2 + \mu_j^2\right) \tag{15.253}$$

or for large N

$$\tilde{P}_j(k/N) \approx \exp\left(\frac{ik}{N}\mu_j - \frac{k^2}{2N^2}\sigma_j^2\right). \tag{15.254}$$

Thus as $N \to \infty$, the characteristic function (15.249) for the variable y converges to

$$\begin{aligned} \tilde{P}^{(N)}(k) &= \prod_{j=1}^{N} \tilde{P}_j(k/N) = \prod_{j=1}^{N} \exp\left(\frac{ik}{N}\mu_j - \frac{k^2}{2N^2}\sigma_j^2\right) \\ &= \exp\left[\sum_{j=1}^{N}\left(\frac{ik}{N}\mu_j - \frac{k^2}{2N^2}\sigma_j^2\right)\right] = \exp\left(i\mu_y k - \frac{1}{2}\sigma_y^2 k^2\right) \end{aligned} \tag{15.255}$$

which is the characteristic function (15.211) of a gaussian (15.210) with mean and variance

$$\mu_y = \frac{1}{N}\sum_{j=1}^{N} \mu_j \quad \text{and} \quad \sigma_y^2 = \frac{1}{N^2}\sum_{j=1}^{N} \sigma_j^2. \tag{15.256}$$

The inverse Fourier transform (15.209) now gives the probability distribution $P^{(N)}(y)$ for the average $y = (x_1 + x_2 + \cdots + x_N)/N$ as

$$P^{(N)}(y) = \int_{-\infty}^{\infty} e^{-iky}\, \tilde{P}^{(N)}(k)\, \frac{dk}{2\pi} \tag{15.257}$$

which in view of (15.255) and (15.211) tends as $N \to \infty$ to Gauss's distribution $P_G(y, \mu_y, \sigma_y)$

$$\begin{aligned} \lim_{N\to\infty} P^{(N)}(y) &= \int_{-\infty}^{\infty} e^{-iky} \lim_{N\to\infty} \tilde{P}^{(N)}(k)\, \frac{dk}{2\pi} \\ &= \int_{-\infty}^{\infty} e^{-iky} \exp\left(i\mu_y k - \frac{1}{2}\sigma_y^2 k^2\right) \frac{dk}{2\pi} \\ &= P_G(y, \mu_y, \sigma_y) = \frac{1}{\sigma_y\sqrt{2\pi}} \exp\left[-\frac{(y-\mu_y)^2}{2\sigma_y^2}\right] \end{aligned} \tag{15.258}$$

with mean μ_y and variance σ_y^2 as given by (15.256). The sense in which the exact distribution $P^{(N)}(y)$ converges to $P_G(y, \mu_y, \sigma_y)$ is that for all a and b the probability $\Pr_N(a < y < b)$ that y lies between a and b as determined by the exact $P^{(N)}(y)$ converges as $N \to \infty$ to the probability that y lies between a and b as determined by the gaussian $P_G(y, \mu_y, \sigma_y)$

$$\lim_{N\to\infty} \Pr_N(a < y < b) = \lim_{N\to\infty} \int_a^b P^{(N)}(y)\, dy = \int_a^b P_G(y, \mu_y, \sigma_y)\, dy. \tag{15.259}$$

This type of convergence is called **convergence in probability** (Feller, 1966, pp. 231, 241–248).

For the special case in which all the means and variances are the same, with $\mu_j = \mu$ and $\sigma_j^2 = \sigma^2$, the definitions in (15.256) imply that $\mu_y = \mu$ and $\sigma_y^2 = \sigma^2/N$. In this case, one may show (Exercise 15.30) that in terms of the variable

$$u \equiv \frac{\sqrt{N}(y-\mu)}{\sigma} = \frac{\left(\sum_{n=1}^N x_j\right) - N\mu}{\sqrt{N}\,\sigma} \tag{15.260}$$

$P^{(N)}(y)$ converges to a distribution that is normal

$$\lim_{N\to\infty} P^{(N)}(y)\, dy = \frac{1}{\sqrt{2\pi}} e^{-u^2/2}\, du. \tag{15.261}$$

To get a clearer idea of when the **central limit theorem** holds, let us write the sum of the N variances as

$$S_N \equiv \sum_{j=1}^N \sigma_j^2 = \sum_{j=1}^N \int_{-\infty}^{\infty} (x_j - \mu_j)^2\, P_j(x_j)\, dx_j \tag{15.262}$$

and the part of this sum due to the regions within δ of the means μ_j as

$$S_N(\delta) \equiv \sum_{j=1}^{N} \int_{\mu_j-\delta}^{\mu_j+\delta} (x_j - \mu_j)^2 \, P_j(x_j) \, dx_j. \tag{15.263}$$

In these terms, Jarl Lindeberg (1876–1932) showed that the exact distribution $P^{(N)}(y)$ converges (in probability) to the gaussian (15.258) as long as the part $S_N(\delta)$ is most of S_N in the sense that for every $\epsilon > 0$

$$\lim_{N\to\infty} \frac{S_N\left(\epsilon\sqrt{S_N}\right)}{S_N} = 1. \tag{15.264}$$

This is **Lindeberg's condition** (Feller 1968, p. 254; Feller 1966, pp. 252–259; Gnedenko 1968, p. 304).

Because we dropped all but the first three terms of the series (15.251) for the characteristic functions $\tilde{P}_j(k/N)$, we may infer that the convergence of the distribution $P^{(N)}(y)$ to a gaussian is quickest near its mean μ_y. If the higher moments μ_{nj} are big, then for finite N the distribution $P^{(N)}(y)$ can have tails that are fatter than those of the limiting gaussian $P_G(y, \mu_y, \sigma_y)$.

Example 15.23 (Illustration of the central limit theorem) The simplest probability distribution is a random number x uniformly distributed on the interval $(0, 1)$. The probability distribution $P^{(2)}(y)$ of the mean of two such random numbers is the integral

$$P^{(2)}(y) = \int_0^1 dx_1 \int_0^1 dx_2 \; \delta((x_1 + x_2)/2 - y). \tag{15.265}$$

Letting $u_1 = x_1/2$, we find

$$P^{(2)}(y) = 4 \int_{\max(0,y-\frac{1}{2})}^{\min(y,\frac{1}{2})} \theta\left(\frac{1}{2} + u_1 - y\right) du_1 = 4y\,\theta\left(\frac{1}{2} - y\right) + 4(1-y)\,\theta\left(y - \frac{1}{2}\right) \tag{15.266}$$

which is the dot-dashed triangle in Fig. 15.5. The probability distribution $P^{(4)}(y)$ is the dashed somewhat gaussian curve in the figure, while $P^{(8)}(y)$ is the solid, nearly gaussian curve.

To work through a more complicated example of the central limit theorem, we first need to learn how to generate random numbers that follow an arbitrary distribution.

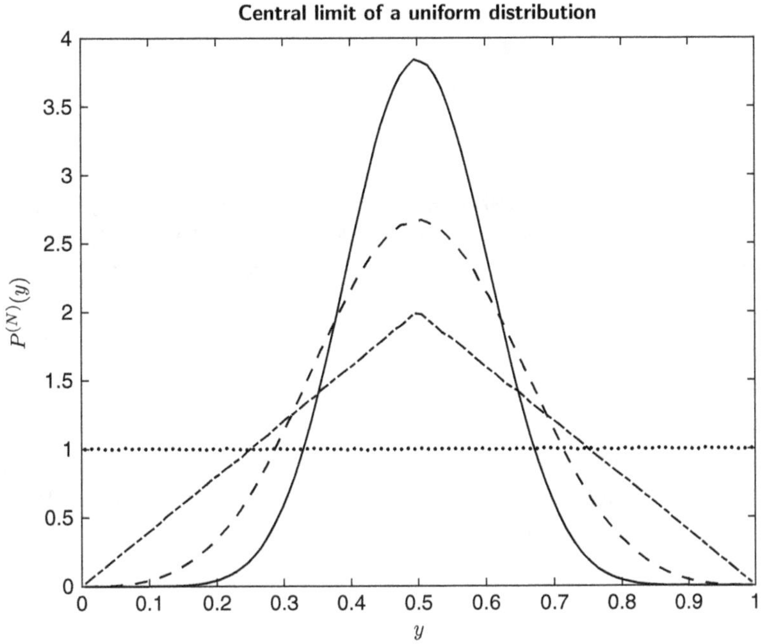

Figure 15.5 The probability distributions $P^{(N)}(y)$ (15.247) for the mean $y = (x_1 + \cdots + x_N)/N$ of N random variables drawn from the uniform distribution are plotted for $N = 1$ (dots), 2 (dot dash), 4 (dashes), and 8 (solid). The distributions $P^{(N)}(y)$ rapidly approach gaussians with the same mean $\mu_y = 1/2$ but with shrinking variances $\sigma^2 = 1/(12N)$.

15.19 Random-Number Generators

To generate truly random numbers, one might use decaying nuclei or an electronic device that makes white noise. But people usually settle for **pseudorandom numbers** computed by a mathematical algorithm. Such algorithms are deterministic, so the numbers they generate are not truly random. But for most purposes, they are random enough.

The easiest way to generate pseudorandom numbers is to use a random-number algorithm that is part of one's favorite FORTRAN, C, or C++ compiler. To run it, one first gives it a random starting point called a **seed**. When using the Intel FORTRAN90 compiler, one can set a random seed by using the line `call random_seed()`. When using the GNU FORTRAN compiler, one can use the statement `call init_random_seed()` to call the subroutine `init_random_seed()` defined in the program clt.f95, which is in the repository Probability_and_statistics at `github.com/kevinecahill`. Once one has set a random seed, the line `call random_number(x)` generates a random number x uniformly distributed on the interval $0 < x < 1$, or an array of such random numbers.

Some applications require random numbers of very high quality. For such applications, one might use Lüscher's RANLUX (Lüscher, 1994; James, 1994).

Most random-number generators are periodic with long periods. The website `www.math.sci.hiroshima-u.ac.jp/~m-mat/MT/emt.html` has codes for the **Mersenne Twister** (Saito and Matsumoto, 2007), which has the very long period of $2^{19937} - 1 \gtrsim 4.3 \times 10^{6001}$. Matlab uses it. The GNU FORTRAN compiler uses George Marsaglia's KISS generator `random_number` which has a period in excess of 2^{59}.

For most purposes, **quasirandom numbers** (Section 16.3) are better than pseudorandom ones.

Random-number generators distribute random numbers u uniformly on the interval $(0, 1)$. How do we make them follow an arbitrary distribution $P(r)$? If the distribution is strictly positive $P(r) > 0$ on the relevant interval (a, b), then its integral

$$F(x) = \int_a^x P(r)\, dr \tag{15.267}$$

is a strictly increasing function on (a, b), that is, $a < x < y < b$ implies $F(x) < F(y)$. Moreover, the function $F(x)$ rises from $F(a) = 0$ to $F(b) = 1$ and takes on every value $0 < y < 1$ for exactly one x in the interval (a, b). Thus the inverse function $F^{-1}(y)$

$$x = F^{-1}(y) \quad \text{if and only if} \quad y = F(x) \tag{15.268}$$

is well defined on the interval $(0, 1)$.

Our random-number generator gives us random numbers u that are uniform on $(0, 1)$. We want a random variable r whose probability $\Pr(r < x)$ of being less than any x is $F(x)$. The trick (Knuth, 1981, p. 116) is to generate a uniformly distributed random number u and then replace it with

$$r = F^{-1}(u). \tag{15.269}$$

For then, since $F(x)$ is one-to-one (15.268), the statements $F^{-1}(u) < x$ and $u < F(x)$ are equivalent, and therefore

$$\Pr(r < x) = \Pr(F^{-1}(u) < x) = \Pr(u < F(x)). \tag{15.270}$$

Example 15.24 ($P(r) = 3r^2$) To turn a distribution of random numbers u uniform on $(0, 1)$ into a distribution $P(r) = 3r^2$ of random numbers r, we integrate and find

$$F(x) = \int_0^x P(r)\, dr = \int_0^x 3r^2\, dr = x^3. \tag{15.271}$$

We then set $r = F^{-1}(u) = u^{1/3}$.

Example 15.25 ($P(r) = 12\,(r - 1/2)^2$) To turn a distribution of random numbers u uniform on $(0, 1)$ into a distribution $P(r) = 12\,(r - 1/2)^2$, we integrate and find

$$F(x) = \int_0^x P(r)\,dr = \int_0^x 12\,(r - 1/2)^2\,dr = 4x^3 - 6x^2 + 3x. \tag{15.272}$$

We set $u = 4r^3 - 6r^2 + 3r$ and solve this cubic equation for r

$$r = \frac{1}{2}\left[1 - (1 - 2u)^{1/3}\right] \tag{15.273}$$

(or we ask Wolfram Alpha for the inverse function to $F(x)$).

15.20 Illustration of the Central Limit Theorem

To make things simple, we'll take all the probability distributions $P_j(x)$ to be the same and equal to $P_j(x_j) = 3x_j^2$ on the interval $(0, 1)$ and zero elsewhere. Our random-number generator gives us random numbers u that are uniformly distributed on $(0, 1)$, so by the example (15.24) the variable $r = u^{1/3}$ is distributed as $P_j(x) = 3x^2$.

The central limit theorem tells us that the distribution

$$P^{(N)}(y) = \int 3x_1^2\, 3x_2^2 \cdots 3x_N^2\, \delta((x_1 + x_2 + \cdots + x_N)/N - y)\, d^N x \tag{15.274}$$

of the mean $y = (x_1 + \cdots + x_N)/N$ tends as $N \to \infty$ to Gauss's distribution

$$\lim_{N\to\infty} P^{(N)}(y) = \frac{1}{\sigma_y\sqrt{2\pi}} \exp\left(-\frac{(x - \mu_y)^2}{2\sigma_y^2}\right) \tag{15.275}$$

with mean μ_y and variance σ_y^2 given by (15.256). Since the P_j's are all the same, they all have the same mean

$$\mu_y = \mu_j = \int_0^1 3x^3 dx = \frac{3}{4} \tag{15.276}$$

and the same variance

$$\sigma_j^2 = \int_0^1 3x^4 dx - \left(\frac{3}{4}\right)^2 = \frac{3}{5} - \frac{9}{16} = \frac{3}{80}. \tag{15.277}$$

By (15.256), the variance of the mean y is then $\sigma_y^2 = 3/80N$. Thus as N increases, the mean y tends to a gaussian with mean $\mu_y = 3/4$ and ever narrower peaks.

For $N = 1$, the probability distribution $P^{(1)}(y)$ is

$$P^{(1)}(y) = \int 3x_1^2\, \delta(x_1 - y)\, dx_1 = 3y^2 \tag{15.278}$$

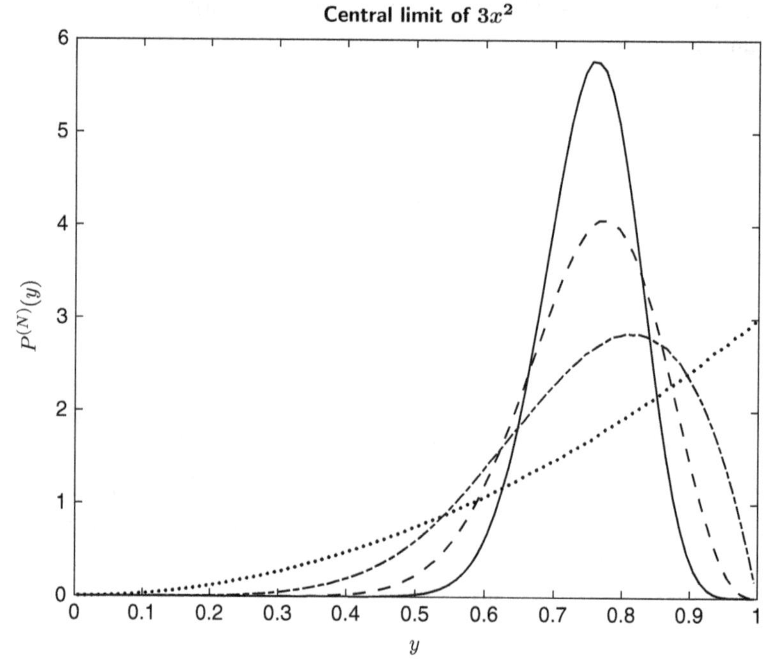

Figure 15.6 The probability distributions $P^{(N)}(y)$ (15.274) for the mean $y = (x_1 + \cdots + x_N)/N$ of N random variables drawn from the quadratic distribution $P(x) = 3x^2$ are plotted for $N = 1$ (dots), 2 (dot dash), 4 (dashes), and 8 (solid). The four distributions $P^{(N)}(y)$ rapidly approach gaussians with the same mean $\mu_y = 3/4$ but with shrinking variances $\sigma_y^2 = 3/(80N)$.

which is the probability distribution we started with. In Fig. 15.6, this is the quadratic, dotted curve.

For $N = 2$, the probability distribution $P^{(1)}(y)$ is (Exercise 15.29)

$$
\begin{aligned}
P^{(2)}(y) &= \int 3x_1^2 \, 3x_2^2 \, \delta((x_1 + x_2)/2 - y) \, dx_1 \, dx_2 \qquad (15.279)\\
&= \theta(\frac{1}{2} - y) \, \frac{96}{5} \, y^5 + \theta(y - \frac{1}{2}) \left(\frac{36}{5} - \frac{96}{5} \, y^5 + 48y^2 - 36y \right).
\end{aligned}
$$

You can get the probability distributions $P^{(N)}(y)$ for $N = 2^j$ by running the Fortran or C++ version of the program central_limit_of_3x^2 both of which are in Probability_and_statistics at `github.com/kevinecahill`.

The distributions $P^{(N)}(y)$ for $N = 1, 2, 4,$ and 8 are plotted in Fig. 15.6. $P^{(1)}(y) = 3y^2$ is the original distribution. $P^{(2)}(y)$ is trying to be a gaussian, while $P^{(4)}(y)$ and $P^{(8)}(y)$ have almost succeeded. The variance $\sigma_y^2 = 3/80N$ shrinks with N.

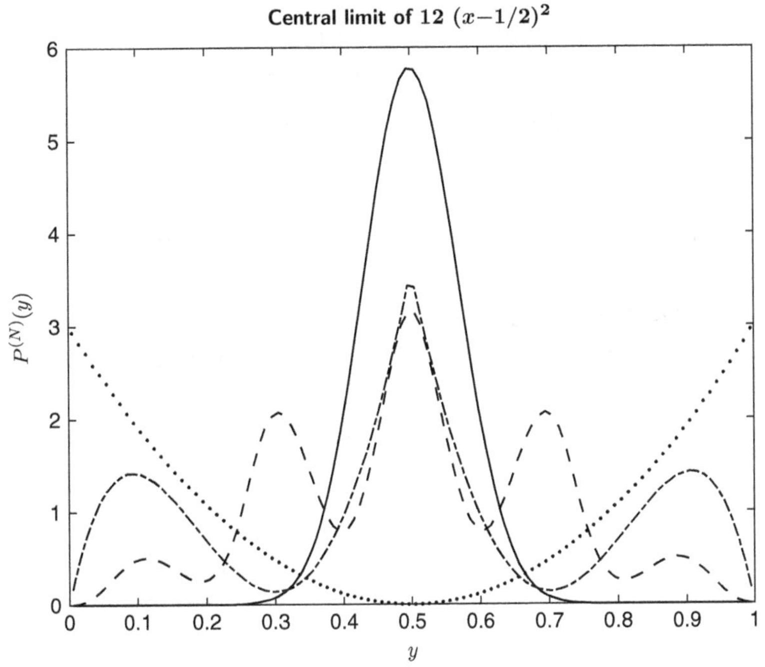

Figure 15.7 The probability distributions $P^{(N)}(y)$ (15.247) for the mean $y = (x_1 + \cdots + x_N)/N$ of N random variables drawn from the quadratic distribution $P(x) = 12\,(x - 1/2)^2$ of Example 15.25 are plotted for $N = 1$ (dots), 2 (dot dash), 4 (dashes), and 32 (solid). The distributions have the same mean $\mu_y = 1/2$ and shrinking variances $\sigma_y^2 = 3/(20N)$.

The quadratic distribution $P(x) = 12\,(x - 1/2)^2$ of Example 15.25 is very different from a gaussian centered at $x = 1/2$. Yet we see in Fig. 15.7 that the probability distributions $P^{(N)}(y)$ (15.247) for the mean $y = (x_1 + \cdots + x_N)/N$ of N random variables drawn from it do converge to such a gaussian.

Although FORTRAN95 is an ideal language for computation, C++ is more versatile and more modular, and Java is easier to use.

15.21 Measurements, Estimators, and Friedrich Bessel

The exact, physical probability distribution $P(x; \boldsymbol{\theta})$ for a stochastic variable x may depend upon one or more unknown parameters $\boldsymbol{\theta} = (\theta_1, \ldots, \theta_m)$. Experimenters seek to determine the unknown parameters $\boldsymbol{\theta}$, such as the mean μ and the variance σ^2, by collecting data in the form of observed values $\boldsymbol{x} = x_1, \ldots, x_N$ of the stochastic variable x. They assume that the probability distribution for the sequence $\boldsymbol{x} = (x_1, \ldots, x_N)$ is the product of N factors of the physical distribution $P(x; \boldsymbol{\theta})$

$$P(\boldsymbol{x};\boldsymbol{\theta}) = \prod_{j=1}^{N} P(x_j;\boldsymbol{\theta}). \tag{15.280}$$

They approximate the unknown value of a parameter θ_ℓ as the mean value of its **estimator** $u_\ell^{(N)}(\boldsymbol{x})$

$$E[u_\ell^{(N)}] = \int u_\ell^{(N)}(\boldsymbol{x})\, P(\boldsymbol{x};\boldsymbol{\theta})\, d^N x = \theta_\ell + b_\ell^{(N)}(\boldsymbol{\theta}). \tag{15.281}$$

If as $N \to \infty$, the **bias** $b_\ell^{(N)}(\boldsymbol{\theta}) \to 0$, then the estimator $u_\ell^{(N)}(\boldsymbol{x})$ is **consistent**.

Inasmuch as the mean (15.29) is the integral of the physical distribution

$$\mu = \int x\, P(x;\boldsymbol{\theta})\, dx \tag{15.282}$$

a natural estimator for the mean is

$$u_\mu^{(N)}(\boldsymbol{x}) = (x_1 + \cdots + x_N)/N. \tag{15.283}$$

Its expected value is

$$\begin{aligned} E[u_\mu^{(N)}] &= \int u_\mu^{(N)}(\boldsymbol{x})\, P(\boldsymbol{x};\boldsymbol{\theta})\, d^N x = \int \frac{x_1 + \cdots + x_N}{N}\, P(\boldsymbol{x};\boldsymbol{\theta})\, d^N x \qquad (15.284)\\ &= \frac{1}{N}\sum_{k=1}^{N} \int x_k\, P(x_k;\boldsymbol{\theta})\, dx_k \prod_{k\neq j=1}^{N} \int P(x_j;\boldsymbol{\theta})\, dx_j = \frac{1}{N}\sum_{k=1}^{N} \mu = \mu. \end{aligned}$$

Thus the natural estimator $u_\mu^{(N)}(\boldsymbol{x})$ of the mean (15.283) has $b_\ell^{(N)} = 0$, and so it is a consistent and unbiased estimator for the mean.

Since the variance (15.32) of the probability distribution $P(x;\boldsymbol{\theta})$ is the integral

$$\sigma^2 = \int (x-\mu)^2\, P(x;\boldsymbol{\theta})\, dx \tag{15.285}$$

the variance of the estimator u_μ^N is

$$\begin{aligned} V[u_\mu^{(N)}] &= \int \left(u_\mu^{(N)}(\boldsymbol{x}) - \mu\right)^2 P(\boldsymbol{x};\boldsymbol{\theta})\, d^N x = \int \left[\frac{1}{N}\sum_{j=1}^{N}(x_j - \mu)\right]^2 P(\boldsymbol{x};\boldsymbol{\theta})\, d^N x\\ &= \frac{1}{N^2}\sum_{j,k=1}^{N} \int (x_j-\mu)\,(x_k-\mu)\, P(\boldsymbol{x};\boldsymbol{\theta})\, d^N x \qquad (15.286)\\ &= \frac{1}{N^2}\sum_{j,k=1}^{N} \delta_{jk} \int (x_j-\mu)^2\, P(\boldsymbol{x};\boldsymbol{\theta})\, d^N x = \frac{1}{N^2}\sum_{k=1}^{N}\sigma^2 = \frac{\sigma^2}{N} \end{aligned}$$

in which σ^2 is the variance (15.285) of the physical distribution $P(x;\boldsymbol{\theta})$. We'll learn in the next section that no estimator of the mean can have a lower variance than this.

A natural estimator for the variance of the probability distribution $P(x;\boldsymbol{\theta})$ is

$$u_{\sigma^2}^{(N)}(x) = B \sum_{j=1}^{N} \left(x_j - u_\mu^{(N)}(\boldsymbol{x})\right)^2 \tag{15.287}$$

in which $B = B(N)$ is a constant of proportionality. The naive choice $B(N) = 1/N$ leads to a biased estimator. To find the correct value of B, we set the expected value $E[u_{\sigma^2}^{(N)}]$ equal to σ^2

$$E[u_{\sigma^2}^{(N)}] = \int B \sum_{j=1}^{N} \left(x_j - u_\mu^{(N)}(\boldsymbol{x})\right)^2 P(\boldsymbol{x};\boldsymbol{\theta})\, d^N x = \sigma^2 \tag{15.288}$$

and solve for B. Subtracting the mean μ from both x_j and $u_\mu^{(N)}(\boldsymbol{x})$, we express σ^2/B as the sum of three terms

$$\frac{\sigma^2}{B} = \sum_{j=1}^{N} \int \left[x_j - \mu - \left(u_\mu^{(N)}(\boldsymbol{x}) - \mu\right)\right]^2 P(\boldsymbol{x};\boldsymbol{\theta})\, d^N x = S_{jj} + S_{j\mu} + S_{\mu\mu} \tag{15.289}$$

the first of which is

$$S_{jj} = \sum_{j=1}^{N} \int \left(x_j - \mu\right)^2 P(\boldsymbol{x};\boldsymbol{\theta})\, d^N x = N\sigma^2. \tag{15.290}$$

The cross-term $S_{j\mu}$ is

$$\begin{aligned} S_{j\mu} &= -2 \sum_{j=1}^{N} \int \left(x_j - \mu\right)\left(u_\mu^{(N)}(\boldsymbol{x}) - \mu\right) P(\boldsymbol{x};\boldsymbol{\theta})\, d^N x \\ &= -\frac{2}{N} \sum_{j=1}^{N} \int \left(x_j - \mu\right) \sum_{k=1}^{N} (x_k - \mu)\, P(\boldsymbol{x};\boldsymbol{\theta})\, d^N x = -2\sigma^2. \end{aligned} \tag{15.291}$$

The third term is the variance (15.286) multiplied by N

$$S_{\mu\mu} = \sum_{j=1}^{N} \int \left(u_\mu^{(N)}(\boldsymbol{x}) - \mu\right)^2 P(\boldsymbol{x};\boldsymbol{\theta})\, d^N x = N V[u_\mu^N] = \sigma^2. \tag{15.292}$$

Thus the factor B must satisfy

$$\sigma^2/B = N\sigma^2 - 2\sigma^2 + \sigma^2 = (N-1)\sigma^2 \tag{15.293}$$

which tells us that $B = 1/(N-1)$, which is **Bessel's correction**. Our estimator for the variance $\sigma^2 = E[u_{\sigma^2}^{(N)}]$ of the probability distribution $P(x;\boldsymbol{\theta})$ then is

$$u_{\sigma^2}^{(N)}(\boldsymbol{x}) = \frac{1}{N-1}\sum_{j=1}^{N}\left(x_j - u_{\mu}^{(N)}(\boldsymbol{x})\right)^2 = \frac{1}{N-1}\sum_{j=1}^{N}\left(x_j - \frac{1}{N}\sum_{k=1}^{N}x_k\right)^2. \tag{15.294}$$

It is consistent and unbiased since $E[u_{\sigma^2}^{(N)}] = \sigma^2$ by construction (15.288). It gives for the variance σ^2 of a single measurement the undefined ratio 0/0, as it should, whereas the naive choice $B = 1/N$ absurdly gives zero.

On the basis of N measurements $x_1, \ldots, x_N$ we can estimate the mean of the unknown probability distribution $P(x;\boldsymbol{\theta})$ as $\mu_N = (x_1 + \cdots + x_N)/N$. And we can use Bessel's formula (15.294) to estimate the variance σ^2 of the unknown distribution $P(x;\boldsymbol{\theta})$. Our formula (15.286) for the variance $\sigma^2(\mu_N)$ of the mean μ_N then gives

$$\sigma^2(\mu_N) = \frac{\sigma^2}{N} = \frac{1}{N(N-1)}\sum_{j=1}^{N}\left(x_j - \frac{1}{N}\sum_{k=1}^{N}x_k\right)^2. \tag{15.295}$$

Thus we can use N measurements x_j to estimate the mean μ to within a standard error or standard deviation of

$$\sigma(\mu_N) = \sqrt{\frac{\sigma^2}{N}} = \sqrt{\frac{1}{N(N-1)}\sum_{j=1}^{N}\left(x_j - \frac{1}{N}\sum_{k=1}^{N}x_k\right)^2}. \tag{15.296}$$

Few formulas have seen so much use.

15.22 Information and Ronald Fisher

The **Fisher information matrix** of a distribution $P(\boldsymbol{x};\boldsymbol{\theta})$ is the mean of products of its partial logarithmic derivatives

$$\begin{aligned} F_{k\ell}(\boldsymbol{\theta}) &\equiv E\left[\frac{\partial \ln P(\boldsymbol{x};\boldsymbol{\theta})}{\partial\theta_k}\frac{\partial \ln P(\boldsymbol{x};\boldsymbol{\theta})}{\partial\theta_\ell}\right] \\ &= \int \frac{\partial \ln P(\boldsymbol{x};\boldsymbol{\theta})}{\partial\theta_k}\frac{\partial \ln P(\boldsymbol{x};\boldsymbol{\theta})}{\partial\theta_\ell}\,P(\boldsymbol{x};\boldsymbol{\theta})\,d^N x \end{aligned} \tag{15.297}$$

(Ronald Fisher, 1890–1962). Fisher's matrix (Exercise 15.31) is symmetric $F_{k\ell} = F_{\ell k}$ and nonnegative (1.39), and when it is positive (1.40), it has an inverse. By differentiating the normalization condition

$$\int P(\boldsymbol{x};\boldsymbol{\theta})\,d^N x = 1 \tag{15.298}$$

we have

$$0 = \int \frac{\partial P(\boldsymbol{x};\boldsymbol{\theta})}{\partial \theta_k}\, d^N x = \int \frac{\partial \ln P(\boldsymbol{x};\boldsymbol{\theta})}{\partial \theta_k}\, P(\boldsymbol{x};\boldsymbol{\theta})\, d^N x \tag{15.299}$$

which says that the **score**, which is the θ derivative of the logarithm of the probability distribution, has a mean value that vanishes. Using commas to denote θ derivatives as in

$$(\ln P)_{,k} \equiv \frac{\partial \ln P}{\partial \theta_k} \quad \text{and} \quad (\ln P)_{,k\ell} \equiv \frac{\partial^2 \ln P}{\partial \theta_k \partial \theta_\ell} \tag{15.300}$$

and differentiating the identity (15.299), one has (Exercise 15.32)

$$0 = \int (\ln P)_{,k}\, (\ln P)_{,\ell}\, P\, d^N x + \int (\ln P)_{,k\ell}\, P\, d^N x \tag{15.301}$$

so that another form of Fisher's information matrix is

$$F_{k\ell}(\boldsymbol{\theta}) = -\, E\left[(\ln P)_{,k\ell}\right] = -\int (\ln P)_{,k\ell}\, P\, d^N x. \tag{15.302}$$

Cramér and Rao used Fisher's information matrix to form a lower bound on the covariance (15.43) matrix $C[u_k, u_\ell]$ of any two estimators. To see how this works, we use the vanishing (15.299) of the mean of the score to write the covariance of the ℓth estimator $u_\ell(\boldsymbol{x})$ with the kth score $(\ln P(\boldsymbol{x};\boldsymbol{\theta}))_{,k}$ as the θ_k-derivative $\langle u_\ell \rangle_{,k}$ of the mean $\langle u_\ell \rangle$

$$\begin{aligned} C[u_\ell, (\ln P)_{,k}] &= \int (u_\ell - \theta_\ell - b_\ell)\, (\ln P)_{,k}\, P\, d^N x \\ &= \int u_\ell\, (\ln P)_{,k}\, P\, d^N x = \int u_\ell\, P_{,k}\, d^N x \\ &= \langle u_\ell \rangle_{,k} = (\theta_\ell + b_\ell)_{,k} = \delta_{\ell k} + b_{\ell,k}. \end{aligned} \tag{15.303}$$

Thus for any *constants* $y_1, y_2, \ldots, y_J$ and $w_1, w_2, \ldots, w_J$, where J is the number of estimators u_k, we have

$$\int \sum_{\ell,k=1}^{J} y_\ell\, (u_\ell - \theta_\ell - b_\ell)\, \sqrt{P}\, (\ln P)_{,k}\, \sqrt{P}\, w_k\, d^N x = \sum_{\ell,k=1}^{J} y_\ell\, \langle u_\ell \rangle_{,k}\, w_k. \tag{15.304}$$

In matrix notation with $u'_{\ell k} = \langle u_\ell \rangle_{,k}$ and $(\ln P)'_k = (\ln P)_{,k}$, this equation is

$$\int y \cdot (u - \theta - b)\, \sqrt{P}\, \sqrt{P}\, (\ln P)' \cdot w\, d^N x = y^{\mathsf{T}}\, u'\, w. \tag{15.305}$$

Squaring and applying Schwarz's inequality (7.418), we get

$$\int (y \cdot (u - \theta - b))^2\, P\, d^N x \int \left((\ln P)' \cdot w\right)^2\, P\, d^N x \ge \left(y^{\mathsf{T}} u' w\right)^2. \tag{15.306}$$

On the left-hand side of this equation, the first term is $y^{\top}Cy$ in which $C_{\ell\ell'} = C[u_\ell, u_{\ell'}]$ is the covariance (15.43) of the estimators $u_\ell(\boldsymbol{x})$ and $u_{\ell'}(\boldsymbol{x})$, and the second term is $w^{\top}Fw$ in which $F_{kk'}$ is Fisher's information matrix (15.297):

$$y^{\top}Cy \;\; w^{\top}Fw \geq \left(y^{\top}u'w\right)^2. \tag{15.307}$$

The Fisher information matrix F is real and symmetric, and its eigenvalues are nonnegative. If all its eigenvalues are positive (as they are unless P is independent of one or more of the θ_k's), then F has an inverse F^{-1}, and we can set $w = F^{-1}u'^{\top}y$. The inequality (15.307) then becomes

$$y^{\top}Cy \;\; y^{\top}u'F^{-1}FF^{-1}u'^{\top}y \geq y^{\top}u'F^{-1}u'^{\top}y \;\; y^{\top}u'F^{-1}u'^{\top}y. \tag{15.308}$$

Setting $FF^{-1} = I$ and cancelling the common factor $y^{\top}u'F^{-1}u'^{\top}y$, we arrive at the **Cramér–Rao inequality**

$$y^{\top}Cy \geq y^{\top}u'F^{-1}u'^{\top}y. \tag{15.309}$$

Recalling the formula (15.303) which expresses u' as $u'_{\ell k} = \langle u_\ell \rangle_{,k} = \delta_{\ell k} + b_{\ell,k}$, we have

$$y_\ell \, C[u_\ell, u_k] \, y_k \geq y_\ell \left(\delta_{\ell k} + b_{\ell,k}\right) F^{-1}_{km} \left(\delta_{mn} + b_{n,m}\right) y_n \tag{15.310}$$

or more succinctly

$$C \geq (I + b')F^{-1}(I + b'^{\top}). \tag{15.311}$$

In these inequalities, the y_ℓ's are arbitrary numbers. Thus setting $y_\ell = \delta_{\ell k}$ and using the symmetry $F_{k\ell} = F_{\ell k}$, we can write the Cramér–Rao inequality (15.310) in terms of the variance $V[u_k] = C[u_k, u_k]$ as

$$V[u_k] = C[u_k, u_k] \geq F^{-1}_{kk} + 2F^{-1}_{k\ell}\, b_{k,\ell} + b_{k,\ell} F^{-1}_{\ell m} b_{k,m}. \tag{15.312}$$

If the estimator u_k is unbiased, this lower bound simplifies to

$$V[u_k] \geq F^{-1}_{kk}. \tag{15.313}$$

Example 15.26 (Cramér-Rao bound for a gaussian) The elements of Fisher's information matrix for the mean μ and variance σ^2 of Gauss's distribution for N data points $x_1, \ldots, x_N$

$$P_G^{(N)}(\boldsymbol{x}, \mu, \sigma) = \prod_{j=1}^{N} P_G(x_j; \mu, \sigma) = \left(\frac{1}{\sigma\sqrt{2\pi}}\right)^N \exp\left(-\sum_{j=1}^{N} \frac{(x_j - \mu)^2}{2\sigma^2}\right) \tag{15.314}$$

are

$$
\begin{aligned}
F_{\mu\mu} &= \int \left[\left(\ln P_G^{(N)}(\boldsymbol{x}, \mu, \sigma) \right)_{,\mu} \right]^2 P_G^{(N)}(\boldsymbol{x}, \mu, \sigma)\, d^N x \\
&= \sum_{i,j=1}^{N} \int \left(\frac{x_i - \mu}{\sigma^2} \right) \left(\frac{x_j - \mu}{\sigma^2} \right) P_G^{(N)}(\boldsymbol{x}, \mu, \sigma)\, d^N x \\
&= \sum_{i=1}^{N} \int \left(\frac{x_i - \mu}{\sigma^2} \right)^2 P_G^{(N)}(\boldsymbol{x}, \mu, \sigma)\, d^N x = \frac{N}{\sigma^2} \qquad (15.315) \\
F_{\mu\sigma^2} &= \int (\ln P_G^{(N)}(\boldsymbol{x}, \mu, \sigma))_{,\mu} (\ln P_G^{(N)}(\boldsymbol{x}, \mu, \sigma))_{,\sigma^2}\, P_G^{(N)}(\boldsymbol{x}, \mu, \sigma)\, d^N x \\
&= \sum_{i,j=1}^{N} \int \left[\frac{x_i - \mu}{\sigma^2} \right] \left[\frac{(x_j - \mu)^2}{2\sigma^4} - \frac{1}{2\sigma^2} \right] P_G^{(N)}(\boldsymbol{x}, \mu, \sigma)\, d^N x = 0
\end{aligned}
$$

$F_{\sigma^2\mu} = F_{\mu\sigma^2} = 0$, and

$$
\begin{aligned}
F_{\sigma^2\sigma^2} &= \int \left[(\ln P_G^{(N)}(\boldsymbol{x}, \mu, \sigma))_{,\sigma^2} \right]^2 P_G^{(N)}(\boldsymbol{x}, \mu, \sigma)\, d^N x \\
&= \sum_{i,j=1}^{N} \int \left[\frac{(x_i - \mu)^2}{2\sigma^4} - \frac{1}{2\sigma^2} \right] \left[\frac{(x_j - \mu)^2}{2\sigma^4} - \frac{1}{2\sigma^2} \right] P_G^{(N)}(\boldsymbol{x}, \mu, \sigma)\, d^N x \\
&= \frac{N}{2\sigma^4}. \qquad (15.316)
\end{aligned}
$$

The inverse of Fisher's matrix then is diagonal with $(F^{-1})_{\mu\mu} = \sigma^2/N$ and $(F^{-1})_{\sigma^2\sigma^2} = 2\sigma^4/N$.

The variance of any unbiased estimator $u_\mu(x)$ of the mean must exceed its Cramér–Rao lower bound (15.313), and so $V[u_\mu] \geq (F^{-1})_{\mu\mu} = \sigma^2/N$. The variance $V[u_\mu^{(N)}]$ of the natural estimator of the mean $u_\mu^{(N)}(\boldsymbol{x}) = (x_1 + \cdots + x_N)/N$ is σ^2/N by (15.286), and so it respects and saturates the lower bound (15.313)

$$
V[u_\mu^{(N)}] = E[(u_\mu^{(N)} - \mu)^2] = \sigma^2/N = (F^{-1})_{\mu\mu}. \qquad (15.317)
$$

One may show (Exercise 15.33) that the variance $V[u_{\sigma^2}^{(N)}]$ of Bessel's estimator (15.294) of the variance is (Riley et al., 2006, p. 1248)

$$
V[u_{\sigma^2}^{(N)}] = \frac{1}{N} \left(\nu_4 - \frac{N-3}{N-1} \sigma^4 \right) \qquad (15.318)
$$

where ν_4 is the fourth central moment (15.31) of the probability distribution. For the gaussian $P_G(\boldsymbol{x}; \mu, \sigma)$ one may show (Exercise 15.34) that this moment is $\nu_4 = 3\sigma^4$, and so for it

$$
V_G[u_{\sigma^2}^{(N)}] = \frac{2}{N-1} \sigma^4. \qquad (15.319)
$$

Thus the variance of Bessel's estimator of the variance respects but does not saturate its Cramér–Rao lower bound (15.313, 15.316)

$$V_G[u^{(N)}_{\sigma^2}] = \frac{2}{N-1}\,\sigma^4 > \frac{2}{N}\,\sigma^4. \tag{15.320}$$

Estimators that saturate their Cramér–Rao lower bounds are **efficient**. The natural estimator $u^{(N)}_{\mu}(\boldsymbol{x})$ of the mean is efficient as well as consistent and unbiased, and Bessel's estimator $u^{(N)}_{\sigma^2}(\boldsymbol{x})$ of the variance is consistent and unbiased but not efficient.

15.23 Maximum Likelihood

Suppose we measure some quantity x at various values of another variable t and find the values $x_1, x_2, \ldots, x_N$ at the known points $t_1, t_2, \ldots, t_N$. We might want to fit these measurements to a curve $x = f(t; \boldsymbol{\alpha})$ where $\boldsymbol{\alpha} = \alpha_1, \ldots, \alpha_M$ is a set of $M < N$ parameters. In view of the central limit theorem, we'll assume that the points x_j fall in Gauss's distribution about the values $x_j = f(t_j; \boldsymbol{\alpha})$ with some known variance σ^2. The probability of getting the N values $x_1, \ldots, x_N$ then is

$$P(\boldsymbol{x}) = \prod_{j=1}^{N} P(x_j, t_j, \sigma) = \left(\frac{1}{\sigma\sqrt{2\pi}}\right)^N \exp\left(-\sum_{j=1}^{N} \frac{(x_j - f(t_j; \boldsymbol{\alpha}))^2}{2\sigma^2}\right). \tag{15.321}$$

To find the M parameters $\boldsymbol{\alpha}$, we maximize the likelihood $P(\boldsymbol{x})$ by minimizing the argument of its exponential

$$0 = \frac{\partial}{\partial \alpha_\ell} \sum_{j=1}^{N} \left(x_j - f(t_j; \boldsymbol{\alpha})\right)^2 = -2 \sum_{j=1}^{N} \left(x_j - f(t_j; \boldsymbol{\alpha})\right) \frac{\partial f(t_j; \boldsymbol{\alpha})}{\partial \alpha_\ell}. \tag{15.322}$$

If the function $f(t; \boldsymbol{\alpha})$ depends nonlinearly upon the parameters $\boldsymbol{\alpha}$, then we may need to use numerical methods to solve this **least-squares** problem.

But if the function $f(t; \boldsymbol{\alpha})$ depends **linearly** upon the M parameters $\boldsymbol{\alpha}$

$$f(t; \boldsymbol{\alpha}) = \sum_{k=1}^{M} g_k(t)\, \alpha_k \tag{15.323}$$

then the equations (15.322) that determine these parameters $\boldsymbol{\alpha}$ are linear

$$0 = \sum_{j=1}^{N} \left(x_j - \sum_{k=1}^{M} g_k(t_j)\, \alpha_k\right) g_\ell(t_j). \tag{15.324}$$

In matrix notation with G the $N \times M$ rectangular matrix with entries $G_{jk} = g_k(t_j)$, they are

$$G^{\mathsf{T}} \boldsymbol{x} = G^{\mathsf{T}} G \boldsymbol{\alpha}. \tag{15.325}$$

The basis functions $g_k(t)$ may depend nonlinearly upon the independent variable t. If one chooses them to be sufficiently different that the columns of G are linearly independent, then the rank of G is M, and the nonnegative matrix $G^{\mathsf{T}} G$ has an inverse. The matrix G then has a pseudoinverse (1.432)

$$G^{+} = \left(G^{\mathsf{T}} G\right)^{-1} G^{\mathsf{T}} \tag{15.326}$$

and it maps the N-vector $\boldsymbol{x}$ into our parameters $\boldsymbol{\alpha}$

$$\boldsymbol{\alpha} = G^{+} \boldsymbol{x}. \tag{15.327}$$

The product $G^{+} G = I_M$ is the $M \times M$ identity matrix, while

$$G G^{+} = P \tag{15.328}$$

is an $N \times N$ projection operator (exercise 15.35) onto the $M \times M$ subspace for which $G^{+}G = I_M$ is the identity operator. Like all projection operators, P satisfies $P^2 = P$.

15.24 Karl Pearson's Chi-Squared Statistic

The argument of the exponential (15.321) in $P(\boldsymbol{x})$ is (the negative of) Karl Pearson's chi-squared statistic (Pearson, 1900)

$$\chi^2 \equiv \sum_{j=1}^{N} \frac{(x_j - f(t_j;\boldsymbol{\alpha}))^2}{2\sigma^2}. \tag{15.329}$$

When the function $f(t;\boldsymbol{\alpha})$ is linear (15.323) in $\boldsymbol{\alpha}$, the N-vector $f(t_j;\boldsymbol{\alpha})$ is $f = G\,\boldsymbol{\alpha}$. Pearson's χ^2 then is

$$\chi^2 = (\boldsymbol{x} - G\,\boldsymbol{\alpha})^2/2\sigma^2. \tag{15.330}$$

Now (15.327) tells us that $\boldsymbol{\alpha} = G^{+} \boldsymbol{x}$, and so in terms of the projection operator $P = G\,G^{+}$, the vector $\boldsymbol{x} - G\,\boldsymbol{\alpha}$ is

$$\boldsymbol{x} - G\,\boldsymbol{\alpha} = \boldsymbol{x} - G\,G^{+} \boldsymbol{x} = \left(I - G\,G^{+}\right) \boldsymbol{x} = (I - P)\,\boldsymbol{x}. \tag{15.331}$$

So χ^2 is proportional to the squared length

$$\chi^2 = \tilde{\boldsymbol{x}}^2/2\sigma^2 \tag{15.332}$$

of the vector

$$\tilde{\boldsymbol{x}} \equiv (I - P)\,\boldsymbol{x}. \tag{15.333}$$

Thus if the matrix G has rank M, and the vector $\boldsymbol{x}$ has N independent components, then the vector $\tilde{\boldsymbol{x}}$ has only $N - M$ independent components.

Example 15.27 (Two position measurements) Suppose we measure a position twice with error σ, get x_1 and x_2, and choose $G^{\mathsf{T}} = (1, 1)$. Then the single parameter α is their average $\alpha = (x_1 + x_2)/2$, and χ^2 is

$$\begin{aligned}\chi^2 &= \left\{\left[x_1 - (x_1 + x_2)/2\right]^2 + \left[x_2 - (x_1 + x_2)/2\right]^2\right\} \Big/ 2\sigma^2 \\ &= \left\{\left[(x_1 - x_2)/2\right]^2 + \left[(x_2 - x_1)/2\right]^2\right\} \Big/ 2\sigma^2 \\ &= \left[(x_1 - x_2)/\sqrt{2}\right]^2 /2\sigma^2. \end{aligned} \tag{15.334}$$

Thus instead of having two independent components x_1 and x_2, χ^2 just has one $(x_1 - x_2)/\sqrt{2}$.

We can see how this happens more generally if we use as basis vectors the $N - M$ orthonormal vectors $|j\rangle$ in the kernel of P (that is, the $|j\rangle$'s annihilated by P)

$$P|j\rangle = 0 \quad 1 \le j \le N - M \tag{15.335}$$

and the M that lie in the range of the projection operator P

$$P|k\rangle = |k\rangle \quad N - M + 1 \le k \le N. \tag{15.336}$$

In terms of these basis vectors, the N-vector $\boldsymbol{x}$ is

$$\boldsymbol{x} = \sum_{j=1}^{N-M} x_j |j\rangle + \sum_{k=N-M+1}^{N} x_k |k\rangle \tag{15.337}$$

and the last M components of the vector $\tilde{\boldsymbol{x}}$ vanish

$$\tilde{\boldsymbol{x}} = (I - P)\boldsymbol{x} = \sum_{j=1}^{N-M} x_j |j\rangle. \tag{15.338}$$

Example 15.28 (N position measurements) Suppose the N values of x_j are the measured values of the position $f(t_j; \alpha) = x_j$ of some object. Then $M = 1$, and we choose $G_{j1} = g_1(t_j) = 1$ for $j = 1, \ldots, N$. Now $G^{\mathsf{T}} G = N$ is a 1×1 matrix, the number N, and the parameter α is the mean $\overline{x}$

$$\alpha = G^{+}\boldsymbol{x} = \left(G^{\mathsf{T}} G\right)^{-1} G^{\mathsf{T}} \boldsymbol{x} = \frac{1}{N} \sum_{j=1}^{N} x_j = \overline{x} \tag{15.339}$$

of the N position measurements x_j. So the vector $\tilde{\boldsymbol{x}}$ has components $\tilde{x}_j = x_j - \overline{x}$ and is orthogonal to $G^{\mathsf{T}} = (1, 1, \ldots, 1)$

$$G^{\mathsf{T}} \tilde{\boldsymbol{x}} = \left(\sum_{j=1}^{N} x_j \right) - N\overline{x} = 0. \tag{15.340}$$

The matrix G^{T} has rank 1, and the vector $\tilde{\boldsymbol{x}}$ has $N - 1$ independent components.

Suppose now that we have determined our M parameters $\boldsymbol{\alpha}$ and have a theoretical fit

$$x = f(t; \boldsymbol{\alpha}) = \sum_{k=1}^{M} g_k(t)\, \alpha_k \tag{15.341}$$

which when we apply it to N measurements x_j gives χ^2 as

$$\chi^2 = (\tilde{\boldsymbol{x}})^2 / 2\sigma^2. \tag{15.342}$$

How good is our fit?

A χ^2 distribution with $N - M$ **degrees of freedom** has by (15.238) mean

$$E[\chi^2] = N - M \tag{15.343}$$

and variance

$$V[\chi^2] = 2(N - M). \tag{15.344}$$

So our χ^2 should be about

$$\chi^2 \approx N - M \pm \sqrt{2(N - M)}. \tag{15.345}$$

If it lies within this range, then (15.341) is a good fit to the data. But if it exceeds $N - M + \sqrt{2(N - M)}$, then the fit isn't so good. On the other hand, if χ^2 is less than $N - M - \sqrt{2(N - M)}$, then we may have used too many parameters or overestimated σ. Indeed, by using N parameters with $G\, G^{+} = I_N$, we could get $\chi^2 = 0$ every time.

The probability that χ^2 exceeds χ_0^2 is the integral (15.237)

$$\Pr_n(\chi^2 > \chi_0^2) = \int_{\chi_0^2}^{\infty} P_n(\chi^2/2)\, d\chi^2 = \int_{\chi_0^2}^{\infty} \frac{1}{2\Gamma(n/2)} \left(\frac{\chi^2}{2} \right)^{n/2-1} e^{-\chi^2/2} d\chi^2 \tag{15.346}$$

in which $n = N - M$ is the number of data points minus the number of parameters, and $\Gamma(n/2)$ is the gamma function (6.108, 5.68). So an M-parameter fit to N data points has only a chance of ϵ of being good if its χ^2 is greater than a χ_0^2 for which $\Pr_{N-M}(\chi^2 > \chi_0^2) = \epsilon$. These probabilities $\Pr_{N-M}(\chi^2 > \chi_0^2)$ are plotted in

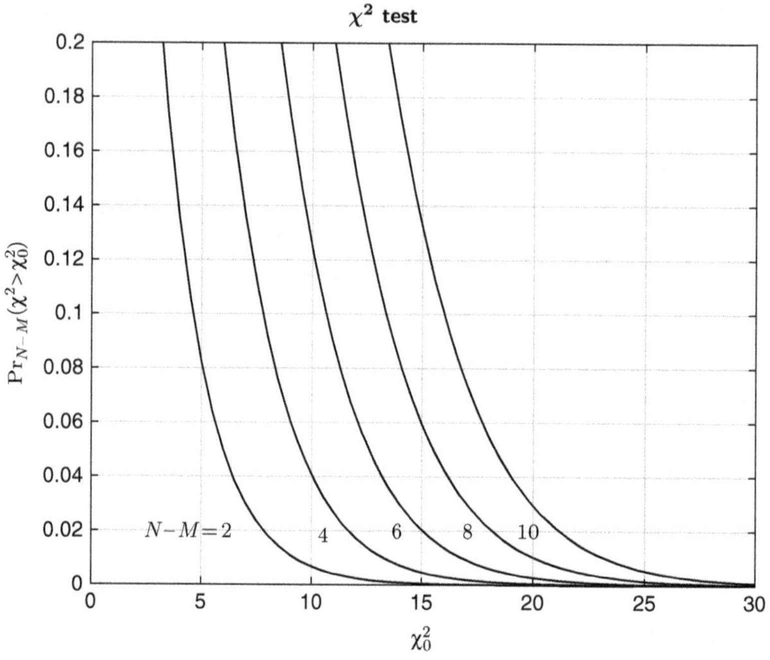

Figure 15.8 The probabilities $\Pr_{N-M}(\chi^2 > \chi_0^2)$ are plotted from left to right for $N - M = 2, 4, 6, 8$, and 10 degrees of freedom as functions of χ_0^2.

Fig. 15.8 for $N - M = 2, 4, 6, 8$, and 10. In particular, the probability of a value of χ^2 greater than $\chi_0^2 = 20$ respectively is 0.000045, 0.000499, 0.00277, 0.010336, and 0.029253 for $N - M = 2, 4, 6, 8$, and 10.

15.25 Kolmogorov's Test

Suppose we want to use a sequence of N measurements x_j to determine the probability distribution that they come from. Our empirical probability distribution is

$$P_e^{(N)}(x) = \frac{1}{N}\sum_{j=1}^{N} \delta(x - x_j). \tag{15.347}$$

Our cumulative probability for events less than x then is

$$\Pr_e^{(N)}(-\infty, x) = \int_{-\infty}^{x} P_e^{(N)}(x')\, dx' = \int_{-\infty}^{x} \frac{1}{N}\sum_{j=1}^{N} \delta(x' - x_j)\, dx'. \tag{15.348}$$

So if we label our events in increasing order $x_1 \le x_2 \le \cdots \le x_N$, then the probability of an event less than x is a staircase

$$\Pr_e^{(N)}(-\infty, x) = \frac{j}{N} \quad \text{for} \quad x_j < x < x_{j+1}. \tag{15.349}$$

Having approximately and experimentally determined our empirical cumulative probability distribution $\Pr_e^{(N)}(-\infty, x)$, we might want to know whether it comes from some hypothetical, theoretical cumulative probability distribution $\Pr_t(-\infty, x)$. One way to do this is to compute the distance D_N between the two cumulative probability distributions

$$D_N = \sup_{-\infty<x<\infty} \left|\Pr_e^{(N)}(-\infty, x) - \Pr_t(-\infty, x)\right| \tag{15.350}$$

in which **sup** stands for ***supremum*** and means **least upper bound**. Since cumulative probabilities lie between zero and one, it follows (Exercise 15.36) that the Kolmogorov distance is bounded by $0 \le D_N \le 1$.

In general, as the number N of data points increases, we expect that our empirical distribution $\Pr_e^{(N)}(-\infty, x)$ should approach the actual empirical distribution $\Pr_e(-\infty, x)$ from which the events x_j came. In this case, the Kolmogorov distance D_N should converge to a limiting value D_∞

$$\lim_{N\to\infty} D_N = D_\infty = \sup_{-\infty<x<\infty} |\Pr_e(-\infty, x) - \Pr_t(-\infty, x)| \in [0, 1]. \tag{15.351}$$

If the empirical distribution $\Pr_e(-\infty, x)$ is the same as the theoretical distribution $\Pr_t(-\infty, x)$, then we expect that $D_\infty = 0$. This expectation is confirmed by a theorem due to Glivenko (Glivenko, 1933; Cantelli, 1933) according to which the probability that the Kolmogorov distance D_N should go to zero as $N \to \infty$ is unity, $\Pr(D_\infty = 0) = 1$.

The real issue is how fast D_N should decrease with N if our events x_j do come from $\Pr_t(-\infty, x)$. This question was answered by Kolmogorov who showed (Kolmogorov, 1933) that if the events x_j of the empirical distribution $\Pr_e(-\infty, x)$ do come from the theoretical distribution $\Pr_t(-\infty, x)$, and if $\Pr_t(-\infty, x)$ is continuous, then for large N the probability that $\sqrt{N}\, D_N$ (D_N being the Kolmogorov distance between the empirical and theoretical cumulative distributions) is less than u is given (for $u > 0$) by the **Kolmogorov function** $K(u)$

$$\lim_{N\to\infty} \Pr(\sqrt{N}\, D_N < u) = K(u) \equiv 1 + 2\sum_{k=1}^{\infty}(-1)^k e^{-2k^2u^2}. \tag{15.352}$$

Amazingly, this upper bound is **universal and independent of the particular probability distributions** $\Pr_e(-\infty, x)$ and $\Pr_t(-\infty, x)$.

On the other hand, if the events x_j of the empirical distribution $\Pr_e(-\infty, x)$ come from a probability distribution that is different from $\Pr_t(-\infty, x)$, then as $N \to \infty$ we should expect that $\Pr_e^{(N)}(-\infty, x) \to \Pr_e(-\infty, x)$, and so that D_N should converge to a positive constant $D_\infty \in (0, 1]$. In this case, we expect that as $N \to \infty$ the quantity $\sqrt{N}\, D_N$ should grow with N as $\sqrt{N}\, D_\infty$.

Example 15.29 (Kolmogorov's test) How do we use (15.352)? As illustrated in Fig. 15.9, Kolmogorov's distribution $K(u)$ rises from zero to unity on $(0, \infty)$, reaching 0.9993 already at $u = 2$. So if our points x_j come from the theoretical distribution, then Kolmogorov's theorem (15.352) tells us that as $N \to \infty$, the probability that $\sqrt{N}\, D_N$ is less than 2 is more than 99.9%. But if the experimental points x_j do not come from the theoretical distribution, then the quantity $\sqrt{N}\, D_N$ should grow as $\sqrt{N}\, D_\infty$ as $N \to \infty$.

To see what this means in practice, I took as the theoretical distribution $P_t(x) = P_G(x, 0, 1)$ which has the cumulative probability distribution (15.94)

$$\Pr_t(-\infty, x) = \frac{1}{2}\left[\operatorname{erf}\left(x/\sqrt{2}\right) + 1\right]. \tag{15.353}$$

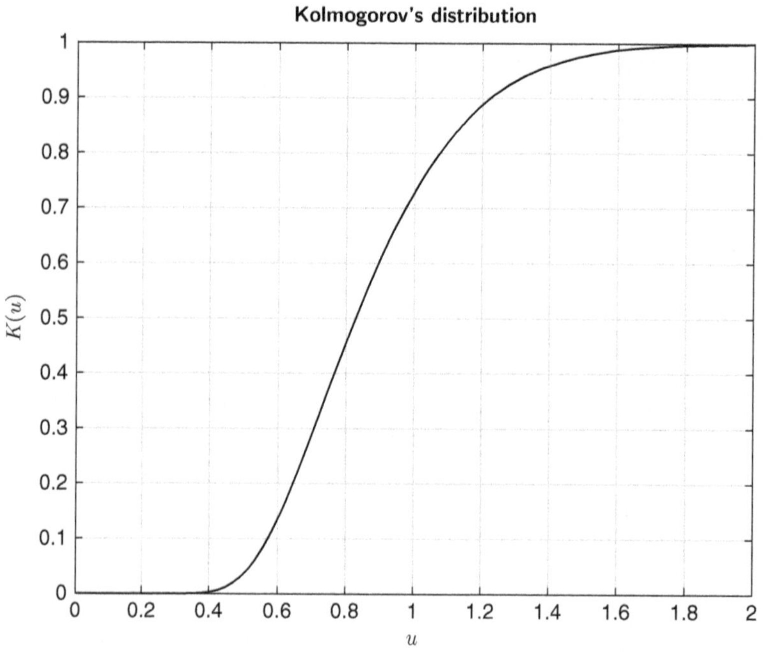

Figure 15.9 Kolmogorov's cumulative probability distribution $K(u)$ defined by (15.352) rises from zero to unity as u runs from zero to about two.

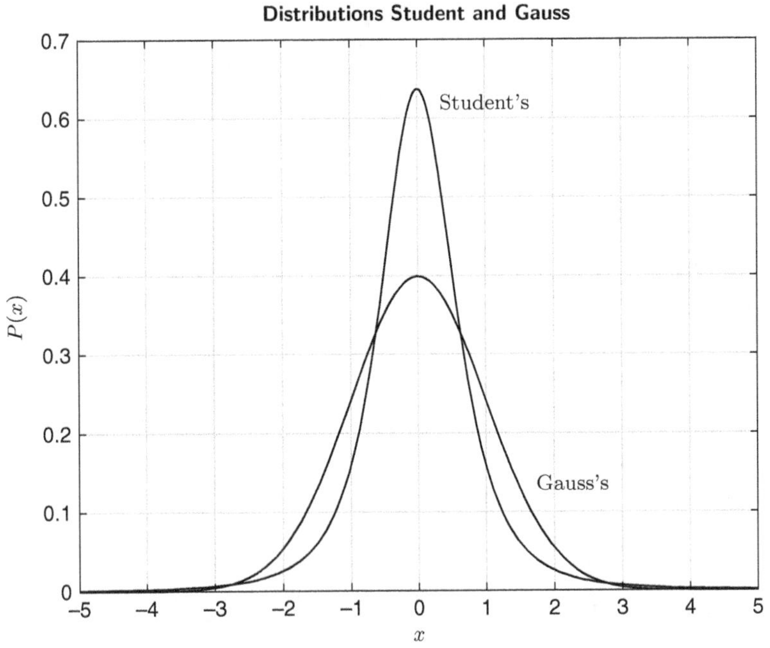

Figure 15.10 The probability distributions of Gauss $P_G(x, 0, 1)$ and Gosset/Student $P_S(x, 3, 1)$ with zero mean and unit variance.

I generated $N = 10^m$ points x_j for $m = 1, 2, 3, 4, 5$, and 6 from this theoretical distribution $P_t(x) = P_G(x, 0, 1)$ and computed $u_N = \sqrt{10^m}\, D_{10^m}$ for these points. I found $\sqrt{10^m}\, D_{10^m} = 0.6928, 0.7074, 1.2000, 0.7356, 1.2260$, and 1.0683. All were less than 2, as expected since I had taken the experimental points x_j from the theoretical distribution.

To see what happens when the experimental points do not come from the theoretical distribution $P_t(x) = P_G(x, 0, 1)$, I generated $N = 10^m$ points x_j for $m = 1, 2, 3, 4, 5$, and 6 from Gosset's Student's distribution $P_S(x, 3, 1)$ defined by (15.227) with $\nu = 3$ and $a = 1$. Both $P_t(x) = P_G(x, 0, 1)$ and $P_S(x, 3, 1)$ have the same mean $\mu = 0$ and standard deviation $\sigma = 1$, as illustrated in Fig. 15.10. For these points, I computed $u_N = \sqrt{N}\, D_N$ and found $\sqrt{10^m}\, D_{10^m} = 0.7741, 1.4522, 3.3837, 9.0478, 27.6414$, and 87.8147. Only the first two are less than 2, and the last four grow as $\sqrt{N}$, indicating that the x_j had not come from the theoretical distribution. In fact, we can approximate the limiting value of D_N as $D_\infty \approx u_{10^6}/\sqrt{10^6} = 0.0878$. The exact value is (Exercise 15.39) $D_\infty = 0.0868552356$.

At the risk of overemphasizing this example, I carried it one step further. I generated $\ell = 1, 2, \ldots 100$ sets of $N = 10^m$ points $x_j^{(\ell)}$ for $m = 2, 3$, and 4 drawn from $P_G(x, 0, 1)$ and from $P_S(x, 3, 1)$ and used them to form 100 empirical cumulative probabilities $\mathrm{Pr}_{e,G}^{(\ell,10^m)}(-\infty, x)$ and $\mathrm{Pr}_{e,S}^{(\ell,10^m)}(-\infty, x)$ as defined by (15.347–15.349).

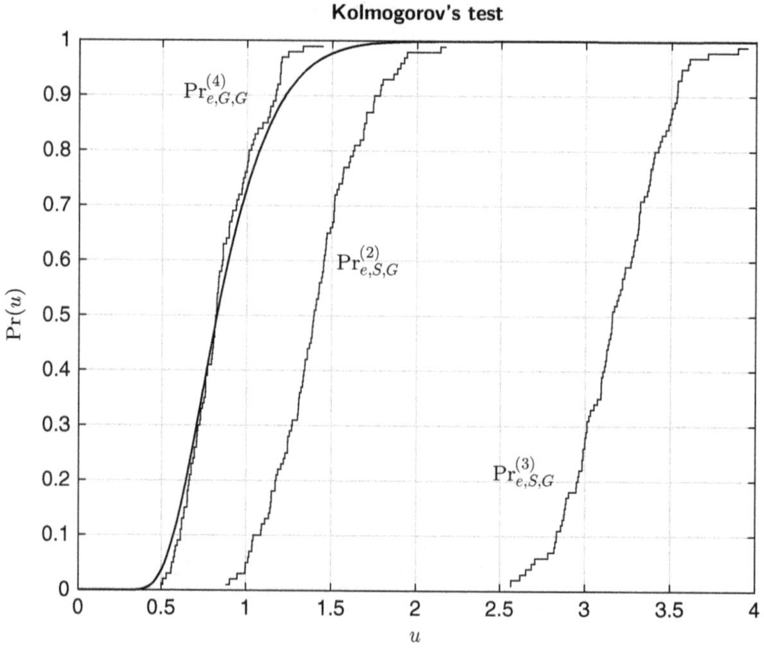

Figure 15.11 Kolmogorov's test is applied to points x_j taken from Gauss's distribution $P_G(x, 0, 1)$ and from Gosset's Student's distribution $P_S(x, 3, 1)$ to see whether the x_j came from $P_G(x, 0, 1)$. The thick smooth curve is Kolmogorov's universal cumulative probability distribution $K(u)$ defined by (15.352). The thin jagged curve that clings to $K(u)$ is the cumulative probability distribution $\Pr^{(4)}_{e,G,G}(-\infty, u)$ made (15.354) from points taken from $P_G(x, 0, 1)$. The other curves $\Pr^{(m)}_{e,S,G}(-\infty, u)$ for $m = 2$ and 3 are made (15.355) from 10^m points taken from $P_S(x, 3, 1)$.

Next, I computed the distances $D^{(\ell)}_{G,G,10^m}$ and $D^{(\ell)}_{S,G,10^m}$ of each of these cumulative probabilities from the gaussian distribution $P_G(x, 0, 1)$. I labeled the two sets of 100 quantities $u^{(\ell,m)}_{G,G} = \sqrt{10^m}\, D^{(\ell)}_{G,G,10^m}$ and $u^{(\ell,m)}_{S,G} = \sqrt{10^m}\, D^{(\ell)}_{S,G,10^m}$ in increasing order as $u^{(m)}_{G,G,1} \le u^{(m)}_{G,G,2} \le \cdots \le u^{(m)}_{G,G,100}$ and $u^{(m)}_{S,G,1} \le u^{(m)}_{S,G,2} \le \cdots \le u^{(m)}_{S,G,100}$. I then used (15.347–15.349) to form the cumulative probabilities

$$\Pr^{(m)}_{e,G,G}(-\infty, u) = \frac{j}{N_s} \quad \text{for} \quad u^{(m)}_{G,G,j} < u < u^{(m)}_{G,G,j+1} \tag{15.354}$$

and

$$\Pr^{(m)}_{e,S,G}(-\infty, u) = \frac{j}{N_s} \quad \text{for} \quad u^{(m)}_{S,G,j} < u < u^{(m)}_{S,G,j+1} \tag{15.355}$$

for $N_s = 100$ sets of 10^m points.

I plotted these cumulative probabilities in Fig. 15.11. The thick smooth curve is Kolmogorov's universal cumulative probability distribution $K(u)$ defined by (15.352). The thin jagged curve that clings to $K(u)$ is the cumulative probability distribution $\text{Pr}^{(4)}_{e,G,G}(-\infty, u)$ made from 100 sets of 10^4 points taken from $P_G(x, 0, 1)$. As the number of sets increases beyond 100 and the number of points 10^m rises further, the probability distributions $\text{Pr}^{(m)}_{e,G,G}(-\infty, u)$ converge to the universal cumulative probability distribution $K(u)$ and provide a numerical verification of Kolmogorov's theorem. Such curves make poor figures, however, because they hide beneath $K(u)$. The curves labeled $\text{Pr}^{(m)}_{e,S,G}(-\infty, u)$ for $m = 2$ and 3 are made from 100 sets of $N = 10^m$ points taken from $P_S(x, 3, 1)$ and tested as to whether they instead come from $P_G(x, 0, 1)$. Note that as $N = 10^m$ increases from 100 to 1000, the cumulative probability distribution $\text{Pr}^{(m)}_{e,S,G}(-\infty, u)$ moves farther from Kolmogorov's universal cumulative probability distribution $K(u)$. In fact, the curve $\text{Pr}^{(4)}_{e,S,G}(-\infty, u)$ made from 100 sets of 10^4 points lies beyond $u > 8$, too far to the right to fit in the figure. Kolmogorov's test gets more conclusive as the number of points $N \to \infty$.

Warning, mathematical hazard: While binned data are ideal for chi-squared fits, they ruin Kolmogorov tests. The reason is that if the data are in bins of width w, then the empirical cumulative probability distribution $\text{Pr}^{(N)}_e(-\infty, x)$ is a staircase function with steps as wide as the bin-width w even in the limit $N \to \infty$. Thus **even if the data come from the theoretical distribution**, the limiting value D_∞ of the Kolmogorov distance will be positive. In fact, one may show (exercise 15.40) that when the data do come from the theoretical probability distribution $P_t(x)$ assumed to be continuous, then the value of D_∞ is

$$D_\infty \approx \sup_{-\infty<x<\infty} \frac{w\, P_t(x)}{2}. \tag{15.356}$$

Thus in this case, the quantity $\sqrt{N}\, D_N$ would diverge as $\sqrt{N}\, D_\infty$ and lead one to believe that the data had not come from $P_t(x)$.

Suppose we have made some changes in our experimental apparatus and our software, and we want to see whether the new data $x'_1, x'_2, \ldots, x'_{N'}$ we took after the changes are consistent with the old data $x_1, x_2, \ldots, x_N$ we took before the changes. Then following equations (15.347–15.349), we can make two empirical cumulative probability distributions – one $\text{Pr}^{(N)}_e(-\infty, x)$ made from the N old points x_j and the other $\text{Pr}^{(N')}_e(-\infty, x)$ made from the N' new points x'_j. Next, we compute the distances

$$\begin{aligned} D^+_{N,N'} &= \sup_{-\infty<x<\infty} \left(\text{Pr}^{(N)}_e(-\infty, x) - \text{Pr}^{(N')}_e(-\infty, x)\right) \\ D_{N,N'} &= \sup_{-\infty<x<\infty} \left|\text{Pr}^{(N)}_e(-\infty, x) - \text{Pr}^{(N')}_e(-\infty, x)\right|. \end{aligned} \tag{15.357}$$

Smirnov (Smirnov 1939; Gnedenko 1968, p. 453) has shown that as $N, N' \to \infty$ the probabilities that

$$u^+_{N,N'} = \sqrt{\frac{NN'}{N+N'}}\, D^+_{N,N'} \quad \text{and} \quad u_{N,N'} = \sqrt{\frac{NN'}{N+N'}}\, D_{N,N'} \tag{15.358}$$

are less than u are

$$\begin{aligned} \lim_{N,N'\to\infty} \Pr(u^+_{N,N'} < u) &= 1 - e^{-2u^2} \\ \lim_{N,N'\to\infty} \Pr(u_{N,N'} < u) &= K(u) \end{aligned} \tag{15.359}$$

in which $K(u)$ is Kolmogorov's distribution (15.352).

Further Reading

Students can learn about quantum probability and statistics in the book *Quantum Detection and Estimation Theory* (Helstrom, 1976). They can learn more about classical probability and statistics in these books: *Mathematical Methods for Physics and Engineering* (Riley et al., 2006), *An Introduction to Probability Theory and Its Applications I, II* (Feller, 1968, 1966), *Theory of Financial Risk and Derivative Pricing* (Bouchaud and Potters, 2003), and *Probability and Statistics in Experimental Physics* (Roe, 2001).

Exercises

15.1 Find the probabilities that two thrown fair dice give 4, 5, or 6.

15.2 Redo the three-door example for the case in which there are 100 doors, and 98 are opened to reveal empty rooms after one picks a door. Should one switch? What are the odds?

15.3 Show that the zeroth moment μ_0 and the zeroth central moment ν_0 always are unity, and that the first central moment ν_1 always vanishes.

15.4 Compute the variance of the uniform distribution on (0, 1).

15.5 In the formulas (15.25 and 15.32) for the variances of discrete and continuous distributions, show that $E[(x - \langle x\rangle)^2] = \mu_2 - \mu^2$.

15.6 (a) Show that the covariance $\langle (x - \overline{x})(y - \overline{y})\rangle$ is equal to $\langle x\, y\rangle - \langle x\rangle\langle y\rangle$ as asserted in (15.43). (b) Derive (15.47) for the variance $V[ax + by]$.

15.7 Derive expression (15.48) for the variance of a sum of N variables.

15.8 Find the range of $pq = p(1-p)$ for $0 \le p \le 1$.

15.9 Show that the variance of the binomial distribution (15.51) is given by (15.55).

15.10 Redo the polling example (15.18–15.20) for the case of a slightly better poll in which 16 likely voters were asked and 13 said they'd vote for Nancy Pelosi. What's the probability that she'll win the election? (You may use Maple or some other program to do the tedious integral.)

15.11 Without using the fact that the Poisson distribution is a limiting form of the binomial distribution, show from its definition (15.66) and its mean (15.68) that its variance is equal to its mean, as in (15.70).

15.12 Show that Gauss's approximation (15.83) to the binomial distribution is a normalized probability distribution with mean $\langle x \rangle = \mu = pN$ and variance $V[x] = pqN$.

15.13 Derive the approximations (15.97 and 15.98) for binomial probabilities for large N.

15.14 Compute the central moments (15.31) of the gaussian (15.84).

15.15 Derive formula (15.93) for the probability that a gaussian random variable falls within an interval.

15.16 Show that the expression (15.100) for $P(y|600)$ is negligible on the interval (0, 1) except for y near 3/5.

15.17 Determine the constant A of the homogeneous solution $\langle \boldsymbol{v}(t) \rangle_{gh}$ and derive expression (15.167) for the general solution $\langle \boldsymbol{v}(t) \rangle$ to (15.165).

15.18 Derive Equation (15.168) for the variance of the position $\boldsymbol{r}$ about its mean $\langle \boldsymbol{r}(t) \rangle$. You may assume that $\langle \boldsymbol{r}(0) \rangle = \langle \boldsymbol{v}(0) \rangle = 0$ and that $\langle (\boldsymbol{v} - \langle \boldsymbol{v}(t) \rangle)^2 \rangle = 3kT/m$.

15.19 Derive Equation (15.197) for the ensemble average $\langle \boldsymbol{r}^2(t) \rangle$ for the case in which $\langle \boldsymbol{r}^2(0) \rangle = 0$ and $d\langle \boldsymbol{r}^2(0) \rangle / dt = 0$.

15.20 Use (15.220) to derive the lower moments (15.222) of the binomial distribution and those of Gauss and Poisson.

15.21 Find the third and fourth moments μ_3 and μ_4 for the distributions of Poisson (15.215) and Gauss (15.210).

15.22 Derive formula (15.226) for the first five cumulants of an arbitrary probability distribution.

15.23 Show that like the characteristic function, the moment-generating function $M(t)$ for an average of several independent random variables factorizes $M(t) = M_1(t/N)\, M_2(t/N) \cdots M_N(t/N)$.

15.24 Derive formula (15.233) for the moments of the log-normal probability distribution (15.232).

15.25 Why doesn't the log-normal probability distribution (15.232) have a sensible power series about $x = 0$? What are its derivatives there?

15.26 Compute the mean and variance of the exponential distribution (15.234).

15.27 Show that the chi-square distribution $P_{3,G}(v, \sigma)$ with variance $\sigma^2 = kT/m$ is the Maxwell–Boltzmann distribution (15.118).

15.28 Compute the inverse Fourier transform (15.209) of the characteristic function (15.239) of the symmetric Lévy distribution for $\nu = 1$ and 2.

15.29 Show that the integral that defines $P^{(2)}(y)$ gives formula (15.279) with two Heaviside functions. Hint: keep x_1 and x_2 in the interval (0, 1).

15.30 Derive the normal distribution (15.261) in the variable (15.260) from the central limit theorem (15.258) for the case in which all the means and variances are the same.

15.31 Show that Fisher's matrix (15.297) is symmetric $F_{k\ell} = F_{\ell k}$ and nonnegative (1.39), and that when it is positive (1.40), it has an inverse.

15.32 Derive the integral equations (15.299 and 15.301) from the normalization condition $\int P(\boldsymbol{x}; \boldsymbol{\theta})\, d^N x = 1$.

15.33 Show that the variance $V[u^{(N)}_{\sigma^2}]$ of Bessel's estimator (15.294) is given by (15.318).

15.34 Compute the fourth central moment (15.31) of Gauss's probability distribution $P_G(x; \mu, \sigma^2)$.

15.35 Show that when the real $N \times M$ matrix G has rank M, the matrices $P = G\,G^+$ and $P_\perp = 1 - P$ are projection operators that are mutually orthogonal $P(I - P) = (I - P)P = 0$.

15.36 Show that Kolmogorov's distance D_N is bounded, $0 \le D_N \le 1$.

15.37 Show that Kolmogorov's distance D_N is the greater of the two Smirnov distances

$$\begin{aligned} D_N^+ &= \sup_{-\infty<x<\infty} \left(\Pr_e^{(N)}(-\infty, x) - \Pr_t(-\infty, x) \right) \\ D_N^- &= \sup_{-\infty<x<\infty} \left(\Pr_t(-\infty, x) - \Pr_e^{(N)}(-\infty, x) \right). \end{aligned} \tag{15.360}$$

15.38 Derive the formulas

$$\begin{aligned} D_N^+ &= \sup_{1\le j\le N} \left(\frac{j}{N} - \Pr_t(-\infty, x_j) \right) \\ D_N^- &= \sup_{1\le j\le N} \left(\Pr_t(-\infty, x_j) - \frac{j-1}{N} \right) \end{aligned} \tag{15.361}$$

for D_N^+ and D_N^-.

15.39 Compute the exact limiting value D_∞ of the Kolmogorov distance between $P_G(x, 0, 1)$ and $P_S(x, 3, 1)$. Use the cumulative probabilities (15.353 and 15.230) to find the value of x that maximizes their difference. Using Maple or some other program, you should find $x = 0.6276952185$ and then $D_\infty = 0.0868552356$.

15.40 Show that when the data do come from the theoretical probability distribution (assumed to be continuous) but are in bins of width w, then the limiting value D_∞ of the Kolmogorov distance is given by (15.356).

15.41 Suppose in a poll of 1000 likely voters, 510 have said they would vote for Nancy Pelosi. Redo Example 15.16.

16

Monte Carlo Methods

16.1 The Monte Carlo Method

The Monte Carlo method is simple, robust, and useful. It has many applications. It is used, for instance, in numerical integration, data analysis, statistical mechanics, lattice gauge theory, chemical physics, biophysics, and finance.

16.2 Numerical Integration

Suppose one wants to numerically integrate a function $f(x)$ of a vector $x = (x_1, \ldots, x_n)$ over a region $\mathcal{R}$. One generates a large number N of pseudorandom values for the n coordinates x within a hyperrectangle of length L that contains the region $\mathcal{R}$, keeps the $N_{\mathcal{R}}$ points $x_k = (x_{1k}, \ldots, x_{nk})$ that fall within the region $\mathcal{R}$, computes the average $\langle f(x_k) \rangle$, and multiplies by the hypervolume $V_{\mathcal{R}}$ of the region

$$\int_{\mathcal{R}} f(x)\, d^n x \approx \frac{V_{\mathcal{R}}}{N_{\mathcal{R}}} \sum_{k=1}^{N_{\mathcal{R}}} f(x_k). \tag{16.1}$$

If the hypervolume $V_{\mathcal{R}}$ is hard to compute, you can have the Monte Carlo code compute it for you. The hypervolume $V_{\mathcal{R}}$ is the volume L^n of the enclosing hypercube multiplied by the number $N_{\mathcal{R}}$ of times the N points fall within the region $\mathcal{R}$

$$V_{\mathcal{R}} = \frac{N_{\mathcal{R}}}{N} L^n. \tag{16.2}$$

The integral formula (16.1) then becomes

$$\int_{\mathcal{R}} f(x)\, d^n x \approx \frac{L^n}{N} \sum_{k=1}^{N_{\mathcal{R}}} f(x_k). \tag{16.3}$$

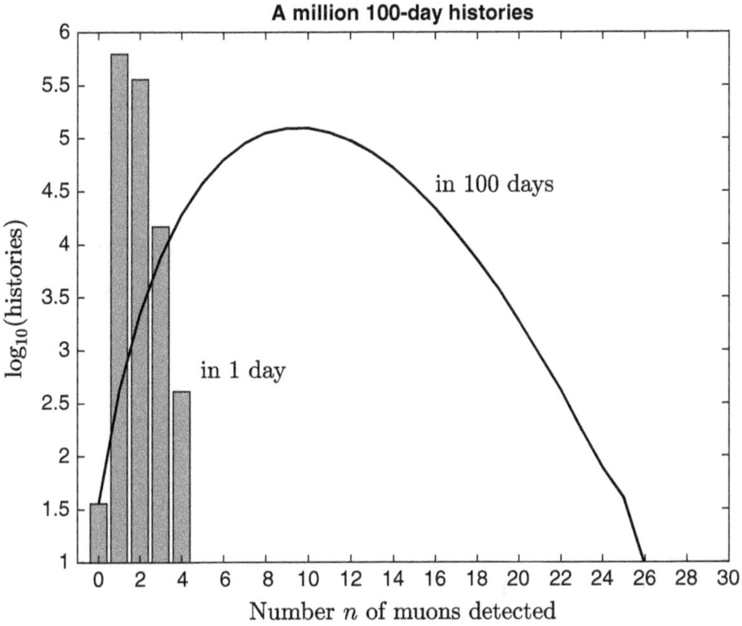

Figure 16.1 The number of histories of 100 days (out of 1,000,000 histories) in which a maximum of n muons is detected on a single day (boxes) and in 100 days (curve). A Matlab script for this figure is in Monte_Carlo_methods at `github.com/kevinecahill`.

For instance, if the system is a protein, the vector x might be the $3N$ spatial coordinates of the N atoms of the protein. A protein composed of 200 amino acids has about 4000 atoms, and so the vector x would have some 12,000 components. Suppose $E(x)$ is the energy of configuration x of the protein in its cellular environment of salty water crowded with macromolecules. How do we generate a sequence of "native states" of the protein at temperature T?

We start with some random or artificial initial configuration x^0 and then make random changes δx in successive configurations x. One way to do this is to make a small, random change δx_i in coordinate x_i and then to test whether to accept this change by comparing the energies $E(x)$ and $E(x')$ of the two configurations x and x', which differ by δx_i in coordinate x_i. (Estimating these energies is not trivial; Gromacs and TINKER can help.)

It is important that the random changes be symmetric, that is, the probability of choosing to test whether to go from x to x' when one is at x should be equal to the probability of choosing to test whether to go from x' to x when one is at x'. A simple way to ensure this symmetry is to define x_i' in terms of x_i, a suitable step size δx, and a random number r (uniformly distributed between 0 and 1) as

$$x_i' = x_i + \left(r - \frac{1}{2}\right) \delta x. \tag{16.7}$$

Also, the sequences of configurations should be **ergodic**; that is, from any configuration x, one should be able to get to any other configuration x' by a sequence of changes $\delta x_i = x_i' - x_i$.

How do we decide whether to accept or reject δx_i? We use the following **Metropolis step**: If the energy $E' = E(x')$ of the new configuration x' is less than the energy $E(x)$ of the current configuration x, then we accept the new configuration x'. But if $E' > E$, then we accept x' with probability

$$P(x \to x') = e^{-(E'-E)/kT} \tag{16.8}$$

by generating a random number $r \in [0, 1]$ and accepting x' if

$$r < e^{-(E'-E)/kT}. \tag{16.9}$$

If one does not accept x', then the system remains in configuration x.

In FORTRAN90, the Metropolis step might be

```
if ( newE <= oldE ) then ! accept
   x(i) = x(i) + dx
else ! accept conditionally
   call random_number(r)
   if ( r <= exp(- (newE - oldE)/(k*T)) ) then ! accept
      x(i) = x(i) + dx
   end if
end if
```

The next step is to vary another coordinate, such as x_{i+1}. Once one has varied all of the coordinates, one has finished a **sweep** through the system. After thousands or millions of such sweeps, the protein is said to be **thermalized**. Once the protein has thermalized, one can start measuring its properties, such as its shape. One computes a physical quantity every hundred or thousand sweeps and takes the average of these measurements. That average is the mean value of the physical quantity at temperature T.

Why does this work? Consider two configurations x and x' that respectively have energies $E = E(x)$ and $E' = E(x')$ and are occupied with probabilities $P_t(x)$ and $P_t(x')$ as the system is thermalizing. If $E > E'$, then the rate $R(x \to x')$ of going from x to x' is the rate v of choosing to test x' when one is at x times the probability $P_t(x)$ of being at x, that is, $R(x \to x') = v\, P_t(x)$. The reverse rate $R(x' \to x)$ is $R(x' \to x) = v\, P_t(x')\, e^{-(E-E')/kT}$ with the same v since the random walk is symmetric. The net rate from $x \to x'$ then is

$$R(x \to x') - R(x' \to x) = v \left(P_t(x) - P_t(x')\, e^{-(E-E')/kT} \right). \tag{16.10}$$

This net flow of probability from $x' \to x$ is positive if and only if

$$P_t(x)/P_t(x') > e^{-(E-E')/kT}. \tag{16.11}$$

The probability distribution $P_t(x)$ therefore flows with each sweep toward the Boltzmann distribution $\exp(-E(x)/kT)$. The flow slows and stops when the two rates are equal $R(x' \to x) = R(x \to x')$ a condition called **detailed balance**. At this equilibrium, the distribution $P_t(x)$ satisfies

$$P_t(x) = P_t(x')\, e^{-(E-E')/kT} \tag{16.12}$$

in which $P_t(x')\, e^{E'/kT}$ is independent of x. So the thermalizing distribution $P_t(x)$ approaches the distribution $P(x) = c\, e^{-E/kT}$ in which c is independent of x. Since the sum of these probabilities must be unity, we have

$$\sum_x P(x) = c \sum_x e^{-E(x)/kT} = 1 \tag{16.13}$$

which means that the constant c is the inverse of the **partition function**

$$Z(T) = \sum_x e^{-E(x)/kT}. \tag{16.14}$$

The thermalizing distribution approaches Boltzmann's distribution (1.392)

$$P_t(x) \to P_B(x) = e^{-E(x)/kT}/Z(T). \tag{16.15}$$

Example 16.2 (Z_2 lattice gauge theory) To simulate Z_2 gauge theory on a lattice, one represents spacetime as a lattice of points in d dimensions. Two nearest-neighbor points are separated by the lattice spacing a and joined by a link. One puts an element $U = \pm 1$ of the group Z_2 on each link. One then assigns an action $S_\Box$ to each elementary square or *plaquette* of the lattice. For the Z_2 gauge group (Example 11.6), the action $S_\Box$ of a square with vertices 1, 2, 3, and 4 is

$$S_\Box = 1 - U_{1,2}\, U_{2,3}\, U_{3,4}\, U_{4,1} \tag{16.16}$$

where each $U = \pm 1$. Then, one replaces $E(x)/kT$ with βS in which the action S is a sum of all the plaquette actions S_p.

You can study Z_2 lattice gauge theory by using the program puregauge.cc available at Michael Creutz's website (`latticeguy.net/lattice.html`). By running it on a 6^4 lattice from low temperature $\beta = 1$ to high temperature $\beta = 0$ and back again, you can exhibit hysteresis as in Fig. 16.2.

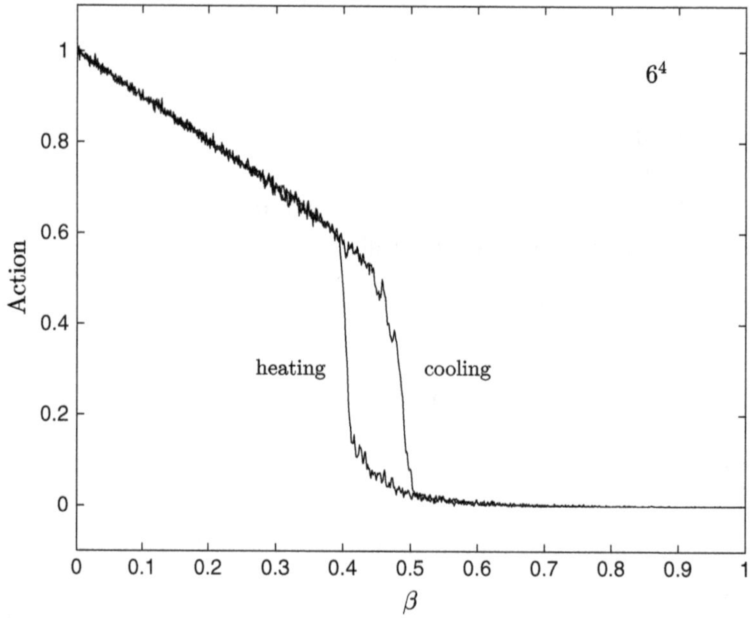

Figure 16.2 The hysteresis loop is a sign of a first-order phase transition.

Example 16.3 ($SU(3)$ lattice gauge theory) For each elementary square of the lattice, the plaquette variable U_p is the product of elements U of the gauge group $SU(3)$ around the square, $U_p = U_{1,2}\, U_{2,3}\, U_{3,4}\, U_{4,1}$. The euclidian action of the theory is then the sum over all the plaquettes of the lattice of the traces

$$S = \beta \sum_p \left[1 - \frac{1}{6} \text{Tr} \left(U_p + U_p^\dagger \right) \right] \tag{16.17}$$

in which $\beta = 6/g^2$ is inversely proportional to the square of the coupling constant g.

Although the generation of configurations distributed according to the Boltzmann probability distribution (1.392) is one of its most useful applications, the Monte Carlo method is much more general. It can generate configurations x distributed according to any probability distribution $P(x)$.

To generate configurations distributed according to $P(x)$, we accept any new configuration x' if $P(x') > P(x)$ and also accept x' with probability

$$P(x \to x') = P(x')/P(x) \tag{16.18}$$

if $P(x) > P(x')$.

This works for the same reason that the Boltzmann version works. Consider two configurations x and x'. After the system has thermalized, the probabilities $P_t(x)$ and $P_t(x')$ have reached equilibrium, and so the rate $R(x \to x')$ from $x \to x'$ must equal the rate $R(x' \to x)$ from $x' \to x$. If $P(x') > P(x)$, then $R(x \to x')$ is

$$R(x \to x') = v\, P_t(x) \tag{16.19}$$

in which v is the rate of choosing $\delta x = x' - x$, while the rate $R(x' \to x)$ is

$$R(x' \to x) = v\, P_t(x')\, P(x)/P(x') \tag{16.20}$$

with the same v since the random walk is symmetric. Equating the two rates $R(x \to x') = R(x' \to x)$, we find that after thermalization

$$P_t(x) = P(x)\, P_t(x')/P(x') = c\, P(x) \tag{16.21}$$

in which c is independent of x. Thus $P_t(x)$ converges to $P(x)$ at equilibrium.

So far we have assumed that the rate of choosing $x \to x'$ is the same as the rate of choosing $x' \to x$. In **Smart Monte Carlo** schemes, physicists arrange the rates $v_{x \to x'}$ and $v_{x' \to x}$ so as to steer the flow and speed-up thermalization. To compensate for this asymmetry, they change the second part of the Metropolis step from $x \to x'$ when $P(x) > P(x')$ to accept conditionally with probability

$$P(x \to x') = P(x')\, v_{x' \to x} / \left[P(x)\, v_{x \to x'} \right]. \tag{16.22}$$

Now if $P(x) > P(x')$, then $R(x' \to x)$ is

$$R(x' \to x) = v_{x' \to x}\, P_t(x') \tag{16.23}$$

while the rate $R(x \to x')$ is

$$R(x \to x') = v_{x \to x'}\, P_t(x)\, P(x')\, v_{x' \to x} / \left[P(x)\, v_{x \to x'} \right]. \tag{16.24}$$

Equating the two rates $R(x' \to x) = R(x \to x')$, we find

$$P_t(x) = P(x)\, P_t(x')/P(x') \tag{16.25}$$

which implies that $P_t(x)$ converges to

$$P_t(x) = P(x)/Z \tag{16.26}$$

in which $Z = \int P(x)\, dx$.

Example 16.4 (Highly multiple integration) You can use the general Metropolis method (16.18–16.21) to integrate a function $f(x)$ of many variables $x = (x_1, \ldots, x_n)$ if you can find a positive function $g(x)$ similar to $f(x)$ whose integral $I[g]$ you know. You just use the probability distribution $P(x) = g(x)/I[g]$ to find the mean value of $I[g] f(x)/g(x)$:

$$\int f(x)\, d^n x = I[g] \int \frac{f(x)}{g(x)} \frac{g(x)}{I[g]}\, d^n x = I[g] \int \frac{f(x)}{g(x)}\, P(x)\, d^n x. \tag{16.27}$$

16.6 Simulated Annealing

One can use the Monte Carlo method to find the absolute minimum (or maximum) of a function $E(x)$ of many variables $x = (x_1, x_2, \ldots, x_N)$. To avoid being trapped in a local minimum, one starts with a sequence of Metropolis steps (16.8–16.9) at a high value of kT and then gradually lowers the value of kT to zero.

16.7 Solving Arbitrary Problems

If you know how to generate a suitably large space of trial solutions to a problem, and you also know how to compare the quality of any two of your solutions, then you can use a Monte Carlo method to solve the problem. The hard parts of this seemingly magical method are characterizing a big enough space of solutions s and constructing a quality function or functional that assigns a number $Q(s)$ to every solution in such a way that if s is a better solution than s', then

$$Q(s) > Q(s'). \tag{16.28}$$

But once one has characterized the space of possible solutions s and has constructed the quality function $Q(s)$, then one simply generates zillions of random solutions and selects the one that maximizes the function $Q(s)$ over the space of all solutions.

If one can characterize the solutions as vectors of a certain dimension, $s = (x_1, \ldots, x_n)$, then one may use the Monte Carlo method of the previous section (16.5) by setting $P(s) = Q(s)$.

16.8 Evolution

The reader may think that the use of Monte Carlo methods to solve arbitrary problems is quite a stretch. Yet nature has applied them to the problem of evolving species that survive. As a measure of the quality $Q(s)$ of a given solution s, nature used the time derivative of the logarithm of its population $\dot{P}(t)/P(t)$. The space of solutions is the set of possible genomes. Leaving aside DNA methylation, histone acetylation, and other epigenetic changes, we may idealize each solution or genome as a sequence of nucleotides $s = b_1 b_2 \ldots b_N$ some thousands or billions of bases long, each base b_k being adenine, cytosine, guanine, or thymine (A, C, G, or T). Since there are four choices for each base, the set of solutions is huge. The genome for *homo sapiens* has some 3 billion bases (or base pairs, DNA being double stranded), and so the solution space is a set with

$$\mathcal{N} = 4^{3\times 10^9} = 10^{1.8\times 10^9} \tag{16.29}$$

elements. By comparison, a googol is only 10^{100}.

In evolution, a Metropolis step begins with a random change in the sequence of bases; changes in a germ-line cell can change a new individual. Some of these changes are due to errors in the normal mechanisms by which genomes are copied and repaired; the holoenzyme DNA polymerase copies DNA with remarkable fidelity, but it makes one error per billion base pairs. Sexual reproduction makes bigger random changes in genomes. In meiosis, the paternal and maternal versions of each of our 23 chromosomes are duplicated, and the four versions swap segments of DNA in a process called genetic recombination or crossing-over. The cell then divides twice producing four haploid germ cells each with a single paternal, maternal, or mixed version of each chromosome. Two haploid cells, one from each parent, join to start a new individual. Sexual reproduction makes evolution more ergodic, which is why most complex modern organisms use it. The second part of the evolutionary Metropolis step is done by the newly born individual: if he or she survives and multiplies, the change is accepted; if he or she dies without progeny, it is rejected. In 4 billion years, evolution has turned simple molecules into human beings.

John Holland and others have used analogs of these Metropolis steps to write genetic algorithms that can solve wide classes of problems (Holland, 1975; Vose, 1999; Schmitt, 2001).

Further Reading

The classic *Quarks, Gluons, and Lattices* (Creutz, 1983) is a marvelous introduction to the subject; his website (latticeguy.net/lattice.html) is an extraordinary resource, as is Rubinstein's *Simulation and the Monte Carlo Method* (Rubinstein and Kroese, 2007). *Molecular Biology of the Cell* (Alberts et al., 2015) is one of the best textbooks ever written.

Exercises

16.1 Go to Michael Creutz's website `latticeguy.net/lattice.html` and get his C-code for Z_2 lattice gauge theory. Compile and run it, and make a graph like Fig. 16.2 which exhibits hysteresis.

16.2 Modify his code and produce a graph showing the coexistence of two phases at the critical coupling $\beta_t = 0.5 \ln(1+\sqrt{2})$. Hint: Do a cold start and then 100 updates at β_t, then do a random start and do 100 updates at β_t. Plot the values of the action against the update number 1, 2, 3, ..., 100.

16.3 Modify Creutz's C code for Z_2 lattice gauge theory so as to be able to vary the dimension d of spacetime. Show that for $d = 2$, there's no hysteresis loop (there's no phase transition). For $d = 3$, show that any hysteresis loop is minimal (there's a second-order phase transition).

16.4 What happens when $d = 5$?

16.5 Use Example 16.4 to compute the ten-dimensional integral

$$\mathcal{I} = \int \exp\left[-\left(x^2 + (x^2)^2\right)\right] d^{10}x \tag{16.30}$$

over $\mathbb{R}^{10}$ where $x^2 = x_1^2 + \cdots + x_{10}^2$.

17

Artificial Intelligence

17.1 Steps Toward Artificial Intelligence

Alan Turing proposed the Turing test in 1950. In 1960, Marvin Minsky wrote "Steps toward artificial intelligence" (Minsky, 1961). In 1961, James Slagle wrote a program that does symbolic integration. The years 1964–1966 saw Joseph Weizenbaum's Eliza: a program that simulates a psychiatrist (`tex-edit.com`), programs that can beat some intelligence tests, and programs that can analyze forms and learn what an arch is. In 1975, Edward Shortliffe, Bruce Buchanan, and Stanley Cohen developed Mycin, an expert system that chooses antibiotics for patients (Shortliffe and Buchanan, 1975). After 1975, came modern rule-based expert systems that perform many useful functions, such as parking airplanes at airports.

17.2 Slagle's Symbolic Automatic Integrator

Slagle's LISP program SAINT is a rule-based expert system. It first does problem reduction by applying the rules: (1) $\int c\, f(x)\, dx = c \int f(x)\, dx$; (2) $\int \sum_i f_i(x)\, dx = \sum_i \int f_i(x)\, dx$; and (3) if in the integral $\int (P(x)/Q(x))\, dx$ the degree of the polynomial $P(x)$ exceeds that of $Q(x)$, then divide. The program then looks in its small table of 26 integrals to see if it's done. If it's not done, it tries trig substitutions such as these:

$$\int f(\tan x)\, dx = \int \frac{f(y)}{1+y^2}\, dy, \quad \int f(1-x^2)\, dx = \int f(\cos^2 y)\, \cos y\, dy,$$

$$\text{and} \quad \int f(1+x^2)\, dx = \int f(\sec^2 y)\, (1+\tan^2 y)\, dy.$$

Example 17.1 (Slagle's SAINT does an integral) SAINT sets $x = \sin y$

$$\int \frac{x^4}{(1-x^2)^{5/2}}\, dx = \int \frac{\sin^4 y}{\cos^5 y} \cos y\, dy = \int \frac{\sin^4 y}{\cos^4 y}\, dy = \int \tan^4 y\, dy.$$

It then sets $z = \tan y$, applies rule 3, and divides:

$$\int \tan^4 y\, dy = \int \frac{z^4}{1+z^2}\, dz = \int \left(z^2 - 1 + \frac{1}{z^2+1}\right) dz.$$

Finally, using rule (2), it does each integral separately and gets

$$\int \frac{x^4}{(1-x^2)^{5/2}}\, dx = \frac{1}{3}\tan^3(\arcsin x) - \tan(\arcsin x) + \arcsin x.$$

17.3 Neural Networks

As of this writing, early 2018, most things that a human can do in one second can be done by artificial intelligence. The main technique is the training and use of neural networks. The more data one has, the better one can train a neural network. And the more data one has, the more neurons one needs to optimally use the training data. So the trend is toward huge data sets and large neural networks. A three-year-old child has 10^{15} synapses.

A single neuron takes input signals a_i from other neurons $i = 1, \ldots, n$, giving weight w_i to signal a_i. It fires if the sum of the n weighted inputs $w_i a_i$ exceeds its bias b. The signal the neuron sends out is then

$$a = \frac{1}{2}\left(w_1 a_1 + \cdots + w_n a_n - b + |w_1 a_1 + \cdots + w_n a_n - b|\right) \tag{17.1}$$

which often is written in terms of the function $\mathrm{ReLU}(x) = (x + |x|)/2$ as $a = \mathrm{ReLU}\,(w_1 a_1 + \cdots + w_n a_n - b)$.

The neurons of a typical network are organized into layers $\ell = 1, \ldots, m$ of neurons with n_ℓ neurons labeled by the index $j = 1, \ldots, n_\ell$. In this notation, the signal emitted by the jth neuron of layer ℓ is

$$a_j^\ell = \mathrm{ReLU}\left(w_{j1}^\ell a_1^{\ell-1} + \cdots + w_{jn_{\ell-1}}^\ell a_{n_{\ell-1}}^{\ell-1} - b_j^\ell\right). \tag{17.2}$$

If there are m layers of neurons, then the prediction of the network is the n_m nonnegative numbers a_j^m. If the task of the network is to classify vectors x into C categories, then the probability the network assigns to vector x's being in the ith category is

$$p(i|x) = \frac{a_i^m}{\sum_{k=1}^C a_k^m}. \tag{17.3}$$

For example, a vector x might represent the darkness of the pixels of an image of the number 2 handwritten on a white sheet of paper. A perfect network would give $p(i|x) = \delta_{i2}$.

If the neural network assigns probability $0 \le p(i|x) \le 1$ to image x's belonging to the ith category while the correct category is $\ell(x)$ then the squared error made by the neural network is

$$E^2(x) = \sum_{i=1}^{C} |p(i|x) - \delta_{i\ell(x)}|^2 \tag{17.4}$$

summed over the C categories. The squared error made by the neural network on N images $\{x\} = \{x_1, x_2, \ldots, x_N\}$ would be

$$E^2(\{x\}, \{v\}) = \frac{1}{N} \sum_{k=1}^{N} \sum_{i=1}^{C} |p(i|x_k) - \delta_{i\ell(x_k)}|^2. \tag{17.5}$$

One trains a neural network by adjusting its parameters w^ℓ_{jk} and b^ℓ_j so as to lower its error $E(\{x\})$. If the network has m layers of n neurons, then the w^ℓ_{jk} and the b^ℓ_j constitute $M = mn(n+1)$ adjustable parameters, or 3,003,000 if $n = 10^3$ and $m = 3$. We can number the parameters w^ℓ_{jk} and b^ℓ_j with a single index ℓ, setting $v_1 = w^1_{11}$, $v_2 = w^1_{12}, \ldots, v_M = b^m_n$. The error $E(\{x\}, \{v\})$ of the network depend upon these parameters.

One can use the Monte Carlo method (Section 16.6) of simulated annealing to find the parameters $\{v\}$ that minimize the error $E(\{x\}, \{v\})$. Another procedure is to successively compute the partial derivatives $\partial E^2/\partial v_i$ of the squared error $E^2(\{x\}, \{v\})$ with respect to the parameters $\{v_1, \ldots, v_M\}$ and to successively change these parameters by a suitably small negative multiple of these partial derivatives, setting $v'_i = v_i - \epsilon\, \partial E^2/\partial v_i$.

17.4 A Linear Unbiased Neural Network

If we simplify our neural network by replacing the function ReLU(x) by x and setting all the biases to zero, then the most elaborate neural network reduces to a linear map, $y = A\,x$, in which the real matrix A maps an unknown vector x into a category y.

Suppose X is a matrix that represents the training set of vectors, so that the ith element of its kth column is the ith element of the kth training vector $x^{(k)}$, that is, $X_{ik} = x^{(k)}_i$. Let T be the matrix of correct assignments of the training vectors X. That is, T_{ik} is the correct assignment of the training vector $x^{(k)}$. Ideally, we then should like to have $A\,X = T$. If the training matrix X were a square nonsingular matrix, we could set $A = T\,X^{-1}$, but if we have lots of training vectors, more than there are categories to which we seek to assign them, then X has more columns than rows. The rows of X typically are long and linearly independent. The matrix $X\,X^{\mathsf{T}}$ then has an inverse, and we may use the form (1.433) of the Moore–Penrose pseudomatrix $X^{+} = X^{\mathsf{T}}\,(X X^{\mathsf{T}})^{-1}$. Our best guess for the matrix A then is

$$A = T\,X^{+} = T\,X^{\mathsf{T}}\,(X X^{\mathsf{T}})^{-1} \tag{17.6}$$

as long as the matrix $X X^{\mathsf{T}}$ is nonsingular.

Example 17.2 (Reading handwritten numbers) The MNIST website `yann.lecun.com/exdb/mnist/` lists four (high endian) files that one can use to train and test a neural network. The gzipped file train-images-idx3-ubyte.gz contains 60,000 images $x(i)$ of handwritten numbers. Each image $x(i)$ is a real 28-by-28 matrix, which is equivalent to a real vector in a space of 784 dimensions. The file train-labels-idx1-ubyte.gz contains the 60,000 labels that the 60,000 handwritten numbers of the train-images file represent. The files t10k-images-idx3-ubyte.gz and t10k-labels-idx1-ubyte.gz are similar files of 10,000 different handwritten integers and their labels.

We seek a matrix A_{ik} with 10 rows, $i = 0, \dots, 9$, and 784 columns, $k = 1, \dots, 784$. The singular-value decomposition of this matrix should be like

$$A = \sum_{\ell=0}^{9} |\ell\rangle\langle\bar{\ell}| \tag{17.7}$$

in which the ith element of the vector $|\ell\rangle$ is $\delta_{i\ell}$, and the vector $|\bar{\ell}\rangle$ is the normalized sum of all the 60,000 training vectors $x(i,\ell)$ that represent the integer ℓ. We make it in two steps

$$|\tilde{\ell}\rangle = \sum_{i=1}^{60{,}000} |x(i,\ell)\rangle \qquad \text{and} \qquad |\bar{\ell}\rangle = \frac{|\tilde{\ell}\rangle}{\sqrt{\langle\tilde{\ell}|\tilde{\ell}\rangle}}. \tag{17.8}$$

This unbiased linear neural network correctly identifies 82.16% of the handwritten test images of the MNIST website. The Fortran programs for this neural network are in Artificial_intelligence at `github.com/kevinecahill`.

Further Reading

`Playground.tensorflow.org`; `Tensorflow.org`; `Deeplearningbook.org`; `Ocw.mit.edu/6-034F10`; `Neuralnetworksanddeeplearning.com`; `ocw.mit.edu/6-034F10`; *The Elements of Statistical Learning* by Trevor Hastie, Robert Tibshirani, and Jerome Friedman (Hastie et al., 2016); and the A.I. novel *Coding Lina* by Sean Cahill (Cahill, 2018).

18

Order, Chaos, and Fractals

18.1 Hamilton Systems

A **Hamilton system** of n **degrees of freedom** has n coordinates q_i and n momenta p_i whose time derivatives are the partial derivatives

$$\dot{q}_i = \frac{\partial H}{\partial p_i} \quad \text{and} \quad \dot{p}_i = -\frac{\partial H}{\partial q_i} \tag{18.1}$$

of a hamiltonian H, which is a function of the $2n$ q's and p's and possibly of the time t. The time derivative of any function $F(q, p)$ of the q's and p's is then

$$\begin{aligned}\frac{dF}{dt} &= \frac{\partial F}{\partial t} + \sum_{i=1}^{n} \frac{\partial F}{\partial q_i}\dot{q}_i + \frac{\partial F}{\partial p_i}\dot{p}_i = \frac{\partial F}{\partial t} + \sum_{i=1}^{n} \frac{\partial F}{\partial q_i}\frac{\partial H}{\partial p_i} - \frac{\partial F}{\partial p_i}\frac{\partial H}{\partial q_i} \\ &= \frac{\partial F}{\partial t} + [F, H]_{\text{Pb}}\end{aligned} \tag{18.2}$$

in which the last term is the **Poisson bracket** $[F, H]_{\text{Pb}}$. In quantum mechanics, F and H become operators and their Poisson bracket is replaced by their commutator divided by $i\hbar$

$$\sum_{i=1}^{n} \frac{\partial F}{\partial q_i}\frac{\partial H}{\partial p_i} - \frac{\partial F}{\partial p_i}\frac{\partial H}{\partial q_i} = [F, H]_{\text{Pb}} \to \frac{1}{i\hbar}[F, H] = \frac{FH - HF}{i\hbar}. \tag{18.3}$$

A Hamilton system with n symmetries has n conserved quantities C_i. If time translation invariance is one of the symmetries, and if the hamiltonian H and the conserved quantities are time independent, then the system is **autonomous**, the hamiltonian is one of the conserved quantites, and its Poisson bracket with each of the conserved quantities vanishes $[C_i, H] = 0$. If the conserved quantities have vanishing Poisson brackets $[C_i, C_j] = 0$, then they are **in involution**. An autonomous Hamilton system of degree n that has n independent conserved quantities that are in involution is Liouville **integrable**. In principle, one can integrate Hamilton's equations (18.1) for such a system.

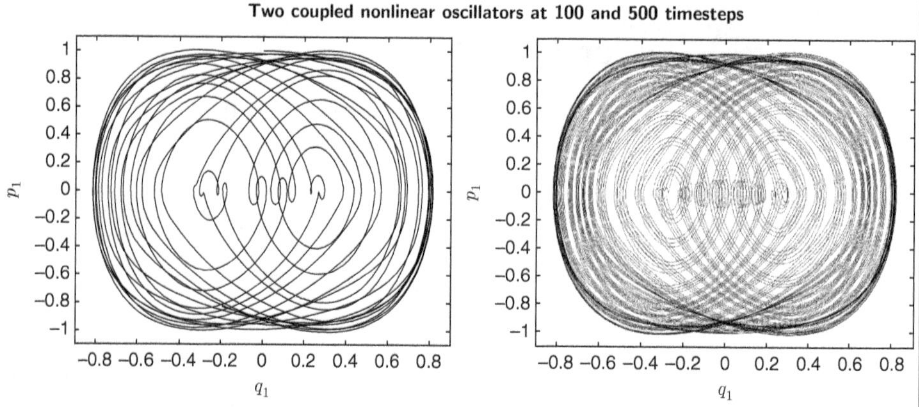

Figure 18.1 The $q_1(t)$, $p_1(t)$ coordinates of a harmonic oscillator coupled to another harmonic oscillator by the hamiltonian (18.5) for 100 (left) and 500 (right) timesteps. The initial conditions are $q_1 = q_2 = 0$ and $p_1 = p_2 = 1$, with $m = 1$, $\omega_2 = \omega_1/2 = 1/2$, and $\lambda = 1/2$. The Matlab scripts and Fortran programs that generate the figures of this chapter are in the repository Order_chaos_and_fractals at `github.com/kevinecahill`.

Example 18.1 (Harmonic oscillators) The coordinates q_i and momenta p_i of a Hamilton system of n independent harmonic oscillators with hamiltonian

$$H = \sum_{i=1}^{n} \frac{p_i^2}{2m_i} + \frac{m_i \omega_i^2 q_i^2}{2} \tag{18.4}$$

obey Hamilton's equations $\dot{q}_i = p_i/m_i$ and $\dot{p}_i = -m_i\omega_i^2 q_i$. This system is Liouville integrable. The energy $E_i = p_i^2/2m_i + m_i\omega_i q_i^2/2$ of each oscillator is conserved. The coordinates $q_i(t) = a_i \sin(\omega_i(t + \tau_i))$ and momenta $p_i = a_i m_i \omega_i \cos(\omega_i(t + \tau_i))$ run over the n-dimensional surface of an n-torus in a q, p phase space of $2n$ dimensions.

Few systems are integrable. Two coupled harmonic oscillators with

$$H = \frac{p_1^2 + p_2^2}{2m} + \frac{m(\omega_1^2 q_1^2 + \omega_2^2 q_2^2)}{2} + m\omega_1\omega_2(q_1 - q_2)^2 + \lambda\, q_1^2 q_2^2 \left(q_1^2 + q_2^2\right) \tag{18.5}$$

display complicated behavior as illustrated in Fig. 18.1 for equal masses but different frequencies. This system is not integrable.

Example 18.2 (The two-body problem) Two interacting particles moving in empty 3-dimensional space are a system of 6 degrees of freedom with 7 conserved quantities – the energy $E_1 + E_2$, the momentum $\vec{p}_1 + \vec{p}_2$, and the angular momentum $\vec{r}_1 \times \vec{p}_1 + \vec{r}_2 \times \vec{p}_2$ – which are in involution and independent. The system is integrable. Its motion is ordered, and if bounded, lies on the surface of a 6-torus.

Hamilton systems are special: The $2n$ time derivatives $\dot{q}_i$ and $\dot{p}_i$ satisfy (Exercise 18.1) the identities

$$\frac{\partial \dot{q}_i}{\partial q_j} = -\frac{\partial \dot{p}_j}{\partial p_i}, \quad \frac{\partial \dot{q}_i}{\partial p_j} = \frac{\partial \dot{q}_j}{\partial p_i}, \quad \text{and} \quad \frac{\partial \dot{p}_i}{\partial q_j} = \frac{\partial \dot{p}_j}{\partial q_i}. \tag{18.6}$$

The integral $A = \oint p_1\, dq_1 + \cdots + p_n dq_n$ over a closed trajectory in phase space is a time-independent Poincaré invariant (Example 14.9). Areas of phase space are constant in time (Example 14.11) along Hamilton trajectories (18.1)

$$\frac{d}{dt}\left(\sum_{i=1}^{n} \begin{vmatrix} \delta p_i & \delta q_i \\ \Delta p_i & \Delta q_i \end{vmatrix}\right) = 0. \tag{18.7}$$

Although they are special, few Hamilton systems are integrable. Three interacting particles moving in empty space are a system of 9 degrees of freedom with only 7 independent conserved quantities. The three-body problem is not integrable.

18.2 Autonomous Systems of Ordinary Differential Equations

An autonomous system of n first-order ordinary differential equations

$$\dot{x}_1 = F_1(x_1, x_2, \ldots, x_n), \quad \ldots, \quad \dot{x}_n = F_n(x_1, x_2, \ldots, x_n). \tag{18.8}$$

is more general than it may seem at first sight. For a nonautonomous system of n equations with functions $F_i(x_1, \ldots, x_n, t)$ is equivalent to an autonomous system of $n+1$ equations with $t = x_{n+1}$ and $F_{n+1}(x_1, \ldots, x_{n+1}) = 1$. And a system of n higher-order ordinary differential equations is equivalent to an autonomous first-order system of more than n first-order ordinary differential equations.

Example 18.3 (Forced van der Pol oscillator) The second-order, time-dependent differential equation $\ddot{y} + \mu(y^2 - 1)\dot{y} + y = a\sin(\omega t)$ describes a forced van der Pol oscillator. Setting $x_1 = \dot{y}$, $x_2 = y$, and $x_3 = t$, we may write it as the first-order autonomous system

$$\dot{x}_1 = -x_2 - \mu(x_2^2 - 1)x_1 + a\sin(\omega x_3), \quad \dot{x}_2 = x_1, \quad \text{and} \quad \dot{x}_3 = 1 \tag{18.9}$$

which exhibits chaos for certain values of its parameters μ, a, and ω. The unforced oscillator ($\omega = a = 0$) has trajectories that converge to limit cycles as illustrated in Fig. 18.2. The outward spiral starts from $y(0) = 0.01$ and $\dot{y}(0) = 0$ with $\mu = 1/8$ (left); the inward spiral starts from $y(0) = 6$ and $\dot{y}(0) = 0$ with $\mu = 1/64$ (right).

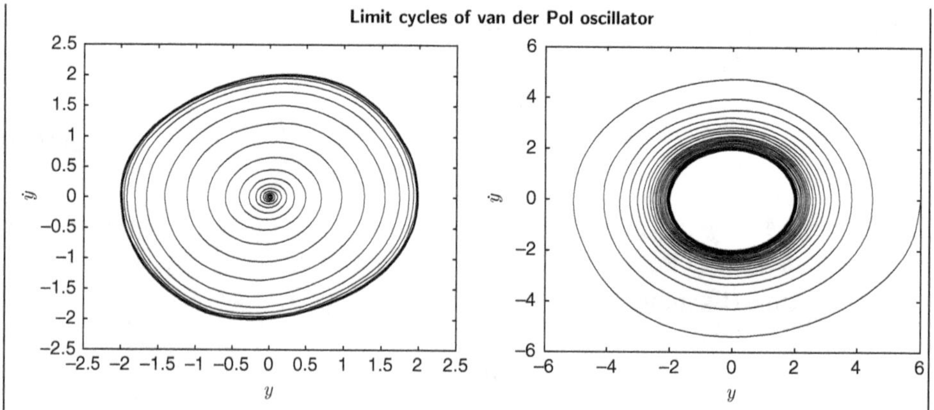

Figure 18.2 Trajectories of the unforced ($\omega = a = 0$) van der Pol oscillator (18.9) converge to their attractors which are limit cycles. The outward spiral starts from $y(0) = 0.01$ and $\dot{y}(0) = 0$ with $\mu = 1/8$ (left); the inward spiral starts from $y(0) = 6$ and $\dot{y}(0) = 0$ with $\mu = 1/64$ (right).

18.3 Attractors

Hamilton systems evolve in ways that conserve their areas in phase space (18.7), but arbitrary autonomous systems evolve more generally. Phase-space areas of **dissipative** systems typically shrink. If they converge to a point or to a set of points, that point or set is an **attractor**. An attractor may be a point of dimension zero, a loop or **limit cycle** of dimension one, a surface of integral dimension, or a **fractal** – a set whose dimension is not an integer (Section 18.6). A fractal attractor is a **strange attractor**.

Example 18.4 (Lorenz butterfly) The Lorenz system is three first-order differential equations

$$\dot{x} = \sigma\,(y - x), \qquad \dot{y} = r\,x - y - x\,z, \qquad \dot{z} = x\,y - b\,z \tag{18.10}$$

in which $\dot{y}$ and $\dot{z}$ have the nonlinear terms $-x\,z$ and $x\,y$, and the Prandtl number σ, the Rayleigh number r, and the parameter b are all positive. The script lorenz.m in Order_chaos_and_fractals at `github.com/kevinecahill` generates the plot of $x = x(1)$ and $z = x(3)$ in Fig. 18.3 for initial conditions $x = z = 0$ and $y = 8$.

Example 18.5 (Rössler system) The autonomous system

$$\dot{x} = -\,y - z, \quad \dot{y} = x + a\,y, \quad \text{and} \quad \dot{z} = b + z(x - c) \tag{18.11}$$

with $a = b = 0.2$ and initial conditions $x(0) = y(0) = z(0) = 0$ displays a simple limit cycle for $c = 2$, a period-two limit cycle for $c = 3$, a period-four limit cycle for

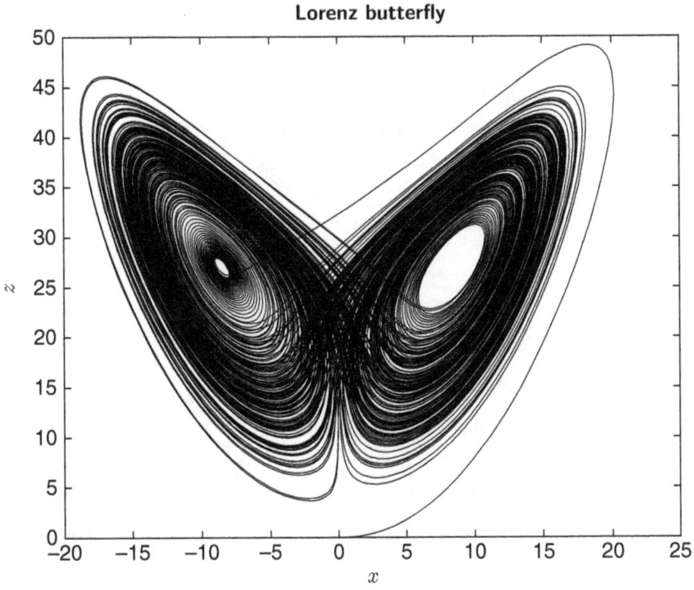

Figure 18.3 The trajectory of the Lorenz system (18.10) for $0 \le t \le 300$ approaches a strange attractor of dimension $D_{KY} = 2.06215$.

$c = 4$, a period-eight limit cycle for $c = 4.15$, and a strange attractor for $c = 5.7$ as shown in Fig. 18.4.

18.4 Chaos

Early in the last century, Henri Poincaré studied the three-body problem and found very complicated orbits. In this and many other systems, he found that after a transient period, classical motion assumes one of four forms:

1. periodic (a limit cycle);
2. steady or damped or stopped;
3. quasi-periodic (more than one frequency); or
4. chaotic.

Chaos takes different forms in different dynamical systems, and no single definition of chaos fits all of them. Most are **extremely sensitive to initial conditions**. For instance, two trajectories $x(t) = (x_1(t), \ldots, x_n(t))$ and $x'(t) = (x'_1(t), \ldots, x'_n(t))$ of an autonomous system (18.8) may diverge from each other exponentially

$$|||x'(t) - x(t)||| = e^{\lambda t} \, |||x'(0) - x(0)|||. \qquad (18.12)$$

in which λ is a **Lyapunov exponent**.

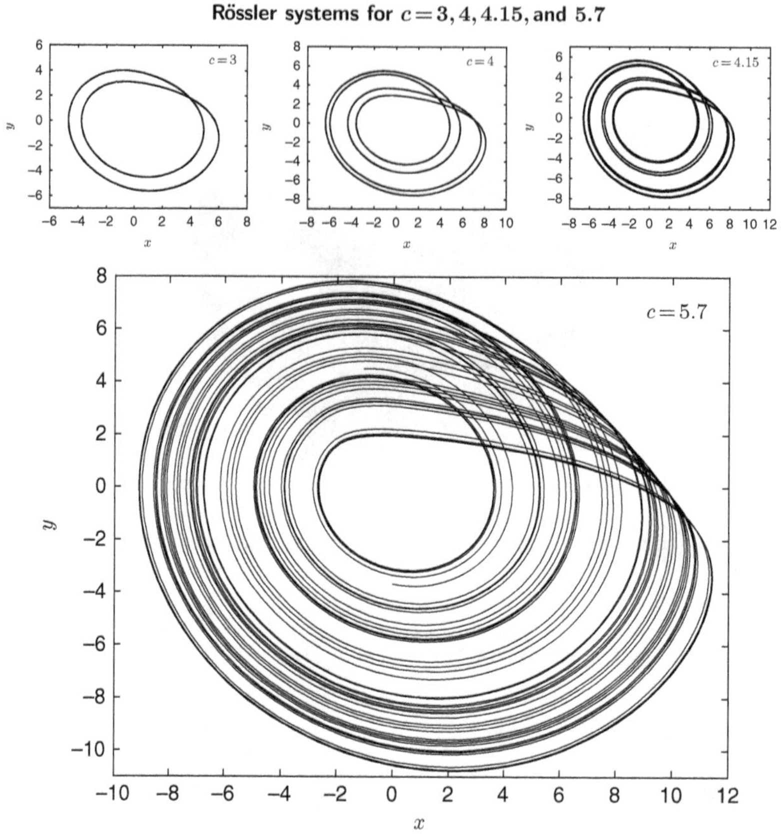

Figure 18.4 The last 10,000 points $(x(i), y(i))$ of the solution of the Rössler equations (18.11) trace the attractor for initial conditions $x(0) = y(0) = z(0) = 0$ and various values of the parameter c. For $c = 2$, the attractor is a simple loop (not shown); for $c = 3$, it is two loops; for $c = 4$, it is four loops; for $c = 4.15$, it is eight loops; and for $c = 5.7$, it is an infinite set of loops of dimension $D_{KY} = 2.0132$.

A first-order, autonomous dynamical system can be chaotic only it has at least $n = 3$ dimensions. The driven, damped pendulum (Example 18.9), the Lorenz system (Example 18.4), and the Rössler system (Example 18.5) all have $n = 3$ and evolve chaotically for certain values of their parameters. Here are three more examples:

Example 18.6 (Duffing's equation) If one attaches a thin piece of iron to the end of a rod that moves sinusoidally in the x-direction at frequency ω near two magnets, then the x coordinate is described by the forced Duffing equation $\ddot{x} + a\dot{x} + bx^3 + cx = g \sin(\omega t + \phi)$ and varies chaotically for suitable values of a, b, c, g, ω, and ϕ.

Example 18.7 (Dripping faucet) Drops from a slowly dripping faucet tend to fall regularly at times t_n separated by a constant interval $\Delta t = t_{n+1} - t_n$. At a slightly higher flow rate, the drops fall separated by intervals that alternate in their durations Δt, δt, Δt, δt, Δt, δt in a **period-two** sequence. At some higher flow rates, no regularity is apparent.

Example 18.8 (Rayleigh–Benard convection) Consider a fluid in a gravitational field above a hot plate and below a cold one. If the difference ΔT is small enough, then steady convective cellular flow occurs. But if ΔT is above the chaotic threshold, the fluid boils chaotically.

18.5 Maps

Successive crossings $x_j = (x_{1j}, \ldots, x_{nj})$ from one side to the other of a suitably oriented surface by an n-dimensional trajectory form a **Poincaré map**

$$x_{j+1} = M(x_j) \tag{18.13}$$

in a space of $n-1$ dimensions. Poincaré maps are invertible $x_j = M^{-1}(x_{j+1})$. An invertible map can be chaotic only if it has at least 2 dimensions so that it comes from a dynamical system that has at least 3 dimensions. A 1-dimensional map that is not invertible can display chaos.

The Lyapunov exponent of a smooth 1-dimensional map $x_{j+1} = f(x_j)$ is the limit

$$h(x_1) = \lim_{j \to \infty} \frac{1}{j} \left[\ln |f'(x_1)| + \cdots + \ln |f'(x_j)| \right]. \tag{18.14}$$

A bounded sequence that has a positive Lyapunov exponent and that does not converge to a periodic sequence is **chaotic** (Alligood et al., 1996, p. 110). Other aspects of chaos lead to other definitions.

Example 18.9 (Driven, damped pendulum) The angle θ of a sinusoidally driven, damped pendulum obeys the differential equation

$$\ddot{\theta} + b\dot{\theta} + \sin\theta = F \cos t \tag{18.15}$$

which is second order and nonautonomous. We put it into autonomous form by defining $x_1 = \dot{\theta}$, $x_2 = \theta$, and $x_3 = t$. In these variables, the pendulum equation (18.15) is the first-order autonomous system

$$\dot{x}_1 = F \cos x_3 - \sin x_2 - b\,x_1, \quad \dot{x}_2 = x_1, \quad \text{and} \quad \dot{x}_3 = 1 \tag{18.16}$$

with $n = 3$ dependent variables. Figure 18.5 displays a Poincaré map of the trajectory of the damped driven pendulum (18.15) with $b = 0.22$ and $F = 2.7$. This map is a

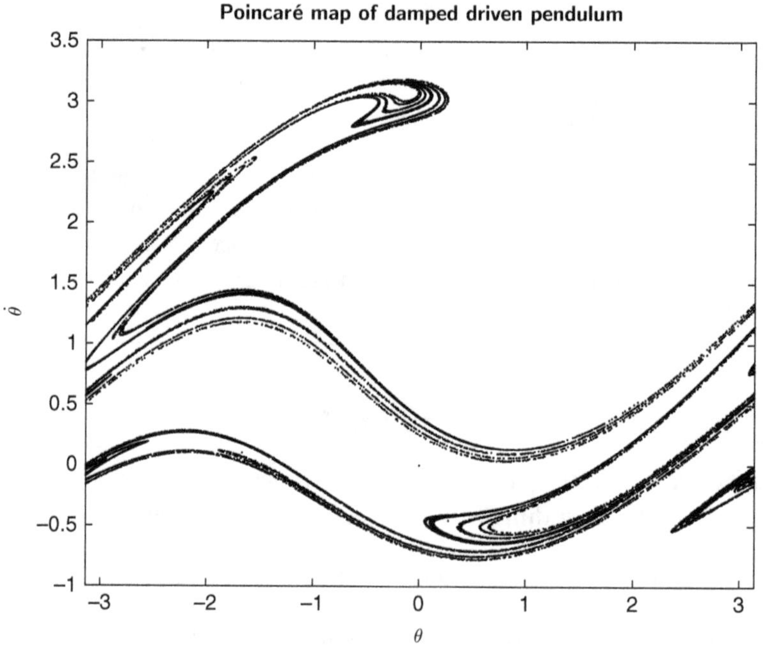

Figure 18.5 Poincaré map of the first million crossings of the surface $x_3 = 0$ (mod 2π) by the trajectory $(\theta(t), \dot\theta(t))$ of the damped driven pendulum (18.15) with $b = 0.22$ and $F = 2.7$. The points form a Cantor-set-like strange attractor of dimension $D_b \simeq 1.38$ (Grebogi et al., 1987). The initial conditions were $\theta(0) = \dot\theta(0) = 0$.

strange attractor of dimension $D_b \simeq 1.38$ (Grebogi et al., 1987). The horizontal axis is the angle $\theta(2\pi j)$ modulo 2π [that is, sign(θ) mod($|\theta|, 2\pi$)] for $j = 0, 1, 2, \ldots, 10^6$ or the first million crossings of the surface $x_3 = 0$ mod 2π.

Example 18.10 (Logistic map) For $0 < r < 4$, the 1-dimensional **logistic map** $x_{j+1} = r\, x_j(1 - x_j)$ describes a population with a limited food supply as does the differential equation (7.63). Because the quadratic equation for x_j in terms of x_{j+1} has two solutions, the logistic map is not invertible. For $0 < r < 1$, sequences of x_i's starting from any $0 < x_0 < 1$ converge to 0. This **attractor** begins to rise at $r = 1$ and reaches 2/3 at $r = 3$ where it bifurcates into two attractors as shown in Fig. 18.6. These attractors split again at $r_2 = 1 + \sqrt{6} \approx 3.4495$, and again at $r_3 \approx 3.54409$, and again at $r_4 \approx 3.5644$. By $r_\infty \approx 3.569946$, the attractors have split an infinite number of times. Chaos appears in increasingly striking forms as r exceeds r_∞. At $r = 3.8$, two sequences respectively starting from $x_0 = 0.2$ and $x_0' = 0.20001$ differ after 19 iterations by seemingly random amounts: 0.218 at $n = 21$, 0.623 at $n = 23$, and 0.723 at $n = 74$. At $r = 4$, the logistic map is totally chaotic and equivalent to the **tent map** $x_{j+1} = 1 - 2|x_j - 1|$. The **2x-modulo-1 map** $x_{j+1} = 2x_j$ mod 1 is similarly chaotic.

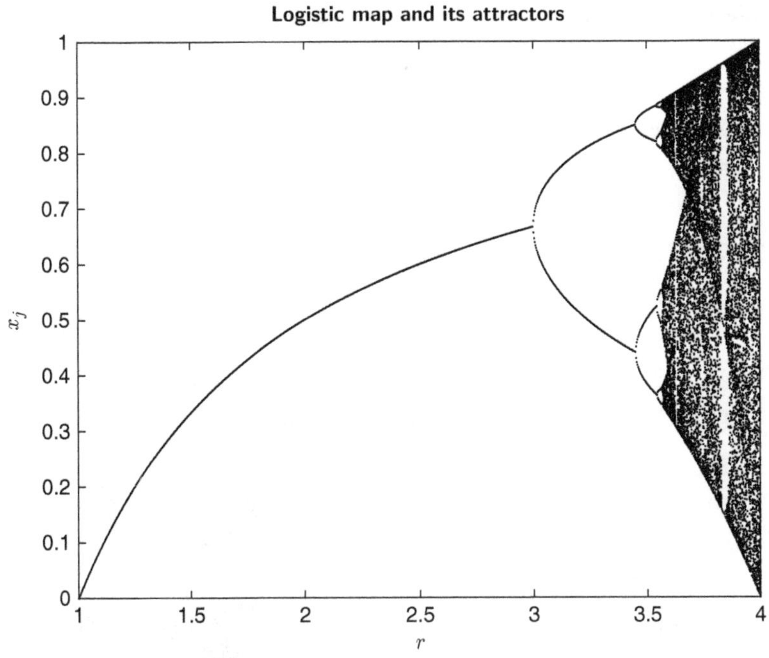

Figure 18.6 Points x_j of the logistic map for $29990 \leq n \leq 30000$ and 10 random starting points $0 < x_0 < 1$ for $1 \leq r \leq 4$. From $r = 1$ to $r = 3$, the attractor rises from 0 to 2/3 where it splits into two attractors.

Example 18.11 (Population map) The map $x_{j+1} = r\, x_j(x_j - 1)$ for $0 < r < 2$ and $x_j > 0$ describes a population whose rates of reproduction and death respectively are proportional to $r\, x_j^2$ and $(r+1)\, x_j$. For $r < 1$, sequences starting from any $0 < x_0 < 1$ approach the attractor $x = 0$. As shown in Fig. 18.7, that attractor splits at $r = 1$ into two attractors which become four at $r_2 = \sqrt{2}$. Somewhat above $r_3 \approx 1.544$, the four attractors split into eight. They split again at $r_4 \approx 1.565$. Chaos begins as r exceeds 1.57.

Example 18.12 (Bernoulli shift) The simplest chaotic map is the **Bernoulli shift** in which the initial point x_0 is an arbitrary number between 0 and 1 with the binary-decimal expansion

$$x_0 = \sum_{k=1}^{\infty} 2^{-k} a_k = 0.a_1 a_2 a_3 a_4 \ldots \tag{18.17}$$

and successive points lack a_1, then a_2, and so forth:

$$x_1 = 0.a_2 a_3 a_4 a_5 \ldots, \quad x_2 = 0.a_3 a_4 a_5 a_6 \ldots, \quad x_3 = 0.a_4 a_5 a_6 a_7 \ldots. \tag{18.18}$$

Two unequal irrational numbers x_0 and x_0' no matter how close generate sequences that roam independently, irregularly, and ergotically over the interval (0, 1).

Example 18.13 (Hénon's map) The 2-dimensional map

$$x_{j+1} = f(x_j) + B\, y_j \quad \text{and} \quad y_{j+1} = x_j \tag{18.19}$$

for $B \neq 0$ is invertible. If $f(x_j) = A - x_j^2$, it is **Hénon's map**, which for $A = 1.4$ and $B = 0.3$ is chaotic and converges to the strange attractor in Fig. 18.10.

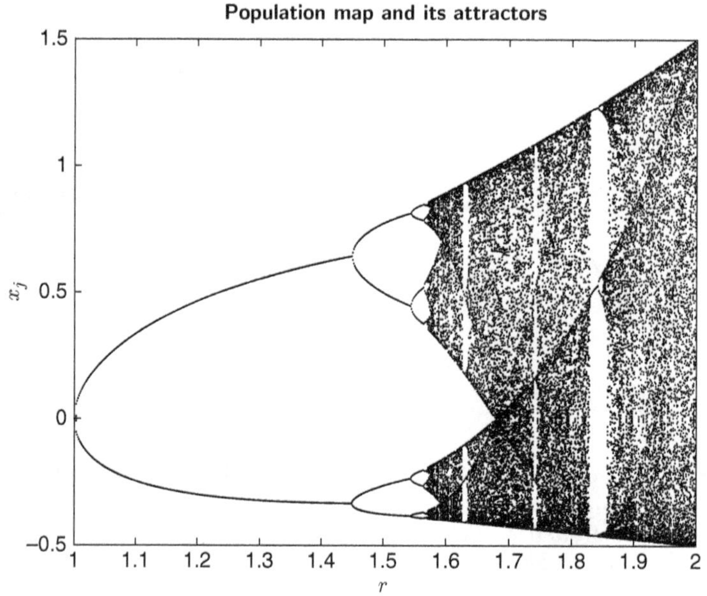

Figure 18.7 Points x_n for $29990 \le n \le 30000$ of the population map (Example 18.11) for 10 random starting points $0 < x_0 < 1$ and $1 \le r \le 2$. For $0 \le r \le 1$, the attractor of the population map is $x = 0$. The attractor splits at $r = 1$ into two attractors until $r_2 = \sqrt{2}$ where it splits into four. Slightly above $r_3 \approx 1.544$, the four attractors split into eight.

18.6 Fractals

A fractal set has a dimension that is not an integer. How can that be?

Felix Hausdorff and Abram Besicovitch have shown how to define the dimension of a weird set of points. To compute their **box-counting** dimension of a set, we cover it with line segments, squares, cubes, or n-dimensional "boxes" of side ϵ. If we need $N(\epsilon)$ boxes, then the fractal dimension D is the limit as $\epsilon \to 0$

$$D_b = \lim_{\epsilon \to 0} \frac{\ln(N(\epsilon))}{\ln(1/\epsilon)}. \tag{18.20}$$

For instance, we can cover the interval $[a, b]$ with $N(\epsilon) = (b-a)/\epsilon$ line segments of length ϵ, so the dimension of the segment $[a, b]$ is

$$D_b = \lim_{\epsilon\to 0} \frac{\ln(N(\epsilon))}{\ln(1/\epsilon)} = \lim_{\epsilon\to 0} \frac{\ln((b-a)/\epsilon)}{\ln(1/\epsilon)} = 1 + \frac{\ln(b-a)}{\ln(1/\epsilon)} = 1 \tag{18.21}$$

as it should be.

Example 18.14 (Cantor set) The Cantor set is defined by a limiting process in which the set at the nth stage consists of 2^n line segments each of length $1/3^n$. The first five approximations to the **Cantor set** are drawn in the figure (18.8). We can cover the nth approximation with $N(\epsilon) = 2^n$ line segments each of length $\epsilon_n = 1/3^n$, and so the fractal dimension is

$$D_b = \lim_{\epsilon\to 0} \frac{\ln(N(\epsilon))}{\ln(1/\epsilon)} = \lim_{n\to\infty} \frac{\ln(N(\epsilon_n))}{\ln(1/\epsilon_n)} = \lim_{n\to\infty} \frac{\ln(2^n)}{\ln(3^n)} = \frac{\ln 2}{\ln 3} = 0.6309297\ldots \tag{18.22}$$

which is not an integer or even a rational number.

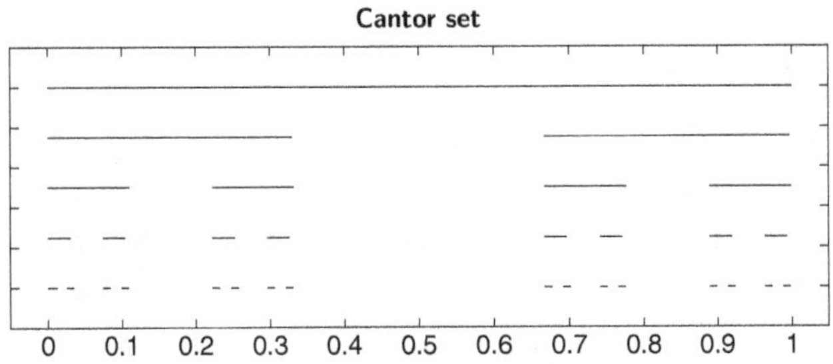

Figure 18.8 The first five approximations to the Cantor set.

Example 18.15 (Koch snowflake) In 1904, the Swedish mathematician Helge von Koch described the Koch curve (or the Koch snowflake), whose construction is shown in Fig. 18.9. With each step, there are 4 times as many line segments, each one being 3 times smaller. The length L of the curve at step n is thus $L = (4/3)^n$ which grows without limit as $n \to \infty$. Its box dimension is

$$D_b = \lim_{n\to\infty} \frac{\ln(N(\epsilon_n))}{\ln(1/\epsilon_n)} = \lim_{n\to\infty} \frac{\ln(4^n)}{\ln(3^n)} = \frac{\ln 4}{\ln 3} = 1.2618595\ldots. \tag{18.23}$$

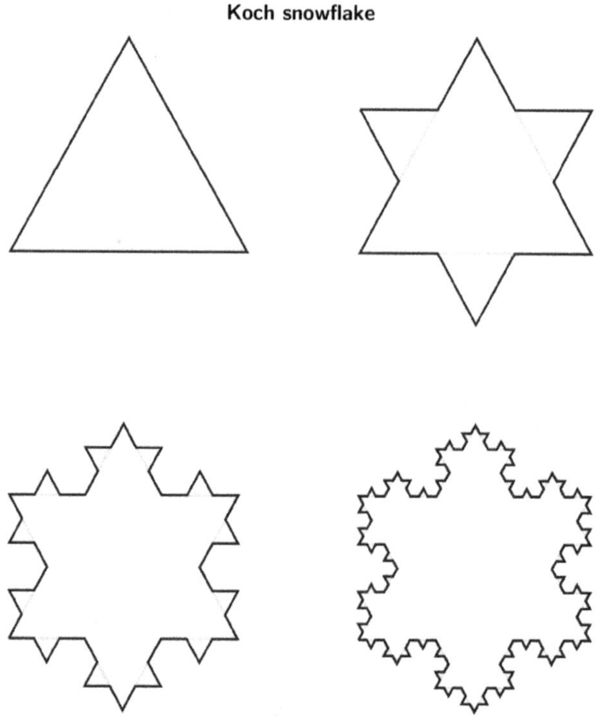

Figure 18.9 First four steps in making a Koch snowflake. (Source: https://en.wikipedia.org/wiki/Koch_snowflake#/media/File:KochFlake.svg)

Closely related to the box-counting dimension is the **self-similar dimension** D_s. To define it, we consider the number of self-similar structures of linear size x needed to cover the figure after n steps and take the limit

$$D_s = \lim_{x\to 0} \frac{\ln N(x)}{\ln 1/x}. \tag{18.24}$$

In the case of von Koch's curve, $x = 1/3^n$ and $N(x) = 4^n$. So the self-similar dimension of von Koch's curve is

$$D_s = \lim_{x\to 0} \frac{\ln N(x)}{\ln 1/x} = \lim_{n\to\infty} \frac{\ln 4^n}{\ln 3^n} = \frac{n \ln 4}{n \ln 3} = \frac{\ln 4}{\ln 3} = 1.2618595\ldots \tag{18.25}$$

which is equal to its box dimension $D_{b,K}$ given by (18.23).

Other definitions of the dimension of a set, such as the correlation dimension D_2 (Grassberger and Procaccia, 1983) and the Kaplan–Yorke dimension D_{KY} (Kaplan and Yorke, 1979), can be easier to measure than the box-counting and self-similar dimensions.

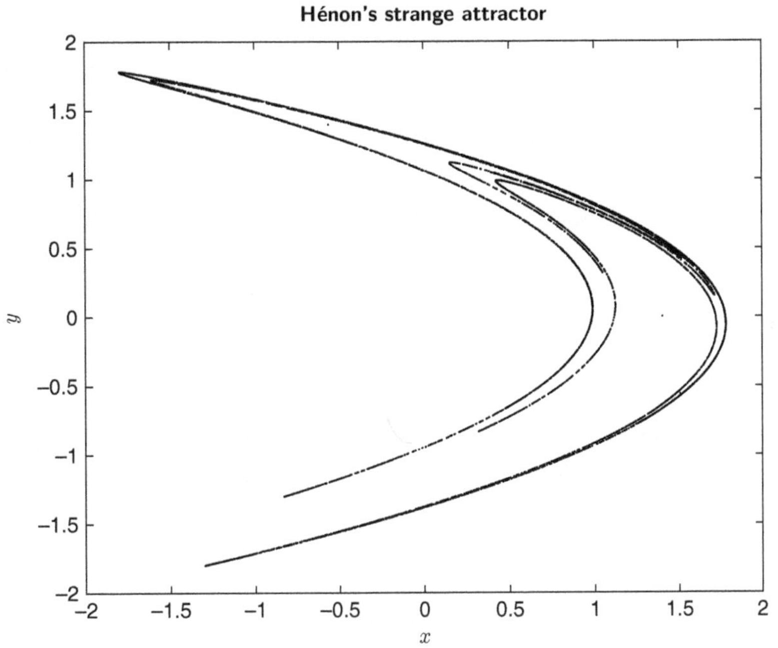

Figure 18.10 The first 10^4 points of the strange attractor of Hénon's map (18.13) with $A = 1.4$ and $B = 0.3$ and $(x_0, y_0) = (0, 0)$.

Attractors of fractal dimension are **strange**. The Lorenz butterfly (Example 18.4, Fig. 18.3) is a strange attractor of Kaplan–Yorke dimension $D_{KY} = 2.06215$, and the loops of the Rössler system for $c = 5.7$ (Example 18.5, Fig. 18.4) form a strange attractor of dimension $D_{KY} = 2.0132$ (Sprott, 2003). The Poincaré map of the damped driven pendulum (18.15, Fig. 18.5) is a strange attractor of box-counting dimension $D_b \approx 1.38$ (Grebogi et al., 1987). Hénon's map (18.19, Fig. 18.10) with $A = 1.4$ and $B = 0.3$ is chaotic with a strange attractor of dimension $D_b = 1.261 \pm 0.003$ (Russell et al., 1980). Chaotic systems often have strange attractors; but chaotic systems can have nonfractal attractors, and nonchaotic systems can have strange attractors.

Further Reading

The books *Nonlinear Dynamics and Chaos: With Applications to Physics, Biology, Chemistry, and Engineering* (Strogatz, 2014), *CHAOS: An Introduction to Dynamical Systems* (Alligood et al., 1996), and *Chaos and Time-Series Analysis* (Sprott, 2003) are superb.

Exercises

18.1 Use Hamilton's equations to derive the special relations (18.6) that the time derivatives $\dot{q}_i$ and $\dot{p}_i$ satisfy.

18.2 A period-one sequence of a map $x_{i+1} = f(x_i)$ is a point p for which $p = f(p)$. Find the period-one sequences of $x_{i+1} = rx_i(1 - x_i/K)$.

18.3 A period-two sequence of a map $x_{i+1} = f(x_i)$ is two different points p and q for which $q = f(p)$ and $p = f(q)$. Estimate the period-two sequences of the logistic map $f(x) = ax(1 - x)$ for $a = 1$, 2, and 3. Hint: Graph the functions $f(f(x))$ and $I(x) = x$ on the interval [0, 1].

18.4 A period-three sequence of a map $x_{i+1} = f(x_i)$ is three different points p, q, and r for which $q = f(p)$, $r = f(q)$, and $p = f(r)$. Li and Yorke have shown that a map with a period-three sequence is chaotic. Estimate the period-three sequences of the map $f(x) = 4x(1 - x)$.

19

Functional Derivatives

19.1 Functionals

A **functional** $G[f]$ is a map from a space of functions to a set of numbers. For instance, the **action** functional $S[q]$ for a particle in 1 dimension maps the coordinate $q(t)$, which is a function of the time t, into a number – the action of the process. If the particle has mass m and is moving slowly and freely, then for the interval (t_1, t_2) its action is

$$S_0[q] = \int_{t_1}^{t_2} dt \, \frac{m}{2} \left(\frac{dq(t)}{dt} \right)^2 . \tag{19.1}$$

If the particle is moving in a potential $V(q(t))$, then its action is

$$S[q] = \int_{t_1}^{t_2} dt \left[\frac{m}{2} \left(\frac{dq(t)}{dt} \right)^2 - V(q(t)) \right] . \tag{19.2}$$

19.2 Functional Derivatives

A **functional derivative** is a functional

$$\delta G[f][h] = \frac{d}{d\epsilon} G[f + \epsilon h] \bigg|_{\epsilon=0} \tag{19.3}$$

of a functional. For instance, if $G_n[f]$ is the functional

$$G_n[f] = \int dx \, f^n(x) \tag{19.4}$$

then its functional derivative is the functional that maps the pair of functions f, h to the number

$$\begin{aligned} \delta G_n[f][h] &= \frac{d}{d\epsilon} G_n[f + \epsilon h] \bigg|_{\epsilon=0} = \frac{d}{d\epsilon} \int dx \, (f(x) + \epsilon h(x))^n \bigg|_{\epsilon=0} \\ &= \int dx \, n f^{n-1}(x) h(x). \end{aligned} \tag{19.5}$$

Physicists often use the less elaborate notation

$$\frac{\delta G[f]}{\delta f(y)} = \delta G[f][\delta_y] \tag{19.6}$$

in which the function $h(x)$ is $\delta_y(x) = \delta(x - y)$. Thus in the preceding example

$$\frac{\delta G[f]}{\delta f(y)} = \int dx\, n f^{n-1}(x) \delta(x - y) = n f^{n-1}(y). \tag{19.7}$$

Functional derivatives of functionals that involve powers of derivatives also are easily dealt with. Suppose that the functional involves the square of the derivative $f'(x)$

$$G[f] = \int dx\, \left(f'(x)\right)^2. \tag{19.8}$$

Then its functional derivative is

$$\begin{aligned} \delta G[f][h] &= \frac{d}{d\epsilon} G[f + \epsilon h]\bigg|_{\epsilon=0} = \frac{d}{d\epsilon} \int dx\, \left(f'(x) + \epsilon h'(x)\right)^2 \bigg|_{\epsilon=0} \\ &= \int dx\, 2 f'(x) h'(x) = -2 \int dx\, f''(x) h(x) \end{aligned} \tag{19.9}$$

in which we have integrated by parts and used suitable boundary conditions on $h(x)$ to drop the surface terms. In physics notation, $h(x) = \delta(x - y)$, and

$$\frac{\delta G[f]}{\delta f(y)} = -2 \int dx\, f''(x) \delta(x - y) = -2 f''(y). \tag{19.10}$$

Let's now compute the functional derivative of the action (19.2), which involves the square of the time-derivative $\dot{q}(t)$ and the potential energy $V(q(t))$

$$\begin{aligned} \delta S[q][h] &= \frac{d}{d\epsilon} S[q + \epsilon h]\bigg|_{\epsilon=0} \\ &= \frac{d}{d\epsilon} \int dt \left[\frac{m}{2} \left(\dot{q}(t) + \epsilon \dot{h}(t)\right)^2 - V(q(t) + \epsilon h(t)) \right] \bigg|_{\epsilon=0} \\ &= \int dt \left[m \dot{q}(t) \dot{h}(t) - V'(q(t)) h(t) \right] \\ &= \int dt \left[-m \ddot{q}(t) - V'(q(t)) \right] h(t) \end{aligned} \tag{19.11}$$

where we once again have integrated by parts and used suitable boundary conditions to drop the surface terms. In physics notation, this is

$$\frac{\delta S[q]}{\delta q(t)} = \int dt' \left[-m \ddot{q}(t') - V'(q(t')) \right] \delta(t' - t) = -m \ddot{q}(t) - V'(q(t)). \tag{19.12}$$

In these terms, the stationarity of the action $S[q]$ is the vanishing of its functional derivative either in the form

$$\delta S[q][h] = 0 \tag{19.13}$$

for arbitrary functions $h(t)$ (that vanish at the endpoints of the interval) or equivalently in the form

$$\frac{\delta S[q]}{\delta q(t)} = 0 \tag{19.14}$$

which is Lagrange's equation of motion

$$m\ddot{q}(t) = -V'(q(t)). \tag{19.15}$$

Physicists also use the compact notation

$$\frac{\delta^2 Z[j]}{\delta j(y)\delta j(z)} \equiv \left.\frac{\partial^2 Z[j+\epsilon\delta_y+\epsilon'\delta_z]}{\partial\epsilon\,\partial\epsilon'}\right|_{\epsilon=\epsilon'=0} \tag{19.16}$$

in which $\delta_y(x) = \delta(x-y)$ and $\delta_z(x) = \delta(x-z)$.

Example 19.1 (Shortest path is a straight line) On a plane, the length of the path $(x, y(x))$ from (x_0, y_0) to (x_1, y_1) is

$$L[y] = \int_{x_0}^{x_1} \sqrt{dx^2 + dy^2} = \int_{x_0}^{x_1} \sqrt{1 + y'^2}\, dx. \tag{19.17}$$

The shortest path $y(x)$ minimizes this length $L[y]$, so

$$\begin{aligned}\delta L[y][h] &= \left.\frac{d}{d\epsilon} L[y+\epsilon h]\right|_{\epsilon=0} = \left.\frac{d}{d\epsilon}\int_{x_0}^{x_1}\sqrt{1+(y'+\epsilon h')^2}\,dx\right|_{\epsilon=0}\\ &= \int_{x_0}^{x_1}\frac{y'h'}{\sqrt{1+y'^2}}\,dx = -\int_{x_0}^{x_1} h\frac{d}{dx}\frac{y'}{\sqrt{1+y'^2}}\,dx = 0\end{aligned} \tag{19.18}$$

since $h(x_0) = h(x_1) = 0$. This can vanish for arbitrary $h(x)$ only if

$$\frac{d}{dx}\frac{y'}{\sqrt{1+y'^2}} = 0 \tag{19.19}$$

which implies $y'' = 0$. Thus $y(x)$ is a straight line, $y = mx + b$.

19.3 Higher-Order Functional Derivatives

The second functional derivative is

$$\delta^2 G[f][h] = \frac{d^2}{d\epsilon^2}\, G[f+\epsilon h]|_{\epsilon=0}\,. \tag{19.20}$$

So if $G_n[f]$ is the functional

$$G_n[f] = \int f^n(x)dx \tag{19.21}$$

then

$$\begin{aligned}
\delta^2 G_n[f][h] &= \frac{d^2}{d\epsilon^2}\, G_n[f+\epsilon h]|_{\epsilon=0} = \frac{d^2}{d\epsilon^2} \int \left(f(x) + \epsilon h(x)\right)^n dx \bigg|_{\epsilon=0} \\
&= \frac{d^2}{d\epsilon^2} \int \binom{n}{2} \epsilon^2 h^2(x) f^{n-2}(x)\, dx \bigg|_{\epsilon=0} \\
&= n(n-1) \int f^{n-2}(x) h^2(x) dx.
\end{aligned} \tag{19.22}$$

Example 19.2 ($\delta^2 S_0$) The second functional derivative of the action $S_0[q]$ (19.1) is

$$\begin{aligned}
\delta^2 S_0[q][h] &= \frac{d^2}{d\epsilon^2} \int_{t_1}^{t_2} dt\, \frac{m}{2} \left(\frac{dq(t)}{dt} + \epsilon \frac{dh(t)}{dt} \right)^2 \bigg|_{\epsilon=0} \\
&= \int_{t_1}^{t_2} dt\, m \left(\frac{dh(t)}{dt} \right)^2 \ge 0
\end{aligned} \tag{19.23}$$

and is positive for all functions $h(t)$. Thus the stationary classical trajectory

$$q(t) = \frac{t - t_1}{t_2 - t_1}\, q(t_2) + \frac{t_2 - t}{t_2 - t_1}\, q(t_1) \tag{19.24}$$

is a **minimum** of the action $S_0[q]$.

The second functional derivative of the action $S[q]$ (19.2) is

$$\begin{aligned}
\delta^2 S[q][h] &= \frac{d^2}{d\epsilon^2} \int_{t_1}^{t_2} dt \left[\frac{m}{2} \left(\frac{dq(t)}{dt} + \epsilon \frac{dh(t)}{dt} \right)^2 - V(q(t) + \epsilon h(t)) \right] \bigg|_{\epsilon=0} \\
&= \int_{t_1}^{t_2} dt \left[m \left(\frac{dh(t)}{dt} \right)^2 - \frac{\partial^2 V(q(t))}{\partial q^2(t)} h^2(t) \right]
\end{aligned} \tag{19.25}$$

and it can be positive, zero, or negative. Chaos sometimes arises in systems of several particles when the second variation of $S[q]$ about a stationary path is negative, $\delta^2 S[q][h] < 0$ while $\delta S[q][h] = 0$.

The nth functional derivative is defined as

$$\delta^n G[f][h] = \frac{d^n}{d\epsilon^n}\, G[f + \epsilon h]|_{\epsilon=0}\,. \tag{19.26}$$

The nth functional derivative of the functional (19.21) is

$$\delta^n G_N[f][h] = \frac{N!}{(N-n)!} \int f^{N-n}(x) h^n(x) dx. \tag{19.27}$$

19.4 Functional Taylor Series

It follows from the Taylor-series theorem (5.7) that

$$e^{\delta} G[f][h] = \sum_{n=0}^{\infty} \frac{\delta^n}{n!} G[f][h] = \sum_{n=0}^{\infty} \frac{1}{n!} \frac{d^n}{d\epsilon^n} G[f+\epsilon h]\bigg|_{\epsilon=0} = G[f+h] \tag{19.28}$$

which illustrates an advantage of the present mathematical notation.

The functional $S_0[q]$ of Eq. (19.1) provides a simple example of the functional Taylor series (19.28):

$$\begin{aligned} e^{\delta} S_0[q][h] &= \left(1 + \frac{d}{d\epsilon} + \frac{1}{2}\frac{d^2}{d\epsilon^2}\right) S_0[q+\epsilon h]\bigg|_{\epsilon=0} \\ &= \frac{m}{2}\int_{t_1}^{t_2} \left(1 + \frac{d}{d\epsilon} + \frac{1}{2}\frac{d^2}{d\epsilon^2}\right) \left(\dot q(t) + \epsilon \dot h(t)\right)^2 dt \bigg|_{\epsilon=0} \\ &= \frac{m}{2}\int_{t_1}^{t_2} \left(\dot q^2(t) + 2\dot q(t)\dot h(t) + \dot h^2(t)\right) dt \\ &= \frac{m}{2}\int_{t_1}^{t_2} \left(\dot q(t) + \dot h(t)\right)^2 dt = S_0[q+h]. \end{aligned} \tag{19.29}$$

If the function $q(t)$ makes the action $S_0[q]$ stationary, and if $h(t)$ is smooth and vanishes at the endpoints of the time interval, then

$$S_0[q+h] = S_0[q] + S_0[h]. \tag{19.30}$$

More generally, if $q(t)$ makes the action $S[q]$ stationary, and $h(t)$ is any loop from and to the origin, then

$$S[q+h] = e^{\delta} S[q][h] = S[q] + \sum_{n=2}^{\infty} \frac{1}{n!} \frac{d^n}{d\epsilon^n} S[q+\epsilon h]|_{\epsilon=0}\,. \tag{19.31}$$

If $S_2[q]$ also is quadratic in q and $\dot q$, then

$$S_2[q+h] = S_2[q] + S_2[h]. \tag{19.32}$$

19.5 Functional Differential Equations

In inner products like $\langle q'|f\rangle$, we represent the momentum operator as

$$p = \frac{\hbar}{i} \frac{d}{dq'} \tag{19.33}$$

because then

$$\langle q'|p\,q|f\rangle = \frac{\hbar}{i}\frac{d}{dq'}\langle q'|q|f\rangle = \frac{\hbar}{i}\frac{d}{dq'}\Big(q'\langle q'|f\rangle\Big) = \left(\frac{\hbar}{i} + q'\frac{\hbar}{i}\frac{d}{dq'}\right)\langle q'|f\rangle \tag{19.34}$$

which respects the commutation relation $[q, p] = i\hbar$.

So too in inner products $\langle\phi'|f\rangle$ of eigenstates $|\phi'\rangle$ of $\phi(\boldsymbol{x}, t)$

$$\phi(\boldsymbol{x}, t)|\phi'\rangle = \phi'(\boldsymbol{x})|\phi'\rangle \tag{19.35}$$

we can represent the momentum $\pi(\boldsymbol{x}, t)$ canonically conjugate to the field $\phi(\boldsymbol{x}, t)$ as the functional derivative

$$\pi(\boldsymbol{x}, t) = \frac{\hbar}{i}\frac{\delta}{\delta\phi'(\boldsymbol{x})} \tag{19.36}$$

because then

$$\begin{aligned}\langle\phi'|\pi(\boldsymbol{x}', t)\phi(\boldsymbol{x}, t)|f\rangle &= \frac{\hbar}{i}\frac{\delta\langle\phi'|\phi(\boldsymbol{x}, t)|f\rangle}{\delta\phi'(\boldsymbol{x}')} = \frac{\hbar}{i}\frac{\delta\big(\phi'(\boldsymbol{x})\langle\phi'|f\rangle\big)}{\delta\phi'(\boldsymbol{x}')} \\ &= \frac{\hbar}{i}\frac{\delta}{\delta\phi'(\boldsymbol{x}')}\left(\int \delta(\boldsymbol{x}-\boldsymbol{x}')\,\phi'(\boldsymbol{x}')\,d^3x'\,\langle\phi'|f\rangle\right) \\ &= \frac{\hbar}{i}\left(\delta(\boldsymbol{x}-\boldsymbol{x}') + \phi'(\boldsymbol{x})\frac{\delta}{\delta\phi'(\boldsymbol{x}')}\right)\langle\phi'|f\rangle \\ &= \langle\phi'| - i\hbar\delta(\boldsymbol{x}-\boldsymbol{x}') + \phi(\boldsymbol{x}, t)\,\pi(\boldsymbol{x}', t)|f\rangle\end{aligned} \tag{19.37}$$

which respects the equal-time commutation relation

$$[\phi(\boldsymbol{x}, t), \pi(\boldsymbol{x}', t)] = i\,\hbar\,\delta(\boldsymbol{x}-\boldsymbol{x}'). \tag{19.38}$$

We can use the representation (19.36) for $\pi(x)$ to find the wave function of the ground state $|0\rangle$ of the hamiltonian

$$H = \frac{1}{2}\int\left[\pi^2 + (\nabla\phi)^2 + m^2\phi^2\right]d^3x \tag{19.39}$$

where we have set $\hbar = c = 1$. We will use the trick we used in Section 3.12 to find the ground state $|0\rangle$ of the harmonic-oscillator hamiltonian

$$H_0 = \frac{p^2}{2m} + \frac{m\omega^2 q^2}{2}. \tag{19.40}$$

In that trick, one writes

$$\begin{aligned}H_0 &= \frac{1}{2m}(m\omega q - ip)(m\omega q + ip) + \frac{i\omega}{2}[p, q] \\ &= \frac{1}{2m}(m\omega q - ip)(m\omega q + ip) + \frac{1}{2}\hbar\omega\end{aligned} \tag{19.41}$$

and seeks a state $|0\rangle$ that is annihilated by $m\omega q + ip$

$$\langle q'|m\omega q + ip|0\rangle = \left(m\omega q' + \hbar\frac{d}{dq'}\right)\langle q'|0\rangle = 0. \tag{19.42}$$

The solution to this differential equation

$$\frac{d}{dq'}\langle q'|0\rangle = -\frac{m\omega q'}{\hbar}\langle q'|0\rangle \tag{19.43}$$

is

$$\langle q'|0\rangle = \left(\frac{m\omega}{\pi\hbar}\right)^{1/4}\exp\left(-\frac{m\omega q'^2}{2\hbar}\right) \tag{19.44}$$

in which the prefactor is a constant of normalization.

Extending that trick to the hamiltonian (19.39), we factor H

$$H = \frac{1}{2}\int\left[\sqrt{-\nabla^2 + m^2}\,\phi - i\pi\right]\left[\sqrt{-\nabla^2 + m^2}\,\phi + i\pi\right]d^3x + C \tag{19.45}$$

in which C is the (infinite) constant

$$C = \frac{i}{2}\int\left[\pi, \sqrt{-\Delta + m^2}\,\phi\right]d^3x. \tag{19.46}$$

The ground state $|0\rangle$ of H therefore must satisfy the functional differential equation $\langle\phi'|\sqrt{-\nabla^2 + m^2}\,\phi + i\pi|0\rangle = 0$ or

$$\frac{\delta\langle\phi'|0\rangle}{\delta\phi'(\boldsymbol{x})} = -\sqrt{-\nabla^2 + m^2}\,\phi'(\boldsymbol{x})\,\langle\phi'|0\rangle. \tag{19.47}$$

The solution to this equation is

$$\langle\phi|0\rangle = N\exp\left(-\frac{1}{2}\int\phi(\boldsymbol{x})\sqrt{-\nabla^2 + m^2}\,\phi(\boldsymbol{x})\,d^3x\right) \tag{19.48}$$

in which N is a normalization constant. To see that this functional does satisfy equation (19.47), we compute the derivative

$$\frac{d\langle\phi + \epsilon h|0\rangle}{d\epsilon} = N\frac{d}{d\epsilon}\exp\left[-\frac{1}{2}\int(\phi + \epsilon h)\sqrt{-\nabla^2 + m^2}\,(\phi + \epsilon h)\,d^3x\right] \tag{19.49}$$

which at $\epsilon = 0$ is

$$\left.\frac{d\langle\phi + \epsilon h|0\rangle}{d\epsilon}\right|_{\epsilon=0} = -\frac{1}{2}\left[\int h(\boldsymbol{x})\sqrt{-\nabla^2 + m^2}\,\phi(\boldsymbol{x})\,\delta^3x \right. \\ \left. + \int\phi(\boldsymbol{x})\sqrt{-\nabla^2 + m^2}\,h(\boldsymbol{x})\,d^3x\right]\langle\phi|0\rangle. \tag{19.50}$$

We integrate the second term by parts and drop the surface terms because the smooth function h goes to zero quickly as its arguments go to infinity. We then have

$$\left.\frac{d\langle\phi+\epsilon h|0\rangle}{d\epsilon}\right|_{\epsilon=0} = -\int h(\boldsymbol{x}')\sqrt{-\nabla^2+m^2}\,\phi(\boldsymbol{x}')\,d^3x'\,\langle\phi|0\rangle. \tag{19.51}$$

Letting $h(\boldsymbol{x}') = \delta^{(3)}(\boldsymbol{x}'-\boldsymbol{x})$, we arrive at (19.47).

Since $\phi(x)$ is real, its spatial Fourier transform

$$\tilde{\phi}(\boldsymbol{p}) = \int e^{-i\boldsymbol{p}\cdot\boldsymbol{x}}\,\phi(\boldsymbol{x})\,\frac{d^3x}{(2\pi)^{3/2}} \tag{19.52}$$

satisfies $\tilde{\phi}(-\boldsymbol{p}) = \tilde{\phi}^*(\boldsymbol{p})$. In terms of it, the ground-state wave function is

$$\langle\phi|0\rangle = N\,\exp\left(-\frac{1}{2}\int|\tilde{\phi}(\boldsymbol{p})|^2\sqrt{\boldsymbol{p}^2+m^2}\,d^3p\right). \tag{19.53}$$

Example 19.3 (Other theories, other vacua) We can find exact ground states for interacting theories with hamiltonians like

$$H = \frac{1}{2}\int\left[\sqrt{-\nabla^2+m^2}\,\phi - ic_n\phi^n - i\pi\right]\left[\sqrt{-\nabla^2+m^2}\,\phi + ic_n\phi^n + i\pi\right]d^3x. \tag{19.54}$$

The state $|\Omega\rangle$ will be an eigenstate of H with eigenvalue zero if

$$\frac{\delta\langle\phi|\Omega\rangle}{\delta\phi(\boldsymbol{x})} = -\left[\sqrt{-\nabla^2+m^2}\,\phi(\boldsymbol{x}) + ic_n\phi^n\right]\langle\phi|\Omega\rangle. \tag{19.55}$$

By extending the argument of equations (19.45–19.51), one may show (Exercise 19.4) that the wave functional of the vacuum is

$$\langle\phi|\Omega\rangle = N\,\exp\left[-\int\left(\frac{1}{2}\phi\sqrt{-\nabla^2+m^2}\,\phi + \frac{ic_n}{n+1}\phi^{n+1}\right)d^3x\right]. \tag{19.56}$$

Exercises

19.1 Compute the action $S_0[q]$ (19.1) for the classical path (19.24).

19.2 Use (19.25) to find a formula for the second functional derivative of the action (19.2) of the harmonic oscillator for which $V(q) = m\omega^2q^2/2$.

19.3 Derive (19.53) from Equations (19.48 and 19.52).

19.4 Show that (19.56) satisfies (19.55).

20

Path Integrals

20.1 Path Integrals and Richard Feynman

Since Richard Feynman invented them over 70 years ago, path integrals have been used with increasing frequency in high-energy and condensed-matter physics, in optics and biophysics, and even in finance. Feynman used them to express the amplitude for a process as a sum of all the ways the process could occur each weighted by an exponential of its classical action $\exp(iS/\hbar)$. Others have used them to compute partition functions and to study the QCD vacuum (Richard Feynman, 1918–1988).

20.2 Gaussian Integrals and Trotter's Formula

Path integrals are based upon the gaussian integral (6.179) which holds for real $a \neq 0$ and real b

$$\int_{-\infty}^{\infty} e^{iax^2+2ibx}\, dx = \sqrt{\frac{i\pi}{a}}\; e^{-ib^2/a} \tag{20.1}$$

and upon the gaussian integral (6.133)

$$\int_{-\infty}^{\infty} e^{-ax^2+2ibx}\, dx = \sqrt{\frac{\pi}{a}}\; e^{-b^2/a} \tag{20.2}$$

which holds both for $\operatorname{Re} a > 0$ and also for $\operatorname{Re} a = 0$ with b real and $\operatorname{Im} a \neq 0$.

The extension of the integral formula (20.1) to any $n \times n$ real symmetric nonsingular matrix s_{jk} and any real vector c_j is (Exercises 20.1 & 20.2)

$$\int_{-\infty}^{\infty} e^{is_{jk}x_jx_k+2ic_jx_j}\, dx_1 \dots dx_n = \sqrt{\frac{(i\pi)^n}{\det s}}\; e^{-ic_j(s^{-1})_{jk}c_k} \tag{20.3}$$

in which $\det a$ is the determinant of the matrix a, a^{-1} is its inverse, and sums over the repeated indices j and k from 1 to n are understood. One may similarly extend the gaussian integral (20.2) to any positive symmetric $n \times n$ matrix s_{jk} and any vector c_j (Exercises 20.3 & 20.4)

$$\int_{-\infty}^{\infty} e^{-s_{jk}x_j x_k + 2ic_j x_j}\, dx_1 \cdots dx_n = \sqrt{\frac{\pi^n}{\det s}}\; e^{-c_j (s^{-1})_{jk} c_k}. \tag{20.4}$$

Path integrals also are based upon Trotter's product formula (Trotter, 1959; Kato, 1978)

$$e^{a+b} = \lim_{n\to\infty} \left(e^{a/n}\, e^{b/n}\right)^n \tag{20.5}$$

both sides of which are symmetrically ordered and obviously equal when $ab = ba$.

Separating a given hamiltonian $H = K + V$ into a kinetic part K and a potential part V, we can use Trotter's formula to write the time-evolution operator $e^{-itH/\hbar}$ as

$$e^{-it(K+V)/\hbar} = \lim_{n\to\infty} \left(e^{-itK/(n\hbar)}\, e^{-itV/(n\hbar)}\right)^n \tag{20.6}$$

and the Boltzmann operator $e^{-\beta H}$ as

$$e^{-\beta(K+V)} = \lim_{n\to\infty} \left(e^{-\beta K/n}\, e^{-\beta V/n}\right)^n. \tag{20.7}$$

20.3 Path Integrals in Quantum Mechanics

Path integrals can represent matrix elements of the time-evolution operator $\exp(-i(t_b - t_a)H/\hbar)$ in which H is the hamiltonian. For a particle of mass m moving nonrelativistically in 1 dimension in a potential $V(q)$, the hamiltonian is

$$H = \frac{p^2}{2m} + V(q). \tag{20.8}$$

The position and momentum operators q and p obey the commutation relation $[q, p] = i\hbar$. Their eigenstates $|q'\rangle$ and $|p'\rangle$ have eigenvalues q' and p' for all real numbers q' and p'

$$q\,|q'\rangle = q'\,|q'\rangle \quad \text{and} \quad p\,|p'\rangle = p'\,|p'\rangle. \tag{20.9}$$

These eigenstates are complete. Their outer products $|q'\rangle\langle q'|$ and $|p'\rangle\langle p'|$ provide expansions for the identity operator I and have inner products (4.69) that are phases

$$I = \int_{-\infty}^{\infty} |q'\rangle\langle q'|\, dq' = \int_{-\infty}^{\infty} |p'\rangle\langle p'|\, dp' \quad \text{and} \quad \langle q'|p'\rangle = \frac{e^{iq'p'/\hbar}}{\sqrt{2\pi\hbar}}. \tag{20.10}$$

Setting $\epsilon = (t_b - t_a)/n$ and writing the hamiltonian (20.8) over $\hbar$ as $H/\hbar = p^2/(2m\hbar) + V/\hbar = k + v$, we can write Trotter's formula (20.6) for the time-evolution operator as the limit as $n \to \infty$ of n factors of $e^{-i\epsilon k} e^{-i\epsilon v}$

$$e^{-i(t_b-t_a)(k+v)} = e^{-i\epsilon k}\, e^{-i\epsilon v}\, e^{-i\epsilon k}\, e^{-i\epsilon v} \cdots e^{-i\epsilon k}\, e^{-i\epsilon v}\, e^{-i\epsilon k}\, e^{-i\epsilon v}. \tag{20.11}$$

The advantage of using Trotter's formula is that we now can evaluate the matrix element $\langle q_1|e^{-i\epsilon k}\, e^{-i\epsilon v}|q_a\rangle$ between eigenstates $|q_a\rangle$ and $|q_1\rangle$ of the position operator q by inserting the momentum-state expansion (20.10) of the identity operator I between these two exponentials

$$\langle q_1|e^{-i\epsilon k}\, e^{-i\epsilon v}|q_a\rangle = \langle q_1|\, e^{-i\epsilon p^2/(2m\hbar)} \int_{-\infty}^{\infty} |p'\rangle\langle p'|\, dp'\; e^{-i\epsilon\, V(q)/\hbar}|q_a\rangle \tag{20.12}$$

and using the eigenvalue formulas (20.9)

$$\langle q_1|e^{-i\epsilon k}\, e^{-i\epsilon v}|q_a\rangle = \int_{-\infty}^{\infty} e^{-i\epsilon p'^2/(2m\hbar)}\, \langle q_1|p'\rangle\, e^{-i\epsilon V(q_a)/\hbar}\, \langle p'|q_a\rangle\, dp'. \tag{20.13}$$

Now using the formula (20.10) for the inner product $\langle q_1|p'\rangle$ and the complex conjugate of that formula for $\langle p'|q_a\rangle$, we get

$$\langle q_1|e^{-i\epsilon k}\, e^{-i\epsilon v}|q_a\rangle = e^{-i\epsilon\, V(q_a)/\hbar} \int_{-\infty}^{\infty} e^{-i\epsilon p'^2/(2m\hbar)}\, e^{i(q_1-q_a)p'/\hbar}\, \frac{dp'}{2\pi\hbar}. \tag{20.14}$$

In this integral, the momenta that are important are very high, being of order $\sqrt{m\hbar/\epsilon}$ which diverges as $\epsilon \to 0$; nonetheless, the integral converges.

If we adopt the suggestive notation $q_1 - q_a = \epsilon\, \dot{q}_a$ and use the gaussian integral (20.1) with $a = -\epsilon/(2m\hbar)$, $x = p$, and $b = \epsilon\dot{q}/(2\hbar)$

$$\int_{-\infty}^{\infty} \exp\left(-i\,\epsilon\,\frac{p^2}{2m\hbar} + i\,\epsilon\,\frac{\dot{q}\,p}{\hbar}\right)\frac{dp}{2\pi\hbar} = \sqrt{\frac{m}{2\pi i\epsilon\hbar}}\, \exp\left(i\,\frac{\epsilon}{\hbar}\,\frac{m\dot{q}^2}{2}\right), \tag{20.15}$$

then we find

$$\begin{aligned}\langle q_1|e^{-i\epsilon k}\, e^{-i\epsilon v}|q_a\rangle &= \frac{1}{2\pi\hbar}\, e^{-\epsilon\, V(q_a)} \int_{-\infty}^{\infty} \exp\left(-i\,\frac{\epsilon p'^2}{2m\hbar} + i\frac{\epsilon\,\dot{q}_a\,p'}{\hbar}\right) dp' \\ &= \left(\frac{m}{2\pi i\hbar\epsilon}\right)^{1/2} \exp\left[i\,\frac{\epsilon}{\hbar}\left(\frac{m\dot{q}_a^2}{2} - V(q_a)\right)\right].\end{aligned} \tag{20.16}$$

The dependence of the amplitude $\langle q_1|e^{-i\epsilon k}\, e^{-i\epsilon v}|q_a\rangle$ upon q_1 is hidden in the formula $\dot{q}_a = (q_1 - q_a)/\epsilon$.

The next step is to use the position-state expansion (20.10) of the identity operator to link two of these matrix elements together

$$\begin{aligned}\langle q_2|\left(e^{-i\epsilon k}\, e^{-i\epsilon v}\right)^2|q_a\rangle &= \int_{-\infty}^{\infty} \langle q_2|\, e^{-i\epsilon k}\, e^{-i\epsilon v}\, |q_1\rangle\langle q_1|\, e^{-i\epsilon k}\, e^{-i\epsilon v}\, |q_a\rangle\, dq_1 \\ &= \frac{m}{2\pi i\hbar\epsilon} \int_{-\infty}^{\infty} \exp\left[i\frac{\epsilon}{\hbar}\left(\frac{m\dot{q}_1^2}{2} - V(q_1) + \frac{m\dot{q}_a^2}{2} - V(q_a)\right)\right] dq_1\end{aligned}$$

where now $\dot{q}_1 = (q_2 - q_1)/\epsilon$.

By stitching together $n = (t_b - t_a)/\epsilon$ time intervals each of length ϵ and letting $n \to \infty$, we get

$$\begin{aligned}\langle q_b|e^{-ni\epsilon H/\hbar}|q_a\rangle &= \int \langle q_b|e^{-i\epsilon k}\, e^{-i\epsilon v}|q_{n-1}\rangle \cdots \langle q_1|e^{-i\epsilon k}\, e^{-i\epsilon v}|q_a\rangle\, dq_{n-1}\cdots dq_1 \\ &= \left(\frac{m}{2\pi i\hbar\epsilon}\right)^{n/2} \int \exp\left[i\frac{\epsilon}{\hbar}\sum_{j=0}^{n-1} \frac{m\dot{q}_j^2}{2} - V(q_j)\right] dq_{n-1}\cdots dq_1 \\ &= \left(\frac{m}{2\pi i\hbar\epsilon}\right)^{n/2} \int \exp\left(i\frac{\epsilon}{\hbar}\sum_{j=0}^{n-1} L_j\right) dq_{n-1}\cdots dq_1 \qquad (20.17)\end{aligned}$$

in which $L_j = m\dot{q}_j^2/2 - V(q_j)$ is the lagrangian of the jth interval, and the q_j integrals run from $-\infty$ to ∞. In the limit $\epsilon \to 0$ with $n\epsilon = (t_b - t_a)/\epsilon$, this multiple integral is an integral over all paths $q(t)$ that go from q_a, t_a to q_b, t_b

$$\langle q_b|e^{-i\,(t_b-t_a)\,H/\hbar}|q_a\rangle = \int e^{iS[q]/\hbar}\, Dq \qquad (20.18)$$

in which each path is weighted by the phase of its classical action

$$S[q] = \int_{t_a}^{t_b} L(\dot{q}, q)\, dt = \int_{t_a}^{t_b} \left(\frac{m\dot{q}(t)^2}{2} - V(q(t))\right) dt \qquad (20.19)$$

in units of $\hbar$ and $Dq = (mn/(2\pi i\hbar(t_b - t_a)))^{n/2}\, dq_{n-1}\cdots dq_1$.

If we multiply the path-integral (20.18) for $\langle q_b|e^{-i\,(t_b-t_a)\,H/\hbar}|q_a\rangle$ from the left by $|q_b\rangle$ and from the right by $\langle q_a|$ and integrate over q_a and q_b as in the resolution (20.10) of the identity operator, then we can write the time-evolution operator as an integral over all paths from t_a to t_b

$$e^{-i\,(t_b-t_a)\,H/\hbar} = \int |q_b\rangle\, e^{iS[q]/\hbar}\, \langle q_a|\, Dq\, dq_a\, dq_b \qquad (20.20)$$

with $Dq = (mn/(2\pi i\hbar(t_b - t_a)))^{n/2}\, dq_{n-1}\cdots dq_1$ and $S[q]$ the action (20.19).

The path integral for a particle moving in 3-dimensional space is

$$\langle \boldsymbol{q}_b|e^{-i(t_b-t_a)H/\hbar}|\boldsymbol{q}_a\rangle = \int \exp\left[\frac{i}{\hbar}\int_{t_a}^{t_b} \frac{1}{2}m\,\dot{\boldsymbol{q}}^2(t) - V(\boldsymbol{q}(t))\, dt\right] D\boldsymbol{q} \qquad (20.21)$$

where $D\boldsymbol{q} = (mn/(2\pi i\hbar(t_b - t_a)))^{3n/2}\, d\boldsymbol{q}_{n-1}\cdots d\boldsymbol{q}_1$.

Let us first consider macroscopic processes whose actions are large compared to $\hbar$. Apart from the factor $D\boldsymbol{q}$, the amplitude (20.21) is a sum of phases $e^{iS[\boldsymbol{q}]/\hbar}$ one for each path from $\boldsymbol{q}_a, t_a$ to $\boldsymbol{q}_b, t_b$. When is this amplitude big? When is it small? Suppose there is a path $\boldsymbol{q}_c(t)$ from $\boldsymbol{q}_a, t_a$ to $\boldsymbol{q}_b, t_b$ that obeys the classical equation of motion (19.14–19.15)

$$\frac{\delta S[\boldsymbol{q}_c]}{\delta \boldsymbol{q}_{jc}} = m\ddot{\boldsymbol{q}}_{jc} + V'(\boldsymbol{q}_c) = 0. \tag{20.22}$$

Its action may be minimal. It certainly is stationary: a path $\boldsymbol{q}_c(t)+\delta\boldsymbol{q}(t)$ that differs from $\boldsymbol{q}_c(t)$ by a small detour $\delta\boldsymbol{q}(t)$ has an action $S[\boldsymbol{q}_c+\delta\boldsymbol{q}]$ that differs from $S[\boldsymbol{q}_c]$ only by terms of second order and higher in $\delta\boldsymbol{q}$. Thus a classical path has infinitely many neighboring paths whose actions differ only by integrals of $(\delta\boldsymbol{q})^n$, $n \geq 2$, and so have the same action to within a small fraction of $\hbar$. These paths add with nearly the same phase to the path integral (20.21) and so make a huge contribution to the amplitude $\langle \boldsymbol{q}_b|e^{-i(t_b-t_a)H/\hbar}|\boldsymbol{q}_a\rangle$. But if no classical path goes from $\boldsymbol{q}_a, t_a$ to $\boldsymbol{q}_b, t_b$, then the nonclassical, nonstationary paths that go from $\boldsymbol{q}_a, t_a$ to $\boldsymbol{q}_b, t_b$ have actions that differ from each other by large multiples of $\hbar$. These amplitudes cancel each other, and their sum, which is the amplitude for going from $\boldsymbol{q}_a, t_a$ to $\boldsymbol{q}_b, t_b$, is small. **Thus the path-integral formula for an amplitude in quantum mechanics explains why macroscopic processes are described by the principle of stationary action** (Section 7.13).

What about microscopic processes whose actions are tiny compared to $\hbar$? The path integral (20.21) gives large amplitudes for all microscopic processes. On very small scales, anything can happen that doesn't break a conservation law.

The path integral for two or more particles $\{\boldsymbol{q}\} = \{\boldsymbol{q}_1, \ldots, \boldsymbol{q}_k\}$ interacting with a potential $V(\{\boldsymbol{q}\})$ is

$$\langle \{\boldsymbol{q}\}_b|e^{-i(t_b-t_a)H/\hbar}|\{\boldsymbol{q}\}_a\rangle = \int e^{iS[\{q\}]/\hbar}\, D\{\boldsymbol{q}\} \tag{20.23}$$

where

$$S[\{\boldsymbol{q}\}] = \int_{t_a}^{t_b} \left[\frac{m_1\dot{\boldsymbol{q}}_1^2(t)}{2} + \cdots + \frac{m_k\dot{\boldsymbol{q}}_k^2(t)}{2} - V(\{\boldsymbol{q}(t)\})\right] dt \tag{20.24}$$

and $D\{\boldsymbol{q}\} = D\boldsymbol{q}_1 \cdots D\boldsymbol{q}_k$.

Example 20.1 (A free particle) For a free particle, the potential is zero, and the path integral (20.18, 20.19) is the $\epsilon \to 0$, $n \to \infty$ limit of

$$\langle q_b|e^{-itH/\hbar}|q_a\rangle = \left(\frac{m}{2\pi i\hbar\epsilon}\right)^{n/2} \times \int \exp\left[i\frac{m\epsilon}{2\hbar}\left(\frac{(q_b-q_{n-1})^2}{\epsilon^2} + \cdots + \frac{(q_1-q_a)^2}{\epsilon^2}\right)\right] dq_{n-1}\cdots dq_1. \tag{20.25}$$

The q_1 integral is by the gaussian formula (20.1)

$$\frac{m}{2\pi i\hbar\epsilon}\int e^{im[(q_2-q_1)^2+(q_1-q_a)^2]/(2\hbar\epsilon)}dq_1 = \sqrt{\frac{m}{2\pi i\hbar 2\epsilon}}\, e^{im(q_2-q_a)^2/(2\hbar 2\epsilon)}. \tag{20.26}$$

The q_2 integral is (Exercise 20.5)

$$\frac{m}{2\sqrt{2}\pi i\hbar\epsilon}\int e^{im[(q_3-q_2)^2+(q_2-q_a)^2/2]/(2\hbar\epsilon)}dq_2 = \sqrt{\frac{m}{2\pi i\hbar 3\epsilon}}\, e^{im(q_3-q_a)^2/(2\hbar 3\epsilon)}. \tag{20.27}$$

Doing all $n-1$ integrals (20.25) in this way and setting $n\epsilon = t_b - t_a$, we get

$$\begin{aligned}\langle q_b|\, e^{-i\,(t_b-t_a)\,H/\hbar}\,|q_a\rangle &= \sqrt{\frac{m}{2\pi i\hbar n\epsilon}}\,\exp\left[\frac{im(q_b-q_a)^2}{2\hbar n\epsilon}\right] \\ &= \sqrt{\frac{m}{2\pi i\hbar(t_b-t_a)}}\,\exp\left[\frac{im(q_b-q_a)^2}{2\hbar(t_b-t_a)}\right].\end{aligned} \tag{20.28}$$

The path integral (20.25) is perfectly convergent even though the velocities $\dot{q}_j = (q_{j+1}-q_j)/\epsilon$ that are important are very high, being of order $\sqrt{\hbar/(m\epsilon)}$.

It is easier to compute this amplitude (20.28) by using the outer products (20.10) (Exercise 20.6).

In 3 dimensions, the amplitude to go from $\boldsymbol{q}_a, t_a$ to $\boldsymbol{q}_b, t_b$ is

$$\langle \boldsymbol{q}_t|e^{-i(t_b-t_a)H/\hbar}|\boldsymbol{q}_0\rangle = \left(\frac{m}{2\pi i\hbar(t_b-t_a)}\right)^{3/2}\exp\left[\frac{im(\boldsymbol{q}_b-\boldsymbol{q}_a)^2}{2\hbar(t_b-t_a)}\right]. \tag{20.29}$$

20.4 Path Integrals for Quadratic Actions

If a path $q(t) = q_c(t)+x(t)$ differs from a classical path $q_c(t)$ by a detour $x(t)$ that vanishes at the endpoints $x(t_a) = 0 = x(t_b)$ so that both paths go from q_a, t_a to q_b, t_b, then the difference $S[q_c+x] - S[q_c]$ in their actions vanishes to first order in the detour $x(t)$ (Section 7.13). Thus the actions of the two paths differ by a time integral of quadratic and higher powers of the detour $x(t)$

$$\begin{aligned}S[q_c+x] &= \int_{t_a}^{t_b}\frac{1}{2}m\dot{q}(t)^2 - V(q(t))\,dt \\ &= \int_{t_a}^{t_b}\frac{1}{2}m\left(\dot{q}_c(t)+\dot{x}(t)\right)^2 - V(q_c(t)+x(t))\,dt \\ &= \int_{t_a}^{t_b}\left[\frac{m}{2}\dot{q}_c^2 + m\dot{q}_c\dot{x} + \frac{m}{2}\dot{x}^2 - V(q_c) - V'(q_c)x - \frac{V''(q_c)}{2}x^2\right. \\ &\qquad\left. - \frac{V'''(q_c)}{6}x^3 - \frac{V''''(q_c)}{24}x^4 - \cdots\right]dt\end{aligned} \tag{20.30}$$

$$= \int_{t_a}^{t_b} \left[\frac{m}{2}\dot{q}_c^2 - V(q_c)\right] dt + \int_{t_a}^{t_b} \left[\frac{m}{2}\dot{x}^2 - \frac{V''(q_c)}{2}x^2 - \frac{V'''(q_c)}{6}x^3 - \frac{V''''(q_c)}{24}x^4 - \cdots\right] dt$$

$$= S[q_c] + \Delta S[q_c, x]$$

in which $S[q_c]$ is the action of the classical path, and the detour $x(t)$ is a loop that goes from $x(t_a) = 0$ to $x(t_b) = 0$.

If the potential $V(q)$ is quadratic in the position q, then the third V''' and higher derivatives of the potential vanish, and the second derivative is a constant $V''(q_c(t)) = V''$. In this quadratic case, the correction $\Delta S[q_c, x]$ depends only on the time interval $t_b - t_a$ and on $\hbar$, m, and V''

$$\Delta S[q_c, x] = \Delta S[x] = \int_{t_a}^{t_b} \left[\frac{1}{2}m\,\dot{x}^2(t) - \frac{1}{2}V''\,x^2(t)\right] dt. \tag{20.31}$$

It is independent of the classical path.

Thus for quadratic actions, the path integral (20.18) is an exponential of the action $S[q_c]$ of the classical path multiplied by a function $f(t_b - t_a, \hbar, m, V'')$ of the time interval $t_b - t_a$ and of $\hbar$, m, and V''

$$\begin{aligned} \langle q_b|e^{-i(t_b-t_a)H/\hbar}|q_a\rangle &= \int e^{iS[q]/\hbar} Dq = \int e^{i(S[q_c]+\Delta S[x])/\hbar} Dq \\ &= e^{iS[q_c]/\hbar} \int e^{i\Delta S[x])/\hbar}\, Dx \\ &= f(t_b - t_a, \hbar, m, V'')\, e^{iS[q_c]/\hbar}. \end{aligned} \tag{20.32}$$

The function $f = f(t_b - t_a, \hbar, m, V'')$ is the limit as $n \to \infty$ of the $(n-1)$-dimensional integral

$$f = \left[\frac{mn}{2\pi i\hbar(t_b - t_a)}\right]^{n/2} \int e^{i\Delta S[x])/\hbar} dx_{n-1} \cdots dx_1 \tag{20.33}$$

where

$$\Delta S[x] = \frac{t_b - t_a}{n} \sum_{j=1}^{n} \frac{1}{2}m\frac{(x_j - x_{j-1})^2}{[(t_b - t_a)/n]^2} - \frac{1}{2}V''\,x_j^2 \tag{20.34}$$

and $x_n = 0 = x_0$.

More generally, the path integral for any quadratic action of the form

$$S[q] = \int_{t_a}^{t_b} u\,\dot{q}^2(t) + v\,q(t)\dot{q}(t) + w\,q^2(t) + s(t)\,\dot{q}(t) + j(t)\,q(t)\,dt \tag{20.35}$$

is (Exercise 20.7)

$$\langle q_b|e^{-i(t_b-t_a)H/\hbar}|q_a\rangle = f(t_a, t_b, \hbar, u, v, w)\, e^{iS[q_c]/\hbar}. \tag{20.36}$$

The dependence of the amplitude upon $s(t)$ and $j(t)$ is contained in the classical action $S[q_c]$ of the classical path q_c.

These formulas (20.32–20.36) may be generalized to any number of particles with coordinates $\{q\} = \{q^1, \ldots, q^k\}$ moving nonrelativistically in a space of multiple dimensions as long as the action is quadratic in the $\{q\}$'s and their velocities $\{\dot{q}\}$. The amplitude is then an exponential of the action $S[\{q\}_c]$ of the classical path multiplied by a function $f(t_a, t_b, \hbar, \ldots)$ that is independent of the classical path q_c

$$\langle \{q\}_b | e^{-i(t_b - t_a)H/\hbar} | \{q\}_a \rangle = f(t_a, t_b, \hbar, \ldots)\, e^{iS[\{q\}_c]/\hbar}. \tag{20.37}$$

Example 20.2 (Free particle) The classical path of a free particle going from $\boldsymbol{q}_a$ at time t_a to $\boldsymbol{q}_b$ at time t_b is

$$\boldsymbol{q}_c(t) = \boldsymbol{q}_a + \frac{t - t_a}{t_b - t_a}(\boldsymbol{q}_b - \boldsymbol{q}_a). \tag{20.38}$$

Its action is

$$S[\boldsymbol{q}_c] = \int_{t_a}^{t_b} \frac{1}{2} m\, \dot{\boldsymbol{q}}_c^2\, dt = \frac{m(\boldsymbol{q}_b - \boldsymbol{q}_a)^2}{2(t_b - t_a)} \tag{20.39}$$

and for this case our quadratic-potential formula (20.37) is

$$\langle \boldsymbol{q}_b | e^{-i(t_b - t_a)H/\hbar} | \boldsymbol{q}_a \rangle = f(t_b - t_a, \hbar, m)\, \exp\left[i \frac{m(\boldsymbol{q}_b - \boldsymbol{q}_a)^2}{2\hbar(t_b - t_a)} \right] \tag{20.40}$$

which agrees with our explicit calculation (20.29) when $f(t_b - t_a, \hbar, m) = [m/(2\pi i \hbar(t_b - t_a))]^{3/2}$.

Example 20.3 (Bohm–Aharonov effect) From the formula (12.55) for the action of a relativistic particle of mass m and charge e, it follows (Exercise 20.18) that the action of a nonrelativistic particle in an electromagnetic field with no scalar potential is

$$S = \int_{t_a}^{t_b} \left[\frac{1}{2} m \dot{\boldsymbol{q}}^2 + e\boldsymbol{A} \cdot \dot{\boldsymbol{q}} \right] dt = \int_{\boldsymbol{q}_a}^{\boldsymbol{q}_b} \left[\frac{1}{2} m \dot{\boldsymbol{q}} + e\, \boldsymbol{A} \right] \cdot d\boldsymbol{q}\,. \tag{20.41}$$

Since this action is quadratic in $\dot{\boldsymbol{q}}$, the amplitude for a particle to go from $\boldsymbol{q}_a$ at t_a to $\boldsymbol{q}_b$ at t_b is an exponential of the classical action

$$\langle \boldsymbol{q}_b | e^{-i(t_b - t_a)H/\hbar} | \boldsymbol{q}_a \rangle = f(t_b - t_a, \hbar, m, e)\, e^{iS[\boldsymbol{q}_c]/\hbar} \tag{20.42}$$

multiplied by a function $f(t_b - t_a, \hbar, m, e)$ that is independent of the path $\boldsymbol{q}_c$. A beam of such particles goes horizontally past but not through a vertical pipe in which a vertical magnetic field is confined. The particles can go both ways around the pipe of cross-sectional area S but do not enter it. The difference in the phases of the amplitudes for the two paths is a loop integral from the source to the detector around the pipe and back to the source

$$\oint \left[\frac{m\dot{\boldsymbol{q}}}{2} + e\,\boldsymbol{A} \right] \cdot \frac{d\boldsymbol{q}}{\hbar} = \oint \frac{m\dot{\boldsymbol{q}} \cdot d\boldsymbol{q}}{2\hbar} + \frac{e}{\hbar} \int_S \boldsymbol{B} \cdot d\boldsymbol{S} = \oint \frac{m\dot{\boldsymbol{q}} \cdot d\boldsymbol{q}}{2\hbar} + \frac{e\Phi}{\hbar} \quad (20.43)$$

in which Φ is the magnetic flux through the cylinder.

Example 20.4 (Harmonic oscillator) The action

$$S = \int_{t_a}^{t_b} \frac{1}{2} m \dot{q}^2(t) - \frac{1}{2} m \omega^2 q^2(t)\, dt \quad (20.44)$$

of a harmonic oscillator is quadratic in q and $\dot{q}$. So apart from a factor f, its path integral (20.32–20.34) is an exponential

$$\langle q_b | e^{-i(t_b - t_a)H/\hbar} | q_a \rangle = f\, e^{iS[q_c]/\hbar} \quad (20.45)$$

of the action $S[q_c]$ (Exercise 20.8)

$$S[q_c] = \frac{m\omega \left[(q_a^2 + q_b^2) \cos(\omega(t_b - t_a)) - 2 q_a q_b \right]}{2 \sin(\omega(t_b - t_a))} \quad (20.46)$$

of the classical path

$$q_c(t) = q_a \cos \omega(t - t_a) + \frac{q_b - q_a \cos \omega(t_b - t_a)}{\sin \omega(t_b - t_a)} \sin \omega(t - t_a) \quad (20.47)$$

that runs from q_a, t_a to q_b, t_b and obeys the classical equation of motion $m\ddot{q}_c(t) = -\,\omega^2 q_c(t)$.

The factor f is a function $f(t_b - t_a, \hbar, m, m\omega^2)$ of the time interval and the parameters of the oscillator. It is the $n \to \infty$ limit of the $(n-1)$-dimensional integral (20.33)

$$f = \left[\frac{mn}{2\pi i \hbar (t_b - t_a)} \right]^{n/2} \int e^{i \Delta S[x])/\hbar}\, dx_{n-1} \cdots dx_1 \quad (20.48)$$

over all loops that run from 0 to 0 in time $t_b - t_a$ in which the quadratic correction to the classical action is (20.34)

$$\Delta S[x] = \frac{t_b - t_a}{n} \sum_{j=1}^{n} \frac{1}{2} m \frac{(x_j - x_{j-1})^2}{[(t_b - t_a)/n]^2} - \frac{1}{2} m \omega^2 x_j^2, \quad (20.49)$$

and $x_n = 0 = x_0$.

Setting $t_b - t_a = T$, we use the many-variable imaginary gaussian integral (20.3) to write f as

$$f = \left[\frac{mn}{2\pi i \hbar T} \right]^{n/2} \int e^{i a_{jk} x_j x_k} dx_{n-1} \cdots dx_1 = \left[\frac{mn}{2\pi i \hbar T} \right]^{n/2} \sqrt{\frac{(i\pi)^{n-1}}{\det a}} \quad (20.50)$$

in which the quadratic form $a_{jk} x_j x_k$ is

$$\frac{nm}{\hbar T} \sum_{j=1}^{n} \left[-\, x_j x_{j-1} + \frac{1}{2}(x_j^2 + x_{j-1}^2) - \frac{(\omega T)^2}{2n^2} x_j^2 \right] \quad (20.51)$$

which has no linear term because $x_0 = x_n = 0$.

The $(n-1)$-dimensional square matrix a is a tridiagonal Toeplitz matrix

$$a = \frac{nm}{2\hbar T} \begin{pmatrix} y & -1 & 0 & 0 & \cdots \\ -1 & y & -1 & 0 & \cdots \\ 0 & -1 & y & -1 & \cdots \\ 0 & 0 & -1 & y & \cdots \\ \vdots & \vdots & \vdots & \ddots & \ddots \end{pmatrix}. \tag{20.52}$$

Apart from the factor $nm/(2\hbar T)$, the matrix $a = (nm/(2\hbar T))\, C_{n-1}(y)$ is a tridiagonal matrix $C_{n-1}(y)$ whose off-diagonal elements are -1 and whose diagonal elements are $y = 2 - (\omega T)^2/n^2$. Their determinants $|C_n(y)| = \det C_n(y)$ obey (Exercise 20.9) the recursion relation

$$|C_{n+1}(y)| = y\,|C_n(y)| - |C_{n-1}(y)| \tag{20.53}$$

and have the initial values $|C_1(y)| = y$ and $|C_2(y)| = y^2 - 1$. The trigonometric functions $S_n(y) = \sin[(n+1)\theta]/\sin\theta$ with $y = 2\cos\theta$ obey the same recursion relation and have the same initial values (Exercise 20.10), so

$$|C_n(y)| = \frac{\sin(n+1)\theta}{\sin\theta}. \tag{20.54}$$

Since for large n

$$\theta = \arccos(y/2) = \arccos\left(1 - \frac{\omega^2 t^2}{2n^2}\right) \approx \frac{\omega T}{n}, \tag{20.55}$$

the determinant of the matrix a is

$$\begin{aligned} \det a &= \left(\frac{nm}{2\hbar T}\right)^{n-1} |C_{n-1}(y)| = \left(\frac{nm}{2\hbar T}\right)^{n-1} \frac{\sin n\theta}{\sin\theta} \\ &\approx \left(\frac{nm}{2\hbar T}\right)^{n-1} \frac{\sin(\omega T)}{\sin(\omega T/n)} \approx \left(\frac{nm}{2\hbar T}\right)^{n-1} \frac{n \sin\omega T}{\omega T}. \end{aligned} \tag{20.56}$$

Thus the factor f is

$$\begin{aligned} f &= \left[\frac{mn}{2\pi i\hbar T}\right]^{n/2} \sqrt{\frac{(i\pi)^{n-1}}{\det a}} = \left[\frac{mn}{2\pi i\hbar T}\right]^{n/2} \sqrt{\left(\frac{2\pi i\hbar T}{nm}\right)^{n-1} \frac{\omega T}{n \sin\omega T}} \\ &= \sqrt{\frac{m\omega}{2\pi i\hbar \sin\omega T}}. \end{aligned} \tag{20.57}$$

The amplitude (20.45) is then an exponential of the action $S[q_c]$ (20.46) of the classical path (20.47) multiplied by this factor f

$$\begin{aligned} \langle q_b | e^{-i(t_b - t_a)H/\hbar} | q_a \rangle &= \sqrt{\frac{m\omega}{2\pi i\hbar \sin\omega(t_b - t_a)}} \\ &\times \exp\left\{ \frac{i}{\hbar} \frac{m\omega\left[(q_a^2 + q_b^2)\cos(\omega(t_b - t_a)) - 2q_a q_b\right]}{2\sin(\omega(t_b - t_a))} \right\}. \end{aligned} \tag{20.58}$$

As these examples (20.1 and 20.4) suggest, path integrals are well defined.

20.5 Path Integrals in Statistical Mechanics

At the imaginary time $t = -i\hbar\beta = -i\hbar/(kT)$, the time-evolution operator $e^{-itH/\hbar}$ becomes the **Boltzmann operator** $e^{-\beta H}$ whose trace is the **partition function** $Z(\beta)$ at inverse energy $\beta = 1/(kT)$

$$Z(\beta) = \mathrm{Tr}\big(e^{-\beta H}\big) = \sum_n \langle n|e^{-\beta H}|n\rangle \tag{20.59}$$

in which the states $|n\rangle$ form a complete orthonormal set, $k = 8.617 \times 10^{-5}$ eV/K is Boltzmann's constant, and T is the absolute temperature. Partition functions are important in statistical mechanics and quantum field theory.

Since the Boltzmann operator $e^{-\beta H}$ is the time-evolution operator $e^{-itH/\hbar}$ at the imaginary time $t = -i\hbar\beta$, we can write it as a path integral by imitating the derivation of the preceding section (20.3). We will use the same hamiltonian $H = p^2/(2m) + V(q)$ and the operators q and p which have complete sets of eigenstates (20.9) that satisfy (20.10).

Changing our definitions of ϵ, k, and v to $\epsilon = \beta/n$, $k = \beta p^2/(2m)$, and $v = \beta V(q)$, we can write Trotter's formula (20.7) for the Boltzmann operator as the $n \to \infty$ limit of n factors of $e^{-\epsilon k}\, e^{-\epsilon v}$

$$e^{-\beta H} = e^{-\epsilon k}\, e^{-\epsilon v}\, e^{-\epsilon k}\, e^{-\epsilon v} \cdots e^{-\epsilon k}\, e^{-\epsilon v}\, e^{-\epsilon k}\, e^{-\epsilon v}. \tag{20.60}$$

To evaluate the matrix element $\langle q_1|e^{-\epsilon k}\, e^{-\epsilon v}|q_a\rangle$, we insert the identity operator $\langle q_1|e^{-\epsilon k}\, I\, e^{-\epsilon v}|q_a\rangle$ as an integral (20.10) over outer products $|p'\rangle\langle p'|$ of momentum eigenstates and use the inner products $\langle q_1|p'\rangle = e^{iq_1 p'/\hbar}/\sqrt{2\pi\hbar}$ and $\langle p'|q_a\rangle = e^{-iq_a p'/\hbar}/\sqrt{2\pi\hbar}$

$$\begin{aligned}\langle q_1|e^{-\epsilon k}\, e^{-\epsilon v}|q_a\rangle &= \int_{-\infty}^{\infty} e^{-\epsilon p^2/(2m)}|p'\rangle\langle p'|e^{-\epsilon V(q)}|q_a\rangle\, dp' \\ &= e^{-\epsilon\, V(q_a)} \int_{-\infty}^{\infty} e^{-\epsilon p'^2/(2m)}\, e^{ip'(q_1-q_a)/\hbar}\, \frac{dp'}{2\pi\hbar}.\end{aligned} \tag{20.61}$$

If we adopt the suggestive notation $q_1 - q_a = \epsilon\hbar\,\dot q_a$ and use the gaussian integral (20.2) with $a = \epsilon/(2m)$, $x = p$, and $b = \epsilon\dot q/2$

$$\int_{-\infty}^{\infty} \exp\left(-\,\epsilon\,\frac{p^2}{2m} + i\epsilon\,\dot q\, p\right)\frac{dp}{2\pi\hbar} = \sqrt{\frac{m}{2\pi\epsilon\hbar^2}}\, \exp\left(-\,\epsilon\,\frac{m\dot q^2}{2}\right), \tag{20.62}$$

then we find

$$\begin{aligned}\langle q_1|e^{-\epsilon k}\, e^{-\epsilon v}|q_a\rangle &= e^{-\epsilon\, V(q_a)} \int_{-\infty}^{\infty} \exp\left(-\epsilon\,\frac{p'^2}{2m} + i\,\epsilon\, p'\,\dot q_a\right)\frac{dp'}{2\pi\hbar} \\ &= \left(\frac{m}{2\pi\hbar^2\epsilon}\right)^{1/2} \exp\left[-\,\epsilon\left(\frac{m\,\dot q_a^2}{2} + V(q_a)\right)\right]\end{aligned} \tag{20.63}$$

in which q_1 is hidden in the formula $q_1 - q_a = \hbar\epsilon\,\dot{q}_a$.

The next step is to link two of these matrix elements together

$$\begin{aligned}\langle q_2|\big(e^{-\epsilon k}\,e^{-\epsilon v}\big)^2|q_a\rangle &= \int_{-\infty}^{\infty}\langle q_2|e^{-\epsilon k}\,e^{-\epsilon v}|q_1\rangle\langle q_1|e^{-\epsilon k}\,e^{-\epsilon v}|q_a\rangle dq_1\\ &= \frac{m}{2\pi\hbar^2\epsilon}\int_{-\infty}^{\infty}\exp\left\{-\epsilon\left[\frac{m\dot{q}_1^2}{2}+V(q_1)+\frac{m\dot{q}_0^2}{2}+V(q_a)\right]\right\}dq_1.\end{aligned}$$

Passing from 2 to n and suppressing some integral signs, we get

$$\begin{aligned}\langle q_b|e^{-n\epsilon H}|q_a\rangle &= \iiint_{-\infty}^{\infty}\langle q_b|e^{-\epsilon k}\,e^{-\epsilon v}|q_{n-1}\rangle\cdots\langle q_1|e^{-\epsilon k}\,e^{-\epsilon v}|q_a\rangle\,dq_{n-1}\cdots dq_1\\ &= \left(\frac{m}{2\pi\hbar^2\epsilon}\right)^{n/2}\iiint_{-\infty}^{\infty}\exp\left[-\epsilon\sum_{j=0}^{n-1}\left(\frac{m\,\dot{q}_j^2}{2}+V(q_j)\right)\right]dq_{n-1}\cdots dq_1.\end{aligned}$$

Setting $du = \hbar\epsilon = \hbar\beta/n$ and taking the limit $n \to \infty$, we find that the matrix element $\langle q_b|e^{-\beta H}|q_a\rangle$ is the path integral

$$\langle q_b|e^{-\beta H}|q_a\rangle = \int e^{-S_e[q]/\hbar}\,Dq \tag{20.64}$$

in which each path is weighted by its euclidian action

$$S_e[q] = \int_0^{\hbar\beta}\frac{m\dot{q}^2(u)}{2} + V(q(u))\,du, \tag{20.65}$$

$\dot{q}$ is the derivative of the coordinate $q(u)$ with respect to euclidian time $u = \hbar\beta$, and $Dq \equiv (n\,m/2\pi\,\hbar^2\beta)^{n/2}\,dq_{n-1}\cdots dq_1$.

A derivation identical to the one that led from (20.60) to (20.65) leads in a more elaborate notation to

$$\langle q_b|e^{-(\beta_b-\beta_a)H}|q_a\rangle = \int e^{-S_e[q]/\hbar}\,Dq \tag{20.66}$$

in which each path is weighted by its euclidian action

$$S_e[q] = \int_{\hbar\beta_a}^{\hbar\beta_b}\frac{m\dot{q}^2(u)}{2} + V(q(u))\,du, \tag{20.67}$$

and $\dot{q}$ and Dq are the same as in (20.65).

If we multiply the path integral (20.66) from the left by $|q_b\rangle$ and from the right by $\langle q_a|$ and integrate over q_a and q_b as in the resolution (20.10) of the identity operator, then we can write the Boltzmann operator as an integral over all paths from t_a to t_b

$$e^{-(\beta_b-\beta_a)\,H} = \int |q_b\rangle\,e^{-S_e[q]/\hbar}\,\langle q_a|\,Dq\,dq_a\,dq_b \tag{20.68}$$

with $Dq = (mn/(2\pi i\hbar(t_b - t_a)))^{n/2}\, dq_{n-1} \cdots dq_1$ and $S_e[q]$ the action (20.67).

To get the partition function $Z(\beta)$, we set $q_b = q_a \equiv q_n$ and integrate over all n q's letting $n \to \infty$

$$\begin{aligned} Z(\beta) = \operatorname{Tr} e^{-\beta H} &= \int \langle q_n | e^{-\beta H} | q_n \rangle \, dq_n \\ &= \int \exp\left[-\frac{1}{\hbar} \int_0^{\hbar\beta} \frac{m\dot{q}^2(u)}{2} + V(q(u))\, du \right] Dq \end{aligned} \tag{20.69}$$

where $Dq \equiv (n\,m/2\pi\,\hbar^2\,\beta)^{n/2}\, dq_n \cdots dq_1$. We sum over all loops $q(u)$ that go from $q(0) = q_n$ at euclidian time 0 to $q(\hbar\beta) = q_n$ at euclidian time $\hbar\beta$.

In the low-temperature limit, $T \to 0$ and $\beta \to \infty$, the Boltzmann operator $\exp(-\beta H)$ projects out the ground state $|E_0\rangle$ of the system

$$\lim_{\beta\to\infty} e^{-\beta H} = \lim_{\beta\to\infty} \sum_n e^{-\beta E_n} |E_n\rangle\langle E_n| = e^{-\beta E_0} |E_0\rangle\langle E_0|. \tag{20.70}$$

The maximum-entropy **density operator** (Section 1.35, Example 1.54) is the Boltzmann operator $e^{-\beta H}$ divided by its trace $Z(\beta)$

$$\rho = \frac{e^{-\beta H}}{\operatorname{Tr}(e^{-\beta H})} = \frac{e^{-\beta H}}{Z(\beta)}. \tag{20.71}$$

Its matrix elements are matrix elements of the Boltzmann operator (20.65) divided by the partition function (20.69)

$$\langle q_b | \rho | q_a \rangle = \frac{\langle q_b | e^{-\beta H} | q_a \rangle}{Z(\beta)}. \tag{20.72}$$

In 3 dimensions with $\dot{\boldsymbol{q}}(u) = d\boldsymbol{q}(u)/du$, the $\boldsymbol{q}_a, \boldsymbol{q}_b$ matrix element of the Boltzmann operator is the analog of Equation (20.65) (Exercise 20.36)

$$\langle \boldsymbol{q}_b | e^{-\beta H} | \boldsymbol{q}_a \rangle = \int \exp\left[-\frac{1}{\hbar} \int_0^{\hbar\beta} \frac{m\dot{\boldsymbol{q}}^2(u)}{2} + V(\boldsymbol{q}(u))\, du \right] D\boldsymbol{q} \tag{20.73}$$

where $D\boldsymbol{q} \equiv (n\,m/2\pi\,\hbar^2\beta)^{3n/2}\, d\boldsymbol{q}_{n-1} \cdots d\boldsymbol{q}_1$, and the partition function is the integral over all loops that go from $q_0 = q_n$ to q_n in time $\hbar\beta$

$$Z(\beta) = \int \exp\left[-\frac{1}{\hbar} \int_0^{\hbar\beta} \frac{m\dot{\boldsymbol{q}}^2(u)}{2} + V(\boldsymbol{q}(u))\, du \right] D\boldsymbol{q} \tag{20.74}$$

where now $D\boldsymbol{q} \equiv (n\,m/2\pi\,\hbar^2\,\beta)^{3n/2}\, d\boldsymbol{q}_n \cdots d\boldsymbol{q}_1$.

Because the Boltzmann operator $e^{-\beta H}$ is the time-evolution operator $e^{-itH/\hbar}$ at the imaginary time $t = -iu = -i\hbar\beta = -i\hbar/(kT)$, the path integrals of statistical mechanics are called **euclidian path integrals**.

Example 20.5 (Density operator for a free particle) For a free particle, the matrix element of the Boltzmann operator $e^{-\beta H}$ is the $n = \beta/\epsilon \to \infty$ limit of the integral of n factors of the integral (20.63) with $V = 0$

$$\langle q_b|e^{-\beta H}|q_a\rangle = \left(\frac{m}{2\pi\hbar^2\epsilon}\right)^{n/2} \times \int \exp\left[-\frac{m(q_b - q_{n-1})^2}{2\hbar^2\epsilon} \cdots - \frac{(q_1 - q_a)^2}{2\hbar^2\epsilon}\right] dq_{n-1}\cdots dq_1.$$

The formula (20.2) gives for the q_1 integral

$$\left(\frac{m}{2\pi\hbar^2\epsilon}\right)^{1/2} \int e^{-[m(q_2-q_1)^2+m(q_1-q_a)^2]/(2\hbar^2\epsilon)} dq_1 = \frac{e^{-m(q_2-q_a)^2/(2\hbar^2 2\epsilon)}}{\sqrt{2}}.$$

The q_2 integral is (Exercise 20.11)

$$\left(\frac{m}{4\pi\hbar^2\epsilon}\right)^{1/2} \int e^{-m(q_3-q_2)^2/(2\hbar^2\epsilon)-m(q_2-q_a)^2/(4\hbar^2\epsilon)} dq_2 = \frac{e^{-m(q_3-q_a)^2/(2\hbar^2 3\epsilon)}}{\sqrt{3}}. \tag{20.75}$$

All $n-1$ integrations give

$$\langle q_b|e^{-\beta H}|q_a\rangle = \sqrt{\frac{m}{2\pi\hbar^2\epsilon}} \frac{e^{-m(q_b-q_a)^2/(2\hbar^2 n\epsilon)}}{\sqrt{n}} = \sqrt{\frac{m}{2\pi\hbar^2\beta}}\, e^{-m(q_b-q_a)^2/(2\hbar^2\beta)}.$$

The partition function is the integral of this matrix element over $q_a = q_b$

$$Z(\beta) = \left(\frac{m}{2\pi\hbar^2\beta}\right)^{1/2} \int dq_a = \left(\frac{m}{2\pi\hbar^2\beta}\right)^{1/2} L \tag{20.76}$$

where L is the (infinite) 1-dimensional volume of the system. The q_b, q_a matrix element of the maximum-entropy density operator is

$$\langle q_b|\rho|q_a\rangle = \frac{e^{-m(q_b-q_a)^2/(2\hbar^2\beta)}}{L}. \tag{20.77}$$

In 3 dimensions, Equations (20.5 and 20.76) are

$$\langle \boldsymbol{q}_b|e^{-\beta H}|\boldsymbol{q}_a\rangle = \left(\frac{mkT}{2\pi\hbar^2}\right)^{3/2} e^{-m(\boldsymbol{q}_b-\boldsymbol{q}_a)^2/(2\hbar^2\beta)} \quad \text{and} \quad Z(\beta) = \left(\frac{mkT}{2\pi\hbar^2}\right)^{3/2} L^3.$$

Example 20.6 (Partition function at high temperatures) At high temperatures, the time $\hbar\beta = \hbar/(kT)$ is very short, and the density operator (20.5) for a free particle shows that free paths are damped and limited to distances of order $\hbar/\sqrt{mkT}$. We thus can approximate the path integral (20.74) for the partition function by replacing the potential $V(q(u))$ by $V(q_n)$ and then using the free-particle matrix element (20.5)

$$Z(\beta) \approx \int e^{-\beta V(q_n)} \exp\left[-\frac{1}{\hbar}\int_0^{\hbar\beta} \frac{m\dot{\boldsymbol{q}}^2(u)}{2}\, du\right] D\boldsymbol{q} \tag{20.78}$$
$$= \int e^{-\beta V(q_n)} \langle \boldsymbol{q}_n | e^{-\beta H} | \boldsymbol{q}_n \rangle\, dq_n = \left(\frac{mkT}{2\pi\hbar^2}\right)^{3/2} \int e^{-\beta V(q_n)}\, dq_n.$$

20.6 Boltzmann Path Integrals for Quadratic Actions

Apart from the factor $Dq \equiv (n\,m/2\pi\,\hbar^2\beta)^{n/2}\,dq_{n-1}\cdots dq_1$, the euclidian path integral

$$\langle q_b|e^{-\beta H}|q_a\rangle = \int \exp\left[-\frac{1}{\hbar}\int_0^{\hbar\beta} \frac{m\dot{q}^2(u)}{2} + V(q(u))\, du\right] Dq \tag{20.79}$$

is a sum of positive terms $e^{-S_e[q]/\hbar}$ one for each path from q_a, 0 to q_b, β. If a path from q_a, 0 to q_b, β obeys the euclidian classical equation of motion

$$m\,\frac{d^2 q_{ec}}{du^2} = m\,\ddot{q}_{ec} = V'(q_{ec}) \tag{20.80}$$

then its euclidian action

$$S_e[q] = \int_0^{\hbar\beta} \frac{m\,\dot{q}^2(u)}{2} + V(u)\, du \tag{20.81}$$

is stationary and may be minimal. So we can approximate the euclidian action $S_e[q_{ec}+x]$ as we approximated the action $S[q_c+x]$ in Section 20.4. The euclidian action $S_e[q_{ec}+x]$ of an arbitrary path from q_a, 0 to q_b, β is the stationary euclidian action $S_e[q_{ec}]$ plus a u-integral of quadratic and higher powers of the detour x which goes from $x(0)=0$ to $x(\hbar\beta)=0$

$$\begin{aligned} S_e[q_{ec}+x] &= \int_0^{\hbar\beta} \left[\frac{m}{2}\dot{q}_{ec}^2 + V(q_{ec})\right] du \\ &+ \int_0^{\hbar\beta} \left[\frac{m}{2}\dot{x}^2 + \frac{V''(q_{ec})}{2}x^2 + \frac{V'''(q_{ec})}{6}x^3 + \frac{V''''(q_{ec})}{24}x^4 + \cdots\right] du \\ &= S_e[q_{ec}] + \Delta S_e[q_{ec}, x], \end{aligned} \tag{20.82}$$

and the path integral for the matrix element $\langle q_b|e^{-\beta H}|q_a\rangle$ is

$$\langle q_b|e^{-\beta H}|q_a\rangle = e^{-S_e[q_{ec}]/\hbar} \int e^{-\Delta S_e[q_{ec},x]/\hbar}\, Dx \tag{20.83}$$

as $n\to\infty$ where $Dx = (n\,m/2\pi\,\hbar^2\beta)^{n/2}\,dq_{n-1}\cdots dq_1$ in the limit $n\to\infty$.

If the action is quadratic in q and $\dot{q}$, then the action $\Delta S_e[q_{ec}, x]$ of the detour x is independent of the euclidian classical path q_{ec}, and so the path integral over x is a function f only of the parameters β, m, $\hbar$, and V''

$$\langle q_b | e^{-\beta H} | q_a \rangle = e^{-S_e[q_{ec}]/\hbar} \int e^{-\Delta S_e[x]/\hbar} Dx = f(\beta, \hbar, m, V'')\, e^{-S_e[q_{ec}]/\hbar} \quad (20.84)$$

in which with $x_n = 0 = x_0$ the function f is

$$f(\beta, \hbar, m, V'') = \left[\frac{mn}{2\pi\hbar^2\beta}\right]^{n/2} \int e^{-\Delta S_e[x]/\hbar} dx_{n-1} \cdots dx_1,$$
$$\Delta S_e[x] = \frac{\hbar\beta}{n} \sum_{j=1}^{n} \frac{m}{2\hbar^2} \frac{(x_j - x_{j-1})^2}{(\beta/n)^2} + \frac{1}{2} V'' x_j^2. \quad (20.85)$$

Example 20.7 (Density operator for the harmonic oscillator) The path $q_{ec}(\beta)$ that satisfies the euclidian classical equation of motion (20.80)

$$\ddot{q}_{ec}(u) = \frac{d^2 q_{ec}(u)}{du^2} = \omega^2 q_{ec}(u) \quad (20.86)$$

and goes from $q_a, 0$ to $q_b, \hbar\beta$ is

$$q_{ec}(u) = \frac{\sinh(\omega u)\, q_b + \sinh[\omega(\hbar\beta - u)]\, q_a}{\sinh(\hbar\omega\beta)}. \quad (20.87)$$

Its euclidian action is (Exercise 20.20)

$$S_e[q_{ec}] = \int_0^{\hbar\beta} \frac{m\dot{q}_{ec}^2(u)}{2} + \frac{m\omega^2 q_{ec}^2(u))}{2}\, du$$
$$= \frac{m\omega}{2\hbar \sinh(\hbar\omega\beta)} \left[\cosh(\hbar\omega\beta)(q_a^2 + q_b^2) - 2q_a q_b\right]. \quad (20.88)$$

Since $V'' = m\omega^2$, our formulas (20.84 and 20.85) for quadratic actions give as the matrix element

$$\langle q_b | e^{-\beta H} | q_a \rangle = f(\beta, \hbar, m, m\omega^2)\, e^{-S_e[q_{ec}]/\hbar} \quad (20.89)$$

in which

$$f(\beta, \hbar, m, m\omega^2) = \left[\frac{mn}{2\pi\hbar^2\beta}\right]^{n/2} \int e^{-\Delta S_e[x]/\hbar)}\, dx_{n-1} \cdots dx_1,$$
$$\Delta S_e[x] = \frac{\hbar\beta}{n} \sum_{j=1}^{n} \frac{m}{2\hbar^2} \frac{(x_j - x_{j-1})^2}{(\beta/n)^2} + \frac{m\omega^2 x_j^2}{2}, \quad (20.90)$$

and $x_n = 0 = x_0$. We can do this integral by using the formula (20.4) for a many variable real gaussian integral

$$f = \left[\frac{mn}{2\pi\hbar^2 B}\right]^{n/2} \int e^{-a_{jk} x_j x_k} dx_{n-1} \cdots dx_1 = \left[\frac{mn}{2\pi\hbar^2 B}\right]^{n/2} \sqrt{\frac{(\pi)^{n-1}}{\det a}} \quad (20.91)$$

in which the positive quadratic form $a_{jk}x_j x_k$ is

$$\frac{nm}{2\hbar^2 B}\sum_{j=1}^{n}\left[-2x_j x_{j-1}+x_j^2+x_{j-1}^2+\frac{(\hbar\omega B)^2}{n^2}x_j^2\right] \tag{20.92}$$

which has no linear term because $x_0 = x_n = 0$.

The matrix a is $(nm/(2\hbar^2 B))\,C_{n-1}(y)$ in which $C_{n-1}(y)$ is a square, tridiagonal, $(n-1)$-dimensional matrix whose off-diagonal elements are -1 and whose diagonal elements are $y = 2+(\hbar\omega\beta)^2/n^2$. The determinants $|C_n(y)|$ obey the recursion relation $|C_{n+1}(y)| = y\,|C_n(y)| - |C_{n-1}(y)|$ and have the initial values $C_1(y) = y$ and $C_2(y) = y^2 - 1$. So do the hyperbolic functions $\sinh(n+1)\theta/\sinh\theta$ with $y = 2\cosh\theta$. So we set $C_n(y) = \sinh(n+1)\theta/\sinh\theta$ with $\theta = \operatorname{arccosh}(y/2)$. We then get as the matrix element (20.89)

$$\langle q_b|e^{-\beta H}|q_a\rangle = \sqrt{\frac{m\omega}{2\pi\hbar\sinh(\hbar\omega\beta)}}\exp\left[-\frac{m\omega[\cosh(\hbar\omega\beta)(q_a^2+q_b^2)-2q_a q_b]}{2\hbar\sinh(\hbar\omega\beta)}\right]. \tag{20.93}$$

The partition function is the integral over q_a of this matrix element for $q_b = q_a$

$$\begin{aligned} Z(\beta) &= \sqrt{\frac{m\omega}{2\pi\hbar\sinh(\hbar\omega\beta)}}\int\exp\left[-\frac{m\omega[\cosh(\hbar\omega\beta)-1]q_a^2}{\hbar\sinh(\hbar\omega\beta)}\right]dq_a \\ &= \frac{1}{\sqrt{2[\cosh(\hbar\omega\beta)-1]}}. \end{aligned} \tag{20.94}$$

The matrix elements of the maximum-entropy density operator (20.71) are

$$\begin{aligned} \langle q_b|\rho|q_a\rangle &= \frac{\langle q_b|e^{-\beta H}|q_a\rangle}{Z(\beta)} \\ &= \sqrt{\frac{m\omega[\cosh(\hbar\omega\beta)-1]}{\pi\hbar\sinh(\hbar\omega\beta)}}\exp\left[-\frac{m\omega[\cosh(\hbar\omega\beta)(q_a^2+q_b^2)-2q_a q_b]}{2\hbar\sinh(\hbar\omega\beta)}\right] \end{aligned} \tag{20.95}$$

which reveals the ground-state wave functions

$$\lim_{\beta\to\infty}\langle q_b|\rho|q_a\rangle = \langle q_b|0\rangle\langle 0|q_a\rangle = \sqrt{\frac{m\omega}{\pi\hbar}}\,e^{-m\omega(q_a^2+q_b^2)/(2\hbar)}. \tag{20.96}$$

The partition function gives us the ground-state energy

$$\lim_{\beta\to\infty} Z(\beta) = \lim_{\beta\to\infty}\frac{1}{\sqrt{2[\cosh(\hbar\omega\beta)-1]}} = e^{-\beta E_0} = e^{-\beta\hbar\omega/2}. \tag{20.97}$$

20.7 Mean Values of Time-Ordered Products

In the Heisenberg picture, the position operator at time t is

$$q(t) = e^{itH/\hbar}q\,e^{-itH/\hbar} \tag{20.98}$$

in which $q = q(0)$ is the position operator at time $t = 0$ or equivalently the position operator in the Schrödinger picture. The position operator q at the imaginary time $t = -iu = -i\hbar\beta = -i\hbar/(kT)$ is the euclidian position operator

$$q_e(u) = q_e(\hbar\beta) = e^{uH/\hbar} q \, e^{-uH/\hbar}. \tag{20.99}$$

The time-ordered product of two position operators is

$$\mathcal{T}[q(t_1)q(t_2)] = \left\{ \begin{array}{ll} q(t_1)\,q(t_2) & \text{if } t_1 \ge t_2 \\ q(t_2)\,q(t_1) & \text{if } t_2 \ge t_1 \end{array} \right\} = q(t_>)\,q(t_<) \tag{20.100}$$

in which $t_>$ is the later and $t_<$ the earlier of the two times t_1 and t_2. Similarly, the time-ordered product of two euclidian position operators at euclidian times $u_1 = \hbar\beta_1$ and $u_2 = \hbar\beta_2$ is

$$\mathcal{T}[q_e(u_1)q_e(u_2)] = \left\{ \begin{array}{ll} q_e(u_1)\,q_e(u_2) & \text{if } u_1 \ge u_2 \\ q_e(u_2)\,q_e(u_1) & \text{if } u_2 \ge u_1 \end{array} \right\} = q_e(u_>)\,q_e(u_<). \tag{20.101}$$

The matrix element of the time-ordered product (20.100) of two position operators and two exponentials $e^{-itH/\hbar}$ between states $|a\rangle$ and $|b\rangle$ is

$$\begin{aligned} \langle b|e^{-itH/\hbar}\mathcal{T}[q(t_1)q(t_2)]e^{-itH/\hbar}|a\rangle &= \langle b|e^{-itH/\hbar}q(t_>)q(t_<)e^{-itH/\hbar}|a\rangle \qquad (20.102)\\ &= \langle b|e^{-i(t-t_>)H/\hbar}q\, e^{-i(t_>-t_<)H/\hbar}q\, e^{-i(t+t_<)H/\hbar}|a\rangle. \end{aligned}$$

We use the path-integral formula (20.20) for each of the exponentials on the right-hand side of this equation and find (Exercise 20.13)

$$\langle b|e^{-itH/\hbar}\mathcal{T}[q(t_1)q(t_2)]e^{-itH/\hbar}|a\rangle = \int \langle b|q_b\rangle q(t_1)q(t_2)e^{iS[q]/\hbar}\langle q_a|a\rangle\, Dq \tag{20.103}$$

in which the integral is over all paths that run from $-t$ to t. This equation simplifies if the states $|a\rangle$ and $|b\rangle$ are eigenstates of H with eigenvalues E_m and E_n

$$e^{-it(E_n+E_m)/\hbar}\langle n|\mathcal{T}[q(t_1)q(t_2)]|m\rangle = \int \langle n|q_b\rangle q(t_1)q(t_2)e^{iS[q]/\hbar}\langle q_a|m\rangle\, Dq. \tag{20.104}$$

By setting $n = m$ and omitting the time-ordered product, we get

$$e^{-2itE_n/\hbar} = \int \langle n|q_b\rangle e^{iS[q]/\hbar}\langle q_a|n\rangle\, Dq. \tag{20.105}$$

The ratio of (20.104) with $n = m$ to (20.105) is

$$\langle n|\mathcal{T}[q(t_1)q(t_2)]|n\rangle = \frac{\displaystyle\int \langle n|q_b\rangle q(t_1)q(t_2)e^{iS[q]/\hbar}\langle q_a|n\rangle\, Dq}{\displaystyle\int \langle n|q_b\rangle e^{iS[q]/\hbar}\langle q_a|n\rangle\, Dq} \tag{20.106}$$

in which the integrations are over all paths that go from $-t \le t_<$ to $t \ge t_>$. The mean value of the time-ordered product of k position operators is

$$\langle n|\mathcal{T}[q(t_1)\cdots q(t_k)]|n\rangle = \frac{\int \langle n|q_b\rangle q(t_1)\cdots q(t_k)e^{iS[q]/\hbar}\langle q_a|n\rangle\, Dq}{\int \langle n|q_b\rangle e^{iS[q]/\hbar}\langle q_a|n\rangle\, Dq} \tag{20.107}$$

in which the integrations are over all paths that go from some time before $t_1, \ldots, t_k$ to some time after them.

We may perform the same operations on the euclidian position operators by replacing t by $-iu = -i\hbar\beta$. A matrix element of the euclidian time-ordered product (20.101) between two states is

$$\begin{aligned}\langle b|e^{-uH/\hbar}\mathcal{T}[q_e(u_1)q_e(u_2)]e^{-uH/\hbar}|a\rangle &= \langle b|e^{-uH/\hbar}q_e(u_>)q_e(u_<)e^{-uH/\hbar}|a\rangle \\ &= \langle b|e^{-(u-u_>)H/\hbar}q\, e^{-(u_>-u_<)H/\hbar}q\, e^{-(u+u_<)H/\hbar}|a\rangle.\end{aligned} \tag{20.108}$$

As $u \to \infty$, the exponential $e^{-uH/\hbar}$ projects (20.70) states in onto the ground state $|0\rangle$ which is an eigenstate of H with energy E_0. So we replace the arbitrary states in (20.108) with the ground state and use the path-integral formula (20.68) for the last three exponentials of (20.108)

$$e^{-2uE_0/\hbar}\langle 0|\mathcal{T}[q_e(u_1)q_e(u_2)]|0\rangle = \int \langle 0|q_b\rangle q(u_1)q(u_2)e^{-S_e[q]/\hbar}\langle q_a|0\rangle\, Dq. \tag{20.109}$$

The same equation without the time-ordered product is

$$e^{-2uE_0/\hbar}\langle 0|0\rangle = e^{-2uE_0/\hbar} = \int \langle 0|q_b\rangle e^{-S_e[q]/\hbar}\langle q_a|0\rangle\, Dq. \tag{20.110}$$

The ratio of the last two equations is

$$\langle 0|\mathcal{T}[q_e(u_1)q_e(u_2)]|0\rangle = \frac{\int \langle 0|q_b\rangle q(u_1)q(u_2)e^{-S_e[q]/\hbar}\langle q_a|0\rangle\, Dq}{\int \langle 0|q_b\rangle e^{-S_e[q]/\hbar}\langle q_a|0\rangle\, Dq} \tag{20.111}$$

in which the integration is over all paths from $u = -\infty$ to $u = \infty$. The mean value in the ground state of the time-ordered product of k euclidian position operators is

$$\langle 0|\mathcal{T}[q_e(u_1)\cdots q_e(u_k)]|0\rangle = \frac{\int \langle 0|q_b\rangle\, q(u_1)\cdots q(u_k)\, e^{-S_e[q]/\hbar}\langle q_a|0\rangle\, Dq}{\int \langle 0|q_b\rangle e^{-S_e[q]/\hbar}\langle q_a|0\rangle\, Dq}. \tag{20.112}$$

20.8 Quantum Field Theory on a Lattice

Quantum mechanics imposes upon n coordinates q_i and conjugate momenta p_k the equal-time commutation relations

$$[q_i, p_k] = i\,\hbar\,\delta_{i,k} \quad \text{and} \quad [q_i, q_k] = [p_i, p_k] = 0. \tag{20.113}$$

In a theory of a single spinless quantum field, a coordinate $q_{\boldsymbol{x}} \equiv \phi(\boldsymbol{x})$ and a conjugate momentum $p_{\boldsymbol{x}} \equiv \pi(\boldsymbol{x})$ are associated with each point $\boldsymbol{x}$ of space. The operators $\phi(\boldsymbol{x})$ and $\pi(\boldsymbol{x})$ obey the commutation relations

$$[\phi(\boldsymbol{x}), \pi(\boldsymbol{x}')] = i\,\hbar\,\delta(\boldsymbol{x}-\boldsymbol{x}') \quad \text{and} \quad [\phi(\boldsymbol{x}), \phi(\boldsymbol{x}')] = [\pi(\boldsymbol{x}), \pi(\boldsymbol{x}')] = 0 \tag{20.114}$$

inherited from quantum mechanics.

To make path integrals, we replace space by a 3-dimensional lattice of points $\boldsymbol{x} = a(i, j, k) = (ai, aj, ak)$ and eventually let the distance a between adjacent points go to zero. On this lattice and at equal times $t = 0$, the field operators obey discrete forms of the commutation relations (20.114)

$$\begin{aligned} [\phi(a(i, j, k)), \pi(a(\ell, m, n))] &= i\,\frac{\hbar}{a^3}\,\delta_{i,\ell}\,\delta_{j,m}\,\delta_{k,n} \\ [\phi(a(i, j, k)), \phi(a(\ell, m, n))] &= [\pi(a(i, j, k)), \pi(a(\ell, m, n))] = 0. \end{aligned} \tag{20.115}$$

The vanishing commutators imply that the field and the momenta have compatible eigenvalues for all lattice points $a(i, j, k)$

$$\phi(a(i, j, k))|\phi'\rangle = \phi'(a(i, j, k))|\phi'\rangle \text{ and } \pi(a(i, j, k))|\pi'\rangle = \pi'(a(i, j, k))|\pi'\rangle. \tag{20.116}$$

Their inner products are

$$\langle\phi'|\pi'\rangle = \prod_{i,j,k} \sqrt{\frac{a^3}{2\pi\hbar}}\, e^{ia^3\phi'(a(i,j,k))\pi'(a(i,j,k))/\hbar}. \tag{20.117}$$

These states are complete

$$\int |\phi'\rangle\langle\phi'| \prod_{i,j,k} d\phi'(a(i, j, k)) = I = \int |\pi'\rangle\langle\pi'| \prod_{i,j,k} d\pi'(a(i, j, k)) \tag{20.118}$$

and orthonormal

$$\langle\phi'|\phi''\rangle = \prod_{i,j,k} \delta(\phi'(a(i, j, k)) - \phi''(a(i, j, k))) \tag{20.119}$$

with a similar equation for $\langle\pi'|\pi''\rangle$.

The hamiltonian for a free field of mass m is

$$H = \frac{1}{2} \int \pi^2 + c^2(\nabla\phi)^2 + \frac{m^2c^4}{\hbar^2}\phi^2 \, d^3x = \frac{a^3}{2} \sum_v \pi_v^2 + c^2(\nabla\phi_v)^2 + \frac{m^2c^4}{\hbar^2}\phi_v^2 \tag{20.120}$$

where $v = a(i, j, k)$, $\pi_v = \pi(a(i, j, k))$, $\phi_v = \phi(a(i, j, k))$, and the square of the lattice gradient $(\nabla\phi_v)^2$ is

$$\begin{aligned}\big[(\phi(a(i+1, j, k)) - \phi(a(i, j, k)))^2 + (\phi(a(i, j+1, k)) - \phi(a(i, j, k)))^2 \\ + (\phi(a(i, j, k+1)) - \phi(a(i, j, k)))^2\big]/a^2.\end{aligned} \tag{20.121}$$

Other interactions, such as $c^3\phi^4/\hbar$, can be added to this hamiltonian.

To simplify the appearance of the equations in the rest of this chapter, I will often use **natural units** in which $\hbar = c = 1$. To convert the value of a physical quantity from natural units to universal units, one multiplies or divides its natural-unit value by suitable factors of $\hbar$ and c until one gets the right dimensions. For example, if $V = 1/m$ is a time in natural units, where m is a mass, then the time in arbitrary units is $T = \hbar/(mc^2)$. If $V = 1/m$ is a length in natural units, then the length in universal units is $L = \hbar/(mc)$.

We set $K = a^3 \sum_v \pi_v^2/2$ and $V = (a^3/2) \sum_v (\nabla\phi_v)^2 + m^2\phi_v^2 + P(\phi_v)$ in which $P(\phi_v)$ represents the self-interactions of the field. With $\epsilon = (t_b - t_a)/n$, Trotter's product formula (20.6) is the $n \to \infty$ limit of

$$e^{-i(t_b-t_a)(K+V)} = \left(e^{-i(t_b-t_a)K/n} e^{-i(t_b-t_a)V/n}\right)^n = \left(e^{-i\epsilon K} e^{-i\epsilon V}\right)^n. \tag{20.122}$$

We insert I in the form (20.118) between $e^{-i\epsilon K}$ and $e^{-i\epsilon V}$

$$\langle\phi_1|e^{-i\epsilon K} \, e^{-i\epsilon V}|\phi_a\rangle = \langle\phi_1|e^{-i\epsilon K} \int |\pi'\rangle\langle\pi'| \prod_v d\pi'_v e^{-i\epsilon V}|\phi_a\rangle \tag{20.123}$$

and use the eigenstate formula (20.116)

$$\langle\phi_1|e^{-i\epsilon K} \, e^{-i\epsilon V}|\phi_a\rangle = e^{-i\epsilon V(\phi_a)} \int e^{-i\epsilon K(\pi')} \langle\phi_1|\pi'\rangle\langle\pi'|\phi_a\rangle \prod_v d\pi'_v \tag{20.124}$$

and the inner-product formula (20.117)

$$\langle\phi_1|e^{-i\epsilon K} \, e^{-i\epsilon V}|\phi_a\rangle = e^{-i\epsilon V(\phi_a)} \prod_v \left[\int \frac{a^3 d\pi'_v}{2\pi} e^{a^3[-i\epsilon\pi_v^2/2 + i(\phi_{1v}-\phi_{av})\pi'_v]}\right]. \tag{20.125}$$

Using the gaussian integral (20.1), we set $\dot\phi_a = (\phi_1 - \phi_a)/\epsilon$ and get

$$\langle\phi_1|e^{-i\epsilon K} \, e^{-i\epsilon V}|\phi_a\rangle = \prod_v \left[\left(\frac{a^3}{2\pi i\epsilon}\right)^{1/2} e^{i\epsilon a^3[\dot\phi_{av}^2 - (\nabla\phi_{av})^2 - m^2\phi_{av}^2 - P(\phi_v)]/2}\right]. \tag{20.126}$$

The product of $n = (t_b - t_a)/\epsilon$ such time intervals is

$$\langle\phi_b|e^{-i(t_b-t_a)H}|\phi_a\rangle = \prod_v \left[\left(\frac{a^3 n}{2\pi i(t_b - t_a)}\right)^{n/2} \int e^{iS_v}\, D\phi_v\right] \tag{20.127}$$

in which

$$S_v = \frac{t_b - t_a}{n}\frac{a^3}{2}\sum_{j=0}^{n-1}\left[\dot\phi_{jv}^2 - (\nabla\phi_{jv})^2 - m^2\phi_{jv}^2 - P(\phi_v)\right], \tag{20.128}$$

$\dot\phi_{jv} = n(\phi_{j+1,v} - \phi_{j,v})/(t_b - t_a)$, and $D\phi_v = d\phi_{n-1,v}\cdots d\phi_{1,v}$.

The amplitude $\langle\phi_b|e^{-i(t_b-t_a)H}|\phi_a\rangle$ is the integral over all fields that go from $\phi_a(\boldsymbol{x})$ at t_a to $\phi_b(\boldsymbol{x})$ at t_b each weighted by an exponential

$$\langle\phi_b|e^{-i(t_b-t_a)H}|\phi_a\rangle = \int e^{iS[\phi]}\, D\phi \tag{20.129}$$

of its action

$$S[\phi] = \int_{t_a}^{t_b} dt \int d^3x\, \frac{1}{2}\left[\dot\phi^2 - (\nabla\phi)^2 - m^2\phi^2 - P(\phi)\right] \tag{20.130}$$

in which $D\phi$ is the $n \to \infty$ limit of the product over all spatial vertices v

$$D\phi = \prod_v \left[\left(\frac{a^3 n}{2\pi i(t_b - t_a)}\right)^{n/2} d\phi_{n-1,v}\cdots d\phi_{1,v}\right]. \tag{20.131}$$

Equivalently, the time-evolution operator is

$$e^{-i(t_b-t_a)H} = \int |\phi_b\rangle e^{iS[\phi]}\langle\phi_a|\, D\phi\, D\phi_a D\phi_b \tag{20.132}$$

in which $D\phi_a D\phi_b = \prod_v d\phi_{a,v} d\phi_{b,v}$ is an integral over the initial and final states.

As in quantum mechanics (Section 20.4), the path integral for an action that is quadratic in the fields is an exponential of the action of a classical process $S[\phi_c]$ times a function of the times t_a, t_b and of other parameters

$$\langle\phi_b|e^{-i(t_b-t_a)H}|\phi_a\rangle = \int e^{iS[\phi]}\, D\phi = f(t_a, t_b, \ldots)\, e^{iS[\phi_c]} \tag{20.133}$$

in which $S[\phi_c]$ is the action of the process that goes from $\phi(\boldsymbol{x}, t_a) = \phi_a(\boldsymbol{x})$ to $\phi(\boldsymbol{x}, t_b) = \phi_b(\boldsymbol{x})$ and obeys the classical equations of motion, and the function f is a path integral over all fields that go from $\phi(\boldsymbol{x}, t_a) = 0$ to $\phi(\boldsymbol{x}, t_b) = 0$.

Example 20.8 (Classical processes) The field

$$\phi(\boldsymbol{x}, t) = \int e^{i\boldsymbol{k}\cdot\boldsymbol{x}} [a(\boldsymbol{k}) \cos \omega t + b(\boldsymbol{k}) \sin \omega t] \, d^3k \tag{20.134}$$

with $\omega = \sqrt{\boldsymbol{k}^2 + m^2}$ makes the action (20.130) for $P = 0$ stationary because it is a solution of the equation of motion $\nabla^2\phi - \ddot{\phi} - m^2\phi = 0$. In terms of the Fourier transforms

$$\tilde{\phi}(\boldsymbol{k}, t_a) = \int e^{-i\boldsymbol{k}\cdot\boldsymbol{x}} \, \phi(\boldsymbol{x}, t_a) \frac{d^3x}{(2\pi)^3} \quad \text{and} \quad \tilde{\phi}(\boldsymbol{k}, t_b) = \int e^{-i\boldsymbol{k}\cdot\boldsymbol{x}} \, \phi(\boldsymbol{x}, t_b) \frac{d^3x}{(2\pi)^3}, \tag{20.135}$$

the solution that goes from $\phi(\boldsymbol{x}, t_a)$ to $\phi(\boldsymbol{x}, t_b)$ is

$$\phi(\boldsymbol{x}, t) = \int e^{i\boldsymbol{k}\cdot\boldsymbol{x}} \, \frac{\sin \omega(t_b - t) \, \tilde{\phi}(\boldsymbol{k}, t_a) + \sin \omega(t - t_a) \, \tilde{\phi}(\boldsymbol{k}, t_b)}{\sin \omega(t_b - t_a)} \, d^3k. \tag{20.136}$$

The solution that evolves from $\phi(\boldsymbol{x}, t_a)$ and $\dot{\phi}(\boldsymbol{x}, t_a)$ is

$$\phi(\boldsymbol{x}, t) = \int e^{i\boldsymbol{k}\cdot\boldsymbol{x}} \left[\cos \omega(t - t_a) \, \tilde{\phi}(\boldsymbol{k}, t_a) + \frac{\sin \omega(t - t_a)}{\omega} \, \tilde{\dot{\phi}}(\boldsymbol{k}, t_a) \right] d^3k \tag{20.137}$$

in which the Fourier transform $\tilde{\dot{\phi}}(\boldsymbol{k}, t_a)$ is defined as in (20.135).

Like a position operator (20.98), a field at time t is defined as

$$\phi(\boldsymbol{x}, t) = e^{itH/\hbar} \phi(\boldsymbol{x}, 0) e^{-itH/\hbar} \tag{20.138}$$

in which $\phi(\boldsymbol{x}) = \phi(\boldsymbol{x}, 0)$ is the field at time zero, which obeys the commutation relations (20.114). The time-ordered product of several fields is their product with newer (later time) fields standing to the left of older (earlier time) fields as in the definition (20.100). The logic (20.102–20.106) of the derivation of the path-formulas for time-ordered products of position operators also applies to field operators. One finds (Exercise 20.14) for the mean value of the time-ordered product of two fields in an energy eigenstate $|n\rangle$

$$\langle n|\mathcal{T}[\phi(x_1)\phi(x_2)]|n\rangle = \frac{\int \langle n|\phi_b\rangle \phi(x_1)\phi(x_2) e^{iS[\phi]/\hbar} \langle \phi_a|n\rangle \, D\phi}{\int \langle n|\phi_b\rangle e^{iS[\phi]/\hbar} \langle \phi_a|n\rangle \, D\phi} \tag{20.139}$$

in which the integrations are over all paths that go from before t_1 and t_2 to after both times. The analogous result for several fields is (Exercise 20.15)

$$\langle n|\mathcal{T}[\phi(x_1)\cdots\phi(x_k)]|n\rangle = \frac{\displaystyle\int \langle n|\phi_b\rangle \phi(x_1)\cdots\phi(x_k) e^{iS[\phi]/\hbar}\langle\phi_a|n\rangle\, D\phi}{\displaystyle\int \langle n|\phi_b\rangle e^{iS[\phi]/\hbar}\langle\phi_a|n\rangle\, D\phi} \tag{20.140}$$

in which the integrations are over all paths that go from before the times $t_1, \ldots, t_k$ to after them.

20.9 Finite-Temperature Field Theory

Since the Boltzmann operator $e^{-\beta H} = e^{-H/(kT)}$ is the time evolution operator $e^{-itH/\hbar}$ at the imaginary time $t = -i\hbar\beta = -i\hbar/(kT)$, the formulas of finite-temperature field theory are those of quantum field theory with t replaced by $-iu = -i\hbar\beta = -i\hbar/(kT)$.

As in Section 20.8, we use as our hamiltonian $H = K + V$ where K and V are sums over all lattice vertices $v = a(i, j, k) = (ai, aj, ak)$ of the cubes of volume a^3 times the squared momentum and the potential energy

$$H = K + V = \frac{a^3}{2}\sum_v \pi_v^2 + \frac{a^3}{2}\sum_v (\nabla\phi_v)^2 + m^2\phi_v^2 + P(\phi_v). \tag{20.141}$$

A matrix element of the first term of the Trotter product formula (20.7)

$$e^{-\beta(K+V)} = \lim_{n\to\infty} \left(e^{-\beta K/n}\, e^{-\beta V/n}\right)^n \tag{20.142}$$

is the imaginary-time version of (20.125) with $\epsilon = \hbar\beta/n$

$$\langle\phi_1|e^{-\epsilon K}\, e^{-\epsilon V}|\phi_a\rangle = e^{-\epsilon V(\phi_a)} \prod_v \left[\int \frac{a^3 d\pi'_v}{2\pi} e^{a^3[-\epsilon\pi_v^2/2 + i(\phi_{1v}-\phi_{av})\pi'_v]}\right]. \tag{20.143}$$

Setting $\dot\phi_{av} = (\phi_{1v} - \phi_{av})/\epsilon$, we find, instead of (20.126)

$$\langle\phi_1|e^{-\epsilon K}\, e^{-\epsilon V}|\phi_a\rangle = \prod_v \left[\left(\frac{a^3}{2\pi\epsilon}\right)^{1/2} e^{-\epsilon a^3[\dot\phi_{av}^2 + (\nabla\phi_{av})^2 + m^2\phi_{av}^2 + P(\phi_v)]/2}\right]. \tag{20.144}$$

The product of $n = \hbar\beta/\epsilon$ such inverse-temperature intervals is

$$\langle\phi_b|e^{-\beta H}|\phi_a\rangle = \prod_v \left[\left(\frac{a^3 n}{2\pi\beta}\right)^{n/2} \int e^{-S_{ev}}\, D\phi_v\right] \tag{20.145}$$

in which the euclidian action is

$$S_{ev} = \frac{\beta}{n}\frac{a^3}{2}\sum_{j=0}^{n-1} \left[\dot\phi_{jv}^2 + (\nabla\phi_{jv})^2 + m^2\phi_{jv}^2 + P(\phi_v)\right] \tag{20.146}$$

where $\dot\phi_{jv} = n(\phi_{j+1,v} - \phi_{j,v})/\beta$ and $D\phi_v = d\phi_{n-1,v}\cdots d\phi_{1,v}$.

The amplitude $\langle\phi_b|e^{-(\beta_b-\beta_a)H}|\phi_a\rangle$ is the integral over all fields that go from $\phi_a(\boldsymbol{x})$ at β_a to $\phi_b(\boldsymbol{x})$ at β_b each weighted by an exponential

$$\langle\phi_b|e^{-(\beta_b-\beta_a)H}|\phi_a\rangle = \int e^{-S_e[\phi]}\,D\phi \tag{20.147}$$

of its euclidian action

$$S_e[\phi] = \int_{\beta_a}^{\beta_b} du \int d^3x\, \frac{1}{2}\left[\dot{\phi}^2 + (\nabla\phi)^2 + m^2\phi^2 + P(\phi)\right] \tag{20.148}$$

in which $D\phi$ is the $n \to \infty$ limit of the product over all spatial vertices v

$$D\phi = \prod_v \left[\left(\frac{a^3 n}{2\pi(\beta_b - \beta_a)}\right)^{n/2} d\phi_{n-1,v}\cdots d\phi_{1,v}\right]. \tag{20.149}$$

Equivalently, the Boltzmann operator is

$$e^{-(\beta_b-\beta_a)H} = \int |\phi_b\rangle e^{-S_e[\phi]}\,\langle\phi_a|\,D\phi\,D\phi_a D\phi_b \tag{20.150}$$

in which $D\phi_a D\phi_b = \prod_v d\phi_{a,v} d\phi_{b,v}$ is an integral over the initial and final states.

The trace of the Boltzmann operator is the partition function

$$Z(\beta) = \mathrm{Tr}(e^{-\beta H}) = \int e^{-S_e[\phi]}\,\langle\phi_a|\phi_b\rangle\,D\phi\,D\phi_a D\phi_b = \int e^{-S_e[\phi]}\,D\phi\,D\phi_a \tag{20.151}$$

which is an integral over all fields that go back to themselves in euclidian time β.

Like a position operator (20.99), a field at an imaginary time $t = -iu = -i\hbar\beta$ is defined as

$$\phi_e(\boldsymbol{x}, u) = \phi_e(\boldsymbol{x}, \hbar\beta) = e^{uH/\hbar}\phi(\boldsymbol{x}, 0)\,e^{-uH/\hbar} \tag{20.152}$$

in which $\phi(\boldsymbol{x}) = \phi(\boldsymbol{x}, 0) = \phi_e(\boldsymbol{x}, 0)$ is the field at time zero, which obeys the commutation relations (20.114). The euclidian-time-ordered product of several fields is their product with newer (higher $u = \hbar\beta$) fields standing to the left of older (lower $u = \hbar\beta$) fields as in the definition (20.101).

The euclidian path integrals for the mean values of euclidian-time-ordered-products of fields are similar to those (20.139 and 20.140) for ordinary time-ordered-products. The euclidian-time-ordered-product of the fields $\phi(x_j) = \phi(\boldsymbol{x}_j, u_j)$ is the path integral

$$\langle n|\mathcal{T}[\phi_e(x_1)\phi_e(x_2)]|n\rangle = \frac{\displaystyle\int \langle n|\phi_b\rangle\phi(x_1)\phi(x_2)e^{-S_e[\phi]/\hbar}\langle\phi_a|n\rangle\,D\phi}{\displaystyle\int \langle n|\phi_b\rangle e^{-S_e[\phi]/\hbar}\langle\phi_a|n\rangle\,D\phi} \tag{20.153}$$

in which the integrations are over all paths that go from before u_1 and u_2 to after both euclidian times. The analogous result for several fields is

$$\langle n|\mathcal{T}[\phi_e(x_1)\cdots\phi_e(x_k)]|n\rangle = \frac{\int \langle n|\phi_b\rangle\phi(x_1)\cdots\phi(x_k)e^{-S_e[\phi]/\hbar}\langle\phi_a|n\rangle\,D\phi}{\int \langle n|\phi_b\rangle e^{-S_e[\phi]/\hbar}\langle\phi_a|n\rangle\,D\phi} \tag{20.154}$$

in which the integrations are over all paths that go from before the times $u_1, \ldots, u_k$ to after them.

In the low-temperature $\beta = 1/(kT) \to \infty$ limit, the Boltzmann operator is proportional to the outer product $|0\rangle\langle 0|$ of the ground-state kets, $e^{-\beta H} \to e^{-\beta E_0}|0\rangle\langle 0|$. In this limit, the integrations are over all fields that run from $u = -\infty$ to $u = \infty$ and the only energy eigenstate that contributes is the ground state of the theory

$$\langle 0|\mathcal{T}[\phi_e(x_1)\cdots\phi_e(x_k)]|0\rangle = \frac{\int \langle 0|\phi_b\rangle\phi(x_1)\cdots\phi(x_k)e^{-S_e[q]/\hbar}\langle\phi_a|0\rangle\,D\phi}{\int \langle 0|\phi_b\rangle e^{-S_e[q]/\hbar}\langle\phi_a|0\rangle\,D\phi}. \tag{20.155}$$

Formulas like this one are used in lattice gauge theory.

20.10 Perturbation Theory

Field theories with hamiltonians that are quadratic in their fields like

$$H_0 = \int \frac{1}{2}\left[\pi^2(x) + (\nabla\phi(x))^2 + m^2\phi^2(x)\right] d^3x \tag{20.156}$$

are soluble. Their fields evolve in time as

$$\phi(\boldsymbol{x}, t) = e^{itH_0}\phi(\boldsymbol{x}, 0)e^{-itH_0}. \tag{20.157}$$

The mean value in the ground state of H_0 of a time-ordered product of these fields is a ratio (20.140) of path integrals

$$\langle 0|\mathcal{T}\left[\phi(x_1)\cdots\phi(x_k)\right]|0\rangle = \frac{\int \langle 0|\phi_b\rangle\,\phi(x_1)\cdots\phi(x_n)\,e^{iS_0[\phi]}\langle\phi_a|0\rangle\,D\phi}{\int \langle 0|\phi_b\rangle\,e^{iS_0[\phi]}\langle\phi_a|0\rangle\,D\phi} \tag{20.158}$$

in which the action $S_0[\phi]$ is quadratic in the field ϕ

$$S_0[\phi] = \frac{1}{2}\int -\partial_a\phi(x)\partial^a\phi(x) - m^2\phi^2(x)\,d^4x. \tag{20.159}$$

Here $-\partial_a \phi \partial^a \phi = \dot{\phi}^2 - (\nabla\phi)^2$, and the integrations are over all fields that run from ϕ_a at a time before the times $t_1, \ldots, t_k$ to ϕ_b at a time after $t_1, \ldots, t_k$. The path integrals in the ratio (20.158) are gaussian and doable.

The Fourier transforms

$$\tilde{\phi}(p) = \int e^{-ipx}\phi(x)\, d^4x \quad \text{and} \quad \phi(x) = \int e^{ipx}\tilde{\phi}(p)\, \frac{d^4p}{(2\pi)^4} \tag{20.160}$$

turn the spacetime derivatives in the action into a quadratic form

$$S_0[\phi] = -\frac{1}{2}\int |\tilde{\phi}(p)|^2\,(p^2 + m^2)\,\frac{d^4p}{(2\pi)^4} \tag{20.161}$$

in which $p^2 = \boldsymbol{p}^2 - p^{02}$ and $\tilde{\phi}(-p) = \tilde{\phi}^*(p)$ by (4.25) since the field ϕ is real.

The initial $\langle\phi_a|0\rangle$ and final $\langle 0|\phi_b\rangle$ wave functions produce the $i\epsilon$ in the Feynman propagator (6.260). Although its exact form doesn't matter here, the wave function $\langle\phi|0\rangle$ of the ground state of H_0 is the exponential (19.53)

$$\langle\phi|0\rangle = c\exp\left[-\frac{1}{2}\int |\tilde{\phi}(\boldsymbol{p})|^2\sqrt{\boldsymbol{p}^2 + m^2}\,\frac{d^3p}{(2\pi)^3}\right] \tag{20.162}$$

in which $\tilde{\phi}(\boldsymbol{p})$ is the spatial Fourier transform of the eigenvalue $\phi(\boldsymbol{x})$

$$\tilde{\phi}(\boldsymbol{p}) = \int e^{-i\boldsymbol{p}\cdot\boldsymbol{x}}\,\phi(\boldsymbol{x})\,d^3x \tag{20.163}$$

and c is a normalization factor that will cancel in ratios of path integrals.

Apart from $-2i\ln c$ which we will not keep track of, the wave functions $\langle\phi_a|0\rangle$ and $\langle 0|\phi_b\rangle$ add to the action $S_0[\phi]$ the term

$$\Delta S_0[\phi] = \frac{i}{2}\int\sqrt{\boldsymbol{p}^2 + m^2}\left(|\tilde{\phi}(\boldsymbol{p}, t)|^2 + |\tilde{\phi}(\boldsymbol{p}, -t)|^2\right)\frac{d^3p}{(2\pi)^3} \tag{20.164}$$

in which we envision taking the limit $t \to \infty$ with $\phi(\boldsymbol{x}, t) = \phi_b(\boldsymbol{x})$ and $\phi(\boldsymbol{x}, -t) = \phi_a(\boldsymbol{x})$. The identity (Weinberg, 1995, pp. 386–388)

$$f(+\infty) + f(-\infty) = \lim_{\epsilon\to 0+}\epsilon\int_{-\infty}^{\infty} f(t)\,e^{-\epsilon|t|}\,dt \tag{20.165}$$

(Exercise 20.22) allows us to write $\Delta S_0[\phi]$ as

$$\Delta S_0[\phi] = \lim_{\epsilon\to 0+}\frac{i\epsilon}{2}\int\sqrt{\boldsymbol{p}^2 + m^2}\int_{-\infty}^{\infty}|\tilde{\phi}(\boldsymbol{p}, t)|^2\,e^{-\epsilon|t|}\,dt\,\frac{d^3p}{(2\pi)^3}. \tag{20.166}$$

So to first order in ϵ, the change in the action is (Exercise 20.23)

$$\Delta S_0[\phi] = \lim_{\epsilon\to 0+} \frac{i\epsilon}{2} \int \sqrt{\boldsymbol{p}^2 + m^2} \int_{-\infty}^{\infty} |\tilde{\phi}(\boldsymbol{p}, t)|^2 \, dt \, \frac{d^3 p}{(2\pi)^3}$$
$$= \lim_{\epsilon\to 0+} \frac{i\epsilon}{2} \int \sqrt{\boldsymbol{p}^2 + m^2} \, |\tilde{\phi}(p)|^2 \, \frac{d^4 p}{(2\pi)^4}. \tag{20.167}$$

Thus the modified action is

$$S_0[\phi, \epsilon] = S_0[\phi] + \Delta S_0[\phi] = -\frac{1}{2} \int |\tilde{\phi}(p)|^2 \left(p^2 + m^2 - i\epsilon\sqrt{\boldsymbol{p}^2 + m^2}\right) \frac{d^4 p}{(2\pi)^4}$$
$$= -\frac{1}{2} \int |\tilde{\phi}(p)|^2 \left(p^2 + m^2 - i\epsilon\right) \frac{d^4 p}{(2\pi)^4} \tag{20.168}$$

since the square root is positive. In terms of the modified action, our formula (20.158) for the time-ordered product is the ratio

$$\langle 0|\mathcal{T}\left[\phi(x_1)\cdots\phi(x_n)\right]|0\rangle = \frac{\displaystyle\int \phi(x_1)\cdots\phi(x_n)\, e^{iS_0[\phi,\epsilon]}\, D\phi}{\displaystyle\int e^{iS_0[\phi,\epsilon]}\, D\phi}. \tag{20.169}$$

We can use this formula (20.169) to express the mean value in the vacuum $|0\rangle$ of the time-ordered exponential of a spacetime integral of $j(x)\phi(x)$, in which $j(x)$ is a classical (c-number, external) current, as the ratio

$$Z_0[j] \equiv \langle 0|\,\mathcal{T}\left\{\exp\left[i\int j(x)\,\phi(x)\,d^4x\right]\right\}|0\rangle$$
$$= \frac{\displaystyle\int \exp\left[i\int j(x)\,\phi(x)\,d^4x\right] e^{iS_0[\phi,\epsilon]}\, D\phi}{\displaystyle\int e^{iS_0[\phi,\epsilon]}\, D\phi}. \tag{20.170}$$

Since the state $|0\rangle$ is normalized, the mean value $Z_0[0]$ is unity, $Z_0[0] = 1$. If we absorb the current into the action

$$S_0[\phi, \epsilon, j] = S_0[\phi, \epsilon] + \int j(x)\,\phi(x)\,d^4x \tag{20.171}$$

then in terms of the current's Fourier transform

$$\tilde{j}(p) = \int e^{-ipx}\, j(x)\, d^4x \tag{20.172}$$

the modified action $S_0[\phi, \epsilon, j]$ is (Exercise 20.24)

$$S_0[\phi, \epsilon, j] = -\frac{1}{2} \int \left[|\tilde{\phi}(p)|^2 \left(p^2 + m^2 - i\epsilon\right) - \tilde{j}^*(p)\tilde{\phi}(p) - \tilde{\phi}^*(p)\tilde{j}(p)\right] \frac{d^4 p}{(2\pi)^4}. \tag{20.173}$$

Changing variables to $\tilde{\psi}(p) = \tilde{\phi}(p) - \tilde{j}(p)/(p^2 + m^2 - i\epsilon)$, we can write the action $S_0[\phi, \epsilon, j]$ as (Exercise 20.25)

$$S_0[\phi, \epsilon, j] = -\frac{1}{2}\int \left[|\tilde{\psi}(p)|^2 \left(p^2 + m^2 - i\epsilon\right) - \frac{\tilde{j}^*(p)\tilde{j}(p)}{(p^2 + m^2 - i\epsilon)} \right] \frac{d^4p}{(2\pi)^4}$$

$$= S_0[\psi, \epsilon] + \frac{1}{2}\int \left[\frac{\tilde{j}^*(p)\tilde{j}(p)}{(p^2 + m^2 - i\epsilon)} \right] \frac{d^4p}{(2\pi)^4}. \tag{20.174}$$

And since $D\phi = D\psi$, our formula (20.170) gives simply (Exercise 20.26)

$$Z_0[j] = \exp\left(\frac{i}{2}\int \frac{|\tilde{j}(p)|^2}{p^2 + m^2 - i\epsilon} \frac{d^4p}{(2\pi)^4} \right). \tag{20.175}$$

Going back to position space, one finds (Exercise 20.27)

$$Z_0[j] = \exp\left[\frac{i}{2}\int j(x)\, \Delta(x - x')\, j(x')\, d^4x\, d^4x' \right] \tag{20.176}$$

in which $\Delta(x - x')$ is Feynman's **propagator** (6.260)

$$\Delta(x - x') = \Delta_F(x - x') = \int \frac{e^{ip(x-x')}}{p^2 + m^2 - i\epsilon} \frac{d^4p}{(2\pi)^4}. \tag{20.177}$$

The functional derivative (Chapter 19) of $Z_0[j]$, defined by (20.170), is

$$\frac{1}{i}\frac{\delta Z_0[j]}{\delta j(x)} = \langle 0|\, \mathcal{T}\left[\phi(x) \exp\left(i\int j(x')\phi(x') d^4x' \right) \right] |0\rangle \tag{20.178}$$

while that of equation (20.176) is

$$\frac{1}{i}\frac{\delta Z_0[j]}{\delta j(x)} = Z_0[j] \int \Delta(x - x')\, j(x')\, d^4x'. \tag{20.179}$$

Thus the second functional derivative of $Z_0[j]$ evaluated at $j = 0$ gives

$$\langle 0|\, \mathcal{T}\left[\phi(x)\phi(x') \right] |0\rangle = \frac{1}{i^2}\frac{\delta^2 Z_0[j]}{\delta j(x)\delta j(x')}\bigg|_{j=0} = -i\, \Delta(x - x'). \tag{20.180}$$

Similarly, one may show (Exercise 20.28) that

$$\langle 0|\, \mathcal{T}\big[\phi(x_1)\phi(x_2)\phi(x_3)\phi(x_4) \big] |0\rangle = \frac{1}{i^4}\frac{\delta^4 Z_0[j]}{\delta j(x_1)\delta j(x_2)\delta j(x_3)\delta j(x_4)}\bigg|_{j=0}$$

$$= -\Delta(x_1 - x_2)\Delta(x_3 - x_4) - \Delta(x_1 - x_3)\Delta(x_2 - x_4)$$

$$- \Delta(x_1 - x_4)\Delta(x_2 - x_3). \tag{20.181}$$

Suppose now that we add a potential $V(\phi)$ to the free hamiltonian (20.156). Scattering amplitudes are matrix elements of the time-ordered exponential $\mathcal{T}\exp\left[-i\int V(\phi)\,d^4x\right]$ (Weinberg, 1995, p. 260). Our formula (20.169) for the mean value in the ground state $|0\rangle$ of the free hamiltonian H_0 of any time-ordered product of fields leads us to

$$\langle 0|\mathcal{T}\left\{\exp\left[-i\int V(\phi)\,d^4x\right]\right\}|0\rangle = \frac{\displaystyle\int \exp\left[-i\int V(\phi)\,d^4x\right] e^{iS_0[\phi,\epsilon]}\,D\phi}{\displaystyle\int e^{iS_0[\phi,\epsilon]}\,D\phi}. \tag{20.182}$$

Using (20.180 and 20.181), we can cast this expression into the magical form

$$\langle 0|\mathcal{T}\left\{\exp\left[-i\int V(\phi)\,d^4x\right]\right\}|0\rangle = \exp\left[-i\int V\left(\frac{\delta}{i\delta j(x)}\right)d^4x\right] Z_0[j]\Bigg|_{j=0}. \tag{20.183}$$

The generalization of the path-integral formula (20.169) to the ground state $|\Omega\rangle$ of an interacting theory with action S is

$$\langle\Omega|\mathcal{T}\left[\phi(x_1)\cdots\phi(x_n)\right]|\Omega\rangle = \frac{\displaystyle\int \phi(x_1)\cdots\phi(x_n)\,e^{iS[\phi,\epsilon]}\,D\phi}{\displaystyle\int e^{iS[\phi,\epsilon]}\,D\phi} \tag{20.184}$$

in which a term like $i\epsilon\phi^2$ is added to make the modified action $S[\phi,\epsilon]$.

These are some of the techniques one uses to make states of incoming and outgoing particles and to compute scattering amplitudes (Weinberg, 1995, 1996; Srednicki, 2007; Zee, 2010).

20.11 Application to Quantum Electrodynamics

In the Coulomb gauge $\nabla\cdot\boldsymbol{A}=0$, the QED hamiltonian is

$$H = H_m + \int\left[\frac{1}{2}\boldsymbol{\pi}^2 + \frac{1}{2}(\nabla\times\boldsymbol{A})^2 - \boldsymbol{A}\cdot\boldsymbol{j}\right]d^3x + V_C \tag{20.185}$$

in which H_m is the matter hamiltonian, and V_C is the Coulomb term

$$V_C = \frac{1}{2}\int\frac{j^0(\boldsymbol{x},t)\,j^0(\boldsymbol{y},t)}{4\pi|\boldsymbol{x}-\boldsymbol{y}|}\,d^3x\,d^3y. \tag{20.186}$$

The operators $\boldsymbol{A}$ and $\boldsymbol{\pi}$ are canonically conjugate, but they satisfy the Coulomb-gauge conditions $\nabla\cdot\boldsymbol{A}=0$ and $\nabla\cdot\boldsymbol{\pi}=0$.

One may show (Weinberg, 1995, pp. 413–418) that in this theory, the analog of Equation (20.184) is

$$\langle\Omega|\mathcal{T}\left[\mathcal{O}_1\cdots\mathcal{O}_n\right]|\Omega\rangle = \frac{\int \mathcal{O}_1\cdots\mathcal{O}_n\, e^{iS_C}\,\delta[\nabla\cdot\boldsymbol{A}]\,DA\,D\psi}{\int e^{iS_C}\,\delta[\nabla\cdot\boldsymbol{A}]\,DA\,D\psi} \tag{20.187}$$

in which the Coulomb-gauge action is

$$S_C = \int \frac{1}{2}\dot{\boldsymbol{A}}^2 - \frac{1}{2}(\nabla\times\boldsymbol{A})^2 + \boldsymbol{A}\cdot\boldsymbol{j} + \mathcal{L}_m\, d^4x - \int V_C\, dt \tag{20.188}$$

and the functional delta function

$$\delta[\nabla\cdot\boldsymbol{A}] = \prod_x \delta(\nabla\cdot\boldsymbol{A}(x)) \tag{20.189}$$

enforces the Coulomb-gauge condition. The term $\mathcal{L}_m$ is the action density of the matter field ψ.

Tricks are available. We introduce a new field $A^0(x)$ and consider the factor

$$F = \int \exp\left[i\int \frac{1}{2}\left(\nabla A^0 + \nabla\triangle^{-1}j^0\right)^2 d^4x\right] DA^0 \tag{20.190}$$

which is just a *number* independent of the charge density j^0 since we can cancel the j^0 term by shifting A^0. By $\triangle^{-1}$, we mean $-\ 1/4\pi|\boldsymbol{x}-\boldsymbol{y}|$. By integrating by parts, we can write the number F as (Exercise 20.29)

$$\begin{aligned} F &= \int \exp\left[i\int \frac{1}{2}\left(\nabla A^0\right)^2 - A^0 j^0 - \frac{1}{2}\,j^0\triangle^{-1}j^0\, d^4x\right] DA^0 \\ &= \int \exp\left[i\int \frac{1}{2}\left(\nabla A^0\right)^2 - A^0 j^0\, d^4x + i\int V_C\, dt\right] DA^0. \end{aligned} \tag{20.191}$$

So when we multiply the numerator and denominator of the amplitude (20.187) by F, the awkward Coulomb term V_C cancels, and we get

$$\langle\Omega|\mathcal{T}\left[\mathcal{O}_1\cdots\mathcal{O}_n\right]|\Omega\rangle = \frac{\int \mathcal{O}_1\cdots\mathcal{O}_n\, e^{iS'}\,\delta[\nabla\cdot\boldsymbol{A}]\,DA\,D\psi}{\int e^{iS'}\,\delta[\nabla\cdot\boldsymbol{A}]\,DA\,D\psi} \tag{20.192}$$

where now DA includes all four components A^μ and

$$S' = \int \frac{1}{2}\dot{\boldsymbol{A}}^2 - \frac{1}{2}(\nabla\times\boldsymbol{A})^2 + \frac{1}{2}\left(\nabla A^0\right)^2 + \boldsymbol{A}\cdot\boldsymbol{j} - A^0 j^0 + \mathcal{L}_m\, d^4x. \tag{20.193}$$

Since the delta-functional $\delta[\nabla \cdot A]$ enforces the Coulomb-gauge condition, we can add to the action S' the term $(\nabla \cdot \dot{A})\, A^0$ which is $-\dot{A} \cdot \nabla A^0$ after we integrate by parts and drop the surface term. This extra term makes the action gauge invariant

$$\begin{aligned} S &= \int \frac{1}{2}(\dot{A} - \nabla A^0)^2 - \frac{1}{2}(\nabla \times A)^2 + A \cdot j - A^0 j^0 + \mathcal{L}_m \, d^4x \\ &= \int -\tfrac{1}{4} F_{ab}\, F^{ab} + A^b j_b + \mathcal{L}_m \; d^4x. \end{aligned} \tag{20.194}$$

Thus at this point we have

$$\langle \Omega | \mathcal{T}\, [\mathcal{O}_1 \cdots \mathcal{O}_n]\, | \Omega \rangle = \frac{\int \mathcal{O}_1 \cdots \mathcal{O}_n \, e^{iS}\, \delta[\nabla \cdot A]\, DA\, D\psi}{\int e^{iS}\, \delta[\nabla \cdot A]\, DA\, D\psi} \tag{20.195}$$

in which S is the gauge-invariant action (20.194), and the integral is over all fields. The only relic of the Coulomb gauge is the gauge-fixing delta functional $\delta[\nabla \cdot A]$.

We now make the gauge transformations $A'_b(x) = A_b(x) + \partial_b \Lambda(x)$ and $\psi'(x) = e^{iq\Lambda(x)} \psi(x)$ in the numerator and also, using a different gauge transformation Λ', in the denominator of the ratio (20.195) of path integrals. Since we are integrating over all gauge fields, these gauge transformations merely change the order of integration in the numerator and denominator of that ratio. They are like replacing $\int_{-\infty}^{\infty} f(x)\, dx$ by $\int_{-\infty}^{\infty} f(y)\, dy$. They change nothing, and so $\langle \Omega | \mathcal{T}\, [\mathcal{O}_1 \cdots \mathcal{O}_n]\, | \Omega \rangle = \langle \Omega | \mathcal{T}\, [\mathcal{O}_1 \cdots \mathcal{O}_n]\, | \Omega \rangle'$ in which the prime refers to the gauge transformations Λ and Λ'.

We've seen that the action S is gauge invariant. So is the measure $DA\, D\psi$. We now restrict ourselves to operators $\mathcal{O}_1 \cdots \mathcal{O}_n$ that are **gauge invariant**. So in $\langle \Omega | \mathcal{T}\, [\mathcal{O}_1 \cdots \mathcal{O}_n]\, | \Omega \rangle'$, the replacement of the fields by their gauge transforms affects only the Coulomb-gauge term $\delta[\nabla \cdot A]$

$$\langle \Omega | \mathcal{T}\, [\mathcal{O}_1 \cdots \mathcal{O}_n]\, | \Omega \rangle = \frac{\int \mathcal{O}_1 \cdots \mathcal{O}_n \, e^{iS}\, \delta[\nabla \cdot A + \triangle \Lambda]\, DA\, D\psi}{\int e^{iS}\, \delta[\nabla \cdot A + \triangle \Lambda']\, DA\, D\psi}. \tag{20.196}$$

We now have two choices. If we integrate over all gauge functions $\Lambda(x)$ and $\Lambda'(x)$ in both the numerator and the denominator of this ratio (20.196), then apart from over-all constants that cancel, the mean value in the vacuum of the time-ordered product is the ratio

$$\langle \Omega | \mathcal{T}\, [\mathcal{O}_1 \cdots \mathcal{O}_n]\, | \Omega \rangle = \frac{\int \mathcal{O}_1 \cdots \mathcal{O}_n \, e^{iS}\, DA\, D\psi}{\int e^{iS}\, DA\, D\psi} \tag{20.197}$$

in which we integrate over all matter fields, gauge fields, and gauges. That is, **we do not fix the gauge**.

The analogous formula for the euclidian time-ordered product is

$$\langle\Omega|\mathcal{T}\left[\mathcal{O}_{e,1}\cdots\mathcal{O}_{e,n}\right]|\Omega\rangle = \frac{\displaystyle\int \mathcal{O}_1\cdots\mathcal{O}_n\, e^{-S_e}\, DA\, D\psi}{\displaystyle\int e^{-S_e}\, DA\, D\psi} \tag{20.198}$$

in which the euclidian action S_e is the spacetime integral of the energy density. This formula is quite general; it holds in nonabelian gauge theories and is important in lattice gauge theory.

Our second choice is to multiply the numerator and the denominator of the ratio (20.196) by the exponential $\exp[-i\frac{1}{2}\alpha\int(\triangle\Lambda)^2\, d^4x]$ and then integrate over $\Lambda(x)$ in the numerator and over $\Lambda'(x)$ in the denominator. This operation just multiplies the numerator and denominator by the same constant factor, which cancels. But if before integrating over all gauge transformations, we shift Λ so that $\triangle\Lambda$ changes to $\triangle\Lambda - \dot{A}^0$, then the exponential factor is $\exp[-i\frac{1}{2}\alpha\int(\dot{A}^0 - \triangle\Lambda)^2\, d^4x]$. Now when we integrate over $\Lambda(x)$, the delta function $\delta(\nabla\cdot\boldsymbol{A}+\triangle\Lambda)$ replaces $\triangle\Lambda$ by $-\nabla\cdot\boldsymbol{A}$ in the inserted exponential, converting it to $\exp[-i\frac{1}{2}\alpha\int(\dot{A}^0 + \nabla\cdot\boldsymbol{A})^2\, d^4x]$. This term changes the gauge-invariant action (20.194) to the gauge-fixed action

$$S_\alpha = \int -\frac{1}{4}F_{ab}\,F^{ab} - \frac{\alpha}{2}(\partial_b A^b)^2 + A^b j_b + \mathcal{L}_m\; d^4x. \tag{20.199}$$

This Lorentz-invariant, gauge-fixed action is much easier to use than the Coulomb-gauge action (20.188) with the Coulomb potential (20.186). We can use it to compute scattering amplitudes perturbatively. The mean value of a time-ordered product of operators in the ground state $|0\rangle$ of the free theory is

$$\langle 0|\mathcal{T}\left[\mathcal{O}_1\cdots\mathcal{O}_n\right]|0\rangle = \frac{\displaystyle\int \mathcal{O}_1\cdots\mathcal{O}_n\, e^{iS_\alpha}\, DA\, D\psi}{\displaystyle\int e^{iS_\alpha}\, DA\, D\psi}. \tag{20.200}$$

By following steps analogous to those that led to (20.177), one may show (Exercise 20.30) that in Feynman's gauge, $\alpha = 1$, the photon propagator is

$$\langle 0|\mathcal{T}\left[A_\mu(x)A_\nu(y)\right]|0\rangle = -\,i\Delta_{\mu\nu}(x-y) = -\,i\int \frac{\eta_{\mu\nu}}{q^2 - i\epsilon}\; e^{iq\cdot(x-y)}\,\frac{d^4q}{(2\pi)^4}. \tag{20.201}$$

20.12 Fermionic Path Integrals

In our brief introduction (1.11–1.12) and (1.44–1.46), to Grassmann variables, we learned that because $\theta^2 = 0$ the most general function $f(\theta)$ of a single Grassmann variable θ is $f(\theta) = a + b\,\theta$. So a complete integral table consists of the integral of this linear function

$$\int f(\theta)\, d\theta = \int a + b\,\theta\, d\theta = a \int d\theta + b \int \theta\, d\theta. \tag{20.202}$$

This equation has two unknowns, the integral $\int d\theta$ of unity and the integral $\int \theta\, d\theta$ of θ. We choose them so that the integral of $f(\theta + \zeta)$

$$\int f(\theta + \zeta)\, d\theta = \int a + b\,(\theta + \zeta)\, d\theta = (a + b\,\zeta) \int d\theta + b \int \theta\, d\theta \tag{20.203}$$

is the same as the integral (20.202) of $f(\theta)$. Thus the integral $\int d\theta$ of unity must vanish, while the integral $\int \theta\, d\theta$ of θ can be any constant, which we choose to be unity. Our complete table of integrals is then

$$\int d\theta = 0 \quad \text{and} \quad \int \theta\, d\theta = 1. \tag{20.204}$$

The anticommutation relations for a fermionic degree of freedom ψ are

$$\{\psi, \psi^\dagger\} \equiv \psi\,\psi^\dagger + \psi^\dagger \psi = 1 \quad \text{and} \quad \{\psi, \psi\} = \{\psi^\dagger, \psi^\dagger\} = 0. \tag{20.205}$$

Because ψ has $\psi^\dagger$, it is conventional to introduce a variable $\theta^* = \theta^\dagger$ that anticommutes with itself and with θ

$$\{\theta^*, \theta^*\} = \{\theta^*, \theta\} = \{\theta, \theta\} = 0. \tag{20.206}$$

The logic that led to (20.204) now gives

$$\int d\theta^* = 0 \quad \text{and} \quad \int \theta^*\, d\theta^* = 1. \tag{20.207}$$

We define the reference state $|0\rangle$ as $|0\rangle \equiv \psi|s\rangle$ for a state $|s\rangle$ that is not annihilated by ψ. Since $\psi^2 = 0$, the operator ψ annihilates the state $|0\rangle$

$$\psi|0\rangle = \psi^2|s\rangle = 0. \tag{20.208}$$

The effect of the operator ψ on the state

$$|\theta\rangle = \exp\left(\psi^\dagger\theta - \frac{1}{2}\theta^*\theta\right)|0\rangle = \left(1 + \psi^\dagger\theta - \frac{1}{2}\theta^*\theta\right)|0\rangle \tag{20.209}$$

is

$$\psi|\theta\rangle = \psi\left(1 + \psi^\dagger\theta - \frac{1}{2}\theta^*\theta\right)|0\rangle = \psi\psi^\dagger\theta|0\rangle = (1 - \psi^\dagger\psi)\theta|0\rangle = \theta|0\rangle \tag{20.210}$$

while that of θ on $|\theta\rangle$ is

$$\theta|\theta\rangle = \theta\left(1 + \psi^\dagger\theta - \frac{1}{2}\theta^*\theta\right)|0\rangle = \theta|0\rangle. \tag{20.211}$$

The state $|\theta\rangle$ therefore is an eigenstate of ψ with eigenvalue θ

$$\psi|\theta\rangle = \theta|\theta\rangle. \tag{20.212}$$

The bra corresponding to the ket $|\zeta\rangle$ is

$$\langle\zeta| = \langle 0|\left(1 + \zeta^*\psi - \frac{1}{2}\zeta^*\zeta\right) \tag{20.213}$$

and the inner product $\langle\zeta|\theta\rangle$ is (Exercise 20.31)

$$\begin{aligned}
\langle\zeta|\theta\rangle &= \langle 0|\left(1 + \zeta^*\psi - \frac{1}{2}\zeta^*\zeta\right)\left(1 + \psi^\dagger\theta - \frac{1}{2}\theta^*\theta\right)|0\rangle \\
&= \langle 0|1 + \zeta^*\psi\psi^\dagger\theta - \frac{1}{2}\zeta^*\zeta - \frac{1}{2}\theta^*\theta + \frac{1}{4}\zeta^*\zeta\theta^*\theta|0\rangle \\
&= \langle 0|1 + \zeta^*\theta - \frac{1}{2}\zeta^*\zeta - \frac{1}{2}\theta^*\theta + \frac{1}{4}\zeta^*\zeta\theta^*\theta|0\rangle \\
&= \exp\left[\zeta^*\theta - \frac{1}{2}\left(\zeta^*\zeta + \theta^*\theta\right)\right].
\end{aligned} \tag{20.214}$$

Example 20.9 (A gaussian integral) For any number c, we can compute the integral of $\exp(c\,\theta^*\theta)$ by expanding the exponential

$$\int e^{c\,\theta^*\theta}\,d\theta^* d\theta = \int (1 + c\,\theta^*\theta)\,d\theta^* d\theta = \int (1 - c\,\theta\,\theta^*)\,d\theta^* d\theta = -c. \tag{20.215}$$

The identity operator for the space of states

$$c|0\rangle + d|1\rangle \equiv c|0\rangle + d\psi^\dagger|0\rangle \tag{20.216}$$

is (Exercise 20.32) the integral

$$I = \int |\theta\rangle\langle\theta|\,d\theta^* d\theta = |0\rangle\langle 0| + |1\rangle\langle 1| \tag{20.217}$$

in which the differentials anticommute with each other and with other fermionic variables: $\{d\theta, d\theta^*\} = 0$, $\{d\theta, \theta\} = 0$, $\{d\theta, \psi\} = 0$, and so forth.

The case of several Grassmann variables $\theta_1, \theta_2, \ldots, \theta_n$ and several Fermi operators $\psi_1, \psi_2, \ldots, \psi_n$ is similar. The θ_k anticommute among themselves and with the Fermi operators

$$\{\theta_i, \theta_j\} = \{\theta_i, \theta_j^*\} = \{\theta_i^*, \theta_j^*\} = 0 \quad \text{and} \quad \{\theta_i, \psi_k\} = \{\theta_i^*, \psi_k\} = 0 \tag{20.218}$$

while the ψ_k satisfy

$$\{\psi_k, \psi_\ell^\dagger\} = \delta_{k\ell} \quad \text{and} \quad \{\psi_k, \psi_l\} = \{\psi_k^\dagger, \psi_\ell^\dagger\} = 0. \tag{20.219}$$

The reference state $|0\rangle$ is

$$|0\rangle = \left(\prod_{k=1}^{n} \psi_k\right) |s\rangle \tag{20.220}$$

in which $|s\rangle$ is any state not annihilated by any ψ_k (so the resulting $|0\rangle$ isn't zero). The direct-product state

$$|\theta\rangle \equiv \exp\left(\sum_{k=1}^{n} \psi_k^\dagger \theta_k - \frac{1}{2}\theta_k^* \theta_k\right) |0\rangle = \left[\prod_{k=1}^{n}\left(1 + \psi_k^\dagger \theta_k - \frac{1}{2}\theta_k^*\theta_k\right)\right] |0\rangle \tag{20.221}$$

is (Exercise 20.33) a simultaneous eigenstate $\psi_k|\theta\rangle = \theta_k|\theta\rangle$ of each ψ_k. It follows that

$$\psi_\ell \psi_k |\theta\rangle = \psi_\ell \theta_k |\theta\rangle = -\theta_k \psi_\ell |\theta\rangle = -\theta_k \theta_\ell |\theta\rangle = \theta_\ell \theta_k |\theta\rangle \tag{20.222}$$

and so too $\psi_k \psi_\ell |\theta\rangle = \theta_k \theta_\ell |\theta\rangle$. Since the ψ's anticommute, their eigenvalues must also

$$\theta_\ell \theta_k |\theta\rangle = \psi_\ell \psi_k |\theta\rangle = -\psi_k \psi_\ell |\theta\rangle = -\theta_k \theta_\ell |\theta\rangle. \tag{20.223}$$

The inner product $\langle \zeta|\theta\rangle$ is

$$\begin{aligned} \langle \zeta|\theta\rangle &= \langle 0| \left[\prod_{k=1}^{n}(1 + \zeta_k^* \psi_k - \frac{1}{2}\zeta_k^*\zeta_k)\right]\left[\prod_{\ell=1}^{n}(1 + \psi_\ell^\dagger \theta_\ell - \frac{1}{2}\theta_\ell^*\theta_\ell)\right] |0\rangle \\ &= \exp\left[\sum_{k=1}^{n} \zeta_k^* \theta_k - \frac{1}{2}\left(\zeta_k^*\zeta_k + \theta_k^*\theta_k\right)\right] = e^{\zeta^\dagger\theta - (\zeta^\dagger\zeta + \theta^\dagger\theta)/2}. \end{aligned} \tag{20.224}$$

The identity operator is

$$I = \int |\theta\rangle\langle\theta| \prod_{k=1}^{n} d\theta_k^* d\theta_k. \tag{20.225}$$

Example 20.10 (Gaussian Grassmann integral) For any 2×2 matrix A, we may compute the gaussian integral

$$g(A) = \int e^{-\theta^\dagger A \theta}\, d\theta_1^* d\theta_1 d\theta_2^* d\theta_2 \tag{20.226}$$

by expanding the exponential. The only terms that survive are the ones that have exactly one of each of the four variables θ_1, θ_2, θ_1^*, and θ_2^*. Thus the integral is the determinant of the matrix A

$$g(A) = \int \frac{1}{2} \left(\theta_k^* A_{k\ell} \theta_\ell\right)^2 \, d\theta_1^* d\theta_1 d\theta_2^* d\theta_2$$
$$= \int \left(\theta_1^* A_{11} \theta_1 \, \theta_2^* A_{22} \theta_2 + \theta_1^* A_{12} \theta_2 \, \theta_2^* A_{21} \theta_1\right) \, d\theta_1^* d\theta_1 d\theta_2^* d\theta_2$$
$$= A_{11} A_{22} - A_{12} A_{21} = \det A. \tag{20.227}$$

The natural generalization to n dimensions is

$$\int e^{-\theta^\dagger A \theta} \prod_{k=1}^{n} d\theta_k^* d\theta_k = \det A \tag{20.228}$$

and is true for any $n \times n$ matrix A. If A is invertible, then the invariance of Grassmann integrals under translations implies that

$$\int e^{-\theta^\dagger A \theta + \theta^\dagger \zeta + \zeta^\dagger \theta} \prod_{k=1}^{n} d\theta_k^* d\theta_k = \int e^{-\theta^\dagger A (\theta + A^{-1}\zeta) + \theta^\dagger \zeta + \zeta^\dagger (\theta + A^{-1}\zeta)} \prod_{k=1}^{n} d\theta_k^* d\theta_k$$
$$= \int e^{-\theta^\dagger A \theta + \zeta^\dagger \theta + \zeta^\dagger A^{-1} \zeta} \prod_{k=1}^{n} d\theta_k^* d\theta_k$$
$$= \int e^{-(\theta^\dagger + \zeta^\dagger A^{-1}) A \theta + \zeta^\dagger \theta + \zeta^\dagger A^{-1} \zeta} \prod_{k=1}^{n} d\theta_k^* d\theta_k$$
$$= \int e^{-\theta^\dagger A \theta + \zeta^\dagger A^{-1} \zeta} \prod_{k=1}^{n} d\theta_k^* d\theta_k$$
$$= \det A \; e^{\zeta^\dagger A^{-1} \zeta}. \tag{20.229}$$

The values of θ and $\theta^\dagger$ that make the argument $-\theta^\dagger A \theta + \theta^\dagger \zeta + \zeta^\dagger \theta$ of the exponential stationary are $\overline{\theta} = A^{-1}\zeta$ and $\overline{\theta^\dagger} = \zeta^\dagger A^{-1}$. So a gaussian Grassmann integral is equal to its exponential evaluated at its stationary point, apart from a prefactor involving the determinant $\det A$. Exercises (20.2 and 20.4) are about the bosonic versions (20.3 and 20.4) of this result.

One may further extend these definitions to a Grassmann field $\chi_m(x)$ and an associated Dirac field $\psi_m(x)$. The $\chi_m(x)$'s anticommute among themselves and with all fermionic variables at all points of spacetime

$$\{\chi_m(x), \chi_n(x')\} = \{\chi_m^*(x), \chi_n(x')\} = \{\chi_m^*(x), \chi_n^*(x')\} = 0 \tag{20.230}$$

and the Dirac field $\psi_m(x)$ obeys the equal-time anticommutation relations

$$\begin{aligned} \{\psi_m(\boldsymbol{x}, t), \psi_n^\dagger(\boldsymbol{x}', t)\} &= \delta_{mn} \, \delta(\boldsymbol{x} - \boldsymbol{x}') \quad (n, m = 1, \ldots, 4) \\ \{\psi_m(\boldsymbol{x}, t), \psi_n(\boldsymbol{x}', t)\} &= \{\psi_m^\dagger(\boldsymbol{x}, t), \psi_n^\dagger(\boldsymbol{x}', t)\} = 0. \end{aligned} \tag{20.231}$$

As in (20.220), we use eigenstates of the field ψ at $t = 0$. If $|0\rangle$ is defined in terms of a state $|s\rangle$ that is not annihilated by any $\psi_m(\boldsymbol{x}, 0)$ as

$$|0\rangle = \left[\prod_{m,\boldsymbol{x}} \psi_m(\boldsymbol{x},0)\right] |s\rangle \tag{20.232}$$

then (Exercise 20.34) the state

$$\begin{aligned} |\chi\rangle &= \exp\left(\int \sum_m \psi_m^\dagger(\boldsymbol{x},0)\, \chi_m(\boldsymbol{x}) - \frac{1}{2}\chi_m^*(\boldsymbol{x})\chi_m(\boldsymbol{x})\, d^3x\right) |0\rangle \\ &= \exp\left(\int \psi^\dagger \chi - \frac{1}{2}\chi^\dagger \chi\, d^3x\right) |0\rangle \end{aligned} \tag{20.233}$$

is an eigenstate of the operator $\psi_m(\boldsymbol{x},0)$ with eigenvalue $\chi_m(\boldsymbol{x})$

$$\psi_m(\boldsymbol{x},0)|\chi\rangle = \chi_m(\boldsymbol{x})|\chi\rangle. \tag{20.234}$$

The inner product of two such states is (Exercise 20.35)

$$\langle\chi'|\chi\rangle = \exp\left[\int \chi'^\dagger \chi - \frac{1}{2}\chi'^\dagger \chi' - \frac{1}{2}\chi^\dagger \chi\; d^3x\right]. \tag{20.235}$$

The identity operator is the integral

$$I = \int |\chi\rangle\langle\chi|\, D\chi^* D\chi \tag{20.236}$$

in which

$$D\chi^* D\chi \equiv \prod_{m,\boldsymbol{x}} d\chi_m^*(\boldsymbol{x}) d\chi_m(\boldsymbol{x}). \tag{20.237}$$

The hamiltonian for a free Dirac field ψ of mass m is the spatial integral

$$H_0 = \int \overline{\psi}\, (\boldsymbol{\gamma}\cdot\nabla + m)\, \psi\; d^3x \tag{20.238}$$

in which $\overline{\psi} \equiv i\psi^\dagger\gamma^0$ and the gamma matrices (11.327) satisfy

$$\{\gamma^a, \gamma^b\} = 2\,\eta^{ab} \tag{20.239}$$

where η is the 4×4 diagonal matrix with entries $(-1, 1, 1, 1)$. Since $\psi|\chi\rangle = \chi|\chi\rangle$ and $\langle\chi'|\psi^\dagger = \langle\chi'|\chi'^\dagger$, the quantity $\langle\chi'|\exp(-i\epsilon H_0)|\chi\rangle$ is by (20.235)

$$\begin{aligned} \langle\chi'|e^{-i\epsilon H_0}|\chi\rangle &= \langle\chi'|\chi\rangle \exp\left[-i\epsilon \int \overline{\chi}'\, (\boldsymbol{\gamma}\cdot\nabla + m)\, \chi\; d^3x\right] \\ &= \exp\left[\int \frac{1}{2}(\chi'^\dagger - \chi^\dagger)\chi - \frac{1}{2}\chi'^\dagger(\chi' - \chi) - i\epsilon\overline{\chi}'(\boldsymbol{\gamma}\cdot\nabla + m)\chi d^3x\right] \\ &= \exp\left\{\epsilon\int\left[\frac{1}{2}\dot{\chi}^\dagger\chi - \frac{1}{2}\chi'^\dagger\dot{\chi} - i\overline{\chi}'\,(\boldsymbol{\gamma}\cdot\nabla + m)\,\chi\right]\, d^3x\right\} \end{aligned} \tag{20.240}$$

in which $\chi'^{\dagger} - \chi^{\dagger} = \epsilon \dot{\chi}^{\dagger}$ and $\chi' - \chi = \epsilon \dot{\chi}$. Everything within the square brackets is multiplied by ϵ, so we may replace $\chi'^{\dagger}$ by $\chi^{\dagger}$ and $\overline{\chi}'$ by $\overline{\chi}$ so as to write to first order in ϵ

$$\langle \chi' | e^{-i\epsilon H_0} | \chi \rangle = \exp\left[\epsilon \int \frac{1}{2} \dot{\chi}^{\dagger} \chi - \frac{1}{2} \chi^{\dagger} \dot{\chi} - i \overline{\chi} \left(\boldsymbol{\gamma} \cdot \nabla + m \right) \chi \, d^3 x \right] \quad (20.241)$$

in which the dependence upon χ' is through the time derivatives.

Putting together $n = 2t/\epsilon$ such matrix elements, integrating over all intermediate-state dyadics $|\chi\rangle\langle\chi|$, and using our formula (20.236), we find

$$\langle \chi_t | e^{-2itH_0} | \chi_{-t} \rangle = \int \exp\left[\int \frac{1}{2} \dot{\chi}^{\dagger} \chi - \frac{1}{2} \chi^{\dagger} \dot{\chi} - i \overline{\chi} \left(\boldsymbol{\gamma} \cdot \nabla + m \right) \chi d^4 x \right] D\chi^* D\chi . \quad (20.242)$$

Integrating $\dot{\chi}^{\dagger} \chi$ by parts and dropping the surface term, we get

$$\langle \chi_t | e^{-2itH_0} | \chi_{-t} \rangle = \int \exp\left[\int - \chi^{\dagger} \dot{\chi} - i \overline{\chi} \left(\boldsymbol{\gamma} \cdot \nabla + m \right) \chi \, d^4 x \right] D\chi^* D\chi . \quad (20.243)$$

Since $-\chi^{\dagger} \dot{\chi} = -i \overline{\chi} \gamma^0 \dot{\chi}$, the argument of the exponential is

$$i \int - \overline{\chi} \gamma^0 \dot{\chi} - \overline{\chi} \left(\boldsymbol{\gamma} \cdot \nabla + m \right) \chi \, d^4 x = i \int - \overline{\chi} \left(\gamma^\mu \partial_\mu + m \right) \chi \, d^4 x . \quad (20.244)$$

We then have

$$\langle \chi_t | e^{-2itH_0} | \chi_{-t} \rangle = \int \exp\left(i \int \mathcal{L}_0(\chi) \, d^4 x \right) D\chi^* D\chi \quad (20.245)$$

in which $\mathcal{L}_0(\chi) = -\overline{\chi} \left(\gamma^\mu \partial_\mu + m \right) \chi$ is the action density (11.329) for a free Dirac field. Thus the amplitude is a path integral with phases given by the classical action $S_0[\chi]$

$$\langle \chi_t | e^{-2itH_0} | \chi_{-t} \rangle = \int e^{i \int \mathcal{L}_0(\chi) \, d^4 x} D\chi^* D\chi = \int e^{i S_0[\chi]} D\chi^* D\chi \quad (20.246)$$

and the integral is over all fields that go from $\chi(\boldsymbol{x}, -t) = \chi_{-t}(\boldsymbol{x})$ to $\chi(\boldsymbol{x}, t) = \chi_t(\boldsymbol{x})$. Any normalization factor will cancel in ratios of such integrals.

Since Fermi fields anticommute, their time-ordered product has an extra minus sign

$$\mathcal{T}\left[\overline{\psi}(x_1) \psi(x_2) \right] = \theta(x_1^0 - x_2^0) \, \overline{\psi}(x_1) \, \psi(x_2) - \theta(x_2^0 - x_1^0) \, \psi(x_2) \, \overline{\psi}(x_1) . \quad (20.247)$$

The logic behind our formulas (20.140) and (20.158) for the time-ordered product of bosonic fields now leads to an expression for the time-ordered product of $2n$ Dirac fields (with $D\chi''$ and $D\chi'$ and so forth suppressed)

$$\langle 0|\mathcal{T}\left[\overline{\psi}(x_1)\cdots\psi(x_{2n})\right]|0\rangle = \frac{\displaystyle\int \langle 0|\chi''\rangle\, \overline{\chi}(x_1)\cdots\chi(x_{2n})\, e^{iS_0[\chi]}\langle\chi'|0\rangle\, D\chi^* D\chi}{\displaystyle\int \langle 0|\chi''\rangle\, e^{iS_0[\chi]}\langle\chi'|0\rangle\, D\chi^* D\chi}. \tag{20.248}$$

As in (20.169), the effect of the inner products $\langle 0|\chi''\rangle$ and $\langle\chi'|0\rangle$ is to insert ϵ-terms which modify the Dirac propagators

$$\langle 0|\mathcal{T}\left[\overline{\psi}(x_1)\cdots\psi(x_{2n})\right]|0\rangle = \frac{\displaystyle\int \overline{\chi}(x_1)\cdots\chi(x_{2n})\, e^{iS_0[\chi,\epsilon]}\, D\chi^* D\chi}{\displaystyle\int e^{iS_0[\chi,\epsilon]}\, D\chi^* D\chi}. \tag{20.249}$$

Imitating (20.170), we introduce a Grassmann external current $\zeta(x)$ and define a fermionic analog of $Z_0[j]$

$$Z_0[\zeta] \equiv \langle 0|\,\mathcal{T}\left[e^{\int \overline{\zeta}\psi+\overline{\psi}\zeta\, d^4x}\right]|0\rangle = \frac{\displaystyle\int e^{\int \overline{\zeta}\chi+\overline{\chi}\zeta\, d^4x}\, e^{iS_0[\chi,\epsilon]} D\chi^* D\chi}{\displaystyle\int e^{iS_0[\chi,\epsilon]} D\chi^* D\chi}. \tag{20.250}$$

Example 20.11 (Feynman's fermion propagator) Since

$$\begin{aligned} i\left(\gamma^\mu\partial_\mu + m\right)\Delta(x-y) &\equiv i\left(\gamma^\mu\partial_\mu + m\right)\int \frac{d^4p}{(2\pi)^4} e^{ip(x-y)}\frac{-i\left(-i\gamma^\nu p_\nu + m\right)}{p^2+m^2-i\epsilon} \\ &= \int \frac{d^4p}{(2\pi)^4} e^{ip(x-y)}\left(i\gamma^\mu p_\mu + m\right)\frac{\left(-i\gamma^\nu p_\nu + m\right)}{p^2+m^2-i\epsilon} \\ &= \int \frac{d^4p}{(2\pi)^4} e^{ip(x-y)}\frac{p^2+m^2}{p^2+m^2-i\epsilon} = \delta^4(x-y), \end{aligned} \tag{20.251}$$

the function $\Delta(x-y)$ is the inverse of the differential operator $i(\gamma^\mu\partial_\mu + m)$. Thus the Grassmann identity (20.229) implies that $Z_0[\zeta]$ is

$$\begin{aligned} \langle 0|\,\mathcal{T}\left[e^{\int \overline{\zeta}\psi+\overline{\psi}\zeta\, d^4x}\right]|0\rangle &= \frac{\displaystyle\int e^{\int [\overline{\zeta}\chi+\overline{\chi}\zeta - i\overline{\chi}(\gamma^\mu\partial_\mu+m)\chi] d^4x} D\chi^* D\chi}{\displaystyle\int e^{iS_0[\chi,\epsilon]} D\chi^* D\chi} \\ &= \exp\left[\int \overline{\zeta}(x)\Delta(x-y)\zeta(y)\, d^4x d^4y\right]. \end{aligned} \tag{20.252}$$

Differentiating we get

$$\langle 0|\mathcal{T}\left[\psi(x)\overline{\psi}(y)\right]|0\rangle = \Delta(x-y) = -i\int \frac{d^4p}{(2\pi)^4} e^{ip(x-y)}\frac{-i\gamma^\nu p_\nu + m}{p^2+m^2-i\epsilon}. \tag{20.253}$$

20.13 Application to Nonabelian Gauge Theories

The action of a generic non-abelian gauge theory is

$$S = \int - \tfrac{1}{4} F_{a\mu\nu} F_a^{\mu\nu} - \overline{\psi} \left(\gamma^\mu D_\mu + m\right) \psi \; d^4x \tag{20.254}$$

in which the Maxwell field is

$$F_{a\mu\nu} \equiv \partial_\mu A_{a\nu} - \partial_\nu A_{a\mu} + g\, f_{abc}\, A_{b\mu}\, A_{c\nu} \tag{20.255}$$

and the covariant derivative is

$$D_\mu \psi \equiv \partial_\mu \psi - i g\, t_a\, A_{a\mu}\, \psi. \tag{20.256}$$

Here g is a coupling constant, f_{abc} is a structure constant (11.68), and t_a is a generator (11.57) of the Lie algebra (Section 11.16) of the gauge group.

One may show (Weinberg, 1996, pp. 14–18) that the analog of equation (20.195) for quantum electrodynamics is

$$\langle \Omega | \mathcal{T} \left[\mathcal{O}_1 \cdots \mathcal{O}_n \right] | \Omega \rangle = \frac{\displaystyle\int \mathcal{O}_1 \cdots \mathcal{O}_n \, e^{iS} \, \delta[A_{a3}] \, DA \, D\psi}{\displaystyle\int e^{iS} \, \delta[A_{a3}] \, DA \, D\psi} \tag{20.257}$$

in which the functional delta function

$$\delta[A_{a3}] \equiv \prod_x \delta(A_{a3}(x)) \tag{20.258}$$

enforces the axial-gauge condition, and $D\psi$ stands for $D\psi^* D\psi$.

Initially, physicists had trouble computing nonabelian amplitudes beyond the lowest order of perturbation theory. Then DeWitt showed how to compute to second order (DeWitt, 1967), and Faddeev and Popov, using path integrals, showed how to compute to all orders (Faddeev and Popov, 1967).

20.14 Faddeev–Popov Trick

The path-integral tricks of Faddeev and Popov are described in (Weinberg, 1996, pp. 19–27). We will use gauge-fixing functions $G_a(x)$ to impose a gauge condition on our non-abelian gauge fields $A^a_\mu(x)$. For instance, we can use $G_a(x) = A^3_a(x)$ to impose an axial gauge or $G_a(x) = i\,\partial_\mu A^\mu_a(x)$ to impose a Lorentz-invariant gauge.

Under an infinitesimal gauge transformation (13.371)

$$A^\lambda_{a\mu} = A_{a\mu} - \partial_\mu \lambda_a - g\, f_{abc}\, A_{b\mu}\, \lambda_c \tag{20.259}$$

the gauge fields change, and so the gauge-fixing functions $G_b(x)$, which depend upon them, also change. The jacobian J of that change at $\lambda = 0$ is

$$J = \det\left(\frac{\delta G_a^\lambda(x)}{\delta\lambda_b(y)}\right)\Bigg|_{\lambda=0} \equiv \frac{DG^\lambda}{D\lambda}\Bigg|_{\lambda=0} \tag{20.260}$$

and it typically involves the delta function $\delta^4(x-y)$.

Let $B[G]$ be any functional of the gauge-fixing functions $G_b(x)$ such as

$$B[G] = \prod_{x,a} \delta(G_a(x)) = \prod_{x,a} \delta(A_a^3(x)) \tag{20.261}$$

in an axial gauge or

$$B[G] = \exp\left[\frac{i}{2}\int (G_a(x))^2\, d^4x\right] = \exp\left[-\frac{i}{2}\int \left(\partial_\mu A_a^\mu(x)\right)^2 d^4x\right] \tag{20.262}$$

in a Lorentz-invariant gauge.

We want to understand functional integrals like (20.257)

$$\langle\Omega|\mathcal{T}\left[\mathcal{O}_1\cdots\mathcal{O}_n\right]|\Omega\rangle = \frac{\int \mathcal{O}_1\cdots\mathcal{O}_n\, e^{iS}\, B[G]\, J\, DA\, D\psi}{\int e^{iS}\, B[G]\, J\, DA\, D\psi} \tag{20.263}$$

in which the operators $\mathcal{O}_k$, the action functional $S[A]$, and the differentials $DAD\psi$ (but not the gauge-fixing functional $B[G]$ or the Jacobian J) are gauge invariant. The axial-gauge formula (20.257) is a simple example in which $B[G] = \delta[A_{a3}]$ enforces the axial-gauge condition $A_{a3}(x) = 0$ and the determinant $J = \det\left(\delta_{ab}\partial_3\delta(x-y)\right)$ is a constant that cancels.

If we translate the gauge fields by gauge transformations Λ and Λ', then the ratio (20.263) does not change

$$\langle\Omega|\mathcal{T}\left[\mathcal{O}_1\cdots\mathcal{O}_n\right]|\Omega\rangle = \frac{\int \mathcal{O}_1^\Lambda\cdots\mathcal{O}_n^\Lambda\, e^{iS^\Lambda}\, B[G^\Lambda]\, J^\Lambda\, DA^\Lambda\, D\psi^\Lambda}{\int e^{iS^{\Lambda'}}\, B[G^{\Lambda'}]\, J^{\Lambda'}\, DA^{\Lambda'}\, D\psi^{\Lambda'}} \tag{20.264}$$

any more than $\int f(y)\,dy$ is different from $\int f(x)\,dx$. Since the operators $\mathcal{O}_k$, the action functional $S[A]$, and the differentials $DAD\psi$ are gauge invariant, most of the Λ-dependence goes away

$$\langle\Omega|\mathcal{T}\left[\mathcal{O}_1\cdots\mathcal{O}_n\right]|\Omega\rangle = \frac{\int \mathcal{O}_1\cdots\mathcal{O}_n\, e^{iS}\, B[G^\Lambda]\, J^\Lambda\, DA\, D\psi}{\int e^{iS}\, B[G^{\Lambda'}]\, J^{\Lambda'}\, DA\, D\psi}. \tag{20.265}$$

Let $\Lambda\lambda$ be a gauge transformation Λ followed by an infinitesimal gauge transformation λ. The jacobian J^Λ is a determinant of a product of matrices which is a product of their determinants

$$
\begin{aligned}
J^\Lambda &= \det\left(\frac{\delta G_a^{\Lambda\lambda}(x)}{\delta\lambda_b(y)}\right)\Big|_{\lambda=0} = \det\left(\int \frac{\delta G_a^{\Lambda\lambda}(x)}{\delta\Lambda\lambda_c(z)}\frac{\delta\Lambda\lambda_c(z)}{\delta\lambda_b(y)}\,d^4z\right)\Big|_{\lambda=0} \\
&= \det\left(\frac{\delta G_a^{\Lambda\lambda}(x)}{\delta\Lambda\lambda_c(z)}\right)\Big|_{\lambda=0} \det\left(\frac{\delta\Lambda\lambda_c(z)}{\delta\lambda_b(y)}\right)\Big|_{\lambda=0} \\
&= \det\left(\frac{\delta G_a^{\Lambda}(x)}{\delta\Lambda_c(z)}\right) \det\left(\frac{\delta\Lambda\lambda_c(z)}{\delta\lambda_b(y)}\right)\Big|_{\lambda=0} \equiv \frac{DG^\Lambda}{D\Lambda}\frac{D\Lambda\lambda}{D\lambda}\Big|_{\lambda=0}.
\end{aligned}
\tag{20.266}
$$

Now we integrate over the gauge transformations Λ (and Λ') with weight function $\rho(\Lambda) = (D\Lambda\lambda/D\lambda|_{\lambda=0})^{-1}$ and find, since the ratio (20.265) is Λ-independent

$$
\begin{aligned}
\langle\Omega|\mathcal{T}\left[\mathcal{O}_1\cdots\mathcal{O}_n\right]|\Omega\rangle &= \frac{\int \mathcal{O}_1\cdots\mathcal{O}_n\, e^{iS}\, B[G^\Lambda]\,\frac{DG^\Lambda}{D\Lambda}\,D\Lambda\, DA\, D\psi}{\int e^{iS}\, B[G^\Lambda]\,\frac{DG^\Lambda}{D\Lambda}\,D\Lambda\, DA\, D\psi} \\
&= \frac{\int \mathcal{O}_1\cdots\mathcal{O}_n\, e^{iS}\, B[G^\Lambda]\,DG^\Lambda\, DA\, D\psi}{\int e^{iS}\, B[G^\Lambda]\,DG^\Lambda\, DA\, D\psi} \\
&= \frac{\int \mathcal{O}_1\cdots\mathcal{O}_n\, e^{iS}\, DA\, D\psi}{\int e^{iS}\, DA\, D\psi}.
\end{aligned}
\tag{20.267}
$$

Thus the mean value in the vacuum of a time-ordered product of gauge-invariant operators is a ratio of path integrals over all gauge fields without any gauge fixing. No matter what gauge condition G or gauge-fixing functional $B[G]$ we use, the resulting gauge-fixed ratio (20.263) is equal to the ratio (20.267) of path integrals over all gauge fields without any gauge fixing. All gauge-fixed ratios (20.263) give the same time-ordered products, and so we can use whatever gauge condition G or gauge-fixing functional $B[G]$ is most convenient.

The analogous formula for the euclidian time-ordered product is

$$
\langle\Omega|\mathcal{T}_e\left[\mathcal{O}_1\cdots\mathcal{O}_n\right]|\Omega\rangle = \frac{\int \mathcal{O}_1\cdots\mathcal{O}_n\, e^{-S_e}\, DA\, D\psi}{\int e^{-S_e}\, DA\, D\psi}
\tag{20.268}
$$

where the euclidian action S_e is the spacetime integral of the energy density. This formula is the basis for lattice gauge theory.

The path-integral formulas (20.197 and 20.198) derived for quantum electrodynamics therefore also apply to nonabelian gauge theories.

20.15 Ghosts

Faddeev and Popov showed how to do perturbative calculations in which one does fix the gauge. To continue our description of their tricks, we return to the gauge-fixed expression (20.263) for the time-ordered product

$$\langle \Omega | \mathcal{T} \left[\mathcal{O}_1 \cdots \mathcal{O}_n \right] | \Omega \rangle = \frac{\int \mathcal{O}_1 \cdots \mathcal{O}_n \, e^{iS} \, B[G] \, J \, DA \, D\psi}{\int e^{iS} \, B[G] \, J \, DA \, D\psi} \tag{20.269}$$

set $G_b(x) = -i \partial_\mu A_b^\mu(x)$ and use (20.262) as the gauge-fixing functional $B[G]$

$$B[G] = \exp\left[\frac{i}{2} \int (G_a(x))^2 \, d^4x \right] = \exp\left[-\frac{i}{2} \int \left(\partial_\mu A_a^\mu(x) \right)^2 d^4x \right]. \tag{20.270}$$

This functional adds to the action density the term $-(\partial_\mu A_a^\mu)^2/2$ which leads to a gauge-field propagator like the photon's (20.201)

$$\langle 0 | \mathcal{T} \left[A_\mu^a(x) A_\nu^b(y) \right] | 0 \rangle = -i \delta_{ab} \Delta_{\mu\nu}(x - y) = -i \int \frac{\eta_{\mu\nu} \delta_{ab}}{q^2 - i\epsilon} \, e^{iq\cdot(x-y)} \, \frac{d^4q}{(2\pi)^4}. \tag{20.271}$$

What about the determinant J? Under an infinitesimal gauge transformation (20.259), the gauge field becomes

$$A_{a\mu}^\lambda = A_{a\mu} - \partial_\mu \lambda_a - g \, f_{abc} \, A_{b\mu} \, \lambda_c \tag{20.272}$$

and so $G_a^\lambda(x) = i \partial^\mu A_{a\mu}^\lambda(x)$ is

$$G_a^\lambda(x) = i \partial^\mu A_{a\mu}(x) + i \partial^\mu \int \left[-\delta_{ac} \partial_\mu - g \, f_{abc} A_{b\mu}(x) \right] \delta^4(x - y) \lambda_c(y) \, d^4y. \tag{20.273}$$

The jacobian J then is the determinant (20.260) of the matrix

$$\left(\frac{\delta G_a^\lambda(x)}{\delta \lambda_c(y)} \right) \Bigg|_{\lambda=0} = -i \delta_{ac} \, \Box \, \delta^4(x - y) - ig \, f_{abc} \frac{\partial}{\partial x^\mu} \left[A_b^\mu(x) \delta^4(x - y) \right] \tag{20.274}$$

that is

$$J = \det \left(-i \delta_{ac} \, \Box \, \delta^4(x - y) - ig \, f_{abc} \frac{\partial}{\partial x^\mu} \left[A_b^\mu(x) \delta^4(x - y) \right] \right). \tag{20.275}$$

But we've seen (20.228) that a determinant can be written as a fermionic path integral

$$\det A = \int e^{-\theta^\dagger A\,\theta} \prod_{k=1}^{n} d\theta_k^* d\theta_k. \tag{20.276}$$

So we can write the jacobian J as

$$J = \int \exp\left[\left[\int i\omega_a^* \Box \omega_a + ig f_{abc}\omega_a^* \partial_\mu (A_b^\mu \omega_c)\, d^4x\right] D\omega^* D\omega \tag{20.277}$$

which contributes the terms $-\partial_\mu \omega_a^* \partial^\mu \omega_a$ and

$$-\,\partial_\mu \omega_a^*\, g\, f_{abc}\, A_b^\mu \omega_c = \partial_\mu \omega_a^*\, g\, f_{abc}\, A_c^\mu \omega_b \tag{20.278}$$

to the action density.

Thus we can do perturbation theory by using the modified action density

$$\mathcal{L}' = -\tfrac{1}{4} F_{a\mu\nu} F_a^{\mu\nu} - \frac{1}{2}\left(\partial_\mu A_a^\mu\right)^2 - \partial_\mu \omega_a^* \partial^\mu \omega_a + \partial_\mu \omega_a^*\, g\, f_{abc}\, A_c^\mu \omega_b - \overline{\psi}\,(\not{D} + m)\,\psi \tag{20.279}$$

in which $\not{D} \equiv \gamma^\mu D_\mu = \gamma^\mu(\partial_\mu - igt^a A_{a\mu})$. The **ghost** field ω is a mathematical device, not a physical field describing real particles, which would be spinless fermions violating the spin-statistics theorem (Example 11.22).

20.16 Effective Field Theories

Suppose a field ϕ whose mass M is huge compared to accessible energies interacts with a field ψ of a low-energy theory such as the standard model

$$L_\phi = -\frac{1}{2}\partial_a\phi(x)\,\partial^a\phi(x) - \frac{1}{2}M^2\phi^2(x) + g\,\overline{\psi}(x)\psi(x)\phi(x). \tag{20.280}$$

Compared to the mass term M^2, the derivative terms $\partial_a\phi\,\partial^a\phi$ contribute little to the low-energy path integral. So we represent the effect of the heavy field ϕ as $L_{\phi 0} = -\frac{1}{2}M^2\phi^2 + g\overline{\psi}\psi\phi$. Completing the square

$$L_{\phi 0} = -\frac{1}{2}M^2\left(\phi - \frac{g}{M^2}\overline{\psi}\psi\right)^2 + \frac{g^2}{2M^2}\,(\overline{\psi}\psi)^2 \tag{20.281}$$

and shifting ϕ by $g\overline{\psi}\psi/M^2$, we see that the gaussian path integral is

$$\int \exp\left[i\int -\frac{1}{2}M^2\phi^2 + \frac{g^2}{2M^2}(\overline{\psi}\psi)^2\, d^4x\right] D\phi = \exp\left[i\int \frac{g^2}{2M^2}(\overline{\psi}\psi)^2\, d^4x\right]$$

apart from a field-independent factor. The net effect of heavy field ϕ is thus to add to the low-energy theory a new interaction

$$L_\psi = \frac{g^2}{2M^2}\,(\overline{\psi}\psi)^2 \tag{20.282}$$

which is small because M^2 is large. If a gauge boson Y_a of huge mass M interacts as $L_{Y0} = -\frac{1}{2}M^2 Y_a Y^a + ig\overline{\psi}\gamma^a\psi Y_a$ with a spin-one-half field ψ, then $L_\psi = -\left(g^2/(2M^2)\right)\overline{\psi}\gamma^a\psi\,\overline{\psi}\gamma_a\psi$ is the new low-energy interaction.

20.17 Complex Path Integrals

In this chapter, it has been tacitly assumed that the action is quadratic in the time derivatives of the fields. This assumption makes the hamiltonian quadratic in the momenta and the path integral over them gaussian. In general, however, the partition function is a path integral over fields and momenta like

$$Z(\beta) = \int \exp\left\{\int_0^\beta \int \left[i\dot{\phi}(x)\pi(x) - H(\phi,\pi)\right]dt d^3x\right\} D\phi\, D\pi \qquad (20.283)$$

in which the exponential is not a probability distribution. To study such theories, one often can numerically integrate over the momentum, make a look-up table for $P[\phi]$, and then apply the usual Monte Carlo methods of Section 16.5 (Amdahl and Cahill, 2016). Programs that do this are in the repository Path_integrals at `github.com/kevinecahill`.

Further Reading

"Space-Time Approach to Non-relativistic Quantum Mechanics" (Feynman, 1948), *Quantum Mechanics and Path Integrals* (Feynman et al., 2010), *Statistical Mechanics* (Feynman, 1972), *The Quantum Theory of Fields I, II, & III* (Weinberg, 1995, 1996, 2005), *Quantum Field Theory in a Nutshell* (Zee, 2010), and *Quantum Field Theory* (Srednicki, 2007) all provide excellent treatments of path integrals. Some applications are described in *Path Integrals in Quantum Mechanics, Statistics, Polymer Physics, and Financial Markets* (Kleinert, 2009).

Exercises

20.1 From (20.1), derive the multiple gaussian integral for real a_j and b_j

$$\int_{-\infty}^{\infty} \exp\left(\sum_{j=1}^{n} ia_j x_j^2 + 2ib_j x_j\right) \prod_{j=1}^{n} dx_j = \prod_{j=1}^{n} \sqrt{\frac{i\pi}{a_j}}\, e^{-ib_j^2/a_j}. \qquad (20.284)$$

20.2 Use (20.284) to derive the multiple imaginary gaussian integral (20.3). Hint: Any real symmetric matrix s can be diagonalized by an orthogonal transformation $a = oso^{\mathsf{T}}$. Let $y = ox$.

20.3 Use (20.2) to show that for positive a_j

$$\int_{-\infty}^{\infty} \exp\left(\sum_j -a_j x_j^2 + 2ib_j x_j\right) \prod_{j=1}^{n} dx_j = \prod_{j=1}^{n} \sqrt{\frac{\pi}{a_j}}\, e^{-b_j^2/a_j}. \quad (20.285)$$

20.4 Use (20.285) to derive the many variable real gaussian integral (20.4). Same hint as for Exercise 20.2.

20.5 Do the q_2 integral (20.27).

20.6 Insert the identity operator in the form of an integral (20.10) of outer products $|p\rangle\langle p|$ of eigenstates of the momentum operator p between the exponential and the state $|q_a\rangle$ in the matrix element (20.25) and so derive for that matrix element $\langle q_b|\exp(-i(t_b - t_a)H/\hbar)|q_a\rangle$ the formula (20.28). Hint: use the inner product $\langle q|p\rangle = \exp(iqp/\hbar)/\sqrt{2\pi\hbar}$, and do the resulting Fourier transform.

20.7 Derive the path-integral formula (20.36) for the quadratic action (20.35).

20.8 Show that for the simple harmonic oscillator (20.44) the action $S[q_c]$ of the classical path from q_a, t_a to q_b, t_b is (20.46).

20.9 Show that the determinants $|C_n(y)| = \det C_n(y)$ of the tridiagonal matrices (20.52) satisfy the recursion relation (20.53) and have the initial values $|C_1(y)| = y$ and $|C_2(y)| = y^2 - 1$. Incidentally, the Chebyshev polynomials (9.65) of the second kind $U_n(y/2)$ obey the same recursion relation and have the same initial values, so $|C_n(y)| = U_n(y/2)$.

20.10 (a) Show that the functions $S_n(y) = \sin(n+1)\theta/\sin\theta$ with $2\cos\theta = y$ satisfy the Toeplitz recursion relation (20.53) which after a cancellation simplifies to $\sin(n+2)\theta = 2\cos\theta\, \sin(n+1)\theta - \sin n\theta$. (b) Derive the initial conditions $S_0(y) = 1$, $S_1(y) = y$, and $S_2(y) = y^2 - 1$.

20.11 Do the q_2 integral (20.75).

20.12 Show that the euclidian action (20.88) is stationary if the path $q_{ec}(u)$ obeys the euclidian equation of motion $\ddot{q}_{ec}(u) = \omega^2 q_{ec}(u)$.

20.13 By using (20.20) for each of the three exponentials in (20.102), derive (20.103) from (20.102). Hint: From (20.20), one has

$$q e^{-i\,(t_b - t_a)\,H/\hbar} q = \int q_b |q_b\rangle\, e^{iS[q]/\hbar}\, \langle q_a| q_a\, Dq\, dq_a\, dq_b \qquad (20.286)$$

in which $q_a = q(t_a)$ and $q_b = q(t_b)$.

20.14 Derive the path-integral formula (20.139) from (20.129–20.132).

20.15 Derive the path-integral formula (20.153) from (20.147–20.150).

20.16 Show that the vector $\overline{Y}$ that makes the argument $-iY^{\mathsf{T}}SY + iD^{\mathsf{T}}Y$ of the multiple gaussian integral

$$\int_{-\infty}^{\infty} \exp\left(-iY^\mathsf{T}SY + iD^\mathsf{T}Y \right) \prod_{i=1}^{n} dy_i = \sqrt{\frac{\pi^n}{\det(iS)}} \exp\left(\frac{i}{4} D^\mathsf{T}S^{-1}D \right) \tag{20.287}$$

stationary is $\overline{Y} = S^{-1}D/2$, and that the multiple gaussian integral (20.287) is equal to its exponential $\exp(-iY^\mathsf{T}SY + iD^\mathsf{T}Y)$ evaluated at its stationary point $Y = \overline{Y}$ apart from a prefactor involving $\det iS$.

20.17 Show that the vector $\overline{Y}$ that makes the argument $-Y^\mathsf{T}SY + D^\mathsf{T}Y$ of the multiple gaussian integral

$$\int_{-\infty}^{\infty} \exp\left(-Y^\mathsf{T}SY + D^\mathsf{T}Y \right) \prod_{i=1}^{n} dy_i = \sqrt{\frac{\pi^n}{\det(S)}} \exp\left(\frac{1}{4} D^\mathsf{T}S^{-1}D \right) \tag{20.288}$$

stationary is $\overline{Y} = S^{-1}D/2$, and that the multiple gaussian integral (20.288) is equal to its exponential $\exp(-Y^\mathsf{T}SY + D^\mathsf{T}Y)$ evaluated at its stationary point $Y = \overline{Y}$ apart from a prefactor involving $\det S$.

20.18 By taking the nonrelativistic limit of the formula (12.55) for the action of a relativistic particle of mass m and charge q, derive the expression (20.41) for the action of a nonrelativistic particle in an electromagnetic field with no scalar potential.

20.19 Work out the path-integral formula for the amplitude for a mass m initially at rest to fall to the ground from height h in a gravitational field of local acceleration g to lowest order and then including loops up to an overall constant. Hint: use the technique of Section 20.4.

20.20 Show that the euclidian action of the stationary solution (20.87) is (20.88).

20.21 Derive formula (20.161) for the action $S_0[\phi]$ from (20.159 and 20.160).

20.22 Derive identity (20.165). Split the time integral at $t = 0$ into two halves, use $\epsilon\, e^{\pm\epsilon t} = \pm\, d\, e^{\pm\epsilon t}/dt$ and then integrate each half by parts.

20.23 Derive the third term in Equation (20.167) from the second term.

20.24 Use (20.171) and the Fourier transform (20.172) of the external current j to derive the formula (20.173) for the modified action $S_0[\phi, \epsilon, j]$.

20.25 Derive Equation (20.174) from Equation (20.173).

20.26 Derive the formula (20.175) for $Z_0[j]$ from the formula for $S_0[\phi, \epsilon, j]$.

20.27 Derive Equations (20.176 and 20.177) from formula (20.175).

20.28 Derive Equation (20.181) from the formula (20.176) for $Z_0[j]$.

20.29 Show that the time integral of the Coulomb term (20.186) is the term that is quadratic in j^0 in the number F defined by (20.190).

20.30 By following steps analogous to those that led to (20.177), derive the formula (20.201) for the photon propagator in Feynman's gauge.

20.31 Derive expression (20.214) for the inner product $\langle \zeta | \theta \rangle$.

20.32 Derive the representation (20.217) of the identity operator I for a single fermionic degree of freedom from the rules (20.204 and 20.207) for Grassmann integration and the anticommutation relations (20.206).

20.33 Derive the eigenvalue equation $\psi_k |\theta\rangle = \theta_k |\theta\rangle$ from the definitions (20.220 and 20.221) of the eigenstate $|\theta\rangle$.

20.34 Derive the eigenvalue relation (20.234) for the Fermi field $\psi_m(\boldsymbol{x}, t)$ from the anticommutation relations (20.230 and 20.231) and the definitions (20.232 and 20.233).

20.35 Derive the formula (20.235) for the inner product $\langle \chi' | \chi \rangle$ from the definition (20.233) of the ket $|\chi\rangle$.

20.36 Imitate the derivation of the path-integral formula (20.66) and derive its 3-dimensional version (20.73).

21

Renormalization Group

21.1 Renormalization and Interpolation

Probably because they describe point particles, quantum field theories are divergent. Unknown physics at very short distance scales, removes these infinities. Since these infinities really are absent, we can cancel them consistently in **renormalizable** theories by a procedure called **renormalization**. One starts with an action that contains infinite, unknown charges and masses, such as $e_0 = e - \delta e$ and $m_0 = m - \delta m$, in which $-e$ is the charge of the electron and m is its mass. One then uses δe and δm to cancel unwanted infinite terms as they appear in perturbation theory.

Because the underlying theory is finite, the value of a divergent scattering amplitude may change by a finite amount when we compute it at two different sets of initial and final momenta. This happens, for example, in the theory of a scalar field ϕ with action density

$$\mathcal{L} = -\frac{1}{2}\partial_i\phi\,\partial^i\phi - \frac{1}{2}m^2\phi^2 - \frac{g}{4!}\phi^4. \tag{21.1}$$

The amplitude for the elastic scattering of two bosons of initial four-momenta p_1 and p_2 into two of final momenta p_1' and p_2' is

$$\begin{aligned} A = g - \frac{g^2}{16\pi^2}\int_0^\infty k^3 dk \int_0^1 dx\Big\{&[k^2 + m^2 - sx(1-x)]^{-2} \\ + [k^2 + m^2 - tx(1-x)]^{-2} + [k^2 + m^2 - ux(1-x)]^{-2}&\Big\} \end{aligned} \tag{21.2}$$

to one-loop order (Weinberg, 1995, section 12.2). In this formula, s, t, and u are the Mandelstam variables $s = -(p_1 + p_2)^2$, $t = -(p_1 - p_1')^2$, and $u = -(p_1 - p_2')^2$, and $k^2 = k_0^2 + k_1^2 + k_2^2 + k_3^2$ after a Wick rotation.

The amplitude $A(s,t,u)$ diverges logarithmically, but the difference between it and its value $A_0 = A(s_0, t_0, u_0)$ at some point (s_0, t_0, u_0) is finite

$$
\begin{aligned}
A(s,t,u) - A_0 = {} & \\
& -\frac{g^2}{16\pi^2}\int_0^\infty k^3 dk \int_0^1 dx \left\{ \frac{x(1-x)(s-s_0)[2k^2+2m^2-(s+s_0)x(1-x)]}{[k^2+m^2-sx(1-x)]^2[k^2+m^2-s_0x(1-x)]^2} \right. \\
& \left. + (s, s_0 \to t, t_0) + (s, s_0 \to u, u_0) \right\}.
\end{aligned} \tag{21.3}
$$

The second and third terms within the big curly brackets are the same as the first term but with s and s_0 replaced by t and t_0 in the second term and by u and u_0 in the third. The k integral is finite as is $A(s,t,u) - A_0$

$$
\begin{aligned}
A(s,t,u) - A_0 = {} & -\frac{g^2}{32\pi^2}\int_0^1 dx \left\{ \ln\left[\frac{m^2 - s_0 x(1-x)}{m^2 - sx(1-x)}\right] \right. \\
& \left. + \ln\left[\frac{m^2 - t_0 x(1-x)}{m^2 - tx(1-x)}\right] + \ln\left[\frac{m^2 - u_0 x(1-x)}{m^2 - ux(1-x)}\right] \right\}.
\end{aligned} \tag{21.4}
$$

If we choose as the renormalization point $s_0 = t_0 = u_0 = -4\mu^2/3$, then we get the usual result (Weinberg, 1995, 1996, sections 12.2, 18.1–2).

21.2 Renormalization Group in Quantum Field Theory

We can use the scattering amplitude (21.4) to define a **running coupling constant** g_μ at energy scale μ as the experimentally measured, finite, physical amplitude A at $s_0 = t_0 = u_0 = -\mu^2$. Then with $g_\mu \equiv A(s_0, t_0, u_0)$, the scattering amplitude (21.4) is finite to order g^2

$$
\begin{aligned}
A(s,t,u) = g_\mu - {} & \frac{g^2}{32\pi^2}\int_0^1 dx \left\{ \ln\left[\frac{1 + (\mu/m)^2 x(1-x)}{1 - s/m^2 x(1-x)}\right] \right. \\
& \left. + \ln\left[\frac{1 + (\mu/m)^2 x(1-x)}{1 - t/m^2 x(1-x)}\right] + \ln\left[\frac{1 + (\mu/m)^2 x(1-x)}{1 - u/m^2 x(1-x)}\right] \right\}.
\end{aligned} \tag{21.5}
$$

Callan (Callan, 1970) and Symanzik (Symanzik, 1970) noticed that this scattering amplitude, like any physical quantity, is **independent of the sliding scale** μ. Thus its derivative with respect to μ vanishes

$$
\begin{aligned}
0 = \frac{\partial A(s,t,u)}{\partial \mu} & = \frac{\partial g_\mu}{\partial \mu} - \frac{3g^2}{32\pi^2}\frac{\partial}{\partial \mu}\int_0^1 dx \, \ln\left[1 + (\mu/m)^2 x(1-x)\right] \\
& = \frac{\partial g_\mu}{\partial \mu} - \frac{3g^2}{32\pi^2}\int_0^1 dx \, \frac{(2\mu/m^2)x(1-x)}{1 + (\mu/m)^2 x(1-x)}.
\end{aligned} \tag{21.6}
$$

For $\mu \gg m$, the integral is $2/\mu$. So at high energies, the running coupling constant obeys the differential equation

$$\mu \frac{\partial g_\mu}{\partial \mu} \equiv \beta(g_\mu) = \frac{3g^2}{16\pi^2} = \frac{3g_\mu^2}{16\pi^2} \tag{21.7}$$

in which the last equality holds to second order in g_μ. Integrating the beta function $\beta(g_\mu)$, we get

$$\ln \frac{E}{M} = \int_M^E \frac{d\mu}{\mu} = \int_{g_M}^{g_E} \frac{dg_\mu}{\beta(g_\mu)} = \frac{16\pi^2}{3} \int_{g_M}^{g_E} \frac{dg_\mu}{g_\mu^2} = \frac{16\pi^2}{3} \left(\frac{1}{g_M} - \frac{1}{g_E} \right). \tag{21.8}$$

So the running coupling constant g_μ at energy $\mu = E$ is

$$g_E = \frac{g_M}{1 - 3\, g_M \ln(E/M)/16\pi^2}. \tag{21.9}$$

As the energy $E = \sqrt{s}$ rises above M, while staying below the singular value $E = M \exp(16\pi^2/3g_M)$, the running coupling constant g_E slowly increases, as does the scattering amplitude, $A \approx g_E$.

Example 21.1 (Quantum electrodynamics) Vacuum polarization makes the one-loop amplitude for the scattering of two electrons proportional to $A(q^2) = e^2 \left[1 + \pi(q^2)\right]$ rather than to e^2 (Gell-Mann and Low, 1954), (Weinberg, 1995, section 11.2). Here e is the renormalized charge, $q = p_1' - p_1$ is the four-momentum transferred to the first electron, and

$$\pi(q^2) = \frac{e^2}{2\pi^2} \int_0^1 x(1-x) \ln \left[1 + \frac{q^2 x(1-x)}{m^2} \right] dx \tag{21.10}$$

represents the polarization of the vacuum. One defines the square of the running coupling constant e_μ^2 to be the amplitude $A(q^2)$ at $q^2 = \mu^2$

$$e_\mu^2 = A(\mu^2) = e^2 \left[1 + \pi(\mu^2)\right]. \tag{21.11}$$

For $q^2 = \mu^2 \gg m^2$, the vacuum-polarization term $\pi(\mu^2)$ is (exercise 21.1)

$$\pi(\mu^2) \approx \frac{e^2}{6\pi^2} \left[\ln \frac{\mu}{m} - \frac{5}{6} \right]. \tag{21.12}$$

The amplitude then is

$$A(q^2) = e_\mu^2 \frac{1 + \pi(q^2)}{1 + \pi(\mu^2)}, \tag{21.13}$$

and since it must be independent of μ, we have

$$0 = \frac{d}{d\mu} \frac{A(q^2)}{1 + \pi(q^2)} = \frac{d}{d\mu} \frac{e_\mu^2}{1 + \pi(\mu^2)} \approx \frac{d}{d\mu} \left\{ e_\mu^2 \left[1 - \pi(\mu^2) \right] \right\}. \tag{21.14}$$

So by differentiating e_μ and the vacuum-polarization term (21.12), we find

$$0 = 2e_\mu \left(\frac{de_\mu}{d\mu}\right)\left[1 - \pi(\mu^2)\right] - e_\mu^2 \frac{d\pi(\mu^2)}{d\mu} = 2e_\mu \left(\frac{de_\mu}{d\mu}\right)\left[1 - \pi(\mu^2)\right] - e_\mu^2 \frac{e^2}{6\pi^2\mu}. \tag{21.15}$$

But by (21.10) the vacuum-polarization term $\pi(\mu^2)$ is of order e^2, which is the same as e_μ^2 to lowest order in e_μ. Thus we arrive at the Callan–Symanzik equation

$$\mu \frac{de_\mu}{d\mu} \equiv \beta(e_\mu) = \frac{e_\mu^3}{12\pi^2} \tag{21.16}$$

which we can integrate

$$\ln \frac{E}{M} = \int_M^E \frac{d\mu}{\mu} = \int_{e_M}^{e_E} \frac{de_\mu}{\beta(e_\mu)} = 12\pi^2 \int_{e_M}^{e_E} \frac{de_\mu}{e_\mu^3} = 6\pi^2 \left(\frac{1}{e_M^2} - \frac{1}{e_E^2}\right)$$

to

$$e_E^2 = \frac{e_M^2}{1 - e_M^2 \ln(E/M)/6\pi^2}. \tag{21.17}$$

The fine-structure constant $e_\mu^2/4\pi$ slowly rises from $\alpha = 1/137.036$ at m_e to

$$\frac{e^2(45.5\text{GeV})}{4\pi} = \frac{\alpha}{1 - 2\alpha \ln(45.5/0.00051)/3\pi} = \frac{1}{134.6} \tag{21.18}$$

at $\sqrt{s} = 91$ GeV. When all light charged particles are included, one finds that the fine-structure constant rises to $\alpha = 1/128.87$ at $E = 91$ GeV.

Example 21.2 (Quantum chromodynamics) The beta functions of scalar field theories and of quantum electrodynamics are positive, and so interactions in these theories become stronger at higher energy scales. But Yang–Mills theories have beta functions that can be negative because of the cubic interactions of the gauge fields and the ghost fields (20.279). If the gauge group is $SU(3)$, then the beta function is

$$\mu \frac{dg_\mu}{d\mu} \equiv \beta(g_\mu) = -\frac{11g^3}{16\pi^2} = -\frac{11g_\mu^3}{16\pi^2} \tag{21.19}$$

to lowest order in g_μ. Integrating, we find

$$\ln \frac{E}{M} = \int_M^E \frac{d\mu}{\mu} = \int_{g_M}^{g_E} \frac{dg_\mu}{\beta(g_\mu)} = -\frac{16\pi^2}{11} \int_{g_M}^{g_E} \frac{dg_\mu}{g_\mu^3} = \frac{8\pi^2}{11}\left(\frac{1}{g_M^2} - \frac{1}{g_E^2}\right) \tag{21.20}$$

and

$$g_E^2 = g_M^2 \left[1 + \frac{11g_M^2}{8\pi^2} \ln \frac{E}{M}\right]^{-1} \tag{21.21}$$

which shows that as the energy E of a scattering process increases, the running coupling slowly **decreases**, going to zero at infinite energy, an effect called **asymptotic freedom** (Gross and Wilczek, 1973; Politzer, 1973).

If the gauge group is $SU(N)$, and the theory has n_f flavors of quarks with masses below μ, then the beta function is

$$\beta(g_\mu) = -\frac{g_\mu^3}{4\pi^2}\left(\frac{11N}{12} - \frac{n_f}{6}\right) \tag{21.22}$$

which is negative as long as $n_f < 11N/2$. Using this beta function with $N = 3$ and again integrating, we get instead of (21.21)

$$g_E^2 = g_M^2\left[1 + \frac{(11 - 2n_f/3)g_M^2}{16\pi^2}\ln\frac{E^2}{M^2}\right]^{-1}. \tag{21.23}$$

So with

$$M^2 \equiv \Lambda^2 \exp\left(\frac{16\pi^2}{(11 - 2n_f/3)g_M^2}\right) \tag{21.24}$$

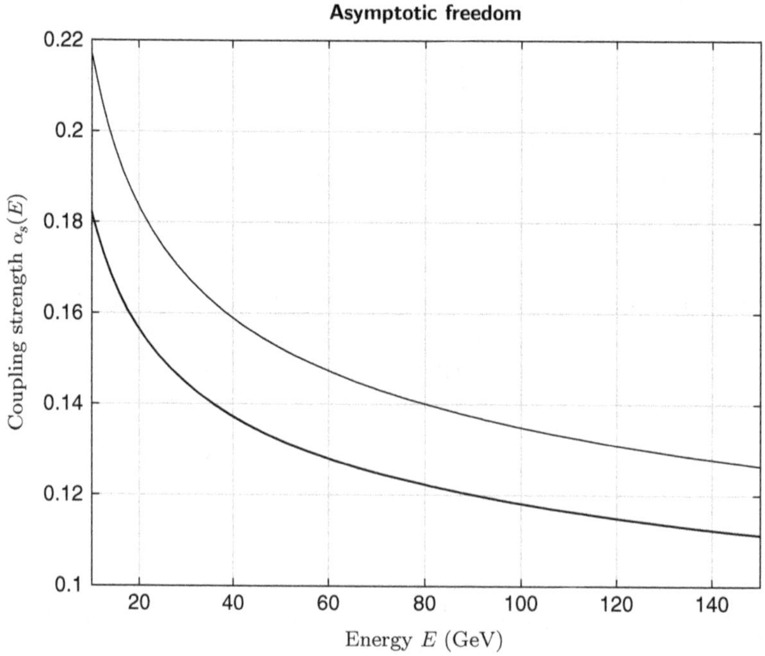

Figure 21.1 The strong-structure constant $\alpha_s(E)$ as given by the one-loop formula (21.25) (thin curve) and by a three-loop formula (thick curve) with $\Lambda = 230$ MeV and $n_f = 5$ is plotted for $m_b \ll E \ll m_t$. This chapter's Matlab scripts are in Renormalization_group at github.com/kevinecahill.

we find (Exercise 21.2)

$$\alpha_s(E) \equiv \frac{g^2(E)}{4\pi} = \frac{12\pi}{(33 - 2n_f)\ln(E^2/\Lambda^2)} \tag{21.25}$$

which expresses the dimensionless QCD coupling constant $\alpha_s(E)$ appropriate to energy E in terms of a parameter Λ that has the dimension of energy. Sidney Coleman called this **dimensional transmutation**. For $\Lambda = 230$ MeV and $n_f = 5$, Fig. 21.1 displays $\alpha_s(E)$ in the range $4.19 = m_b \ll E \ll m_t = 172$ GeV. The thin curve is the one-loop formula (21.25), and the thick curve is a three-loop formula (Weinberg, 1996, p. 156).

21.3 Renormalization Group in Lattice Field Theory

Let us consider a quantum field theory on a lattice (Gattringer and Lang, 2010, chap. 3) in which the strength of the nonlinear interactions depends upon a single dimensionless coupling constant g. The spacing a of the lattice regulates the infinities, which return as $a \to 0$. The value of an observable P computed on this lattice will depend upon the lattice spacing a and on the coupling constant g, and so will be a function $P(a, g)$ of these two parameters. The *right* value of the coupling constant is the value that makes the result of the computation be as close as possible to the physical value P. So the correct coupling constant is not a constant at all, but rather a function $g(a)$ that varies with the lattice spacing or cutoff a. Thus as we vary the lattice spacing and go to the continuum limit $a \to 0$, we must adjust the coupling function $g(a)$ so that what we compute, $P(a, g(a))$, is equal to the physical value P. That is, $g(a)$ must vary with a so as to keep $P(a, g(a))$ constant at $P(a, g(a)) = P$

$$\frac{dP(a, g(a))}{da} = 0. \tag{21.26}$$

Writing this condition as a dimensionless derivative

$$a\,\frac{dP(a, g(a))}{da} = \frac{da}{d\ln a}\frac{dP(a, g(a))}{da} = \frac{dP(a, g(a))}{d\ln a} = 0 \tag{21.27}$$

we arrive at the **Callan–Symanzik equation**

$$0 = \frac{dP(a, g(a))}{d\ln a} = \left(\frac{\partial}{\partial \ln a} + \frac{dg}{d\ln a}\frac{\partial}{\partial g}\right) P(a, g(a)). \tag{21.28}$$

The coefficient of the second partial derivative with a minus sign is the lattice beta function

$$\beta_L(g) \equiv -\frac{dg}{d\ln a}. \tag{21.29}$$

Since the lattice spacing a and the energy scale μ are inversely related, the lattice beta function differs from the continuum beta function by a minus sign.

In $SU(N)$ gauge theory, the first two terms of the lattice beta function for small g are $\beta_L(g) = -\beta_0\, g^3 - \beta_1\, g^5$ where for n_f flavors of light quarks

$$\begin{aligned}\beta_0 &= \frac{1}{(4\pi)^2}\left(\frac{11}{3}N - \frac{2}{3}n_f\right)\\ \beta_1 &= \frac{1}{(4\pi)^4}\left(\frac{34}{3}N^2 - \frac{10}{3}Nn_f - \frac{N^2-1}{N}n_f\right)\end{aligned} \tag{21.30}$$

and $N = 3$ in quantum chromodynamics.

Combining the definition (21.29) of the beta function with its expansion $\beta_L(g) = -\beta_0\, g^3 - \beta_1\, g^5$ for small g, one gets the differential equation

$$\frac{dg}{d\ln a} = \beta_0\, g^3 + \beta_1\, g^5 \tag{21.31}$$

which one may integrate

$$\int d\ln a = \ln a\; - \ln c = \int \frac{dg}{\beta_0 g^3 + \beta_1 g^5} = -\frac{1}{2\beta_0 g^2} + \frac{\beta_1}{2\beta_0^2}\ln\left(\frac{\beta_0 + \beta_1 g^2}{g^2}\right)$$

to find that the lattice spacing has an essential singularity at $g = 0$

$$a(g) = c\left(\frac{\beta_0 + \beta_1 g^2}{g^2}\right)^{\beta_1/2\beta_0^2} e^{-1/(2\beta_0 g^2)} \tag{21.32}$$

in which c is a constant of integration. The term $\beta_1 g^2$ is of higher order in g, and if one drops it and absorbs a power of β_0 into a new constant of integration Λ, then one finds

$$a(g) = \frac{1}{\Lambda}\left(\beta_0 g^2\right)^{-\beta_1/2\beta_0^2} e^{-1/(2\beta_0 g^2)}. \tag{21.33}$$

As $g \to 0$, the lattice spacing $a(g)$ goes to zero *very fast* (as long as $n_f < 17$ for $N = 3$). The inverse of this relation (21.33)

$$g(a) \approx \left[\beta_0 \ln(a^{-2}\Lambda^{-2}) + (\beta_1/\beta_0)\ln\left(\ln(a^{-2}\Lambda^{-2})\right)\right]^{-1/2} \tag{21.34}$$

shows that the coupling constant slowly goes to zero with a, which is a lattice version of **asymptotic freedom**.

21.4 Renormalization Group in Condensed-Matter Physics

In classical statistical mechanics, the partition function $Z(\beta)$ is a sum over all configurations c of $\exp(-\beta H(c))$ in which $H(c)$ is the energy of the configuration c.

A configuration might be a d-dimensional array of spins or a field in d-dimensional space (not spacetime)

$$\phi(\boldsymbol{x}) = \int_\Lambda e^{i\boldsymbol{k}\cdot\boldsymbol{x}}\,\phi(k)\,d^d\boldsymbol{k} \tag{21.35}$$

in which the integral is over Fourier modes $\boldsymbol{k}$ with a cutoff $k < \Lambda$ at an inverse lattice spacing. The hamiltonian might be

$$H[\phi] = \int \left[\frac{1}{2}(\nabla\phi)^2 + g_2\phi^2 + g_4\phi^4 + g_6\phi^6 + \cdots\right] d^d\boldsymbol{x}. \tag{21.36}$$

The partition function $Z(\beta)$ is an integral over all such configurations

$$Z(\beta) = \int \exp\left[-\beta H[\phi]\right] D\phi. \tag{21.37}$$

We change variables from $\phi(\boldsymbol{x})$ to a stretched field $\phi(\boldsymbol{x}/\ell)$ with $\ell > 1$

$$\phi_\ell(\boldsymbol{x}) = a_\ell\,\phi(\boldsymbol{x}/\ell) = a_\ell \int_\Lambda e^{i\boldsymbol{k}\cdot\boldsymbol{x}/\ell}\,\phi(k)\,d^d\boldsymbol{k} \tag{21.38}$$

in which a_ℓ is a factor that keeps the kinetic part of the hamiltonian invariant (Fisher, 1974, 1998; Kosterlitz et al., 1976; Kadanoff, 2009; Wilson, 1971, 1975). To have $H_k[\phi_\ell] = H_k[\phi]$, we need derivative terms to remain invariant

$$\begin{aligned} H_k[\phi_\ell] &= \int \left(\frac{\partial\phi_\ell(\boldsymbol{x})}{\partial x_i}\right)^2 d^d\boldsymbol{x} = \int a_\ell^2 \left(\frac{\partial\phi(\boldsymbol{x}/\ell)}{\partial x_i}\right)^2 d^d\boldsymbol{x} \\ &= \int \ell^d a_\ell^2 \left(\frac{\partial\phi(x/\ell)}{\ell\,\partial x_i/\ell}\right)^2 d^d(\boldsymbol{x}/\ell) = \int \ell^{d-2} a_\ell^2 \left(\frac{\partial\phi(x')}{\partial x_i'}\right)^2 d^d(\boldsymbol{x}') = H_k[\phi]. \end{aligned} \tag{21.39}$$

That is, we need $a_\ell = \ell^{(2-d)/2}$.

How do the various potential-energy terms change? A term $H_n[\phi]$ with ϕ^n changes to

$$\begin{aligned} H_n[\phi_\ell] &= \int g_n\,\phi_\ell^n(\boldsymbol{x})\,d^d\boldsymbol{x} = \int g_n\,a_\ell^n\,\phi^n(\boldsymbol{x}/\ell)\,d^d\boldsymbol{x} = \int g_n\,a_\ell^n\,\phi^n(\boldsymbol{x}/\ell)\,\ell^d\,d^d(\boldsymbol{x}/\ell) \\ &= \ell^d\,a_\ell^n \int g_n\,\phi^n(\boldsymbol{x}')\,d^d\boldsymbol{x}' = \ell^{d+n(2-d)/2}\,H_n[\phi]. \end{aligned} \tag{21.40}$$

In effect, the coupling constant changes to $g_n(\ell) = \ell^{d+n(2-d)/2} g_n$. Similar reasoning shows that the coupling constant $g_{n,p}$ of a term with n factors of the field ϕ and p spatial derivatives of ϕ should vary as $g_{n,p}(\ell) = \ell^{d+n(2-d)/2-p} g_{n,p}$.

Coupling constants with positive exponents $d + n(d-2)/2 - p > 0$ become more important at greater spatial scales and are said to be **relevant**. Those with negative exponents become less important at greater spatial scales and are said to be **irrelevant**. Those with vanishing exponents are **marginal**. The mass term $g_2(\ell)\phi^2 = \ell^2 g_2\phi^2$ is always relevant. The quartic term $g_4(\ell)\phi^4 = \ell^{4-d} g_4\phi^4$ is relevant in fewer than 4 dimensions, marginal in 4 dimensions, and irrelevant in more than 4 dimensions. The term $g_6(\ell)\phi^6 = \ell^{6-2d} g_6\phi^6$ is relevant in fewer than 3 dimensions, marginal in 3 dimensions, and irrelevant in more than 3.

Further Reading

Quantum Field Theory in a Nutshell (Zee, 2010, chaps. III & VI), *An Introduction to Quantum Field Theory* (Peskin and Schroeder, 1995, chap. 12), and *The Quantum Theory of Fields* (Weinberg, 1995, 1996, sections 12.2 & 18.1–2).

Exercises

21.1 Show that for $\mu^2 \gg m^2$, the vacuum polarization term (21.10) reduces to (21.12). Hint: Use $\ln a\, b = \ln a + \ln b$ when integrating.

21.2 Show that by choosing the energy scale Λ according to (21.24), one can derive (21.25) from (21.23).

22
Strings

22.1 The Nambu–Goto String Action

Quantum field theory is plagued with infinities, presumably because it represents particles as points. A more physical approach is to represent elementary particles as objects of finite size. Those that are 1-dimensional are called **strings**.

We'll use $0 \le \sigma \le \sigma_1$ and $\tau_i \le \tau \le \tau_f$ to parametrize the spacetime coordinates $X^\mu(\sigma, \tau)$ of the string. Nambu and Goto suggested using as the action the area (Zwiebach, 2009, chap. 6)

$$S = -\frac{T_0}{c} \int_{\tau_i}^{\tau_f} \int_0^{\sigma_1} \sqrt{\left(\dot{X} \cdot X'\right)^2 - \left(\dot{X}\right)^2 (X')^2} \, d\tau \, d\sigma \tag{22.1}$$

in which

$$\dot{X}^\mu = \frac{\partial X^\mu}{\partial \tau} \quad \text{and} \quad X'^\mu = \frac{\partial X^\mu}{\partial \sigma} \tag{22.2}$$

and a Lorentz metric $\eta_{\mu\nu} = \text{diag}(-1, 1, 1, \dots)$ is used to form the inner products like $\dot{X} \cdot X' = \dot{X}^\mu \eta_{\mu\nu} X^{\nu\prime}$. The action S is the area swept out by a string of length σ_1 in time $\tau_f - \tau_i$.

If $\dot{X} d\tau = dt$ points in the time direction and $X' d\sigma = d\boldsymbol{r}$ points in a spatial direction, then one sees that $\dot{X} \cdot X' = 0$, that $-\left(\dot{X}\right)^2 d\tau^2 = dt^2$, and that $\left(X'\right)^2 d\sigma^2 = d\boldsymbol{r}^2$. So in this simple case, the action (22.1) is

$$S = -\frac{T_0}{c} \int_{t_i}^{t_f} \int_0^{r_1} dt \, dr = -\frac{T_0}{c} (t_f - t_i) r_1 \tag{22.3}$$

which is the area the string sweeps out. The other term $(\dot{X} \cdot X')^2$ within the square root ensures that the action is the area swept out for all $\dot{X}$ and X', and that it is invariant under arbitrary reparametrizations $\sigma \to \sigma'$ and $\tau \to \tau'$.

The equation of motion for the relativistic string follows from the requirement that the action (22.1) be stationary, $\delta S = 0$. Since

$$\delta \dot{X}^\mu = \delta \frac{\partial X^\mu}{\partial \tau} = \frac{\partial \delta X^\mu}{\partial \tau} \quad \text{and} \quad \delta X'^\mu = \delta \frac{\partial X^\mu}{\partial \sigma} = \frac{\partial \delta X^\mu}{\partial \sigma},$$

we may express the change in the action in terms of derivatives of the Lagrange density

$$L = -\frac{T_0}{c} \sqrt{\left(\dot{X} \cdot X'\right)^2 - \left(\dot{X}\right)^2 (X')^2} \tag{22.4}$$

as

$$\delta S = \int_{\tau_i}^{\tau_f} \int_0^{\sigma_1} \left[\frac{\partial L}{\partial \dot{X}^\mu} \frac{\partial \delta X^\mu}{\partial \tau} + \frac{\partial L}{\partial X'^\mu} \frac{\partial \delta X^\mu}{\partial \sigma} \right] d\tau \, d\sigma. \tag{22.5}$$

Its derivatives, which we'll call $\mathcal{P}^\tau_\mu$ and $\mathcal{P}^\sigma_\mu$, are

$$\mathcal{P}^\tau_\mu = \frac{\partial L}{\partial \dot{X}^\mu} = -\frac{T_0}{c} \frac{(\dot{X} \cdot X')X'_\mu - (X')^2 \dot{X}_\mu}{\sqrt{\left(\dot{X} \cdot X'\right)^2 - \left(\dot{X}\right)^2 (X')^2}} \tag{22.6}$$

and

$$\mathcal{P}^\sigma_\mu = \frac{\partial L}{\partial X'^\mu} = -\frac{T_0}{c} \frac{(\dot{X} \cdot X')\dot{X}_\mu - (\dot{X})^2 X'_\mu}{\sqrt{\left(\dot{X} \cdot X'\right)^2 - \left(\dot{X}\right)^2 (X')^2}}. \tag{22.7}$$

In terms of them, the change in the action is

$$\delta S = \int_{\tau_i}^{\tau_f} \int_0^{\sigma_1} \left[\frac{\partial}{\partial \tau} \left(\delta X^\mu \mathcal{P}^\tau_\mu\right) + \frac{\partial}{\partial \sigma} \left(\delta X^\mu \mathcal{P}^\sigma_\mu\right) - \delta X^\mu \left(\frac{\partial \mathcal{P}^\tau_\mu}{\partial \tau} + \frac{\partial \mathcal{P}^\sigma_\mu}{\partial \sigma} \right) \right] d\tau \, d\sigma. \tag{22.8}$$

The total τ-derivative integrates to a term involving the variation δX^μ which we make vanish at the initial and final values of τ. So we drop that term and find that the net change in the action is

$$\delta S = \int_{\tau_i}^{\tau_f} \left[\delta X^\mu \mathcal{P}^\sigma_\mu\right]_0^{\sigma_1} d\tau - \int_{\tau_i}^{\tau_f} \int_0^{\sigma_1} \delta X^\mu \left(\frac{\partial \mathcal{P}^\tau_\mu}{\partial \tau} + \frac{\partial \mathcal{P}^\sigma_\mu}{\partial \sigma} \right) d\tau \, d\sigma. \tag{22.9}$$

Thus the equations of motion for the string are

$$\frac{\partial \mathcal{P}^\tau_\mu}{\partial \tau} + \frac{\partial \mathcal{P}^\sigma_\mu}{\partial \sigma} = 0, \tag{22.10}$$

but the action is stationary only if

$$\int \delta X^\mu(\tau, \sigma_1) \mathcal{P}^\sigma_\mu(\tau, \sigma_1) - \delta X^\mu(\tau, 0) \mathcal{P}^\sigma_\mu(\tau, 0) \, d\tau = 0. \tag{22.11}$$

Closed strings automatically satisfy this condition. Open strings satisfy it if they obey for each end σ_* of the string, each spacetime dimension μ, and all times τ either the **free-endpoint** boundary condition

$$\mathcal{P}^{\sigma}_{\mu}(\tau, \sigma_*) = 0 \tag{22.12}$$

or the **Dirichlet** boundary condition

$$\dot{X}^{\mu}(\tau, \sigma_*) = 0 \tag{22.13}$$

which cannot hold for $\mu = 0$ because $X^0 = ct$ is related to the parameter τ. So the time component of an open string obeys the free-endpoint condition $\mathcal{P}^{\sigma}_{0}(\tau, \sigma_*) = 0$. A **D$n$-brane** is a space of n spatial dimensions to which an end $X^j(\tau, \sigma_*)$ of an open string can attach and on which it can move.

The equation of motion (22.10) shows that the 4-momentum

$$p_\mu = \int \mathcal{P}^{\tau}_{\mu}\, d\sigma \tag{22.14}$$

of both a closed string and an open one moving with free endpoints (22.12) is conserved

$$\frac{dp_\mu}{d\tau} = \int \frac{\partial \mathcal{P}^{\tau}_{\mu}}{\partial \tau}\, d\sigma = -\int \frac{\partial \mathcal{P}^{\sigma}_{\mu}}{\partial \sigma}\, d\sigma = -\left[\mathcal{P}^{\sigma}_{\mu}(\tau, \sigma)\right]_0^{\sigma_1} = 0, \tag{22.15}$$

but the momentum p_ν need not be conserved if X^ν is stuck on a brane. The momentum density of the string is $\mathcal{P}^{\tau}_{\mu}$.

22.2 Static Gauge and Regge Trajectories

The Nambu–Goto action (22.1) remains the same when one changes the parameters σ and τ arbitrarily to $\sigma'(\sigma, \tau)$ and $\tau'(\sigma, \tau)$. This **reparametrization invariance** allows us to choose these parameters so as to simplify any particular calculation without changing the physics.

One such choice is the **static gauge** in which the parameter τ is the time $t = X^0/c$ in some chosen Lorentz frame, and so the σ and τ derivatives of $X = (X^0, \vec{X})$ are $X' = (0, \vec{X}')$ and $\dot{X} = (c, \dot{\vec{X}})$. In this gauge, the time derivative of the 4-momentum p^μ of a freely moving string vanishes because its τ-derivative vanishes (22.15), and the free endpoints of an open string move transversely to the string at the speed of light. We also can chose the parameter σ so that the motion of every point on a string is transverse to the string, $\vec{X}' \cdot \dot{\vec{X}} = X' \cdot \dot{X} = 0$, so that the energy of the string varies as $dE = T_0\, d\sigma$, and so that $X'^2 + \dot{X}^2 = 0$. In this parametrization (Zwiebach, 2009, chaps. 6–8), strings obey the wave equation

$$\ddot{X} = c^2 X'', \tag{22.16}$$

and the free-endpoint boundary condition (22.12) is simply $X'^{\mu}(t, \sigma_*) = 0$.

The angular momentum M_{12} of a string rigidly rotating in the 1–2 plane is

$$M_{12}(\tau) = \int_0^{\sigma_1} X_1(\tau, \sigma)\mathcal{P}_2^\tau(\tau, \sigma) - X_2(\tau, \sigma)\mathcal{P}_1^\tau(\tau, \sigma)\, d\sigma. \tag{22.17}$$

In the above version of the static gauge, the nonzero spatial coordinates of a straight open string rotating in the 1–2 plane are

$$\vec{X}(t, \sigma) = \frac{\sigma_1}{\pi} \cos\frac{\pi\sigma}{\sigma_1}\left(\cos\frac{\pi ct}{\sigma_1},\ \sin\frac{\pi ct}{\sigma_1}\right), \tag{22.18}$$

with momentum densities

$$\vec{\mathcal{P}}^\tau(t, \sigma) = \frac{T_0}{c^2}\frac{\partial \vec{X}}{\partial t} = \frac{T_0}{c}\cos\frac{\pi\sigma}{\sigma_1}\left(-\sin\frac{\pi ct}{\sigma_1},\ \cos\frac{\pi ct}{\sigma_1}\right). \tag{22.19}$$

Its angular momentum (22.17) is

$$M_{12} = \frac{\sigma_1}{\pi}\frac{T_0}{c}\int_0^{\sigma_1} \cos^2\frac{\pi\sigma}{\sigma_1} d\sigma = \frac{\sigma_1^2\, T_0}{2\pi c}. \tag{22.20}$$

Since in this static gauge its energy is $E = \sigma_1 T_0$, the angular momentum $J = M_{12}$ of the relativistic open string (22.18) is proportional to the square of its total energy

$$\frac{J}{\hbar} = \frac{E^2}{2\pi\hbar c T_0}. \tag{22.21}$$

The nonzero spatial coordinates of the corresponding closed string are

$$\vec{X} = \frac{\sigma_1}{4\pi}\left(\sin\frac{2\pi u}{\sigma_1} + \sin\frac{2\pi v}{\sigma_1},\ -\cos\frac{2\pi u}{\sigma_1} - \cos\frac{2\pi v}{\sigma_1}\right) \tag{22.22}$$

in which $u = ct + \sigma$ and $v = ct - \sigma$. One may show (Exercise 22.3) that its angular momentum M_{12} obeys a rule similar to that (22.21) of the open string but with half the slope, $J/\hbar = E^2/(4\pi\hbar c T_0)$.

Rules like (22.21) describe many hadrons made of light (u, d, s) quarks and massless gluons. The nucleon and five baryon resonances obey it with nearly the same value of the string tension $T_0 \approx 0.937$ GeV/fm. Their common **Regge trajectory** is plotted in Fig. 22.1.

The energy E and angular momentum J of a Kerr black hole obey a relation $J/\hbar = aGE^2/(\hbar c^5) = E^2/(2\pi\hbar c T)$ like the Regge-trajectory (22.21) but with string tension $T = c^4/(2\pi a G)$ in which G is Newton's constant and $a < 1$. This string tension is higher by at least 37 orders of magnitude since $1/G = 1.4906 \times 10^{38}\, \text{GeV}^2/(\hbar c)^2$.

A string theory of hadrons took off in 1968 when Gabriel Veneziano published his amplitude for $\pi + \pi$ scattering as a sum of three Euler beta functions (Veneziano, 1968). But after eight years of intense work, this effort was

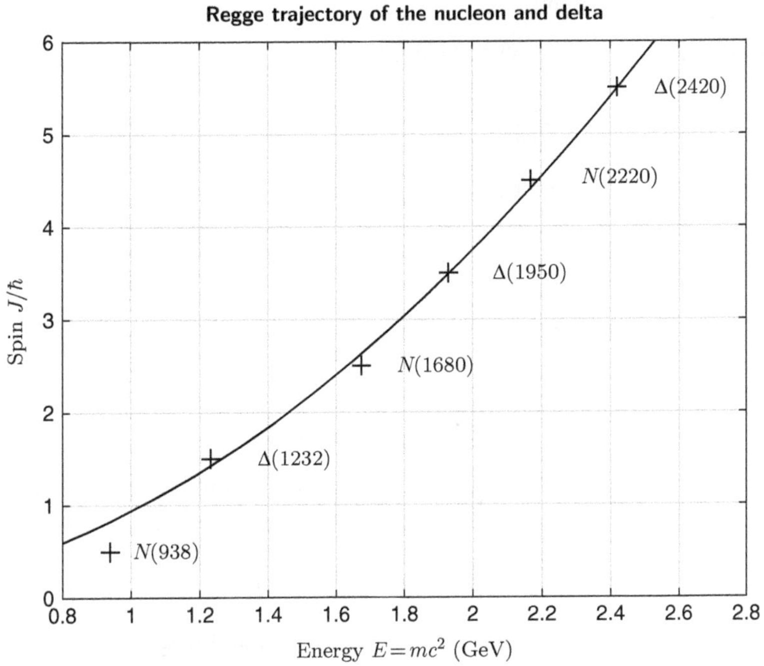

Figure 22.1 The angular momentum and energy of the nucleon and delta resonances approximately fit the curve $J/\hbar = E^2/(2\pi\hbar c\, T_0)$ with string tension $T_0 = 0.94$ GeV/fm. This chapter's Matlab scripts are in Strings at `github.com/kevinecahill`.

largely abandoned with the discovery of quarks at SLAC and the promise of QCD as a theory of the strong interactions. In 1974, Joël Scherk and John Schwarz proposed increasing the string tension by 38 orders of magnitude so as to use strings to make a quantum theory that included gravity (Scherk and Schwarz, 1974). They identified the graviton as an excitation of the closed string.

22.3 Light-Cone Coordinates

Dirac's light-cone coordinates are $x^+ = \left(x^0 + x^1\right)/\sqrt{2}$ in which x^1 is a spatial coordinate and $x^- = \left(x^0 - x^1\right)/\sqrt{2}$. Conventionally x^+ is the light-cone variable corresponding to the time x^0. The light-cone components of the momentum are $p^+ = \left(p^0 + p^1\right)/\sqrt{2}$ and $p^- = \left(p^0 - p^1\right)/\sqrt{2}$.

The squared distance ds^2 is

$$ds^2 = \boldsymbol{dx} \cdot \boldsymbol{dx} - (dx^0)^2 = (dx^2)^2 + (dx^3)^2 - dx^- dx^+ - dx^+ dx^-, \quad (22.23)$$

and $p \cdot x = \boldsymbol{p} \cdot \boldsymbol{x} - p^0 x^0 = p^2 x^2 + p^3 x^3 - p^- x^+ - p^+ x^-$. Just as in quantum mechanics we identify $i\hbar\, \partial_t$ with the energy E so that

$$i\hbar \frac{\partial}{\partial x^0} e^{i(\boldsymbol{p}\cdot\boldsymbol{x}-Et)/\hbar} = \frac{E}{c} e^{i(\boldsymbol{p}\cdot\boldsymbol{x}-Et)/\hbar}, \tag{22.24}$$

so too, in light-cone coordinates, we have

$$i\hbar \frac{\partial}{\partial x^+} e^{i(p^2x^2+p^3x^3-p^-x^+-p^+x^-)/\hbar} = p^- \, e^{i(p^2x^2+p^3x^3-p^-x^+-p^+x^-)/\hbar}. \tag{22.25}$$

So p^- is the light-cone version of E/c.

22.4 Light-Cone Gauge

The equations of motion (22.10) and boundary conditions (22.12 and 22.13) simplify when the string parameters σ and τ obey the light-cone gauge conditions

$$X^+ = \beta\, \alpha'\, p^+ \tau \quad \text{and} \quad p^+ = (2\pi/\beta)\, \mathcal{P}^{\tau +} \tag{22.26}$$

in natural units ($\hbar = c = 1$) with $\alpha' = 1/(2\pi\, \hbar\, c\, T_0)$ and $0 \le \sigma \le \pi$ where $\beta = 2$ for open strings, and $\beta = 1$ for closed strings (Zwiebach, 2009, chaps. 9 & 12). This gauge maintains the wave equation (22.16) and constraints of the static gauge of Section 22.2

$$\ddot{X} = X'' \quad \text{and} \quad \dot{X} \cdot X' = 0 = \dot{X}^2 + X'^2 \tag{22.27}$$

as well as the $X'^\mu(\tau, \sigma_*) = 0$ form of the free-endpoint condition (22.12) while adding the rules

$$\mathcal{P}^{\sigma\mu} = -\frac{1}{2\pi\alpha'} X'^\mu \quad \text{and} \quad \mathcal{P}^{\tau\mu} = \frac{1}{2\pi\alpha'} \dot{X}^\mu \tag{22.28}$$

which make it easier to show (Exercise 22.4) that the angular momentum

$$M_{\mu\nu}(\tau) = \int_0^{\sigma_1} X_\mu(\tau,\sigma)\, \mathcal{P}^\tau_\nu(\tau,\sigma) - X_\nu(\tau,\sigma)\, \mathcal{P}^\tau_\mu(\tau,\sigma)\, d\sigma \tag{22.29}$$

of a freely moving string is conserved, $\dot{M}_{\mu\nu}(\tau) = 0$.

22.5 Quantized Open Strings

In light-cone gauge (22.26), the action is a sum over the transverse coordinates $\ell = 2, 3, \ldots, d$

$$S = \frac{1}{4\pi\alpha'} \int d\tau \int_0^\pi d\sigma \left(\dot{X}^\ell \dot{X}^\ell - X'^\ell X'^\ell \right). \tag{22.30}$$

An expansion for an open string with free endpoints is

$$X^\mu(\tau,\sigma) = x_0^\mu + \sqrt{2\alpha'}\,\alpha_0^\mu\,\tau + i\sqrt{2\alpha'}\sum_{n\neq 0}\frac{e^{-in\tau}}{n}\alpha_n^\mu \cos n\sigma. \tag{22.31}$$

It obeys free-endpoint boundary conditions (22.12) because $(\cos n\sigma)' = n\sin\sigma$ vanishes at $\sigma = 0$ and π. Closed strings have left- and right-moving waves indicated by barred and unbarred amplitudes

$$X^\mu(\tau,\sigma) = x_0^\mu + \sqrt{2\alpha'}\,\alpha_0^\mu\,\tau + i\sqrt{\frac{\alpha'}{2}}\sum_{n\neq 0}\frac{e^{-in\tau}}{n}\left(\alpha_n^\mu e^{in\sigma} + \bar{\alpha}_n^\mu e^{-in\sigma}\right). \tag{22.32}$$

In the light-cone gauge, the transverse coordinates X^ℓ of open strings obey the equal-time commutation relations with the transverse momenta

$$[X^\ell(\tau,\sigma), \mathcal{P}^{\tau\ell'}(\tau,\sigma')] = i\eta^{\ell\ell'}\,\delta(\sigma-\sigma') \quad \text{and} \quad [x_0^-(\tau), p^+(\tau)] = -i, \tag{22.33}$$

and commute with themselves at equal times: $[X^\ell(\tau,\sigma), X^{\ell'}(\tau,\sigma')] = 0$ and $[\mathcal{P}^{\tau\ell}(\tau,\sigma), \mathcal{P}^{\tau\ell'}(\tau,\sigma')] = 0$. The amplitude operators α_n^ℓ obey the commutation relations

$$[\alpha_m^\ell, \alpha_n^{\ell'}] = m\,\eta^{\ell\ell'}\delta_{m+n,0} \quad \text{and} \quad [x_0^\ell, p^{\ell'}] = i\eta^{\ell\ell'}. \tag{22.34}$$

Similar rules apply to closed strings (Zwiebach, 2009, chap. 13).

String theory is a quantum field theory in 2 dimensions τ and σ. Its ground-state energy involves a sum over the positive integers which string theorists interpret by using Ser's series expansion (6.111) of Riemann's zeta function to say that

$$\begin{aligned}\sum_{n=1}^{\infty} n = \zeta(-1) &= -\frac{1}{2}\sum_{n=0}^{\infty}\frac{1}{n+1}\sum_{k=0}^{n}(-1)^k(k+1)^2\binom{n}{k} \\ &= -\frac{1}{2}\sum_{n=0}^{\infty}\frac{1}{n+1}\frac{d}{dx}\left\{x\frac{d}{dx}\left[x(1-x)^n\right]\right\}\bigg|_{x=1} = -\frac{1}{12}.\end{aligned} \tag{22.35}$$

The relation $1 = -\frac{1}{2}(D-2)\sum_n n$ then says that open bosonic strings make sense in $D = 26$ dimensions. The theory also has a tachyon, that is, a particle that moves faster than light.

22.6 Superstrings

In light-cone gauge, one can add fermionic variables $\psi_1^\mu(\tau,\sigma)$ and $\psi_2^\mu(\tau,\sigma)$ to the action in a supersymmetric way

$$\begin{aligned} S = & \frac{1}{4\pi\alpha'}\int d\tau \int_0^\pi d\sigma\, (\dot{X}^\ell \dot{X}^\ell - X'^\ell X'^\ell) \\ & + \frac{1}{4\pi\alpha'}\int d\tau \int_0^\pi d\sigma\, \left[\psi_1^\ell(\partial_\tau + \partial_\sigma)\psi_1^\ell + \psi_2^\ell(\partial_\tau - \partial_\sigma)\psi_2^\ell\right]. \end{aligned} \tag{22.36}$$

The tachyon then goes away, and the number of spacetime dimensions drops from 26 to 10. Although string theory requires renormalization, it does give finite scattering amplitudes.

There are five distinct superstring theories – types I, IIA, and IIB; $E_8 \otimes E_8$ heterotic; and $SO(32)$ heterotic. All five may be related to a single theory in 11 dimensions called **M-theory**, which is not a string theory. M-theory contains membranes (2-branes) and 5-branes, which are not D-branes.

22.7 Covariant and Polyakov Actions

Other actions offer other advantages. The covariant action

$$S = \frac{1}{4\pi\alpha'}\int d\tau d\sigma\, \left(\partial_\tau X^\mu \partial_\tau X_\mu - \partial_\sigma X^\mu \partial_\sigma X_\mu\right) \tag{22.37}$$

offers manifest Lorentz invariance and simple momentum operators

$$\mathcal{P}_\mu = \frac{\partial \mathcal{L}}{\partial \dot{X}^\mu} = \frac{1}{2\pi\alpha'}\dot{X}_\mu. \tag{22.38}$$

The commutation relations are $[X^\mu(\tau,\sigma), \mathcal{P}^\nu(\tau,\sigma')] = i\eta^{\mu\nu}\,\delta(\sigma-\sigma')$, but the 00 commutation relation has a minus sign that requires constraints on the physical states.

Polyakov's action is

$$S = -\frac{1}{4\pi\alpha'}\int d\tau d\sigma\, \sqrt{-h}\, h^{\alpha\beta}\, \partial_\alpha X^\mu\, \partial_\beta X^\nu\, \eta_{\mu\nu} \tag{22.39}$$

in which $h = \det(h_{\alpha\beta})$ and $h^{\alpha\beta}$ is the inverse of the 2×2 matrix $h_{\alpha\beta}$. The Minkowski metric $\eta_{\mu\nu}$ is fixed and diagonal, but the metric $h_{\alpha\beta}$ is dynamical and plays in the 2 dimensions τ, σ a role like that of $g_{\mu\nu}$ in general relativity.

22.8 D-branes or P-branes

One may satisfy Dirichlet boundary conditions (22.13) by requiring the ends of a string to touch but be free to move along a spatial manifold, called a **D-brane** after

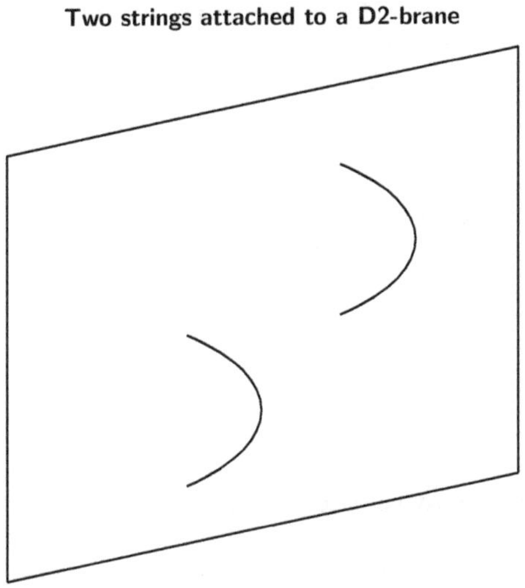

Figure 22.2 Two strings stuck on a D2-brane.

Dirichlet. These branes should be called P-branes after Polchinski. If the manifold to which the string is stuck has p dimensions, then it's called a **Dp-brane**. Figure 22.2 shows two strings whose ends are free to move only within a D2-brane.

Dp-branes offer a natural way to explain the extra six dimensions required in a universe of superstrings. One imagines that the ends of all the strings are free to move only in our 4-dimensional spacetime; the strings are stuck on a D3-brane, which is the 3-dimensional space of our physical universe. The tension of the superstring then keeps it from wandering far enough into the extra six spatial dimensions for us ever to have noticed.

22.9 String–String Scattering

Strings interact by joining and by breaking. The left side of Fig. 22.3 shows two open strings joining to form one open string and then breaking into two open strings; the right side shows two closed strings joining to form one closed string and then breaking into two closed strings. The interactions of strings do not occur at points. Because strings are extended objects, their scattering amplitudes are finite.

22.10 Riemann Surfaces and Moduli

A **homeomorphism** is a map that is one to one and continuous with a continuous inverse. A **Riemann surface** is a 2-dimensional real manifold whose open sets U_α are mapped onto open sets of the complex plane $\mathbb{C}$ by **homeomorphisms** z_α whose

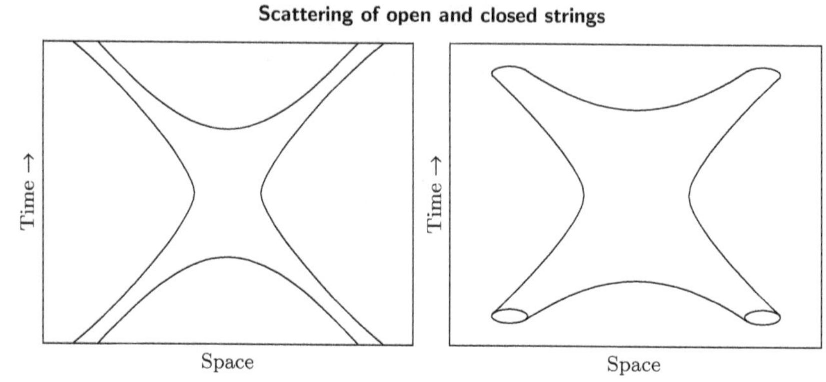

Figure 22.3 Spacetime diagrams of the scattering of two open strings into two open strings (left) and of two closed strings into two closed strings (right).

transition functions $z_\alpha \circ z_\beta^{-1}$ are analytic on the images of the intersections $U_\alpha \cap U_\beta$. Two Riemann surfaces are **equivalent** if they are related by a continuous analytic map that is one to one and onto.

A parameter that labels the geometry of a manifold or that identifies a vacuum in quantum field theory is called a **modulus**. Some Riemann surfaces have several moduli; others have one modulus; others none at all. Some moduli are continuous parameters; others are discrete.

Further Reading

A First Course in String Theory (Zwiebach, 2009).

Exercises

22.1 Derive formulas (22.6) and (22.7).

22.2 Derive Equation (22.8) from (22.5–22.7).

22.3 Use the formulas for the momentum $\vec{\mathcal{P}}^\tau(t,\sigma) = T_0\dot{\vec{X}}/c^2$ and angular momentum (22.17) to derive the Regge relation $J/\hbar = E^2/(4\pi\hbar c T_0)$ for the angular momentum of the closed string (22.22).

22.4 Use the equation of motion (22.10) and the light-cone gauge conditions (22.28) and $X^{\mu\prime}(\tau,\sigma_*) = 0$ to show that the angular momentum (22.29) of a freely moving string is conserved, $\dot{M}_{\mu\nu} = 0$.

References

Aghanim, N., et al. 2018. Planck 2018 results. VI. Cosmological parameters. *arXiv*, 1807.06209.

Aitken, A. C. 1959. *Determinants and Matrices*. Oliver and Boyd.

Aké, Luis, Flores, José L., and Sánchez, Miguel. 2018. Structure of globally hyperbolic spacetimes with timelike boundary. *arXiv*, **1808**, 04412.

Akrami, Y., et al. 2018. Planck 2018 results. X. Constraints on inflation. *arXiv*, 1807.06211.

Alberts, Bruce, Johnson, Alexander, Lewis, Julian, Morgan, David, Raff, Martin, Roberts, Keith, and Walter, Peter. 2015. *Molecular Biology of the Cell*. 6th edn. Garland Science.

Alligood, Kathleen T., Sauer, Tim D., and Yorke, James A. 1996. *CHAOS: An Introduction to Dynamical Systems*. Springer-Verlag.

Alsing, Paul M. and Milonni, Peter W. 2004. Simplified derivation of the Hawking-Unruh temperature for an accelerated observer in vacuum. *Am. J. Phys.*, **72**, 1524–1529.

Amdahl, David and Cahill, Kevin. 2016. Path integrals for awkward actions. *arXiv*, 1611.06685.

Arnold, V. I. 1989. *Mathematical Methods of Classical Mechanics*. 2nd edn. Springer. 5th printing. Chap. 7.

Autonne, L. 1915. *Sur les Matrices Hypohermitiennes et sur les Matrices Unitaires. Ann. Univ. Lyon, Nouvelle Série I*, **Fasc. 38**, 1–77.

Bailey, J., Borer, K., Combley, F., Drumm, H., Krienen, F., Lange, F., Picasso, E., von Ruden, W., Farley, F. J. M., Field, J. H., Flegel, W., and Hattersley, P. M. 1977. Measurements of relativistic time dilatation for positive and negative muons in a circular orbit. *Nature*, **268**, 301–305.

Bekenstein, Jacob D. 1973. Black holes and entropy. *Phys. Rev.*, **D7**, 2333–2346.

Bender, Carl M, and Orszag, Steven A. 1978. *Advanced Mathematical Methods for Scientists and Engineers*. McGraw-Hill.

Bigelow, Matthew S., Lepeshkin, Nick N., and Boyd, Robert W. 2003. Superluminal and Slow Light Propagation in a Room-Temperature Solid. *Science*, **301(5630)**, 200–202.

Bouchaud, Jean-Philippe and Potters, Marc. 2003. *Theory of Financial Risk and Derivative Pricing*. 2nd edn. Cambridge University Press.

Bowman, Judd D., Rogers, Alan E. E., Monsalve, Raul A., Mozdzen, Thomas J., and Mahesh, Nivedita. 2018. An absorption profile centred at 78 megahertz in the sky-averaged spectrum. *Nature*, **555**, 67.

Boyd, Robert W. 2000. *Nonlinear Optics*. 2nd edn. Academic Press.

Brillouin, L. 1960. *Wave Propagation and Group Velocity*. Academic Press.

Brunner, N, Scarani, V, Wegmüller, M, Legré, M, and Gisin, N. 2004. Direct measurement of superluminal group velocity and signal velocity in an optical fiber. *Phys. Rev. Lett.*, **93(20)**, 203902.

Cahill, Sean. 2018. *Coding Lina*. Amazon.

Callan, Jr., Curtis G. 1970. Broken scale invariance in scalar field theory. *Phys. Rev.*, **D2**, 1541–1547.

Cantelli, F. P. 1933. Sulla determinazione empirica di una legge di distribuzione. *Giornale dell'Instituto Italiano degli Attuari*, **4**, 221–424.

Carroll, Sean. 2003. *Spacetime and Geometry: An Introduction to General Relativity*. Benjamin Cummings.

Chandrasekhar, S. 1943. Stochastic problems in physics and astronomy. *Rev. Mod. Phys.*, **15(1)**, 1–89.

Cheng, Ta-Pei. 2010. *Relativity, Gravitation and Cosmology: A Basic Introduction*. Oxford University Press.

Clarke, C. J. S. 1970. On the global isometric embedding of pseudo-Riemannian manifolds. *Proc. Roy. Soc. London Ser. A*, **314**, 417–428.

Cohen-Tannoudji, Claude, Diu, Bernard, and Laloë, Frank. 1977. *Quantum Mechanics*. Hermann & John Wiley.

Cotton, F. Albert. 1990. *Chemical Applications of Group Theory*. 3rd edn. Wiley-Interscience.

Courant, Richard. 1937. *Differential and Integral Calculus, Vol. I*. Interscience.

Courant, Richard and Hilbert, David. 1955. *Methods of Mathematical Physics, Vol. I*. Interscience.

Creutz, Michael. 1983. *Quarks, Gluons and Lattices*. Cambridge University Press.

Darden, Tom, York, Darrin, and Pedersen, Lee. 1993. Particle mesh Ewald: An Nlog(N) method for Ewald sums in large systems. *J. Chem. Phys.*, **98(12)**.

DeWitt, Bryce S. 1967. Quantum theory of gravity. II. The manifestly covariant Theory. *Phys. Rev.*, **162(5)**, 1195–1239.

Dirac, P. A. M. 1967. *The Principles of Quantum Mechanics*. 4th edn. Oxford University Press.

Dirac, P. A. M. 1996. *General Theory of Relativity*. Princeton University Press.

Dodelson, Scott. 2003. *Modern Cosmology*. 1st edn. Academic Press.

Faddeev, L. D. and Popov, V. N. 1967. Feynman diagrams for the Yang-Mills field. *Phys. Lett. B*, **25(1)**, 29–30.

Feller, William. 1966. *An Introduction to Probability Theory and Its Applications*. Vol. II. Wiley.

Feller, William. 1968. *An Introduction to Probability Theory and Its Applications*. 3rd edn. Vol. I. Wiley.

Feynman, Richard P. 1948. Space-time approach to non-relativistic quantum mechanics. *Rev. Mod. Phys.*, **20**, 367–387.

Feynman, Richard P. 1972. *Statistical Mechanics*. Basic Books.

Feynman, Richard P., Hibbs, Albert R., and Styer, Daniel F.(ed.). 2010. *Quantum Mechanics and Path Integrals*. Dover Publications.

Fisher, M. E. 1974. The renormalization group in the theory of critical behavior. *Rev. Mod. Phys.*, **46(4)**, 597–616.

Fisher, M. E. 1998. Renormalization group theory: Its basis and formulation in statistical physics. *Rev. Mod. Phys.*, **70**, 653–681.

Fixsen, D. J. 2009. The temperature of the cosmic microwave background. *Astrophys. J.*, **707**, 916–920.

Gattringer, Christof and Lang, Christian B. 2010. *Quantum Chromodynamics on the Lattice: An Introductory Presentation*. Springer's (Lecture Notes in Physics).

Gehring, G. M., Schweinsberg, A., Barsi, C., Kostinski, N., and Boyd, R. W. 2006. Observation of backwards pulse propagation through a medium with a negative group velocity. *Science*, **312**, 895–897.

Gelfand, Israel M. 1961. *Lectures on Linear Algebra*. Interscience.

Gell-Mann, Murray. 1994. *The Quark and the Jaguar*. W. H. Freeman.

Gell-Mann, Murray. 2008. *Plectics*. Lectures at the University of New Mexico.

Gell-Mann, Murray and Low, F. E. 1954. Quantum electrodynamics at small distances. *Phys. Rev.*, **95**, 1300–1312.

Georgi, H. 1999. *Lie Algebras in Particle Physics*. 2rd edn. Perseus Books.

Ghosh, Ritesh, Dewangan, Gulab C., Mallick, Labani, and Raychaudhuri, Biplab. 2018. Broadband spectral study of the jet-disc emission in the radio-loud narrow-line Seyfert 1 galaxy 1H 0323+342. *MNRAS, arxiv:1806.04089*, **479**, 2464.

Glauber, Roy J. 1963a. Coherent and incoherent states of the radiation field. *Phys. Rev.*, **131**(6), 2766–2788.

Glauber, Roy J. 1963b. The quantum theory of optical coherence. *Phys. Rev.*, **130**(6), 2529–2539.

Glivenko, V. 1933. Sulla determinazione empirica di una legge di distribuzione. *Giornale dell'Instituto Italiano degli Attuari*, **4**, 92–99.

Gnedenko, B. V. 1968. *The Theory of Probability*. Chelsea Publishing Co.

Grassberger, Peter and Procaccia, Itamar. 1983. Measuring the strangeness of strange attractors. *Physica D*, **9**, 189–208.

Grebogi, Celso, Ott, Edward, and Yorke, James A. 1987. Chaos, strange attractors, and fractal basin boundaries in nonlinear dynamics. *Science*, **238**, 632–638.

Greene, Robert E. 1970. *Isometric embeddings of Riemannian and pseudo-Riemannian manifolds*. Memoirs of the American Mathematical Society, No. 97. American Mathematical Society.

Griffiths, Jerry B. and Podolsky, Jiri. 2009. *Exact Space-Times in Einstein's General Relativity*. Cambridge Monographs on Mathematical Physics. Cambridge University Press.

Gross, David J., and Wilczek, Frank. 1973. Ultraviolet behavior of nonabelian gauge theories. *Phys. Rev. Lett.*, **30**, 1343–1346. [,271(1973)].

Günther, M. 1989. Zum Einbettungssatz von J. Nash. *Math. Nachr.*, **144**, 165–187.

Guth, Alan H. 1981. Inflationary universe: A possible solution to the horizon and flatness problems. *Phys. Rev.*, **D23**, 347–356.

Gutzwiller, Martin C. 1990. *Chaos in Classical and Quantum Mechanics*. Springer.

Halmos, Paul R. 1958. *Finite-Dimensional Vector Spaces*. 2nd edn. Van Nostrand.

Hardy, Godfrey Harold and Rogosinski, W. W. 1944. *Fourier Series*. 1st edn. Cambridge University Press.

Hartle, James B. 2003. *Gravity: An Introduction to Einstein's General Relativity*. Pearson.

Hastie, Trevor, Tibshirani, Roberti, and Friedman, Jerome. 2016. *The Elements of Statistical Leraning: Data Mining, Inference, and Prediction*. Springer.

Hau, L. V., Harris, S. E., Dutton, Z, and Behroozi, C. H. 1999. Light speed reduction to 17 metres per second in an ultracold atomic gas. *Nature*, **397**, 594.

Hawking, S. W. 1975. Particle creation by black holes. *Commun. Math. Phys.*, **43**, 199–220.

Helstrom, Carl W. 1976. *Quantum Detection and Estimation Theory*. Academic Press.

Hobson, M. P., Efstathiou, G. P., and Lasenby, A. N. 2006. *General Relativity: An Introduction for Physicists*. Cambridge University Press.

Holland, John H. 1975. *Adaptation in Natural and Artificial Systems*. University of Michigan Press.

Huang, Fang, Schwartz, Samantha L., Byars, Jason M., and Lidke, Keith A. 2011. Simultaneous multiple-emitter fitting for single molecule super-resolution imaging. *Biomed. Opt. Express*, **2**(5), 1377–1393.

Ijjas, Anna and Steinhardt, Paul J. 2018. Bouncing Cosmology made simple. *Class. Quant. Grav.*, **35**(13), 135004.

Ince, E. L. 1956. *Integration of Ordinary Differential Equations*. 7th edn. Oliver and Boyd, Ltd. Chap. 1.

James, F. 1994. RANLUX: A Fortran implementation of the high-quality pseudorandom number generator of Lüscher. *Comp. Phys. Comm.* , **79**, 110.

Kadanoff, Leo P. 2009. More is the same: Phase transitions and mean field theories. *J. Statist. Phys.*, **137**, 777.

Kahneman, Daniel. 2011. *Thinking, Fast and Slow*. Farrar, Straus, and Giroux.

Kaplan, J. and Yorke, J. 1979. *Chaotic Behavior of Multidimensional Difference Equations*. Springer Lecture Notes in Mathematics, vol. 730. Springer. Pages 228–237.

Kato, T. 1978. *Topics in Functional Analysis*. Academic Press. Pages 185–195.

Kleinert, Hagen. 2009. *Path Integrals in Quantum Mechanics, Statistics, Polymer Physics, and Financial Markets*. World Scientific.

Knuth, Donald E. 1981. *The Art of Computer Programming, Volume 2: Seminumerical Algorithms*. 2nd edn. Addison-Wesley.

Kolmogorov, Andrei Nikolaevich. 1933. Sulla determinazione empirica di una legge di distribuzione. *Giornale dell'Instituto Italiano degli Attuari*, **4**, 83–91.

Kosterlitz, J. M., Nelson, David R., and Fisher, Michael E. 1976. Bicritical and tetracritical points in anisotropic antiferromagnetic systems. *Phys. Rev.*, **B13**, 412–432.

Langevin, Paul. 1908. *Sur la théorie du mouvement brownien. Comptes Rend. Acad. Sci. Paris*, **146**, 530–533.

Lichtenberg, Donald B. 1978. *Unitary Symmetry and Elementary Particles*. 2nd edn. Academic Press.

Lin, I-Hsiung. 2011. *Classic Complex Analysis*. World Scientific.

Linde, Andrei D. 1983. Chaotic inflation. *Phys. Lett.*, **129B**, 177–181.

Lüscher, M. 1994. A portable high-quality random number generator for lattice field theory simulations. *Comp. Phys. Comm.* , **79**, 100.

Lyth, David H. and Liddle, Andrew R. 2009. *The Primordial Density Perturbation: Cosmology, Inflation and the Origin of Structure*. Cambridge University Press.

Martincorena, Iñigo, Raine, Keiran M., Gerstung, Moritz, Dawson, Kevin J., Haase, Kerstin, Loo, Peter Van, Davies, Helen, Stratton, Michael R., and Campbell, Peter J. 2017. Universal patterns of selection in cancer and somatic tissues. *Cell*, **171**(5), 1029–1041.

Matzner, Richard A. and Shepley, Lawrence C. 1991. *Classical Mechanics*. Prentice Hall.

Minsky, Marvin. 1961. Steps toward Artificial Intelligence. *Proceedings of the IEEE*, **49**, 8–30.

Misner, Charles W., Thorne, Kip S., and Wheeler, John Archibald. 1973. *Gravitation*. W. H. Freeman.

Morse, Philip M. and Feshbach, Herman. 1953. *Methods of Theoretical Physics*. Vol. I. McGraw-Hill.

Müller, O., and Sánchez, M. 2011. Lorentzian manifolds isometrically embeddable in $\mathbb{L}^N$. *Trans. Amer. Math. Soc.*, **363**(10), 5367–5379.

Nash, John. 1956. The imbedding problem for Riemannian manifolds. *Ann. Math.*, **63(1)**, 20–63.

Nishi, C. C. 2005. Simple derivation of general Fierz-like identities. *Am. J. Phys.*, **73**, 1160–1163.

Olver, Frank W. J., Lozier, Daniel W., Boisvert, Ronald F., and Clark, Charles W. 2010. *NIST Handbook of Mathematical Functions*. Cambridge University Press.

Padmanabhan, Thanu. 2010. *Gravitation: Foundations and Frontiers*. Cambridge University Press.

Parsegian, Adrian. 1969. Energy of an ion crossing a low dielectric membrane: Solutions to four relevant electrostatic problems. *Nature*, **221**, 844–846.

Pathria, R. K. 1972. *Statistical Mechanics*. Pergamon. Ch. 13.

Pearson, Karl. 1900. On the criterion that a given system of deviations from the probable in the case of correlated system of variables is such that it can be reasonably supposed to have arisen from random sampling. *Phil. Mag.*, **50(5)**, 157–175.

Peebles, P. J. E. 1982. Large scale background temperature and mass fluctuations due to scale invariant primeval perturbations. *Astrophys. J.*, **263**, L1–L5. [,85(1982)].

Peskin, Michael Edward, and Schroeder, Daniel V. 1995. *An Introduction to Quantum Field Theory*. Advanced book program. Westview Press. Reprinted in 1997.

Phillips, Robert Brooks, Kondev, Jane, Theriot, Julie, and Orme, Nigel. 2012. *Physical Biology of the Cell*. 2nd edn. Garland Science.

Politzer, H. David. 1973. Reliable perturbative results for strong interactions? *Phys. Rev. Lett.*, **30**, 1346–1349. [,274(1973)].

Riess, A. G., Casertano, S., Yuan, W., Macri, L., Anderson, J., MacKenty, J. W., Bowers, J. B., Clubb, K. I., Filippenko, A. V., Jones, D. O., and Tucker, B. E. 2018. New parallaxes of galactic cepheids from spatially scanning the Hubble Space Telescope: Implications for the Hubble constant. *ApJ*, **855**(Mar.), 136.

Riley, Ken, Hobson, Mike, and Bence, Stephen. 2006. *Mathematical Methods for Physics and Engineering*. 3d edn. Cambridge University Press.

Ritt, Robert K. 1970. *Fourier Series*. 1st edn. McGraw-Hill.

Roe, Byron P. 2001. *Probability and Statistics in Experimental Physics*. Springer.

Roos, C. E., Marraffino, J., Reucroft, S., Waters, J., Webster, M. S., Williams, E. G. H., Manz, A., Settles, R., and Wolf, G. 1980. $\Sigma^{\pm}$ lifetimes and longitudinal acceleration. *Nature*, **286**, 244–245.

Rubinstein, Reuven Y. and Kroese, Dirk P. 2007. *Simulation and the Monte Carlo Method*. 2nd edn. Wiley.

Russell, D. A., Hanson, J. D., and Ott, E. 1980. Dimension of strange attractors. *Phys. Rev. Lett.*, **45(14)**, 1175.

Saito, Mutsuo and Matsumoto, Makoto. 2007. www.math.sci.hiroshima-u.ac.jp/ m-mat/MT /emt.html.

Sakurai, J. J. 1982. *Advanced Quantum Mechanics*. 1st edn. Addison Wesley. 9th printing. Pages 62–63.

Scherk, Joël, and Schwarz, John H. 1974. Dual models for non-hadrons. *Nucl. Phys.*, **B81**, 118.

Schmitt, Lothar M. 2001. Theory of genetic algorithms. *Theor. Comput. Sci.*, **259**, 1–61.

Schutz, Bernard F. 1980. *Geometrical Methods of Mathematical Physics*. Cambridge University Press.

Schutz, Bernard F. 2009. *A First Course in General Relativity*. 2nd edn. Cambridge University Press.

Schwinger, Julian, Deraad, Lester, Milton, Kimball A., and Tsai, Wu-yang. 1998. *Classical Electrodynamics*. Westview Press.

Shortliffe, E. H. and Buchanan, B.G. 1975. A model of inexact reasoning in medicine. *Math. Biosci.*, **23**, 351–379.

Smirnov, N. V. 1939. Estimation of the deviation between empirical distribution curves for two independent random samples. *Bull. Moscow State Univ.*, **2(2)**, 3–14.

Smith, Carlas S, Joseph, Nikolai, Rieger, Bernd, and Lidke, Keith A. 2010. Fast, single-molecule localization that achieves theoretically minimum uncertainty. *Nature Methods*, **7(5)**, 373.

Sprott, Julien C. 2003. *Chaos and Time-Series Analysis*. 1st edn. Oxford University Press.

Srednicki, Mark. 2007. *Quantum Field Theory*. Cambridge University Press.

Stakgold, Ivar. 1967. *Boundary Value Problems of Mathematical Physics, Vol. I*. Macmillan.

Steinberg, A. M., Kwiat, P. G., and Chiao, R. Y. 1993. Measurement of the Single-Photon Tunneling Time. *Phys. Rev. Lett.*, **71(5)**, 708–711.

Steinhardt, Paul J., Turok, Neil, and Turok, N. 2002. A Cyclic model of the universe. *Science*, **296**, 1436–1439.

Stenner, Michael D., Gauthier, Daniel J., and Neifeld, Mark A. 2003. The speed of information in a "fast-light" optical medium. *Nature*, **425**, 695–698.

Strogatz, Steven H. 2014. *Nonlinear Dynamics and Chaos: With Applications to Physics, Biology, Chemistry, and Engineering*. 2nd edn. Westview Press.

Symanzik, K. 1970. Small distance behavior in field theory and power counting. *Commun. Math. Phys.*, **18**, 227–246.

Tinkham, Michael. 2003. *Group Theory and Quantum Mechanics*. Dover Publications.

Titulaer, U. M. and Glauber, R. J. 1965. Correlation functions for coherent fields. *Phys. Rev.*, **140**(3B), B676–682.

Trotter, H. F. 1959. On the product of semi-groups of operators. *Proc. Ann. Math.*, **10**, 545–551.

Veneziano, Gabriel. 1968. Construction of a crossing-symmetric Regge-behaved amplitude for linearly rising Regge trajectories. *Nuovo Cim.*, **57A**, 190.

von Foerster, Heinz, Patricia M. Mora, Patricia M., and Amiot, Lawrence W. 1960. Doomsday: Friday, 13 November, A.D. 2026. *Science*, **132**, 1291–1295.

Vose, Michael D. 1999. *The Simple Genetic Algorithm: Foundations and Theory*. MIT Press.

Wang, Yun-ping and Zhang, Dian-lin. 1995. Reshaping, path uncertainty, and superluminal traveling. *Phys. Rev. A*, **52(4)**, 2597–2600.

Watson, George Neville. 1995. *A Treatise on the Theory of Bessel Functions*. Cambridge University Press.

Waxman, David and Peck, Joel R. 1998. Pleiotropy and the preservation of perfection. *Science*, **279**.

Weinberg, Steven. 1972. *Gravitation and Cosmology*. John Wiley & Sons.

Weinberg, Steven. 1988. *The First Three Minutes*. Basic Books.

Weinberg, Steven. 1995. *The Quantum Theory of Fields*. Vol. I Foundations. Cambridge University Press.

Weinberg, Steven. 1996. *The Quantum Theory of Fields*. Vol. II Modern Applications. Cambridge University Press.

Weinberg, Steven. 2005. *The Quantum Theory of Fields*. Vol. III Supersymmetry. Cambridge University Press.

Weinberg, Steven. 2010. *Cosmology*. Oxford University Press.

Whittaker, E. T. and Watson, G. N. 1927. *A Course of Modern Analysis*. 4th edn. Cambridge University Press.

Wigner, Eugene P. 1964. *Group Theory and Its Application to the Quantum Mechanics of Atomic Spectra*. Revised edn. Academic Press.

Wilson, Kenneth G. 1971. Renormalization group and critical phenomena. 1. Renormalization group and the Kadanoff scaling picture. *Phys. Rev.*, **B4**, 3174–3183.
Wilson, Kenneth G. 1975. The renormalization group: Critical phenomena and the Kondo problem. *Rev. Mod. Phys.*, **47**(Oct), 773–840.
Zee, Anthony. 2010. *Quantum Field Theory in a Nutshell*. 2nd edn. Princeton University Press.
Zee, Anthony. 2013. *Einstein Gravity in a Nutshell*. Princeton University Press.
Zee, Anthony. 2016. *Group Theory in a Nutshell for Physicists*. Princeton University Press.
Zwiebach, Barton. 2009. *A First Course in String Theory*. 2nd edn. Cambridge University Press.

Index

www.ingramcontent.com/pod-product-compliance
Lightning Source LLC
Chambersburg PA
CBHW082138150125
20468CB00005B/118
* 9 7 8 1 1 0 8 4 7 0 0 3 2 *